**V. W. Cairns, P.** ...
Contaminant Effe... ...me 16 1984

**J. O. Nriagu and C. I. Davidson**
Toxic Metals in the Atmosphere Volume 17 1986

**Allan H. Legge and Sagar V. Krupa**
Air Pollutants and Their Effects on the Terrestrial Ecosystem Volume 18 1986

**J. O. Nriagu and J. B. Sprague**
Cadmium in the Aquatic Environment Volume 19 1987

**J. O. Nriagu and E. Nieboer**
Chromium in the Natural and Human Environments Volume 20 1988

**Marlene S. Evans**
Toxic Contaminants and Ecosystem Health: A Great Lakes Focus Volume 21 1988

# TOXIC CONTAMINANTS AND ECOSYSTEM HEALTH: A GREAT LAKES FOCUS

Volume

21

Wiley Series

# Advances in Environmental Science and Technology

JEROME O. NRIAGU, Series Editor

# TOXIC CONTAMINANTS AND ECOSYSTEM HEALTH: A GREAT LAKES FOCUS

Edited by
**Marlene S. Evans**
Great Lakes Research Division
The University of Michigan
Ann Arbor, Michigan

Assisted by
John E. Gannon
Donald C. McNaught
Carlos M. Fetterolf, Jr.
Health of Aquatic Communities Task Force

WILEY
A WILEY-INTERSCIENCE PUBLICATION
**JOHN WILEY & SONS**
**New York • Chichester • Brisbane • Toronto • Singapore**

***Library of Congress Cataloging in Publication Data:***

Toxic contaminants and ecosystem health.

(Wiley series in Advances in environmental science and technology; v. 21)
"A Wiley-Interscience publication."
Includes index.
1. Water—Pollution—Great Lakes Watershed—Toxicology. I. Evans, Marlene S. II. Series: Advances in environmental science and technology; v. 21.
TD180.A38 vol. 21 628 s 87-22996
[TD223.3] [574.5'26322'0977]
ISBN 0-471-85556-1

Printed in the United States of America

10 9 8 7 6 5 4 3 2 1

# CONTRIBUTORS

BARTELL, STEVEN M., Environmental Sciences Division, Oak Ridge National Laboratory, Oak Ridge, Tennessee

BINDER,* ROBERT L., Department of Pharmacology and Toxicology, Medical College of Wisconsin, Milwaukee, Wisconsin

BRECK, JAMES E., Environmental Sciences Division, Oak Ridge National Laboratory, Oak Ridge, Tennessee

CAIRNS, JR., JOHN, Department of Biology and University Center for Environmental Studies, Virginia Polytechnic Institute and State University, Blacksburg, Virginia

DI TORO, DOMINIC M., HydroQual, Inc., Mahwah, New Jersey

ESTERBY, S. R., Aquatic Ecology Division, National Water Research Institute, Burlington, Ontario, Canada

EVANS, MARLENE S., Great Lakes Research Division, The University of Michigan, Ann Arbor, Michigan

FONTAINE III, THOMAS D., Great Lakes Environmental Research Laboratory, National Oceanic and Atmospheric Administration, Ann Arbor, Michigan

GANNON, JOHN E., International Joint Commission, Windsor, Ontario, Canada

GIESY, JOHN P., Pesticide Research Center, Department of Fisheries and Wildlife, Center for Environmental Toxicology, Michigan State University, East Lansing, Michigan

GILBERTSON, MICHAEL, Fish Habitat Management Branch, Department of Fisheries and Oceans, Ottawa, Ontario, Canada

GRANEY, ROBERT L., Pesticide Research Center, Department of Fisheries and Wildlife, Center for Environmental Toxicology, Michigan State University, East Lansing, Michigan

HALLDEN, JAMES A., HydroQual, Inc., Mahwah, New Jersey

HALLETT, DOUGLAS J., ELI-Eco Logic Inc., Rockwood, Ontario, Canada

HARRIS, H. J., University of Wisconsin—Green Bay, Green Bay, Wisconsin

*Present affiliation*: Human and Environmental Safety Division, The Procter and Gamble Company, Miami Valley Laboratories, Cincinnati, Ohio.

HARTMAN, WILBUR L., United States Fish and Wildlife Service, National Fisheries Center—Great Lakes, Ann Arbor, Michigan

HOBBIE, JOHN E., Marine Biological Laboratory and Woods Hole Oceanographic Institution, Woods Hole, Massachusetts

HUMPHREY, HAROLD E. B., Center for Environmental Health Sciences, Michigan Department of Public Health, Lansing, Michigan

JACOBSON, JOSEPH L., Wayne State University, Detroit, Michigan

JACOBSON, SANDRA W., Wayne State University, Detroit, Michigan

LAMBERSON, J. O., U.S. Environmental Protection Agency, Mark O. Hatfield Marine Science Center, Newport, Oregon

LANDRUM, PETER F., National Oceanic and Atmospheric Administration, Great Lakes Environmental Research Laboratory, Ann Arbor, Michigan

LECH, JOHN J., Department of Pharmacology and Toxicology, Medical College of Wisconsin, Milwaukee, Wisconsin

LEE, RICHARD F., Skidaway Institute of Oceanography, Savannah, Georgia

MAC, MICHAEL J., United States Fish and Wildlife Service, National Fisheries Center—Great Lakes, Ann Arbor, Michigan

MCNAUGHT, DONALD C., Department of Ecology and Behavioral Biology, University of Minnesota, St. Paul, Minnesota

MELANCON, MARK J., Department of Pharmacology and Toxicology, Medical College of Wisconsin, Milwaukee, Wisconsin

MOUNT, DONALD I., U.S. Environmental Protection Agency, Environmental Research Laboratory—Duluth, Duluth, Minnesota

NALEPA, THOMAS F., National Oceanic and Atmospheric Administration, Great Lakes Environmental Research Laboratory, Ann Arbor, Michigan

PARSONS, T. R., Department of Oceanography, University of British Columbia, Vancouver, British Columbia, Canada

PLAFKIN, JAMES L., U.S. Environmental Protection Agency, Monitoring and Data Support Division Office of Water Service, Washington, D.C.

RHEE, G-YULL, Wadsworth Center for Labs and Research, New York State Department of Health and School of Public Health Sciences, State University of New York at Albany, Albany, New York

RITACCO, P. J., Marine Ecosystems Research Laboratory, University of Rhode Island, Narragansett, Rhode Island

ROBERTSON, ANDREW, National Marine Pollution Program Office, National Oceanic and Atmospheric Administration, Rockville, Maryland

SICKO-GOAD, LINDA, Great Lakes-Research Division, The University of Michigan, Ann Arbor, Michigan

SPRAGUE, J. B., Department of Zoology, University of Guelph, Guelph, Ontario, Canada

STOERMER, EUGENE F., Great Lakes Research Division, The University of Michigan, Ann Arbor, Michigan

SULLIVAN, BARBARA K., Marine Ecosystems Research Laboratory, University of Rhode Island, Narragansett, Rhode Island

SWARTZ, R. C., U.S. Environmental Protection Agency, Mark O. Hatfield Marine Science Center, Newport, Oregon

VERSTEEG, DONALD J., Pesticide Research Center, Department of Fisheries and Wildlife, Center for Environmental Toxicology, Michigan State University, East Lansing, Michigan

WAKEHAM, STUART G., Marine Biological Laboratory and Woods Hole Oceanographic Institution, Woods Hole, Massachusetts

WARWICK*, W. F., National Water Research Institute, Winnipeg, Manitoba, Canada

WHITE, DAVID S., Great Lakes Research Division, University of Michigan, Ann Arbor, Michigan

WILLFORD**, WAYNE A., United States Fish and Wildlife Service, Ann Arbor, Michigan

**Present affiliation*: National Hydrological Research Centre, Saskatoon, Saskatchewan, Canada
**U.S. Environmental Protection Agency, Great Lakes National Program Office, Chicago, Illinois

# INTRODUCTION TO THE SERIES

The deterioration of environmental quality, which began when mankind first congregated into villages, has existed as a serious problem since the industrial revolution. In the second half of the twentieth century, under the ever-increasing impacts of exponentially growing population and of industrializing society, environmental contamination of the air, water, soil, and food has become a threat to the continued existence of many plant and animal communities of various ecosystems and may ultimately threaten the very survival of the human race. Understandably, many scientific, industrial, and governmental communities have recently committed large resources of money and human power to the problems of environmental pollution and pollution abatement by effective control measures.

*Advances in Environmental Sciences and Technology* deals with creative reviews and critical assessments of all studies pertaining to the quality of the environment and to the technology of its conservation. The volumes published in the series are expected to service several objectives: (1) stimulate interdisciplinary cooperation and understanding among the environmental scientists; (2) provide the scientists with a periodic overview of environmental developments that are of general concern or that are of relevance to their own work or interests; (3) provide the graduate student with a critical assessment of past accomplishment which may help stimulate him or her toward the career opportunities in this vital area; and (4) provide the research manager and the legislative or administrative official with an assured awareness of newly developing research work on the critical pollutants, and with the background information important to their responsibility.

As the skills and techniques of many scientific disciplines are brought to bear on the fundamental and applied aspects of the environmental issues, there is a heightened need to draw together the numerous threads and to present a coherent picture of the various research endeavors. This need and the recent tremendous growth in the field of environmental studies have clearly made some editorial adjustments necessary. Apart from the changes in style and format, each future volume in the series will focus on one particular theme or timely topic, starting with Volume 12. The author(s) of each pertinent section will be expected to critically review the literature and the most important recent developments in the particular field; to critically evaluate new concepts, methods, and data; and to focus attention on important unresolved or controversial questions and on probable future trends. Monographs embodying the results of unusually extensive and well-rounded investigations will also be published in the series. The net result of the new editorial policy should be more integrative and comprehensive volumes on key environmental issues and pollutants. Indeed, the development of realistic standards of environmental quality for many pollutants often entails such a holistic treatment.

JEROME O. NRIAGU, Series Editor

# PREFACE

The Laurentian Great Lakes constitute the world's largest single supply of fresh water. Twenty percent of the world's surface freshwater resource is held in these lakes. Approximately 20 million of the 37 million people living in the Great Lakes drainage basin obtain their drinking water from these lakes. Because of the abundance of water and essential minerals (e.g., iron ore, limestone, gypsum, salt), a major proportion of Canadian (50%) and U.S. (20%) industries are located in the drainage basin. Agriculture is an important component of the region's economy: fertile soils and a mild climate make the region ideally suited for food production. Recreational use is another significant source of income to the region, with over a billion dollars generated annually by people utilizing the lakes for swimming, boating, and fishing. The Great Lakes support a valuable commercial fishery in addition to the sport fishery.

As early as the mid-1800s, the character of the lakes began to change as overfishing and intensifying human activity began to affect various components of the Great Lakes ecosystem. In 1909, Canada and the United States signed the Boundary Water Treaty to prevent disputes between the two countries and to settle questions regarding the chemical, biological, and physical integrity of the lakes. Despite such good intentions, environmental degradation of the lakes continued. By the 1960s, the Great Lakes were in serious trouble as a consequence of eutrophication, overfishing, and the widespread presence of toxic substances such as DDT and mercury. To restore the lakes, Canada and the United States negotiated the Great Lakes Water Quality Agreement, signing it in 1972. Many remedial actions were implemented to improve the quality of Great Lakes waters. In 1978, Canada and the United States signed a revised and improved Great Lakes Water Quality Agreement pledging to "restore and maintain the chemical, physical, and biological integrity of the Great Lakes ecosystem."

Although many goals of this agreement have been met, the presence of toxic substances in the Great Lakes continues to be of concern to the International Joint Commission (including its boards), scientists, lake managers, and the general public. More than 1000 chemicals of potential concern have been detected in Great Lakes waters. Some of these chemicals are by-products of industrial activity and enter the lakes from the various industries located along the shoreline. Other chemicals are used in agriculture and enter the lakes with river inflow and runoff. Urban areas and the atmosphere are two other major sources of toxic substances. Concentrations of toxic substances are especially high in harbor and industrial areas: 38 such locations have been identified as areas of concern by the Water Quality Board of the International Joint Commission (IJC). Fish consumption advisories for the protection of human health are in effect for 31 of these areas.

There is widespread concern about the potential threat toxic substances pose to organisms living in the lakes and to terrestrial organisms, including humans, which utilize Great Lakes resources. Some toxic substances interfere with the reproductive success of and contribute to congenital abnormalities in birds. Other toxic substances have been implicated in the development of neoplastic tumors in bottom-dwelling fish. The failure of lake trout to reproduce in most regions of the Great Lakes is thought by many to be due to toxic substances. Studies have shown that the human population living within the Great Lakes basin is exposed to and accumulates appreciably more toxic-chemical body burdens than populations from other similarly large regions of Canada and the United States. Such differences appear to be due to the fact that toxic contaminant concentrations are comparatively high in Great Lakes food products, including those from the lakes themselves and from the drainage basin. There is concern that some of these toxic substances may adversely affect brain development in human infants and promote cancers in juveniles and adults.

In 1982, the Science Advisory Board (SAB) of the IJC concluded that research efforts were insufficient to adequately address the potential effects of persistent toxic substances on the health of Great Lakes aquatic communities: similar conclusions were made by the Water Quality Board of the IJC (1985), the National Research Council of the United States, and the Royal Council of Canada (1985). In 1983, the SAB established the Health of Aquatic Communities Task Force to synthesize information on the effects of toxic substances on the Great Lakes aquatic community, to explore improved methods for assessing effects of toxic substances on the biota, and to develop recommendations for further research. Members of the Task Force included M. S. Evans, Chairperson, Great Lakes Research Division, The University of Michigan, Ann Arbor, Michigan; J. E. Gannon, Secretary, International Joint Commission, Windsor, Ontario; D. C. McNaught, Department of Ecology and Behavioral Biology, University of Minnesota, St. Paul, Minnesota; and C. M. Fetterolf, Jr., Great Lakes Fishery Commission, Ann Arbor, Michigan.

The culmination of the Task Force's activities was a symposium and workshop held June 18–21, 1985 as part of the annual meeting of the American Society of Limnology and Oceanography. After this meeting, participants prepared the papers that form the basis of this book.

The book begins with a commentary on the protection of aquatic ecosystems. Such considerations include politics and economics in addition to science.

Part 1 consists of six chapters dealing with Great Lakes biota (phytoplankton, zooplankton, benthos, fish, birds and mammals, and humans) and the effects of toxic substances on their well-being. This is an exceedingly difficult issue. These papers show that a multitude of experimental studies have been conducted investigating the effects of toxic substances on various components of aquatic ecosystems. Furthermore, many components (especially fish) of the Great Lakes aquatic community have been intensively studied over the last several decades. However, despite this apparent surfeit of knowledge, researchers have a very poor understanding of how toxic substances have affected these six basic components of Great Lakes communities. There are several reasons for this.

1. Bioassays typically consider only one toxicant and are conducted under otherwise ideal conditions for the test organism. However, in the Great Lakes, organisms typically are exposed to a multitude of toxic substances and to various environmental stresses (e.g., temperature, food limitation). Only a limited number of species have been tested in the laboratory: many of these species are not components of the Great Lakes aquatic community. Thus, it currently is not possible to extrapolate the results of most experimental studies conducted in the laboratory to natural Great Lakes environments where organisms of varying sensitivities are exposed to a variety of toxic substances and environmental stresses.

2. The Great Lakes are highly perturbed systems, having been adversely affected by eutrophication, overfishing, and habitat alterations. Such factors mask changes in aquatic communities resulting from toxic substance stresses. This is most clearly illustrated for fish, the group of organisms for which there is the most detailed and long-term record of species abundance and composition. However, even these data are limited, having, for the most part, been collected in offshore waters and away from regions where toxic concentrations are relatively high.

3. Environmental monitoring studies have not been sufficiently well designed to investigate the effects of toxic substances on aquatic communities. Most components of the aquatic community have been sampled on an infrequent basis. Consistent sampling methods and locations have not been used, and concentrations of toxic substances have not been determined. Relatively few studies have been conducted in areas where toxic concentrations are relatively high.

4. Even where toxic substances have been implicated as having had significant effects on the biota, the specific causal factor has been difficult to ascertain. The confusion in this matter is well illustrated by the early bird and mammal studies.

Despite these uncertainties, it is generally thought that toxic substances have the potential to adversely affect organisms in areas where toxic chemical concentrations are relatively high and those long-lived organisms (including humans) that bioaccumulate certain toxic substances. Impaired reproduction in birds and lake trout, congenital deformities in birds and tumors in fish, and the bioaccumulation of substances known to promote tumors and impair learning are all issues of concern for both the aquatic and human communities.

Part 2 discusses better approaches for investigating the effects of toxic substances on Great Lakes communities. Current approaches, based in large measure on single-species–single-substance bioassays, are insufficient for ensuring the protection of ecosystem structure and function. The toxics issue is complex, and a variety of approaches must be used to investigate effects on ecosystem structure and function.

Bioassays using naturally contaminated water or sediments allow researchers to investigate the toxicity of natural systems. A number of organisms and test designs can be used. For example, the zooplankter *Ceriodaphnia* is being used to investigate toxicity along spatial gradients, e.g., a river. Amphipods are being used to investigate the toxicity of sediments: if indigenous amphipod species are used, field studies can be used to validate the results of laboratory bioassays. Finally, the failure of lake trout to successfully reproduce in the Great Lakes can be investigated through bioassay experiments. Survival of eggs and larvae is quantified using water collected from various regions of the Great Lakes and at various times.

Bioassay experiments have some limitations: only species that can be maintained in the laboratory can be tested, and such studies are time-consuming to conduct. Biochemical approaches are another way to investigate stress resulting from exposure to toxic substances. In addition, biochemical approaches provide insight into how toxic substances affect organism metabolism.

Morphological deformities in chironomids, as an indicator, have great potential for investigating toxicity. For example, by investigating the incidence of deformities over a given region of a lake or river, scientists may be able to delineate the spatial extent of toxicity. Moreover, by conducting paleolimnological studies, researchers can investigate the historical record of toxic stress in a region.

More information is required on how toxic substances can affect ecosystem structure and function. Bioassays and biochemical approaches alone will not provide information on this issue. One useful approach for investigating such interactions is microcosms or mesocosm studies. Three chapters in Part 2 show how such approaches can provide useful information on the complex effects of organic loading, nutrients, hydrocarbons, and metals on the different components of marine and aquatic ecosystems. Such studies illustrate the difficulties in evaluating the effects of toxic substances on aquatic communities. Modeling approaches, which make use of field and experimental data, provide insight into the fate and effect of toxic substances on aquatic ecosystems. Modeling enables researchers to test ideas and develop new hypotheses.

Good statistical design is an important component of all studies investigating the effects and causes of toxic stress. This fact is not sufficiently appreciated during the design of field and laboratory studies.

Although this book focuses primarily on aquatic organisms (including birds and mammals), human health is considered in Parts 1 (bioaccumulation of toxic compounds) and 2 (the effect of toxic substances on infants). Part 2 contains a chapter discussing new methodologies for assessing the effects of prenatal toxic exposure on cognitive functioning in humans.

Part 2 also contains a chapter dealing with ecosystem (rather than chemical) objectives for the Great Lakes. Both the 1972 and 1978 Great Lakes Water Quality Agreements contain specific objectives on the discharge and concentration of various pollutants in Great Lakes waters. More recently, the Aquatic Ecosystem Objective Committee has begun to develop ecosystem objectives based on biological indicators of ecosystem health. The first ecosystem objective is based on the lake trout, an indicator for cold-water, oligotrophic communities in the Great Lakes.

Following the formal symposium, attendees participated in a one-day workshop organized by aquatic group, that is, phytoplankton, zooplankton, benthos, fish, birds and mammals; human health formed a sixth group while modeling and statistics formed a seventh group. Participants were asked to address three questions:

1. What changes have occurred in the five basic components of the Great Lakes aquatic community?
2. What role have toxic substances had in effecting such changes?
3. How can researchers better investigate the effects of toxic substances on these five components of the aquatic community?

These three questions were not appropriate to the human health group and the modeling and statistics group. Accordingly, the latter group modified these questions and addressed eight basic questions regarding toxic contaminants in the Great Lakes.

Part 3 presents the results of these deliberations with the exception of the human health group: recommendations from this group are reported as part of Humphrey's chapter in Part 1. All groups suggest a broader-based, more integrative approach for investigating toxic stress effects on aquatic communities. Many of the approaches are described in Part 2. Each group identified specific areas of research need and general approaches for investigating the effects of toxic substances on the health of the Great Lakes ecosystem. Researchers recognized the fact that the toxic issue is complex and requires a stronger research effort. Since many toxic compounds persist in the Great Lakes long after their inputs are reduced, it is important that researchers obtain a better understanding of their ecosystem effects.

MARLENE S. EVANS

*Ann Arbor, Michigan*
*January 1988*

# ACKNOWLEDGMENTS

This book is based on a symposium workshop organized by the Health of Aquatic Communities Task Force and sponsored by the Science Advisory Board of the International Joint Commission (SAB, IJC) and the American Society of Limnology and Oceanography (ASLO). Sponsorship of the symposium/workshop does not necessarily imply that the findings and conclusions reached by the participants are endorsed by the sponsors.

Many people provided special assistance to the task force over its 1983–1986 term including D. Mount (SAB and U.S. Environmental Protection Agency), F. K. Fahmy (IJC, Windsor), R. Thomas (SAB and Canada Centre for Inland Waters), and G. LaRoche (SAB and Triple-V, Vit-Vitaille-Viron, Inc.). J. Fitchko (IEC Beak Consultants, Ltd.) prepared a report which assisted the task force in developing the symposium and workshop. W. Glooschenko (Canada Centre for Inland Waters), J. Hobbie (Marine Biological Laboratory), P. Landrum (Great Lakes Environmental Research Laboratory, National Oceanic and Atmospheric Administration), A. Robertson (National Oceanic and Atmospheric Administration), and J. Sprague (University of Guelph) assisted the task force in organizing the symposium and workshop. C. Cooper (Michigan State University) provided the task force and workshop leaders with useful information on organizing and conducting a workshop. P. Brezonik (Chairman, ASLO program, University of Minnesota) endorsed the inclusion of the symposium and workshop in the summer meeting of the American Society of Limnology and Oceanography.

Symposium manuscripts received two peer reviews. The contributions of D. Bellinger, J. A. Bowers, U. Borgmann, T. Fontaine, W. Gardner, A. Gilman, M. James, J. Lamberson, R. A. Laskowski-Hoke, P. Landrum, R. F. Lee, J. P. Ludwig, D. M. Passino, J. F. Payne, R. Lee, A. Robertson, R. Ryder, J. Skea, R. B. Spies, E. F. Stoermer, F. B. Taub, and several anonymous reviewers are gratefully acknowledged. Workshop manuscripts were reviewed by workshop participants: the names and affiliations of these participants are given in the various workshop papers and are not repeated here. In addition, the contributions of G. Becking (World Health Organization), N. Chernoff (Health Effects Research Laboratory), E. Nestmann (Environmental Health Directorate), and D. Rapport (Statistics Canada) are acknowledged.

Special appreciation is extended to J. O. Nriagu (National Water Research Institute), editor of the *Advances in Science and Technology* series, for the support and guidance which he provided during the development of this book.

M.S.E.

# CONTENTS

## PART THREE SUMMARY AND RECOMMENDATIONS FOR FUTURE RESEARCH

# 1

# POLITICS, ECONOMICS, SCIENCE—GOING BEYOND DISCIPLINARY BOUNDARIES TO PROTECT AQUATIC ECOSYSTEMS

***John Cairns, Jr.***

*Department of Biology and University Center for Environmental Studies, Virginia Polytechnic Institute and State University, Blacksburg, Virginia 24061*

## 1. INTRODUCTION

> It is the mark of an instructed mind to rest satisfied with the degree of precision which the nature of the subject permits and not to seek an exactness where only an approximation of the truth is possible.
>
> —Aristotle

> Ecological awareness leads to questioning of goals: This threatens the performance of some old established roles. So to raise the human species from the level of subsistence We have to overcome Covert Political Resistance. So we should be propagating, without shadow of apology, A Scientific Discipline of Poleconecology.
>
> —Boulding, 1972

The responsibility of the scientific community is to determine the probability and degree of harm to Great Lakes ecosystems from various stresses, both anthropogenic and natural and present and potential, but not to make a determination of what is safe and what is not. Hazard evaluation or risk analysis is a probablistic determination requiring scientific evidence; such a determination is a perfectly appropriate activity for scientists. The estimation of safety, however, is a political/social determination in which a cost–benefit analysis is made, either formally or informally, but nearly always with a highly subjective nonquantifiable component. This is not an appropriate activity for scientists, although individuals may wish to participate as citizens in the determination of safety. Scientists have often failed to make a sharp distinction between those two processes, and, as a consequence, they have diminished their credibility. On the other hand, scientists must give serious attention to the ways their information will be used so that it can be intelligible to nonscientists and readily communicated to them. It is also important to remember (as the quote from Aristotle is intended to emphasize) that the level of detail and the amount of evidence should be related to the nature of the decision being made and that precision for its own sake is not as well received by the general public and the decision makers as it is likely to be among the scientific community. This is particularly true if gathering additional evidence or increasing the precision of an estimate takes substantial amounts of time and thus delays the decision or encourages making a decision based on less appropriate information.

The splendid chapter by Regier and MacKay (1984) on the technical basis for the goals of the Canada–U.S. Water Quality Agreement and the prospects of its effectiveness in the future, published in the *Review of the Great Lakes Water Quality Agreement,* provides a superb overview of the general problems with a technical basis. The section on toxic chemicals, particularly persistent ones, and the section on ecosystem approach deserve particular attention, as do the issues for the future and final comments. One comment deserves special attention:

> The ecosystem approach should not be confused with a simplified macroscopic overview, nor with a nonsystemic compendium of all relevant data, nor with old-fashioned comprehensive planning, nor with mysticism. In practice the ecosystem approach as it relates to research, is coming to mean a type of study that explains phenomena of the Great Lakes at the level of interest of, and with some direct relevance to, the combined mandates of the binational governance institutions. From this perspective, we may expect to have a sufficiently developed technical understanding of the Great Lakes ecosystems when we can explicitly interrelate the social demands and ecological effects of uses that are administered under the various binational treaties, conventions, agreements, and protocols.

Although I will not dwell on this theme elsewhere in this discussion, it should be a primary concern that the assumptions, hypotheses, and evidence of specialists (e.g., Great Lakes limnologists) be intelligible to those in other disciplines. It is equally important for each discipline to understand what other disciplines expect from it. Section 4.1, ''Just Give Me a Freeze-Dried, Talking Fish on a Stick,'' is intended to illustrate this point. Obviously, no rational person would take this request seriously. Nevertheless, the desire for a reliable, simple, inexpensive, uncomplicated method deserves attention. The complexities that delight specialists are annoying to laymen. It is important to remember that specialists in one field are laymen in all others. Yet, on large problems, all must work together effectively, which means communication without jargon.

## 2. POLITICAL PROBLEMS IN THE PROTECTION OF AQUATIC ECOSYSTEMS

### 2.1. Prescriptive Laws and Regulations

One result of the failure in communication among disciplines is the detailed prescriptive way in which laws and regulations are written. Both state and federal laws and regulations are often so detailed as to obviate sincere and well-intended efforts by both the regulated and the regulator to solve even simple problems in obvious ways. Adversarial confrontations pitting the regulated and the regulators against each other in ways that prevent or obstruct cooperation toward obvious and effective solutions are common in the United States. As a consequence, both factions become embroiled in details that are generally unproductive, and time that should be spent on the basics and the important issues is lost or so diminished that it is worthless. Some case histories of problems with prescriptive legislation are cited and discussed in Cairns and Pratt (1986).

How we got into this unfortunate situation is relatively easy to understand, but it is extremely difficult to see a way out. Under optimal circumstances, all parties would be working vigorously and effectively toward the common goal. It is essential that there be a reasonable degree of mutual trust and respect to do this. In the absence of mutual trust and respect, the regulators produce highly specific tasks to be done in specific ways at specific times, and so on. Because the basis for the working relationship is law and not science, the specifications are drawn by lawyers; the scientists, engineers, and other environmental professionals struggle with the regulations as best they can.

Different disciplines not accustomed to working together can probably do this best on ''neutral ground,'' through an organization not dominated by a particular discipline, such as a consortium. The function of this organization is to furnish information that will influence decisions but not do so in a predetermined or biased way. In some cases, these institutes or consortia could be supported by assessing the industries causing the environmental problems a fee for each unit of production. The Florida Phosphate Institute is financed by a fee assessed by the State of Florida. Despite this, the institute should not be dominated by representatives of the industry furnishing the funding, as is all too often the case. The assimilative-capacity fee is discussed in more detail later.

My recommendations for more cooperation are made with full recognition that environmental protection might well be diminished. This will almost surely happen if the scientific integrity of the research is markedly diminished. This can happen even if the consortium, institute, and so on appears to be under state or some other form of ''outside'' control. A few warning signals that scientific objectivity has been lost are:

1. The board of directors is dominated by members from industry.
2. The organization suppresses ''unfavorable'' results. The most effective way to do this is to pay only for work accomplished and ''approved.'' That is, no money is advanced, and payment is only for expenditures incurred. Since conditions of awarding grants and contracts are often a matter of public record, this important point is easily determined.
3. The organization does not use a peer-review system for research proposals that would place major reliance on outside review by nationally known scientists. ''Reviews'' are carried out by organization scientists under strong organizational control.
4. Most of the ''research'' is ''published'' in ''in-house'' documents rather than peer-reviewed professional journals with national stature.
5. The grant or contract contains restrictions on the freedom of the principal investigator (and others) to publish the results of the research.
6. There is clearly a ''good old boy'' relationship between members of the

organization and some of those submitting proposals. That is, the organization need not fear that unfavorable research results will be obtained.

7. Final approval of funding for scientific investigations is entirely in the hands of nonscientists. The caveat should be added that it can be equally dangerous to leave the decision entirely to scientists who may pursue their own theoretical objectives without regard to the type and amount of information needed to make an informed decision.

The major problem with such organizations is that they give the public the impression that the research is scientific when it is not. When they purport to "protect the environment," one can be nearly certain that this is not happening. There are many examples of this, but stating the names would surely elicit lawsuits, and documenting the assertions would require more space than is available in this discussion.

## 2.2. Political Barriers in Academe to the Study of Systems

A large ecosystem, such as any one of the Great Lakes, cannot be effectively studied *as a system* by any individual, however skillful and hard-working. Only a team can effectively make such a study, as is abundantly evident from the Hubbard Brook studies that were carefully orchestrated by Borman and Likens (1979). However, despite repeated calls by Odum and others (Odum, 1981), the reductionist approach usually prevails over the integrative approach. The many benefits of the reductionist approach are so evident that they need no defense in this discussion. However, the integrative approach, although frequently supported by the academic community in words, is not well supported by actions. Consider the problems of an untenured assistant professor in most biology departments that require publications for tenure and promotion to associate professor. Single-authored papers are given substantial weight by most tenure and promotion committees, whereas publications with multiple authors are generally discounted substantially, even if the candidate is first author. Should the candidate be in any other position of authorship, the publication is better than none at all, but not much. Physicists seem to have less problems with multiple-authored papers than biologists, perhaps because there is abundant evidence that in many areas of physics no one would believe that a single author had sufficient competence to carry out all the necessary data gathering. Why, then, is there such a stigma in ecology when clearly the same principle applies?

It is also worth noting that the size of grants affects the number of people who can work together. Most agencies do not support large multidisciplinary studies.

## 2.3. Episodic Events

Very often episodic events, such as severe droughts or major catastrophic spills, are so widely spaced in time that administrative personnel in a particular region may change between events. Therefore, no bureaucratic memory or experience remains to ensure an appropriate response. As a consequence, even though the scientific database may increase and the ability to predict the probable degree of harm to an ecosystem with precision may improve markedly over a period of time, the organizational response will probably not improve proportionately because those with direct experience in coping with severe episodic events will have left the organization. There are ways to offset this both in hiring and in simulating response to these events on a regular basis so that personnel are better prepared to cope with them.

## 3. ECONOMIC PROBLEMS IN THE PROTECTION OF AQUATIC ECOSYSTEMS

Substantial amounts of research are still needed, including validation of some hypotheses on the environmental fate and effects of persistent chemicals in the Great Lakes. Financing this research means substantial new allocations for this from both the United States and Canada (e.g., Sadar, 1984). These appropriations appear unlikely, given the present attempts to reduce the budget deficit in the United States and the recent cutting of environmental programs in Canada. Potential sources of substantial amounts of money are industry, including agrobusiness, and, to a lesser extent, municipalities that are either point-source or non-point-source dischargers into the Great Lakes or tributaries. A substantial fraction of the money available for environmental matters is now spent in nonproductive ways, such as legal battles and standard waste treatment requirements that are very expensive but often produce negligible or nondemonstrable biological benefits. It is difficult for industry to use alternative forms of waste treatment that might produce equally beneficial results at a lower cost. If some of these nonproductive expenditures were made available to ecologists, much could be accomplished. In short, the scientific and engineering communities in academe and in the private sector would be encouraged to work cooperatively toward a common objective, such as developing a scientifically justifiable quality control system for the Great Lakes ecosystems.

Although Sabine (1985) demonstrates serious errors in the scientific literature and suggests the possibility of a significantly greater unreported error, it is quite clear that he believes the number of errors in articles, reports, and so on that have not been given an impartial objective review are clearly much higher than in those that have. Therefore, merely getting the academic community involved in a much larger way in the process of environmental quality control should have remarkable benefits in the quality and reliability of the information.

Elsewhere (Cairns, 1985), I have called attention to the parallels between the agricultural revolution and the presently occurring environmental revolution. The agricultural revolution began because the unmanaged environment could not produce food in either the quality or quantity that society desired. While there are still hunger problems on a global basis, it is an indisputable fact that in many countries a relatively few farmers are feeding very large numbers of people. In short, managing the environment has not solved all the food problems, but it has made possible an enormous improvement in human nutrition. The environmental revolution is developing because the unmanaged environment is incapable of assimilating societal wastes without being placed in disequilibrium. Although the words "managed environment" will undoubtedly raise the hackles of a wide range of lay and scientific people, *management* should not be regarded as a pejorative term. The prime objective of the environmental revolution will be to manage societal wastes so that the receiving systems are not overloaded and placed into disequilibrium. This means developing a predictive algorithm on the environmental response threshold at all levels of biological organization and simultaneously establishing an error-control feedback loop to show when the predictive model is in error. A striking difference between the two revolutions, not noted in the earlier paper (Cairns, 1985), is that the agricultural revolution occurred primarily on private land whereas the environmental revolution will occur primarily on *the common grounds,* as defined by Hardin (1968). This will make a vast difference in the financing of the two revolutions. In the agricultural revolution on private holdings, skillful investment of funds produced clearly discernible benefits to the investor. In contrast, the environmental revolution will benefit all society.

For all society to benefit, integrated resource management is mandatory. The primary goal should be to protect the health of the resource (e.g., the Great Lakes). However, regulatory measures often place primary emphasis on the performance of waste treatment systems rather than directly on the ecosystem itself. If industries were charged according to the percent of receiving-system assimilative-capacity use, gross adverse biological effects would probably be eliminated quickly (for a substantive debate on this subject, see Campbell, 1981, and response by Cairns, 1981). Discharges producing no measurable adverse effects would not be eliminated. However, they would be reduced because the use charge would be based on the nondegrading portion of the assimilative capacity used. Of course, zero discharge will probably never be cost-effective for an industry, and certainly is not now. It is often cheaper to hire lawyers under the present system than to improve waste treatment. The alternative of technology-based standards has also not proven effective (Graber, 1983).

The idea of environmental-use taxes is not new. An example of such a scheme is mentioned in Dales (1970). The idea of setting limits, but using both renewable and nonrenewable natural resources, is discussed in Boulding (1968). An early discussion of some of the difficulties of calculating the total value of services performed by the environment can be found in Kneese et al. (1970). However, the fields of hazard evaluation and ecological risk analysis provide a fairly accurate, verifiable estimate of the conditions under which the probability of harm is very low. Perhaps having unsuccessfully tried technology-based standards (e.g., Best Applicable Technology), we may wish to reexamine hazard evaluation and ecological risk analysis.

## 4. SCIENTIFIC PROBLEMS IN THE PROTECTION OF AQUATIC ECOSYSTEMS

### 4.1. Just Give Me a Freeze-Dried Talking Fish on a Stick

Some months ago, I attended an interdisciplinary meeting where problems of pollution were discussed. One of the nonecologists present plaintively but humorously asked, "Why can't you environmental toxicologists just give us a freeze-dried talking fish on a stick?" This fish could be inserted in any aquatic ecosystem where pollutional problems were suspected, where it would immediately inflate, determine the biological condition of the water, and give the person holding the stick the answer verbally. If one vigorously suppresses the automatic impulse to explain the complexities of ecology and toxicology to this "ignorant" individual, it becomes abundantly clear that those two simple sentences contain a beautiful description of the kinds of methods the rest of the world wants from ecologists and environmental toxicologists. The following are my interpretations of these characteristics:

1. The method is simple and the apparatus portable.
2. The all-purpose method will work well in any aquatic ecosystem.
3. The method can be used by a variety of individuals with little or no professional training.
4. Results are obtained immediately and communicated in generally understood terms.
5. Because of the stick, investigators need not expose themselves unnecessarily to the potentially harmful material.
6. The device is inexpensive and readily available. (This is less apparent from the two-sentence hypothesis, but, on checking with the originator of the statement, I found that my assumption is correct.)

Although the possibility always exists that some creative individual who does not know it "can't be done" will develop the freeze-dried, talking fish on a stick, for the foreseeable future we must live with the existing methodology. When developing or improving methods to assess the probable or actual damage caused by persistent chemicals, ecologists would do well to keep the above criteria in mind, particularly in terms of communicability. Scientists could, of course, adopt the "trust us" attitude of some professionals, which assumes that the matter is too complicated to explain to laymen and therefore no attempt should be made to do so. This has not been a smashing success in the past for scientists dealing with environmental problems and is unlikely to be so in the future. What will probably happen, if the "trust us" attitude prevails, is that unqualified persons with simplistic solutions will take over. Of course, meeting the needs (which are not well understood) of others is just as difficult as meeting the needs within a discipline, even though the requirements are different. It demands an investment of time and energy in a period when it is impossible to keep up with all the literature within a single discipline. Nevertheless, it is abundantly clear that successful quality control for the Great Lakes, particularly with regard to persistent chemicals, must involve politics, economics, society, and science.

## 4.2. Toxicity Testing for Great Lakes Ecosystems

It seems quite reasonable to follow the usual sequence of toxicity testing for ecosystems in general when determining risks to the Great Lakes ecosystems. The Great Lakes are unique ecosystems and, for a variety of reasons, deserve more than the usual level of protection. However, the evaluatory process itself is no different for the Great Lakes than for other aquatic ecosystems: (1) screening or range-finding tests, (2) predictive tests, (3) confirmatory or validation tests, and (4) monitoring. Specimens from the Great Lakes should be stored as recommended by Lewis (1984).

### *4.2.1. Screening or Range-Finding Tests*

The purpose of the screening or range-finding tests is to determine in a very crude way the short-term toxicological properties of the chemical and the priority that should be given to testing it. It is worth emphasizing that biological testing should be accompanied by chemical/physical testing that would provide information on the probable persistence of the chemical, whether or not it has a high octanol/water partition coefficient, whether it is likely to volatilize or be quickly transformed into another compound, and a variety of other information of this type. Screening tests are virtually always carried out in the laboratory.

### *4.2.2. Predictive Tests*

Predictive tests are also usually carried out in the laboratory but differ from the screening tests in a number of respects. They generally have a higher degree of environmental realism and they determine the response threshold with greater precision. While most screening tests use lethality as the sole end point, predictive tests are more likely to involve more subtle endpoints, such as growth, reproductive success, and behavior. The duration of exposure is likely to be longer, and the number of life history stages tested for a single species is likely to be greater. In addition, the number of species tested will almost certainly be greater than in screening or range-finding tests. In the predictive tests, renewal of the test solution either through continuous flow or replacement in batch processes is likely to occur more frequently than in screening tests. A number of additional differences are discussed in a variety of publications (e.g., Cairns et al., 1982; Westman, 1985).

### *4.2.3. Confirmatory or Validation Tests*

Confirmatory or validation tests should be carried out in the ecosystem at risk or in a surrogate of the ecosystem such as a microcosm or mesocosm (e.g., Odum, 1984). It seems almost platitudinous to state that all predictions regarding the toxicity of potentially hazardous chemicals based on laboratory testing should be confirmed in the ecosystem at risk or a surrogate. However, as recently as 1981, the Committee for Testing the Effects of Chemicals on Ecosystems of the National Research Council called attention to this need because validation tests were so uncommon (National Research Council, 1981). A description of some of the means of validating the predictions based on laboratory evidence can be found in the National Research Council volume or in Odum (1984).

#### *4.2.4. Monitoring*

Monitoring can be carried out with biological, chemical, or physical endpoints. The purpose of monitoring is to take quick corrective action when established quality control conditions are not being met. A number of methods are available (e.g., Cairns et al., 1982; Cairns and van der Schalie, 1980). Hellawell (1978) uses the word *monitoring* in the sense of obligation to take corrective action if quality control conditions are not being met. He uses the word *surveillance* for systematic or routine testing in which no corrective action is mandatory and *survey* as an exercise in which a set of standardized observations is taken from a site. I frequently hear the term *biological monitoring* used, particularly in the United States, to cover all three types of activities defined by Hellawell. It is unfortunate that we fail to make the distinctions that Hellawell has made so well. His definitions follow.

*Survey.* An exercise in which a set of standardized observations (or replicate samples) is taken from a station (or stations) within a short period of time to furnish qualitative or quantitative descriptive data.

*Surveillance.* A continued program of surveys systematically undertaken to provide a series of observations in time.

*Monitoring.* Surveillance undertaken to ensure that previously formulated standards are being met.

### 4.3. Single Species versus Multispecies Toxicity Testing

Since they came into use as management tools in the late 1940s and early 1950s, almost all toxicity tests have been based on single species. Although this is now changing, only rarely does one encounter a multispecies toxicity test being used by an industry or a regulatory agency. The professional literature, however, which only a few years ago rarely had an article on multispecies testing, now has many such articles (e.g., Bishop et al., 1983). The theoretical importance of multispecies aggregations has been advanced by Odum (1953) for many years. At each level of biological organization, proceeding from subcellular through cell, tissue, whole organism, population, community, and ecosystem, new properties are added that cannot be effectively studied at lower levels of organization.

Even when multispecies tests are used, they are often at the end of a protocol (Dickson et al., 1979). There are compelling economic, as well as scientific, reasons for carrying out multispecies toxicity testing (Cairns, 1983). It is worth noting that as of the time this is being prepared, no formally endorsed standard methods (e.g., those of the American Society for Testing and Materials or in *Standard Methods for the Evaluation of Water and Wastewater*) for multispecies toxicity testing exist. However, a number of provisional methods appear in the symposium jointly sponsored by the Ecological Society of America and the Society for Environmental Toxicology and Chemistry on *Multispecies Toxicity Testing* (Cairns, 1985). Additional provisional methods can be found in Cairns et al. (1985) and Cairns and Pratt (1985).

## 4.4. Natural-System Validation of Predictions Based on Laboratory Evidence

Although the field of aquatic toxicity testing might be traced as far back as Aristotle (who transferred fish from one kind of water to another and observed them), its real development began shortly after World War II. Even so, it might appear rather odd that scientifically justifiable validation of predictions based on single-species laboratory tests has not been a major activity in the history of the field and is not carried out to any significant extent today, although much more attention is now being given to this problem. One reason may be that the assumptions of the field of toxicity testing are not explicitly stated, and another reason is that the people in the field have been so busy fighting emergency situations and disasters that the philosophical underpinnings did not receive the attention they deserved. In addition, checks could be made on the accuracy of laboratory toxicity tests in the sense of the ability to replicate them within a single laboratory or among a group of laboratories. Presumably, testers thought that if the numbers generated were accurate in the sense of being repeatable and the error quantifiable, the predictions based on them, even for much more complex systems, were sound. Furthermore, people who were comfortable making the laboratory tests were generally not comfortable in the field and vice versa.

The National Research Council (1981), reviewing methods for ecotoxicology, mentions that single-species tests would be of greatest value if used in combination with tests that can provide data on population interactions and ecosystem processes. More important, the executive summary states, ''Research and development should be directed toward designing and validating test systems and procedures that will detect changes in ecosystem and population attributes.'' Since the National Research Council intended only to examine the adequacy of the present strategy for testing for effects of chemicals on ecosystems, it did not provide a detailed discussion of the validation process. Since I consider toxicity tests to be one of the key elements in predicting the hazard to Great Lakes ecosystems of various chemicals appearing in them (or likely to be introduced into them), the question of the form validation should take deserves considerable attention. It is worth emphasizing that despite the drawbacks of toxicity tests, there is no instrument devised by man that will directly measure toxicity! Knowing chemical concentrations is essential, but these measurements, however precise, are not of themselves adequate predictors of toxicity. A series of commonly, and not so commonly, used predictors of toxicity follow with some consideration as to how they might be validated.

#### 4.4.1. **Ceriodaphnia** *(Mount and Norberg, 1984) and* **Daphnia** *(Ingersoll and Winner, 1982) Toxicity Tests Used in a Situation Where the Test Species Is Not Resident in the Ecosystem into Which the Waste Is Being Discharged or Introduced*

*Level 1.* The degree of biological organization is the same (i.e., single species in both field and laboratory), but the test species is not resident in the receiving system. Validation is carried out with only a single indigenous species. Level 1 validation should, at the very least, demonstrate that an important indigenous species in the receiving system has a response threshold reasonably close to or more tolerant than the test species. The importance could be determined on an ecological, recreational, or commercial basis. If the waste discharge is from an industry, validation might take the form of field studies of an important indigenous species. If the test is for a chemical of more widespread use, such as agricultural chemicals likely to enter as non-point-source discharges, one might carry out the validation in microcosms, mesocosms, or field enclosures with a reasonable array of species likely to be exposed. Presumably, the results would be more persuasive if endpoints other than lethality, such as reproductive behavior and success, growth, or respiration, were used.

*Level 2.* This is the same level of biological organization as Level 1, but more than one indigenous species is used for validating predictions based on toxicity tests carried out with a species not resident in the receiving system. Level 2 validation should demonstrate that an array of species tested individually from a variety of taxonomic groups and trophic levels indigenous to the receiving system do not suffer deleterious effects (e.g., lethality, reproductive, or growth) at no-effect concentrations predicted by the laboratory test. One might use a primary producer, a detritus processor, a grazer, and a carnivore for this purpose. If a point-source discharge is available, the validation in a natural system should not be difficult. If not, a surrogate system should be used.

*Level 3.* This level uses a nonresident test species, but confirmation of predictions of hazard involves attributes of either a community or ecosystem (i.e., a higher level of biological organization than single species). Level 3 validation should involve a process, such as energy flow, nutrient cycling, or some other process involving more than one species, as confirmation that the protection included processes at the higher degrees of biological organization than single species. Such a test will be more expensive than a single-species acute-toxicity test with lethality as an endpoint, but the information value should be proportionately higher since it applies to an entire system instead of a single species.

#### 4.4.2. *Use of an Indigenous Toxicity Testing Species Thought to Be "Representative" or "Sensitive"*

*Level 1.* Determination (in the natural system or a surrogate of the natural system) that the species responds the same way in a more complex natural system as it does under laboratory conditions. If there is an existing point-source discharge, the validation should be relatively simple (e.g., Cairns and Cherry, 1983). If no natural study site is available, a surrogate system should be used with adequate ecological complexity.

*Levels 2 and 3.* Demonstration similar to that required for the nonindigenous test species as described above.

### *4.4.3. Validation of Results from a Microcosm or Mesocosm Designed to Simulate Ecologically Important Characteristics of a Particular Ecosystem*

*Level 1.* Presumably, each microcosm or mesocosm would simulate one or more important attributes of one of the Great Lakes. Therefore, direct validation of a laboratory microcosm could easily be carried out in the area of a point-source discharge by using the same endpoints used in the laboratory microcosm tests. This is a determination of correspondence very similar to that used for single species, the only difference being that this is a higher level of biological organization or a more complex test system. If there is no contaminant entering this system and this has to be simulated, a field enclosure could be used in the particular lake for which the prediction has been made. As was the case in the earlier illustrative validations, if the correspondence between the laboratory results and the field results is good, the results of the laboratory tests can be considered validated. If there is an ecologically, as well as statistically, significant difference, the factors causing this difference should be identified. These factors should be included in the microcosm tests if at all possible, and new predictive tests should be run. At the very least, there should be some evidence that the explanation for the differences in results between the laboratory and natural system is credible.

If a field enclosure in a natural system is used and there is a point-source discharge or an area where the non-point-source runoff is higher than the rest of the lake, validation can be carried out as already described using the same endpoints in both the natural and the artificial systems. No control is needed when the point of the exercise is to compare results in an enclosed system with that of a natural system. If there is no entry of the chemical into the lake, provisional validation can probably be achieved by using a series of field enclosures in another part of the ecosystem. As was the case in earlier illustrations, if the correspondence is high, one can consider that an acceptable provisional validation has been produced. If the correspondence is not particularly good, some explanation for the cause of the differences should be developed, and some means of experimentally confirming that the factors identified were in fact causing the differences in results would be most helpful.

*Level 2.* As was the case in earlier illustrations of Level 2, endpoints characteristic of whatever level of biological organization is being studied, in addition to the ones actually used as endpoints in the laboratory system or field enclosure, should be used. In short, the prediction of ''safety'' for attributes other than the ones directly measured should be determined. Level 1 validation merely determines that the attributes measured are responding as predicted in a natural system.

For Level 2 validation, one should determine that attributes at the same level of biological organization other than those measured directly in the test system do not suffer deleterious effects in the natural system (see chart for explanation).

**LEVEL 2**

| Test system | Natural system |
|---|---|
| Microcosm | Community |
| Attributes | Attributes |
| Detritus processing | Detritus processing |
| Energy transfer | Energy transfer |

**LEVEL 3**

| Test system | Natural system |
|---|---|
| Microcosm | Single species<br>Community |
| Attributes | Attributes |
| Detritus processing | Reproductive success, tissue damage |
| Energy transfer | Rarefaction index, trophic relationships |

*Level 3.* As was the case with the preceding illustrations, Level 3 validation should show that levels of biological organization other than those for which endpoints were developed in the test system are protected. In a sense, this validation is intended to confirm or falsify the assumption or hypothesis that the endpoint chosen (or endpoints chosen) was such a key attribute of the natural system that all other important attributes, or a high percentage of them, were also protected.

## 5. CONCLUSIONS

The purpose of this chapter was to examine the effects of persistent chemicals on the Great Lakes ecosystems and the methods of measurement. This in itself is a very large task indeed.

Although thinking at the systems level has become increasingly popular and a holistic view is often espoused, actions do not always support the public statements. The reductionist approach, although mandatory in some situations, is counterproductive in others. Unless the right questions are asked, it is unlikely that either the information gathered or the protection of the system itself will prove adequate.

Even the most modest plan for protecting the Great Lakes ecosystems will involve considerable sums of money over an indefinite period of time. Of course, these amounts are trivial compared to the value of the resource being protected or the cost of rectifying errors. However, in an era of enormous federal budget deficits and other economic problems, it is unlikely that policy makers will pay for measurements just because a group of professionals accustomed to making these measurements think they are a good idea. What is interesting academically may not be useful for management purposes, and we must not attempt to supplant lost research dollars for purely theoretical work with management dollars designed for a specific goal. This is not to say that research funding will not become available to help in the management and protection of the Great Lakes, but rather that it will be focused on management needs rather than purely theoretical purposes. Boulding (1972) says it well:

> It's nice to be the drafter of a well-constructed plan,
> For spending lots of money for the betterment of Man,
> But Audits are a threat, for it is neither games nor fun
> To look at plans of yesteryear and ask, "What have we done?"
> And learning is unpleasant when we have to do it fast,
> So it's pleasanter to contemplate the future than the past.

## ACKNOWLEDGEMENTS

I am indebted to John Sprague and Carlos Fetterolf for useful comments and suggestions on the first draft of this manuscript. Darla Donald, Editorial Assistant in the University Center for Environmental Studies, prepared the final draft in the form requested by the publisher. The final draft of the manuscript was written at the Rocky Mountain Biological Laboratory during the summer of 1985 and was typed by Angela Miller. Revisions were made at Rocky Mountain Biological Laboratory during the summer of 1986 and were typed by Angela Miller.

## REFERENCES

Bishop, W. E., Caldwell, R. D., and Heidolph, B. B. (Eds.) (1983). *Aquatic Toxicology and Hazard Assessment,* STP 802. Am. Soc. Test. Mater., Philadelphia, PA.

Borman, F. H., and Likens, G. E. (1979). *Pattern and Process in a Forested Ecosystem.* Springer-Verlag, New York.

Boulding, K. E. (1968). *Beyond Economics.* Univ. of Michigan Press, Ann Arbor.

Boulding, K. E. (1972). A ballad of ecological awareness. In M. T. Farrar and J. P. Milton (Eds.). *The Careless Technology,* Nat. Hist. Press, Doubleday, New York, pp. 3, 955.

Cairns, J., Jr. (1981). Discussion of: A critique of assimilative capacity. *J. Water Pollut. Control Fed.* **53**(11), 1653–1655.

Cairns, J., Jr. (1983). The case for simultaneous toxicity testing at different levels of biological organization. *ASTM Spec. Tech. Publ.* STP 802, 111–127.

Cairns, J., Jr. (1985). *Multispecies Toxicity Testing.* Pergamon, New York.

Cairns, J., Jr., and Cherry, D. S. (1983). A site-specific field and laboratory evaluation of fish and Asiatic clam population responses to coal fired discharges. *Water Sci. Technol.* **15,** 31–58.

Cairns, J., Jr., and Pratt, J. R. (1985). Multispecies toxicity testing using indigenous organisms—A new, cost-effective approach to ecosystem protection. In 1985 Environmental Conference, TAPPI Press, Atlanta, GA, pp. 149–159.

Cairns, J., Jr., and Pratt, J. R. (1986). Determining microbial community equilibrium in disturbed wetland ecosystems. In F. G. Webb (Ed.). 12th Annual Conference on Wetlands Restoration and Creation. Hillsborough Community College, Tampa, FL, pp. 201–209.

Cairns, J., Jr., and van der Schalie, W. H. (1980). Biological monitoring. I. Early warning systems. *Water Res.* **14,** 1179–1196.

Cairns, J., Jr., Buikema, A. L., Jr., Cherry, D. S., Herricks, E. E., van der Schalie, W. H., Matthews, R., Niederlehner, B. R., and Rodgers, J. H., Jr. (1982). *Biological Monitoring.* Pergamon, Oxford.

Cairns, J., Jr., Pratt, J. R., and Niederlehner, B. R. (1985). A provisional multispecies test using indigenous organisms. *J. Test. Eval.* **13,** 316–319.

Campbell, I. C. (1981). Communication: A critique of assimilative capacity. *J. Water Pollut. Control Fed.* **53**(5), 604–607.

Dales, J. H. (1970). *Pollution, Property, and Prices.* Univ. of Toronto Press, Toronto.

Dickson, K. L., Cairns, J., Jr., and Maki, A. W. (Eds.) (1979). *Analyzing the Hazard Evaluation Process.* Am. Fish. Soc., Washington, DC.

Graber, R. C. (Ed.) (1983). *Proceedings of the Seminar on Development and Assessment of Environmental Standards.* Am. Acad. Environ. Eng., Annapolis, MD.

Hardin, G. (1968). The tragedy of the commons. *Science* **162,** 1243–1248.

Hellawell, J. M. (1978). *Biological Surveillance of Rivers.* Water Research Centre, Stevenage, UK.

Ingersoll, C. G., and Winner, K. W. (1982). Effect on *Daphia pulex* (deGeer) of daily pulse exposures to copper or cadmium. *Environ. Toxicol. Chem.* **1,** 321–327.

Kneese, A. V., Ayres, R. V., and D'Arge, R. C. (1970). *Economics and the Environment.* Resources for the Future, Washington, DC.

Lewis, R. A. (1984). *Richtlinen fur den einsatz einer umweltprobenbank in der Bundesrepublik Deutschland auf okologischer grundlage.* Biogeographisches Institut der Universitat des Saarlandes im Auftrag des Umweltbundesamtes.

Mount, D. I., and Norberg, T. J. (1984). A seven-day life-cycle cladoceran test. *Environ. Toxicol. Chem.* **3,** 425–434.

National Research Council (1981). *Testing for Effects of Chemicals on Ecosystems,* Report of the Committee to Review Methods for Ecotoxicology. Natl. Acad. Press, Washington, DC.

Odum, E. P. (1953). *Fundamentals of Ecology.* Saunders, Philadelphia, PA.

Odum, E. P. (1981). Revival of non-laboratory science. *Assoc. Southeast. Biol. Bull.* **28,** 119–121.

Odum, E. P. (1984). The mesocosm. *BioScience* **34,** 558–562.

Regier, H., and MacKay, D. (1984). Technical basis for the goals of the agreement and prospects of its effectiveness in the future. In *Review of the Great Lakes Water Quality Agreement.* Natl. Acad. Press, Washington, DC, pp. 4–12.

Sabine, J. R. (1985). The error rate in biological publication: A preliminary survey. *BioScience* **35,** 358–363.

Sadar, M. H. (1984). The importance of establishing a permanent trust fund for Great Lakes research. *J. Great Lakes Res.* **10,** 461–462.

Westman, W. E. (1985). *Ecology, Impact Assessment, and Environmental Planning.* Wiley, New York.

# PART ONE

# GREAT LAKES COMMUNITIES AND TOXIC SUBSTANCES

# 2

# EFFECTS OF TOXICANTS ON PHYTOPLANKTON WITH SPECIAL REFERENCE TO THE LAURENTIAN GREAT LAKES

***Linda Sicko-Goad and Eugene F. Stoermer***

*Great Lakes Research Division, The University of Michigan, Ann Arbor, Michigan 48109*

## 1. INTRODUCTION

Phytoplankton, as a collective group, generally are sensitive to toxicants. Because of their rapid physiological response and short generation time, it is possible to detect deleterious effects of toxicants in relatively short periods of time. The importance of the effects of toxicants on phytoplankton should not be underestimated since this group forms the most important group of primary producers in aquatic foodwebs. Although the effects of toxicants, including both metals and organics, on phytoplankton species have been well documented, the literature is quite diverse and encompasses experiments that range from single species laboratory bioassays to large in situ mesocosm studies. While some physiological effects of toxicants seem to be ubiquitous, the ability to predict the effects of a toxicant on natural phytoplankton communities is confounded by a variety of physical and biological factors that make data interpretation difficult.

Phytoplankton composition and abundance in natural bodies of water are controlled by factors such as temperature, light regime, nutrients, competition, and predation. However, composition and abundance may be modified by toxicant exposure. It is difficult to extrapolate laboratory results of toxicant exposure to natural assemblages for several reasons. First, an assemblage consists of many algal species growing together. Not all species grow optimally at the same time. In view of the fact that certain life cycle stages are known to be more sensitive to toxicants than other stages, a toxicant stress may not reproducibly alter composition or abundance because the physiological state of the populations tested may not be the same. Consequently, the timing of toxicant addition may influence its ultimate effect.

The second most important reason is that other environmental parameters also may alter the cell physiology. For example, cells that are stressed by insufficient nutrients may be more susceptible to toxicants. Consequently, certain species, under different environmental conditions, may demonstrate either tolerance or susceptibility.

Interpreting or predicting the effects of toxicants on natural phytoplankton species or assemblages is further complicated by the fact that most sources of toxicants, both organic and inorganic, are often complex effluents that may also contain nutrients. These toxicants and nutrients may act either synergistically or antagonistically on the phytoplankton assemblage, resulting in unpredictable changes in assemblage composition.

The final important consideration in toxicant studies on natural assemblages is that certain algal species are known to be more tolerant of pollutants. In fact, different clones of the same species may react quite differently to toxicant burdens so that, although the algae appear to be the same species upon microscopic identification based on cell morphology, physiologically expressed genetic differences allow some clones to survive adverse exposure to toxicants. It seems reasonable to assume that algae growing under conditions of prolonged sublethal toxicant exposure have developed mechanisms to withstand the chemical insult, by either excluding or detoxifying the chemical pollutant, or by altering their physiology to compensate for the stress.

This chapter reviews the current literature pertaining to toxicant effects on phytoplankton species. It also contains a brief history of the historical changes in phytoplankton composition in the Laurentian Great Lakes, which we will use as an example of the difficulty of assessing lakewide toxicant stresses.

## 2. HISTORICAL PERSPECTIVE OF PHYTOPLANKTON ASSEMBLAGES IN THE LAURENTIAN GREAT LAKES

Although the successional history of certain components of the Great Lakes algal flora is reasonably well known, the scope and timing of modifications of indigenous communities have only recently been realized. Currently available information indicates that both the quantitative and qualitative aspects of the phytoplankton flora of all of the Great Lakes have been extensively modified. Available information further suggests a common pattern of quantitative and qualitative changes.

Among the earliest recoverable algal samples from the Great Lakes are collections taken by J. W. Bailey in 1839 and sent to the German diatomist C. G. Ehrenberg. From these collections Ehrenberg (1843, 1854) described several species common in the Great Lakes. Somewhat later, Briggs (1872) published on collections for the Straits of Mackinac area, which are still present in American herbaria. These were soon followed by numerous accounts, particularly of plankton collections derived from municipal water supplies. Although there is little purpose in attempting to recapitulate them all here, most can be found in Van Oosten's (1957) bibliography. It is interesting to note that many of these early studies were in response to perceived pollution crises of the day. The problems ranged from sewage contamination (Vorce, 1882; Thomas and Chase, 1887) to declines in valued fish stocks (Ward, 1896). Pollution problems became the pervasive theme driving most studies of the Great Lakes algal flora from 1880 to the present.

Following the early period of interest, the practice of depositing archival samples of collections was substantially abandoned by investigators working on the Great Lakes. Although numerous reports of algae from the various areas of the Great Lakes are available (Van Oosten, 1957; Gannon, 1969; Davis, 1966), they are generally difficult to interpret due to changes in nomenclature and other taxonomic uncertainties. This is regrettable, since these reports contain the only records of certain parts of the algal flora prior to extensive human intervention in the Great Lakes system. Recoverable collections, or at least collections that have been recovered so far, are mainly diatom samples. This early literature, and possible archival samples resulting from the studies, has not been extensively investigated, and further information may become available.

The early literature conveys the impression of a sparse but very diverse flora, particularly in benthic communities, in most parts of the lakes. Although few studies contain reliable chemical and physical evidence, the floristic evidence suggests that diversity was promoted by a very stable external milieu. Taft and Kishler's (1973) review presents evidence that certain regions maintained high primary-producer biomass, again particularly in benthic communities. They argue that physical disruption of the integrated structure of these communities was a major factor in subsequent modification of the entire biotic component of the ecosystem. Indeed, it is difficult for modern aquatic ecologists to conceptualize the structural and functional integration of the indigenous plant communities of the Great Lakes in the same way it is difficult for modern terrestrial ecologists to conceptualize relationships in the unmodified terrestrial floristic formations of North America. While the available literature furnishes a meager inventory of the composition of the indigenous flora, it provides virtually no information regarding its microscale distribution or structural and functional relationships.

The need for some sort of historical perspective on the quasi-steady-state condition of the system was keenly felt by investigators in the early part of the modern era of limnological studies (e.g., Burkholder, 1929; Gottschall, 1930; Gottschall and Jennings, 1933; Daily, 1938; Damann, 1940, 1941; Chandler, 1940, 1942, 1944; Chandler and Weeks, 1945; Langford, 1946). Their attempts at quantitative analysis of phytoplankton communities showed a clear need for extended records of qualitative and quantitative changes in such communities. The source of this information was recapitulation of records of algal abundance developed at municipal water filtration plants at several localities around the lakes. Records from Lake Michigan at Chicago (Damann, 1945), Lake Erie at Cleveland (Davis, 1964), and Lake Ontario at Toronto (Schenk and Thompson, 1965) provided a first-order approximation of trends in phytoplankton composition and abundance. Although such records, having been developed for an entirely practical purpose only secondarily related to phycological ecology, are neither precise nor accurate, they have the saving quality of a certain degree of internal consistency. This allowed development of a reasonably consistent picture of the trends of phytoplankton abundance over the period of record at specific sites. This semiquantitative information, combined with reconstructive studies of qualitative changes from the available preserved samples (e.g., Hohn, 1969; Stoermer and Yang, 1969) provided rather clear evidence of major changes in the phytoplankton flora during the past 30–50 years. The nature of the change was a succession from low and fairly stable abundance of perennial species associated with oligotrophic lakes to high and unstable abundance of populations associated with eutrophic and contaminated habitats. Because of the pattern of settlement and population growth in the Great Lakes drainage basin, this pattern was displayed both sequentially at a given locality and geographically in different regions receiving different degrees of impact. Its implications, together with publicly perceived problems, such as obnoxious tastes and odors in municipal water supplies (Vaughn, 1961; Bierman et al., 1980) and massive overgrowth of attached algae, particularly *Cladophora* (Wolfe and Sweeney, 1982), led to extensive studies designed to understand the causes of this undesirable perturbation and to point out effective remedial measures.

As a result of these studies, we now have a fair idea of the comparative limnology of the lakes and associated levels of phytoplankton standing crop and productivity (Vollenweider et al., 1974). More detailed studies of species composition and distribution are available for Lake Ontario (Munawar and Nauwerck, 1971; Stoermer et al., 1974), Lake Erie (Munawar and Munawar, 1976), Lake Huron (Stoermer and Kreis, 1980; Stoermer et al., 1983a,b; Stoermer and Theriot, 1985), Lake Michigan (Rockwell et al., 1980; Stoermer and Stevenson, 1980; Stoermer, 1984), and Lake Superior (Munawar and Munawar, 1978). Although not well summarized in the journal literature, these studies show huge differences in community composition and seasonal succession between the different lakes and between different regions within the same lake (Stoermer, 1978). They also show the great importance of physical factors, such as the spring thermal bar (Stoermer, 1968; Lorifice and Munawar, 1974), stratification and development of the so-called deep chlorophyll community (Brooks and Torke, 1977), upwelling (Schelske et al., 1971; Haffner et al., 1984), and advective processes (Stoermer and Kreis, 1980; Sell et al., 1984; Stoermer and Theriot, 1985).

Most work on attached algae has concentrated on *Cladophora* communities because of their obvious nuisance potential (Herbst, 1969). Much of the recent work has been summarized in a symposium (Auer, 1982) that contains a review of the literature (Wolfe and Sweeney, 1982) pertaining to the Great Lakes. The history of succession in benthic communities is less well known than it is in the phytoplankton, although the invasion of certain exotic populations, such as *Bangia,* is well documented (Kishler and Taft, 1970; Lin and Blum, 1977). Microscopic benthic algal communities in Lake Superior are extremely diverse (Stoermer, 1975). Diversity appears to decrease and community differentiation with depth appears to increase in Lake Michigan (Stevenson and Stoermer, 1981). Kingston et al. (1983) have shown that a diverse and seasonally stable community is present at depth in Grand Traverse Bay of Lake Michigan and that seasonal succession of eutrophication-tolerant species is characteristic of epilimnetic communities.

The major result of all these studies has been to document the overwhelming importance of phosphorus loading in driving algal succession in the Great Lakes. Both field (Schelske and Stoermer, 1972; Schelske et al., 1972) and laboratory studies (Schelske et al., 1974; Kilham, 1984; Kilham et al., 1977; Tilman and Kilham, 1976; Tilman, 1977; Stoermer et al., 1978; Mechling and Kilham, 1982) have clearly shown that phosphorus is the nutrient controlling production in the Great Lakes. Further, it has become clear that the unique hydrological (Schelske and Stoermer, 1971) and physical (Moll and Stoermer, 1982) characteristics of the Great Lakes make their algal communities particularly susceptible to large perturbations from relatively small increases in phosphorus loading. This understanding has led to development and implementation of management practices designed to limit phosphorus loadings to the lakes. Current evidence (Dobson, 1981; Bierman and Dolan, 1981) suggests that these practices are becoming effective.

At the same time, additional new evidence suggests that the phosphorus enrichment problem is, in terms of perturbation of the entire ecosystem, much more serious and long-standing than generally realized. Qualitative studies of diatom assemblages preserved in sediments of the lower Great Lakes (Duthie and Sreenivasa, 1971; Frederick, 1981; Harris and Vollenweider, 1982) showed community succession beginning with early European settlement and progressing to replacement of substantial portions of the indigenous flora with species characteristic of highly eutrophied habitats. Based on quantitative paleolimnological studies, Schelske et al. (1983) suggested that silica limitation induced by increased phosphorus loadings began very soon after European settlement of the Lake Erie and Lake Ontario basins and that these events are analogous to more recent silica depletion in Lake Michigan. Recent studies show that the diatom flora preserved in Bay of Quinte sediments responded to very low levels of European population pressure and that gross community modification was associated with early forest clearance and agricultural activities (Stoermer et al., 1985a). Parallel studies of Lake Ontario show that silica limitation began by 1850 and the plankton diatom flora was significantly modified by 1900. These effects are evident in both changes in the community (Stoermer et al., 1985b) and changes in the morphology of some persistent species (Stoermer et al., 1985c). The presence of similar effects in the sedimentary history of Lake Superior (Stoermer et al., 1985) indicates that it is an inappropriate model of processes in a quasi-steady-state system.

It is somewhat incongruous that, just as we are beginning to understand the scale and timing of changes in the primary-producer component of the Great Lakes ecosystem, the forcing functions are beginning to be modified. During the past 200 years the Great Lakes have faced ever-increased phosphorus loadings, and the effects of this particular perturbation have been so massive and pervasive as to submerge the concurrent effects of toxicant loadings and biotic interactions. There is now clear evidence that phosphorus-loading limitation strategies are becoming effective (Dobson, 1981) and that reduced loadings are producing the forecast effect (Bierman and Dolan, 1981). While it is certainly debatable whether phosphorus-loading reductions presently achieved, or even those achievable with current management strategies, are sufficient to restore the ecosystem to a balanced and stable condition, it is clear that other effects will become relatively more important in the future.

Discovery and quantification of toxic effects on primary producer communities in the Great Lakes are presently in a primitive stage, and there is every reason to believe that achieving a clear understanding of this problem will be much more difficult than the eutrophication issue. Toxic insults may result from a multitude of exotic compounds, whereas eutrophication is primarily driven by a single factor. The physiological basis of phosphorus effects was clearly understood well in advance of the realization that eutrophication was a serious problem in the Great Lakes, whereas the physiological basis of the effects of many toxic compounds is understood poorly, if at all.

## 3. METHODS OF ASSESSING THE IMPACT OF TOXICANTS ON PHYTOPLANKTON

The methods used to assess toxicant effects on algal species vary from unialgal, single-toxicant bioassays to complex micro- or mesocosms subjected to complex chemical mixtures. There are equally valid, but different reasons for the varied approaches that have been reported in the literature. Single-species, single-toxicant bioassays were developed as a method to quickly assess the toxicity of chemical compounds or suspected contaminated sources such as water or sediments. Microcosm or mesocosm experiments attempt to simulate complex ecosystem interactions that may take place in natural waters.

### 3.1. *Selenastrum* Bioassay

The most commonly used bioassay is the *Selenastrum capricornutum* Printz algal assay bottle test (Miller et al., 1978). *Selenastrum* has been used in a variety of bioassays for the simple reason that it is quite ubiquitous and is able to tolerate a wide range of environmental conditions (Rodhe, 1978). A single-species bioassay with one easily obtainable organism in a standard culture medium is useful for a variety of reasons. First, the growth of an organism is interrelated with the combined effects of all nutrients and growth conditions (Odum, 1971). Use of a standard bioassay protocol circumvents unusual physiological responses that may be species-specific effects. Second, detection of algal growth response becomes more precise as information on the physiology of the test organism accumulates. Third, if algal growth remains limited in the standard bioassay under unlimited nutrient supply and optimal light and temperature conditions, the presence of a toxicant or an inhibitory substance is indicated (Miller et al., 1974; Greene et al., 1975, 1978).

The *Selenastrum* bioassay is useful in determining standard physiological responses to known toxicants and the growth response to suspected toxicants. However, many investigators have extended the concept of single-species, single-toxicant assays to other organisms.

### 3.2. Single-Species Assays

Much of the earlier literature on heavy metal effects on phytoplankton has been thoroughly reviewed by Davies (1978). Most of the reported physiological responses of algae to toxicants have been observed in unialgal cultures of species from a variety of divisions, although the test organisms most often used are green or blue-green algae or diatoms. The repeated use of these divisions of algae is probably related to the availability of axenic or unialgal cultures and the comparative ease of culture. They are also ubiquitously distributed, making them logical candidates for testing. There are numerous reports in the literature on the effects of both organic and inorganic toxicants. Some of the more recent papers are summarized in Tables 2.1 and 2.2. Although these lists are certainly not exhaustive, several trends emerge with respect to physiological effects resulting from exposure to toxicants (also see Boyle, 1984).

Table 2.1 is a summary of effects of metals on a variety of freshwater and marine algal species. The table, for convenience, is divided into two sections: photosynthesis, and growth and metabolism. Several trends emerge that appear to be consistent with all metal ions. The most easily observed effect is reduction in photosynthesis, generally measured by reduced $^{14}C$ uptake. The second most com-

**Table 2.1 Physiological Responses to Metals**

| Organism | Toxicant | Response | Reference |
|---|---|---|---|
| | | *Photosynthesis* | |
| *Anabaena flos-aquae* | Cd | Reduction of thylakoid surface area | Rachlin et al. (1984) |
| *Anabaena inequalis* | Ni, Hg | Photosynthesis inhibited; pigment bleaching | Stratton and Corke (1979); Stratton et al. (1979) |
| *Anacystis, Fragilaria, Melosira* | Cu, Pb | Photosynthetic apparatus and mitochondrial degradation | Sicko-Goad (1982) |
| *Asterionella formosa* | Cd | Reduces photosynthetic activity | Conway (1978) |
| *Asterionella japonica* | Cu, Zn | Uncoupling of photosynthesis from cell division | Fisher et al. (1981) |
| *Chlorella* | Cd, Zn | Reduction in pigment content | De Fillipis and Pallaghy (1976a,b); De Fillipis (1979) |
| *Chorella* | Cr | Stimulates photosynthetic oxygen evolution; depression of chlorophyll synthesis | Miesch and Schmitt-Beckmann (1979) |
| *Chlorella pyrenoidosa* | Cd | Reduced $CO_2$ fixation | Hart and Scaife (1977) |
| Corn | Cd | Inhibits Photosystem II activity | Bazzaz and Govindjee (1974) |
| *Plectonema boryanum* | Mn, Pb, Co, Ni, Cd | Increase in thylakoid surface area | Rachlin et al. (1982a) |
| *Scenedesmus acutus* | Cu | Chlorophyll, lipid, and membrane degradation | Sandmann and Böger (1980) |
| *Skeletonema costatum* | Pb | Increased photosynthetic rate; decreased catabolism; increased carbon excretion | Rivkin (1979) |
| *Stigeoclonium tenue* | Pb | Chloroplast membrane reduction; increased chloroplast lipids | Silverberg (1975) |
| *Thalassiosira aestivalis* | 11 diff. metals | Disruption of chloroplast integrity; bleached cells | Thomas et al. (1980a) |
| | | *Growth and Metabolism* | |
| *Anabaena inequalis* | Ni, Hg | Growth and acetylene reduction inhibited | Stratton and Corke (1979); Stratton et al. (1979) |
| *Anabaena, Ankistrodesmus* | Cd, Cu | Multinucleate growth; cell lysis | Massalski et al. (1981) |
| *Anabaena, Scenedesmus* | Se | Growth inhibition; depressed phosphate uptake | Moede et al. (1980) |
| *Ankistrodesmus, Chlorococcum, Scenedesmus* | Cd, Cu | Stimulate growth at low concentrations Inhibit growth at high concentrations | Devi Prasad and Devi Prasad (1982) |

| | | | |
|---|---|---|---|
| *Ankistrodesmus, Euglena* | Hg | Inhibition of lipid biosynthesis, esp. galactolipids and chlorophylls | Matson et al. (1972) |
| *Ankistrodesmus, Scenedesmus* | Ni | Decreased growth | Spencer and Greene (1981) |
| *Asterionella japonica* | Cu, Zn | Depressed cell division rates; inability to utilize or excrete carbon | Fisher et al. (1981) |
| *Chlorella* | Hg, Zn | Reduction in cell division; no glycollate excretion | De Filippis and Pallaghy (1976c) |
| *Chlorella* | Zn, Cd, Hg | General metabolic activity reduced; ATP reduced | De Fillipis and Pallaghy (1976a,b); De Filippis et al. (1981) |
| *Chorella* | Hg, Cd | Decreased productivity and growth | Gerhards and Weller (1977) |
| *Chlorella saccharophila* | Cd, Cu, Pb, Zn | Growth reduction (rate and days per division cycle) | Rachlin et al. (1982b) |
| *Chlorella vulgaris* | Zn | Growth reduction | Rachlin and Farran (1974) |
| *Chlorella, Navicula, Nitzschia* | Cd | Growth reduction | Rachlin et al. (1982c) |
| *Coelastrum microporum* | Hg | Depressed growth; starch accumulation | Holderness et al. (1975) |
| *Coscinodiscus granii* | Cd, Zn | 20 μg Cd/L inhibits growth; cell cycle sensitivity | Rabsch and Elbrächter (1980) |
| *Cricosphaera carterae* | Cd | Cell division and calcification inhibited | Stillwell and Corum (1982) |
| *Diatoma tenue* | Cu, Pb | Mitochondrial degradation; increased autolytic activity; Cu inhibits polyphosphate formation | Sicko-Goad and Stoermer (1979) |
| *Ditylum brightwellii* | Cu | Osmotic swelling at $10^{-5}M$ | Bentley-Mowat and Reid (1977) |
| *Ditylum brightwellii* | Cu, Zn, Hg, Ag, Tl | Formation of "pseudo" resting spores | Canterford (1980) |
| *Isochrysis galbana* | Hg | Cell division inhibited | Davies (1974) |
| *Poterioochromonas malhamensis* | Pb | Mitochondrial degradation; nuclear deformation increased; autolytic activity | Roderer (1984) |
| *Prorocentrum micans* | Cd | Decreased cell division rates; "brittleness" of cell walls | Kayser and Sperling (1980) |
| *Skeletonema costatum* | Pb | Growth rate, yield, and respiration decreased | Rivkin (1979) |
| *Skeletonema costatum* | Hg, Zn | Plasma membrane damage; osmotic disorganization | Smith (1983) |
| *Tetrahymena pyriformis* | Cu | Increase in refractile granules; stimulation of phagocytosis | Nilsson (1981) |
| *Thalassiosira aestivalis* | Hg, Cd, Pb | Cell clumps, abnormal divisions | Thomas et al. (1980a) |
| *Thalassiosira pseudonana* | Cr | Depressed growth | Riedel (1984) |

**Table 2.2 Physiological Responses to Organics: Growth and Metabolism and Photosynthesis**

| Organism | Toxicant | Response | Reference |
|---|---|---|---|
| *Acrosiphonia* | Crude oils | Photosynthetic inhibition | Kusk (1980) |
| *Agmenellum, Cylindrotheca, Chlorella* | Crude oil; fuel oil | Fuel oils generally more toxic; inhibiting P.S. and growth | Batterton et al. (1978) |
| *Amphidinium, Cyclotella cryptica* | North Sea crude oil and fractions | Low concentrations stimulate growth; high concentrations inhibit; reduction in photosynthesis and chlorophyll content | Karydis and Fogg (1980); Karydis (1982) |
| *Anabaena, Anacystis, Chlorella, Navicula* | Diquat | Growth inhibition; prokaryotes more sensitive | Birmingham and Colman (1983) |
| *Anabaena, Aphanocapsa, Nostoc* | Atrazine, Diuron, Paraquat | Growth inhibition | Hutber et al. (1979) |
| *Anabaena, Chlorella, Scenedesmus* | Atrazine | Reduction in $^{14}C$ uptake and growth | Stratton (1984) |
| *Anacystis nidulans* | DDT | Inhibition of sodium chloride tolerance | Batterton et al. (1972) |
| *Ankistrodesmus, Chlamydomonas, Chlorella, Scenedesmus* | PCB (Aroclor 1242) | Inhibition of $^{14}C$; heterotrophic growth in presence of PCB | Conner (1981); Conner and Mahanty (1979) |
| *Ankistrodesmus, Fragilaria, Microcystis* | Hexachlorobiphenyl | *Fragilaria* growth inhibited; reduction in chlorophyll | Lederman and Rhee (1982) |
| *Ankistrodesmus, Scenedesmus, Synedra* | PCB | Growth inhibition, depressed $^{14}C$ uptake; membrane degeneration and vacuolization | Glooschenko and Glooschenko (1975) |
| *Chlamydomonas, Chlorella* | PCB (Aroclor 1242) | Growth inhibition, but variable with species and strain | Mahanty and Gresshoff (1978) |
| *Chlamydomonas, Scenedesmus* | Chlordane | Low concentrations stimulate division and respiration | Glooschenko and Lott (1977) |
| *Chlamydomonas angulosa* | Crude oil extract | Chloroplast degradation; increase in lipid content | Soto et al. (1979a) |
| *Chlamydomonas angulosa* | Naphthalene | Chloroplast abnormalities; increased starch and lipid content | Soto et al. (1979b) |

| | | | |
|---|---|---|---|
| *Chlamydomonas angulosa* | Naphthalene | Change in plasmalemma permeability; loss of Mn and K; increased cell lipids and carbohydrates; decreased protein | Hutchinson et al. (1981); Soto et al. (1977) |
| *Chlorella*, *Lyngbya* | Herbicides (picloram, prometryn, dinoseb, fluometuron) | Growth inhibition with prometryn and fluometuron | Tubea et al. (1981) |
| *Chlorella*, *Scenedesmus* | PCB (Aroclor 1242) | Disruption of chloroplast membranes; cytokinesis failure | Mahanty et al. (1983) |
| *Chlorella vulgaris* | Atrazine | Growth inhibition | Véber et al. (1981) |
| *Cylindrotheca closterium* | PCB | Growth inhibition, reduction in RNA and chlorophyll | Keil et al. (1971) |
| *Ditylum*, *Lithodesmium* | PCB | Reduction in photosynthesis | Harding and Phillips (1978) |
| *Dunaliella bioculata* | Crude oil photodecomposition products | Reduced growth rate; cell death due to membrane disruption | Heldal et al. (1984) |
| *Exuviella baltica* | Chlordane, Heptachlor | Reduced growth, chlorophylls, $^{14}C$ uptake, and fixation | Magnani et al. (1978) |
| *Fucus*, *Laminaria* | Fuel oils | Change in plasmalemma permeability | Reddin and Prendeville (1981) |
| *Microcystis*, *Scenedesmus* | Coal liquefaction products; shale oil, petroleum products | Photosynthesis inhibition | Giddings and Washington (1981) |
| *Nitzschia*, *Phaeodactylum* | Crude oil; aromatic hydrocarbons; naphthalene | Photosynthesis inhibited | Kusk (1980, 1981a,b) |
| *Nitzschia delicatissima* | DDT | Reduced carbon fixation and chlorophyll; chloroplast degradation | MacFarlane et al. (1972) |
| *Nitzschia palea* | Crude oil | Decreased photosynthesis | Kusk (1978) |

**Table 2.2 (*Continued*)**

| Organism | Toxicant | Response | Reference |
|---|---|---|---|
| *Pavlova lutheri* | Naphthalene; crude oil; diesel fuel | Changes in patterns of cell motility | Vandermeulen et al. (1983) |
| *Phaeodactylum* | Naphthalene, phenanthrene | Photosynthetic inhibition; growth inhibition | Kusk (1981a,b,c) |
| *Scenedesmus* | Diquat | Lipid peroxidation and chlorophyll degradation | Rensen (1975) |
| *Scenedesmus quadricauda* | 3,4-Benzopyrene | Ultrastructural modifications and degradation of nucleus, thylakoids, cell wall, and cytoplasm | Abarzua et al. (1983) |
| *Scenedesmus quadricauda* | 3,4-Benzopyrene | Decreased PNA and protein synthesis; chlorophyll synthesis inhibition; chlorophyll degradation acceleration | Zachleder et al. (1983) |
| *Selanastrum* | PAH | Decreased growth with light less than 550 nm | Cody et al. (1984) |
| *Spirodela oligorrhiza* | PCB (Aroclor 1242) | Chloroplast degradation | Mahanty and Fineran (1976) |
| *Thalassiosira* | PCB | Suppressed growth and photosynthesis | Powers et al. (1977) |
| *Thalassiosira pseudonana* | PCB | Inhibition of membrane-bound enzymes of nitrogen uptake or metabolism | Fisher et al. (1976) |

monly observed response is decrease in pigment content, thought to result from pigment degradation, rather than a loss correlated with increased biomass, hence pigment dilution. The third most commonly observed effect of metals on the photosynthetic apparatus is disruption of morphological integrity, particularly disruption of individual thylakoid membranes. Alteration of membranous organelles or of the plasmalemma is also a common effect of exposure to organic contaminants.

There have been a variety of reported effects of metals on metabolic processes. The simplest test, as with the *Selenastrum* bioassay, is merely a measurement of the effect on absolute growth or growth rate. A general response in growth reduction has been observed for a variety of metals and organisms (Table 2.1). The second most commonly observed phenomenon is a reduction or inhibition of cell division. While the cells may continue to expand and biomass increases, cytokinesis cannot be completed, resulting in abnormally large or deformed cells. A third common pattern is the appearance of characteristics similar to those found in cells that are senescent or dormant. For example, starch or lipid storage products may accumulate, and there may be increased autolytic activity. In one case (Canterford, 1980) the formation of "pseudo" resting spores has been reported. Membrane damage resulting from metal exposure may also be evident in other membrane systems, such as the plasmalemma and the mitochondria.

Table 2.2 summarizes physiological responses to organic toxicants. Again, although the list is not exhaustive, and although the term "organic toxicants" encompasses a wide variety of compounds including herbicides, chlorinated hydrocarbons, and oils, a pattern of response emerges that is similar to the effect of metals. Many of the reports on the varied algal species deal with cessation of growth, although in some reports it is obvious that the effects are rather species-specific. A second general pattern of physiological response is reduction in photosynthesis. The decrease in photosynthesis may be characterized by one of three general mechanisms: (1) inhibition of $^{14}C$ uptake, (2) chlorophyll destruction, or (3) inhibition of chlorophyll synthesis.

A third common mechanism of toxicity appears to be destruction of membranes and membranous organelles such as the mitochondria and chloroplasts. It has been speculated that the effects of several of the organics may result from lipid peroxidation (e.g., Rensen, 1975). If lipid peroxidation occurs, it could lead to changes in membrane structure, resulting in a variety of physiological and morphological effects that are measurable and visible. Changes in membrane composition could account for reported disturbances in membrane permeability (e.g., Hutchinson et al., 1981; Reddin and Prendeville, 1981) or physiological processes that are associated with membrane-bound enzymes (e.g., Fisher et al., 1976; Gavis et al., 1981). This would also explain the variety of cytological abnormalities associated with toxicant exposure.

### 3.3. Multiple-Species–Multiple-Toxicant Studies

Although there are numerous reports on the morphological and physiological effects of toxicants or contaminants on algae, the ability to extrapolate these findings to species in the natural environment has not been well developed. In 1978, Davies concluded, "This review of our current understanding of the effects of metals upon marine plankton has highlighted the fact that although, at present, the data obtained from laboratory experiments cannot usually be used to assess the degree to which natural populations are at risk due to metal pollution, it may now be possible to begin to rectify the situation." Similarly, in 1984, Boyle stated, "The secondary or indirect effects of contaminants on algae, that is the change in the structure and function of algal communities resulting from toxic effects on other organisms in the aquatic community, have received less attention."

The problem of assessing contaminant stress in aquatic ecosystems has been addressed in several different ways in an attempt to mimic the natural environment, both in complexity of community structure and in complexity of contaminant exposure. The following section summarizes some current findings.

### 3.3.1. Metal Mixtures

The need to study the toxic effects of metal mixtures was realized in the 1970s when researchers began to feel uneasy about toxicity bioassays conducted with single metals (Christensen et al., 1979; Wong et al., 1978; Braek et al., 1976; Hutchinson and Stokes, 1975). These studies emphasized the fact that single metals rarely dominate in an environment, and that other environmental variables may act along with the metal mixtures in either a synergistic or antagonistic manner that is dependent upon the metal and organisms involved.

The simplest test of the toxicity of metal mixtures still remains with cultured algae. Bartlett and Rabe (1974) demonstrated that combinations of copper, zinc, and cadmium were similar in toxicity to equal concentrations of zinc in cultures of *Selenastrum*. Their work also suggested that cadmium inhibited copper toxicity. Wong and Beaver (1980) found that metal mixtures were toxic to *Ankistrodesmus* but not to *Chlorella*. Similarly, Wong et al. (1982) demonstrated that metal mixtures reduced the primary productivity of four cultured freshwater algae as well as natural phytoplankton assemblages. Growth inhibition as a function of exposure to metal mixtures was found in both cultures of *Thalassiosira* and natural diatom assemblages by Thomas et al. (1980b). They attributed inhibition to Cu and Hg since deletion of these elements resulted in growth that was equivalent to control cultures containing no metals.

### 3.3.2. Natural Phytoplankton Assemblages

Studies of the effects of contaminant stress on natural phytoplankton assemblages are much more complex. Schelske (1984) has summarized the types of laboratory and in situ experiments that can be conducted with natural assemblages. The range of experimental variables is quite large and the use of enclosure experiments is not without risk.

The use of natural phytoplankton assemblages in laboratory experiments more closely simulates populations that have undergone succession and are adapted to living at lower nutrient concentrations. However, laboratory and larger in situ enclosures cannot duplicate the natural irradiance and turbulence patterns found in the field, so interpretation of results obtained from such studies must take these factors into consideration. The second difficulty encountered with comparing results of toxicant perturbations on natural assemblages is the method of exposure. Schelske (1984) outlines five basic types of experiments that may be conducted: (1) spike, (2) fixed level or fixed supply, (3) dilution, (4) chemostat, and (5) turbidostat. Any or all of these exposure methods may be justified for experiments on natural phytoplankton assemblages depending on the source of material.

A third problem associated with studies of natural phytoplankton assemblages is the inability to obtain species-specific physiological effects. Most studies are concerned with changes in abundance and distribution. With the information already available on the sensitivity of certain life cycle stages (e.g., Rabsch and Elbrächter, 1980; Butler, 1977; Cain et al., 1980) and with the physiological status of cells (Fisher and Wurster, 1973), it is not unreasonable to suspect that different populations of the same species (as with clones, e.g., Butler et al., 1980) may be more sensitive to contaminants than others.

A final problem associated with studies of natural assemblages is that a variety of environmental parameters are known to either mitigate or amplify the effects of toxicants on phytoplankton. For example, pH, temperature, and the presence of chelators or concentration of nutrients may affect metal toxicity by reducing free ion activity of metals (e.g., Rabsch and Elbrächter, 1980; Thomas et al., 1980b; Gipps and Coller, 1980, 1982; Canterford and Canterford, 1980; Hallas et al., 1982; Crist et al., 1981; Ahlf et al., 1980; Harding and Whitton, 1977; Allen et al., 1980; Davies, 1978; Gadd and Griffiths, 1978). Consequently, it is difficult to duplicate chemical and environmental parameters from experiment to experiment.

Reports of effects of contaminants on natural assemblages range from simple laboratory enclosure perturbations to the more elaborate, large-volume, multidisciplinary studies of CEPEX (controlled ecosystem pollution experiments) and MELIMEX (MEtal LIMnological EXperiment). Although results vary with assemblages, it is apparent that community changes occur as a result of differential sensitivity to contaminants (Hollibaugh et al., 1980; Thomas and Seibert, 1977; Harrison et al., 1977; Gachter and Mares, 1979; Throndsen, 1982; Vargo et al., 1982). The change in species composition is toward selection of more tolerant species. Sanders et al. (1981) suggested that tolerant species that are not dominant in the natural assemblage are able to compete successfully with the usual dominant species only when those species are stressed. Blinn et al. (1977) demonstrated that summer phytoplankton assemblages in Lake Powell, Arizona, were more sensitive to Hg exposure than spring assemblages. They suggested that the differential sensitivity was a result of the difference in the assemblages acting in combination with temperature. Inhibition of carbon fixation in natural assemblages as a result of metal exposure has also been reported (Davies and Sleep, 1979, 1980; Woiter et al., 1984), as well as a reduction in respiration as a result of oil spills (De La Cruz, 1982).

## 4. BIOACCUMULATION AND TOLERANCE

In addition to the wealth of information available on the physiological and morphological effects of toxicants on single-species cultures as well as the community changes associated with toxicant exposure, presumably a result of differential sensitivity (Blanck et al., 1984) and change in the ability to compete, another aspect of toxicant exposure that affects species survival is bioaccumulation and tolerance. For purposes of this chapter, the term bioaccumulation will be used to include both adsorption and absorption of toxicants, for the simple reason that in terms of food-web dynamics and transfer, in some cases it is irrelevant whether the toxicant is adsorbed to the cell or is bound to internal cellular components. We have chosen to include both accumulation and tolerance in the same section since, in many instances, tolerance is a direct result of internal binding, thus removing the toxicant from the metabolic pool, although chemical analysis would indicate that the toxicant is associated with the cell.

### 4.1. Metal Accumulation

Metal accumulation is generally believed to be biphasic. That is, there is a rapid uptake that is believed to be related to the adsorption of the metal at the cell surface, followed by a slower passage through the cell membrane (Rai et al., 1981; Wong et al., 1984; Sakaguchi et al., 1979). Geisweid and Urbach (1983) demonstrated that sorption of cadmium ions by three unicellular green algae was divided into a fast and a slow phase. The fast sorption of cadmium, which was completed in less than 5 minutes, was described as a Freundlich or Langmuir sorption isotherm, with cadmium binding rapidly to acidic groups in the cell wall. They also demonstrated that 85% of the cadmium sorbed was taken into the interior of the cell after 3 hours.

Les and Walker (1984) found that metal binding increased significantly with a pH rise from 4 to 7. They also found that most of the accumulated metal was EDTA extractable, indicating that the adsorption or "fast" uptake was most significant for the blue-green. Similar results were reported by Sakaguchi et al. (1979), who also found that heat-killed *Chlorella* cells accumulated more cadmium than living cells. Thus, they concluded that cadmium "uptake" was not directly mediated by metabolic processes, but rather by physicochemical adsorption on the cell surface. Wihlidal and Broda (1978) found that uptake of zinc into the cell interior was directly related to a breakdown of the cell membrane as a permeability barrier in *Chlorella fusca*.

Although bioaccumulation of metals has been reported for a variety of both metals and organisms (Canterford et al., 1978; Dongmann and Nurnberg, 1982; Fisher et al., 1983, 1984; Fujita et al., 1977; Gavis, 1983; Gerhards and Weller, 1977; Gutknecht, 1963; Jennings and Rainbow, 1979; Li, 1980; Pawlaczyk-Szpilowa and Slowik, 1981; Pribil and Marvan, 1976; Schulz-Baldes and Lewin, 1976; Weber et al., 1978), several mechanisms of binding and accumulation sites deserve special mention.

Fisher and Jones (1981) found that six metals were toxic to the diatom *Asterionella japonica* and that the common basis for toxicity was the ability to bind to sulfhydryl groups. Similar results have been reported by Davies (1976, 1978), DeFilippis (1978), and Fisher et al. (1984). Heavy metal binding to polyphosphate, an osmotically inert inorganic phosphate polymer, has been reported in several blue-green and diatoms (Jensen et al., 1982a,b; Baxter and Jensen, 1980a,b; Sicko-Goad and Stoermer, 1979; Daniel and Chamberlain, 1981). Polyphosphate is believed to be a phosphate storage form and is generally not metabolized under phosphate-sufficient conditions (Sicko, 1974). Immobilization of metals by binding or trapping during polymerization serves to remove the metal from the metabolic pool, minimizing immediate cellular damage caused by the intracellular accumulation of metal (Sicko-Goad and Lazinsky, 1982a,b, 1983). Presumably metals so sequestered may be remobilized if the external medium subsequently becomes phosphorus deficient (Sicko-Goad and Lazinsky, 1986).

Several other metal-binding sites have been suggested for algae. Silverberg (1976) found that dense intramitochondrial granules were formed in three green algae upon exposure to cadmium. These granules contained detectable quantities of the metal. However, concentration of the metal into these granules did not render the metal harmless. McLean and Williamson (1977) found that cadmium was accumulated in the red alga *Porphyra* in the nucleus, and that a de novo protein synthesis was a prerequisite for cadmium accumulation. A third binding site that has been reported for *Tetrahymena,* which may also be significant for a number of phototrophic flagellates, is the digestive vacuole and lysosomes (Nilsson, 1978, 1979). Although these organelles are not commonly thought of as components of algae, there is evidence that metal exposure may increase membranous aggregates within algal cells (Sicko-Goad and Stoermer, 1979; Sicko-Goad, 1982), and much of the current thinking on the structure of plant cells indicates that the vacuole may act as the lytic compartment which would be comparable to the digestive vacuole (Matile, 1975).

## 4.2. Organics Accumulation

The literature on organic toxicant accumulation is much more diverse and encompasses a wide variety of chemical compounds. Several reviews have indicated that a wide variety of organics are bioconcentrated by phytoplankton (e.g., Ware and Roan, 1970; Butler, 1977). It has been demonstrated that for some organics, there is also a rapid initial uptake, followed by a slower "metabolic" uptake (Butler, 1977) which resembles the biphasic uptake of metals.

The relationship between water or lipid solubility of the compounds and bioaccumulation potential is unclear and deserves further study. For example, Geyer et al. (1981) found that there was a simple inverse relationship between water solubility and bioconcentration of 34 chemical compounds in the alga *Chlorella fusca.* However, they also point out that more study is needed on organic compounds that have low water solubility and other physicochemical properties that may make the compounds more lipid soluble.

Baluja et al. (1982) found that polychlorinated biphenyls were accumulated by the yeast *Saccharomyces cerevisiae.* The uptake was rapid, being almost complete in the first hours of the exponential growth phase. Lipid biosynthesis decreased as a result of this accumulation, and the biosynthesis levels of two monounsaturated acids C 16:1 and C 18:1 were depressed by approximately 40%. Lederman and Rhee (1982) found that hexachlorobiphenyl (HCB) was bioaccumulated by three different species of Great Lakes algae. The presence of a mucilage layer in *Fragilaria crotonensis* was associated with a twofold increase in bioconcentration of HCB. The same phenomenon was reported for uptake of lindane by *Chlorella* (Hansen, 1979) and is important ecologically because many algal species have extracellular material associated with them.

Bioaccumulation or concentration is a real phenomenon that exists for a wide variety of organic and inorganic toxicants. Although passage through a differentially permeable membrane may result in deleterious physiological effects, adsorption may be equally responsible for bioamplification of the toxicant in the food web.

## 4.3. Tolerance Mechanisms

Heavy metal tolerance mechanisms were reviewed thoroughly by Antonovics et al. in 1971 and that information, coupled with more recent studies, leads to the conclusion that several mechanisms may be responsible for algal adaptation to toxicant stress. These may be summarized as follows:

1. A toxicant exerts no effect on a cell because it is excluded. That is, either a physical or physiological barrier exists that prevents the toxicant from entering the cell.
2. A toxicant exerts no effect because, although it enters a cell, it is transformed or immobilized in such a way that it does not interfere with cell metabolism.

While these observations may be simplistic in definition, they encompass a variety of complex physical, physiological, biochemical, and genetic adaptations that allow certain algal species to grow in areas where they are exposed to a variety of chemical compounds.

The simplest demonstrations of the ability to withstand exposure to a host of chemical compounds are observations of community changes along toxicant gradients coupled with reports of known pollution-tolerant species. For example, *Stigeoclonium tenue* has been reported to withstand both metals and organic pollutants (McLean, 1974; McLean and Benson-Evans, 1974). Austin and Deniseger (1985) used periphyton communities to assess the impact of a heavy metal gradient in Buttle Lake on Vancouver Island, British Columbia. They found that near the source of a lead–zinc–copper mine, species diversity was low but increased with distance from the mine. They also found that diatoms known to be sensitive to metals were rare or absent near the mine, but increased in abundance with distance from the mine. Other reports have demonstrated the utility of periphyton communities established on artificial surfaces to assess contamination by heavy metals (Bell, 1977; Say and Whitton, 1980; Schmidt and Christensen, 1975). It has been suggested (Bruno et al., 1982) that periphyton may represent a significant sink for benzo[a]pyrene in stream systems. These reports demonstrate quite clearly that differences in pollution tolerance exist in natural populations (Stokes et al., 1973; Rushforth et al., 1981). However, the mechanisms of tolerance have not been elucidated by laboratory studies.

The simplest mechanism of tolerance is exclusion. That is, the pollutant is not able to pass through the cell wall or cell membrane. Foster (1977) found that copper was excluded from a copper-tolerant strain of *Chlorella vulgaris.* The metal was not associated with either the cell wall or the cell contents. Similarly, Butler et al. (1980) and Hall et al. (1979) found that copper tolerance in two different algal species was a result of quite different exclusion mechanisms. In *C. vulgaris* (Butler et al., 1980), more extracellular organic material was produced by copper-tolerant strains than nontolerant strains. This organic material formed highly stable organocopper complexes. Copper complexation occurred at the cell wall, with the complex eventually released to the medium. Similar findings of exclusion by the release of extracellular material have been reported by Morris (1972) and Antonovics et al. (1971).

Changes in cell membrane permeability that alter copper uptake have been demonstrated in bacteria (Beppu and Arima, 1964; Chopra, 1971) and in a marine fouling alga, *Ectocarpus siliculosus* (Hall et al., 1979). Tolerance in these organisms was not associated with production of extracellular material or associated organisms, nor was it attributable to cell wall binding.

The role of dissolved organic compounds (DOC) in mitigating the effects of metals on phytoplankton is discussed by Fisher and Frood (1980), who found that there was an apparent lack of selection for metal resistance in "dirty" water clones. They speculate that this lack of selection may be related to the fact that the algae normally are found in water that contains elevated levels of DOC, and the algae probably never "see" high concentrations of ionic metals. The algal strains from areas of high DOC were sensitive to metals when cultured under low ambient DOC. Consequently, metal tolerance by exclusion may be a combination of environmental binding as well as cellular (or extracellular) binding. This factor (dissolved organic carbon) may become important in the Great Lakes as eutrophication is brought under control (Dobson, 1981).

The second general mechanism of tolerance is the immobilization or detoxification process. This mechanism operates in a number of distinct ways. The first most common method is the production of metallothionein-like heavy metal binding proteins. These have been reported in algae and yeast (e.g., Gerhards and Weller, 1977; Naiki and Yamagata, 1976). Immobilization of metals as dense cytoplasmic deposits or in polyphosphate bodies has been proposed as a mechanism to reduce metal toxicity and has been discussed previously. A third, more recent mechanism postulated as a mechanism of tolerance is induction of the enzyme superoxide dismutase. Naiki (1980) has demonstrated that in the yeast *Sac-*

*charomyces cerevisiae,* copper toxicity has two distinct aspects. One is due to the toxicity of superoxide radicals produced by the reaction of copper and reduced sulfur. This toxicity can be decreased by inducing the formation of superoxide dismutase. The second aspect of toxicity is due to the copper ion itself, which only some organisms may eliminate. Sandmann and Böger (1980) have demonstrated that Cu(II) ions induce the light-driven peroxidation of membrane lipids. They found that copper ions inhibit photosynthetic electron transport at the oxidizing side of PS II, providing $O_2$ species that are precursors of the highly reactive hydroxyl radical.

This general overview of tolerance mechanisms demonstrates the fact that a variety of mechanisms exist that allow certain phytoplankton species to adapt to toxicant exposure. Species that are tolerant have most likely been selected in their native environment because they possess physiological mechanisms to compensate for the chemical insult.

The problem of tolerance and bioaccumulation is further complicated by the fact that, aside from the direct effects of toxicants, indirect effects may substantially alter the environment. Indirect (or secondary) effects are changes in an ecosystem that follow and result from primary effects on growth, survival, or reproduction (Hurlbert, 1975). For example, complete or partial removal of an algal species from an ecosystem is followed by changes in algal species composition as well as changes in predators and prey.

## 5. SUMMARY

Phytoplankton are sensitive indicators of environmental perturbations, including toxic inputs. Review of the literature makes it clear that toxic materials have numerous effects on algal species. It appears that many of the effects noted are highly species-specific and, in many cases, are synergistic. Thus, both community composition and the immediate chemical and physical conditions may modify the effects of toxic inputs. Likewise, small changes in any number of physical, chemical, or biological parameters may result in significant changes in species composition and numbers in natural assemblages.

From an ecosystem perspective, three approaches to the study of the impact of toxicants on phytoplankton are necessary. A *mechanistic* approach involves acquiring knowledge of physiological, biochemical, or morphological changes that result from chemical exposure. A second approach is the use of *mesocosms* or *enclosures,* which allow for analysis of many parameters simultaneously and may be used in a predictive fashion for larger masses of water. The third approach is *historical* (paleolimnological). Detailed chemical and biological analyses (for species that have remains preserved in the sediment) allow for interpreting effects that have already occurred in impacted areas.

Although considerable information is available on mechanistic effects on phytoplankton, there is a clear need for further basic experimental studies, particularly on populations which are numerically important in the Great Lakes, and in which the design realistically simulates physical and chemical conditions that occur in the Great Lakes. It is particularly discouraging that there is virtually no information available on phytoplankton species that, based on their known distribution, might be particularly sensitive to toxic effects. It should be emphasized that, because primary producer communities in the Great Lakes contain such a physiologically diverse assemblage of organisms, extrapolation of effects from experiments on some of the commonly available laboratory strains of algae may be very misleading. Well designed experiments on potentially sensitive algal populations from the Great Lakes should be a very high research priority.

Assembling an ecologically useful primary experimental basis for predicting toxic effects on primary producer communities will be a long and difficult task. In this problem we are dealing with multiple compounds with multiple modes of action that are poorly known, if known at all. In contrast, in the eutrophication problem, where some management success has been obtained, undesirable effects were driven by a single element (phosphorus) whose physiological mode of action was very well known. Despite the relative simplicity, at least in retrospect, of the eutrophication problem, it is fair to say that there are still specific experimental problems that need attention before our predictive capability is complete. Full development of a sound, experimentally based, predictive capability for toxic substances will be a much longer and more difficult task.

If we may take example from the eutrophication problem, there is an approach to this problem that may prove useful in the management context. Some of the most convincing and telling evidence brought to bear on the eutrophication problem was derived from gross experiments on entire small-lake ecosystems or on large enclosures in the Great Lakes. While such experiments lack the rigor of carefully controlled experiments on individual populations, they provide a realistic demonstration of community effects and a reasonable basis for management action. It is somewhat surprising that this approach has not been more widely applied to toxic contaminant problems in the Great Lakes to this point. Implementation of such studies should be a high priority.

The other element of problem solution that seems lacking in our current approach to toxic contaminant problems in the Great Lakes is perspective. Since the lakes are greatly modified systems, it is necessary to know something of the history of this perturbation in order to set realistic goals for rehabilitation. Returning to the model provided by the eutrophication issue, this necessary perspective was furnished by retrospective studies of communities sampled previously and, most recently, by paleolimnological studies. The latter approach would appear to hold great promise for providing a useful historical context on contaminant problems. By careful choice of sites, and comparison of records from several sites, it should be possible to gain at least a first-order understanding of the effects of particular contaminants and the interactions between contaminant and other types of loadings. With the substantial recent advances in techniques and methods in paleolimnology, such studies would seem a highly useful approach to the toxic contamination problem and should be tested in the Great Lakes.

## ACKNOWLEDGMENT

Contribution number 474 of the Great Lakes Research Division, The University of Michigan.

## REFERENCES

Abarzua, VA-S., Jonas, L., Putzke, H-P., and Wittenburg, E. (1983). Die wirkung von 3,4-benzpyren auf die ultrastruktur der grunalge *Scenedesmus quadricauda. Arch. Protistenkd.* **127,** 103–113.

Ahlf, W., Irmer, U., and Weber, A. (1980). Uber die Anreicherung von Blei durch Susswattergrunalgen unter Berucksichtigung verschiedener aubenfaktoren. *Z. Pflanzenphysiol.* **100,** 197–207.

Allen, H. E., Hall, R. H., and Brisbin, T. D. (1980). Metal speciation. Effects on aquatic toxicity. *Environ. Sci. Technol.* **14,** 441–443.

Antonovics, J., Bradshaw, A. D., and Turner, R. G. (1971). Heavy metal tolerance in plants. *Adv. Ecol. Res.* **7,** 1–85.

Auer, M. T. (Ed.) (1982). Ecology of filamentous algae. *J. Great Lakes Res.* **8,** 1–237.

Austin, A., and Deniseger, J. (1985). Periphyton community changes along a heavy metals gradient in a long narrow lake. *Environ. Exp. Bot.* **25,** 41–52.

Baluja, G., Hernandez, L. M., and Gonzalez, M. J. (1982). Accumulation of polychlorinated biphenyls by *Saccharomyces cerevisiae* and their effects on lipid biosynthesis. *Bull. Environ. Contam. Toxicol.* **28,** 105–112.

Bartlett, L., and Rabe, F. W. (1974). Effects of copper, zinc, and cadmium on *Selenastrum capricornutum. Water Res.* **8,** 179–185.

Batterton, J. C., Bouse, G. M., and Matsumura, F. (1972). DDT: Inhibition of sodium chloride tolerance by the blue green alga *Anacystis nidulans. Science* **176,** 1141–1143.

Batterton, J. C., Winters, K., and Van Baalen, C. (1978). Sensitivity of three microalgae to crude oils and fuel oils. *Mar. Environ. Res.* **1,** 31–41.

Baxter, M., and Jensen, T. E. (1980a). Uptake of magnesium, strontium, barium and manganese by *Plectonema boryanum* (Cyanophyceae) with special reference to polyphosphate bodies. *Protoplasma* **104,** 81–89.

Baxter, M., and Jensen, T. E. (1980b). A study of methods for in situ X-ray energy dispersive analysis of polyphosphate bodies in *Plectonema boryanum. Arch. Microbiol.* **126,** 213–215.

Bazzaz, M. B., and Govindjee (1974). Effects of cadmium nitrate on spectral characteristics and light reactions of chloroplasts. *Environ. Lett.* **6,** 1–12.

Bell, D. M. (1977). Effects of zinc mine groundwater effluent on a stream diatom community. *Proc. Pa. Acad. Sci.* **51,** 51–53.

Bentley-Mowat, J. A., and Reid, S. M. (1977). Survival of marine phytoplankton in high concentrations of heavy metals, and uptake of copper. *J. Exp. Mar. Biol. Ecol.* **26,** 249–264.

Beppu, M. L., and Arima, K. (1964). Decreased permeability as the mechanism of arsenite resistance in *Pseudomonas pseudomallei. J. Bacteriol.* **88,** 151–157.

Bierman, V. J., Jr., and Dolan, D. M. (1981). Modeling of phytoplankton-nutrient dynamics in Saginaw Bay, Lake Huron. *J. Great Lakes Res.* **7,** 409–439.

Bierman, V. J., Jr., Dolan, D. M., Stoermer, E. F., Gannon, J. E., and Smith, V. E. (1980). *The Development and Calibration of a Spatially Simplified Multi-class Phytoplankton Model for Saginaw Bay, Lake Huron,* Great Lakes Environ. Plann. Study Contrib. No. 33. Great Lakes Basin Comm., Ann Arbor, MI.

Birmingham, B. C., and Colman, B. (1983). Potential phytotoxicity of diquat accumulated by aquatic plants and sediments. *Water, Air, Soil Pollut.* **19,** 123–131.

Blanck, H., Wallin, G., and Wangberg, S.-A. (1984). Species-dependent variation in algal sensitivity to chemical compounds. *Ecotoxicol. Environ. Saf.* **8,** 339–351.

Blinn, D. W., Tompkins, T., and Zaleski, L. (1977). Mercury inhibition on primary productivity using large volume plastic chambers in situ. *J. Phycol.* **13,** 58–61.

Boyle, T. P. (1984). The effect of environmental contaminants on aquatic algae. In L. E. Shubert (Ed.) *Algae As Ecological Indicators,* Academic Press, New York, pp. 237–256.

Braek, G. S., Jensen, A., and Mohus, A. (1976). Heavy metal tolerance of marine phytoplankton. III. Combined effect of copper and zinc ions in cultures of four common species. *J. Exp. Mar. Biol. Ecol.* **25,** 37–55.

Briggs, S. A. (1872). The diatomaceae of Lake Michigan. *Lens* 1, 41–44.

Brooks, A. S., and Torke, B. G. (1977). Vertical and seasonal distribution of chlorophyll *a* in Lake Michigan. *J. Fish. Res. Board Can.* **34,** 2280–2287.

Bruno, M. G., Fannin, T. E., and Leversee, G. J. (1982). The disposition of benzo[a]pyrene in the periphyton communities of two South Carolina streams: Uptake and biotransformation. *Can. J. Bot.* **60,** 2084–2091.

Burkholder, P. R. (1929). Microplankton studies of Lake Erie. *Bull. Buffalo Soc. Nat. Sci.* **14,** 73–93.

Butler, G. L. (1977). Algae and pesticides. *Residue Rev.* **66,** 19–62.

Butler, M., Haskew, A. E. J., and Young, M. M. (1980). Copper tolerance in the green alga, *Chlorella vulgaris. Plant, Cell Environ.* **3,** 118–126.

Cain, J. R., Paschal, D. C., and Hayden, C. M. (1980). Toxicity and bioaccumulation of cadmium in the colonial green alga *Scenedesmus obliquus. Arch. Environ. Contam. Toxicol.* **9,** 9–16.

Canterford, G. S. (1980). Formation and regeneration of abnormal cells of the marine diatom *Ditylum brightwellii* (West) Grunow. *J. Mar. Biol. Assoc. U.K.* **60,** 243–253.

Canterford, G. S., and Canterford, D. R. (1980). Toxicity of heavy metals to the marine diatom *Ditylum brightwellii* (West) Grunow: Correlation between toxicity and metal speciation. *J. Mar. Biol. Assoc. U.K.* **60,** 227–242.

Canterford, G. S., Buchanan, A. S., and Ducker, S. C. (1978). Accumulation of heavy metals by the marine diatom *Ditylum brightwellii* (West) Grunow. *Aust. J. Mar. Freshwater Res.* **29,** 613–622.

Chandler, D. C. (1940). Limnological studies of western Lake Erie. I. Plankton and certain physical-chemical data of the Bass Islands region, from September, 1938 to November, 1939. *Ohio J. Sci.* **40,** 291–336.

Chandler, D. C. (1942). Limnological studies of western Lake Erie. III. Phytoplankton and physical-chemical data from November, 1939 to November, 1940. *Ohio J. Sci.* **42,** 24–44.

Chandler, D. C. (1944). Limnological studies of western Lake Erie. IV. Relation of limnological and climatic factors to the phytoplankton of 1941. *Trans. Am. Microsc. Soc.* **63,** 203–236.

Chandler, D. C., and Weeks, O. B. (1945). Limnological studies of western Lake Erie. V. Relation of limnological and meteorological conditions to the production of phytoplankton in 1942. *Ecol. Monogr.* **15,** 435–457.

Chopra, I. (1971). Decreased uptake of cadmium by a resistant strain of *Staphylococcus aureus. J. Gen. Microbiol.* **63,** 265–267.

Christensen, E. R., Scherfig, J., and Dixon, P. S. (1979). Effects of manganese, copper and lead on *Selenastrum capricornutum* and *Chlorella stigmatophora. Water Res.* **13,** 79–92.

Cody, T. E., Radike, M. J., and Warshawsky, D. (1984). The phototoxicity of benzo[a]pyrene in the green alga *Selenastrum capricornutum. Environ. Res.* **35,** 122–132.

Conner, A. J. (1981). The differential sensitivity of phytoplankton to polychlorinated biphenyls when cultured heterotrophically and photoautotrophically. *Environ. Exp. Bot.* **21,** 241–247.

Conner, A. J., and Mahanty, H. K. (1979). Growth responses of unicellular algae to polychlorinated biphenyls: New evidence for photosynthetic inhibition. *Mauri Ora* **7,** 3–17.

Conway, H. L. (1978). Sorption of arsenic and cadmium and their effects on growth, micronutrient utilization, and photosynthetic pigment composition of *Asterionella formosa*. *J. Fish. Res. Board Can.* **35,** 286–294.

Crist, R. H., Oberholser, K., Shank, N., and Nguyen, M. (1981). Nature of bonding between metallic ions and algal cell walls. *Environ. Sci. Technol.* **15,** 1212–1217.

Daily, W. A. (1938). A quantitative study of the phytoplankton of Lake Michigan collected in the vicinity of Evanston, Illinois. *Butler Univ. Bot. Stud.* **4,** 65–83.

Damann, K. E. (1940). Phytoplankton studies of Lake Michigan at Evanston, Illinois. *Trans Ill. State Acad. Sci.* **33,** 68–70.

Damann, K. E. (1941). Quantitative study of the phytoplankton of Lake Michigan at Evanston, Illinois. *Butler Univ. Bot. Stud.* **5,** 27–44.

Damann, K. E. (1945). Plankton studies of Lake Michigan. I. Seventeen years of plankton data collected at Chicago, Illinois. *Am. Midl. Nat.* **34,** 769–796.

Daniel, G. F., and Chamberlain, A. H. L. (1981). Copper immobilization in fouling diatoms. *Bot. Mar.* **24,** 229–243.

Davies, A. G. (1974). The growth kinetics of *Isochrysis galbana* in cultures containing sublethal concentrations of mercuric chloride. *J. Mar. Biol. Assoc. U.K.* **54,** 157–169.

Davies, A. G. (1976). As assessment of the basis of mercury tolerance in *Dunaliella tertiolecta*. *J. Mar. Biol. Assoc. U.K.* **56,** 39–57.

Davies, A. G. (1978). Pollution studies with marine plankton. Part II. Heavy metals. *Adv. Mar. Biol.* **15,** 381–508.

Davies, A. G., and Sleep, J. A. (1979). Inhibition of carbon fixation as a function of zinc uptake in natural phytoplankton assemblages. *J. Mar. Biol. Assoc. U.K.* **59,** 937–949.

Davies, A. G., and Sleep, J. A. (1980). Copper inhibition of carbon fixation in coastal phytoplankton assemblages. *J. Mar. Biol. Assoc. U.K.* **60,** 841–850.

Davis, C. C. (1964). Evidence of the eutrophication of Lake Erie from phytoplankton records. *Limnol. Oceanogr.* **9,** 275–283.

Davis, C. C. (1966). Plankton studies in the largest Great Lakes of the world, with special reference to the St. Lawrence Great Lakes of North America. Great Lakes Res. Div. Pub. No. 14, Univ. Mich., pp. 1–36.

De Filippis, L. F. (1978). The effect of sub-lethal concentrations of mercury and zinc on *Chlorella*. IV. Characteristics of a general reducing enzyme system for metallic ions. *Z. Pflanzenphysiol.* **86,** 339–352.

De Filippis, L. F. (1979). The effect of heavy metals on the absorption spectra of *Chlorella* cells and chlorophyll solutions. *Z. Pflanzenphysiol.* **93,** 129–137.

De Filippis, L. F., and Pallaghy, C. K. (1976a). The effect of sublethal concentrations of mercury and zinc on *Chlorella*. I. Growth characteristics and uptake of metals. *Z. Pflanzenphysiol.* **78,** 197–207.

De Filippis, L. F., and Pallaghy, C. K. (1976b). The effect of sublethal concentrations of mercury and zinc on *Chlorella*. II. Photosynthesis and pigment composition. *Z. Pflanzenphysiol.* **78,** 314–332.

De Filippis, L. F., and Pallaghy, C. K. (1976c). The effect of sublethal concentrations of mercury and zinc on *Chlorella*. III. Development and possible mechanisms of resistance to metals. *Z. Pflanzenphysiol.* **79,** 323–335.

De Filippis, L. F., Hampp, R., and Ziegler, H. (1981). The effect of sublethal concentrations of zinc, cadmium, and mercury on *Euglena*. Adenylates and energy charge. *Z. Pflanzenphysiol.* **103,** 1–7.

De La Cruz, A. A. (1982). Effects of oil on phytoplankton metabolism in natural and experimental estuarine ponds. *Mar. Environ. Res.* **7,** 257–263.

Devi Prasad, P. V., and Devi Prasad, P. S. (1982). Effect of cadmium, lead and nickel on three freshwater green algae. *Water, Air, Soil Pollut.* **17,** 263–268.

Dobson, H. H. (1981). Trophic conditions and trends in the Laurentian Great Lakes. *WHO Water Qual. Bull.* **6,** 158–160.

Dongmann, G., and Nurnberg, H. W. (1982). Observations with *Thalassiosira rotula* (Meunier) on the toxicity and accumulation of cadmium and nickel. *Ecotoxicol. Environ. Saf.* **6,** 535–544.

Duthie, H. C., and Sreenivasa, M. R. (1971). Evidence for eutrophication of Lake Ontario from the sedimentary diatom succession. In Proc. 14th Conf. Great Lakes Res., pp. 1-3.

Ehrenberg, C. G. (1843). Verbreitung und Einfluss des mikroskopischen lebens in Sud-und Nord-Amerika. *Abh. K. Akad. Wiss. Berlin,* pp. 291–445.

Ehrenberg, C. G. (1854). *Mikrogeologie.* Leopold Voss, Leipzig.

Fisher, N. S., and Frood, D. (1980). Heavy metals and marine diatoms: Influence of dissolved organic compounds on toxicity and selection for metal tolerance among four species. *Mar. Biol.* (*Berlin*) **59,** 85–93.

Fisher, N. S., and Jones, G. J. (1981). Heavy metals and marine phytoplankton: Correlation of toxicity and sulfhydryl-binding. *J. Phycol.* **17,** 108–111.

Fisher, N. S., and Wurster, C. F. (1973). Individual and combined effects of temperature and polychlorinated biphenyls on the growth of three species of phytoplankton. *Environ. Pollut.* **5,** 205–212.

Fisher, N. S., Guillard, R. R. L., and Wurster, C. F. (1976). Effects of a chlorinated hydrocarbon pollutant on the growth kinetics of a marine diatom. In R. P. Canale (Ed.). *Modeling Biochemical Processes in Aquatic Ecosystems.* Ann Arbor Sci. Publ., Ann Arbor, MI, pp. 305–317.

Fisher, N. S., Jones, G. J., and Nelson, D. M. (1981). Effects of copper and zinc on growth, morphology, and metabolism of *Asterionella japonica* (Cleve). *J. Exp. Mar. Biol. Ecol.* **51,** 37–56.

Fisher, N. S., Burns, K. A., Cherry, R. D., and Heyraud, M. (1983). Accumulation and cellular distribution of 241-Am, 210-Po, and 210-Pb in two marine algae. *Mar. Ecol.: Prog. Ser.* **11,** 233–237.

Fisher, N. S., Bohe, M., and Teyssie, J.-L. (1984). Accumulation and toxicity of Cd, Zn, Ag, and Hg in four marine phytoplankters. *Mar. Ecol.: Prog. Ser.* **18,** 201–213.

Foster, P. L. (1977). Copper exclusion as a mechanism of heavy metal tolerance in a green alga. *Nature (London)* **269,** 322–323.

Frederick, V. R. (1981). Preliminary investigation of the algal flora in the sediments of Lake Erie. *J. Great Lakes Res.* **7,** 404–408.

Fujita, M., Iwasaki, K., and Takabatake, E. (1977). Intracellular distribution of mercury in freshwater diatom, *Synedra* cells. *Environ. Res.* **14,** 1–13.

Gachter, R., and Mares, A. (1979). Effects of increased heavy metal loads on phytoplankton communities. *Swiss J. Hydrol.* **41,** 228–246.

Gadd, G. M., and Griffiths, A. J. (1978). Microorganisms and heavy metal toxicity. *Microb. Ecol.* **4,** 303–317.

Gannon, J. E. (1969). *Great Lakes Plankton Investigations: A Bibliography,* Spec. Rep. No. 7. Cent. Great Lakes Stud., University of Wisconsin, Milwaukee.

Gavis, J. (1983). Toxic binding of cupric ion by marine phytoplankton. *J. Mar. Res.* **41,** 53–63.

Gavis, J., Guillard, R. R. L., and Woodward, B. L. (1981). Cupric ion activity and the growth of phytoplankton clones isolated from different marine environments. *J. Mar. Res.* **39,** 315–333.

Geisweid, H. J., and Urbach, E. (1983). Sorption of cadmium by the green microalgae *Chlorella vulgaris, Ankistrodesmus braunii,* and *Eremosphaera viridis. Z. Pflanzenphysiol.* **109,** 127–141.

Gerhards, U., and Weller, H. (1977). Die Aufnahme von Quecksilber, Cadmium und Nickel durch *Chlorella pyrenoidosa. Z. Pflanzenphysiol.* **82,** 292–300.

Geyer, H., Viswanathan, R., Freitag, D., and Korte, F. (1981). Relationship between water solubility of organic chemicals and their bioaccumulation by the alga *Chlorella. Chemosphere* **10,** 1307–1313.

Giddings, J. M., and Washington, J. N. (1981). Coal liquefaction products, shale oil, and petroleum. Acute toxicity to freshwater algae. *Environ. Sci. Technol.* **15,** 106–108.

Gipps, J. F., and Coller, B. A. W. (1980). Effect of physical and culture conditions on uptake of cadmium by *Chlorella pyrenoidosa. Aust. J. Mar. Freshwater Res.* **31,** 747–755.

Gipps, J. F., and Coller, B. A. W. (1982). Effect of some nutrient cations on uptake of cadmium by *Chlorella pyrenoidosa. Aust. J. Mar. Freshwater Res.* **33,** 979–987.

Glooschenko, V., and Glooschenko, W. (1975). Effect of polychlorinated biphenyl compounds on growth of Great Lakes phytoplankton. *Can. J. Bot.* **53,** 653–659.

Glooschenko, V., and Lott, J. N. A. (1977). The effects of chlordane on the green algae *Scenedesmus quadricauda* and *Chlamydomonas* sp. *Can. J. Bot.* **55,** 2866–2872.

Gottschall, R. Y. (1930). Preliminary report on the phytoplankton and pollution in Presque Isle Bay, Lake Erie. *Proc. Pa. Acad. Sci.* **4,** 1–11.

Gottschall, R. Y., and Jennings, O. E. (1933). Limnological studies at Erie, Pennsylvania. *Trans. Am. Microsc. Soc.* **52,** 181–191.

Greene, J. C., Miller, W. E. Shiroyama, T., and Maloney, T. E. (1975). Utilization of algal assays to assess the effects of municipal, industrial, and agricultural waste water effluents upon phytoplankton production in the Snake River system. *Water, Air, Soil Pollut.* **4,** 415–434.

Greene, J. C., Miller, W. E., Shiroyama, T., Soltero, R. A., and Putnam, K. (1978). Use of laboratory cultures of *Selenastrum, Anabaena* and the indigenous isolate *Sphaerocystis* to predict effects of nutrient and zinc interactions upon phytoplankton growth in Long Lake, Washington. *Mitt. Int. Ver. Limnol.* **21,** 372–384.

Gutknecht, J. (1963). Zn-65 uptake by benthic marine algae. *Limnol. Oceanogr.* **8,** 31–38.

Haffner, D. G., Griffith, M., and Herbert, P. D. N. (1984). Phytoplankton community structure and distribution in the nearshore zone of Lake Ontario. *Hydrobiologia* **114,** 51–66.

Hall, A., Fielding, A. H., and Butler, M. (1979). Mechanisms of copper tolerance in the marine fouling alga *Ectocarpus siliculosus*—Evidence for an exclusion mechanism. *Mar. Biol.* (*Berlin*) **54,** 195–199.

Hallas, L. E., Thayer, J. S., and Cooney, J. J. (1982). Factors affecting the toxic effect of tin on estuarine microorganisms. *Appl. Environ. Microbiol.* **44,** 193–197.

Hansen, P-D. (1979). Experiments on the accumulation of lindane (-BHC) by primary producers *Chlorella* spec. and *Chlorella pyrenoidosa. Arch. Environ. Contam. Toxicol.* **8,** 721–731.

Harding, J. P. C., and Whitton, B. A. (1977). Environmental factors reducing the toxicity of zinc to *Stigeoclonium tenue. Br. Phycol. J.* **12,** 17–21.

Harding, L. W., Jr., and Phillips, J. H., Jr. (1978). Polychlorinated biphenyls: Transfer from microparticulates to marine phytoplankton and the effects on photosynthesis. *Science* **202,** 1189–1192.

Harris, G. P., and Vollenweider, R. A. (1982). Paleolimnological evidence of early eutrophication in Lake Erie. *Can. J. Fish. Aquat. Sci.* **39,** 618–626.

Harrison, W. G., Eppley, R. W., and Renger, E. H. (1977). Phytoplankton nitrogen metabolism, nitrogen budgets, and observations on copper toxicity: Controlled ecosystem pollution experiment. *Bull. Mar. Sci.* **27,** 44–57.

Hart, B. A., and Scaife, B. D. (1977). Toxicity and bioaccumulation of cadmium in *Chlorella pyrenoidosa. Environ. Res.* **14,** 401–413.

Heldal, M., Norland, S., Lien, T., and Knutsen, G. (1984). Toxic responses of the green alga *Dunaliella bioculata* (Chlorophyceae, Volvocales) to selected oxidised hydrocarbons. *Environ. Pollut., Ser. A* **34,** 119–132.

Herbst, R. P. (1969). Ecological factors and the distribution of *Cladophora glomerata* in the Great Lakes. *Am. Midl. Nat.* **82,** 90–98.

Hohn, M. H. (1969). Qualitative and quantitative analysis of plankton diatoms, Bass Island area, Lake Erie. *Bull. Ohio Biol. Surv.* [N.S.] **3,** 1–211.

Holderness, J., Fenwick, M. G., and Lynch, D. L. (1975). The effect of methyl mercury on the growth of the green alga *Coelastrum microporum* Naeg. strain 280. *Bull. Environ. Contam. Toxicol.* **13,** 348–350.

Hollibaugh, J. T., Seibert, D. L. R., and Thomas, W. H. (1980). A comparison of the acute toxicities of ten heavy metals to phytoplankton from Saanich Inlet, B. C., Canada. *Estuarine Coastal Mar. Sci.* **10,** 93–105.

Hurlbert, S. H. (1975). Secondary effects of pesticides on aquatic ecosystems. *Residue Rev.* **57,** 82–148.

Hutber, G. N., Rogers, L. J., and Smith, A. J. (1979). Influence of pesticides on the growth of cyanobacteria. *Z. Allg. Mikrobiol.* **19,** 397–402.

Hutchinson, T. C., and Stokes, P. M. (1975). Heavy metal toxicity and algal bioassays. *ASTM Spec. Tech. Publ.* **STP 573,** 321–343.

Hutchinson, T. C., Hellebust, J. A., and Soto, C. (1981). Effect of napthalene and aqueous crude oil extracts on the green flagellate *Chlamydomonas angulosa.* IV. Decreases in cellular manganese and potassium. *Can. J. Bot.* **59,** 742–749.

Jennings, J. R., and Rainbow, P. S. (1979). Accumulation of cadmium by *Dunaliella tertiolecta* Butcher. *J. Plankton Res.* **1,** 67–74.

Jensen, T. E., Baxter, M., Rachlin, J. W., and Jani, V. (1982a). Uptake of heavy metals by *Plectonema boryanum* (Cyanophyceae) into cellular components, especially polyphosphate bodies: An X-ray energy dispersive study. *Environ. Pollut., Ser. A.* **27,** 119–127.

Jensen, T. E., Rachlin, J. W., Jani, V., and Warkentine, B. (1982b). An X-ray energy dispersive study of cellular compartmentalization of lead and zinc in *Chlorella saccharophila* (Chlorophyta), *Navicula incerta* and *Nitzschia closterium* (Bacillariophyta). *Environ. Exp. Bot.* **22,** 219–239.

Karydis, M. (1982). Toxicity of a photooxidised crude oil on two marine microalgae. *Bot. Mar.* **25,** 25–29.

Karydis, M., and Fogg, G. E. (1980). Physiological effects of hydrocarbons on the marine diatom *Cyclotella cryptica. Microb. Ecol.* **6,** 281–290.

Kayser, H., and Sperling, K-R. (1980). Cadmium effects and accumulation in cultures of *Prorocentrum micans* (Dinophyta). *Helgol. Wiss. Meeresunters.* **33,** 89–102.

Keil, J. E., Priester, L. E., and Sandifer, S. H. (1971). Polychlorinated biphenyl (Aroclor 1242): Effects of uptake on growth, nucleic acids, and chlorophyll of a marine diatom. *Bull. Environ. Contam. Toxicol.* **6,** 156–159.

Kilham, S. S. (1984). Silicon and phosphorus growth kinetics and competitive interactions between *Stephanodiscus minutus* and *Synedra* sp. *Verh. Int. Ver. Limnol.* **22,** 435–439.

Kilham, S. S., Knott, C. L., and Tilman, D. (1977). Phosphorus and silicate kinetics for the Lake Michigan diatom *Diatoma elongatum. J. Great Lakes Res.* **3,** 93–99.

Kingston, J. C., Lowe, R. L., Stoermer, E. F., and Ladewski, T. B. (1983). Spatial and temporal distribution of benthic diatoms in northern Lake Michigan. *Ecology* **64,** 1566–1580.

Kishler, J., and Taft, C. E. (1970). *Bangia atropurpurea* (Roth) in Western Lake Erie. *Ohio J. Sci.* **70,** 56–57.

Kusk, K. O. (1978). Effects of crude oil and aromatic hydrocarbons on the photosynthesis of the diatom *Nitzschia palea. Physiol. Plant.* **43,** 1–6.

Kusk, K. O. (1980). Effects of crude oils and aromatic hydrocarbons on the photosynthesis of three species of *Acrosiphonia* grown in the laboratory. *Bot. Mar.* **23,** 587–593.

Kusk, K. O. (1981a). Effects of hydrocarbons on respiration, photosynthesis and growth of the diatom *Phaeodactylum tricornutum. Bot. Mar.* **24,** 413–418.

Kusk, K. O. (1981b). Effects of naphthalene on the diatom *Phaeodactylum tricornutum* grown under varied conditions. *Bot. Mar.* **24,** 485–487.

Kusk, K. O. (1981c). Comparison of the effects of aromatic hydrocarbons on a laboratory alga and natural phytoplankton. *Bot. Mar.* **24,** 611–613.

Langford, R. R. (1946). The study of seasonal and annual plankton production in the eastern end of Lake Ontario. Proc. 9th Meet. Natl. Comm. Fish. Cult., Appendix D.

Lederman, T. C., and Rhee, G-Y. (1982). Influence of a hexachlorobiphenyl on Great Lakes phytoplankton in continuous culture. *Can. J. Fish. Aquat. Sci.* **39,** 388–394.

Les, A., and Walker, R. W. (1984). Toxicity and binding of copper, zinc, and cadmium by the blue-green alga, *Chroococcus paris. Water, Air, Soil Pollut.* **23,** 129–139.

Li, W. K. W. (1980). Cellular accumulation and distribution of cadmium in *Isochrysis galbana* during growth inhibition and recovery. *J. Plankton Res.* **2,** 283–294.

Lin, C. K., and Blum, J. L. (1977). Recent invasion of a red alga (*Bangia atropurpurea*) in Lake Michigan. *J. Fish. Res. Board Can.* **34,** 2413–2416.

Lorifice, G. J., and Munawar, M. (1974). The abundance of diatoms in the southwestern nearshore region of Lake Ontario during the spring thermal bar period. In Proc. 17th Conf. Great Lakes Res., pp. 619–628.

MacFarlane, R. B., Glooschenko, W. A., and Harriss, R. C. (1972). The interaction of light intensity and DDT concentration upon the marine diatom, *Nitzschia delicatissima* Cleve. *Hydrobiologia* **39,** 373–382.

Magnani, B., Powers, C. D., Wurster, C. F., and O'Connors, H. B. (1978). Effects of chlordane and heptachlor on the marine dinoflagellate, *Exuviella baltica,* Lohmann. *Bull. Environ. Contam. Toxicol.* **20,** 1–8.

Mahanty, H. K., and Fineran, B. A. (1976). Effects of a polychlorinated biphenyl (Aroclor 1242) on the ultrastructure of frond cells in the aquatic plant *Spirodela oligorrhiza* (Kurz) Hegelm. *N. Z. J. Bot.* **14,** 13–18.

Mahanty, H. K., and Gresshoff, P. M. (1978). Influence of polychlorinated biphenyls (PCBs) on growth of freshwater algae. *Bot. Gaz.* (*Chicago*) **139,** 202–206.

Mahanty, H. K., Fineran, B. A., and Gresshoff, P. M. (1983). Effects of polychlorinated biphenyls (Aroclor 1242) on the ultrastructure of certain planktonic algae. *Bot. Gaz.* (*Chicago*) **144,** 56–61.

Massalski, A., Laube, V. M., and Kushner, D. J. (1981). Effects of cadmium and copper on the ultrastructure of *Ankistrodesmus braunii* and *Anabaena* 7120. *Microb. Ecol.* **7,** 183–193.

Matile, P. (1975). *The Lytic Compartment of Plant Cells,* Cell Biol. Monogr., Vol. 1. Springer-Verlag, New York.

Matson, R. S., Mustoe, G. E., and Chang, S. B. (1972). Mercury inhibition on lipid biosynthesis in freshwater algae. *Environ. Sci. Technol.* **6,** 158–160.

McLean, M., and Williamson, F. B. (1977). Cadmium accumulation by the marine red alga *Porphyra umbilicalis. Physiol. Plant.* **41,** 268–272.

McLean, R. O. (1974). The tolerance of *Stigeoclonium tenue* Kutz to heavy metals in South Wales. *Br. Phycol. J.* **9,** 91–95.

McLean, R. O., and Benson-Evans, K. (1974). The distribution of *Stigeoclonium tenue* Kutz in South Wales in relation to its use as an indicator of organic pollution. *Br. Phycol. J.* **9,** 83–89.

Mechling, J. J., and Kilham, S. S. (1982). Temperature effects on silicon limited growth of the Lake Michigan diatom *Stephanodiscus minutus. J. Phycol.* **18,** 199–205.

Miesch, H-U., and Schmitt-Beckmann, I. (1979). Influence of tri- and hexavalent chromium on two *Chlorella* strains. *Z. Pflanzenphysiol.* **94,** 231–239.

Miller, W. E., Maloney, T. E., and Greene, J. C. (1974). Algal productivity in 49 lake waters as determined by algal assays. *Water Res.* **8,** 667–679.

Miller, W. E., Greene, J. C., and Shiroyama, T. (1978). The *Selenastrum capricornutum* Printz algal assay bottle test. Experimental design, application, and data interpretation protocol. U.S. Environ. Prot. Agency, Off. Res. Dev. Rep. EPA-600/9-78-018.

Moede, A., Greene, R. W., and Spencer, D. F. (1980). Effects of selenium on the growth and phosphorus uptake of *Scenedesmus dimorphus* and *Anabaena cylindrica*. *Environ. Exp. Bot.* **20,** 207–212.

Moll, R. A., and Stoermer, E. F. (1982). A hypothesis relating trophic status and subsurface chlorophyll maxima of lakes. *Arch. Hydrobiol.* **94,** 425–440.

Morris, O. P. (1972). Shipfouling properties of *Ectocarpus siliculosus*. Ph.D. Thesis, University of Liverpool.

Munawar, M., and Munawar, I. F. (1976). A lakewide study of phytoplankton biomass and its species composition in Lake Erie, April–December, 1970. *J. Fish. Res. Board Can.* **33,** 581–600.

Munawar, M., and Munawar, I. F. (1978). Phytoplankton of Lake Superior 1973. *J. Great Lakes Res.* **4,** 462–480.

Munawar, M., and Nauwerck, A. (1971). The composition and horizontal distribution of phytoplankton in Lake Ontario during the year 1970. Proc. 14th Conf. Great Lakes Res., pp. 69–78.

Naiki, N. (1980). Role of superoxide dismutase in a copper-resistant strain of yeast. *Plant Cell Physiol.* **21,** 775–783.

Naiki, N., and Yamagata, S. (1976). Isolation and some properties of copper-binding proteins found in a copper-resistant strain of yeast. *Plant Cell Physiol.* **17,** 1281–1295.

Nilsson, J. R. (1978). Retention of lead within the digestive vacuole in *Tetrahymena. Protoplasma* **95,** 163–173.

Nilsson, J. R. (1979). Intracellular distribution of lead in *Tetrahymena* during continuous exposure to the metal. *J. Cell Sci.* **39,** 383–396.

Nilsson, J. R. (1981). Effects of copper on phagocytosis in *Tetrahymena. Protoplasma* **109,** 359–370.

Odum, E. P. (1971). *Fundamentals of Ecology,* 3rd ed. Saunders, Philadelphia, PA.

Pawlaczyk-Szpilowa, M., and Slowik, J. (1981). The bioaccumulation of copper and lead by *Scendesmus obliquus. Acta Microbiol. Pol.* **30,** 79–87.

Powers, C. D., Rowland, R. G., O'Connors, H. B., Jr., and Wurster, C. F. (1977). Response to polychlorinated biphenyls of marine phytoplankton isolated cultured under natural conditions. *Appl. Environ. Microbiol.* **34,** 760–764.

Pribil, S., and Marvan, P. (1976). Accumulation of uranium by the chlorococcal alga *Scenedesmus quadricauda. Arch. Hydrobiol., Suppl.* **49,** 214–225.

Rabsch, U., and Elbrächter, M. (1980). Cadmium and zinc uptake, growth, and primary production in *Coscinodiscus granii* cultures containing low levels of cells and dissolved organic carbon. *Helgol. Wiss. Meeresunters.* **33,** 79–88.

Rachlin, J. W., and Farran, M. (1974). Growth response of the green alga *Chlorella vulgaris* to selective concentrations of zinc. *Water Res.* **8,** 575–577.

Rachlin, J. W., Jensen, T. E., Baxter, M., and Jani, V. (1982a). Utilization of morphometric analysis in evaluating response of *Plectonema boryanum* (Cyanophyceae) to exposure to eight heavy metals. *Arch. Environ. Contam. Toxicol.* **11,** 323–333.

Rachlin, J. W., Jensen, T. E., and Warkentine, B. (1982b). The growth response of the green alga (*Chlorella saccharophila*) to selected concentrations of the heavy metals Cd, Cu, Pb, and Zn. *Trace Subst. Environ. Health* **16,** 145–154.

Rachlin, J. W., Jensen, T. E., and Warkentine, B. (1982c). The growth responses of *Chlorella saccharophila, Navicula incerta,* and *Nitzschia closterium* to selected concentrations of cadmium. *Bull. Torrey Bot. Club* **109,** 129–135.

Rachlin, J. W., Jensen, T. E., and Warkentine, B. (1984). The toxicological response of the alga *Anabaena flos-aquae* (Cyanophyceae) to cadmium. *Arch. Environ. Contam. Toxicol.* **13,** 143–151.

Rai, L. C., Gaur, J. P., and Kumar, H. D. (1981). Phycology and heavy metal pollution. *Biol. Rev. Cambridge Philos. Soc.* **56,** 99–151.

Reddin, A., and Prendeville, G. N. (1981). Effect of oils on cell membrane permeability in *Fucus serratus* and *Laminaria digitata. Mar. Pollut. Bull.* **10,** 339–342.

Rensen, J. J. S. (1975). Lipid peroxidation and chlorophyll destruction caused by diquat during photosynthesis in *Scenedesmus. Physiol. Plant.* **33,** 42–46.

Riedel, G. F. (1984). Influence of salinity and sulfate on the toxicity of chromium (VI) to the estuarine diatom *Thalassiosira pseudonana. J. Phycol.* **20,** 496–500.

Rivkin, R. B. (1979). Effects of lead on growth of the marine diatom *Skeletonema costatum. Mar. Biol. (Berlin)* **50,** 239–247.

Rockwell, D. C., Devault, D. S., Palmer, M. F., Marion, C. V., and Bowden, R. J. (1980). Lake Michigan intensive survey 1976–1977. U.S. Environ. Prot. Agency Rep. EPA-905/4-80-003-A.

Roderer, G. (1984). On the toxic effects of tetraethyl lead and its derivatives on the chrysophyte *Poterioochromonas malhamensis.* V. Electron microscopical studies. *Environ. Exp. Bot.* **24,** 17–30.

Rodhe, W. (1978). Algae in culture and nature. *Mitt. Int. Ver. Limnol.* **21,** 7–20.

Rushforth, S. R., Brotherson, J. D., Fungladda, N., and Evenson, W. E. (1981). The effects of dissolved heavy metals on attached diatoms in the Unintah Basin of Utah. *Hydrobiologia* **83,** 313–323.

Sakaguchi, T., Tsuji, T., Nakajima, A., and Horikoshi, T. (1979). Accumulation of cadmium by green microalgae. *Eur. J. Appl. Microbiol. Biotechnol.* **8,** 207–215.

Sanders, J. G., Batchelder, J. H., and Ryther, J. H. (1981). Dominance of a stressed marine phytoplankton assemblage by a copper-tolerant pennate diatom. *Bot. Mar.* **24,** 39–41.

Sandmann, G., and Böger, P. (1980). Copper deficiency and toxicity in *Scenedesmus. Z. Pflanzenphysiol.* **98,** 53–59.

Say, P. J., and Whitton, B. A. (1980). Changes in flora down a stream showing a zinc gradient. *Hydrobiologia* **76,** 255–262.

Schelske, C. L. (1984). *In situ* and natural phytoplankton assemblage bioassays. In L. E. Shubert (Ed.), *Algae as Ecological Indicators.* Academic Press, New York, pp. 15–48.

Schelske, C. L., and Stoermer, E. F. (1971). Eutrophication, silica, and predicted changes in algal quality in Lake Michigan. *Science* **173,** 423–424.

Schelske, C. L., and Stoermer, E. F. (1972). Phosphorus, silica, and eutrophication of Lake Michigan. *Spec. Symp.—Am. Soc. Limnol. Oceanogr.* **1,** 157–171.

Schelske, C. L., Stoermer, E. F., and Feldt, L. E. (1971). Nutrients, phytoplankton productivity and species composition as influenced by upwelling in Lake Michigan. Proc. 14th Conf. Great Lakes Res., pp. 102–113.

Schelske, C. L., Feldt, L. E., Santiago, M. A., and Stoermer, E. F. (1972). Nutrient enrichment and its effects on phytoplankton production and species composition in Lake Superior. Proc. 15th Conf. Great Lakes Res., pp. 149–165.

Schelske, C. L., Rothman, E. D., Stoermer, E. F., and Simmons, M. A. (1974). Responses of phosphorus limited Lake Michigan phytoplankton to factorial enrichments with nitrogen and phosphorus. *Limnol. Oceanogr.* **19,** 409–419.

Schelske, C. L., Stoermer, E. F., Conley, D. J., Robbins, J. A., and Glover, R. M. (1983). Early eutrophication in the lower Great Lakes: New evidence from biogenic silica in sediments. *Science* **222,** 320–322.

Schenk, C. F., and Thompson, R. E. (1965). Long-term changes in water chemistry and abundance of plankton at a single sampling location in Lake Ontario. Great Lakes Res. Div. Publ. 13, Univ. Mich., pp. 197–208.

Schmidt, D. J., and Christensen, C. L. (1975). Diatoms as water quality indicators. I. *Phycologia* **30,** 369–374.

Schulz-Baldes, M., and Lewin, R. A. (1976). Lead uptake in two marine phytoplankton organisms. *Biol. Bull. (Woods Hole, Mass.)* **150,** 118–127.

Sell, D. W., Carney, H. J., and Fahnenstiel, G. L. (1984). Inferring competition between natural phytoplankton populations: The Lake Michigan example reexamined. *Ecology* **65,** 325–328.

Sicko, L. M. (1974). Physiological and cytological aspects of phosphate metabolism in *Plectonema boryanum.* Ph.D. Dissertation, City University of New York, New York.

Sicko-Goad, L. (1982). A morphometric analysis of algal response to low dose, short-term heavy metal exposure. *Protoplasma* **110,** 75–86.

Sicko-Goad, L., and Lazinsky, D. (1982a). Polyphosphate body formation and degradation in *Plectonema boryanum* (Cyanophyceae). *Micron* **13,** 459–460.

Sicko-Goad, L., and Lazinsky, D. (1982b). Polyphosphate in natural phytoplankton assemblages and its formation and degradation in cultured algal cells. In P. Archibald (Ed.), *State-of-the-Art Seminar in Environmental Biology,* Environ. Prot. Agency EPA-600/9-82-007. Washington, DC.

Sicko-Goad, L., and Lazinsky, D. (1983). Synergistic effects of phosphorus nutrient status and lead exposure in three algae. *Micron Microsc. Acta* **14,** 261–262.

Sicko-Goad, L., and Lazinsky, D. (1986). Quantitative ultrastructural changes associated with lead-coupled luxury phosphate uptake and polyphosphate utilization. *Arch. Environ. Contam. Toxicol.* **15,** 617–627.

Sicko-Goad, L., and Stoermer, E. F. (1979). A morphometric study of lead and copper effects on *Diatoma tenue* var. *elongatum* (Bacillariophyta). *J. Phycol.* **15,** 316–321.

Silverberg, B. A. (1975). Ultrastructural localization of lead in *Stigeoclonium tenue* (Chlorophyceae, Ulotrichales) as demonstrated by cytochemical and X-ray microanalysis. *Phycologia* **14,** 265–274.

Silverberg, B. A. (1976). Cadmium-induced ultrastructural changes in mitochondria of freshwater green algae. *Phycologia* **15,** 155–159.

Smith, M. A. (1983). The effect of heavy metals on the cytoplasmic fine structure of *Skeletonema costatum* (Bacillariophyta). *Protoplasma* **116,** 14–23.

Soto, C., Hellebust, J. A., and Hutchinson, T. C. (1977). Effect of naphthalene and aqueous crude oil extracts on the green flagellate *Chlamydomonas angulosa,* III. Changes in cellular composition. *Can. J. Bot.* **55,** 2765–2777.

Soto, C., Hutchinson, T. C., Hellebust, J. A., and Sheath, R. G. (1979a). The effect of crude oil on the morphology of the green flagellate *Chlamydomonas angulosa. Can. J. Bot.* **57,** 2717–2728.

Soto, C., Hellebust, J. A., Hutchinson, T. C., and Sheath, R. G. (1979b). The effect of the hydrocarbon naphthalene on the morphology of the green flagellate *Chlamydomonas angulosa. Can. J. Bot.* **57,** 2729–2739.

Spencer, D. F., and Greene, R. W. (1981). Effects of nickel on seven species of freshwater algae. *Environ. Pollut., Ser. A* **25,** 241–247.

Stevenson, R. J., and Stoermer, E. F. (1981). Quantitative differences between benthic algal communities along a depth gradient in Lake Michigan. *J. Phycol.* **17,** 29–36.

Stillwell, E. F., and Corum, K. E. (1982). Effects of cadmium and low magnesium on cell division and calcification in the marine coccolithophorid *Cricosphaera* (*Hymenomonas*) *carterae. Mar. Biol.* (*Berlin*) **66,** 227–230.

Stoermer, E. F. (1968). Nearshore phytoplankton populations in the Grand Haven, Michigan, vicinity during thermal bar conditions. Proc. 11th Conf. Great Lakes Res., pp. 137–150.

Stoermer, E. F. (1975). Comparison of benthic diatom communities in Lake Michigan and Lake Superior. *Verh. Int. Verh. Limnol.* **19,** 932–938.

Stoermer, E. F. (1978). Phytoplankton as indicators of water quality in the Laurentian Great Lakes. *Trans. Am. Microsc. Soc.* **97,** 2–16.

Stoermer, E. F. (1984). Research on Great Lakes algal communities: Problems from the past, lessons for the future. *J. Great Lakes Res.* **10,** 179–188.

Stoermer, E. F., and Kreis, R. G., Jr. (1980). Phytoplankton composition and abundance in southern Lake Huron. Great Lakes Res. Div. Spec. Rep. 65, Univ. Mich., pp. 1–382.

Stoermer, E. F., and Stevenson, R. J. (1980). *Green Bay Phytoplankton Composition, Abundance, and Distribution.* Great Lakes Nat. Program Off., U.S. Environ. Prot. Agency, Chicago, IL.

Stoermer, E. F., and Theriot, E. C. (1985). Phytoplankton distribution in Saginaw Bay. *J. Great Lakes Res.* **11,** 132–142.

Stoermer, E. F., and Yang, J. J. (1969). Plankton diatom assemblages in Lake Michigan. Great Lakes Res. Div. Spec. Rep. 47, Univ. Mich., pp. 1–169.

Stoermer, E. F., Bowman, M. M. Kingston, J. C., and Schaedel, A. L. (1974). Phytoplankton composition and abundance in Lake Ontario during IFYGL. Great Lakes Res. Div. Spec. Rep. 53, Univ. Mich., pp. 1–373.

Stoermer, E. F., Ladewski, B. G., and Schelske, C. L. (1978). Population responses of Lake Michigan phytoplankton to nitrogen and phosphorus enrichment. *Hydrobiologia* **57,** 249–265.

Stoermer, E. F., Sicko-Goad, L., and Frey, L. C. (1983a). *Effects of Phosphorus Loading on Phytoplankton Distribution and Certain Aspects of Cytology in Saginaw Bay, Lake Huron.* U.S. Environ. Prot. Agency, Duluth, MN.

Stoermer, E. F., Kreis, R. G., Jr., and Ladewski, T. B. (1983b). *Phytoplankton Species Composition, Abundance and Distribution in Southern Lake Huron, 1980.* U.S. Environ. Prot. Agency, Duluth, MN.

Stoermer, E. F., Wolin, J. A., Schelske, C. L., and Conley, D. J. (1985a). Post settlement diatom succession in the Bay of Quinte, Lake Ontario. *Can. J. Fish. Aquat. Sci.* **42,** 754–767.

Stoermer, E. F., Wolin, J. A., Schelske, C. L., and Conley, D. J. (1985b). An assessment of changes during the recent history of Lake Ontario based on siliceous microfossils preserved in the sediments. *J. Phycol.* **21,** 257–276.

Stoermer, E. F., Wolin, J. A., Schelske, C. L., and Conley, D. J. (1985c). Variations in *Melosira islandica* valve morphology related to eutrophication and silica depletion. *Limnol. Oceanogr.* **30,** 414–418.

Stoermer, E. F., Kociolek, J. P., Schelske, C. L., and Conley, D. J. (1985). *Siliceous Microfossil Succession in the Recent History of Lake Superior.* Acad. Nat. Sci., Philadelphia, PA, pp. 137, 106–118.

Stokes, P. M., Hutchinson, T. C., and Krauter, K. (1973). Heavy-metal tolerance in algae isolated from contaminated lakes near Sudbury, Ontario. *Can. J. Bot.* **51,** 2155–2168.

Stratton, G. W. (1984). Effects of the herbicide atrazine and its degradation products, alone and in combination, on phototrophic microorganisms. *Arch. Environ. Contam. Toxicol.* **13,** 35–42.

Stratton, G. W., and Corke, C. T. (1979). The effect of nickel on the growth, photosynthesis, and nitrogenase activity of *Anabaena inaequalis. Can. J. Microbiol.* **25,** 1094–1099.

Stratton, G. W., Huber, A. L., and Corke, C. T. (1979). Effect of mercuric ion on the growth, photosynthesis, and nitrogenase activity of *Anabaena inaequalis. Appl. Environ. Microbiol.* **38,** 537–543.

Taft, C. E., and Kishler, W. J. (1973). *Cladophora as Related to Pollution and Eutrophication in Western Lake Erie.* Water Resour. Cent., Ohio State University, Columbus.

Thomas, B. W., and Chase, H. H. (1887). Diatomaceae of Lake Michigan as collected during the past 16 years from the water supply of the city of Chicago. *Notarisia* **2,** 328–330.

Thomas, W. H., and Seibert, D. L. R. (1977). Effects of copper on the dominance and the diversity of algae: Controlled ecosystem pollution experiment. *Bull. Mar. Sci.* **27,** 23–33.

Thomas, W. H., Hollibaugh, J. T., and Seibert, D. L. R. (1980a). Effects of heavy metals on the morphology of some marine phytoplankton. *Phycologia* **19,** 202–209.

Thomas, W. H., Hollibaugh, J. T., Seibert, D. L. R., and Wallace, G. T., Jr. (1980b). Toxicity of a mixture of ten metals to phytoplankton. *Mar. Ecol.: Prog. Ser.* **2,** 213–220.

Throndsen, J. (1982). Oil pollution and plankton dynamics, III. Effects on flagellate communities in controlled ecosystem experiments in Lindaspollene, Norway, June 1980 and 1981. *Sarsia* **67,** 163–169.

Tilman, D. (1977). Resource competition between planktonic algae: An experimental and theoretical approach. *Ecology* **58,** 338–348.

Tilman, D., and Kilham, S. S. (1976). Phosphate and silicate growth kinetics of the diatoms *Asterionella formosa* and *Cyclotella meneghiniana* in batch and semi-continuous culture. *J. Phycol.* **12,** 375–383.

Tubea, B., Hawxby, K., and Mehta, R. (1981). The effects of nutrient, pH, and herbicide levels on algal growth. *Hydrobiologia* **79,** 221–227.

Vandermeulen, J. H., Silvert, W., and Foda, A. (1983). Sublethal hydrocarbon phytotoxicity in the marine unicellular alga *Pavlova lutheri* Droop. *Aquat. Toxicol.* **4,** 31–49.

Van Oosten, J. (1957). *Great Lakes Fauna, Flora, and Their Environment—A Bibliography.* Great Lakes Comm., Ann Arbor, MI.

Vargo, G. A., Hutchins, M., and Almquist, G. (1982). The effect of low, chronic levels of No. 2 fuel oil on natural phytoplankton assemblages in microcosms. 1. Species composition and seasonal succession. *Mar. Environ. Res.* **6,** 245–264.

Vaughn, J. C. (1961). Coagulative difficulties of the south district filtration plant. *Pure Water* **14,** 34–39.

Véber, K., Zahradnik, J., Breyl, I., and Kredl, F. (1981). Toxic effect and accumulation of atrazine in algae. *Bull. Environ. Contam. Toxicol.* **27,** 872–876.

Vollenweider, R. A., Munawar, M., and Stadelmann, P. (1974). A comparative review of phytoplankton and primary production in the Laurentian Great Lakes. *J. Fish. Res. Board Can.* **31,** 739–762.

Vorce, C. M. (1882). Microscopic forms observed in the waters of Lake Erie. *Proc. Am. Soc. Microsc.* **5,** 187–196.

Ward, H. B. (1896). A biological examination of Lake Michigan in the Traverse Bay region. *Bull. Mich. Fish Comm.* **6,** 1–71.

Ware, G. W., and Roan, C. C. (1970). Interaction of pesticides with aquatic microorganisms and plankton. *Residue Rev.* **33,** 15–45.

Weber, A., Melkonian, M., Lorch, D. W., and Wettern, M. (1978). Accumulation of lead by several green algae. *Mitt. Int. Vern. Limnol.* **21,** 254–260.

Wihlidal, H., and Broda, E. (1978). Aufnahme von markiertem Zink durch *Chlorella* unterschiedlichen Zinkgehalts. *Z. Allg. Mikrobiol.* **18,** 447–451.

Woiter, K., Rabsch, U., Krischker, P., and Davies, A. G. (1984). Influence of low concentrations of cadmium, copper and zinc on phytoplankton of natural water samples. *Mar. Ecol.: Prog. Ser.* **19,** 167–173.

Wolfe, T. L., and Sweeney, R. A. (1982). Laurentian Great Lakes *Cladophora* bibliography. *J. Great Lakes Res.* **8,** 201–237.

Wong, P. T. S., Chau, Y. K., and Luxon, P. L. (1978). Toxicity of a mixture of metals on freshwater algae. *J. Fish. Res. Board Can.* **35,** 479–481.

Wong, P. T. S., Chau, Y. K., and Pate, D. (1982). Physiological and biochemical responses of several freshwater algae to a mixture of metals. *Chemosphere* **11,** 367–376.

Wong, P. T. S., Maguire, R. J., Chau, Y. K., and Kramar, O. (1984). Uptake and accumulation of inorganic tin by a freshwater alga, *Ankistrodesmus falcatus. Can. J. Fish. Aquat. Sci.* **41,** 1570–1574.

Wong, S. L., and Beaver, J. L. (1980). Algal bioassays to determine toxicity of metal mixtures. *Hydrobiologia* **74,** 199–208.

Zachleder, V., Abarzua, S., and Wittenburg, E. (1983). Effects of 3,4-benzopyrene on the course of cell cycle events in the chlorococcal alga *Scenedesmus quadricauda. Planta* **157,** 432–440.

# 3

# THE EFFECTS OF TOXIC SUBSTANCES ON ZOOPLANKTON POPULATIONS: A GREAT LAKES PERSPECTIVE

***Marlene S. Evans***

*Great Lakes Research Division, The University of Michigan, Ann Arbor, Michigan 48109*

***and***

***Donald C. McNaught***

*Department of Ecology and Behavioral Biology, University of Minnesota, Minneapolis, Minnesota 55455*

---

## 1. INTRODUCTION

Over the last several decades, Great Lakes zooplankton communities have been exposed to a variety of environmental pertubations including eutrophication, alterations in indigenous fish communities, thermal pollution, and persistent toxic substances. Understanding the effects of these pertubations on zooplankton populations requires (1) information on the structural and functional features of zooplankton communities in the absence of the pertubation, (2) information on how the pertubation affects the structural and functional characteristics of zooplankton communities, and (3) environmental studies designed to demonstrate cause and effect relationships.

The focus of Great Lakes zooplankton studies has changed over the decades from simple studies describing species composition and abundance to more complex ecosystem studies. In the late 1960s and early 1970s, the major environmental issue was eutrophication and surveillance programs were specifically designed to investigate this issue. While alterations in planktivorous fish communities were known to affect zooplankton community structure in small lakes, such considerations were not incorporated into the design of eutrophication surveillance studies. Thus, changes in zooplankton community structure resulting from eutrophication cannot easily be separated from changes in structure resulting from alterations in fish communities. A second issue of concern in the early 1970s was the effects of power plant operation on the various components of Great Lakes food webs. Numerous monitoring studies were conducted at power plants located along the Great Lakes shorelines. Toxics also emerged as an environmental issue during the 1970s. The primary foci of toxicological studies were the determination of (1) ambient levels of toxic substances in water, (2) concentrations in fish and sediments, and (3) the transport and fate of toxic substances. In contrast to eutrophication and thermal pollution studies, no monitoring studies were specifically designed to investigate the effects of toxic substances on Great Lakes zooplankton communities. Thus, due to insufficient research, our understanding of how toxic substances affect Great Lakes zooplankton communities is limited.

In this chapter, we discuss (1) how toxic substances can affect zooplankton, both at the species and community level, (2) modifying factors affecting toxicity, (3) the effects of various environmental pertubations, including toxic substances, on Great Lakes zooplankton, and (4) the role of zooplankton in the transport, persistence, and biomagnification of toxic substances.

## 2. INDICES OF TOXICITY

Toxic substances, as metabolic poisons, can affect a variety of biochemical and physiological functions. These effects can, in turn, become manifest at the community level as longevity, fecundity, growth, and so on, of the component zooplankton species are altered. Such effects have been investigated through four basic levels of organization:

1. Functional responses of single zooplankton species to single toxic substances
2. Functional responses of single zooplankton species to multiple toxic substances, including mixed effluents and sediment extracts
3. Functional and structural responses of natural assemblages of zooplankton to a single toxic compound
4. Functional and structural responses of natural assemblages of zooplankton to multiple toxic substances, including mixed effluents.

Functional indices include mortality, longevity, behavior, and physiological and biochemical responses. Structural indices include estimates of numbers of zooplankton species (or genera), abundance of various developmental stages, and percentage composition within the community. Less commonly used are diversity indices such as the Shannon-Wiener diversity index, percentage similarity, community similarity, and evenness.

## 3. EFFECTS OF TOXIC SUBSTANCES ON ZOOPLANKTON

The following paragraphs provide a broad overview of the effects of toxic substances on zooplankton. This overview is not comprehensive but illustrates the complexity of toxicity responses.

### 3.1. Single Species Responses to Single Toxic Substances

#### *3.1.1. Test Responses*

Most toxicological studies involving single species and single compounds have focused on acute effects. Such studies provide insight into the relative toxicity of various compounds, differential sensitivity among species, and factors affecting species sensitivity (Biesinger and Christensen, 1972; Baudouin and Scoppa, 1974; Nebeker and Puglisi, 1974; Passino and Novak, 1984).

Mortality measures are most commonly used in short-term (hours) studies while physiological measures are commonly used in longer-term studies. Physiological responses most commonly measured are reproduction and growth, parameters that should have a direct effect on zooplankton abundance. Ingestion and respiration are less commonly measured, possibly because these rates are less easily determined. Excretion and egestion are the least common measures used. Neither function is easily measured in the laboratory. Furthermore, the ecological significance of impaired excretion rates is more difficult to evaluate than other functional indices such as impaired reproduction or shortened longevity.

Toxicity studies based on growth and reproduction (Maki and Johnson, 1975; Nebeker and Puglisi, 1974) are time-consuming, requiring information on animal size, egg numbers, abundance of neonates, and so forth. However, such studies are warranted since reproductive measures appear to be more sensitive indicators of chronic stress than mortality (Marshall, 1966, 1978; Biesinger and Christensen, 1972). Moreover, reproductive fitness (a measure of the number of progeny successfully produced) clearly is strongly related to such processes. The most commonly tested zooplankton are members of the genus *Daphnia,* species that have relatively rapid reproductive rates and are readily cultured in the laboratory.

Although mortality, growth, and reproduction are the three most commonly studied responses of zooplankton to toxic substances, a smaller number of studies have investigated the effects of toxic substances on grazing. McNaught et al. (1982) found that 25 μg/L 2,2′-PCB had no effect on grazing by *Cyclops* and nauplii, but that the degradation product of this congener reduced grazing rates by 43% at 25 μg/L. In the laboratory, toxaphene at experimental concentrations of 0.15–15 μg/

L (in Lake Superior water) reduced *Diaptomus sicilis* grazing by 8–65% and *Limnocalanus macrurus* grazing by 3–57% (McNaught et al., 1984).

Reduced grazing in the presence of toxic compounds has one important implication: if grazing rates are sufficiently reduced, phytoplankton standing stocks may increase. This is a relevant Great Lakes issue since many regions of relatively high concentrations of toxic substances also are regions where phytoplankton standing stocks are large. Although excessive nutrient loading is clearly a major factor affecting phytoplankton abundance in such regions, toxic substances, by reducing zooplankton grazing, may also be a contributing factor.

### *3.1.2. Differential Species Sensitivity*

The results of several studies suggest that zooplankton species may differ in their sensitivity to toxic substances. A good demonstration of differential sensitivity is provided by an investigation of the acute toxicity of 13 metals to three zooplankton species (the cyclopoid *Cyclops abyssorum prealpinus,* the calanoid *Eudiaptomus padunus padunus,* and the cladoceran *Daphnia hyalina*) (Baudouin and Scoppa, 1974). Zooplankton were most sensitive to copper, cadmium, zinc, and mercury and were least sensitive to cesium, magnesium, strontium, and calcium: similar results were obtained testing *Daphnia magna* (Biesinger and Christensen, 1972). The three zooplankton species differed in their tolerance to metal toxicity. In general, *Cyclops* had the highest resistance to metals while *Daphnia* had the lowest resistance. Although zooplankton differed in their sensitivity to metals, the mode of toxic action appeared to be similar. This was inferred from the observation that toxicity curves (time for 50% mortality versus metal concentration) were parallel for the three zooplankton species tested. However, in some cases (e.g., cesium, cadmium, and copper for *Cyclops,* and mercury and calcium for *Daphnia*), the three species regression lines were not parallel, suggesting that the mode of action of these metal differed qualitatively and quantitatively among zooplankton species.

Differential sensitivities to toxic substances have been detected not only in acute tests (Baudouin and Scoppa, 1974) but in chronic tests. The cladoceran *Daphnia pulex* is less sensitive to short- and long-term exposures to dieldrin than the calanoid copepod *Eurytemora affinis* (Daniels and Allen, 1981), contrasting with the results of another study (Baudouin and Scoppa, 1974) where the cladoceran species tested was more sensitive than the calanoid species tested. The greater tolerance of *D. pulex* to dieldrin was related to its simpler life cycle and its larger body size (with the concurrent smaller surface:volume ratio).

Differences in sensitivity to toxic substances have also been detected within taxonomic orders of zooplankton such as the Cladocera. For example, *Bosmina longirostris* (a small cladoceran species) is more sensitive than *Daphnia pulex* to arsenate and DDT (Passino and Novak, 1984).

Although differences in taxonomic sensitivity to inorganic and organic toxic compounds clearly exist, too few laboratory studies have been conducted to classify species on the basis of their tolerance to toxic substances. This is in constrast to other laboratory studies that have allowed researchers to classify organisms on the basis of their tolerance to oxygen, pH, salinity, and temperature.

### *3.1.3. Modifying Factors*

Many environmental factors may modify toxicity. Although limited research has been conducted in this area, there has been a sufficient amount to illustrate the difficulties in extrapolating the results of laboratory studies to real-world aquatic environments.

*3.1.3.1. Water Hardness.* Chemical factors such as water hardness can affect the sensitivity of zooplankton to toxic substances. *Bosmina longirostris* is more sensitive to DDT in hard than soft water (Passino and Novak, 1984). Similarly, water hardness appears to be an important factor modifying the sensitivity of *Daphnia magna* to DDT stress. Animals reared in soft water had a 48-hour $EC_{50}$* (immobilization concentration) of 1.1 $\mu$g DDT/L while animals reared in hard water had a 48-hour $EC_{50}$ of 0.68 $\mu$g DDT/L: tests were conducted in hard water (Berglind and Dave, 1984). DDT alters sodium and potassium ion transport across membranes and thus may have a different effect on zooplankton communities in soft versus hard waters. This suggests that DDT may be more toxic to zooplankton in the hard waters of Lakes Erie and Ontario than in the softer waters of Lakes Huron and Superior. However, water hardness does not modify toxicity of all substances. It does not affect the sensitivity of *Daphnia magna* to potassium bichromate, pentachlorophenol, zinc sulfate, and tetraproplenbenzyl (Berglind and Dave, 1984).

*3.1.3.2. Food Quantity and Quality.* Food is another important modifying factor affecting zooplankton sensitivity to toxic stress. *Daphnia magna* is more sensitive to chronic copper stress when fed a trout-granule diet than when fed vitamin-enriched algae (Winner et al., 1977). Measures of stress included survival, mean brood size, and the instantaneous rate of population growth. In contrast to chronic toxicity, no differences in sensitivity were observed in acute experiments conducted over 72 hours. This suggests that the mechanism of toxic action was different at chronic and acute levels. It also suggests that the ratio between chronic and acute toxicity may vary with environmental conditions. Zooplankton may be more sensitive to stress from toxic substances in winter, when algal standing stocks and primary productivity are low, than in spring when food levels are higher.

The sensitivity of zooplankton to toxic substances may depend on food quality in addition to quantity. The presence of unpalatable species such as blue-green algae, indigestable species such as gelatinous greens, or species larger or smaller than the optimal feeding size may all modify toxicity by affecting grazing and ultimately metabolism. Zooplankton may be especially sensitive to toxic substances in eutrophic waters such as Lake Erie, Green Bay, or Saginaw Bay, where food quality may be relatively low due to the dominance of blue-green and gelatinous green algae. Zooplankton species with poor selective feeding capabilities, for example, *Daphnia* may be especially vulnerable. Food effects also may be

*$EC_{50}$ = median effective concentration for immobilization.

indirect. Toxic substances may alter phytoplankton composition and subsequently affect zooplankton fecundity (Sanders, 1986).

Food level, as a modifying factor, may have different effects on different zooplankton species. Food quality and quantity appears to strongly affect competitive interactions between calanoid copepods and *Daphnia* (McNaught, 1975; Richman and Dodson, 1983). Calanoid copepods, because of their selective feeding capabilities, may be better adapted to eutrophic environments than the passive filter-feeder, *Daphnia*. In eutrophic environments, *Daphnia* may pay a high energetic cost in rejecting undesirable algal species collected on the maxillae (Richman and Dodson, 1983; Porter and McDonough, 1984). Thus, in eutrophic environments, calanoid copepods may be better able to withstand stresses from toxic substances than *Daphnia*. Even within the Cladocera, differences in feeding responses may have significant effects on the ability of organisms to withstand stresses from exposure to toxic substances. *Bosmina* is capable of both filter feeding and of selective feeding (DeMott, 1982; Porter and McDonough, 1984). Thus, in oligotrophic and eutrophic environments, it may have a greater potential to withstand stresses induced by toxic substances than *Daphnia*.

*3.1.3.3. Compensation Mechanisms.* Compensation mechanisms can modify the population response to a chronic toxicant stress. For example, toxic compounds may reduce life expectancy. As animal densities decline, more food is available for the survivors and individual fecundity may increase (Marshall, 1978). At intermediate levels of toxicity, increased birth rates may compensate for increased death rates. However, such compensation mechanisms cannot compensate for increased toxic stress after the maximum attainable birth rate has been reached. Thus, at high levels of toxicity, population densities decline (Marshall, 1978).

*3.1.3.4. Exploitation.* Exploitation may also modify the response of zooplankton to environmental stresses. For example, the effect of chronic, sublethal radiation stress on *Daphnia pulex* populations is modified by density-independent, nonselective exploitation (Marshall, 1967). At low exploitation levels (15% of the population per week), standing stock is strongly reduced as radiation level increases. However, at high exploitation levels (90%) per week, differences in radiation levels have less effects on population abundance. Thus, these two stresses act antagonistically on *Daphnia* populations.

Exploitation and environmental stress may act synergistically, for example, when the largest components of the population are removed: such exploitation occurs when planktivorous fish exert size-selective predation pressures on the zooplankton community. As adult zooplankton are consumed, the population is less able to reproductively compensate for mortality, such as from exposure to a toxic compound. In the Great Lakes, zooplankton may be especially sensitive to toxic substances in regions or times when fish predation is relatively high, for example, in nearshore nursery areas, from spring to late summer (Evans et al., 1980), or in Lake Michigan during the mid 1960s, at the height of the alewife population explosion (Wells, 1970).

## 3.2. Single Species Responses to Multiple Toxic Substances

Zooplankton are exposed to multiple contaminants in the Great Lakes, especially in heavily polluted industrial areas. Unfortunately, only a few laboratory studies have investigated the effects of multiple contaminants on zooplankton functional responses. Fewer studies have investigated the response of zooplankton to naturally contaminated waters containing multiple contaminants. Such waters include mixed effluents collected along the course of a river, waters overlying contaminated sediments, and sediment extracts.

### *3.2.1. Two Compounds*

Both synergistic and antagonistic responses have been observed in studies testing the toxicity of two compounds. PCB (Aroclor 1254), while having no effect on *Daphnia magna* production and survival at 12 ppb, increased susceptibility to *p,p'*-DDT by approximately one-third (Maki and Johnson, 1975). DDT at 0.5 ppb increased the toxicity of PCB by a factor of approximately two. Thus, a significant interaction existed between the two organic contaminants but the interactions were less than additive.

Selenium at concentrations of 5 ppb may reduce chronic toxicity in *Daphnia magna* exposed to copper (Winner, 1984). However, at very low concentrations (essentially 0 ppb), a selenium deficiency may aggravate copper stress. One measure of such stress was an increase in antennal damage.

### *3.2.2. Mixed Effluents*

One of the newest approaches for investigating the toxicity of mixed effluents is the Mount-Norberg three-brood, 7-day test which uses the cladoceran *Ceriodaphnia* (Mount and Norberg, 1984). This rapid test allows investigation of relative toxicity along rivers or other environmental gradients. Applications of this test are discussed by Di Toro et al. (Chapter 20).

Mixed effluents may reduce zooplankton grazing in addition to impairing reproduction. *Daphnia magna* grazing was reduced by exposure to polluted waters from the Niagara River (Mirza, in Fitchko, 1986). *Daphnia retrocurva* filtering rate was reduced when animals were exposed to water containing dilute pulp mill effluent (Cooley, 1977).

### *3.2.3. Sediment Extracts*

Relatively few researchers have attempted to identify which toxicants in mixed effluents or contaminated sediments have the major role in producing the observed response. *Daphnia magna* 96-hour mortalities were significantly correlated with cyanide and zinc in contaminated sediments from various Lake Michigan harbors (Laskowski-Hoke and Prater, 1981). In contrast, mayfly (*Hexagenia limbata*) mortalities were significantly correlated with suspended solids, cadmium, chromium, cyanide, mercury, and zinc. In a later study, more sophisticated multivariate analyses were used (Laskowski-Hoke and Prater, 1984). *Daphnia* mortalities due to exposure to contaminated sediments were associated with lead, iron, copper, and chromium, while mortalities due to exposure to elutriate water were associated with nickel, ammonia, sulfate, and total Kjedahl nitrogen. Such studies illustrate the difficulties in attempting to evaluate the toxicity of sediments (including interstitial water) and the overlying water column. These problems are magnified severalfold when a researcher attempts to understand toxicity effects on communities indigenous to the receiving water body

## 3.3. Zooplankton Assemblage Responses to Single Toxic Substances

### *3.3.1. Mesocosm Studies*

Chronic responses of natural zooplankton assemblages to a toxic compound typically have been conducted in enclosures. The results of these studies have been difficult to interpret for several reasons. First, when a toxic substance perturbs one or several components of a test ecosystem, a complex series of interactions occur. Interpreting such results is limited by our poor understanding of ecosystem structure and function and of differences in species sensitivity to toxic substances. Second, because mesocosm experiments seldom have been replicated, the representativeness of experimental results is uncertain. Finally, few intra- and intersystem ecosystem studies have been conducted. Thus, only limited information on broad ecological principles regarding ecosystem response to toxic substances is available.

*3.3.1.1. CEPEX Studies.* CEPEX (controlled ecosystem pollution experiments) studies represent a complex series of enclosure experiments designed to investigate the effects of copper on several trophic levels, including zooplankton. Although many effects were detected, causal factors were often ambiguous. For example, ctenophore and medusa abundances were reduced by copper concentrations of 5, 10, and 50 $\mu$g/L (Gibson and Grice, 1977). However, abundances remained sufficiently high in all test exposures for these carnivores to exert significant predation pressure on macrozooplankton. Thus, although copepod abundances were severely reduced at a copper exposure of 50 $\mu$g/L, the additive effects of copper toxicity and predation could not be separated (Gibson and Grice, 1977). Although microzooplankton populations also were affected by copper, the direct effects of copper toxicity could not be distinguished from indirect effects resulting from modifications to other trophic levels of the contained population (Beers et al., 1977). For example, as copepod standing stocks were reduced at high copper concentrations, phytoplankton standing stocks increased. Such an increase may have been related to reduced zooplankton grazing (Reeve et al., 1977a).

In addition to examining structural responses of the zooplankton community to copper addition, a number of functional responses were investigated. Reeve et al. (1977a) determined that egestion rates and feeding rates were reduced by exposure to high (up to 10 $\mu$g/L) copper concentrations. Reductions in feeding may have been a functional response to environmental conditions rather than sublethal effects since feeding rates recovered when animals were placed in clean water. Excretion and respiration rates were less sensitive indicators of sublethal stress than were fecal pellet and egg production rates (Reeve et al., 1977b).

*3.3.1.2. Loch Ewe Studies.* Results obtained from a less intensive series of experiments conducted in Loch Ewe indicate that copper stress may affect herbivorous zooplankton more than carnivorous zooplankton (Gamble et al., 1977). These results resemble those obtained CEPEX studies (Gibson and Grice, 1977). Variations in herbivorous zooplankton abundances at high copper concentrations probably were due both to direct copper effects and to the relative increase in predator numbers.

*3.3.1.3. Other Freshwater Studies.* The insecticide permethrin was acutely toxic to all crustacean (copepods and cladocerans) zooplankton in limnocorrals at concentrations of 0.5–50 μg/L while rotifers showed acute effects only at 50 μg/L (Kaushik et al., 1985). Despite their initial adverse response to permethrin, rotifers increased dramatically in abundance in all treated enclosures and dominated the zooplankton community in 3–4 weeks. This rapid increase was related to the inherent rapid reproductive capabilities of rotifers and their release from competitive and predation pressures exerted by certain crustacean zooplankton species. A similar increase in rotifer dominance occurred in Experimental Lakes Area cadmium experiments (Marshall and Mellinger, 1980).

More recently, fish zooplankton phytoplankton interactions were investigated in 100-$m^3$ enclosures contaminated with 1, 10, and 100 μg/L selenium (Salki et al., 1985). No obvious changes in zooplankton abundance, composition, and reproduction were noted in the 1- and 10-μg/L enclosures in comparison to the controls. Differences in species dynamics in the 100-μg/L enclosure were related to biotic interactions due to greater numbers of planktivorous fish introduced in this enclosure relative to the other enclosures. Test concentrations of selenium may have been too low to detect any effects since the 48-hour $EC_{50}$ was 2500 μg/L.

### 3.3.2. *Interlake and Intralake Comparisons*

Marshall and his collaborators conducted similar microcosom experiments both within a single lake (Marshall et al., 1981) and between two lakes (Marshall and Mellinger, 1980). Although they did not design their studies to investigate intralake and interlake responses of zooplankton communities to toxics, such comparisons can be made a posteriori.

In their first study, Marshall and Mellinger (1980) investigated the structural and functional responses of zooplankton communities to cadmium in the Experimental Lakes Area (late July–early August 1977) and Lake Michigan (August 1977). Zooplankton density and species diversity were not affected by cadmium concentrations of ≤0.6 and 1.2 μg Cd/L, respectively, but percentage similarity was significantly reduced by concentrations of ≥0.2 μg Cd/L. However, while zooplankton community structure changed in response to cadmium during their 3-week experiments, such changes were not consistent between lakes. At the end of the 3-week experiments conducted in Lake Michigan, cladocerans accounted for 84.4% of the crustacean zooplankton) with increasing cadmium concentration (from 61.9% of the zooplankton in the control enclosure but only 50.5% in the 5.0-μg Cd/L enclosure. However, in the Experimental Lake experiments, cladocerans increased in dominance relative to copepods with increasing cadmium levels: cladocerans increased from 24.8% in the control enclosure to 60.0% in the 30-μg Cd/L enclosure. Rotifers (not studied in the Lake Michigan experiments) increased in dominance (relative to total crustacean zooplankton) with increasing cadmium concentration (from 61.9% in the control experiment to 93.1% in the 30-μg cadmium/L experiment). A similar increase in rotifer dominance was observed in permethrin limnocorral experiments (Kaushik et al., 1985).

Later, the combined effects of cadmium and zinc on Lake Michigan zooplankton communities were investigated (Marshall et al., 1981). In July 1978, cladocerans decreased slightly in dominance (71.4% for the control to 60.9% for the 2 μg Cd/L + 100 μg Zn/L experiments) with increasing cadmium and zinc concentration while, in August 1978, decreases in dominance (75.9% for the control to 34.1% for the 2 μg Cd/L + 100 μg Zn/L experiments) were more pronounced at the end of three weeks. A decrease in cladoceran dominance with increasing cadmium level also occurred in summer 1977 Lake Michigan experiments (Marshall and Mellinger, 1980). This apparent consistency in community response requires further investigation to determine the causal factors and the potential usefulness of this response as an indicator of toxic stress in Lake Michigan.

#### *3.3.3. Seasonality in Toxicity*

Seasonal variations in the sensitivity of the zooplankton community to toxics such as metals may be significant. For example, a seasonal cycle in the toxicity of cadmium, copper, mercury, and lead, but not for arsenic occurred in the biomass production dynamics of nauplii collected from Burlington Canal (Lake Ontario). Nauplii were most sensitive to metals in late fall and winter and most resistant in summer (Borgmann et al., 1980). Since most microcosm and mesocosm experiments have been conducted in summer, the sensitivity of the zooplankton community to toxic stress may have been underestimated during other seasons of the year. In a similar study of metal mixtures on biomass kinetics of copepods, toxicity was usually accounted for by the summation of noninteractive, single metal effects (Borgmann, 1980).

### 3.4. Responses of Natural Assemblages to Multiple Toxic Substances

#### *3.4.1. Mesocosm studies*

Few studies have investigated the combined effects of toxic substances on zooplankton communities. One such study was conducted by Marshall et al. (1981). In summer 1978, cadmium did not effect zinc toxicity, whereas zinc reduced cadmium uptake and toxicity.

Apparently only one mesocosm study has investigated the effects of multicontaminants on zooplankton community structure. Two limnocorrals in Lake Baldigg, a highly eutrophic Swiss Lake, were continuously loaded with low concentrations of mercury, copper, cadmium, zinc, and lead (Urech, 1979). Marked decreases in crustacean zooplankton abundances were noted but the causal factors were not identified. Egg development time and egg production rates apparently were affected by the metal mixture but responses varied with the species considered. Food supply levels were lower in the experimental limnocorrals suggesting that phytoplankton were also adversely affected.

### *3.4.2. Lake Studies Comparing Polluted and Control Areas*

Laboratory studies have shown that zooplankton grazing may be impaired by toxic substances. Similar results have been obtained using natural assemblages of zooplankton: ingestion rates may vary between polluted and clean areas. For example, crustacean zooplankton collected from the polluted waters of Saginaw Bay had ingestion rates only 13.2% that of similarly fed Lake Huron zooplankton (McNaught et al., 1980). Inhibition varied with the zooplankton species. *Diaptomus* spp. and *Chydorus sphaericus* had experimental grazing rates that were less than 10% of that observed for the Lake Huron populations, while *Eubosmina* grazing rate was reduced by only 79%.

### *3.4.3. River Studies*

Little research has been conducted exposing natural assemblages of phytoplankton and zooplankton to mixed effluents. However, in recent years, such studies have been initiated in various Great Lakes rivers, such as the Raisin River, Michigan. In May 1984, upstream and downstream effluents inhibited gross photosynthesis of a natural assemblage of phytoplankton. The cladoceran *Ceriodaphnia* was used to investigate impaired reproduction: reproduction was inhibited in upstream but stimulated in downstream effluents. Finally, two species of herbivorous zooplankton were used in grazing experiments: grazing was inhibited in upstream but stimulated in downstream effluents (D. C. McNaught, S. Bridgham, and C. Meadows, unpublished report).

Interpreting river studies requires several considerations. First, effluent composition along the length of the river must be known. Some regions receive various combinations of nutrients from agricultural lands and urban areas, stimulating photosynthesis, while other regions receive toxic substances, inhibiting photosynthesis and zooplankton growth; some regions receive a mixture of nutrients and toxic compounds, for example, fertilizers and pesticides from agricultural lands. Second, some care also must be exercised in discussing the results of mixed effluent bioassay studies; results may not be representative of a true community response, especially when indigenous species are not used in the bioassays.

## 4. GREAT LAKES ZOOPLANKTON COMMUNITIES

In the preceding paragraphs, we have discussed experimental approaches used to investigate the effects of toxic substances on the functional and structural characteristics of zooplankton populations. Toxicity is a function of several factors including physical–chemical variables such as the nature of the compound, exposure time, water hardness, and the co-occurrence of other contaminants. Important biological variables include intraspecific differences in sensitivity, differences in sensitivity as a function of the age of the organism, food abundance, season, and the actual stress parameter being investigated. For zooplankton, reproductive measures appear to be more sensitive than measures based on mortality or physiological functions such as egestion. At the community level, interactions among competitors, and direct effects on lower (e.g., phytoplankton) and higher (e.g., predators) trophic levels may be important.

Resolving what effects toxic substances have had on Great Lakes zooplankton communities is difficult. Two specific problems are that (1) zooplankton communities in the Great Lakes have been affected by a variety of environmental perturbations, and (2) few studies were designed specifically to investigate the effects of toxic substances on zooplankton communities. These points are further developed in the following paragraphs.

## 4.1. General Features

Great Lakes zooplankton communities are composed primarily of two groups of organisms: rotifers and crustaceans. Crustacean zooplankton can further be subdivided into copepods (cyclopoids and calanoids) and cladocerans. These taxa differ significantly in their physiology (reproductive rates, longevity, etc.), size, and feeding characteristics (Allen, 1976).

Rotifers, the smallest animals, have the most rapid reproductive rates, while copepods have the longest generation times. Thus, rotifers have the greatest potential to respond to increased food levels in the environment, such as as occurs with eutrophication. Furthermore, rotifers may exhibit greater recovery rates following some nonspecific environmental stress, for example, toxic substances (Anderson, 1970; Hurlbert et al., 1972; Papst and Boyer, 1980). Rotifers, because of their small size, are especially vulnerable to invertebrate predators while adult copepods and cladocerans tend to be more vulnerable to fish predators (Allen, 1976; Stemberger and Evans, 1984). While numerous studies have been conducted on Great Lakes zooplankton, rotifers have been investigated in only a few studies with the primary focus either seasonality in community structure or variations in abundance in response to trophic gradients (Stemberger, 1974; Duffy and Liston, 1978; Stemberger et al., 1979; Evans, 1986).

Most research on Great Lakes zooplankton communities has focused on copepods and cladocerans. Studies conducted over the decades (Robertson and Gannon, 1981) show that crustacean zooplankton exhibit significant seasonal and spatial gradients in abundance and community structure. The nature of these gradients vary within and between lakes as a function of trophic status, abundance of predators, season, and physical regime. In general, copepods dominate the winter and early spring zooplankton communities whereas cladocerans are major components during summer and autumn. During thermal stratification, cladocerans tend to be especially abundant in nearshore waters and, in particular, eutrophic regions.

Although the Lake Michigan zooplankton community is dominated by microzooplankton rotifers, cladocerans, and copepods, certain macrobenthic organisms may, on occasion, migrate into the water column, becoming part of the planktonic community, for example, amphipods and mysids (Mozley and Howmiller, 1977). Of these, the mysid *Mysis relicta* is the strongest vertical migrator and an abundant omnivore in the colder, offshore regions of the Great Lakes. As carnivores, mysids feed most intensively on the smaller zooplankton (Bowers and Vanderploeg, 1982). Thus, they may affect zooplankton community structure, especially in the offshore region (Evans and Jude, 1986). Mysids have been poorly studied in the Great Lakes because these highly motile macroinvertebrates are difficult to sample. Since mysids may modify zooplankton community structure, especially in the offshore area, more information is required on these important macroinvertebrates.

The epibenthic and benthic microcrustacean communities require more study. These copepods and cladocerans may reach relatively high abundances at the sediment–water interface, especially in nearshore regions (Napela and Quigley, 1983; Evans, 1984). Many taxa migrate toward the surface at night (Evans and Stewart, 1977).

## 4.2. Eutrophication and Zooplankton Community Structure

Zooplankton abundance in the Great Lakes varies as a function of trophic status. Mean summer zooplankton abundance in Lakes Superior, Huron, Erie, and Ontario is positively correlated with chlorophyll concentration and phosphorus loading (Patalas, 1972). Similar relationships between zooplankton abundance and phosphorus have been developed for other systems (Shortreed and Stockner, 1986). Each zooplankton species is uniquely adapted for a suite of environmental conditions, including food abundance. Some species are found primarily in nutrient-rich waters while others are more typically found in oligotophic waters (Gannon and Stemberger, 1978). While the mechanisms affecting these adaptations are unclear, sufficient information is known about zooplankton community structure in a variety of environments to use the presence of various species as indicators of trophic status.

Many Great Lakes studies have investigated the relationship between the zooplankton abundance and composition and eutrophication. Such studies include interlake comparisons (Patalas, 1972; Watson, 1974; Watson and Carpenter, 1974) and intralake comparisons (Gannon, 1974; Stemberger et al., 1979; Evans, 1986). There is less documentation of the effects of long-term eutrophication on zooplankton community structure. Relatively few zooplankton studies were conducted in the Great Lakes prior to the 1970s when eutrophication became of concern. Differences in sampling methods and locations preclude the ''linking'' of the various disparate studies. The notable pre-1970s exception is Bradshaw's (1964) western Lake Erie study: copepods and cladocerans increased markedly in abundance between 1938–39, 1948–49, and 1959 as phytoplankton standing stocks increased (Davis, 1964). More recently, a strong decrease in zooplankton abundance in

Saginaw Bay occurred between 1974 and 1980 in apparent response to improvements in water quality, that is, a reduction in phosphorus loading (McNaught et al., unpublished manuscript). Finally, reductions in total phosphorus loading and decreases in alewife (*Alosa pseudoharengus*) abundance (which affected zooplankton grazing) have been implicated as factors affecting major improvements in water quality of offshore Lake Michigan waters, especially during summer 1983–84 (Scavia et al., 1986).

Moderate eutrophication of various regions of the Great Lakes potentially could mitigate the effects of toxic compounds. This could occur as zooplankton reproduction increases with increased food level. However, in highly eutrophic regions, the presence of undesirable algal species may act synergistically with toxic substances. Passive filter feeders such as *Daphnia* may be especially vulnerable to such interactions.

## 4.3. Fish and Zooplankton Community Structure

Many fish are planktivores and potentially can cause significant changes in zooplankton community structure. Planktivorous fish exert size-selective predation pressures on the zooplankton community, feeding most intensively on larger zooplankton. Thus, as fish predators such as the alewife increase in abundance, zooplankton community structure is shifted toward small species (Brooks and Dodson, 1965).

Major alterations in Great Lakes fish community structure occurred as eutrophication intensified (Christie, 1974), making it difficult to separate the effects of these two factors on long-term trends in zooplankton community structure (Brooks, 1969). No studies were designed specifically to investigate the effects of changes in fish populations on Great Lakes zooplankton communities.

The first concrete evidence that fish predation could affect long-term changes in Great Lakes zooplankton community structure is the classic study of Wells (1970) in offshore Lake Michigan. Large-bodied zooplankton decreased in abundance between summer 1955 (pre-alewife population explosion) and 1966 (at the height of the alewife population explosion). There was some evidence of recovery in the abundance of large zooplankton in summer 1968 (following a massive dieoff of alewife in spring 1967). More recently, major shifts occurred in the abundance and composition of large-bodied *Daphnia* in the inshore and offshore waters of Lake Michigan. Such shifts appear related to declines in alewife abundance (Evans and Jude, 1986).

Planktivorous fish inhabit specific regions and depths of the lake during various seasons of the year. Thus, predation pressure on the zooplankton community varies seasonally and spatially. Such effects can be inferred through consideration of spatial gradients in zooplankton size (Hawkins and Evans, 1979) and community structure (Evans et al., 1980). Alterations in fish community structure can affect the inshore and offshore region of Lake Michigan differently (Evans and Jude, 1986). In the inshore region of Lake Michigan, a region of intense fish planktivory during much of the year, recent increases in yellow perch (*Perca flavescens*) abundance apparently caused a major decline in summer zooplankton abundance; zooplankton community structure remained unchanged (Evans, 1986).

Planktivorous fish, by consuming the largest zooplankton, may act synergistically with stresses resulting from exposure to toxic substances. As adult zooplankton are removed by planktivorous fish, zooplankton are less able to compensate (through reproduction) for mortality resulting from chemical stresses. The largest zooplankton species and zooplankton inhabiting the nearshore region from spring to autumn should be especially vulnerable to interactions between toxic substances and fish planktivores.

## 4.4. Thermal Pollution and Zooplankton Communities

As power plant construction increased through the late 1960s and early 1970s, the environmental effects of such operations became of concern. As a consequence of Public Laws 316a and 316b, numerous monitoring studies were conducted around Great Lakes power plants during the 1970s. The studies investigated (1) damage inflicted on organisms during plant passage and (2) alterations in community structure in the vicinity of thermal discharges. Power plant operation appears to be a minor source of environmental stress to the Great Lakes zooplankton community under typical operating conditions (Evans, 1981; Evans et al., 1986). Power plant operation appears to have less potential than eutrophication and fish predation to interact with stresses resulting from exposure to toxic substances.

Consideration of the ecological effects of power plants has resulted in beneficial but costly studies. In contrast, substantially less effort has been directed toward investigating the effects of the discharges of persistent toxic contaminants on the functional and structural characteristics of Great Lakes aquatic communities.

## 4.5. Toxic Substances and Zooplankton Community Structure

As shown in Section 3, no basic principles have emerged about the effects of toxic substances on zooplankton community structure. While zooplankton species may differ in their sensitivity to toxic substances, too few studies have been conducted to classify species on the basis of their tolerances to toxic compounds.

Several long-term changes in Great Lakes zooplankton communities have co-occurred with changes in toxic chemical discharges to the Great Lakes. However, toxic substances have not been shown to cause such changes. The increased abundance of zooplankton in Lake Erie (Bradshaw, 1964) appears to be strongly related to eutrophication, while shifts in the abundance of large-bodied zooplankton in the offshore waters of Lake Michigan appear to be related to changes in fish planktivore abundance (Wells, 1970; Evans and Jude, 1986). Neither laboratory nor mesocosm studies have suggested that toxic substances cause increases in zooplankton abundance or shifts toward larger species. Similarly, recent declines in summer zooplankton abundance in the inshore region of Lake Michigan appear to be related to changes in fish planktivore abundance (Evans, 1986).

Numerous studies have investigated spatial gradients in zooplankton community structure in regions where toxic concentrations are relatively high, such as Milwaukee Harbor, Green Bay, Saginaw Bay, and Lake St. Clair (Leach, 1973; Gannon, 1974; Stemberger, 1974; Stemberger et al., 1979). Spatial gradients in zooplankton community structure in these regions have been investigated in terms of eutrophication but not in terms of other sources of pollution. However, these areas are affected by other pertubations including relatively high toxic chemical concentrations, low dissolved oxygen concentrations, the presence of oils and greases, and altered fish populations. Since toxic substances were not investigated in the above studies, these variables cannot be discounted as factors affecting the zooplankton community.

In recent years, improved research strategies have been developed for investigating the effects of toxic substances on Great Lakes zooplankton communities. The *Ceriodaphnia* test, developed by Mount and Norberg (1984), has been used to assess toxicity of effluents along the Raisin River and is now being used in the Detroit River. Such bioassay studies may soon be applied to other regions of the Great Lakes where toxic concentrations are of concern.

## 5. OTHER CONSIDERATIONS OF TOXICS AND ZOOPLANKTON

Zooplankton potentially can affect the cycling of contaminants through the food web. This includes metabolizing, transporting, and biomagnifying contaminants.

### 5.1. Metabolism

Zooplankton can metabolize various toxic organic contaminants into forms that may be more or less toxic (and persistent) than the parent compound. For example, *Daphnia pulex* can metabolize significant amounts of DDT to DDE but only small amounts of DDD; dieldrin is considerably more resistant to biodegradation (Neudorf and Khan, 1975). *Daphnia pulex* metabolizes the polynuclear aromatic hydrocarbon anthracene slowly (Herbes and Risi, 1978). Many compounds such as photodieldrin and nickel (Khan et al., 1975; Hall, 1982) have significant elimination rates. However, where uptake rates exceed elimination rates, accumulation of the toxic substances occurs (Khan et al., 1975; Herbes and Risi, 1978). Such bioaccumulation may lead to biomagnification of the contaminant from one trophic level to another.

### 5.2. Transport

Zooplankton potentially can physically transport toxic substances through the Great Lakes ecosystem. Many toxic substances enter the lakes from the atmosphere, occurring in enriched concentrations in the surface microlayer (Elzerman and Armstrong, 1979). Such contaminants may become associated with organic platelets and aggregates and thereby be consumed by zooplankton (McNaught, 1982). Alternatively, toxic substances are taken up by phytoplankton and consumed by zooplankton. Phytoplankton and zooplankton take up contaminants rapidly relative to fish (Reinert, 1972) and thus may act as biological sponges for newly discharged contaminants into Great Lakes waters.

In the marine environment, zooplankton fecal pellets have a major role in transporting materials, including toxics, from the surface to the sediments (Elder and Fowler, 1977). Fecal pellets, because of their size and relatively high density, sink rapidly. Thus, substances in fecal pellets are more likely to reach the ocean floor than those contained in more slowly sinking particles, which may be consumed or decompose before reaching sedimentary sinks. Zooplankton egestion products may be similarly important in the open waters of the Great Lakes (Evans et al., 1982; M. S. Evans, unpublished manuscript).

### 5.3. Biomagnification

Zooplankton help biomagnify contaminants from one trophic level to another. Some compounds such as polychlorinated biphenyls and DDT are strongly biomagnified from one trophic level to another while others such as lindane, dieldrin, photodieldrin, chlordane, selenium, mercury, nickel, plutonium, copper, cadmium, zinc, and lead are only weakly biomagnified (Khan and Khan, 1974; Adams and Johnson, 1977; Gachter and Geiger, 1979; Thomann, 1981; Borgmann and Whittle, 1983; Biddinger and Gloss, 1984; Watras et al., 1985).

## 6. CONCLUSIONS

### 6.1. Experimental Studies

1. Tests involving single species provide useful information on acute and chronic toxicity levels of contaminants and allow a number of variables to be investigated. These include comparative toxicity of compounds, interactive effects of multiple compounds, differential species sensitivity, the effects of food level, water hardness, and nonspecific and specific exploitation pressures in modifying a stress response. More information is required on the mode of action of toxic substances and modifying factors.

2. Tests involving single species and exposures to mixed effluents and sediment extracts allow for toxicity evaluation of real-world waters and sediments. Such tests are useful to evaluate the severity of toxic loadings in various areas. Additional research is required to identify the more toxic compounds in mixed effluents and sediments.

3. Tests involving natural zooplankton assemblages and exposure to single or multiple compounds provide information on community response. Such tests are time-consuming and consequently expensive. The primary value of these tests is

to illustrate the complexity of community response. The nature of such responses will vary from community to community, with season, and probably with location.

4. A limited amount of research has been conducted on the response of natural assemblages of zooplankton to mixed effluents. The results of such tests are the most difficult to interpret.

5. Functional responses based on reproduction appear to provide one of the most sensitive and ecologically appropriate measures of stress resulting from exposure to toxic substances. Such responses are best studied on organisms with short generation times and rapid reproductive rates, for example, cladocerans such as *Daphnia* and *Ceriodaphnia*. Although not commonly studied, rotifers also may be useful test organisms, especially those that carry their eggs.

6. There is some evidence that toxic substances may reduce zooplankton grazing, possibly resulting in an increase in phytoplankton standing stocks. This has important implications to those regions of the Great Lakes in which relatively high concentrations of toxic compounds co-occur with large standing stocks of phytoplankton.

## 6.2. Community Structure Studies

1. Very little information exists on the effects of toxic substances on the community structure of Great Lakes zooplankton. Toxic substances may have and continue to affect spatial and temporal changes in zooplankton communities. However, the detection of these effects is difficult because (1) toxic substances have complex modes of action on natural ecosystems, (2) unlike eutrophication and fish predation, no general principles regarding the effects of toxic substances on zooplankton community structure exist, (3) several toxic substances generally co-occur in polluted areas, and (4) other co-occurring pertubations (eutrophication, alterations in fish communities) may mask the effects of toxic substances. More information is needed to define general modes of action of toxic substances on zooplankton communities. This includes information on the relative sensitivity of macro- versus microzooplankton to toxics, and differential sensitivity within species. More information is required on how other factors may modify such sensitivity to toxic substances, including eutrophication and alterations in planktivorous fish abundances.

2. The most sensitive indicators of the mode of action of toxics on zooplankton assemblages are based on species data, such as composition and diversity measures. Change in total abundance is not as sensitive. Species data also provide valuable insight into other environmental changes, especially those related to the fish community and eutrophication.

3. More information is required on the effects of toxic chemical and nutrient discharges on plankton population dynamics, especially along the course of rivers. This would allow for the evaluation of the effects of mixed (toxic chemicals and nutrients) discharges on zooplankton communities at river mouths, in the inshore region, and, eventually, the offshore region of the Great Lakes.

## ACKNOWLEDGMENTS

Drs. W. Gardner and P. Landrum, Great Lakes Environmental Research Laboratory, Ann Arbor, and Dr. James Bowers, Great Lakes Research Division, University of Michigan, provided valuable suggestions for improvement of the manuscript. Contribution number 472, Great Lakes Research Division.

## REFERENCES

Adams, W. J., and Johnson, H. E. (1977). Survey of the selenium content of the aquatic biota of western Lake Erie. *J. Great Lakes Res.* **3,** 10–14.

Allen, J. D. (1976). Life history patterns in zooplankton. *Am. Nat.* **110,** 165–180.

Anderson, R. S. (1970). Effects of rotenone on zooplankton communities and a study of their recovery patterns in two mountain lakes in Alberta. *J. Fish. Res. Board Can.* **27,** 1335–1356.

Baudouin, M. F., and Scoppa, P. (1974). Acute toxicity of various metals to freshwater zooplankton. *Bull. Environ. Control Toxicol.* **12,** 745–751.

Beers, J. R., Stewart, G. L., and Hoskins, K. D. (1977). Dynamics of micro-zooplankton populations treated with copper: Controlled ecosystem pollution experiment. *Bull. Mar. Sci.* **27,** 66–79.

Berglind, R., and Dave, G. (1984). Acute toxicity of chromate, DDT, PCP, TPBS, and zinc to *Daphnia magna* cultured in hard and soft water. *Bull. Environ. Control Toxicol.* **33,** 63–68.

Biddinger, G. R., and Gloss, S. P. (1984). The importance of trophic transfer in the bioaccumulation of chemical contaminants in aquatic ecosystems. *Residue Rev.* **91,** 103–145.

Biesinger, K. E., and Christensen, G. M. (1972). Effects of various metals on survival, growth, reproduction, and metabolism of *Daphnia magna. J. Fish. Res. Board Can.* **29,** 1691–1700.

Borgmann, U. (1980). Interactive effects of metals in mixtures on biomass production kinetics of freshwater copepods. *Can. J. Fish. Aquat. Sci.* **37,** 1295–1302.

Borgmann, U., and Whittle, D. M. (1983). Particle-size-conversion efficiency and contaminant concentrations in Lake Ontario biota. *Can. J. Fish. Aquat. Sci.* **40,** 328–336.

Borgmann, U., Cove, R., and Loveridge, C. (1980). Effects of metals on the biomass production kinetics of freshwater copepods. *Can. J. Fish. Aquat. Sci.* **37,** 567–575.

Bowers, J. A., and Vanderploeg, H. A. (1982). In situ predatory behavior of *Mysis relicta* in Lake Michigan. *Hydrobiologia* **93,** 121–131.

Bradshaw, A. S. (1964). The crustacean zooplankton picture: Lake Erie 1939-49-59; Cayuga 1910-51-61. *Verh. Int. Ver. Limnol.* **15,** 700–706.

Brooks, J. L. (1969). Eutrophication and changes in the composition of the zooplankton. In *Eutrophication: Causes, Consequences, Correctives.* Nat. Acad. Sci. Publ. 1700, pp. 236–255.

Brooks, J. L., and Dodson, S. I. (1965). Predation, body size, and composition of plankton. *Science* **250,** 28–35.

Christie, W. J. (1974). Changes in the fish species composition of the Great Lakes. *J. Fish. Res. Board Can.* **31,** 827–854.

Cooley, J. M. (1977). Filtering rate performance of *Daphnia retrocurva* in pulp mill effluent. *J. Fish. Res. Board Can.* **34,** 863–868.

Daniels, R. E., and Allen, J. D. (1981). Life table evaluation of chronic exposure to a pesticide. *Can. J. Fish. Aquat. Sci.* **38,** 485–494.

Davis, C. C. (1964). Evidence for the eutrophication of Lake Erie from phytoplankton records. *Limnol. Oceanogr.* **9,** 275–283.

DeMott, W. R. (1982). Feeding selectivities and relative ingestion rates of *Daphnia* and *Bosmina. Limnol. Oceanogr.* **27,** 518–527.

Di Toro, D. M., Hallden, J. A., and Plafkin, J. L. (1988). Modeling *Ceriodaphnia* toxicity in receiving waters. In M. S. Evans (Ed.) *Toxicant Contaminants and Ecosystem Health: A Great Lakes Focus.* Wiley, New York.

Duffy, W. G., and Liston, C. R. (1978). Seasonal abundance of planktonic rotifers in a nearshore area of central Lake Michigan. *J. Great Lakes Res.* **4,** 46–49.

Elder, D. L., and Fowler, S. W. (1977). Polychlorinated biphenyls: Penetration into the deep ocean by zooplankton fecal pellet transport. *Science* **197,** 459–461.

Elzerman, A. W., and Armstrong, D. E. (1979). Enrichment of Zn, Cd, Pb, and Cu in the surface microlayer of Lakes Michigan, Ontario, and Mendota. *Limnol. Oceanogr.* **24,** 133–144.

Evans, M. S. (1981). Distributions of zooplankton populations within and adjacent to a thermal plume. *Can. J. Fish. Aquat. Sci.* **38,** 441–448.

Evans, M. S. (1984). Benthic and epibenthic (microcrustaceans, macrobenthos) community structure in the vicinity of a power plant, southeastern Lake Michigan. *Verh. Int. Ver. Limnol.* **22,** 488–494.

Evans, M. S. (1986). Recent declines in zooplankton populations in the inshore region of Lake Michigan: Probable causes and consequences. *Can. J. Fish. Aquat. Sci.* **43,** 154–159.

Evans, M. S. (1986). Lake Huron rotifer and crustacean zooplankton, April–July 1980. *J. Great Lakes Res.* **12,** 281–292.

Evans, M. S., and Jude, D. J. (1986). Recent shifts in *Daphnia* community structure in southeastern Lake Michigan: A comparison of the inshore and offshore regions. *Limnol. Oceanogr.* **31,** 56–67.

Evans, M. S., and Stewart, J. A. (1977). Epibenthic and benthic microcrustaceans (copepods, cladocerans, ostracods) from a nearshore area in southeastern Lake Michigan. *Limnol. Oceanogr.* **22,** 1059–1066.

Evans, M. S., Hawkins, B. E., and Sell, D. W. (1980). Seasonal features of zooplankton assemblages in the nearshore area of southeastern Lake Michigan. *J. Great Lakes Res.* **6,** 275–289.

Evans, M. S., Bathelt, R. W., and Rice, C. P. (1982). PCBs and other toxicants in *Mysis relicta. Hydrobiologia* **93,** 205–215.

Evans, M. S., Warren, D. J., and Page, D. I. (1986). The effects of power plant passage on zooplankton mortalities: Eight years of study at the Donald C. Cook Nuclear Plant. *Water Res.* **20,** 725–734.

Fitchko, J. (1986). *Literature Review of the Effects of Persistent Toxic Substances on Great Lakes Biota,* Health of Aquatic Communities Task Force Report to the Science Advisory Board. International Joint Commission, Windsor, Ontario.

Gachter, R., and Geiger, W. (1979). MELIMNEX, an experimental heavy metal pollution study: Behaviour of heavy metals in an aquatic food chain. *Schweiz. Z. Hydrol.* **41,** 277–290.

Gamble, J. C., Davies, J. M., and Steele, J. H. (1977). Loch Ewe bag experiment, 1974. *Bull. Mar. Sci.* **27,** 146–175.

Gannon, J. E. (1974). The crustacean zooplankton of Green Bay, Lake Michigan. In Proc. 17th Conf. Great Lakes Res., pp. 28–51.

Gannon, J. E., and Stemberger, R. S. (1978). Zooplankton (especially crustaceans and rotifers) as indicators of water quality. *Trans. Am. Microsc. Soc.* **97,** 16–35.

Gibson, V. R., and Grice, G. D. (1977). Response of macro-zooplankton populations to copper: Controlled ecosystem pollution experiment. *Bull. Mar. Sci.* **27,** 85–91.

Hall, T. M. (1982). Free ionic nickel accumulation and localization in the freshwater zooplankton, *Daphnia magna. Limnol. Oceanogr.* **27,** 718–727.

Hawkins, B. E., and Evans, M. S. (1979). Seasonal cycles of zooplankton biomass in southeastern Lake Michigan. *J. Great Lakes Res.* **5,** 256–263.

Herbes, S. E., and Risi, G. F. (1978). Metabolic alteration and excretion of anthracene by *Daphnia pulex*. *Bull. Environ. Contam. Toxicol.* **19,** 147–155.

Hurlbert, S. H., Mulla, M. S., and Wilson, H. R. (1972). The effects of an organophosphorus insecticide on the phytoplankton, zooplankton, and insect populations of fresh-water ponds. *Ecol. Monogr.* **42,** 269–299.

Kaushik, N. K., Stephenson, G. L., Solomon, K. R., and Day, K. E. (1985). Impact of permethrin on zooplankton communities in limnocorrals. *Can. J. Fish. Aquat. Sci.* **42,** 77–85.

Khan, H. M., and Khan, M. A. Q. (1974). Biological magnification of photodieldrin by food chain organisms. *Arch. Environ. Contam. Toxicol.* **2,** 289–301.

Khan, H. M., Neudorf, S., and Khan, M. A. Q. (1975). Absorption and elimination of photodieldrin by *Daphnia* and goldfish. *Bull. Environ. Contam. Toxicol.* **13,** 582–587.

Laskowski-Hoke, R. A., and Prater, B. L. (1981). Relationship of mortality of aquatic biota from 96-hour sediment bioassays and the change in chemical composition of test water. *Bull. Environ. Contam. Toxicol.* **26,** 323–327.

Laskowski-Hoke, R. A., and Prater, B. L. (1984). Multivariate statistical analyses of 96-hour sediment bioassay chemistry data. *Bull. Environ. Contam. Toxicol.* **33,** 400–409.

Leach, J. H. (1973). Seasonal distribution, composition, and abundance of zooplankton in Ontario waters of Lake St. Clair. In Proc. 16th Conf. Great Lakes Res., pp. 54–64.

Maki, A. W., and Johnson, H. E. (1975). Effects of PCB (Aroclor 1254) and *p,p'*-DDT on production and survival of *Daphnia magna* Strauss. *Bull. Environ. Contam. Toxicol.* **13,** 412–419.

Marshall, J. S. (1966). Population dynamics of *Daphnia pulex* as modified by chronic radiation stress. *Ecology* **47,** 561–571.

Marshall, J. S. (1967). Radiation stress in exploited *Daphnia* populations. *Limnol. Oceanogr.* **12,** 154–158.

Marshall, J. S. (1978). Population dynamics of *Daphnia galeata mendotae* as modified by chronic cadmium stress. *J. Fish. Res. Board Can.* **35,** 461–469.

Marshall, J. S., and Mellinger, D. L. (1980). Dynamics of cadmium-stressed plankton communities. *Can. J. Fish. Aquat. Sci.* **37,** 403–414.

Marshall, J. S., Mellinger, D. L., and Parker, J. I. (1981). Combined effects of cadmium and zinc on a Lake Michigan zooplankton community. *J. Great Lakes Res.* **7,** 215–223.

McNaught, D. C. (1975). A hypothesis to explain the succession from calanoids to cladocerans during eutrophication. *Verh. Int. Ver. Limnol.* **19,** 724–731.

McNaught, D. C. (1982). Short cycling of contaminants by zooplankton and their impact on Great Lakes ecosystems. *J. Great Lakes Res.* **8,** 360–366.

McNaught, D. C., Buzzard, M., Griesmer, D., and Kennedy, M. (1980). Zooplankton grazing and population dynamics relative to water quality in Southern Lake Huron. U.S. Environ. Prot. Agency, Off. Res. Dev. Rep. EPA-600/3-80-069.

McNaught, D. C., Griesmer, D., Buzzard, M., and Kennedy, M. (1980). Inhibition of productivity by PCBs and natural products in Saginaw Bay, Lake Huron. U.S. Environ. Prot. Agency, Off. Res. Dev. Data report.

McNaught, D. C., Griesmer, Buzzard, M., and Kennedy, M. (1984). PCBs in Saginaw Bay: Development of functional indices to estimate inhibition of ecosystem fluxes. U.S. Environ. Prot. Agency, Off. Res. Dev. Rep. EPA-600/3-84-008, pp. 1–95.

Mount, D. I., and Norberg, T. J. (1984). A seven-day life-cycle cladoceran test. *Environ. Toxicol. Chem.* **3,** 425–434.

Mozley, S. C., and Howmiller, R. P. (1977). *Environmental Status of the Lake Michigan Region,* Vol. 6. Argonne Natl. Lab., Argonne, IL.

Nalepa, T. F., and Quigley, M. A. (1983). Abundance and biomass of the meiobenthos in nearshore Lake Michigan with comparisons to the macrobenthos. *J. Great Lakes Res.* **9,** 530–549.

Nebeker, A. V., and Puglisi, F. A. (1974). Effect of polychlorinated biphenyls (PCB's) on survival and reproduction of *Daphnia, Gammarus,* and *Tanytarus. Trans. Am. Fish. Soc.* **103,** 722–728.

Neudorf, S., and Khan, M. A. Q. (1975). Pick-up and metabolism of DDT, dieldrin, and photodieldrin by a freshwater alga (*Ankistrodesmus ammalloides*) and a microcrustacean (*Daphnia pulex*). *Bull. Environ. Contam. Toxicol.* **13,** 443–450.

Papst, M. H., and Boyer, M. G. (1980). Effects of two organophosphorus insecticides on the chlorophyll *a* and pheopigment concentrations of standing ponds. *Hydrobiologia* **69,** 245–250.

Passino, D. R. M., and Novak, A. J. (1984). Toxicity of arsenate and DDT to the Cladocera *Bosmina longirostris. Bull. Environ. Contam. Toxicol.* **33,** 325–329.

Patalas, K. (1972). Crustacean plankton and the eutrophication of the St. Lawrence Great Lakes. *J. Fish. Res. Board Can.* **29,** 1451–1462.

Porter, K. G., and McDonough, R. (1984). The energetic cost of response to bluegreen algal filaments by cladocerans. *Limnol. Oceanogr.* **29,** 365–369.

Reeve, M. R., Gamble, J. C., and Walter, M. A. (1977a). Experimental observations on the effects of copper on copepods and other zooplankton: Controlled ecosystem pollution experiment. *Bull. Mar. Sci.* **27,** 92–104.

Reeve, M. R., Walter, M. A., Darcy, K., and Ikeda, K. (1977b). Evaluation of potential indicators of sub-lethal toxic stress on marine zooplankton (feeding, fecundity, respiration, and excretion): Controlled ecosystem pollution experiment. *Bull. Mar. Sci.* **27,** 105–113.

Reinert, R. E. (1972). Accumulation of dieldrin in an algae (*Scenedesmus obliquus*), *Daphnia magna,* and the guppy (*Poecilia reticulata*). *J. Fish. Res. Board Can.* **29,** 1413–1418.

Richman, S., and Dodson, S. I. (1983). The effect of food quality on feeding and respiration by *Daphnia* and *Diaptomus. Limnol. Oceanogr.* **28,** 948–956.

Robertson, A., and Gannon, J. E. (1981). Annotated checklist of the free-living copepods of the Great Lakes. *J. Great Lakes Res.* **7,** 382–393.

Salki, A., Turner, M., Patalas, K., Rudd, J., and Findlay, D. (1985). The influence of fish–zooplankton–phytoplankton interactions on the results of selenium toxicity experiments within large enclosures. *Can. J. Fish. Aquat. Sci.* **42,** 1132–1143.

Sanders, J. G. (1986). Direct and indirect effects of arsenic on the survival and fecundity of estuarine zooplankton. *Can. J. Fish. Aquat. Sci.* **43,** 694–699.

Scavia, D., Fahnensteil, G. L., Evans, M. S., Jude, D. J., and Lehman, J. T. (1986). Influence of salmonid predation and weather on long-term water quality trends in Lake Michigan. *Can. J. Fish. Aquat. Sci.* **43,** 435–443.

Shortreed, K. S., and Stockner, J. G. (1986). Trophic status of 19 subarctic lakes in the Yukon territory. *Can. J. Fish. Aquat. Sci.* **43,** 797–805.

Stemberger, R. S. (1974). Temporal and spatial distributions of planktonic rotifers in Milwaukee Harbor and adjacent Lake Michigan. In Proc. 17th Conf. Great Lakes Res., pp. 120–134.

Stemberger, R. S., and Evans, M. S. (1984). Rotifer seasonal succession and copepod predation in Lake Michigan. *J. Great Lakes Res.* **10,** 417–428.

Stemberger, R. S., Gannon, J. E., and Bricker, F. J. (1979). Spatial and seasonal structure of the rotifer communities in Lake Huron. U. S. Environ. Prot. Agency, Off. Res. Dev. Rep. EPA 600/3-79-05.

Thomann, R. V. (1981). Equilibrium model of fate of microcontaminants in diverse aquatic food chains. *Can. J. Fish. Aquat. Sci.* **38,** 280–296.

Urech, J. (1979). MELIMNEX, an experimental heavy metal pollution study: Effects of increased heavy metal load on crustacea plankton. *Schweiz. Z. Hydrol.* **41,** 247–260.

Watras, C. J., MacFarlane, J., and Morel, F. M. M. (1985). Nickel accumulation by *Scenedesmus* and *Daphnia:* Food-chain transport and geochemical implications. *Can. J. Fish. Aquat. Sci.* **42,** 724–730.

Watson, N. H. F. (1974). Zooplankton of the St. Lawrence Great Lakes-Species composition, distribution, and abundance. *J. Fish. Res. Board Can.* **31,** 783–794.

Watson, N. H. F., and Carpenter, G. F. (1974). Seasonal abundance of crustacean zooplankton and net plankton biomass of Lakes Huron, Erie, and Ontario. *J. Fish. Res. Board Can.* **31,** 309–317.

Wells, L. (1970). Effects of alewife predation on zooplankton populations in Lake Michigan. *Limnol. Oceanogr.* **15,** 556–565.

Winner, R. W. (1984). Selenium effects on antennal integrity and chronic copper toxicity in *Daphnia pulex* (deGreer). *Bull. Environ. Contam. Toxicol.* **33,** 605–611.

Winner, R. W., Keeling, T., Yeager, R., and Farrell, M. P. (1977). Effect of food type on the acute and chronic toxicity of copper to *Daphnia magna. Freshwater Biol.* **7,** 343–349.

4

# BENTHIC INVERTEBRATES AND CONTAMINANT LEVELS IN THE GREAT LAKES: EFFECTS, FATES, AND ROLE IN CYCLING

***Thomas F. Nalepa and Peter F. Landrum***

*National Oceanic and Atmospheric Administration, Great Lakes Environmental Research Laboratory, 2205 Commonwealth Blvd., Ann Arbor, Michigan 48105*

## 1. INTRODUCTION

Most persistent toxic contaminants have a strong affinity for particulate material and, over time, become associated with the bottom sediments. As a result, benthic organisms play an integral role in the effects, fates, and cycling of contaminants throughout the rest of the aquatic system. In general, studies of interactions between the benthos and toxic contaminants can be divided into two categories. The first category examines the response of benthic organisms to changes in contaminant levels. Benthic response may be examined in the field by assessing changes in benthic community structure, or in the laboratory by using acute and chronic toxicity tests. The other category concerns the effect of benthic-mediated processes on contaminant cycling and distribution. Examples of such processes are bioaccumulation, bioturbation, and trophic transfer.

This review will focus on known interactions between benthic organisms and contaminants in the Great Lakes. Benthic organisms are here defined as those invertebrates retained by a 0.5- to 0.6-mm screen, or those commonly referred to as the "macrobenthos." Although other benthic groups, such as bacteria and meiobenthos, are also important in contaminant fates and effects, they will not be considered in this review. Since benthic groups occurring in the Great Lakes are not unique to this system, relevant studies from other freshwater environments will also be presented.

## 2. COMPONENTS OF GREAT LAKES BENTHOS

The typical benthic community in the Great Lakes, excluding the connecting channels and shallow bays and basins, might best be described in terms of broadly defined habitat zones based on gradients in water depth (see Cook and Johnson, 1974, for review). Two important depth-related factors affecting benthic distributions are sediment type and temperature regime. The benthos of the littoral zone (0–30 m) is characterized by low abundances, high diversity, and great variability in both time and space. This variability is a reflection of the physical instability of this zone, particularly at the shallow end of this depth range. Downwellings and oscillating thermoclines cause wide fluctuations in bottom temperatures, and waves and bottom currents cause resuspension of bottom substrates. Organisms with broad environmental tolerances or life cycles suited to these changing conditions do best in this zone; tubificids, naidids, and chironomids are often the dominant forms. As water depth increases within the littoral zone, the bottom environment becomes increasingly stable and the amphipod *Pontoporeia hoyi* and the oligochaete *Stylodrilus heringianus,* both obligate stenotherms, gradually increase in abundance.

In the sublittoral zone (30–70 m), temperatures fluctuate less and the sediments are less influenced by storms and currents. Suspended particles from the littoral begin to settle, providing increased food resources for the benthos. Fewer species are found in this zone than in the littoral, but total biomass reaches a maximum.

Tubificids and sphaeriid clams are abundant, and *P. hoyi* and *S. heringianus* are found at maximum densities.

The profundal (>70 m) is a stable region of uniformly low temperatures, low light regimes, and unconsolidated sediments. *P. hoyi* is clearly the most abundant form in this zone, although the semiplanktonic *Mysis relicta,* because of its large size, is also important when biomass is considered. *S. heringianus* is the dominant oligochaete. Both sphaeriid clams and chironomids are found in the profundal, but abundances are generally low. As depth increases in the profundal, all benthic forms except *Mysis* decrease in abundance. Maximum densities of *Mysis* occur at depths greater than 100 m (Carpenter et al., 1974).

In shallow bays and basins (Green Bay, Saginaw Bay, western Lake Erie), in the connecting channels (St. Mary's River, St. Clair River, Detroit River), and in Lake St. Clair, other benthic forms may also become abundant. The most significant of these are the mayfly larvae *Hexagenia* and unionid bivalves (mussels). *Hexagenia* was once very abundant in the bays and shallow basins. The decline of this species has been well documented and generally attributed to severe oxygen depletion in bottom waters (Britt, 1955; Howmiller and Beeton, 1971; Schneider et al., 1969; Carr and Hiltunen, 1965). In western Lake Erie, *Hexagenia* populations were reduced by 90% after a prolonged period of thermal stratification that resulted in decreased oxygen levels (Britt, 1955). *Hexagenia* is also sensitive to environmental pollutants (Fremling, 1970) and numbers presently remain low in these areas, although recent evidence indicates populations may be increasing in the Detroit River and parts of western Lake Erie (Thornley, 1985).

Mussels are found mostly in the Huron–Erie corridor (St. Clair River, Lake St. Clair, Detroit River) and in the western basin of Lake Erie. This group has generally been considered sensitive to pollutants and, at least in western Lake Erie, abundances have apparently declined over the past few decades (Mackie et al., 1980) and existing populations have showed signs of senescence (Roth and Mozley, 1973). Since mussels are dependent upon fish hosts for the development of their glochidia, shifts in fish populations over the years also may have contributed to these changes.

## 3. EFFECTS OF CONTAMINANTS ON BENTHIC COMMUNITIES

### 3.1. Field Assessments

Whether or not toxic contaminants have altered the structure of Great Lakes benthic communities is difficult to assess. Unless concentrations of a particular contaminant are high and the resulting impact on the benthos is quite dramatic, changes in community structure resulting from chemical contaminants cannot be readily distinguished from other perturbations such as organic pollution, cultural eutrophication, or natural and human-induced ecological changes. For example, with an increase in eutrophication, increased nutrient levels and organic inputs to the bottom cause a shift in the community to species best suited to exploiting these additional food levels. These "opportunistic" species tend to have a larger reproductive capacity and an inherent ability to tolerate changing environmental conditions. These species may also increase in abundance because of a decrease in competition from forms more sensitive to changing conditions. Because the same particles that transport organic material utilized as food also transport toxic contaminants, the relative impact of contaminants cannot be readily distinguished when examining only benthic distributions in the field. Both trophic contaminants (organic carbon, total phosphorus) and chemical contaminants may interact in defining benthic community structure (Lang and Lang-Dobler, 1979).

Throughout the Great Lakes, changes in the benthic community in response to polluting influences are relatively predictable. Species assemblages in polluted areas are distinctive and quite different from assemblages in clean areas. The response of Great Lakes benthos to changes in general pollution levels have been documented in studies of trends through time (Howmiller and Beeton, 1971; Schneider et al., 1969; Carr and Hiltunen, 1965; Robertson and Alley, 1966) and in studies measuring attenuation in space relative to a point discharge or river input (Brinkhurst, 1967; Vander Wal, 1977; Johnson and Brinkhurst, 1971; Nalepa and Thomas, 1976). Typically, tubificid oligochaetes increase in abundance relative to other forms and may achieve great densities, while *S. heringianus, P. hoyi,* and *Hexagenia,* if present initially, become rare and may eventually disappear.

A typical example of pollution-induced changes in the benthos in both time and space has been documented in Green Bay (Howmiller and Beeton, 1970, 1971). Green Bay is a relatively shallow arm of Lake Michigan that receives large amounts of both domestic and industrial wastes (primarily from pulp and papermills). Wastes enter the lower bay via the Fox River and as a result, the environmental quality of the lower bay is severely degraded (Beeton, 1978). The upper bay mixes with cleaner Lake Michigan water and is less influenced by the river. Based on benthic community changes, the quality of the entire bay declined between 1952 and 1969. During this period, the proportion of tubificid oligochaetes (mostly *Limnodrilus hoffmeisteri*) in the total number of benthic organisms increased from 66% to 85% in the lower bay and from 23% to 64% in the middle bay. The sensitive *S. heringianus* was abundant only in the upper bay, while amphipods and sphaeriids declined throughout the area and *Hexagenia* disappeared. In 1969, the area nearest the mouth of the river was devoid of all fauna.

Such evaluations of community changes are based on ecological observations and attributed to "pollution effects" without implying a specific cause and effect relationship. In areas where oxygen concentrations reach low levels [which is the case in many parts of Green Bay (Howmiller and Beeton, 1971)], the cause of community change is rather obvious. However, with toxic contaminants, the specific causative agent or agents are not readily apparent unless concentrations are high enough to entirely eliminate species from a given area. In the Great Lakes, this has been documented in a few instances and has occurred primarily in more shallow areas where the impacts are direct and most immediate. In Lake Superior, *P. hoyi* was not found in an area where concentrations of copper in the sediment exceeded 395 mg/kg (Kraft, 1979) and, in the same general area, *Hexagenia* was not found where copper concentrations exceeded 595 mg/kg (Malueg et al., 1984).

In the St. Mary's River, *Hexagenia* was absent or occurred in reduced numbers in sediments with visible signs of oil (Hiltunen and Schloesser, 1983). Also, in Hamilton Bay, Lake Ontario, oligochaetes and all other benthic forms were absent from sediments with an iron content over 25% (Johnson and Matheson, 1968). Tubificids (mostly *L. hoffmeisteri* and *Tubifex tubifex*) typically increased and *P. hoyi* decreased near a pulpmill effluent in Lake Superior (Vander Wal, 1977). These changes were related to increased organic carbon content of the sediments in areas near the pulpmill discharge. In the area nearest the discharge, tubificid abundances were reduced despite continued high levels of organic carbon. Apparently, some toxic chemical or chemical mixture associated with the high carbon levels was detrimental to the organisms.

The presence or absence of a particularly sensitive species may indeed be useful when assessing the impact of a toxic contaminant in the field. Yet, more often, the effects are not readily discernible from the many other environmental factors that can affect species distributions. For example, the Niagara River has been identified as the primary source of contaminants into Lake Ontario; high contaminant concentrations are found in the sediments east of the river mouth along the southern shoreline (Thomas, 1983). In this area, *P. hoyi* is rare or absent at depths where it is normally found (Nalepa and Thomas, 1976). High contaminant levels may indeed be adversely affecting *P. hoyi,* but low Eh values also characterize the sediments of this area (Thomas et al., 1972), and water temperatures here are higher than in other shoreline areas of Lake Ontario because of frequent downwellings (Lee, 1972). Both of these physical fcatures may also be limiting *P. hoyi* populations.

Various statistical techniques, such as discriminant analysis, principal component analysis, and cluster analyses, may be helpful in defining the relevant physicochemical variables affecting benthic populations (Green and Vascotto, 1978). Thus far, however, these techniques have not been widely used in the Great Lakes. Although useful, the results of such techniques should be interpreted with caution when assessing the impact of toxic contaminants. In the Duluth basin of western Lake Superior, low benthic abundances were attributed to either high metal levels in the sediments or the heavy silt loads in that area (Cook, 1975). A subsequent study used canonical correlation analyses to determine the important factors affecting benthic distributions (Dermott, 1978). Both sphaeriids and *S. heringianus* were negatively associated with levels of Zn in the sediments. Yet, although an overall negative relationship was found, site-specific discrepancies were apparent. Sphaeriids were more abundant in Thunder Bay than in the Duluth basin, even though Zn levels were significantly greater in the former area. Many factors can affect the toxic impact of heavy metals. For instance, the organic fraction of sediments binds heavy metals and reduces their toxicity (Lewis et al., 1973).

## 3.2. Laboratory Assessments and Sublethal Effects

Because of the difficulties associated with field assessments, direct cause and effect relationships are commonly examined using acute toxicity bioassays under controlled laboratory conditions. While bioassays have been useful in establishing water quality criteria and standards for effluents, the results are difficult to apply to field situations because of the multiplicative effects of chemical mixtures, variations in bioavailability, species interactions, long-term low level exposures, and so on. These tests, however, have provided some insights into the complexity of assessing the impact of chemical contaminants on the benthos. The average $LC_{50}$ value of 15 different chemicals to 12 different benthic species was determined and the species were ranked according to mean tolerance levels of all the compounds (Slooff, 1983). The rankings did not always follow traditional lines; that is, species considered generally tolerant to pollution (in the ecological sense of being found in polluted areas) were not necessarily ranked highest in terms of tolerance to the various chemical compounds. In other studies, the ''pollution-tolerant'' tubificids *L. hoffmeisteri* and *T. tubifex* were less tolerant of pentachlorophenol, black liquor, mercury, and cadmium than the ''pollution-sensitive'' *S. heringianus* (Chapman et al., 1982). Interestingly, the same study confirmed the relative tolerances of these species in the traditional sense of organic pollution and eutrophication; *L. hoffmeisteri* and *T. tubifex* were far more tolerant of sewage sludge and low oxygen levels than *S. heringianus.*

Several studies have examined the acute toxicity of Great Lakes harbor sediments and the adverse effects of these sediments on sensitive benthic species were readily apparent. *P. hoyi* and chironomids, when given a choice, preferred clean, unpolluted sediments to sediments from various polluted harbors (Gannon and Beeton, 1969). Also, in 48-hour toxicity tests, mortality for *P. hoyi* in harbor sediments was 70%, and mortality for chironomids was even higher. Prater and Anderson (1977) performed 96-hour toxicity tests with sediments from several areas in Duluth and Superior harbors. The sediments were classified as nonpolluted, moderately polluted, and highly polluted based on mortality of several aquatic organisms including *Hexagenia.* Although highest mortalities (10–50%) occurred in sediments with the greatest concentrations of inorganic and organic contaminants, mortality could not be related to any single compound. The assumption was that either synergistic or antagonistic effects had occurred, or chemical contaminants not measured led to the observed mortalities. Differences in bioavailability from the various sediments may also have contributed to the results. In similar studies on sediments from Lake Michigan harbors, mortality of *Hexagenia* was positively correlated with sediment concentrations of lead, iron, copper and chromium (Laskowski-Hoke and Prater, 1984) and to elutriate concentrations of chromium, cynanide, and zinc (Laskowski-Hoke and Prater, 1981).

In some cases, acute toxicity tests have been useful in interpreting benthic distributions in the field. In an Indiana lake, tubificids were found in sediments with concentrations of cadmium, zinc, and chromium (970, 1400, and 2100 $\mu$g/g, respectively) that eliminated most other forms (Wentsel et al., 1977a), while in an Idaho stream, oligochaete distributions were not affected by zinc concentrations of 1000–7000 $\mu$g/g (Funk et al., 1973). Oligochaete densities in both studies were highest in areas with the highest metal levels, and it was suggested that the elimination of more metal-sensitive predators-competitors such as chironomids may account for this distribution. Subsequent laboratory experiments have shown that chironomids are very sensitive to certain heavy metals (Wentsel et al., 1977b, c). This may explain tubificid distributions in the Detroit River, where some of the highest densities of tubificids were found at stations with the highest heavy metal concentrations (particularly Zn) (Thornley and Hamdy, 1984).

Only one study in the Great Lakes has used acute toxicity assays to interpret benthic community response to a particular contaminant. In the Keewenaw Waterway, Lake Superior, benthic distributions were examined in relation to copper-contaminated sediments (Malueg et al., 1984). Chironomids accounted for 96% of all organisms found in an area with high copper concentrations ($\bar{x}$ = 589 $\mu$g/g), while they comprised only 25% in area with lower copper levels ($\bar{x}$ = 33 $\mu$g/g). *Hexagenia* was not found in the former area, but was the dominant organism in the latter. Acute toxicity tests on *Hexagenia* found no significant difference in mortality using sediments from the two areas. In similar tests using the more sensitive *Daphnia magna* as the test organism, mortality was greater in the sediments with higher copper levels. It was concluded that, although not acutely toxic, high copper levels had a subtle, long-term negative impact on *Hexagenia* populations.

As illustrated in the above study, sublethal levels of contaminants probably have more of an impact on benthic populations than acutely lethal levels simply because such levels occur more frequently and over a much broader area. Laboratory studies have shown that low concentrations of some contaminants have subtle effects on organism behavior and physiology. For instance, heavy metals affected the respiration rates of tubificids (Brkovic-Popovic and Popovic, 1977); phenols, chlorophenol, and styrene affected the swimming behavior of *P. hoyi* (Lindstrom and Lindstrom, 1980); and sediments with heavy metals were actively avoided by both tubificids (McMurtry, 1984) and chironomids (Wentsel et al., 1977b). Often, sublethal effects may not be recognized without employing special techniques. Histopathological studies revealed that low cadmium levels caused rupturing of the body wall and degeneration of amoebocytes in polychaetes (Thompson et al., 1982).

The effects of low, sublethal levels of contaminants are likely to be greatest in depositional areas where contaminants tend to accumulate. However, as noted earlier, sublethal effects are difficult to assess since organic inputs in these areas also are high. Although there have been no studies designed to evaluate the effects of low-level, long-term exposures of contaminants on Great Lakes benthic communities, there is some evidence that such effects may exist. For example, PCB concentrations of over 50 ng/g caused *S. heringianus* to terminate feeding after 58 days (White and Klahr, 1982). This concentration is below the average PCB levels of 85.0 and 115.0 ng/g found in the depositional basins of Lake Ontario and Lake Erie, and below the maximum levels of 75.0 ng/g found in southern Lake Michigan. Also, zinc concentrations of between 58.5 and 123.5 $\mu$g/g significantly lowered swimming activity in *Pontoporeia* (Magnuson et al., 1976). Most depositional areas of Lake Michigan have zinc concentrations of over 60 $\mu$g/g and often higher than 200 $\mu$g/g (Cahill, 1981). If and how such sublethal effects become evident at the population level remains unclear. For instance, despite these seemingly high PCB and zinc levels in southern Lake Michigan, populations of both *S.*

*heringianus* and *Pontoporeia* have increased or remained stable over the years (Nalepa, 1987).

Morphological deformities in organisms occurring in polluted areas may be indicative of sublethal effects. Deformed setae have been noted in oligochaetes from areas with high mercury levels (Milbrink, 1983) and in areas with high concentrations of zinc and copper (Chapman and Brinkhurst, 1984). Deformities have also been noted in the mouthparts of chironomids occurring in areas contaminated with certain heavy metals (Wiederholm, 1984). In the Great Lakes, chironomid deformities have been reported in western Lake Erie near the mouths of the Detroit and Maumee Rivers (Brinkhurst et al., 1968), in the Bay of Quinte (Warwick, 1980), and in Georgian Bay (Hare and Carter, 1976).

## 4. BIOTURBATION AND CONTAMINANT CYCLING

Through their constant burrowing and feeding activities, benthic organisms physically rework and mix the upper layers of sediments. These activities, termed bioturbation, tend to increase the rate of material flux between the sediments and overlying waters and also homogenize many years (often decades) of accumulated sediments. The exchange process is mediated by diffusion or by physical transport. Numerous studies have demonstrated the impact of bioturbation on the direct diffusion of various nutrients from sediments to overlying waters (for reviews, see Petr, 1977; Fisher, 1982). However, considering the sorptive nature of most toxic contaminants, the importance of bioturbation in contaminant cycling lies mainly in the physical transport and resuspension of sediment particles. These processes reintroduce contaminants into the water column.

Each benthic species characteristically alters the physical and chemical nature of the sediments, depending on its particular feeding mode and life habit. Oligochaetes are probably the most important agents of physical transport. These organisms feed at depth and defecate at the sediment surface. Sediments are thus mixed down to the feeding depth, the exact location of which will vary depending on abiotic and biotic factors. In laboratory studies, the feeding depth of *S. heringianus* in Lake Michigan sediments was 4 cm (Krezoski et al., 1984b) and that of *T. tubifex* in Lake Erie sediments was 8 cm (McCall and Fisher, 1980). Field studies of the vertical distribution of cesium-137 and lead-210 have confirmed a zone of constant activity consistent with the vertical distribution of oligochaetes (Robbins, 1982). In addition, several studies in western Lake Erie have shown a mixed zone of heavy metals (Hg, Pb, Cr, and Zn) from 6–20 cm deep (Kemp et al., 1976; Wolery and Walters, 1974; Walters et al., 1974); this uniform distribution was attributed to the reworking activities of the benthos (Wolery and Walters, 1974).

The net effect of sediment reworking is to keep settled contaminants from being completely buried via sedimentation. Thus, even if a contaminant is no longer introduced from external loadings, it will not be readily lost from the system. The reworking rate of oligochaetes (calculated from field densities and laboratory derived egestion rates) has been estimated at 4 times the sedimentation rate in Lake Huron (Krezoski et al., 1978) and over 10 times the sedimentation rate in parts of Lake Erie (Fisher et al., 1980). In addition, since oligochaetes selectively feed on the organic fraction of the sediments (Brinkhurst and Austin, 1979) and persistent contaminants are closely associated with this fraction, the net effect of these feeding activities is to transport contaminants preferentially back to the sediment surface. Release of contaminants from the sediments to the overlying waters was enhanced four- to sixfold in the presence of oligochaetes (Karickhoff and Morris, 1985).

The burrowing activities of oligochaetes increases sediment porosity and decreases tortuosity (Krezoski, 1981; McCall and Fisher, 1980). This is also true of other benthic forms. *P. hoyi* is a surface detritivore that randomly plows through the upper 1–2 cm of sediment. These activities mix the sediments in an "eddy diffusive" manner (Robbins et al., 1979). Similarly, sphaeriid clams live in the upper sediments where they filter feed on the surface or burrow to a depth of several millimeters. Many chironomids and *Hexagenia* live in distinct burrows and filter feed by drawing water in and through these constructions. *Hexagenia* burrows may extend down to 15 cm. Mussels burrow in the upper 2–3 cm and, because of their large size, may alter sediment structure as deep as 10 cm. In western Lake Erie, mussels are estimated to move 2–6% of the upper 10 cm of sediment every week; they also increase water content in the top 4 cm by 10–20% (McCall et al., 1979). This burrowing activity decreases sediment compaction which, in turn, enhances the probability of particle resuspension by physical processes (waves and currents). In addition, both *P. hoyi* and *M. relicta* make excursions into the water column and may resuspend particles directly. In the Baltic, a highly turbid zone extending 1–2 m above the bottom has been observed in areas with high densities of *Pontoporeia* (Ankar, 1977). The importance of sediment resuspension in contaminant cycling has recently been noted (Eadie et al., 1983a; Rosa, 1985). In Lake Michigan, about 14% of particle-associated PCBs in the water column during the winter could be accounted for by resuspension (Eadie et al., 1983a).

By physically mixing the sediments and thereby increasing the exchange of interstitial and overlying waters, the benthos have a profound impact on sediment redox potential and pH. These two parameters, in turn, influence the complexation of heavy metals, altering both bioavailability and toxicity. Free Cu and Zn ions are more toxic than their organically complexed forms, while complexation of Hg, Pb, and Cd enhances their uptake and toxicity. In a study to determine the effect of tubificids and mussels on mercury uptake by fish, mercury was placed at various depths in experimental sediment columns (Jernelov, 1970). In controls (without benthos) the methyl mercury content of fish decreased as the depth of the buried contaminated layer increased. In columns with benthos, the methyl mercury content of fish was higher than that of the controls and remained high until a critical depth was reached. The critical depth was 3 cm for tubificids and 9 cm for mussels. Bioturbation by these organisms likely enhanced mercury methylation by increasing sediment oxidation and microbial processes.

## 5. BIOACCUMULATION OF ORGANIC CONTAMINANTS

An important role of the benthos in contaminant cycling is their tendency to accumulate certain compounds from their immediate environment. Thus, contaminants become concentrated and available to higher trophic levels. Sources of contaminants to the benthos are the overlying waters, the porewaters, and sediment particles. The predominant route for bioaccumulation is considered to be uptake by passive diffusion from the aqueous phase, generally across the integument or respiratory surface (Neely et al., 1974). Absorption through the gut wall will occur when contaminants in the sediments or porewaters are ingested. With the sediments as a source, the contaminants are desorbed from particles as they pass through the digestive tract.

Bioaccumulation of contaminants in aquatic organisms is usually reported as the bioconcentration factor (BCF). This factor is the ratio, at steady state, between the concentration in the organism divided by the concentration in the water or sediment. Since uptake is by passive diffusion, the BCF of a particular organic contaminant is related to its octanol:water partition coefficient ($K_{ow}$) or its ability to partition into the organism. For instance, experiments to determine the uptake of various compounds from Lake Ontario sediments by tubificids showed that the BCF was low for compounds with a $K_{ow}$ of $< 10^5$, increased to a peak for compounds with a $K_{ow}$ of $10^5$–$10^6$, and then decreased for compounds with a $K_{ow}$ of $> 10^6$ (Fig. 4.1) (Oliver, 1984). It was hypothesized that, for compounds with low $K_{ow}$ values, elimination rates by the worms are high and equilibrium is reached very quickly; BCFs are thus mainly governed by the lipophilicity ($K_{ow}$) of the compound. Compounds with intermediate $K_{ow}$ values are eliminated more slowly, and will accumulate in the organisms over time. Compounds with high $K_{ow}$ values are more strongly bound to the sediments and would therefore be less available to the worms. Also, the large molecular size of compounds with a high $K_{ow}$ would

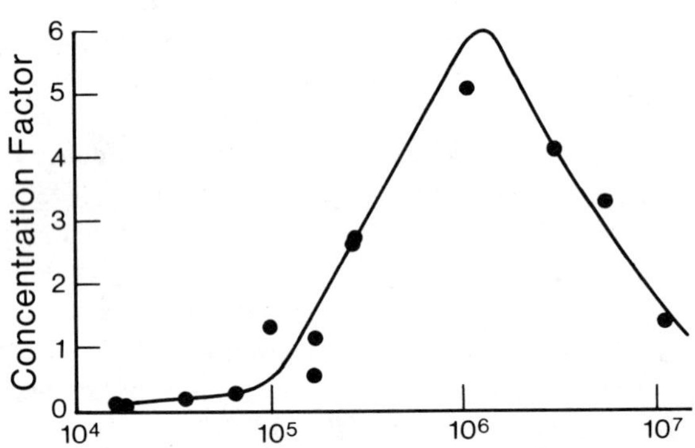


**Figure 4.1.** Worm concentration factor for various chemicals plotted against the chemical's octanol–water partition coefficient (curve fitted to points by eye). From Oliver (1984); with permission.

make chemical transport across worm membranes very difficult. A similar relationship between the $K_{ow}$ of individual PAH compounds and the BCF was noted in *P. hoyi* (Eadie et al., 1983b).

Relatively high concentrations of organic contaminants have been found in Great Lakes benthos; bioaccumulation is evident in most cases although some variation is apparent (Table 4.1). For example, levels of PCBs in oligochaetes from the Detroit River were lower than levels found in the sediments (Smith et al., 1985), but oligochaetes from Lake Ontario had PCB levels that were 2–8 times the levels found in the sediments (Fox et al., 1983). Laboratory experiments have confirmed a BCF of 5 for PCB uptake by oligochaetes from Lake Ontario sediments (Oliver, 1984); similarly, a BCF of 4 was reported for PCB uptake by oligochaetes from Raisin River sediments (Mac et al., 1984).

Obviously, numerous variables may contribute to contaminant uptake by the benthos as related to both the physical nature of the environment and attributes of the organisms themselves. Probably the most important physical factor is the organic content of the sediments and/or the amount of dissolved organics found in the water. Through complexation or adsorption onto organic matter, the amount of contaminant available for biological uptake is dramatically reduced (Lynch and Johnson, 1982; Landrum et al., 1985). Laboratory experiments have shown that benthic organisms in highly organic sediments will have consistently lower BCFs than organisms in sediments with little organic content (Rubinstein et al., 1983; Muir et al., 1985).

Bioaccumulation in a given species is a function of its uptake rate, depuration rate and rate of biotransformation. For the Great Lakes, uptake rate experiments have focused primarily on the uptake of polynuclear aromatic hydrocarbons (PAH) from water by *P. hoyi, S. heringianus,* and *M. relicta* (Table 4.2). Uptake rate constants are highest in *P. hoyi,* and lowest in *M. relicta.* Since the uptake rate from water is directly related to the respiration rate (Murphy and Murphy, 1971; Frank et al., 1986), larger organisms with lower specific metabolic rates such as *Mysis* may be expected to have lower uptake rates. Also, external adsorption through the integument would be less significant in larger organisms since the surface area to volume ratio is lower than in smaller organisms.

The kinetic parameter that has perhaps the most control over the body burden is the conditional depuration rate constant. Depuration rates are estimated by placing exposed organisms into uncontaminated systems and measuring reduction in body burden. Depuration rates for *P. hoyi, S. heringianus* in sediments, and *M. relicta* in water are given in Table 4.3. The rates for *M. relicta* would likely be higher in the presence of uncontaminated food sources (Leversee et al., 1982; Landrum, 1982; Landrum and Scavia, 1983), but they do provide a basis for species comparisons. The conditional depuration rate constant was lowest for *P. hoyi* and highest for *S. heringianus.* This would seem to indicate that lipid levels play a major role in contaminant elimination; lipid levels are highest in *P. hoyi,* and lowest in *S. heringianus* (Gardner et al., 1985). Overall, depuration rates are lower in these three species than rates found in similarly sized organisms in warmer waters (Landrum and Scavia, 1983; Frez and Landrum, 1986; Bahner et al., 1977).

**Table 4.1 Levels of Organic Contaminants Found in Great Lakes Benthos[a]**

| Organism | Organic Contaminant (ng/g dry wt) | | | | | | | Location | Reference |
|---|---|---|---|---|---|---|---|---|---|
| | PCB | DDE | DDD | HEOD | Chlordane | HCB | OCS | | |
| Oligochaetes | 440 (0.7) | 5.9 (4.9) | 1.5 (0.7) | | 4.5 (1.6) | 3.6 (1.2) | | One site, Detroit River | Smith et al. (1985) |
| Oligochaetes | 930–3650 (2.4–7.8) | | | | | 105–735 (0.7–2.3) | | 4 sites, western Lake Ontario | Fox et al. (1983) |
| *Pontoporeia hoyi* | 4215–11,000 (9.0–19.0) | | | | | 230–895 (1.6–6.0) | | 4 sites, western Lake Ontario | Fox et al. (1983) |
| Benthos (mostly *P. hoyi*) | 341 (4.0) | 34 (4.3) | 1.8 (1.2) | 2.9 (3.2) | | | | Lake Ontario near Rochester | Haile et al. (1975) |
| *Pontoporeia hoyi* | 1305 | 154 | 96 | 141 | 252 | | | 1 site, southeastern Lake Michigan | Evans et al. (1982) |
| *Pontoporeia hoyi* | 560 | 58 | 62 | 35 | | | | 1 site, eastern Lake Erie | Whittle and Fitzsimons (1983) |
| *Pontoporeia hoyi* | 1378 | 292 | | 226 | | | | 3 sites, western Lake Ontario | Whittle and Fitzsimons (1983) |
| *Pontoporeia hoyi* | 1849 | 730 | | 376 | | | | 1 site, eastern Lake Ontario | Whittle and Fitzsimons (1983) |
| Unionid bivalves (mussels) | 133 | | | | | | 59 | Huron–Erie Corridor | Pugsley et al. (1985) |

[a]Number in parentheses is the bioconcentration factor (concentration in organism/concentration in sediment).

**Table 4.2 Average Uptake Rate Constants of PAH for Three Great Lakes Benthic Invertebrates[a]**

| Organism | Compound | Uptake Rate Constant ($mL\ g^{-1}\ h^{-1}$) | Reference[b] |
|---|---|---|---|
| *Stylodrilus heringianus* | Anthracene | 86 (10) | 1 |
| | Phenanthrene | 91 (13) | 1 |
| | Pyrene | 112 (1) | 1 |
| | Benzo[*a*]pyrene | 85 (16) | 1 |
| *Pontoporeia hoyi* | Biphenyl | 85 (23) | 2 |
| | Anthracene | 126 (36) | 3,4 |
| | Phenanthrene | 123 (37) | 4 |
| | Pyrene | 208 (52) | 2 |
| | Benzo[*a*]pyrene | 137 (7) | 4 |
| | Benzo[*a*]anthracene | 148 (50) | 2 |
| *Mysis relicta* | Phenanthrene | 32 (12) | 4 |
| | Anthracene | 63 (18) | 4 |
| | Benzo[*a*]pyrene | 112 (35) | 4 |

[a]Constants are given as milliliters of water cleared of compound per gram wet weight of organism per hour. Number in parentheses is standard deviation.
[b]1, Frank et al. (1986); 2, Landrum (unpublished data); 3, Landrum (1982); 4, Frez and Landrum (1986).

Biotransformation has an impact on both the toxicity and overall fate of the contaminant. Although biotransformation has been noted in chironomids (Leversee et al., 1982), attempts to measure biotransformation of PAH in either *P. hoyi* or *S. heringianus* have not yielded any measurable rate (Landrum, 1982; Frank et al., 1986). Thus, rates of biotransformation by these organisms are very slow or

**Table 4.3 Depuration Rate Constant of PAH for Three Great Lakes Benthic Invertebrates**

| Organism | Compound | Depuration Rate Constant[a] ($h^{-1}$) | Reference[b] |
|---|---|---|---|
| *Pontoporeia hoyi* | Benzo[*a*]pyrene | 0.0016 (0.0012) | 1,2 |
| | Anthracene | 0.004 (0.003) | 2 |
| | Phenanthrene | 0.004 (0.001) | 2 |
| *Mysis relicta* | Benzo[*a*]pyrene | 0.013 (0.002) | 2 |
| | Anthracene | 0.015 (0.005) | 2 |
| | Phenanthrene | 0.012 (0.003) | 2 |
| *Stylodrilus heringianus* | Benzo[*a*]pyrene | 0.012 (0.002) | 3 |
| | Anthracene | 0.017 (0.004) | 3 |
| | Phenanthrene | 0.018 (0.003) | 3 |
| | Pyrene | 0.017 (0.001) | 3 |

[a]Number in parentheses is standard deviation.
[b]1, Landrum et al. (1985); 2, Frez and Landrum (1986); 3, Frank et al. (1986).

nonexistent, which may also contribute to their relatively low depuration rate constants.

As previously noted, contaminants can be taken up directly from the water or from the sediments via the food pathway. For PAH uptake, models derived from field data indicate that the relative contribution of each source is dependent upon the relative concentration of contaminant, the physicochemical properties of the specific compound, and the life habit of the organism. When sediment PAH concentrations are low, both *P. hoyi* and oligochaetes obtain a substantial fraction of their PAH from water, but when sediment concentrations are high, much of their PAH is obtained from that source (Eadie et al., 1983b). Soluble PAH compounds such as phenanthrene, fluoranthene, pyrene, and chrysene are taken up to a greater extent from the water, while insoluble compounds such as benzo[*a*]pyrene (BaP) are taken up from the sediments. Laboratory-derived kinetics models of BaP predict that sediments contribute only 10–50% of the body burden in *P. hoyi* (Landrum et al., 1985), but 40–90% in *S. heringianus* (Frank et al., 1986). The reason, of course, is that *P. hoyi* is more readily found at the sediment surface and has more contact with the overlying waters than *S. heringianus*, which occurs deeper in the sediments. In terms of sediment contribution, the proportion of a contaminant passing through the gut that is actually incorporated into the body depends upon the feeding rate of the organism and the assimilation efficiency of the particular compound. Tubificids assimilated 20–30% of sediment-associated hexachlorobiphenyl (Krezoski et al., 1984a); this is considerably lower than the 90% assumed in some models (Thomann, 1981; Thomann and Connolly, 1984). Overall, the contribution of the food pathway to the total body burden of contaminants is an area that has not been systematically studied (see Biddinger and Gloss, 1984, for review).

## 6. BIOACCUMULATION OF INORGANIC CONTAMINANTS

The mechanism for the accumulation of heavy metals is markedly different from that of organic compounds. Most metals require an active or facilitated process for assimilation, while organic compounds accumulate by passive diffusion; the exception is organometalics such as methyl mercury which also accumulate via a passive mode. As a result, the accumulation of metals occurs mostly from the sediments through the food pathway. In addition, although free metals are readily available for uptake, they complex very strongly with the organic fraction in both sediments and water, significantly reducing their bioavailability. This complexation is much stronger for metals than for organic compounds.

Because many factors affect the complexation and thus the bioavailability of heavy metals (geochemical phases of the metal, organic content of sediment, Eh, pH), bioconcentration factors in the benthos tend to be highly variable and strongly influenced by local conditions. For instance, concentrations of nine heavy metals (Cu, Zn, Pb, Fe, Mn, Ni, Co, Hg) were examined in both tubificids and sediments at two stations in the Frazer River system (Chapman et al., 1979a). BCFs were less than 0.6 for all metals except mercury, which had a BCF of 1.7 and 4.0 at the two stations. In contrast, another study in the same river system with the same benthic group found that, of the five metals examined, BCFs were greater than 1.0 for Cu, Zn, Pb, and Mn and less than 1.0 for only Fe (Bindra and Hall, 1978).

Few studies in the Great Lakes have examined metal levels in both the benthos and sediments. Engler (1979) noted that oligochaetes did not bioaccumulate Cd, Cu, Mn, Zn, or Fe from Lake Erie sediments while in Lake St. Clair, average levels of Cr and Pb in mussels were 28.6 and 1.4 times average levels found in the sediments (Pugsley et al., 1984). These latter bioconcentration factors were based on sediment samples taken down to 10 cm, which may not accurately reflect actual exposure concentrations. Bioconcentration factors for mussels in the Illinois River were less than 1.0 for 8 different heavy metals, including Cr and Pb (Mathis and Cummings, 1973).

## 7. TROPHIC TRANSFER

Benthic organisms are an important food item in the diet of most species of Great Lakes fish. The smaller forage species, such as alewife (*Alosa pseudoharengus*), smelt (*Osmerus mordax*), slimy sculpin (*Cottus cognatus*), deepwater sculpin (*Myoxocephalus thompsoni*), trout-perch (*Percopsis omiscomaycus*), and bloater (*Corgonus hoyi*), feed on both *P. hoyi* and *M. relicta* (Janssen and Brandt, 1980; Wells and Beeton, 1963; Heberger and House, 1974; Anderson and Smith, 1971; Wells, 1980). These fish, in turn, serve as food for the large piscivores. These valuable gamefish may also feed directly on benthic species during the early stages of their life cycles. Although laboratory experiments have indicated that fish may obtain most of their body burden of organic contaminants from their food (Macek and Korn, 1970; Weininger, 1978), little is known of the actual proportion attributable to benthic organisms. In comparing PCB concentrations of various size organisms ranging from zooplankton to lake trout (*Salvelinus namaycush*), *P. hoyi*, slimy sculpin, and lake trout had higher PCB concentrations than might be expected from a strict size-dependent relationship (Fig. 4.2) (Borgmann and Whittle, 1983). Lake trout feed on slimy sculpin, which, in turn, feeds heavily on *P. hoyi*. From a series of simulation models, Breck (1984) estimated that 40% of the body burden of PCBs in adult alewife came from *P. hoyi* and that 25% of the PCBs entering the entire alewife population (juvenile and adult) in Lake Michigan could be attributed to uptake from benthic organisms (*P. hoyi* and *M. relicta*).

Various species of ducks from the Great Lakes are also known to feed on the benthos including scaup (*Aythya affinis, Aythya marila*), goldeneyes (*Bucephala clangula*), and oldsquaw (*Clangula hyemalis*) (Rofritz, 1977; Hickey et al., 1966; Drobney et al., 1982). Although contaminant levels are high in waterfowl (Smith et al., 1985), sources of contaminants are difficult to establish because of their seasonal migrations and their great diversity of food items.

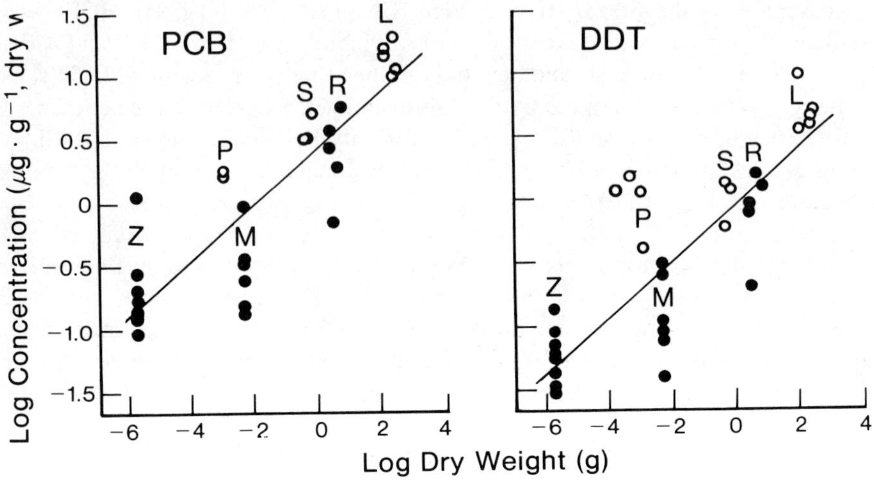


**Figure 4.2.** PCB and DDT concentrations as a function of body size in aquatic organisms from the eastern portion of Lake Ontario. Z, zooplankton; P, *Pontoporeia hoyi;* M, *Mysis relicta;* S, slimy sculpin; R, rainbow smelt; L, lake trout. From Borgmann and Whittle (1983); with permission.

## 8. BENTHIC ORGANISMS AS BIOMONITORING TOOLS

The use of benthic organisms to monitor both heavy metals and organic compounds in the environment has several advantages over simply measuring concentrations in the water or sediment: (1) Concentrations in the benthos provide an indication of the bioavailability of the particular contaminant. The major concern in the whole toxic contaminant issue is not whether contaminants are present, but whether they are available for bioaccumulation. (2) Many chemical and physical features of both water and sediments will affect the bioavailability of a given contaminant. Benthic organisms integrate contaminants from various sources forming a basis for assessing bioavailability. (3) Since most benthic communities are relatively stable in both time and space, contaminant levels in the benthos might be used in comparing concentrations at different sites, and also in comparing concentrations at the same site at different times.

Several benthic groups are potentially useful as biomonitors of contaminants in the Great Lakes. The most frequently analyzed groups have been oligochaetes and *P. hoyi* (Table 4.1). The advantages of using oligochaetes as a monitoring tool for heavy metals have recently been reviewed (Chapman et al., 1979b). Oligochaetes are the most widespread of the Great Lakes benthic groups and, because of the high tolerance levels of many oligochaete species, are often the only benthic group available for contaminant assessment in polluted areas. *P. hoyi* is the dominant species in the profundal waters of the Great Lakes where contaminants accumulate, is an important component in the diet of many fish, and tends to bioaccumulate organic contaminants to a greater extent than oligochaetes. In terms of assessing trends in bioavailable contaminants, life habits of the individual taxonomic groups as well as varying tendencies to bioaccumulate contaminants should be considered. For example, concentrations of two contaminants (hexachlorobenzene and PCB) in amphipods (mostly *P. hoyi*) and oligochaetes were compared to corresponding concentrations in surficial (0–2 cm) Lake Ontario sediments (Fox et al., 1983). Concentrations in oligochaetes were directly related to concentrations in the sediments, but no correlation was observed for amphipods (Fig. 4.3). Oligochaetes integrate contaminants in the upper sediments and thus body burdens will reflect contaminant levels down to the depth of the feeding zone. *P. hoyi,* on the other hand, feeds mostly on particulates recently settled from the water column. These surface particles have contaminant concentrations several times greater than concentrations in sediments integrated over several centimeters (Eadie et al., 1983a). Also, *P. hoyi* tends to bioaccumulate contaminants to a greater extent than oligochaetes; thus, body burdens in *P. hoyi* will not reflect sediment concentrations as closely as body burdens in oligochaetes.

Unionid bivalves are potentially an ideal biomonitoring organism. Mussels filter large quantities of water and thus are exposed to large amounts of contaminants from both the water and from suspended particulates. They are relatively long-lived (10–20 years), which allows the sampling of more than one year class, and are large enough to provide adequate tissue for analysis with a minimum amount of sampling effort. In the Great Lakes, mussels were successfully used to delineate sources of an organic contaminant (octachlorostyrene) in the Huron–Erie corridor (Pugsley et al., 1985). The disadvantage of using mussels as in situ bioindicators is their limited distribution in the Great Lakes. As noted, mussels are found in

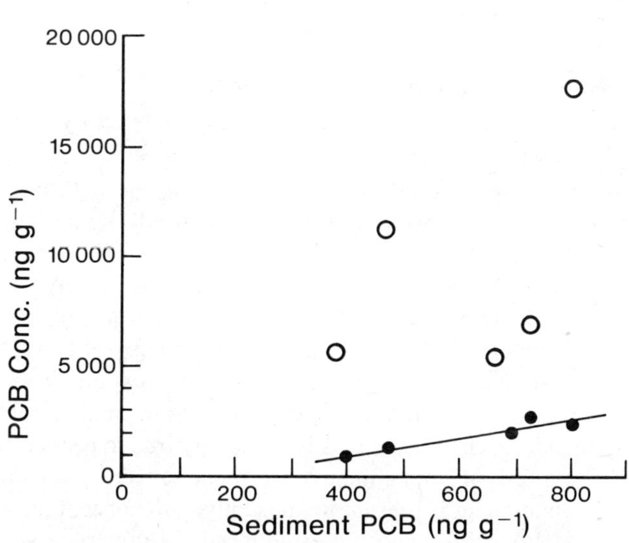


**Figure 4.3.** Concentrations (dry weight) of PCB in oligochaetes and amphipods as a function of the corresponding surficial sediment concentrations at five Lake Ontario sampling sites. ●, oligochaetes; $r = 0.96$. ○, amphipods, $r = 0.47$. Modified from Fox et al. (1983).

abundance only in the Huron–Erie corridor and in western Lake Erie. A possible means of circumventing this problem is to place uncontaminated mussels in cages that are anchored near suspected contaminant sources (Kauss and Hamdy, 1985). Mussels bioaccumulate contaminants very rapidly and maximum concentrations may be attained in 8–16 days (Kauss et al., 1983).

The mayfly nymph *Hexagenia* sp. may also prove to be a useful biomonitoring organism in certain areas of the Great Lakes. It is a rather large organism (up to 20 mg dry wt), is an important item in the diet of many fish, and ingests large amounts of sediment (Dermott, 1981). In the Great Lakes, contaminant levels in this genus have not been closely examined. In the upper Mississippi River, body burdens in *Hexagenia* were used as an indicator of the persistence and widespread distribution of PCB in that system (Clements and Kawatski, 1984).

Biological variables will also affect contaminant levels of benthic organisms. Contaminant uptake/retention by a particular organism will be affected by its lipid content, reproductive condition, age, and weight. For instance, seasonal changes in the body burden of BaP in *P. hoyi* varied by a factor of at least two without any change in contaminant load (Landrum et al., 1985). These changes were related to seasonal variations in lipid levels. As noted by Phillips (1978), the introduction of such biological variability necessitates the maintenance of rigid sampling and interpretative procedures.

## 9. SUMMARY AND RESEARCH NEEDS

Basically three approaches are available for studying the effects of pollutants on biota (Malins et al., 1984):

1. Effects can be inferred from field observations.
2. Effects can be derived from acute and chronic laboratory tests under controlled conditions.
3. Semicontrolled experiments can be conducted in the field in pens or other enclosures, or in model ecosystems under "natural" conditions.

In the Great Lakes, the first approach has been the most widely used method in assessing pollutant effects on the benthos. This approach has been successful in delineating specific zones of impact and in assessing the severity of the response. However, benthic populations are subject to wide fluctuations in both time and space and reflect the combined influences of all environmental variables, whether related to contaminant levels or not. This is especially true in nearshore and harbor areas where the impacts of contaminants are most severe. With this approach, unless a specific contaminant is present in unusually high concentrations, a definite cause-and-effect relationship cannot be established. Multivariate statistical techniques may be useful in discerning the most relevant factors affecting distributions (Dermott, 1978) but, unfortunately, few studies in the Great Lakes have simultaneously examined benthic distributions and levels of organic and inorganic pollutants.

The laboratory approach is useful in making relative toxicity comparisons of individual compounds or polluted sediments, and in evaluating tolerance levels of different benthic species. Only through the controlled conditions of the laboratory can the effects and fates of specific contaminants be separated from other environmental influences. However, the main limitation of laboratory studies is that they do not realistically simulate natural conditions. The complex physical, chemical, and biological factors that may affect a compound's toxicity and bioavailability in the field simply cannot be duplicated in the laboratory. Perhaps a more realistic assessment is to combine the two approaches, that is, to use laboratory toxicity tests in conjunction with detailed field observations. To date, only Malueg et al. (1984) have used both approaches to assess contaminant impact on Great Lakes benthic populations. Yet, this combined approach was successful only because one contaminant (copper) was at issue. Also of concern is the impact of interactions among low, sublethal levels of contaminants in the profundal areas of the Great Lakes. Such effects might best be examined by exposing natural populations (in native sediments) to low contaminant levels over long periods of time. Mesocosms, in conjunction with laboratory studies, may prove useful in this regard (Donaghay, 1984; McIntyre, 1977).

The ultimate goal of contaminant assessment is to develop predictive capabilities using effects models. Such models are still in the initial stages, yet inherent in

Collect sediments and associated benthic species

Analyze sediments chemically; and perform tissue-chemical, gross pathological, and histo-pathological analyses; evaluate community structures.

Codify sediments into groups based on chemical compositions and determine relationships between (1) chemical "clusters" and principal chemical components of sediment and (2) pathological abnormalities in biota and variations in community structures.

Conduct controlled laboratory and in situ field studies to identify chemicals responsible for toxicities and determine bioavailabilities and dose-response relationships.

Incorporate data into appropriate computer based model system.

**Figure 4.4.** Protocol for relating sediment contaminants to biological effects. Modified from Malins et al. (1984).

this approach is a clear understanding of the fate of the contaminant in the organism, and also the environmental and physiological variables that may affect uptake and elimination. Kinetics and fates research should permit the estimation of an exposure scenario for the benthos and, by combining such a scenario with information on community structure and acute and chronic effects, a predictive model of contaminmant effects may be developed. A protocol for combining field studies, laboratory experiments, and model development to examine contaminant effects on the biota was given by Malins et al. (1984) and is illustrated in Fig. 4.4. The concepts presented might certainly be used as a initial guide in planning future efforts to assess contaminant effects on the Great Lakes benthos.

## ACKNOWLEDGMENTS

This is contribution 467 of the Great Lakes Environmental Research Laboratory, National Oceanic and Atmospheric Administration. We thank B. J. Eadie, W. A. Frez, J. A. Robbins, and W. S. Gardner for reviewing the manuscript and providing helpful suggestions.

## REFERENCES

Anderson, E. D., and Smith, L. L., Jr. (1971). A synoptic study of food habits of 30 fish species from western Lake Superior. *Minn. Agric. Exp. Stn., Tech. Bull.* **279,** 1–199.

Ankar, S. (1977). The soft bottom ecosystem of the northern Baltic proper with special reference to the macrofauna. *Contr. Asko Lab. Univ. Stockholm*, No. 19, 1–62.

Bahner, L. H., Wilson, A. J., Jr., Sheppard, J. M., Patrick, J. M., Jr., Goodman, L. R., and Walsh, G. E. (1977). Kepone bioconcentration, accumulation, loss and transfer through estuarine food chains. *Chesapeake Sci.* **18,** 299–308.

Beeton, A. M. (1978). Present status of the Green Bay ecosystem. In H. J. Harris and V. Garsow (Eds.), Green Bay Workshop Proceedings. Univ. Wisconsin Sea Grant, WIS-SG-78-234, pp. 114–125.

Biddinger, G. R., and Gloss, S. P. (1984). The importance of trophic transfer in the bioaccumulation of chemical contaminants in aquatic ecosystems. *Residue Rev.* **91,** 103–146.

Bindra, K., and Hall, K. J. (1978). *Geochemical Partitioning of Trace Metals in Sediments and Factors Affecting Bioaccumulation in Benthic Organisms,* Wastewater Techn. Rep. University of British Columbia.

Borgmann, U., and Whittle, D. M. (1983). Particle-size-conversion efficiency and contaminant concentrations in Lake Ontario biota. *Can. J. Fish. Aquat. Sci.* **40,** 328–336.

Breck, J. E. (1984). Bioenergetics-based computer model for the behavior and fate of polychlorinated biphenyls and dieldrin in Lake Michigan alewives. Abstr., 5th Annu. Meet. Soc. Environ. Toxicol. Chem., Arlington, VA.

Brinkhurst, R. O. (1967). The distribution of aquatic oligochaetes in Saginaw Bay, Lake Huron. *Limnol. Oceanogr.* **12,** 137–143.

Brinkhurst, R. O., and Austin, M. J. (1979). Assimilation by aquatic oligochaeta. *Int. Rev. Gesamten Hydrobiol.* **64,** 245–250.

Brinkhurst, R. O., Hamilton, A. L., and Herrington, H. B. (1968). *Components of the Bottom Fauna of the St. Lawrence Great Lakes,* Prelim. Rep., No. 33. Great Lakes Inst., University of Toronto.

Britt, N. W. (1955). Stratification in western Lake Erie in summer of 1953: Effects on the *Hexagenia* (*Ephemeroptera*) population. *Ecology* **36,** 239–244.

Brković-Popović, I., and Popović, M. (1977). Effects of heavy metals on survival and respiration rate of tubificid worms, I. Effects on survival. *Environ. Pollut.* **13,** 65–72.

Cahill, R. A. (1981). Geochemistry of recent Lake Michigan sediments. Circ. 517 Ill. State Geol. Surv., pp. 1–96.

Carpenter, G. F., Mansey, E. L., and Watson, N. H. F. (1974). Abundance and life history of *Mysis relicta* in the St. Lawrence Great Lakes. *J. Fish. Res. Board Can.* **31,** 319–325.

Carr, J. F., and Hiltunen, J. K. (1965). Changes in the bottom fauna of western Lake Erie from 1930–1961. *Limnol. Oceanogr.* **10,** 551–569.

Chapman, P. M., and Brinkhurst, R. O. (1984). Lethal and sublethal tolerances of aquatic oligochaetes with reference to their use as a biotic index of pollution. *Hydrobiologia* **115,** 139–144.

Chapman, P. M., Churchland, L. M., Thomson, P. A., and Michnowsky, E. (1979a). Heavy metal studies with oligochaetes. In R. O. Brinkhurst and D. G. Cook (Eds.), *Aquatic Oligochaete Biology,* Plenum, New York, pp. 442–502.

Chapman, P. M., Churchland, L. M., Thomson, P. A., and Michnowsky, E. (1979b). Tubificid oligochaetes as monitors of heavy metal pollution. Tech. Rep. 862 Fish. Mar. Serv. (Can.), pp. 278–294.

Chapman, P. M., Farrell, M. A., and Brinkhurst, R. O. (1982). Relative tolerances of selected aquatic oligochaetes to individual pollutants and environment factors. *Aquat. Toxicol.* **2,** 47–67.

Clements, J. R., and Kawatski, J. A. (1984). Occurrence of polychlorinated biphenyls (PCB's) in adult mayflies (*Hexagenia bilineata*) of the upper Mississippi River. *J. Freshwater Biol.* **2,** 611–614.

Cook, D. G. (1975). A preliminary report on the benthic macroinvertebrates of Lake Superior. Tech. Rep. 572 Fish. Mar. Serv. (Can.), pp. 1–44.

Cook, D. G., and Johnson, M. G. (1974). Benthic invertebrates of St. Lawrence Great Lakes. *J. Fish. Res. Board Can.* **31,** 763–782.

Dermott, R. (1978). Benthic diversity and substrate-fauna associations in Lake Superior. *J. Great Lakes Res.* **4,** 505–512.

Dermott, R. (1981). Ingestion rate of burrowing mayfly *Hexagenia limbata* as determined with $^{14}C$. *Hydrobiologia* **83,** 499–503.

Donaghay, P. L. (1984). Utility of mesocosms to assess marine pollution. In H. H. White (Ed.), *Concepts in Marine Pollution Measurements.* Maryland Sea Grant College, University of Maryland, College Park. pp. 589–620.

Drobney, R. D., Jones, J. J., and Noseworthy, S. M. (1982). *The Effects of Winter Navigation on Waterfowl and Benthic Communities.* Final report to the U.S. Fish and Wildlife Service (Grant DOI-C-14-16-009). Patuxent Wildl. Res. Cent., Laurel, MD.

Eadie, B. J., Rice, C. P., and Frez, W. A. (1983a). The role of the benthic boundary in the cycling of PCBs in the Great Lakes. In D. Mackay, S. Peterson, S. J. Eisenreich, and M. S. Simmons (Eds.), *Physical Behavior of PCBs in the Great Lakes.* Ann Arbor Sci. Publ., Ann Arbor, MI pp. 213–228.

Eadie, B. J., Robbins, J. A., Landrum, P. F., Rice, C. P., Simmons, M. S., McCormick, M. J., Eisenreich, S. J., Bell, G. L., Pickett, R. L., Johansen, P., Rossman, R., Hawley, N., and Voice, T. (1983b). *The Cycling of Toxic Organics in the Great Lakes: A Three-year Status Report,* Tech. Memo. No. 45. Natl. Ocean. Atmos. Admin., Environ. Res. Lab., Great Lakes Environ. Res. Lab., Ann Arbor, MI.

Engler, R. M. (1979). Bioaccumulation of toxic substances from contaminated sediments by fish and benthic organisms. In S. A. Peterson and K. K. Randolph (Eds.), *Management of Bottom Sediments Containing Toxic Substances,* (EPA-600/3-79-102). U. S. Environ. Prot. Agency, Washington, DC, pp. 325–354.

Evans, M. S., Bathelt, R. W., and Rice, C. W. (1982). PCBs and other toxicants in *Mysis relicta*. *Hydrobiologia* **93,** 205–215.

Fisher, J. B. (1982). Effects of macrobenthos on the chemical diagenesis of freshwater sediments. In P. L. McCall and M. J. S. Tevesz (Eds.), *Animal–Sediment Relations. The Biogenic Alteration of Sediments.* Plenum, New York, pp. 177–218.

Fisher, J. B., Lick, W. S., McCall, P. L., and Robbins, J. A. (1980). Vertical mixing of lake sediments by tubificid oligochaetes. *J. Geophys. Res.* **85,** 3997–4006.

Fox, M. E., Carey, J. H., and Oliver, B. G. (1983). Compartmental distribution of organochlorine contaminants in the Niagara River and the western basin of Lake Ontario. *J. Great Lakes Res.* **9,** 287–294.

Frank, A. P., Landrum, P. F., and Eadie, B. J. (1986). Polycyclic aromatic hydrocarbon rates of uptake, depuration, and biotransformation by Lake Michigan *Stylodrilus heringianus. Chemosphere* **15,** 317–330.

Fremling, C. R. (1970). Mayfly distribution as a water quality index. U.S. Environmental Protection Agency, Water Quality Office. Water Pollut. Control Res. Ser. 16030 DOH 11/70, 1–39.

Frez, W. A., and Landrum, P. F. (1986). Species-dependent uptake of PAH in Great Lakes invertebrates. In M. Cook and A. J. Dennis (Eds.), *Polynuclear Aromatic Hydrocarbons: Chemistry, Characterization,* and *Carcinogenisis.* Ninth International Symposium, Battelle, Columbus, OH, pp. 291–304.

Funk, W. H., Rabe, R. W., Filby, R., and Parker, J. I. (1973). *An Integrated Study on the Impact of Metallic Trace Element Pollution in the Coeur d'Alene-Spokane Rivers-Lake Drainage System,* Publ. No. 222-946. Natl. Tech. Inf. Serv., Springfield, VA.

Gannon, J. E., and Beeton, A. M. (1969). Studies on the effects of dredged materials from selected Great Lakes harbors on plankton and benthos. Spec. Rep. 8 Univ. Wis.-Milwaukee, Cent. Great Lakes Stud., pp. 1–82.

Gardner, W. G., Nalepa, T. F., Frez, W. A., Cichocki, E. A., and Landrum, P. F. (1985). Seasonal patterns in lipid content of Lake Michigan macroinvertebrates. *Can. J. Fish. Aquat. Sci.* **42,** 1827–1832.

Green, R. H., and Vascotto, G. L. (1978). A method for the analysis of environmental factors controlling patterns of species composition in aquatic communities. *Water Res.* **12,** 583–590.

Haile, C. L., Veith, G. D., Lee, G. F., and Boyle, W. C. (1975). Chlorinated hydrocarbons in the Lake Ontario ecosystem (IFYGL). U. S. Environ. Prot. Agency, Off. Res. Dev. Rep. EPA-660/3-75-022, 1–28.

Hare, L., and Carter, J. C. H. (1976). The distribution of *Chironomus* (s.s.)? *cucini* (salinarius group) larvae (Diptera: Chironomidae) in Parry Sound, Georgian Bay, with particular reference to structural deformities. *Can. J. Zool.* **54,** 2129–2134.

Heberger, R. F., and House, R. (1974). Food of rainbow smelt and alewives in Lake Ontario. Abstr., 17th Conf. Great Lakes Res., Hamilton, Ont.

Hickey, J. J., Keith, J. A., and Coon, F. B. (1966). An exploration of pesticides in a Lake Michigan ecosystem. *J. Appl. Ecol.* **3** (Suppl.), 141–154.

Hiltunen, J. K., and Schloesser, D. W. (1983). The occurrence of oil and the distribution of *Hexagenia* (Ephemeroptera: Ephemeridae) nymphs in the St. Marys River, Michigan and Ontario. *Freshwater Invertebr. Biol.* **2,** 199–203.

Howmiller, R. P., and Beeton, A. M. (1970). The oligochaete fauna of Green Bay, Lake Michigan. Proc. 13th Conf. Great Lakes Res. pp. 15–46.

Howmiller, R. P., and Beeton, A. M. (1971). Biological evaluation of environmental quality, Green Bay, Lake Michigan. *J. Water Pollut. Control Fed.* **43,** 123–133.

Janssen, J., and Brandt, S. B. (1980). Feeding ecology and vertical migration of adult alewives (*Alosa pseudoharengus*) in Lake Michigan. *J. Fish. Res. Board Can.* **37,** 177–184.

Jernelov, A. (1970). Release of methyl mercury from sediments with layers containing inorganic mercury at different depths. *Limnol. Oceanogr.* **15,** 958–960.

Johnson, M. G., and Brinkhurst, R. O. (1971). Associations and species diversity in benthic macroinvertebrates of Bay of Quinte and Lake Ontario. *J. Fish. Res. Board Can.* **28,** 1683–1697.

Johnson, M. G., and Matheson, D. H. (1968). Macroinvertebrate communities of the sediments of Hamilton Bay and adjacent Lake Ontario. *Limnol. Oceanogr.* **13,** 99–111.

Karickhoff, S. W., and Morris, K. R. (1985). Impact of tubificid oligochaetes on pollutant transport in bottom sediments. *Environ. Sci. Technol.* **19,** 51–56.

Kauss, P. B., and Hamdy, Y. (1985). Biological monitoring of organochlorine contaminants in the St. Clair and Detroit Rivers using introduced clams, *Elliptio complanatus. J. Great Lakes Res.* **11,** 247–263.

Kauss, P. B., Suns, K., and Johnson, A. F. (1983). Monitoring of PCBs in water, sediments and biota of the Great Lakes—Some recent examples. In D. Mackay, S. Paterson, S. J. Eisenreich, and M. S. Simmons (Eds.), *Physical Behavior of PCBs in the Great Lakes.* Ann Arbor Sci. Publ., Ann Arbor, MI, pp. 385–409.

Kemp, A. L. W., Thomas, R. L., Dell, C. E., and Jaquet, J. M. (1976). Cultural impact on the geochemistry of sediments in Lake Erie. *J. Fish. Res. Board Can.* **33,** 440–462.

Kraft, K. J. (1979). *Pontoporeia* distribution along the Keweenaw shore of Lake Superior affected by copper tailings. *J. Great Lakes Res.* **5,** 28–35.

Krezoski, J. R. (1981). The influence of zoobenthos on fine-grained particle reworking and benthic solute transport in Great Lakes sediments. Ph.D. Thesis, University of Michigan, Ann Arbor.

Krezoski, J. R., Klump, J. V., and Kaster, J. L. (1984a). Dual radiotracer studies of the assimilation of organic contaminants by deposit feeding benthos. *EOS, Trans. Am. Geophys. Union* **64,** 1088 (Abstr. 42F-14).

Krezoski, J. R., Mozley, S. C., and Robbins, J. A. (1978). Influence of benthic macroinvertebrates on mixing of profundal sediments in southeastern Lake Huron. *Limnol. Oceanogr.* **23,** 1011–1016.

Krezoski, J. R., Robbins, J. A., and White, D. S. (1984b). Dual radiotracer measurement of zoobenthos-mediated solute and particle transport in freshwater sediments. *J. Geophys. Res.* **89,** 7937–7947.

Landrum, P. F. (1982). Uptake, depuration and biotransformation of anthracene by the scud, *Pontoporeia hoyi. Chemosphere* **11,** 1049–1057.

Landrum, P. F., and Scavia, D. (1983). Influence of sediment on anthracene uptake, depuration and biotransformation by the amphipod *Hyallela azteca. Can. J. Fish. Aquat. Sci.* **40,** 298–305.

Landrum, P. F., Eadie, B. J., Faust, W. R., Morehead, N. R., and McCormick, M. J. (1985). Role of sediment in the bioaccumulation of benzo[a]pyrene by the amphipod *Pontoporeia hoyi.* In M. Cooke and A. J. Dennis (Eds.), *Polynuclear Aromatic Hydrocarbons: Mechanisms, Methods, and Metabolism.* Battelle, Columbus, OH, pp. 799–812.

Lang, C., and Lang-Dobler, B. (1979). The chemical environment of tubificid and lumbriculid worms according to the pollution level of the sediment. *Hydrobiologia* **65,** 273–282.

Laskowski-Hoke, R. A., and Prater, B. L. (1981). Relationship of mortality of aquatic biota from 96-hour sediment bioassays and the change in chemical composition of the test water. *Bull. Environ. Contam. Toxicol.* **26,** 323–327.

Laskowski-Hoke, R. A., and Prater, B. L. (1984). Multivariate statistical analysis of 96-hour sediment bioassay and chemistry data. *Bull. Environ. Contam. Toxicol.* **33,** 400–409.

Lee, A. H. (1972). Regional characterizations of the thermal properties of Lake Ontario, Proc. 15th Conf. Great Lakes Res. pp. 625–634.

Leversee, G. J., Giesy, J. P., Landrum, P. F., Gerould, S., Bowling, J. W., Fannin, T. E., Haddock, J. D., and Bartell, S. M. (1982). Kinetics and biotransformation of benzo[a]pyrene in *Chironomus riparus. Arch. Environ. Contam. Toxicol.* **27,** 359–367.

Lewis, A. G., Whitfield, P., and Ramnarine, A. (1973). The reduction of copper toxicity in a marine copepod by sediment extract. *Limnol. Oceanogr.* **18,** 324–326.

Lindstrom, M., and Lindstrom, A. (1980). Changes in the swimming activity of *Pontoporeia affinis* (Crustacea, Amphipoda) after exposure to sublethal concentrations of phenol, 4-chlorophenol and styrene. *Ann. Zool. Fenn.* **17,** 221–231.

Lynch, T. R., and Johnson, H. E. (1982). Availability of a hexachlorobiphenyl isomer to benthic amphipods from experimentally contaminated natural sediments. ASTM Spec. Tech. Publ. STP 766, 273–287.

Mac, M. J., Edsall, C. C., Hesselberg, R. J., and Sayers, R. E., Jr. (1984). Flow-through bioassay for measuring bioaccumulation of toxic substances from sediment. U.S. Environ. Prot. Agency EPA-905/3-84-007, 1–17.

Macek, K. J., and Korn, S. (1970). Significance of the food chain in DDT accumulation by fish. *J. Fish. Res. Board Can.* **27,** 1496–1498.

Mackie, G. L., White, D. S., and Zdeba, T. W. (1980). A guide to freshwater mollusks of the Laurentian Great Lakes with special emphasis on the genus *Pisidium.* U.S. Environ. Prot. Agency, Off. Res. Dev. Rep. EPA-600/3-80-068, 1–144.

Magnuson, J. J., Forbes, A. M., and Hall, R. J. (1976). *An Assessment of the Environmental Effects of Dredged Material Disposal in Lake Superior,* Final Rep., Vol. 3. Duluth-Superior and Keweenaw Study Areas, Mar. Stud. Cent., University of Wisconsin, Madison.

Malins, D. C., Hodgins, H. O., Varanasi, U., MacLeod, W. D., Jr., McCain, B. B., and Chan, S. L. (1984). Chemical measurements of organic pollutants and effects criteria. In H. H. White (Ed.), *Concepts in Marine Pollution Measurements.* Maryland Sea Grant College, University of Maryland, College Park, pp. 405–417.

Malueg, K. W., Schuytema, G. S., Krawczyk, D. F., and Gakstatter, J. H. (1984). Laboratory sediment toxicity tests, sediment chemistry and distribution of benthic macroinvertebrates in sediments from the Keweenaw Waterway, Michigan. *Environ. Toxicol. Chem.* **3,** 233–242.

Mathis, J., and Cummings, T. T. (1973). Selected metals in sediments, water, and biota in the Illinois River. *J. Water Pollut. Control Fed.* **45,** 1573–1583.

McCall, P. L., and Fisher, J. B. (1980). Effects of tubificid oligochaetes on physical and chemical properties of Lake Erie sediments. In R. O. Brinkhurst and D. G. Cook (Eds.), *Aquatic Oligochaete Biology.* Plenum, New York, pp. 253–318.

McCall, P. L., Tevesz, M. J. S., and Schwelgien, S. F. (1979). Sediment mixing by *Lampsilis radiata siliquoidea* (Mollusca) from Western Lake Erie. *J. Great Lakes Res.* **5,** 105–111.

McIntyre, A. D. (1977). Effects of pollution on inshore benthos. In B. C. Coull (Ed.), *Ecology of Marine Benthos.* Univ. of South Carolina Press, Columbia, pp. 301–318.

McMurtry, M. J. (1984). Avoidance of sublethal doses of copper and zinc by tubificid oligochaetes. *J. Great Lakes Res.* **10,** 267–272.

Milbrink, G. (1983). Characteristic deformities in tubificid oligochaetes inhabiting polluted bays of Lake Vanern, Southern Sweden. *Hydrobiologia* **106,** 169–184.

Muir, D. C. G., Rawn, G. P., Townsend, B. E., Lockhart, W. L., and Greenhalgh, R. (1985). Bioaccumulation of cypermethrin, deltamethrin, fenvalerate and permethrin by *Chironomus tentans* larvae in sediment and water. *Environ. Toxicol. Chem.* **4,** 51–61.

Murphy, P. G., and Murphy, J. V. (1971). Correlations between respiration and direct uptake of DDT in the mosquito fish, *Gambusia affinis. Bull. Environ. Contam. Toxicol.* **6,** 581–588.

Nalepa, T. F. (1987). Long-term changes in the macrobenthos of southern Lake Michigan, *Can. J. Fish. Aquat. Sci.* **44,** 515–524.

Nalepa, T. F., and Thomas, N. A. (1976). Distribution of macrobenthic species in Lake Ontario in relation to sources of pollution and sediment parameters. *J. Great Lakes Res.* **2,** 150–163.

Neely, W. B., Branson, D. R., and Blau, G. E. (1974). Partition coefficient to measure bioconcentration potential of organic chemicals in fish. *Environ. Sci. Technol.* **8,** 1113–1115.

Oliver, B. G. (1984). Uptake of chlorinated organics from anthropogenically contaminated sediments by oligochaete worms. *Can. J. Fish. Aquat. Sci.* **41,** 878–883.

Petr, T. (1977). Bioturbation and exchange of chemicals in the mud–water interface. In H. L. Golterman (Ed.), *Interactions between Sediments and Freshwater*. Junk, The Hague, pp. 216–226.

Phillips, D. J. H. (1978). Use of biological indicator organisms to quantitate organochlorine pollutants in aquatic environments—A review. *Environ. Pollut.* **16,** 167–229.

Prater, B. L., and Anderson, M. (1977). A 96-hour sediment bioassay of Duluth and Superior harbor basins using *Hexagenia limbata, Asellus communis, Daphnia magna,* and *Pimephalos promelas* as test organisms. *Bull. Environ. Contam. Toxicol.* **13,** 159–169.

Pugsley, C. W., Hebert, P., and McQuarrie, P. (1984). Patterns of contaminant distribution in clams and sediments taken from the Huron-Erie corridor during 1983. Abstr., 27th Conf. Great Lakes Res., St. Catherines, Ont.

Pugsley, C. W., Hebert, P. D. N., Wood, G. W., Brotea, G., and Obal, T. W. (1985). Distribution of contaminants in clams and sediments from the Huron-Erie corridor. I-PCBs and octachlorostyrene. *J. Great Lakes Res.* **11,** 275–289.

Robbins, J. A. (1982). Stratigraphic and dynamic effects of sediment reworking by Great Lakes zoobenthos. *Hydrobiologia* **92,** 611–622.

Robbins, J. A., McCall, P. L., Fisher, J. B., and Krezoski, J. R. (1979). Effect of deposit feeders on migration of cesium-137 in lake sediments. *Earth Planet. Sci. Lett.* **42,** 277–287.

Robertson, A., and Alley, W. P. (1966). A comparative study of Lake Michigan macrobenthos. *Limnol. Oceanogr.* **11,** 576–583.

Rofritz, D. J. (1977). Oligochaeta as a winter food source for the oldsquaw. *J. Wildl. Manage.* **41,** 590–591.

Rosa, F. (1985). Sedimentation and sediment resuspension in Lake Ontario. *J. Great Lakes Res.* **11,** 13–25.

Roth, J. C., and Mozley, S. C. (1973). A re-examination of the naiades (Unionidae, Pelecypoda) of western Lake Erie, 1972. Abstr., 21st Annu. Meet. Midwest Benthol. Soci., East Lansing, MI.

Rubinstein, N. I., Lores, E., and Gregory, N. R. (1983). Accumulation of PCBs, mercury, and cadmium by *Vereis virens, Mercenaria mercenaria* and *Palaemonetes pugio* from contaminated harbor sediments. *Aquat. Toxicol.* **3,** 249–260.

Schneider, J. C., Hooper, F. F., and Beeton, A. M. (1969). The distribution and abundance of benthic fauna in Saginaw Bay, Lake Huron. Proc. 12th Conf. Great Lakes Res. pp. 80–90.

Slooff, W. (1983). Benthic macroinvertebrates and water quality assessment: Some toxicological considerations. *Aquat. Toxicol.* **4,** 73–82.

Smith, V. E., Spurr, J. M., Filkins, J. C., and Jones, J. J. (1985). Organchlorine contaminants of wintering ducks foraging on Detroit River sediments. *J. Great Lakes Res.* **11,** 231–246.

Thomann, R. V. (1981). Equilibrium model of fate of microcontaminants in diverse food chains. *Can. J. Fish. Aquat. Sci.* **38,** 280–296.

Thomann, R. V., and Connolly, J. P. (1984). Model of PCB in the Lake Michigan lake trout food chain. *Environ. Sci. Technol.* **18,** 65–71.

Thomas, R. L. (1983). Lake Ontario sediments as indicators of the Niagara River as a primary source of contaminants. *J. Great Lakes Res.* **9,** 118–124.

Thomas, R. L., Kemp, A. L. W., and Lewis, C. F. M. (1972). Report on the surficial sediment distribution of the Great Lakes, 1. Lake Ontario. Sci. Ser. 10th Cent. Inland Waters (Can.) pp. 1–52.

Thompson, K. A., Brown, D. A., Chapman, P. M., and Brinkhurst, R. O. (1982). Histopathological effects and cadmium-binding protein synthesis in the marine oligochaete *Monopylephorus cuticulatus* following cadmium exposure. *Trans. Am. Microsc. Soc.* **101,** 10–26.

Thornley, S. (1985). Macrozoobenthos of the Detroit and St. Clair Rivers with comparisons to neighboring waters. *J. Great Lakes Res.* **11,** 290–296.

Thornley, S., and Hamdy, Y. (1984). *An Assessment of the Bottom Fauna and Sediments of the Detroit River* (ISBNO-7743-8474-3). Ontario Ministry of the Environment Report, London, Ontario.

Vander Wal, J. (1977). Relations between Nipigon Bay benthic macroinvertebrates and selected aspects of their habitat. *J. Fish. Res. Board Can.* **34,** 824–829.

Walters, L. J., Wolery, T. J., and Myser, R. O. (1974). Occurrence of As, Cd, Co, Cr, Cu, Fe, Hg, Ni, Sb, and Zn in Lake Erie sediments. Proc. 17th Conf. Great Lakes Res., pp. 219–234.

Warwick, W. F. (1980). Chironomidae (Diptera) responses to 2800 years of cultural influence: A palaeolimnological study with special reference to sedimentation eutrophication and contamination processes. *Can. Entomol.* **112,** 1193–1238.

Weininger, D. (1978). Accumulation of PCBs in lake trout in Lake Michigan. Ph.D. thesis, University of Wisconsin, Milwaukee.

Wells, L. (1980). Food of alewives, yellow perch, spottail shiners, trout-perch, and slimy and fourhorn sculpins in southeastern Lake Michigan. Tech. Pap. 98 U.S. Fish Wildl. Serv., pp. 1–12.

Wells, L., and Beeton, A. M. (1963). Food of the bloater *Coregonus hoyi* in Lake Michigan. *Trans. Am. Fish. Soc.* **92,** 245–255.

Wentsel, R., McIntosh, A., and Anderson, V. (1977a). Sediment contamination and benthic macroinvertebrate distribution in a metal-impacted lake. *Environ. Pollut.* **14,** 187–192.

Wentsel, R., McIntosh, A., McCafferty, W. P., Atchison, G., and Anderson, V. (1977b). Avoidance response of midge larvae (*Chironomus tentans*) to heavy metal contaminated sediment. *Hydrobiologia* **55,** 171–175.

Wentsel, R., McIntosh, A., and Atcheson, G. (1977c). Sublethal effects of metal contaminated sediment on midge larvae (*Chironomus tentans*). *Hydrobiologia* **56,** 153–156.

White, D. S., and Klahr, P. C. (1982). Effect of PCBs on the rate of sediment reworking by *Stylodrilus heringianus* (Oligochaeta). Abstr., 30th Annu. Meet. North Am. Benthol. Soc., Ann Arbor, MI.

Whittle, D. M., and Fitzsimons, J. D. (1983). The influence of the Niagara River on contaminant burdens of Lake Ontario biota. *J. Great Lakes Res.* **9,** 295–302.

Wiederholm, T. (1984). Incidence of deformed chironomid larvae (Diptera: Chironomidae) in Swedish lakes. *Hydrobiologia* **109,** 243–249.

Wolery, T. J., and Walters, L. J., Jr. (1974). Pollutant mercury and sedimentation in the western basin of Lake Erie. Proc. 17th Conf. Great Lakes Res., pp. 235–249.

# 5

# HISTORICAL CHANGES IN THE MAJOR FISH RESOURCES OF THE GREAT LAKES

***Wilbur L. Hartman***

*United States Fish and Wildlife Service, National Fisheries Center–Great Lakes, Ann Arbor, Michigan 48105*

## 1. INTRODUCTION

The largest and most valuable assemblage of freshwater fish resources in the world is in the Laurentian Great Lakes of North America. This is not surprising, since collectively the five lakes contain 95% of the surface area of fresh water in the United States, and among the lakes of the world, Lake Superior ranks 2nd in area (82,100 $km^2$), Lake Huron ranks 5th (59,500 $km^2$), Lake Michigan ranks 6th (57,750 $km^2$), Lake Erie ranks 13th (25,657 $km^2$), and Lake Ontario ranks 17th (19,000 $km^2$) according to Herdendorf (1982). The lakes contain a great variety of habitats, ranging from deep, oligotrophic basins to shallower, mesotrophic basins to the shallowest of eutrophic embayments, and, accordingly, they contain a wide variety of important fish species.

When European missionaries, fur traders, and settlers first reached the Great Lakes, they found bountiful supplies of fish, only small amounts of which were used for subsistence by the relatively few native Americans who lived along the shores. Lake trout (*Salvelinus namaycush*), lake whitefish (*Coregonus clupeaformis*), lake herring (*Coregonus artedii*), yellow perch (*Perca flavescens*), walleye (*Stizostedium vitreum vitreum*), burbot (*Lota lota*), several ciscoes (*Coregonus* spp.), and lake sturgeon (*Acipenser fluvescens*) were plentiful and soon were the targets of subsistence and commercial pursuit by the settlers (Kutkuhn, 1985).

Unfortunately, as markets developed in the midwest during the 1800s, and the need for domestic protein intensified during World Wars I and II, the misconception grew and persisted among fishermen that fish resources in the vast expanses of the Great Lakes were limitless and inexhaustible. So did the misconception that the volume of water contained in the Great Lakes was so enormous, and the watershed so vast, that the system was impervious to physical alterations and to physical and chemical pollution (Kutkuhn, 1985), despite recognition of such problems by the United States and Canada in enactment of the Boundary Waters Treaty in 1909, and periodic visible evidence of habitat degradation in tributaries to and inshore waters of the Great Lakes.

Many changes have occurred in the fishery resources and their critical habitats in the Great Lakes over the past century. Several valuable stocks have become extinct, or severely reduced in abundance. Other species of fish have invaded the Great Lakes or have been purposefully introduced, with mixed results. Water quality and critical habitats have been degraded—seriously in some instances.

My purpose here is to review historic changes in the major fish resources of the five Great Lakes, and to identify the cause or causes for those changes. In some instances it will be clear that intensive fishing was the primary cause of change; in other instances it will be nearly as clear that predation by the sea lamprey played a significant if not dominant role in change; and in still others it will be clear (or at least circumstantial) that exotic species other than the sea lamprey have been implicated. The exotics that have invaded or been accidentally or purposefully released into the Great Lakes system have not only adversely affected indigenous fishes, but some have developed into new and valuable resources. However, when it comes to degradation of water quality and of critical habitat, the linkages to changes in fish populations are neither clear-cut nor quantifiable—their impacts were generally far more subtle and difficult to detect, but not necessarily of less importance. Inasmuch as a comprehensive review of all changes in fishery resources, water quality, and habitat conditions in the Great Lakes would be encyclopedic, I confine discussion here to prime examples related to the most historically important fish resources. One of the values of the approach used in this manuscript is the collation in a single-source document of the conclusions reached by many scientists on reasons for changes in the fish resources across the Great Lakes basin.

Two important points are necessary to clarify the perspective of this paper: (1) traditional fishery research on the Great Lakes has focused on the large offshore and inshore stocks of fish, when the main effects of chemicals and toxic substances on fish populations are likely to occur very nearshore in tributaries, harbors, and connecting waterways; and (2) that being the case, most fishery scientists believe that direct impacts of toxic substances on inshore and offshore stocks of fish would be of less importance than exploitation, physical destruction of or exclusion from riverine or estuarine spawning grounds, and the introduction of exotic species of fish (including the parasitic sea lamprey).

In 1971, under sponsorship of the binational Great Lakes Fishery Commission, an International Symposium on Salmonid Communities in Oligotrophic Lakes was convened at Geneva Park, Ontario. The primary goal of this Symposium was to attempt to identify the separate and joint effects of three major stresses imposed on fish communities by man: exploitation, cultural eutrophication, and species introduction. Papers presented on each of the five Great Lakes—Lake Superior (Lawrie and Rahrer, 1973), Lake Michigan (Wells and McLain, 1973), Lake Huron (Berst and Spangler, 1973), Lake Erie (Hartman, 1973), and Lake Ontario (Christie, 1973)—were very useful source materials for events, from historic times to about 1970, and much of this present account is synopsized from those sources. To extend the scope of this paper to 1983 or 1984, I have consulted more recent formal publications, technical reports, official commercial catch records, and unpublished information.

The use of commercial catch statistics to detect changes in the abundance of fish requires qualification. The reporting of catches by commercial fishermen is often imprecise; most illegal harvests are not recorded; and the sport-fishery component of the harvest is not included. Only in the past 10–15 years has sport fishery become a significant portion of the total catches. However, most general trends are still accurately reflected in the commercial catches.

Finally, the dynamics of stocked fishes, most notably the Pacific salmons, are not addressed here. These resources are for the most part artificially generated and changes in their abundance are usually due to changes in stocking levels and levels of exploitation.

## 2. MAJOR FISH RESOURCES OF LAKE SUPERIOR

### 2.1. The Lake

Lake Superior is the largest, deepest, and coldest of the Great Lakes. Generally regarded as still being in a relatively pristine state (Lawrie and Rahrer, 1973), it remains lowest in concentrations of sodium and potassium, calcium, chloride, sulfate, and total dissolved solids, with no evident trend toward increase since 1890 (Weiler and Chawla, 1969). It is a classic example of a large oligotrophic lake sustaining populations of cold stenotherm fishes in its open waters.

### 2.2. Lake Herring

Since 1900, lake herring have contributed the largest harvests to the commercial fisheries in Lake Superior, yielding between 7 to 20 million pounds annually (Fig. 5.1). But in the late 1950s and early 1960s landings declined severely in Michigan, especially in Wisconsin and Minnesota waters. Anderson and Smith (1971) re-

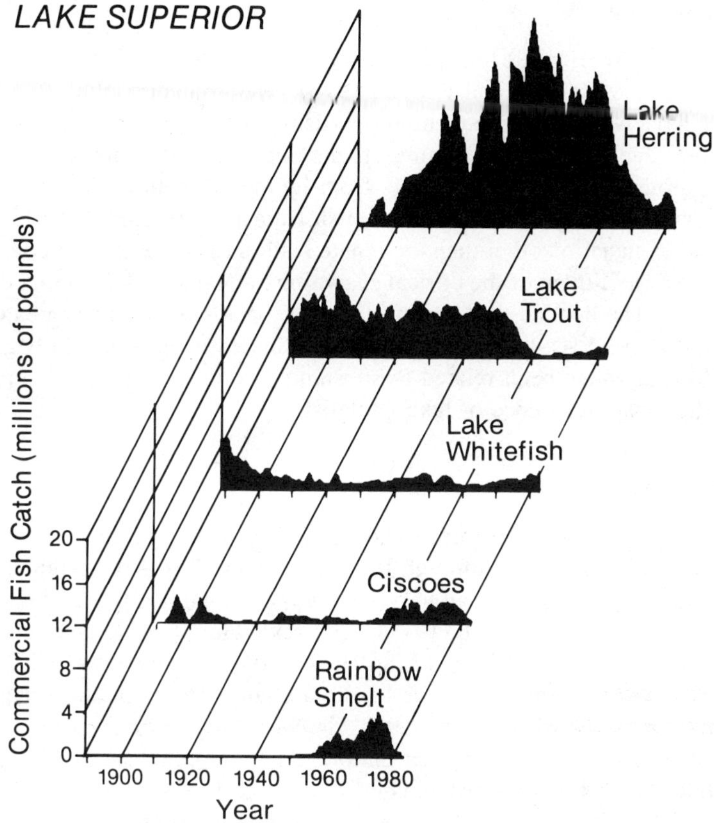


**Figure 5.1.** Commercial harvests of major fish species in Lake Superior, 1889–1983. Sources of data: Baldwin et al. (1979) and National Fisheries Center–Great Lakes annual summaries of commercial harvests of fish from the Great Lakes.

jected overfishing as the cause of the collapse in these areas; finding no deleterious abiotic factor of general significance, they concluded that competition for food with the increasing populations of bloaters (*Coregonus hoyi*), and particularly of the introduced rainbow smelt (*Osmerus mordax*), was the most probable cause (Lawrie and Rahrer, 1973). Selgeby et al. (1978), who studied predation by rainbow smelt on lake herring in Black Bay (in Ontario waters of western Lake Superior), where large and stable populations of rainbow smelt and lake herring coexisted, and in the Apostle Island's region (Wisconsin) where the abundance of lake herring was severely reduced, concluded that predation on herring larvae was not a major factor controlling or suppressing lake herring in either region. Selgeby (1982), in a classic study of the decline of the Wisconsin fishery from 1936 to 1978, demonstrated that sequential overexploitation of six major unit stocks caused the collapse. Unexpectedly, extremely strong year classes were produced in 1978 and 1980 and the population is steadily rebuilding.

## 2.3. Lake Whitefish

Lake whitefish experienced four obvious episodes of change (Fig. 5.1) in the history of the fishery (Lawrie and Rahrer, 1973). The gradual decline from 1891 to 1930 may have been due to sequential overfishing of unit stocks, as well as to the loss of river and estuarine spawning stocks because of the deposition over the spawning grounds of woody materials from lumbering and log-rafting activities. The later partial resurgence of lake whitefish from 1930 through the 1950s has no clear causal explanation other than the innate resiliency of the species coupled with reduction in despoilment of the critical spawning habitat after substantial decreases in lumbering. The decline that began in the late 1950s is positively related to the rise in abundance of sea lampreys. Finally, the increased harvest that began in the early 1970s may have been related to substantial control of the sea lamprey, and again to the innate resilience of lake whitefish.

## 2.4. Lake Trout

Annual harvests of lake trout from Lake Superior held fairly steady between 3 and 6 million pounds from 1889 through the early 1950s (Fig. 5.1). Abundance then declined precipitously and annual harvests dropped to barely 10% of their previous levels by 1960. Pycha and King (1975), after reviewing the available information, concluded that intensified fishing during 1929 to 1949, combined with a 305% rise in effective fishing effort that resulted from a change from cotton to nylon twine in gill nets during the period 1950–1952, depleted and seriously stressed the lake trout populations. The parasitic sea lamprey, first reported in Lake Superior in 1946 (Table 5.1), exploded in abundance during the 1950s; all investigators agree that sea lamprey predation on the already stressed lake trout populations contributed heavily to the precipitous decline.

Control of the sea lamprey began in 1953 with installation of electrical barriers

**Table 5.1 Chronology of First Appearance of Certain Exotic Species in the Great Lakes**

| Lake | Rainbow smelt | Sea lamprey | Alewife[a] | White perch |
|---|---|---|---|---|
| Ontario | 1929[b] | 1850[c] | 1860 | 1950 |
| Erie | 1932[d] | 1921[a] | 1931 | 1953 |
| Huron | 1925[b] | 1932[a] | 1933 | 1980 |
| Michigan | 1923[b] | 1936[a] | 1949 | — |
| Superior | 1930[b] | 1946[a] | 1954 | — |

[a]From Christie (1974).
[b]From Van Oosten (1937).
[c]From Lark (1973).
[d]From Trautman (1957).

and traps on streams where adults enter from the lake to spawn. Treatment of the streams with a larvicide, which began in 1960, proved to be effective in greatly reducing populations of ammocoetes, which usually remain in the fine sediments of the spawning streams for several years before they metamorphose and migrate to the lake to begin their adult parasitic phase.

Some remnant native stocks of lake trout survived (e.g., Gull Island Shoal, Wisconsin; Isle Royale and Stannard Rock, Michigan; deep waters of Thunder Bay, Ontario). Beginning in the 1960s, millions of hatchery-reared fingerling and yearling lake trout were stocked annually in Lake Superior. The slight but steady rise in harvest after the mid 1960s was largely composed of stocked fish. By 1984, stocks of introduced and of wild (lake-produced) lake trout had substantially increased in most areas. Nevertheless, intensive fishing and sea lamprey predation continue to impede rapid and complete restoration of self-sustaining populations of lake trout to their historical levels of abundance.

### 2.5. Ciscoes

At least six species of ciscoes, usually called ''chubs'' by the fishermen, were available to the earliest commercial fisheries. Sequential ''fishing up and fishing out'' proceeded from the larger species to the smaller species, but annual harvests remained dependable, although modest, through the early 1950s (Fig. 5.1). Collapse of the lake trout population then resulted in a shift in effort by the fishery to bloaters, which led to a substantial increase in annual harvest that was sustained for over 20 years until 1978, when a decline began.

### 2.6. Rainbow Smelt

Rainbow smelt were first introduced successfully into the upper Great Lakes by a planting in 1912 in Crystal Lake, Michigan, which drains into Lake Michigan. They were first reported in Lake Superior in 1930 (Table 5.1). The commercial fishery began to harvest them in 1952 and annual harvest varied generally between 1 and 3 million pounds until 1980 (Fig. 5.1). During this period rainbow smelt became very important prey of lake trout, coho (*Oncorhynchus kitsutch*) and chinook (*Oncorhynchus tshawytscha*) salmon, and brook (*Salvelinus fontinalis*), brown (*Salmo trutta*), and rainbow (*Salmo gairdneri*) trout. Heavy mortality among older smelt, for reasons unknown, resulted in markedly reduced spawning stocks and year-class failures during the late 1970s and caused a ninefold drop in standing biomass as well as harvest between 1978 and 1983. Even the unusually strong year class produced in 1983 will unlikely by itself reestablish rainbow smelt to their former levels of abundance. Fortunately, the continued recovery of lake herring populations should again contribute an important component to the community of prey fishes that is critical for sustaining native and introduced salmonid populations.

## 3. MAJOR FISH RESOURCES OF LAKE HURON

### 3.1. The Lake

Deep, cold, oligotrophic Lake Huron receives inflow from Lakes Superior and Michigan, in addition to that from its own watershed. Water quality has changed only slightly in the open waters of the lake since the early 1800s (Beeton, 1965; Berst and Spangler, 1973). Weiler and Chawla (1969) reported that in most aspects concentrations of chemicals in Lake Huron proper were intermediate between those in Lakes Superior and Michigan. The only significant increases have been observed in Saginaw Bay and various harbors and bays in Georgian Bay and the North Channel.

### 3.2. Lake Trout

Historically the harvest of lake trout from Lake Huron fluctuated between 1.8 and 2.7 million pounds annually (Fig. 5.2). Lake trout harvests first declined in the main body of the lake in the early 1940s, and then in Saginaw and Georgian bays. Lake trout production was insignificant in U.S. waters by 1946 and in Ontario waters by 1960. Only two small remnant populations have survived: one in McGregor Bay in the North Channel, and one in Parry Sound in Georgian Bay. The decline of the lake trout in Lake Huron was associated primarily with overfishing. However, predation by the expanding population of sea lampreys, first reported in Lake Huron in 1932 (Table 5.1), was the critical factor in the final decline of the lake trout in Lake Huron (Berst and Spangler, 1973).

Sea lamprey control was begun in the Lake Huron watershed in 1960, and the first round of chemical treatments of lamprey producing streams was completed in 1968. Since lake trout stocking began in 1973, more than 10 million young hatchery-reared lake trout have been stocked in U.S. waters of Lake Huron. Although some lake trout stocked in 1973 reached maturity in 1979, no fertilized eggs were found until 1981 when a few were collected off Rockport, Michigan. In the spring of 1982 a few swim-up fry were captured in the same area. Since then a few unclipped, lake-produced fingerlings and yearlings have been captured off Rockport, in Thunder Bay, and (in 1985) off Port Austin. In Canadian waters, the emphasis over the past decade has been on the stocking of splake hybrids (lake trout × brook trout) in southern Georgian Bay, Owen Sound, and North Channel. Splake were engineered genetically to be relatively invulnerable to sea lamprey—a kind of safety precaution in case control of sea lamprey would not become sufficiently effective to permit survival and recovery of lake trout populations. Splake were selected for their smaller size and earlier age at maturity, and their habit of occupying shallower waters than lake trout (Berst and Spangler, 1970). Initial survival and growth of these hybrids was relatively good, but fishing mortality was high.

Recent increases in annual harvests of lake trout have been the result of an intensified Indian fishery in Treaty-ceded waters of northern Lake Huron in Mich-

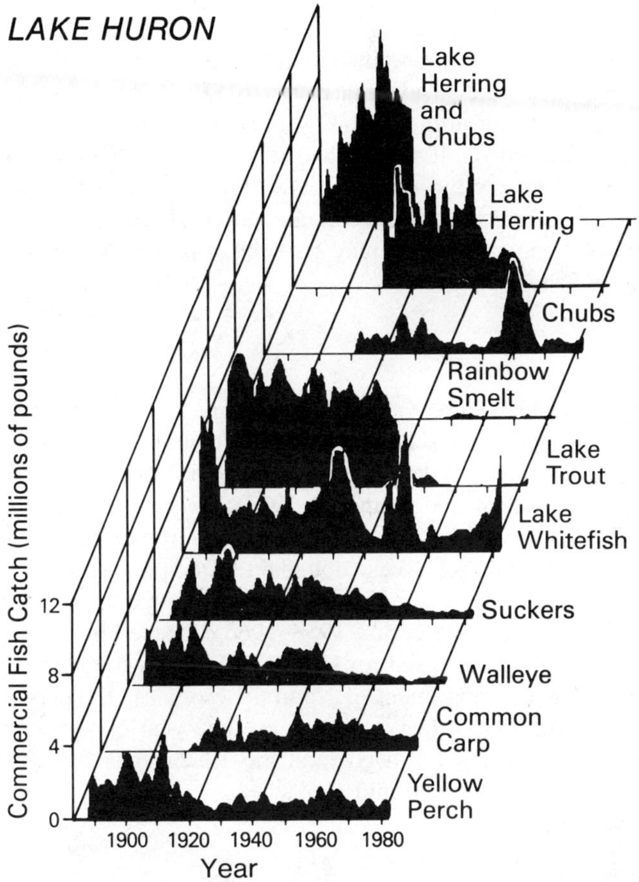


**Figure 5.2.** Commercial harvests of major fish species in Lake Huron, 1883–1983. Sources of data: Baldwin et al. (1979) and National Fisheries Center–Great Lakes annual summaries of commercial harvests of fish from the Great Lakes.

igan. Recent regulations imposed by State of Michigan and tribal fishery managers may reduce fishing mortality on lake trout and increase the potential for the accumulation of brood stocks and ultimately the reestablishment of self-sustaining populations.

### 3.3. Lake Whitefish

Annual harvests of lake whitefish have generally ranged from 1 to 5 million pounds since 1889, with only two short periods of lower harvests—1941–1946 and 1957–1960 (Fig. 5.2). In the most recent years, intensification of the tribal fishery for lake whitefish in the U.S. waters of northern Lake Huron has substantially increased harvest levels, but the populations still appear to be thriving and apparently increasing.

### 3.4. Lake Herring

Historically, the commercial importance of lake herring nearly equaled that of lake trout. Annual landings were steady and generally ranged between 2 and 6 million pounds from the late 1800s until 1940 (Fig. 5.2). They then declined and the lake herring fishery collapsed by 1955 and had not recovered by 1983. Inasmuch as most lake herring were produced in Saginaw Bay, Beeton (1969) suggested that changes in water quality leading to accelerated eutrophication and increases in the concentrations of toxic substances may have been primarily responsible for the initial loss of the fishery.

### 3.5. Bloaters

Bloaters ranked fourth in importance to the commercial fishery in Lake Huron, with annual harvests seldom exceeding 1 million pounds prior to the 1950s (Fig. 5.2). Harvests were somewhat depressed during 1940–1955, the period of the final decline of lake trout (Berst and Spangler, 1973). Perhaps sea lamprey predation was a factor. Exploratory fishing in 1956 by the U.S. Bureau of Commercial Fisheries revealed that bloaters were very abundant (Anonymous, 1958). As a result of this information the fishery for bloaters intensified—perhaps too greatly—and annual harvests throughout the period 1958–1966 exceeded historic records. The populations then abruptly collapsed. In 1970 the Michigan commercial fishery was closed by the Michigan Department of Natural Resources. Bloaters showed little sign of recovery until 1977 when an above-average year class was produced. A series of strong year classes followed, and the 1982 year class was the largest recorded since the collapse of the mid 1960s.

### 3.6. Rainbow Smelt

Rainbow smelt were first reported in Lake Huron in 1925 (Table 5.1), and supported a modest commercial fishery beginning in the early 1950s (Fig. 5.2). Of far greater importance, however, is the rainbow smelt's function as prey for such salmonids as lake trout, coho and chinook salmon, and brown and rainbow trout. Rainbow smelt abundance continued to remain high in the early 1980s whereas alewife abundance sharply declined. Fortunately, the recent marked increase in bloaters has largely filled the prey fish function previously occupied by alewives.

### 3.7. Alewives

Alewives (*Alosa pseudoharengus*) were first recorded in Lake Huron in 1933 (Table 5.1). They became fairly abundant by the 1960s and sustained a small commercial fishery in South Bay, Ontario (Coble, 1967). Like rainbow smelt, alewives became very important prey for some salmonids. A marked decline of alewives began in the early 1970s, however, and continued into the 1980s. A combination of harsh winters and intensified predation by increasing populations of salmonids may have been chiefly responsible (R. Argyle, National Fisheries Center-Great Lakes, Ann Arbor, MI, personal communication).

### 3.8. Suckers, Walleyes, Yellow Perch, and Common Carp

Several warm- and cool-water species characteristically concentrate in the embayments, harbors, and along the shallower shorelines of Lake Huron. Over the years the walleye and yellow perch have contributed markedly to the commercial and sport fisheries (Fig. 5.2). Most suckers, yellow perch, and common carp (*Cyprinus carpio*) harvested in U.S. waters have come from Saginaw Bay. Through 1946, most walleyes commercially harvested in Lake Huron also came from Saginaw Bay. The population collapsed during the late 1940s (Fig. 5.2). Commercial fishermen believed that pollution from Saginaw River was mainly responsible for this decline. Schneider (1977) detailed the circumstantial evidence and his similar conclusion was supported by Schneider and Leach (1979). Recent cooperative efforts between sportsmans clubs and the Michigan Department of Natural Resources to improve the remnant populations of walleyes in Saginaw Bay have been successful.

Yellow perch yielded relatively stable catches from 1930 to the late 1950s (Fig. 5.2); several strong year classes were then produced and abundance increased. But the fisheries soon intensified and abundance declined into the 1970s. Spangler et al. (1977) suggested that walleye stocks in McGregor Bay and near the mouth of the Spanish River very likely were adversely affected by acidification of the watersheds or the release of toxic mill wastes. During the early 1980s the abundance of yellow perch began to increase in Saginaw Bay and southern Lake Huron.

## 4. MAJOR FISH RESOURCES OF LAKE MICHIGAN

### 4.1. The Lake

Lake Michigan is generally oligotrophic, but has a large, shallow, warm-water embayment, Green Bay, on its northwest side. Since 1890, concentrations of various chemicals and total dissolved solids, even in the open lake waters, have increased much more in Lake Michigan than in Lakes Superior and Huron (Weiler and Chawla, 1969). Of particular concern regarding the reproduction by fish has been the appearance of relatively high levels of certain toxic substances (for example, see Willford et al., 1981).

### 4.2. Lake Trout

The lake trout was the most valuable commercial species in Lake Michigan from 1890 until the mid 1940s. As Wells and McLain (1973) noted, lake trout production began declining precipitously in 1945 (Fig. 5.3). The sea lamprey was first reported in Lake Michigan in 1936 (Table 5.1; Applegate, 1950). By the early

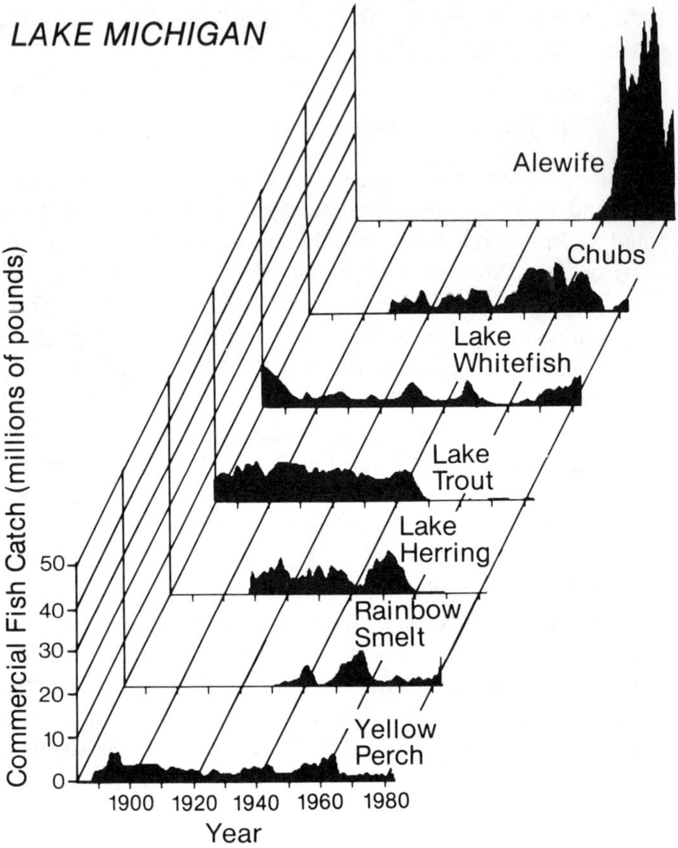


**Figure 5.3.** Commercial harvests of major fish species in Lake Michigan, 1885–1983. Sources of data: Baldwin et al. (1979) and National Fisheries Center–Great Lakes annual summaries of commercial harvests of fish from the Great Lakes.

1940s it was abundant and the lake trout was its preferred host. By 1955, lake trout were extremely rare (Eschmeyer, 1957), and the species probably became extinct in Lake Michigan in 1956. The precise cause of the extinction of lake trout is still debatable. Hile et al. (1951) and Eschmeyer (1957) concluded that sea lampreys were totally responsible. Others (Smith, 1968; Wells and McLain, 1973) believed, as I do, that the disappearance of lake trout resulted from the combined stresses of an intensive fishery and sea lamprey predation.

Sea lamprey control was begun in Lake Michigan in 1953 with electrical barriers and traps on spawning streams. In 1960, a program was started in which a species-selective chemical lampricide was used to eradicate larval sea lampreys in spawning streams. Beginning in 1965, about 2 million hatchery-reared yearling lake trout (plus some fingerlings), marked by the removal of one or two fins, were stocked annually in Lake Michigan. Growth and survival have been good, but very little evidence of natural reproduction has been obtained. Some lake trout fry have been captured in the southern end of the West Arm of Grand Traverse Bay—158 fry in 1978 and 572 fry in 1979—(Wagner, 1981). Sixty lake trout fry were captured in 1980 off Port Sheldon in southeastern Lake Michigan (Jude et al., 1981). In the early 1980s, the percentage of young lake trout with a full complement of fins (i.e., that presumably had not been fin clipped) substantially increased in samples taken in Grand Traverse Bay by biologists of the Michigan Department of Natural Resources.

### 4.3. Lake Whitefish

The lake whitefish was the predominant species in the Lake Michigan fishery in the mid 1800s (Milner, 1874; Smith and Snell, 1891). However, harvests declined slowly through the late 1800s (Fig. 5.3)—perhaps due to a combination of intensive fishing, including wholesale capture of small fish (Koelz, 1929), and degradation of stream and river-mouth spawning areas by sawdust and bark from lumbering and sawmill operations. Increased harvests during 1928–1932 and 1947–1949 were the result of the occurrence of single exceptionally strong year classes (Wells and McLain, 1973). Depressed harvests during 1952–1964 may have been a result of reduced stocks due to sea lamprey predation. After the sea lamprey was controlled, annual harvests of lake whitefish increased steadily and were as high in the early 1980s as in the early 1900s.

### 4.4. Lake Herring

The lake herring was another very important commercial species in the late 1800s, but declined somewhat by 1900 (Fig. 5.3)—presumably due to heavy exploitation, although pollution may have been detrimental in southern Green Bay (Wells and McLain, 1973). A gradual decline during the 1930s and early 1940s seems also to be related to intensive exploitation, but by then rainbow smelt may have been adversely affecting small lake herring by predation or competition. A catastrophic die-off of smelt during 1942–1943 (Van Oosten, 1947) was perhaps not coincidentally followed by an exceptionally strong year class of lake herring in 1943 that supported the fishery for a number of years. In 1957, however, a severe and sudden collapse of the lake herring fishery began. Was it overfishing? Was it the resurgence of rainbow smelt and the explosive colonization by alewives (through competition with young herring for planktonic food, or predation on them)? Was it continued degradation of water quality on the spawning grounds in Green Bay, where about 87% of the annual harvest of lake herring from Lake Michigan had been made, at least during 1936–1955 (Smith, 1956)? These questions remain unanswered.

### 4.5. Ciscoes

The complex community of ciscoes indigenous to Lake Michigan originally contained seven species, listed here in order of decreasing size (from Koelz, 1926):

blackfin cisco (*Coregonus nigripinnis*), deepwater cisco (*Coregonus johannae*), shortjaw cisco (*Coregonus zenithicus*), longjaw cisco (*Coregonus alpenae*), shortnose cisco (*Coregonus reighardi*), kiyi (*Coregonus kiyi*), and bloaters. Over the past 100 years, sequential fishing out the largest species, then the next largest, and so on—easily done by gradually reducing mesh sizes in gill nets—ultimately resulted in the bloater being the only abundant cisco remaining by the early 1960s. Actually, relieved of predation by lake trout and burbot, whose populations had essentially collapsed by 1950, and relieved of competition for food and space with larger species of ciscoes, bloaters had greatly increased in abundance by the mid 1950s (Fig. 5.3). Concomitantly, as the lake trout resource began to collapse in the late 1940s, fishermen shifted to bloaters and annual harvests were increased to substantially higher levels during the 1950s and early 1960s. Bloater abundance then probably declined as a result of an explosive increase in the abundance of alewives, which compete with bloaters for space and possibly food and, as Smith (1968, 1972) suggested, may prey on their young. However, exploitation that included an intense otter-trawl fishery created in 1959 to take small bloaters for animal food may also have played a role in the decline. L. Wells (personal communication) also believes that by the mid 1950s, alewives were depressing bloater reproduction, causing a decline in their abundance that led to faster growth and thus greater abundance of large bloaters for the fishery.

Great concern about the depleted bloater population brought state and federal agencies together to protect and manage the bloater stocks on a lakewide basis. In 1974, the Lake Michigan Lake Committee, under the aegis of the Great Lakes Fishery Commission, established a Lake Michigan Chub Technical Committee (Brown et al., 1985) composed of scientists from all agencies involved. On advice from this Technical Committee, a lakewide ban on bloater fishing was finally achieved in late 1976, after challenges by commercial fishermen in court were denied. With fishing pressure relieved and alewife abundance reduced, the bloater recovered dramatically, after producing a number of strong year classes in the late 1970s and early 1980s. In 1979, the Wisconsin Department of Natural Resources established a total allowable catch of 1 million pounds (dressed weight) and increased this quota to 2 million pounds in 1982. The Michigan Department of Natural Resources, being faced with high concentrations of dieldrin in bloaters from its southern waters and the emergence of the Tribal commercial fishery in waters ceded by the Treaty of 1836 (Brown et al., 1985), has reacted cautiously to the recovery of the bloaters. Only limited harvests by state-licensed commercial fishermen have been permitted.

### 4.6. Rainbow Smelt

Rainbow smelt were first reported in Lake Michigan in 1923 (Table 5.1), and had colonized the entire lake by 1936 (Wells and McLain, 1973). Commercial harvests increased rapidly to a peak in 1941, when 4.8 million pounds were landed (Fig. 5.3). A massive mortality during the winter of 1942–43—apparently caused by disease (Van Oosten, 1947)—decimated the resource. The abundance of rainbow smelt rapidly increased again, peaking in 1958. An abrupt, severe decline in commercial harvest then ensued, probably due to reduced abundance caused by competition with and predation by the exploding population of alewives. Rainbow smelt began a recovery in the late 1960s and by the 1970s became important prey for introduced lake trout, Pacific salmons, and other salmonids, particularly in western and northern Lake Michigan. Despite the predation on them, rainbow smelt remained abundant, although moderately large annual fluctuations in year-class strength continued.

## 4.7. Alewives

Alewives, first recorded in Lake Michigan in 1949 (Table 5.1), were dispersed throughout most of the lake by 1953 (Miller, 1957). Populations of alewives exploded in the late 1950s and early 1960s and an enormous die-off in 1967 (Brown, 1972) left millions of pounds of alewife carcasses clogging water intakes to steel mills, power plants, and municipal water filtration plants, and littering the beaches in windrows. Perhaps 70% of the alewives in Lake Michigan died during the 1967 die-off. The populations recovered to some extent and commercial harvests (Fig. 5.3) and estimated standing stocks (Hatch et al., 1981) remained relatively strong during the 1970s and into the 1980s. However, survey indices of abundance dropped sharply in 1982, and again in 1983 (L. Wells, personal communication), although the harvest was substantial in both years. The abundance had increased slightly by 1985.

As in Lake Huron, the recovery of bloater populations in Lake Michigan restored an indigeneous component of the prey community which, together with rainbow smelt and sculpins (*Cottus sp.*), should sustain salmonids at their elevated levels of abundance despite the reduced alewife population.

## 4.8. Yellow Perch

Evidently yellow perch were not particularly abundant in Lake Michigan before 1882, but then increased spectacularly—first in Green Bay, and then in the southern regions of the lake (Smith and Snell, 1891). High commercial production during 1894–1896 (Fig. 5.3) was probably the result of increased abundance and greater fishing effort as markets for this species rapidly developed (Wells and McLain, 1973). Through the next half-century, annual commercial harvests fluctuated between 1 and 3 million pounds. Then in the early 1960s a short burst of substantially higher harvests occurred when fishing intensity increased, perhaps because the success of the Lake Erie yellow perch fishery had led to the development of lucrative markets (M. Keller, personal communication). An abrupt decline in commercial harvests in 1965 continued through 1983. Wells and McLain (1973), Wells (1977), and Wells and Jorgenson (1983), though acknowledging that the sport and commercial fisheries affected the yellow perch populations to some degree, considered the alewife to be a strong factor in the abrupt decline of yellow perch in the mid 1960s and their failure to recover during the 1970s. They believed that the mechanism was a combination of heavy concentrations of alewives competing for food and space with or preying on yellow perch fry. Sharp declines in abundance of alewives in 1983 and 1984 were accompanied by the production of extraordinarily strong year classes of yellow perch in southern Lake Michigan (L. Wells, personal communication).

## 5. MAJOR FISH RESOURCES OF LAKE ERIE

### 5.1. The Lake

The annual average inflow of water from the upper Great Lakes through the Detroit River represents about 95% of the total inflow into Lake Erie. The theoretical flushing rate of 920 days (Federal Water Pollution Control Administration, 1968) makes water quality in Lake Erie highly vulnerable to change by the inflowing water (Hartman, 1973). This lake is the shallowest, most southern, and warmest of the Laurentian Great Lakes and was by far the most productive in harvests of fish in all except 4 years from 1914 to 1973 (Leach and Nepszy, 1976).

Lake Erie is naturally divided geologically into three basins. The western basin is the shallowest (mean depth, 7.4 m), contains many shoals, islands, and rock reefs, and is generally considered to have the most important spawning and nursery grounds in the lake (Hartman, 1973). The large central basin which makes up 63% of Lake Erie's surface area, is considerably deeper than the western basin with an extensive flat muddy plain 18–24 m deep. This basin stratifies thermally during summer and the hypolimnion has become anoxic in recent decades. The eastern basin has a maximum depth of 64 m, and the hypolimnion has not been detected anoxic during thermal stratification. Several major stresses on fish in Lake Erie (in the approximate order of importance) were identified by Regier and Hartman (1973): the commercial fishery, cultural eutrophication, the introduction of new species, tributary and shoreline restructuring, shoreline and agricultural land erosion and consequent siltation, and the introduction of toxic materials from industrial sources.

### 5.2. Lake Herring

The lake herring was an extremely important commercial species in Lake Erie for many years beginning in the 1880s. Hundreds of millions of pounds were harvested up to the catastrophic collapse in 1925 (Fig. 5.4). One totally unexpected, exceptionally strong year class produced in 1944 sustained a sizable fishery for 2 or 3 years. But the species then again became commercially unimportant. Van Oosten (1930) concluded his analysis of the available data by stating that the lake herring was "commercially exterminated by overfishing." Regier et al. (1969) agreed. Beeton and Edmondson (1972) suggested that dwindling catches in western Lake Erie and increases in central and eastern Lake Erie before the decline were related to degradation of the environment that progressed from west to east,

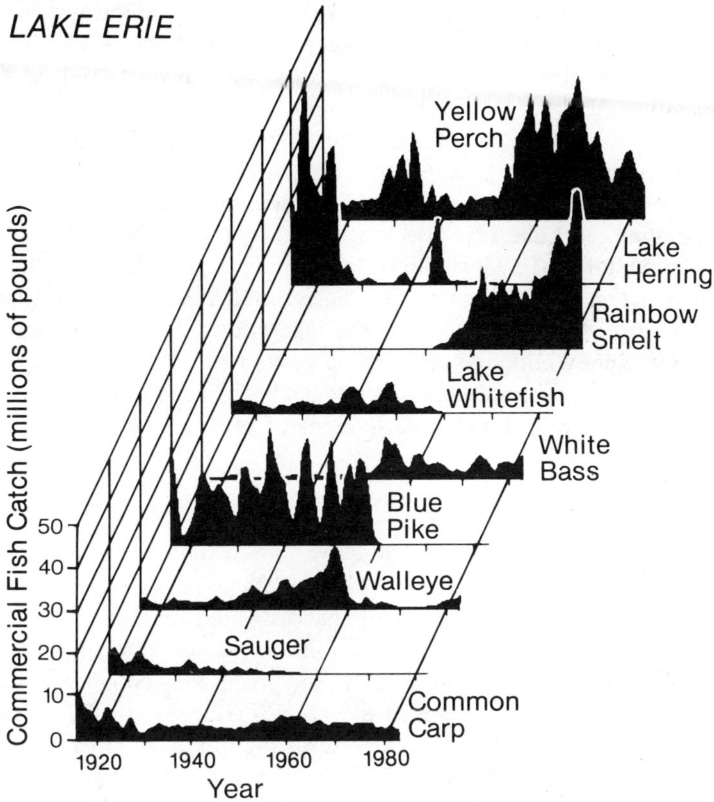


**Figure 5.4.** Commercial harvests of major fish species in Lake Erie, 1915–1983. Sources of data: Baldwin et al. (1979) and National Fisheries Center–Great L..kes annual summaries of commercial harvests of fish from the Great Lakes.

and the concentration of lake herring and the fishery in eastern waters. Leach and Nepszy (1976) argued against this conclusion, stating, "if eutrophication was severe enough in thc 1920s to detrimentally affect herring stocks, it is unlikely that the species would have recovered in the 1940s when environmental conditions had worsened."

## 5.3. Lake Whitefish

Historically, lake whitefish were apparently important to the commercial fishery during the late 1800s, although catch records are scanty for U.S. waters. Since 1915, when annual catch records were first reported lakewide, lake whitefish constituted only a relatively small component of the commercial harvest until their collapse in the mid 1950s (Fig. 5.4). It is, however, an important species in the sense that most investigators have concluded that environmental degradation, to some extent at least, contributed to the commercial extinction of this species (Hartman, 1973; Leach and Nepszy, 1976). Lake whitefish are coldwater stenotherms at the southern edge of their zoogeographical range in Lake Erie. They require cold, adequately oxygenated bottom waters for summer habitat, and silt-free river or lake spawning areas for successful spawning (Hartman, 1973). Pollution in the Detroit River and siltation in the Maumee River destroyed whitefish spawning grounds early in the century (Trautman, 1957). A general climatic warming in the Great Lakes region since 1918 resulted in an increase of 0.8 to 1.1 C° in annual mean temperatures in Lake Erie (Beeton, 1961) by the 1950s, and added further stress on lake whitefish (Lawler, 1965). Loss of the over-summer sanctuary in the central basin, when cultural eutrophication resulted in anoxic conditions in the hypolimnion, placed further stress on the lake whitefish resource, remnants of which are now chiefly confined to the deep eastern basin. In the past few years, 1980 to 1985, stronger year classes of whitefish have been produced in the eastern basin and the resource is showing early signs of recovery (R. Kenyon, personal communication).

## 5.4. Blue Pike

Although the blue pike (*Stizostedion vitreum glaucum*) was extremely important to the commercial fishery for many years, harvests fluctuated markedly after 1915 (Fig. 5.4); annual production exceeded 8 million pounds four times, and dipped below 2 million pounds seven times before the precipitous collapse of the fishery in 1958, and subsequent extinction of the species (Hartman, 1973). Various explanations have been offered for the demise of the blue pike. Fishing intensified during the 1940s after lake herring and lake whitefish populations collapsed and the need for domestically produced protein greatly increased during World War II. In 1948, Canadian fishermen began converting gill nets from cotton to the much more effective nylon, greatly increased the amount of gear fished, and began extensive use of fish-finding sonar and ship-to-ship radio (Regier et al., 1969). Just as for lake whitefish, the over-summer sanctuary for blue pike in the hypolimnion of the central basin was lost in the mid 1950s, when anoxic conditions developed. Leach and Nepszy (1976) also suggested that the exploding populations of smelt must have stressed the blue pike resource through competition or predation, or both. Finally, Regier et al. (1969) suggested that the final phase of the disintegration of the blue pike resource was the ultimate disappearance of the species by genetic introgression with the numerically dominant walleye.

## 5.5. Sauger

Though catch records are quite incomplete, sauger (*Stizostedion canadense*) harvests were relatively high between 1885 and 1924 when annual landings generally ranged between 2 and 6 million pounds (Fig. 5.4). The catch then declined steadily until the species became commercially extinct in the late 1950s. The remarkable regularity of the decline prompted Regier et al. (1969) to suggest that progressively deteriorating environmental conditions, including degradation of spawning reefs and streams in the western basin by pollution, siltation, and damming may have seriously stressed the sauger's reproductive potential. Scott and Crossman (1973) felt that increased fishing from 1930 to 1940, coupled with the biological characteristics of slow growth and late sexual maturity of the species, may have been significant factors. Finally, Regier et al. (1969) suggested that, as in the blue pike, introgressive hybridization with the dominant walleye may have eliminated the pure sauger strain. During the 1970s the Ohio Division of Wildlife reintroduced this species in the western basin by stocking. No self-sustaining populations became established (C. Baker, personal communication).

## 5.6. Walleye

Much has been written about historic changes in walleye populations through about 1974 (Regier et al., 1969; Parsons, 1970; Hartman, 1972, 1973; Regier and Hartman, 1973; Leach and Nepszy, 1976; Nepszy, 1977). Commercial harvests were relatively steady at about 2 million pounds annually from 1915 until the mid 1930s. A sharp and steady increase then began, peaked at over 15 million pounds in 1956, and was followed by a precipitous decline to less than 0.5 million pounds in 1969 (Fig. 5.4). Major reasons suggested for the decline included primarily overfishing, but also degradation of the environment by siltation of spawning areas, destruction of summer oxygen regimes in the western and central basins, major changes in the benthos in those basins (Britt, 1955), and the potentially deleterious effect of proliferating populations of rainbow smelt. Performance of walleye populations in the western basin after 1970 appeared to clarify causal factors for their decline.

In 1970, the discovery in western Lake Erie of walleyes with mercury concentrations that exceeded action levels for human health according to the Federal Food and Drug Administration resulted in closure of commercial fisheries in Ohio and Ontario. Commercial fishing for walleyes in Michigan waters had been closed earlier for other reasons. This moratorium provided the States of Michigan and Ohio, the Province of Ontario, and the U.S. Fish and Wildlife Service, the opportunity to plan a cooperative interagency management strategy that would allow rebuilding of the resource and would control harvest to prevent future overexploitation. Under the aegis of the international Great Lakes Fishery Commission, a "Walleye Scientific Protocol Committee" was established, composed of representatives from the Ohio Department of Natural Resources, the Michigan Department of Natural Resources, the Ontario Ministry of Natural Resources, and the U.S. Fish and Wildlife Service. All historic and current data from all agencies were pooled and analyzed, and a system was developed for establishing annual quota recommendations to be apportioned to the three jurisdictions on the basis of lake surface area: Ontario, 39%; Ohio, 52%; and Michigan, 9%. The interagency management system was successful. Quotas (readjusted each year on the basis of the previous year's catch—sport and commercial—plus survey information on the strength of year classes that will be recruited into the fisheries) have been generally adhered to, and recovery of the walleye population has been phenomenal—beyond anyone's expectations. For example, the estimated fishable stock of walleyes 17

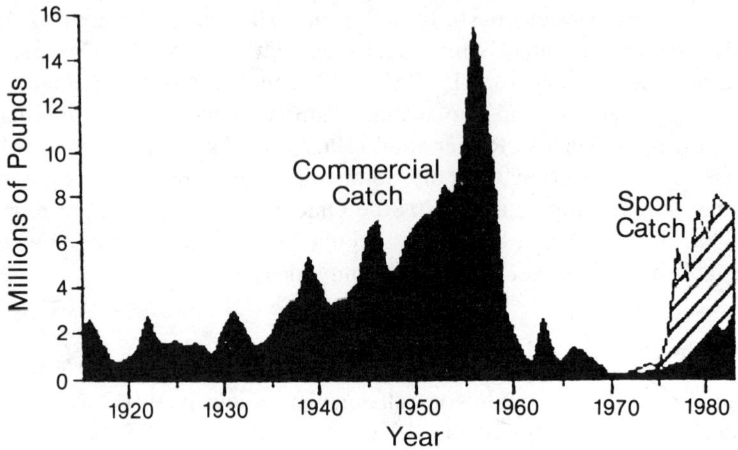


**Figure 5.5.** Commercial harvest and estimated sport harvest of walleyes from Lake Erie, 1915–1983.

inches long or larger in spring 1985 was 21.8 million fish, and the total allowable catch for all fisheries was set at 4.9 million fish. Adding the sport fishery harvest to the commercial harvest (Fig. 5.5) reveals the magnitude of the recovery. Obviously, the environment is still able to sustain a thriving walleye population resource, and overexploitation must have been the single most important factor in the decline of the walleye during the 1960s.

Walleye population that spawned in the Detroit, Maumee, Sandusky, Cuyahoga, and Grand rivers were greatly reduced over the years due to pollution, sedimentation, and damming of the critical spawning grounds and nursery areas (Langlois, 1945; Regier et al., 1969; Busch et al., 1975). As a consequence of improved farmland erosion control and pollution reductions, spawning populations in the Maumee and Sandusky Rivers, for example, substantially increased during the 1970s.

## 5.7. Yellow Perch

Another historically important commercial species, the yellow perch, supported a major fishery during the period 1928–1935, peaking at over 14 million pounds in 1932 (Fig. 5.4). Harvests then returned to former levels at about 5 million pounds annually until they increased sharply in the mid 1950s—a change attributed largely to the decline of the blue pike and lake whitefish, and a shift in effort of especially the U.S. trap-net fishery to yellow perch (Nepszy, 1977). Irregularity in year-class strength later characterized the resource; strong year classes were produced only in 1962, 1965, and 1970 (Hartman, 1973). A steady decline in abundance of yellow perch began in the early 1970s and the catch dropped to barely 6 million pounds in 1983. Furthermore, survey assessments of the 1983 and 1984 year classes in the western basin proved both to be rather weak (K. M. Muth, personal communication). Factors continuing to suppress the yellow perch in Lake Erie are continued intensive fishing, competition with and predation by rainbow smelt, and perhaps some deleterious effects of the expanding populations of a recent invader, the white perch (Table 5.1).

### 5.8. Rainbow Smelt

Rainbow smelt were first reported in Lake Erie in 1932 (Table 5.1), and are believed to have entered Lake Erie from the upper Great Lakes by way of the Detroit River (Van Oosten, 1937). A sizable population was evident in the early 1950s (Kennedy, 1961), and introduction of a trawling technique and the availability of a market generated a fishery that by the mid 1960s was harvesting 13 to 15 million pounds annually (Leach and Nepszy, 1976). The fishery is now primarily a trawl fishery in Canadian waters in the central basin (winter and early spring) and the deeper eastern basin (spring, fall, and winter). A slight and brief depression in harvests in about 1970 (Fig. 5.4) was probably due to stress imposed on rainbow smelt by unusually heavy parasitism by the microsporidan *Glugea hertwigi* in the fall of 1969 (Nepszy et al., 1978). Thereafter, harvests of rainbow smelt increased markedly, peaking at 37 million pounds in 1982, then dropping to 29 million pounds in 1983 and 17 million pounds in 1984. It is too early to predict the velocity and extent of this decline, much less its cause—other than the obvious evidence of intensive fishing.

## 6. MAJOR FISH RESOURCES OF LAKE ONTARIO

### 6.1. The Lake

The open waters of deep, cold, oligotrophic Lake Ontario historically sustained valuable populations of cold-stenotherm fishes such as Atlantic salmon (*Salmo salar*), lake herring, lake whitefish, deepwater ciscoes, and lake trout. Weiler and Chawla (1969) reported that increases in concentrations of sodium, potassium, calcium, chloride, sulfate, and total dissolved solids in the Great Lakes since 1910 were highest by far in Lakes Erie and Ontario. Nevertheless, Christie (1973) believed that the open lake is still suitable habitat for indigenous fish resources, that oxygen regimes are satisfactory, and that none of the chemicals that have been introduced and measured approach levels known to be toxic to fish. In contrast, eutrophication became advanced in the 1970s in such inshore areas as the Bay of Quinte (Hurley, 1970; Hurley and Christie, 1977). In the 1980s, point-source loading of phosphorus was substantially reduced and the process of eutrophication reversed in the Bay of Quinte (J. Christie, personal communication).

### 6.2. Lake Herring and Other Ciscoes

Historically, lake herring and other ciscoes were the most important resources fished commercially until the fishery for lake whitefish boomed in the 1930s. Unfortunately the catch reporting system did not require commercial fishermen to distinguish between lake herring and the three other species of ciscoes until 1952, when few of these fish were being caught. Through much of the period 1884–1944, annual harvests generally exceeded 1 million pounds (Fig. 5.6). A slight depression in the catch during the mid 1920s was attributed by Christie (1973), in part, to a modest diversion of the fishery toward lake trout and lake whitefish. Christie (1973) also suggested the following scenario of community dynamics: cisco abundance increased during the period of low predator abundance in the 1880s and 1890s, decreased when lake trout and burbot abundance peaked in the 1920s, and increased again in the 1930s and early 1940s as predators once more decreased. However, a severe and long-standing decline occurred in the mid 1940s,

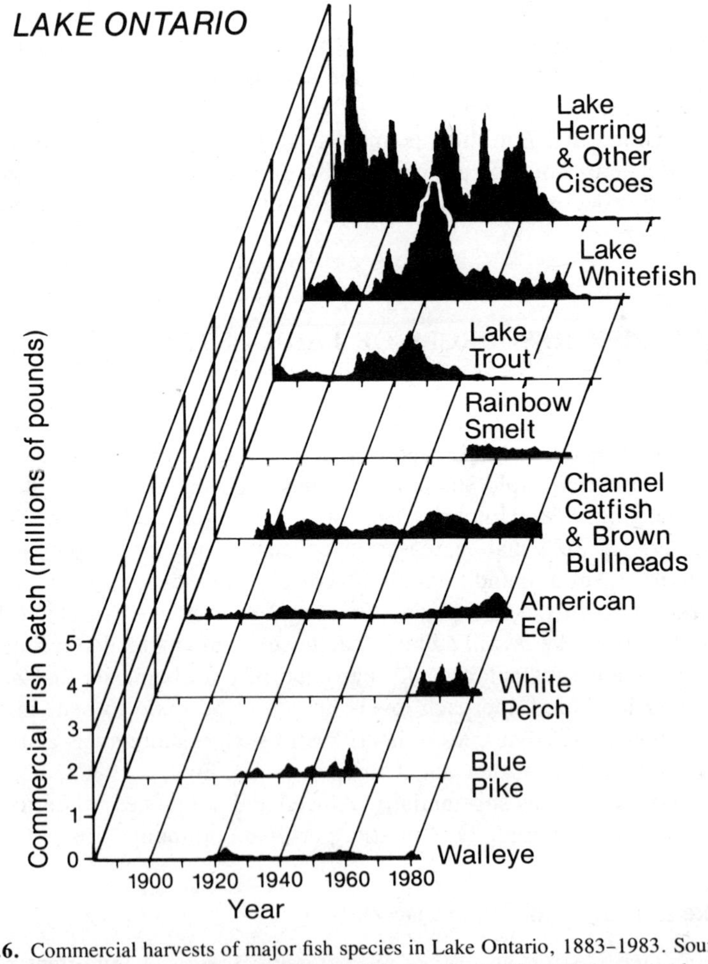


**Figure 5.6.** Commercial harvests of major fish species in Lake Ontario, 1883–1983. Sources of data: Baldwin et al. (1979) and National Fisheries Center–Great Lakes annual summaries of commercial harvests of fish from the Great Lakes.

perhaps as a result of intensive fishing and environmental degradation. Christie (1973) also speculated that the ascendancy of rainbow smelt populations in the 1940s and the potential of their predation on bloater eggs and larvae, or competition for food with young bloaters, may also have been important factors. Others believe (R. O'Gorman, personal communication) that the predatory or competitive impact of the exotic invading alewife and cultural degradation of nearshore spawning grounds also were factors contributing to the decline of deepwater ciscoes.

## 6.3. Lake Whitefish

The commercial fishery for lake whitefish intensified after the turn of the century, peaked in 1924 at nearly 2.8 million pounds, and then declined steadily and dropped below 100,000 pounds in 1966 (Fig. 5.6). Christie (1968) and Christie and Regier (1973) concluded that stress induced by the fishery alone was sufficient to account for the decline of lake whitefish. Conversion from linen to nylon gill nets in 1950 resulted in an almost threefold increase in fishing intensity (Christie, 1963). The final stress imposed on lake whitefish was very likely predation by the sea lamprey during the 1950s, after preferred prey populations of lake trout and burbot had declined.

## 6.4. Lake Trout

Lake trout were most abundant in the eastern waters of Lake Ontario where more spawning reefs were most numerous—Scotch Bonnet Shoal, Nicholson Island, Main Duck Island, Charity Shoal, and Pigeon Island. Although lake trout were harvested since records were first kept in 1867, the fishery did not intensify significantly until 1909, and the peak harvest (1.1 million pounds) came in 1925 (Fig. 5.6). The catch then declined steadily and lake trout became virtually extinct by 1950. Intensive fishing and predation by the sea lamprey (first reported in Lake Ontario in 1850, Table 5.1) are considered the major factors causing the demise of lake trout in Lake Ontario.

A major interagency, binational program of sea lamprey control began in Lake Ontario in the mid 1970s, followed, in 1974, by the annual stocking of hundreds of thousands of hatchery-reared yearling lake trout. In April 1983 one lake-produced lake trout fingerling was captured in the eastern region of the lake, and in spring 1986 seventy-five swim-up fry and one small fingerling were captured on historic lake trout spawning grounds near Stony Island in the eastern region (C. Schneider, New York Department of Environment Conservation, personal communication).

## 6.5. Rainbow Smelt

Rainbow smelt were first recorded from Lake Ontario in 1929 (Table 5.1), but their numbers remained relatively low until substantial increases occurred in the mid 1940s. The commercial fishery for rainbow smelt, primarily Canadian, began in 1952; the harvest reached 296,000 pounds in 1958 and then gradually and steadily declined to less than 100,000 pounds in the 1980s (Fig. 5.6). During the 1970s, rainbow smelt were ''buffered'' from predation to some extent by much larger populations of alewives (R. O'Gorman, personal communication). Only after the major die-offs of alewives in the late 1980s did mortality rates for rainbow smelt increase. Nevertheless, rainbow smelt remain abundant in Lake Ontario and an important component of the diet of lake trout, Pacific salmon, and other salmonids.

## 6.6. American Eels

Always a small component of the commercial fishery, the fishery for American eels (*Anguilla rostrata*) intensified in the early 1960s as reduced eel catches in Europe created a demand for the North American product (Christie, 1973). Harvests increased, peaked in 1971 at over 700,000 pounds, and then declined to 150,000 pounds by 1983 (Fig. 5.6). The population depends on elvers from spawning grounds in the Atlantic Ocean reaching Lake Ontario by way of the St. Lawrence River. The presence of dams that partly obstruct immigration of elvers through the St. Lawrence River may be depressing recruitment of elvers into the Lake Ontario population. Also, New York banned possession of elvers in September 1982 because of unacceptable levels of contaminants in elver flesh, and commercial impacts to some ports in Europe were closed for the same reason.

## 6.7. White Perch

Scott and Christie (1963) believed that the white perch (*Morone americana*) invaded Lake Ontario in about 1950 (Table 5.1) from the Atlantic watershed by way of the Erie Barge Canal. Since then this species has fluctuated in abundance, as have harvests by the commercial fisheries. Most production has come from the eutrophic Bay of Quinte.

## 6.8. Walleyes and Blue Pike

Walleyes have been of only modest importance to the commercial fishery, the annual harvest seldom exceeding 100,000 pounds (Fig. 5.5). A recent program of control on nutrient loading in the Bay of Quinte has been effective and certain depressed fish resources such as walleye have dramatically rebounded in abundance.

The commercial harvest of blue pike was generally unimportant, compared to the landings of lake herring and ciscoes, lake whitefish, and lake trout. Nevertheless, during the total fishery on blue pike starting in 1918 and terminating in 1961, before extinction of the species, the harvest of blue pike totaled 5.5 million pounds (Fig. 5.5). The reason for the disappearance of blue pike seems to center on the large brood stocks in Lake Erie being the source of blue pike migrating into Lake Ontario. When the blue pike populations were decimated and eventually eradicated in Lake Erie the source of blue pike for Lake Ontario was removed.

## 7. A CENTURY OF CHANGE, A DECADE OF RECOVERY

As I have documented above, major changes in the fish resources of the Great Lakes have occurred over the past century. I believe that virtually every major change in the fish resources of the Great Lakes during the past century was directly or indirectly caused by humans, masking any subtle effects that might have been the result of ecological succession.

Intensive fishing was sometimes the major cause of change—as in lake herring in western Lake Superior; bloaters in Lake Huron; lake herring, walleyes, and yellow perch in Lake Erie; lake whitefish in Lake Ontario; and lake trout throughout the Great Lakes. In some instances predation by sea lamprey played a significant if not dominant role in change, as in lake trout in all five Great Lakes, and lake whitefish in Lakes Huron and Ontario. And in some instances, exotic species other than the sea lamprey, such as alewives, smelt, and white perch, have been associated with changes in indigenous fish resources, for example, yellow perch and bloaters in Lake Michigan, and bloaters in Lake Huron.

Linkages between degradation of water quality and critical spawning and nursery habitats, however, have not generally been either clearcut or quantifiable. Perhaps most obvious has been the damming of rivers that prevented fish from reaching spawning grounds, for example, river-spawning walleyes and lake sturgeon along the south shore of Lake Erie. Deforestation of certain watersheds, resulting in siltation and elevated temperatures, and destruction of critical riverine spawning habitats for Atlantic salmon in Lake Ontario were also factors contributing to the disappearance of that species by 1900 (Christie, 1973).

Sedimentation, eutrophication, and the release of toxic substances have also adversely affected river spawners, and nearshore spawners in the shallow embayments around the Great Lakes, for example, percids in Green Bay, Saginaw Bay, and the Bay of Quinte. Eutrophication dramatically altered the benthic community in western Lake Erie, and resulted in anoxic conditions during summer thermal stratification in the hypolimnion of the central basin. As a consequence, the over-summer sanctuary for cold stenotherms such as lake whitefish, lake trout, and blue pike was lost. Elevated levels of mercury in walleyes in western Lake Erie and Lake St. Clair led to a moratorium on fishing for that species in the early 1970s. There is some evidence that the combination of many contaminants in the tissues of lake trout in southeastern Lake Michigan is impairing successful reproduction (Willford et al., 1981; Mac et al., 1985).

Another concern is the apparent relationship between elevated levels of carcinogenic chemicals in sediments in certain harbors, tributaries, and connecting channels of the Great Lakes and elevated levels of tumors in brown bullheads (Baumann et al., 1982), and suckers and carp (Leatherland and Sonstegard, 1984). The impacts of this factor on the size and structure of fish populations has as not yet been determined.

Some changes are likely irreversible. The exotic alewife, rainbow smelt, white perch, and Pacific salmon (some natural production is occurring in tributaries to the Great Lakes) are now virtually permanent members of the fish communities.

Although integrated pest management control has generally reduced the sea lamprey to only 10% of its former abundance, its eradication will be very difficult. In fact, substantially improved water quality and reduced sedimentation provides more suitable habitat for successful spawning by the sea lamprey. Various populations of blue pike, saugers, lake herring, and ciscoes are extinct, and their restoration is unlikely.

However, there are encouraging signs as well, and it appears that we are well into a "decade of recovery" for Great Lakes fish resources and their critical habitats. For example, various populations have demonstrated remarkable recoveries from severely depressed conditions of a decade or two ago: lake herring in Lake Superior, bloater chubs in Lakes Michigan and Huron, yellow perch in Lake Michigan, and walleyes in Lake Erie and the Bay of Quinte (Lake Ontario). Lake whitefish populations in Lakes Superior, Huron, and Michigan are thriving and commercial catches are comparable to historic high levels. Some rivers are cleaner now, and as a result, walleye and white bass populations (for example) are thriving in two tributaries of Lake Erie, the Maumee and Sandusky rivers. Finally, body burdens of certain contaminants are decreasing in fish due to point-source control and improved enforcement of existing regulations.

## ACKNOWLEDGMENT

Contribution 668 of the National Fisheries Center-Great Lakes, U.S. Fish and Wildlife Service, Ann Arbor, MI 48105. This article was written by an employee of the U.S. Government as part of his official duties and therefore cannot be copyrighted.

## REFERENCES

Anderson, E. D., and Smith, L. L., Jr. (1971). Factors affecting abundance of lake herring (*Coregonus artedii* Lesueur) in western Lake Superior. *Trans. Am. Fish. Soc.* **100,** 691–707.

Anonymous. (1958). Great Lakes fishery investigations. U.S. Bureau of Commercial Fisheries. *Comm. Fish. Rev.* **16**(8), 27–29.

Applegate, V. C. (1950). Natural history of the sea lamprey (*Petromyzon marinus*) in Michigan. U.S., Fish Wildl. Serv., Spec. Sci. Rep. Fish 55, pp. 1–237.

Baldwin, N. S., Saalfeld, R. W., Ross, M. A., and Buettner, H. J. (1979). Commercial fish production in the Great Lakes, 1867–1977. Great Lakes Fish. Comm. Tech. Rep. 3, 1–187.

Baumann, P. C., Smith, W. D., and Ribick, M. (1982). Hepatic tumor rates and polynuclear aromatic hydrocarbon levels in two populations of brown bullheads (*Ictalurus nebulosis*). In M. Cooke and A. J. Dennis (Eds.), *Physical and Biological Chemistry.* Battelle, Columbus, OH, pp. 93–102.

Beeton, A. M. (1961). Environmental changes in Lake Erie. *Trans. Am. Fish. Soc.* **90,** 153–159.

Beeton, A. M. (1965). Eutrophication of the St. Lawrence Great Lakes. *Limnol. Oceanogr.* **10**(2), 240–254.

Beeton, A. M. (1969). Changes in the environment and biota of the Great Lakes. In *Eutrophication: Causes, Consequences, Correctives.* Natl. Acad. Sci., Washington, DC, pp. 150–187.

Beeton, A. M., and Edmondson, W. T. (1972). The eutrophication problem. *J. Fish. Res. Board Can.* **29,** 673–682.

Berst, A. H., and Spangler, G. R. (1970). Population dynamics of F. splake (*Salvelinus fontinalis* × *S. namacush*) in Lake Huron. *J. Fish. Res. Board Can.* **27**, 1017–1032.

Berst, A. H., and Spangler, G. R. (1973). Lake Huron: The ecology of the fish community and man's effects on it. Great Lakes Fish. Comm. Tech. Rep. 21, pp. 1–41.

Britt, N. W. (1955). Stratification in western Lake Erie in summer of 1953: Effects on the *Hexagenia* (Ephemeroptera) population. *Ecology* **36,** 239–244.

Brown, E. H., Jr. (1972). Population biology of alewives, *Alosa pseudoharengus,* in Lake Michigan, 1949–70. *J. Fish. Res. Board Can.* **29,** 477–500.

Brown, E. H., Jr., Argyle, R. L., Payne, N. R., and Holey, M. E. (1988). Dynamics of destabilized chub populations in Lakes Michigan and Huron, 1950–84. *Can. J. Fish. Aquat. Sci.* (in press).

Busch, W.-D. N., Scholl, R. L., and Hartman, W. L. (1975). Environmental factors affecting the strength of walleye (*Stizostedion vitreum vitreum*) year-classes in western Lake Erie, 1960–70. *J. Fish. Res. Board Can.* **32,** 1733–1743.

Christie, W. J. (1963). Effects of artificial propagation and the weather on recruitment in the Lake Ontario whitefish fishery. *J. Fish. Res. Board Can.* **20**(3), 597–646.

Christie, W. J. (1968). Possible influences of fishing in the decline of Great Lakes fish stocks. Proc. 11th Conf. Great Lakes Res., pp. 31–38.

Christie, W. J. (1973). A review of the changes in the fish species composition of Lake Ontario. Great Lakes Fish. Comm. Tech. Rep. 23, pp. 1–65.

Christie, W. J. (1974). Changes in the fish species composition of the Great Lakes. *J. Fish. Res. Board Can.* **31,** 827–854.

Christie, W. J., and Regier, H. A. (1973). Temperature as a major factor influencing reproductive success of fish—Two examples. *Rapp. P.-V. Reun., Cons. Int. Explor. Mer* **164,** 208–218.

Coble, D. W. (1967). The white sucker population of South Bay, Lake Huron, and effects of the sea lamprey on it. *J. Fish. Res. Board Can.* **24**(10), 2117–2136.

Eschmeyer, P. H. (1957). The near extinction of lake trout in Lake Michigan. *Trans. Am. Fish. Soc.* **85,** 102–119.

Federal Water Pollution Control Administration (1968). *Lake Erie Environmental Summary, 1963–64.* U.S. Dept. of Interior, Great Lakes Region, Cleveland Program Office, FWPCA, Cleveland, OH.

Hartman, W. L. (1972). Lake Erie: Effects of exploitation, environmental changes, and new species on the fishery resources. *J. Fish. Res. Board Can.* **29,** 899–912.

Hartman, W. L. (1973). Effects of exploitation, environmental changes, and new species on the fish habitats and resources of Lake Erie. Great Lakes Fish. Comm. Tech. Rep. 22, pp. 1–43.

Hatch, R. W., Haack, P. M., and Brown, E. J., Jr. (1981). Estimation of alewife biomass in Lake Michigan, 1967–1978. *Trans. Am. Fish. Soc.* **110,** 575–584.

Herdendorf, C. E. (1982). Large lakes of the world. *J. Great Lakes Res.* **8,** 379–412.

Hile, R., Eschmeyer, P. H., and Lunger, G. F. (1951). Decline of the lake trout fishery in Lake Michigan. *Fish. Bull.* **52,** 77–95.

Hurley, D. A. (1970). Seasonal changes in several biotic and abiotic constituents at three stations in the Bay of Quinte. Ont. Dep. Lands For., Sect. Rep. (Fish.) 73, pp. 1–50.

Hurley, D. A., and Christie, W. J. (1977). Depreciation of the warmwater fish community in the Bay of Quinte, Lake Ontario. *J. Fish. Res. Board Can.* **34,** 1849–1860.

Jude, D. J., Klinger, S. A., and Enk, M. D. (1981). Evidence of natural reproduction of planted lake trout in Lake Michigan. *J. Great Lakes Res.* **7,** 57–61.

Kennedy, W. A. (1961). Daily catch records of Crewe Brothers Fishery, Lake Erie—1904 to 1956. *Fish. Res. Board Can. Manuscr. Rep., Ser. Biol.* **706,** 1–411.

Koelz, W. (1926). Fishing industry of the Great Lakes. Rep. U.S. Comm. Fish., 1925, pp. 553–617.

Koelz, W. (1929). Coregonid fishes of the Great Lakes. Bull. U.S. Bur. Fish. 43, Part 2, pp. 297–643.

Kutkuhn, J. H. (1985). *Fish and Fisheries of the Upper Great Lakes: Past Contribution and Future Potential,* Mimeo. U.S. Fish Wildl. Serv., Washington, DC.

Langlois, T. H. (1945). Water, fishes, and cropland management. *Trans. North Am. Wildl. Conf.* **10,** 190–196.

Lark, J. G. I. (1973). An early record of the sea lamprey (*Petromyzon marinus*) from Lake Ontario. *J. Fish. Res. Board Can.* **30,** 131–133.

Lawler, G. H. (1965). Fluctuations in the success of year-classes of whitefish populations with special reference to Lake Erie. *J. Fish. Res. Board Can.* **22,** 1197–1227.

Lawrie, A. H., and Rahrer, J. F. (1973). Lake Superior: A case history of the lake and its fisheries. Great Lakes Fish. Comm. Tech. Rep. 19, pp. 1–69.

Leach, J. H., and Nepszy, S. J. (1976). The fish community in Lake Erie. *J. Fish. Res. Board Can.* **33,** 622–638.

Leatherland, J. F., and Sonstegard, R. A. (1984). Pathobiological responses of feral teleosts to environmental stressors: Interlake studies on the physiology of Great Lakes salmon. *Adv. Environ. Sci. Technol.* **16,** 115–149.

Mac, J. M., Edsall, C. C., and Seelye, J. G.(1985). Survival of lake trout eggs and fry reared in water from the upper Great Lakes. *J. Great Lakes Res.* **11**(4), 520–529.

Miller, R. R. (1957). Origin and dispersal of the alewife, *Alosa pseudoharengus,* and the gizzard shad, *Dorosoma cepedianum,* in the Great Lakes. *Trans. Am. Fish. Soc.* **86,** 97–111.

Milner, J. W. (1874). Report on the fisheries of the Great Lakes: The results of inquiries prosecuted in 1871 and 1872. Rep. U.S. Comm. Fish Fish., 1872, 1873, pp. 1–78.

Nepszy, S. J. (1977). Changes in percid populations and species interactions in Lake Erie. *J. Fish. Res. Board Can.* **34,** 1861–1868.

Nepszy, S. J., Budd, J., and Dechtiar, A. O. (1978). Mortality of young-of-the-year rainbow smelt (*Osmerus mordax*) in Lake Erie associated with the occurrence of *Glugea hertwigs. J. Wildl. Dis.* **14,** 233–239.

Parsons, J. W. (1970). Walleye fishery of Lake Erie in 1943–62 with emphasis on contributions of the 1942–61 year-classes. *J. Fish. Res. Board Can.* **27,** 1475–1489.

Pycha, R. L., and King, G. R. (1975). Changes in the lake trout population of southern Lake Superior in relation to the fishery, the sea lamprey, and stocking, 1950–70. Great Lakes Fish. Comm. Tech. Rep. 28, pp. 1–34.

Regier, H. A., and Hartman, W. L. (1973). Lake Erie's fish community: 150 years of cultural stresses. *Science* **180,** 1248–1255.

Regier, H. A., Applegate, V. C., and Ryder, R. A. (1969). The ecology and management of the walleye in western Lake Erie. Great Lakes Fish. Comm. Tech. Rep. 15, pp. 1–101.

Schneider, J. C. (1977). History of the walleye fisheries of Saginaw Bay, Lake Huron. Mich., Dep. Nat. Resour., Fish. Res. Rep. 1850, pp. 1–16.

Schneider, J. C., and Leach, J. H. (1979). Walleye stocks in the Great Lakes, 1800–1975: Fluctuations and possible causes. Great Lakes Fish. Comm. Tech. Rep. 31, pp. 1–51.

Scott, W. B., and Christie, W. J. (1963). The invasion of the lower Great Lakes by the white perch, *Roccus americanus* (Gmelin). *J. Fish. Res. Board Can.* **20**(5), 1189–1195.

Scott, W. B., and Crossman, E. J. (1973). Freshwater fishes of Canada. *Bull., Fish. Res. Board Can.* **184,** 1–966.

Selgeby, J. H. (1982). Decline of lake herring (*Coregonus artedii*) in Lake Superior: An analysis of the Wisconsin herring fishery. *Can. J. Fish. Aquat. Sci.* **39,** 554–563.

Selgeby, J. H., MacCallum, W. R., and Swedberg, D. V. (1978). Predation by rainbow (*Osmerus mordax*) on lake herring (*Coregonus artedii*) in western Lake Superior. *J. Fish. Res. Board Can.* **35,** 1457–1463.

Smith, H. M., and Snell, M. (1891). Review of the fisheries of the Great Lakes in 1885, with introduction and description of fishing vessels and boats by J. W. Collins. Rep. U.S. Comm. Fish Fish., 1887, pp. 1-333.

Smith, S. H. (1956). Life history of the lake herring of Green Bay, Lake Michigan. *Fish. Bull.* **57,** 87–138.

Smith, S. H. (1968). Species succession and fishery exploitation in the Great Lakes. *J. Fish. Res. Board Can.* **25,** 667–693.

Smith, S. H. (1972). The future of salmonid communities in the Laurentian Great Lakes. *J. Fish. Res. Board Can.* **29,** 951–957.

Spangler, G. R., Payne, N. R., and Winterton, K. (1977). Percids in the Canadian waters of Lake Huron. *J. Fish. Res. Board Can.* **34,** 1839–1848.

Trautman, M. B. (1957). *The Fishes of Ohio.* Ohio State Univ. Press, Columbus.

Van Oosten, J. (1930). The disappearance of the Lake Erie cisco—A preliminary report. *Trans. Am. Fish. Soc.* **60,** 204–214.

Van Oosten, J. (1937). The dispersal of rainbow smelt, *Osmerus mordax* (Mitchill), in the Great Lakes region. *Trans. Am. Fish. Soc.* **60,** 204–214.

Van Oosten, J. (1947). Mortality of smelt, *Osmerus mordax* (Mitchill), in Lakes Huron and Michigan during the fall and winter of 1942–1943. *Trans. Am. Fish. Soc.* **74,** 310–337.

Wagner, W. C. (1981). Reproduction of planted lake trout in Lake Michigan. *North Am. J. Fish. Manage.* **1,** 159–164.

Weiler, R. R., and Chawla, V. K. (1969). Dissolved mineral quality of Great Lakes waters. Proc. 12th Conf. Great Lakes Res., pp. 801–818.

Wells, L. (1977). Changes in yellow perch (*Perca flavescens*) populations of Lake Michigan, 1954–75. *J. Fish. Res. Board Can.* **34**(10), 1821–1829.

Wells, L., and Jorgenson, S. C. (1983). Population biology of yellow perch in southern Lake Michigan, 1971–79. Tech. Pap. U.S. Fish Wildl. Serv. 109, pp. 1–19.

Wells, L., and McLain, A. L. (1973). Lake Michigan: Man's effects on native fish stocks and other biota. Great Lakes Fish. Comm. Tech. Rep. 20, pp. 1–55.

Willford, W. A., Bergstedt, R. A., Berlin, W. H., Foster, N. R., Hesselberg, R. J., Mac, M. J., Passino, D. R. M., Reinert, R. E., and Rottiers, D. V. (1981). Introduction and summary. In *Chlorinated Hydrocarbons As a Factor in the Reproduction and Survival of Lake Trout (Salvelinus namaycush) in Lake Michigan.* Tech. Pap. U.S. Fish Wildl. Serv. 105, pp. 1–7.

# 6

# EPIDEMICS IN BIRDS AND MAMMALS CAUSED BY CHEMICALS IN THE GREAT LAKES

*Michael Gilbertson*

*Fish Habitat Management Branch, Department of Fisheries and Oceans, Ottawa, Ontario, Canada K1A 0E6**



*Present address: Commercial Chemicals Branch, Environment Canada, Ottawa, Ontario, Canada K1A 0H3

# 1. EPIDEMICS AND CAUSES

During the past twenty years there has been a series of anomalies involving reproductive failure and mortality in populations of mammals and birds in the Great Lakes basin. The common factor between these observed anomalies has been that the organisms affected were primarily fish eaters. Subsequent etiological research has shown that the toxic agents in the fish were industrial and agricultural chemicals and that these were organochlorine compounds.

This review is mostly concerned with case histories of anomalies that were noticed in the mid 1960s and early 1970s and it is valid to ask why these old spoors should be reworked. First, after a period when the levels of organochlorine chemicals declined, recent data show that the levels in fish from some Great Lakes are increasing (International Joint Commission, 1983). Second, researchers from Green Bay, Wisconsin, have shown that there is a new outbreak of reproductive dysfunction and abnormalities in fish-eating birds. Third, various interest groups, such as mink (*Mustela vison*) farmers, need to know whether Great Lakes fish are yet safe to be used as a feedstuff for their mink-ranching operations. Fourth, these case histories could show principles of how to structure research on severely contaminated populations to obtain credible results. Finally, these case histories could be important in negotiating cleanup pursuant to the Great Lakes Water Quality Agreement.

Classically (Table 6.1), epidemiological cases begin with the observation of an anomaly at some level of biological organization. For instance, the observed anomaly may be a missing year class, reduced hatchability, a tumor, an enlarged liver, a deformity, or biochemical or histological lesion. Subsequent to these initial signals, detailed data on the species affected and geographic and temporal variation in the incidence of the anomaly yields information on the scale, severity, and trends of the epidemic. The pathological signs and symptoms are investigated and described. Finally, the question is posed as to what is causing the anomaly and epidemiological approaches give way to etiological methods using experimentation based on the formulation of testable hypotheses of causation. After a causal agent

**6.1 Flow Diagram of Activities for Formal Demonstration of Causation of a Biological Anomaly by a Suspected Agent**

*Initial signals*

Observation of a biological anomaly

*Epidemiology*

Field studies of species affected and of geographic and temporal incidence and severity of the biological anomaly

Description of pathological signs and symptoms

Formulation of hypotheses of causation

*Etiology*

Experimentation with possible causal agent

Experimentation with other causal agents with similar mode of action

Conclusions about causation of the biological anomaly

*Source:* After Gilbertson (1984).

has been demonstrated, alternative causal agents must also be considered. At some point in the procedure the anomaly may acquire a name.

The International Joint Commission has already sponsored a literature review of the effects of chemicals on Great Lakes aquatic populations (Fitchko, 1986). This present review aims at a transformation of the literature review into a series of selected case histories based on the most complete examples of investigations to relate biological anomalies observed in the field to a cause or causes.

## 2. REPRODUCTIVE FAILURE IN RANCH MINK

### 2.1. Initial Report

Twenty years ago, Dr. G. R. Hartsough (1965), Director of Ranch Service, Great Lakes Mink Association, published a short, carefully worded news article implicating Great Lakes fish in a large number of reproductive failures in ranch mink operations in North Central states and in Ontario. The report noted that the evidence was based only on clinical evidence and that it required substantiation. He also noted the dilemma for the decision maker. On the one hand, it was important that no mistake be made in attributing the cause of the problem to Great Lakes fish because removal of this feed source would strain alternative feed sources and increase feed costs. On the other hand, he recognized his obligation to the industry to prevent a widespread problem. Chemical analyses had shown that pesticide residues were present in significant quantities in female mink and in fish, but he also noted that the offending chemical might not be a pesticide but some entirely different compound. One aspect that is not clear from the article is the exact source of the fish, because little was recorded about which of the Great Lakes contained the fish with the toxic factors at that time.

The details of the epidemiology and etiology of the mink reproductive failure have been published in several review articles (Aulerich and Ringer, 1977; Ringer, 1982). There are, however, few references to the geographic extent of the failures or to the economic losses incurred. Without these data it is difficult to gauge the significance of the outbreaks. The disease in the mink was associated with fish caught in Lakes Michigan and Erie fed to mink during the early 1960s. However, all Great Lakes fish became suspect and use was rapidly discontinued. The industry turned to sources of ocean fish. However, no estimate has yet been made of the socioeconomic impact on the Great Lakes fishery or of the increased costs of these alternative feedstuffs.

### 2.2. Symptoms and Hypotheses

In 1967 mink ranch operations, adult animals were not adversely affected by feeding up to 15% of Lake Michigan coho salmon (*Oncorhynchus kisutch*) in their diet (Aulerich et al., 1971). This level of feed even supported normal breeding and whelping. However, depending on the percentage of coho salmon in the diet and the duration of feeding, mortality of the kits reached 80%. Based on these studies, several hypotheses were put forward as possible explanations for the observed reproductive failure including rancidity, mercury contamination, organochlorine pesticides and the possibility of other chemical contaminants (Hartsough, 1965; Aulerich and Ringer, 1970; Aulerich et al., 1971).

In 1968, the first of a series of experiments was undertaken to verify the industry's observations associated with feeding Great Lakes fish to mink and to identify the etiological agent. The experiments not only confirmed (Table 6.2) the presence of a toxic agent in Lake Michigan coho salmon but also showed that the agent was present in other Lake Michigan species and in Lake Erie fish. In addition to significant increases in kit mortality and general failure to whelp when fed Lake Michigan fish species, kits weighed less at whelping and after 4 weeks.

Adult mortality occurred in all experimental mink fed by-products from canning Lake Michigan coho salmon. Symptoms at death were nonspecific but included anorexia, bloody stools, fatty infiltration and degeneration of the livers and kidneys, and hemorrhagic gastric ulcers (Aulerich et al., 1973).

### 2.3. Etiology

No relationship was found between rancidity or mercury content and any of the biological parameters of whelping success or kit mortality. Based on the results of organochlorine pesticide residue analysis of Lake Michigan coho salmon sampled

**Table 6.2 Reproductive Performance of Mink Fed Various Fish**

| Fish Portion of Diet | Number of Females | | Number of Kits | | |
|---|---|---|---|---|---|
| | | | Birth | | |
| | Mated | Whelped | Alive | Dead | Alive at 4 weeks |
| Ocean Whiting | 12 | 10 | 45 (8.2 g) | 10 | 29 (125 g) |
| Lake Michigan | | | | | |
| Coho Salmon | 10 | 1 | 0 | 1 | 0 |
| Bloater Chub | 12 | 7 | 10 (7.0 g)* | 12 | 5 (91 g)** |
| Lake Erie Coho Salmon | 12 | 11 | 32 (7.6 g) | 17 | 7 (86 g)** |

*$P < 0.05$.
**$P < 0.01$.
*Source:* After Aulerich et al. (1973).

in 1969, there was an apparent relationship between the content of organochlorine pesticides (DDT and related isomers or dieldrin) and reproductive failure (Aulerich et al., 1971). However, previous work (Duby, 1970; Aulerich and Ringer, 1970) had shown that mink were able to reproduce normally when exposed to an intake of DDT or dieldrin greater than the calculated DDT or dieldrin intake in these experiments in which fish were fed to mink. In addition, there was no indication of clinical signs of organochlorine pesticide toxicity such as tremors, convulsions, or loss of coordination (Aulerich et al., 1973). Another reproductive experiment in which the acetone hexane extract of Lake Michigan coho salmon was fed to mink showed that the toxic factor was in the lipids (Aulerich and Ringer, 1977).

The next clue to the possible etiology came from analytical chemistry. In addition to DDT and dieldrin found in the Lake Michigan coho salmon sampled in 1969, polychlorinated biphenyls were detected at concentrations between 10 and 15 ppm (Aulerich and Ringer, 1977). Samples of three PCB products (Aroclor 1242, 1248, and 1254) were obtained from the manufacturer (Monsanto) and their toxicity was tested in a subsequent mink-breeding experiment (Aulerich et al., 1973). Mink were fed a basic diet with either (1) ocean fish, (2) Lake Michigan coho salmon, or (3) ocean fish with 30 ppm PCB (10 ppm of each of the three above Aroclors). The reproductive success (Table 6.3) and PCB content of the adults were determined and the gross pathology was determined after death. None of the mated females fed Lake Michigan coho salmon or 30 ppm PCB whelped, whereas the controls produced 35 live kits. There was a striking similarity between the clinical signs and lesions of the mink that died when fed diets containing coho salmon from Lake Michigan and those fed PCB. The levels of PCB in the tissues of mink that died after eating Lake Michigan coho salmon were remarkably similar to those of mink that were fed the PCB (Table 6.3).

**Table 6.3 Reproductive Performance and Average PCB Residues in Tissues of Mink Fed Various Dietary Treatments**

| | Dietary Treatment | | |
|---|---|---|---|
| | Control 30% Ocean Fish Mix | 30% Lake Michigan Coho Salmon | 30% Ocean Fish Mix plus 30 ppm PCB[a] |
| No. females mated | 12 | 12 | 11 |
| No. females whelped | 11 | 0 | 0 |
| No. kits born | | | |
| Alive | 35 | — | — |
| Dead | 19 | — | — |
| No. live kits at 4 weeks | 29 | — | — |
| PCB residue in ppm in | | | |
| Brain | 0.1 | 11.07 | 11.00 |
| Liver | 0.01 | 5.21 | 4.18 |
| Kidney | 0.01 | 6.37 | 4.47 |

[a]PCB consisted of 10 ppm each of Aroclor 1242, 1248, and 1254.

*Source:* After Aulerich and Ringer (1977).

Aulerich and Ringer (1977) used five criteria to "strongly implicate" PCB as the etiological agent responsible for the observed reproductive failure and lesions:

1. The PCB residue content of Lake Michigan coho salmon
2. The sensitivity of mink to PCB
3. The similarity of the reproductive complications in mink fed coho salmon and PCB
4. The similarity of the clinical signs and lesions in mink that died after being fed Lake Michigan coho salmon or PCB
5. Accumulation of comparable quantities of PCB in mink that died after being fed Lake Michigan coho salmon or PCB.

## 2.4. Critique

There is no doubt that this series of observations and experiments was an excellent piece of epidemiological and etiological sleuthing. It represents one of the very few cases in which low-level exposure to organic chemicals has been shown to be the cause of a subtle epidemic in a mammal population, other than humans. Perhaps the most important aspect of this work is that the initial signals of the presence of a toxic factor in Lake Michigan fish were detected 20 years before epidemiological evidence showed effects on human reproduction (Fein et al., 1984; Jacobson et al., 1984).

How has this causal explanation stood up to time? For the mink industry the avoidance of fish products from the Great Lakes has resulted in a cessation of losses attributable to this source of feedstuff. During the past decade research has demonstrated other organic compounds that have a similar mode of action to PCB (Goldstein, 1980) including various extremely toxic isomers of the polychlorinated dibenzo-*p*-dioxins and polychlorinated dibenzofurans. Their concentrations in tissues may also correlate with the levels of PCB and other organochlorine compounds. Could these compounds have contributed significantly to the mink reproductive failure? Recent analytical data on Great Lakes fish have shown the presence of these compounds (Stalling et al., 1983). Various polychlorinated dibenzofurans are contaminants of PCB fluids (Bowes et al., 1975). There are, however, several other sources of these compounds to the Great Lakes environment. Incinerators are sources of both polychlorinated dibenzo-*p*-dioxins and dibenzofurans (Rappe, 1984). The extremely toxic 2,3,7,8-tetrachlorodibenzo-*p*-dioxin is leaching from chemical dumps where wastes from production of 2,4,5-trichlorophenol were buried.

The relative toxicities of many of these compounds have been determined using rapid screening bioassays based on induction of aryl hydrocarbon hydroxylase activity (Bradlaw and Casterline, 1979) or production of a keratinization response in cell cultures (Knutson and Poland, 1980). If the mode of action of these compounds is the same between mammal species, detailed isomer specific analysis of Great Lakes fish and of mink experimentally exposed to these fish could provide an indication of the relative significance of the PCB content compared to other microcontaminants in causing the mink reproductive failure. These data may not be important to the mink ranchers that are no longer dependent on Great Lakes fish as a feedstuff. They are, however, important for administrators of the Great Lakes Water Quality Agreement so that priorities can be decided on a scientific basis for cleanup of sources of the most dangerous pollutants, in order to reduce human exposure through eating contaminated fish.

In addition, recent reports have shown that population densities of wild mink and otters (*Lutra canadensis*) are very low in areas close to Lake Ontario (Robert Foley, New York Department of Environmental Conservation, Gloverville, NY, personal communication). Similarly, trapping data from the Wisconsin Department of Natural Resources (Kohn and Ashbrenner, 1984) showed that no otters were caught in four counties near the Door Peninsula, which divides Green Bay from Lake Michigan. In both instances the presence of organochlorine pollutants in fish was the suspected cause, though historical trapping data, the possible role of human disturbance and the contaminant levels must be considered before a causal connection can be made. Further evidence of the probable role of PCB in the ecology of wild piscivorous mammals was provided by the analysis of a dead mink found in a marsh near Green Bay. The liver of this animal contained 5.7 ppm PCB (T. J. Kubiak, U.S. Fish and Wildlife Service, Green Bay, WI, personal communication) and this corresponds with the levels in mink liver associated with adult mortality in feeding experiments (Aulerich and Ringer, 1977).

Because mink and otter are piscivorous and extremely sensitive to PCB and, likely, to compounds that have a similar mode of action, these species have a high potential to be critically exposed. Though data on mink and otter abundance may be difficult to collect, there is a priority requirement to undertake population surveys to document areas where the species have been extirpated. Without these data it will continue to be difficult to make a scientific case for stringent controls on sources of organochlorine pollutants to the Great Lakes Ecosystem.

## 3. REPRODUCTIVE FAILURE AND MORTALITY IN BIRDS

### 3.1. Initial Reports

Birds, in contrast to mammals, fish, and humans, have given clear, easily observed signals when chemically induced epizootics occurred. These signals were observed in colonies of fish-feeding birds on Lakes Michigan and Huron and on Lake Ontario between 1964 and 1970.

#### *3.1.1. Lake Michigan and Lake Huron*

In 1964, bird banders visiting herring gull (*Larus argentatus*) colonies in Door County, Green Bay, Wisconsin, noted a drastic reduction in the number of bandable chicks compared with previous years (Keith, 1966); the colonies contained a great many disused nests with cold eggs. In 1963, adult herring gulls had been shot for pesticide residue analysis and found to contain more than 2000 ppm in the body fat. In 1964, Keith (1966) undertook his detailed pioneering work on the herring gull colony on Little Sister Island, Green Bay, Wisconsin, to investigate whether reproductive success was affected by this pesticide residue. In the same year, Ludwig and Tomoff (1966) noted unusually low reproductive success of herring gulls nesting on Bellow Island in Grand Traverse Bay, Lake Michigan, and a decrease in the number of nesting pairs at the colony. Subsequent annual observation of colony size showed a precipitous decline in the breeding population (Ludwig and Ludwig, 1969).

In the late 1960s, Postupalsky documented reproductive failure and 30% eggshell thinning among double-crested cormorant (*Phalacrocorax auritus*) populations in Georgian Bay, Lake Huron (cited in Weseloh et al., 1983). During the past four years, 1981–1984, there has been an anomalous incidence of congenital abnormalities among Green Bay, Wisconsin, colonies of Forster's terns (*Sterna forsteri*), black-crowned night herons (*Nycticorax nycticorax*), and double-crested cormorants (T. J. Kubiak, personal communication).

#### *3.1.2. Lake Ontario*

There is anecdotal evidence from two sources that severe reproductive failure in herring gulls occurred in Lake Ontario colonies in 1966 (Gilbertson, 1983) that had been apparently normal in 1965. Several other bird species exhibited reproductive failure including double-crested cormorants in Lake Ontario in 1963 (cited in Gilbertson, 1975) and common terns (*Sterna hirundo*) in Hamilton Harbour in 1970 (Gilbertson and Reynolds, 1972). Bald eagles (*Haliaeetus leucocephalus*) had completely disappeared from this lower Great Lake by the late 1950s.

## 3.2. Upper Lakes

### *3.2.1. Symptoms and Hypotheses*

The detailed surveys of reproductive success of Lake Michigan herring gulls showed (Table 6.4) that eggs in the colonies from Green Bay (Keith, 1966) and Grand Traverse Bay (Ludwig and Tomoff, 1966) had a low hatching success of 41 and 28%, respectively. Twenty percent of the eggs at the Grand Traverse Bay colony were cracked and shell damage was noted in the Green Bay colony. At these two colonies only three or four fledged young were produced per 10 nests. In contrast, at a midlake colony in the Beaver Islands, over 80% of the eggs hatched and an estimated 12 young were produced per 10 nests.

Adult birds at the colony at Grand Traverse Bay were exceptionally aggressive and it was suggested that this behavioral anomaly could account for the high proportion of cracked eggs. Both authors suggested that disturbances in adult behavior might adversely affect incubation of the eggs (Keith, 1966; Ludwig and Tomoff, 1966). The presence of organochlorine chemicals in the eggs and in adults was hypothesized to be the cause of the reproductive failure at the Green Bay and Grand Traverse Bay colonies (Ludwig and Tomoff, 1966).

### *3.2.2. Etiology*

Herring gull eggs from the three Lake Michigan study colonies were analyzed for organochlorine residues. The results (Table 6.5) indicate that eggs from the Green Bay colony were the most contaminated and contained twice the level of DDE compared to the least contaminated eggs from the colony at the Beaver Islands.

There was an apparent inverse relationship between fledged young per pair and the residue levels in the eggs. Both papers concluded, however, that there were difficulties relating the observed reproductive failure to organochlorine residues. Keith (1966) had reservations based on the discrepancy between his findings of low hatchability and experimental feeding studies with quail (*Colinus virginianus*) and pheasant (*Phasianus colchicus*) in which eggs of contaminated hens hatched as frequently as eggs of controls. Ludwig and Tomoff (1966) concluded that because the residues in samples of live and dead eggs from the Grand Traverse Bay colony were similar to the levels in live and dead eggs from the Beaver Islands the residue content could not be a factor in the reproductive failure.

### *3.2.3. Critique*

These studies were among the first in the Great Lakes to try to investigate the role of organochlorine insecticides on avian reproduction. In the Green Bay studies, there were no controls but reliance was placed on comparisons with other published studies of herring gull reproduction. Even in the mid 1960s it was likely

**Table 6.4 Estimated Reproductive Success in Lake Michigan Herring Gull Colonies, 1964 and 1965**

| Location | Year | Number of Nests | Eggs per Nest | Percentage of Eggs | | | Fledged Young per Pair |
|---|---|---|---|---|---|---|---|
| | | | | Hatched | Cracked | Fledged | |
| Little Sister Island, Green Bay, WI (45° 12′N, 87° 10′W) | 1964 | 115 | 2.2 | 41 | 13 | 14 | 0.3 |
| Bellow Island, Grand Traverse Bay, MI (45° 06′N, 85° 34′W) | 1965 | 287 | 2.9 | 28 | 20 | 14 | 0.4 |
| Pismire Island, Beaver Islands, MI (45° 45′N, 85° 31′W) | 1965 | 344 | 2.9 | 80 | 6.5 | 42 | 1.2 |

*Source:* After Keith (1966); Ludwig and Tomoff (1966).

**Table 6.5 Organochlorine Residue Levels in Lake Michigan Herring Gull Eggs, Collected Alive, in Wet Weight Parts per Million**

| Location | Year | DDT | TDE | DDT + TDE | DDE | Dieldrin |
|---|---|---|---|---|---|---|
| Green Bay | 1964 | 19 | 6 | — | 200 | — |
| Grand Traverse Bay | 1965 | — | — | 18 | 140 | 1.5 |
| Beaver Islands | 1965 | — | — | 16 | 100 | 1.0 |

*Source:* After Keith (1966); Ludwig and Tomoff (1966).

difficult to find relatively uncontaminated colonies. There were serious problems in the egg-sampling regime since eggs were not sampled randomly and some eggs were taken at the end of incubation. Thus no generalizations can be drawn with certainty concerning the relative contamination of the colonies. These two studies were undertaken and published before the phenomenon of eggshell thinning had been documented (Ratcliffe, 1967) and before the relationship between DDE and eggshell thinning had been demonstrated on a correlative basis (Hickey and Anderson, 1968; Gilbertson, 1974) or on an experimental basis (Wiemeyer and Porter, 1970).

Organochlorine residue analysis in the mid 1960s was still in its infancy. As Ludwig and Ludwig (1969) later stated, the chemist who analyzed the Lake Michigan eggs suspected that part of the DDT and DDD reported represented interference from polychlorinated biphenyls. Estimation of the levels of PCBs was not included in these analyses and thus their role in the reproductive failure cannot be assessed. More particularly, the subject of ultratrace contaminants such as polychlorinated dibenzo-*p*-dioxins and dibenzofurans was not even being discussed by analytical chemists involved in analysis of environmental samples. The absence of any stored specimens precludes the possibility of reanalyzing the samples.

### 3.2.4. *New Symptoms and Hypotheses*

The research (Ludwig and Tomoff, 1966) did, however, provide a new hypothesis (Ludwig and Ludwig, 1969): the presence of organochlorine insecticides in herring gulls might cause significant mortality in adults, particularly during times of stress such as migration, storms, and moulting. Long-term observation of the colony at Grand Traverse Bay had shown that the population increased during the 20-year period 1944–1964 from 1000 to 2500 breeding pairs. However, in 1964 a slight decline to 2200 was noted. In subsequent annual visits to this island, a rapid population decline was recorded until only 850 breeding pairs remained in 1968 (Ludwig and Ludwig, 1969).

### 3.2.5. *Etiology of Adult Mortality*

To test this new hypothesis, the Ludwigs in 1967 captured 15 nesting adult herring gulls from Bellow Island, Grand Traverse Bay, and transported them to Port Huron to investigate the effects of starvation on the gulls. Even as they were capturing the gulls, they observed an adult bird with symptoms of pesticide poisoning, including uncontrollable eye tremors and violent body spasms (Ludwig and Ludwig, 1969). Previous research with presumably uncontaminated herring gulls captured at Rogers City, Lake Huron, had shown that this species could withstand 30% loss of body weight in 17 days without losing vigor or flight. In contrast, within 3 days the first of the gulls started to exhibit signs of insecticide poisoning, a second bird died in tremors on day 6, and four others died by the tenth day. Chemical analysis of the brains indicated that combined levels of insecticides associated with mortality in this population were about 12 ppm DDT, 6 ppm DDD, 180 ppm DDE, and 2 ppm dieldrin. The authors concluded that the 20% annual decline in the breeding population of herring gulls on Bellow Island at Grand Traverse Bay resulted from increased adult mortality caused by insecticide residues.

Further support for the conclusion of increased adult mortality caused by organochlorine chemicals was forthcoming from subsequent investigations. In Southern Ontario, an anomalous number of emaciated dead or dying ring-billed gulls (*Larus delawarensis*) were found during the 1969 and 1973 fall migrations. Sileo et al. (1977) undertook autopsies and residue analysis on specimens submitted to the University of Guelph. The investigations resulted in an improved method for determining the significance of multiple residues by calculation of an organochlorine index for residues in the brain. Based on residue analysis of wild birds and experimental poisoning of a variety of bird species, Sileo et al. (1977) concluded that the following individual levels of organochlorine compounds in the brain were associated with death: 5 ppm dieldrin, 150 ppm DDE, or 300 ppm PCB. The organochlorine index was calculated by adding the ratios of the levels determined by residue analysis compared with the above levels associated with death. The investigators showed that there was an inverse relationship between body weight and brain organochlorine indices. They concluded that stresses such as post nuptial or post juvenal molt likely contributed to the observed emaciation and mortality.

## 3.3. Lake Ontario

The epidemiology and etiology of the reproductive failures in the lower Great Lakes have been reviewed in detail by Mineau et al. (1984) and Gilbertson (1983).

### 3.3.1. *Epidemiology*

In the 1960s and 1970s Lake Ontario and Lake Michigan were two freshwater locations with high levels of organochlorine pollutants. After the initial findings of the 1970 reproductive failure of common terns in Hamilton Harbour (Gilbertson and Reynolds, 1972), fieldwork was undertaken to determine the geographic extent of the reproductive failure and the species affected. Seven species were studied during the next two years: herring gulls, ring-bill gulls, common terns, caspian terns (*Hydroprogne caspia*), black terns (*Chlidonias niger*), black-crowned night herons, and double-crested cormorants. In addition, contaminated eggs of these species were collected and incubated artificially to determine hatchability under controlled environmental conditions and to investigate signs and symptoms of the reproductive pathology.

### *3.3.2. Signs and Symptoms*

These preliminary studies revealed serious reproductive failure among double-crested cormorants in Lake Ontario. Several of the Lake Ontario species, particularly the common tern, exhibited a high incidence of congenital abnormalities such as crossed beaks, small eyes, or duplicate limbs (Gilbertson et al., 1976). Similarly, low hatchability and fledging rates were evident in several of these species. The artificial incubation procedures indicated that the low hatching success was due, at least in part, to a factor inside the egg. Characteristically, chicks that died after pipping the shell were undersize and had large unresorbed yolk sacs.

After the 1972 field season, the herring gull was chosen (Gilbertson, 1974) as the species for routine monitoring of trends of reproductive success and of levels of organochlorine compounds and for detailed etiological research. Among the most important reasons for the choice was the relatively nonmigratory habits of the adult breeding population in the Great Lakes. Thus the adult gulls tended to lay eggs that reflected the organochlorine contamination near the colony. This routine monitoring subsequently became part of the surveillance plan of the International Joint Commission in 1977.

The data for the number of fledged young per pair are shown in Table 6.6 and indicated that severe reproductive failure occurred in the Scotch Bonnet Island colony in Lake Ontario between 1972 and 1975 but that reproduction was normal in the year 1977. Similarly the Port Colborne colony in eastern Lake Erie exhibited reduced reproductive success between 1974 and 1976 but was normal in 1978. Reproductive success in Lakes Huron and Superior appeared normal between 1975 and 1978. At the same time organochlorine residues in herring gull eggs collected from colonies in the lower Great Lakes declined (Weseloh et al., 1979). Thus interpretation of cause and effect was made more complex because of the shifting baseline of fledging success and residue levels.

In 1972, fledging success in five Lake Ontario colonies was between 0.06 and

**Table 6.6 Fledged Young per Pair of Great Lakes Herring Gulls**

| Location | 1972 | 1973 | 1974 | 1975 | 1976 | 1977 | 1978 |
|---|---|---|---|---|---|---|---|
| Lake Ontario Scotch Bonnet Island | 0.12 | 0.06 | | 0.15 | | 1.10 | 1.01 |
| Lake Erie Port Colborne | | | 0.48 | 0.65 | 0.79 | | 1.45 |
| Lake Huron Chantry Island | | | | 1.48 | | 1.12 | 1.40 |
| Lake Superior Granite Island | | | | 1.32 | 1.55 | | 1.12 |

*Source:* After Weseloh et al. (1979).

0.21 fledged young per pair (Gilbertson, 1974), which represents only about a tenth of the normal fledging success. These studies also demonstrated a relationship between eggshell thinning and DDE: eggshell breakage and flaking was observed in the most contaminated colonies.

Detailed studies of the colony on Scotch Bonnet Island in 1973 showed that the reproductive failure was characterized by failure to lay eggs in 27% of the nests, early embryonic mortality in 20% of the eggs laid, low hatching success (17%), low fledging success (0.06 fledged young per pair), and high chick mortality (74%) (Gilbertson and Hale, 1974a,b). Chemical analysis of eggs from the Scotch Bonnet Island colony showed the presence of about 140 ppm DDE and 550 ppm PCB on a dry weight basis. In addition, the eggs contained dieldrin, heptachlor epoxide, and hexachlorobenzene.

### 3.3.3. *Hypotheses*

1. Based on the small size of chicks that pipped the shell but failed to hatch and the similarity of these findings to those caused by antithyroid compounds, it was hypothesized (Gilbertson and Fox, 1977) that an antithyroid compound was acting on the embryos. No definitive study on this aspect has yet been completed and published.

2. Based on the high levels of PCB present it was postulated that a chick edema active compound such as polychlorinated dibenzofurans could be causing the high embryonic mortality and abnormalities. A study (Gilbertson and Fox, 1977) was then undertaken to look for signs of chick edema disease in herring gull chicks at hatching.

3. Based on the poor hatchability and high rate of disappearance of eggs it was postulated (Fox et al., 1978; Peakall et al., 1980) that adult breeding behavior had been disturbed as had also been noted in the two Lake Michigan studies (Keith, 1966; Ludwig and Tomoff, 1966).

### 3.3.4. *Etiology*

These hypotheses concerned both intrinsic and extrinsic factors affecting the hatchability of the eggs. To separate the relative importance of these factors, a series of egg exchange experiments was conducted between clean and contaminated colonies in 1975, 1976, and 1977 (Peakall et al., 1980). These three years were the period in which herring gull reproduction in the Scotch Bonnet Island colony on Lake Ontario improved from 0.15 to 1.1 fledged young per pair. The design of the egg exchange experiment is shown in Table 6.7.

"Clean" maritime and "dirty" Lake Ontario eggs were incubated by clean and dirty adults and by artificial incubation. The results showed that in 1975 there were both marked intrinsic and extrinsic (adult behavioral) factors affecting hatchability; in 1976 the extrinsic was less evident and in 1977 there was no apparent effect.

To investigate the behavioral aspect, a study was undertaken in 1976, using telemetered eggs. The study showed that Lake Ontario herring gulls were less attentive to their nests and applied less heat to their eggs than uncontaminated

**Table 6.7 Egg Exchange Experiments: Overall Design and Summary of Hatching Success**

| | | | % Hatched | | |
|---|---|---|---|---|---|
| Incubation | Egg | Expected Result | 1975 | 1976 | 1977 |
| Clean adult | Clean | Normal reproduction | 86 | — | 68 |
| Clean adult | Dirty | Intrinsic factors only | 10 | — | 63 |
| Dirty adult | Clean | Extrinsic factors only | 7 | 48 | 81 |
| Dirty adult | Dirty | Both intrinsic and extrinsic factors | 2 | 37 | 70 |
| Artificial | Clean | Normal reproduction | 60 | 57 | — |
| Artificial | Dirty | Intrinsic factors | 37 | 23 | — |
| Conclusion | | | Marked intrinsic and extrinsic factors | Marked intrinsic, some extrinsic factors | No effects seen |

*Source:* After Peakall et al. (1980).

controls on Kent Island, New Brunswick (Fox et al.,1978). In the Lake Ontario colony there was a significant 1 C° decrease in the average temperature of the telemetered eggs in unsuccessful nests compared with successful nests. Unsuccessful nests were left unattended for three times longer than successful nests. The level of organochlorine residues in eggs removed for analysis from telemetered nests was correlated with the length of time that nests were left unattended (Fox et al., 1978).

To investigate whether there was a chick edema factor present in the Lake Ontario and eastern Lake Erie herring gull eggs a bioassay was undertaken in 1974 (reviewed in Gilbertson, 1983). It was postulated that if a chick edema factor were present in these eggs then a suite of signs and symptoms would be evident, including embryonic mortality, growth retardation, abnormalities, subcutaneous edema, pericardial edema, peritonial edema, porphyria, and liver enlargement and necrosis with fatty degeneration (Gilbertson and Fox, 1977).

Eggs were collected from two colonies on Lake Ontario, at the eastern end of Lake Erie, and from control colonies in New Brunswick and northern Alberta, and were incubated artifically. Only 26% of the eggs from Lake Ontario colonies hatched compared with 53% from Lake Erie colonies and 69% for the controls. Chicks from the Great Lakes colonies, particularly those that failed to hatch, showed extensive subcutaneous edema and fragility of the skin and pericardial edema. Lake Ontario chicks exhibited a milky edema in the peritoneal cavity. In addition, there was evidence of growth retardation: Lake Ontario chicks that hatched had tarsus lengths that were 7% smaller than controls, and those that pipped and failed to hatch were 20% smaller and had large unresorbed yolk sacs. Comparison of liver weights to tarsus lengths indicated that Lake Ontario chicks were 19% greater than controls and Lake Erie chicks were 24% larger, indicating liver enlargement. Lake Ontario chicks that pipped but failed to hatch had markedly elevated porphyrin levels and there was evidence of liver pathology in chicks from Lake Erie and Lake Ontario colonies including fatty degeneration and liver necrosis (Gilbertson and Fox, 1977). The rate of abnormalities in several colonies of fish-feeding birds on the lower Great Lakes in the year 1971 to 1973 was about 1%, indicating the presence of a teratogenic compound (Gilbertson et al., 1976). In the incubation experiment one abnormal herring gull embryo was found among the 47 eggs from Lake Ontario.

Based on these findings it was concluded that there was a chick edema active compound in the herring gull eggs from the lower Great Lakes. Subsequent chemical analysis of aliquots of frozen egg samples revealed the presence of 2,3,7,8-tetrachlorodibenzo-*p*-dioxin at a concentration of 500 ppt in 1974 (Canadian Wildlife Service, unpublished data). Other retrospective analyses showed that levels in eggs had been as high was 1200 ppt in the early 1970s.

### *3.3.5. Critique*

The detailed review of the chick edema studies (Gilbertson, 1983) concluded with two questions. How important was chick edema disease in the reproductive failures in fish-feeding colonial birds in the lower Great Lakes? What other chick edema compounds were present that might have contributed to the signs of chick edema disease? The review also concluded that those questions could not be answered without isomer-specific analysis of the chick edema compounds present in the eggs.* These questions might seem academic a decade after the end of the epidemic. However, a new outbreak of chick edema disease has been reported (T. J. Kubiak, personal communication). Investigations at Green Bay have had several advantages over the previous work and this is likely to result in the first comprehensive statement of the causes of an epidemic. These advantages include knowledge of the workable methodologies used in the studies on the lower Great Lakes and access to reliable ultratrace analysis which will yield congener specific analysis.

## 4. CONCLUSIONS

These case histories, concerning reproductive dysfunction in populations of fish-feeding mammals and birds, formally demonstrate the causal interrelationship between the observed epidemics and contamination of Great Lakes fish with organochlorine chemicals. This formal demonstration is based on a consistency of the evidence relating the characteristics of the observed reproductive pathology to the presence of specific chemicals in the fish and in the adults and their progeny and to the experimentally determined toxicological properties of the chemicals. Because these case histories are the essence of the complaints concerning organochlorine pollution of the Great Lakes, it is vital that an unassailable scientific case, based on sound epidemiology and etiology, should be prepared as the basis for cleanup negotiations or litigation (Mortimer, 1984; Gilbertson, 1984). To be unassailable, cases have to be prepared using the methodology and rigor of forensic science.

**Note Added in Press:* Harris, Kubiak, and Trick (1985) have published a detailed account of their studies on Forster's terns nesting at Green Bay, Lake Michigan, and at an inland location. At the Green Bay colony, egg hatchability was significantly reduced and at hatching chicks weighed less and had enlarged livers. Embryonic mortality was associated with subcutaneous edema. Egg-swap experiments and controlled incubation studies showed evidence of aberrant parental behavior in Green Bay terns resulting in an increase of 3.7 days in the incubation period. In addition, Green Bay tern eggs took 4.6 days longer to incubate. Thus eggs hatchability at the Green Bay colony was affected by factors intrinsic and extrinsic to the eggs. Fledging success in the Green Bay colony was markedly reduced.

Organochlorine chemical residue analysis of eggs from the two colonies revealed significantly higher residues of various chick edema active compounds including polychlorinated biphenyls, dibenzo-*p*-dioxins, and dibenzofurans at the Green Bay colony. Transformation of the results of isomer-specific analyses to 2,3,7,8-TCDD equivalents revealed that more than 80% of the chick edema activity was contributed by 2,3,4,3′,4′-pentachlorobiphenyl.

This combination of biological and chemical analysis has produced a sophisticated epidemiological case history. Specific regulatory controls can now be formulated for the industries on the Fox River based on a knowledge of the relative contribution of the various etiological agents present. The case history reemphasizes the powerful versatility of studies of avian reproduction for investigating environmental toxicology.

The case histories of research on the effects of chemicals on fish-eating birds and mammals in the Great Lakes show the effectiveness of using these organisms as indicators of the health of the aquatic environment. In contrast, epidemiological and etiological research on fish and human populations have proved much more difficult to conduct and, to date, much less successful. It is recommended that research and monitoring on birds and mammals should continue to be fostered to provide this essential part of the scientific case for restoring ecosystem health.

The model depicted in Table 6.1 has been used, sometimes perhaps unknowingly, to investigate the incidence and cause of outbreaks of chemically induced diseases. The model has application when the preventive approach to management of chemicals has not succeeded and contamination by persistent, toxic, and bioconcentrated chemicals has occurred on a widespread geographic scale. The successful implementation of the model demands teamwork and careful scientific management because no individual has the requisite skills to solve the toxicological riddles posed by releases of chemicals to the natural environment.

## ACKNOWLEDGMENTS

I am indebted to Tim Kubiak and Robert Foley for information on the situation in Wisconsin and New York, respectively, and to Dr. David Peakall, Dr. Chip Weseloh, and Pierre Mineau for reviewing the manuscript. Dr. Marlene Evans kindly asked me to undertake this assignment. Ms. Brenda Black patiently processed the words in the various drafts.

## REFERENCES

Aulerich, R. J., and Ringer, R. K. (1970). Some effects of chlorinated hydrocarbon pesticides on mink. *Am. Fur Breeder* **43,** 10–11.

Aulerich, R. J., and Ringer, R. K. (1977). Current status of PCB toxicity to mink, and effect on their reproduction. *Arch. Environ. Contam. Toxicol.* **6,** 279–292.

Aulerich, R. J., Ringer, R. K., Seagran, H. L., and Youatt, W. G. (1971). Effects of feeding coho salmon and other Great Lakes fish on mink reproduction. *Can. J. Zool.* **49,** 611–616.

Aulerich, R. J., Ringer, R. K., and Iwamoto, S. (1973). Reproductive failure and mortality in mink fed on Great Lakes fish. *J. Reprod. Fertil., Suppl.* **19,** 365–376.

Bowes, G. W., Mulvihill, M. J., Simoneit, B. R. T., Burlingame, A. L., and Risebrough, R. W. (1975). Identification of chlorinated dibenzofurans in American polychlorinated biphenyls. *Nature (London)*, **256,** 305–307.

Bradlaw, J. A., and Casterline, J. L. (1979). Induction of enzyme activity in cell culture: A rapid screen for detection of planar polychlorinated organic compounds. *J. Assoc. Off. Anal. Chem.* **62,** 904–916.

Duby, R. T. (1970). Pesticides vs reproduction still a puzzle. *Am. Fur Breeder* **43,** 15.

Fein, G. G., Jacobson, J. L., Jacobson, S. W., Schwartz, P. M., and Dowler, J. K. (1984). Prenatal exposure to polychlorinated biphenyls: Effects on birth size and gestational age. *J. Pediatr.* **105,** 315–320.

Fitchko, J. (1986). Literature review of the effects of persistent toxic substances on Great Lakes Biota. Report of the Health of Aquatic Communities Task Force to the Great Lakes Science Advisory Board, International Joint Commission. 256 pp.

Fox, G. A., Gilman, A. P., Peakall, D. B., and Anderka, F. W. (1978). Behavioural abnormalities of nesting Lake Ontario herring gulls. *J. Wildl. Manage.* **42,** 477–483.

Gilbertson, M. (1974). Pollutants in breeding herring gulls in the lower Great Lakes. *Can. Field Nat.* **88,** 273–280.

Gilbertson, M. (1975). A Great Lakes tragedy. *Nature (Canada)* **4,** 22–24.

Gilbertson, M. (1983). Etiology of chick-edema disease in herring gulls in the lower Great Lakes. *Chemosphere* **12,** 357–370.

Gilbertson, M. (1984). Need for development of epidemiology for chemically induced diseases in fish in Canada. *Can. J. Fish. Aquat. Sci.* **41,** 1534–1540.

Gilbertson, M., and Fox, G. A. (1977). Pollutant-associated embryonic mortality of Great Lakes herring gulls. *Environ. Pollut.* **12,** 211–216.

Gilbertson, M., and Hale, R. (1974a). Early embryonic mortality in a herring gull colony in Lake Ontario. *Can. Field Nat.* **88,** 354–356.

Gilbertson, M., and Hale, R. (1974b). Characteristics of the breeding failure of a colony of herring gulls on Lake Ontario. *Can. Field Nat.* **88,** 356–358.

Gilbertson, M., and Reynolds, L. M. (1972). Hexachlorobenzene (HCB) in the eggs of common terns in Hamilton Harbour, Ontario. *Bull. Environ. Contam. Toxicol.* **7,** 371–373.

Gilbertson, M., Morris, R. D., and Hunter, R. A. (1976). Abnormal chicks and PCB residue levels in eggs of colonial birds on the lower Great Lakes (1971–73). *Auk* **93,** 434–442.

Goldstein J. A. (1980). Structure-activity relationship for the biochemical effects of halogenated aromatic hydrocarbons and the relationship to toxicity. In R. Kimbrough (Ed.), *Halogenated Biphenyls, Terphenyls, Naphthalenes, Dibenzodioxins, and Related Products.* Elsevier/North-Holland, Amsterdam, pp. 151–190.

Hartsough, G. R. (1965). Great Lakes fish now suspect as mink food. *Am. Fur Breeder* **38,** 35–27.

Hickey, J. J., and Anderson, D. W. (1968). Chlorinated hydrocarbons and eggshell changes in raptorial and fish-eating birds. *Science* **162,** 271–273.

International Joint Commission (1983). *Report on Great Lakes Water Quality,* Appendix. Great Lakes Surveillance, IJC, Windsor, Ontario.

Jacobson, J. L., Jacobson, S. W., Schwartz, P. M., Fein G. G. and Dowler, J. K. (1984). Prenatal exposure to an environmental toxic: A test of the multiple effects model. *Dev. Psychol.* **20,** 523–532.

Keith, J. A. (1966). Reproduction in a population of herring gulls (*Larus argentatus*) contaminated by DDT. *J. Appl. Ecol.* **3** (Suppl.), 57–70.

Knutson, J. C., and Poland, A. (1980). Keratinization of mouse teratoma cell line XB produced by 2,3,7,8-tetrachlorodibenzo-*p*-dioxin: An *in vitro* model of toxicity. *Cell (Cambridge, Mass.)* **22,** 27–36.

Kohn, B. E., and Ashbrenner, J. E. (1984). *Harvest and Population Status of River Otter in Wisconsin,* Rep. No. 129. Department of Natural Resources, Madison, Wis.

Ludwig, J. P., and Ludwig, C. E. (1969). The effect of starvation on insecticide contaminated herring gulls removed from a Lake Michigan colony. *Proc. 12th Conf. Great Lakes Res.*, 53–60.

Harris, H. J., Kubiak, T. J. and Trick, T. A. (1985). *Microcontaminants and Reproductive Impairment of Forster's Terns on Green Bay,* Final Report to U.S. Fish and Wildlife Service. University of Wisconsin, Sea Grant Inst., Wis. Dept. of Nat. Resour. and Green Bay Metropolitan Sewerage Dist., Sea Grant Office, ES-105, University of Wisconsin, Green Bay.

Ludwig, J. P., and Tomoff, C. S. (1966). Reproductive success and insecticide residues in Lake Michigan herring gulls. *Jack-Pine Warbler* **44,** 77–84.

Mineau, P., Fox, G. A., Norstrom, R. J., Weseloh, D. V., Hallett, D. J., and Ellenton, J. A. (1984). Using the herring gull to monitor levels and effects of organochlorine contamination in the Canadian Great Lakes. In J. O. Nriagu and M. S. Simmons (Eds.), *Toxic Contaminants in the Great Lakes.* Wiley, New York, pp. 425–452.

Mortimer, C. (1984). *The Lake Michigan Pollution Case* (WIS-SG-81-237). Sea Grant Institute and Centre for Great Lakes Studies, University of Wisconsin, Milwaukee.

Peakall, D. B., Fox, G. A., Gilman, A. P., Hallett D. J., and Norstrom, R. J. (1980). Reproductive success of herring gulls as an indicator of Great Lakes Water Quality. In B. K. Afghan and D. Mackay (Eds.), *Hydrocarbons and Halogenated Hydrocarbons in the Aquatic Environment.* Plenum, New York, pp. 337–344.

Rappe, C. (1984). Analysis of polychlorinated dioxins and furans. *Environ. Sci. Technol.* **18,** 78A–90A.

Ratcliffe, D. A. (1967). Decrease in eggshell weight in certain birds of prey. *Nature* (*London*) **215,** 208–210.

Ringer, R. K. (1982). Toxicology of PCBs in mink and ferrets. In J. H. Highland (Ed.), *Hazardous Waste Disposal: Assessing the Problem.* Ann Arbor Sci. Publ., Woburn, MA.

Sileo, L., Karstad, L., Frank, R., Holdrinet, M. V. H., Addison, E., and Braun, H. E. (1977). Organochlorine poisoning of ring-billed gulls in southern Ontario. *J. Wildl. Dis.* **13,** 313–322.

Stalling, D. L., Smith, L. M., Petty, J. D., Hogan, J. W., Johnson, J. L., Rappe, C., and Buser, H. R. (1983). Residues of polychlorinated dibenzo-*p*-dioxins and dibenzofurans in Laurentian Great Lakes fish. In R. E. Tucker, A. L. Young, and A. P. Gray (Eds.), *Human and Environmental Risks of Chlorinated Dioxins and Related Compounds.* Plenum, New York, pp. 221–240.

Weseloh, D. V., Mineau, P., and Hallett, D. J. (1979). Organochlorine contaminants and trends in reproduction in Great Lakes herring gulls, 1974–1978. *Trans. North Am. Wildl. Nat. Resour. Conf.* **44,** 543–557.

Weseloh, D. V., Teeple, S. M., and Gilbertson, M. (1983). Double-crested cormorants of the Great Lakes: Egg laying parameters, reproductive failure, and contaminant residues in eggs, Lake Huron 1972–1973. *Can. J. Zool.* **61,** 427–436.

Wiemeyer, S. N., and Porter, R. D. (1970). DDE thins eggshells of captive American kestrels. *Nature* (*London*) **227,** 737–738.

# 7

# CHEMICAL CONTAMINANTS IN THE GREAT LAKES: THE HUMAN HEALTH ASPECT

*Harold E. B. Humphrey*

*Center for Environmental Health Sciences, Michigan Department of Public Health, Lansing, Michigan 48909*

## 1. INTRODUCTION

North America's Great Lakes define the boundaries of the state of Michigan and play a dominant role in its beauty, recreation, and economy. The lake resource has felt the impact of the activities of a regional society with nearly 40 million inhabitants. The nature of the lakes are such that materials used in terrestrial activities often accumulate and remain in the aquatic environment for decades or longer. Thus, the lakes and their sediments act as a sink or reservoir for terrestrial and atmospheric inputs. Lipophilic pollutants such as PCBs and DDT biomagnify within the aquatic food chain, which makes them available to higher forms of life including humans who consume fish from the lakes. Calculations by Swain (1983) and evaluations by Humphrey (1983b) have shown that exposure from fish far outweighs terrestrial, atmospheric, or water column sources. Humans therefore, can represent multigenerational endpoint receptors for contaminants found in the Great Lakes.

## 2. CHEMICAL CONTAMINANTS

### 2.1. Contaminants in Great Lakes Fish

Nearly 1000 chemical substances have been identified in the Great Lakes aquatic environment (International Joint Commission, 1983). Some of these substances have reported mammalian toxicity (International Joint Commission, 1981), are persistent and are known to bioaccumulate and biomagnify in the aquatic environment. A few of the contaminants have been detected in Great Lakes fish at greater than trace quantities. Of these, PCBs dominate the spectrum of major contaminants likely to reach a person's dinner plate (Table 7.1). Although localized pollution may alter the spectrum of chemical contaminants at specific sites and in certain species, the compounds shown represent those generally encountered at higher concentrations in Great Lakes trout and salmon. Fish, therefore, represent the prin-

**Table 7.1 Concentration of the Most Commonly Detected Contaminants Found in Cooked Lake Michigan Fish Provided by Study Participants**

| | Median Concentration (ppb) | | | |
|---|---|---|---|---|
| Contaminant | Lake Trout | Chinook | Coho | Other[a] |
| PCB | 3012 | 1478 | 807 | 168 |
| DDT | 1505 | 761 | 447 | 137 |
| HCB | 5 | 3 | 2 | 1 |
| Oxychlordane | 25 | — | — | — |
| trans-Nonachlor | 195 | 107 | 55 | — |
| Dieldrin | 75 | 53 | — | — |

[a]Pike, perch, walleye, whitefish.

*Source*. Humphrey (1983a).

cipal link between contaminants in the aquatic environment and humans. Accordingly, persons who consume fish become a part of the aquatic ecosystem and represent potential endpoint receptors for contaminants in the environment.

### 2.2. Potential Risk from Fish Consumption

Concentrations of lipid soluble contaminants are magnified to significant proportions in edible fish through food chain bioconcentration. Fish therefore, can represent a significant risk factor to humans who eat them on a regular basis. Carcinogenic risk from eating fish contaminated with PCBs and other organics such as DDT, dieldrin, toxaphene, and chlordane have been reported to be 10–100 times greater than the risk associated with drinking groundwater contaminated with lower halogenated compounds such as chloroform, 1,1-dichloroethylene, and other commonly encountered organics (Connor, 1984). Greater than average consumption rates and consumption of highly contaminated species contribute to the increased risk reported in these calculations.

Connor (1984) stated that most of the known carcinogenic risk in his calculations was due to PCBs (because of their higher concentration) and that other contaminants, even though some are more potent, provided only minor contributions to this risk because of their low concentrations in fish.

### 2.3. Potential Exposure from Fish Consumption

Laboratory analysis of cooked Lake Michigan fish substantiates the risk emphasis placed on PCBs. Although hundreds of chemicals have been detected in the Great Lakes and many may be present in raw fish at trace levels, only a few represent significant dose factors in cooked fish. Evaluation of specimens collected from the dinner plates of participants in our 1981 cohort study (Humphrey, 1983a) demonstrate that the concentration of PCBs dominate the levels of the six contaminants most often detected (Table 7.1). The data also show that predator fish with higher fat content such as sport-caught trout and salmon contain significantly higher levels of contaminants after cooking than smaller or more lean species such as perch or whitefish.

PCBs are ubiquitous contaminants that can reach humans in a variety of ways. Although levels of environmental contamination are declining, freshwater fish continue to represent a predominant source of exposure in comparison with air, water (Humphrey 1983b), and other food sources (Food and Drug Administration, 1983). It has been estimated that plant material contributes a high proportion of dietary exposure to PCBs (Hallett, Chapter 24). These estimates may be excessive because lipophilic/hydrophobic compounds are neither absorbed into plant tissues nor distributed throughout the plant, but instead are absorbed onto root and leaf surfaces (Chou et al., 1978; Fries and Marrow, 1978) where they may be removed by normal food preparation and washing procedures. This is verified by the infrequent detection of PCB in processed plant materials such as vegetables (Food and Drug Administration, 1983). In contrast, PCBs associated with the fats of fish or animal flesh cannot be removed by washing and are only partially removed by cooking procedures. Thus, contaminant fish or meat products constitute a dietary source of exposure of such magnitude that even occasional consumption provides a dose sufficient to elevate human serum concentrations above background levels in the human population (Humphrey, 1983b; Kreiss et al., 1982; and Tables 7.3 and 7.4 below). The need to evaluate the potential contribution of contaminated fish to dietary exposure prompted the initiation of the epidemilogical studies described in this chapter.

## 3. EPIDEMIOLOGICAL INVESTIGATIONS

### 3.1. Pilot Study

Funds provided by the Food and Drug Administration in 1973 permitted the initiation of a study in 1973 and 1974 (Humphrey, 1976) to quantitate human PCB exposure from fish in order to obtain a better understanding of fish contaminant dose factors. A cohort of 158 persons, 90 of whom regularly ate fish they caught from Lake Michigan, were enrolled. It was found that this group consumed an average of 32 pounds of fish per year and some ate as much as 262 pounds per year. This average was nearly five times the national per capita fish consumption average commonly used in risk estimates. Regular consumption of these meals represents a significant source of exposure to fat-soluble contaminants such as PCBs and DDT (Table 7.1). This was verified by comparison of PCB serum concentrations with fish consumption patterns for members of the study group (Table 7.2). Those who regularly ate Great Lakes salmon and trout, 24 pounds per year or more, had serum concentrations significantly higher ($P < 0.001$) than persons who seldom or never ate such fish. Thus, serum PCB was positively correlated with the quantity of salmon and lake trout consumed. As a result, persons in the upper range of consumption had median serum concentrations approximately four times greater than those of the unexposed group.

### 3.2. Cohort Study

Evaluation of human exposure to environmental contaminants is generally hindered by limited observations on a small number of subjects long after the event

**Table 7.2 Concentration of PCBs Found in the Serum of the Pilot Study Participants**

| Participants | No. People | Consumption (lb/yr) | Serum PCB (ppb) Range | Serum PCB (ppb) Median |
|---|---|---|---|---|
| Nonfisheaters | 29 | 0 | <5–41 | 15 |
| Occasional fisheaters | 39 | 0–6 | <5–41 | 20 |
| Regular fisheaters | 90 | >24 | 25–366 | 56 |

*Source*. Humphrey (1976).

**Table 7.3 Concentration of PCBs Found in the Serum of the Cohort Study Participants**

| Group | No. Participants | Serum PCB (ppb) Median | Range |
|---|---|---|---|
| Comparison (0–6 lb/yr) | 419 | 6.6 | <3–59.5 |
| Exposed (24–270 lb/yr)[a] | 572 | 21.4 | <3–202.7 |

[a]Median fish consumption: 38.5 lb/yr.
*Source*. Humphrey (1983a).

has occurred. The popularity of catching fish from the Great Lakes provides an opportunity to evaluate this relationship while it is occurring. Funds provided by the U.S. Environmental Protection Agency in 1979–1982 permitted expansion of the earlier pilot study group and the establishment of a study cohort of 991 participants. Persons who regularly ate one or more meals of sport-caught fish per week (>26 pounds/year) were recruited from 11 communities along the Michigan shoreline of Lake Michigan. A group of 572 persons who met this criteria were identified and agreed to participate. A matched comparison group of 419 persons who ate little or no sport-caught fish (<6 pounds/year) were randomly selected from the same communities. Annual fish consumption rates were closely monitored over two seasons by field visits, interviews, and dietary logs. For the exposed group these rates were found to be similar to those stated previously. Serum PCB levels were higher in the exposed group than in the comparison group in this study (Table 7.3), as they were in the earlier pilot study (Table 7.2).

## 4. HUMAN EXPOSURE ASSESSMENT

### 4.1. Serum Concentrations

The analytical laboratory protocol developed for the cohort study specifically quantitated PCBs and nine other organochlorine compounds (Table 7.4). Improvements in the analytical quantitation of PCB, as discussed in Section 4.3, accounts for the apparent difference between median serum levels for participants in the pilot study (Table 7.2) and later cohort study (Table 7.3).

PCBs, DDT, and DDE were detectable in nearly every fisheater (Table 7.5). These compounds were less frequently detected in the comparison group whose range of concentrations (Table 7.4) were lower and median concentrations of 6.6, 10.5, and 9 ppb, respectively, were significantly ($P = 0.0001$) lower than the values of 21.4, 26.4, and 21.7 ppb recorded for fisheaters (Table 7.5). The other seven compounds were detectable less than 50% of the time in fisheaters and seldom in the comparisons (Tables 7.4, 7.5). The median concentration of PCBs found in the comparison group is consistent with values reported in other studies of unexposed populations (Humphrey, 1983b; Kreiss et al., 1982). The significant

**Table 7.4 Concentration Range (ppb) of Contaminants Frequently Found in the Serum of the Cohort Study Participants**

| Contaminant | Fisheaters (572) | Comparison (419) |
|---|---|---|
| PCB 1016 | 5–39 | 5–16 |
| PCB 1260 | 3–203 | 3–53 |
| DDE | 1–425 | 1–80 |
| DDT | 1–513 | 1–91 |
| HCB | 1–3 | 0 |
| β-BHC | 1–5 | 1-4 |
| Oxychlordane | 2–6 | 2 |
| Heptachlor Epoxide | 2–3 | 2–3 |
| trans-Nonachlor | 1–12 | 1–3 |
| Dieldrin | 1–6 | 0 |
| Mirex | 1–3 | 0 |

*Source*. Humphrey (1983a).

**Table 7.5 Frequency and Concentration of Chemical Contaminants Found in the Serum of the 572 Fisheaters Participating in the Cohort Study**

| Contaminant | Detection Limit (ppb) | CV%[a] | % of Cohort above Detection | Concentration (ppb) | |
|---|---|---|---|---|---|
| | | | | Range | Median |
| Total PCB | 3 | 9.71 | 99.1 | <3–202.7 | 21.4 |
| DDE | 1 | 7.31 | 99.5 | <1–425.3 | 21.7 |
| Total DDT | 1 | 7.33 | 99.7 | <1–512.6 | 26.4 |
| HCB | 1 | 20.83 | 3.3 | <1– 3.2 | <1 |
| β-BHC | 1 | — | 18.4 | <1– 4.8 | <1 |
| Oxychlordane | 2 | — | 4.2 | <2– 5.5 | <2 |
| Heptachlor Epoxide | 2 | — | 0.5 | <2– 2.6 | <2 |
| trans-Nonachlor | 1 | — | 45.3 | <1– 12.1 | <1 |
| Dieldrin | 1 | — | 19.6 | <1– 6.0 | <1 |
| Mirex | 2 | 6.99 | 0.2 | <2– 3.1 | <2 |

[a]CV%, coefficient of variation.
*Source*. Humphrey (1983a).

(Hotellings T square, $P = 0.0001$) elevation of PCB concentration in Lake Michigan fisheaters confirms our earlier findings (Humphrey, 1976) and correlates ($P < 0.0001$) with fish consumption. The upper range of serum PCB concentration (203 ppb) found in Lake Michigan fisheaters is higher than those reported for fisheaters tested at other geographic sites (Kreiss, 1985).

### 4.2. Influence of Fish Consumption

Ubiquitous sources of PCBs contribute to a gradual increase in body burden over a lifetime. Evaluation of serum PCB levels in a rural population with no specific source of exposure showed that 65- to 70-year-olds had mean levels three times those of 1- to 19-year-olds (Kreiss et al., 1982). Regular exposure to a known source of PCBs (fish) adds to this baseline level as shown by the elevated serum concentrations reported for the fisheaters in our studies (Tables 7.2–7.4). These additional body burdens are retained for decades. The reported correlation between serum concentrations and both quantity of fish eaten and duration (years) of regular sport-caught fish consumption (Humphrey, 1983a) shows that these are significant factors contributing to the increased serum levels and, accordingly, lifetime body burdens.

The influence of fish consumption on human exposure indices is further illustrated by serum values for 90 of the fisheaters who maintained fish consumption above 24 pounds/year and participated in both the pilot and cohort studies. At the time of the pilot study the median total serum PCB concentration for this group was 56 ppb (Table 7.2), which was the sum of the PCBs quantitated by Aroclor 1254 (42 ppb) and Aroclor 1260 (14 ppb) standards. Eight years later, the same group had a median total serum PCB concentration of 47 ppb, again the sum of Aroclor 1254 (23 ppb) and Aroclor 1260 (24 ppb). Thus, the median concentration of the more highly chlorinated homologs of PCB (those associated with the Aroclor 1260 standard) rose from 14 pbb in 1973 to 24 ppb in 1981. Neither the gradual lifetime rise in serum PCBs nor changes in laboratory methodology (which did not impact PCB quantitation using the Aroclor 1260 standard) account for the observed increase in concentration. The data demonstrate that continued ingestion of PCB contaminated fish by these participants resulted in an accumulation of the more persistent forms of PCB in their bodies.

Another example is provided by examination of the serum PCB and DDT levels for participants who live and catch fish from the northern section of Lake Michigan. These people preferred lake trout as a first choice of fish sought, 47% of the time, in contrast to their neighbors to the south who sought this species only 30% of the time. As stated earlier (Table 7.1) lake trout are more highly contaminated than other species of fish. Accordingly, the median serum concentrations of both PCBs and DDT were greater in the northern participants than in the participants from the middle and southern areas of the lake shore (Humphrey, 1983a).

### 4.3. Predominant Forms of PCBs in Serum

Improvements in analytical chemistry have allowed a more precise evaluation of the PCB mixture in serum. For the cohort study, the laboratory quantitated PCBs using a temperature-programmed gas chromatograph technique (Needham et al., 1981) and the Webb and McCall isomer identification method (Webb and McCall, 1973). This approach improved the separation of PCBs and DDT, avoiding the inflated PCB values that were a problem with the laboratory methodology used in the pilot study, and allowed identification and quantitation of the major elution peaks associated with the Aroclor 1016 and 1260 standards. Median concentrations for each of the major PCB elution peaks found in the serum of the fisheater and comparison participants were calculated (Fig. 7.1). Peaks associated with the more highly chlorinated Aroclor 1260 standard dominate the PCB pattern seen in human

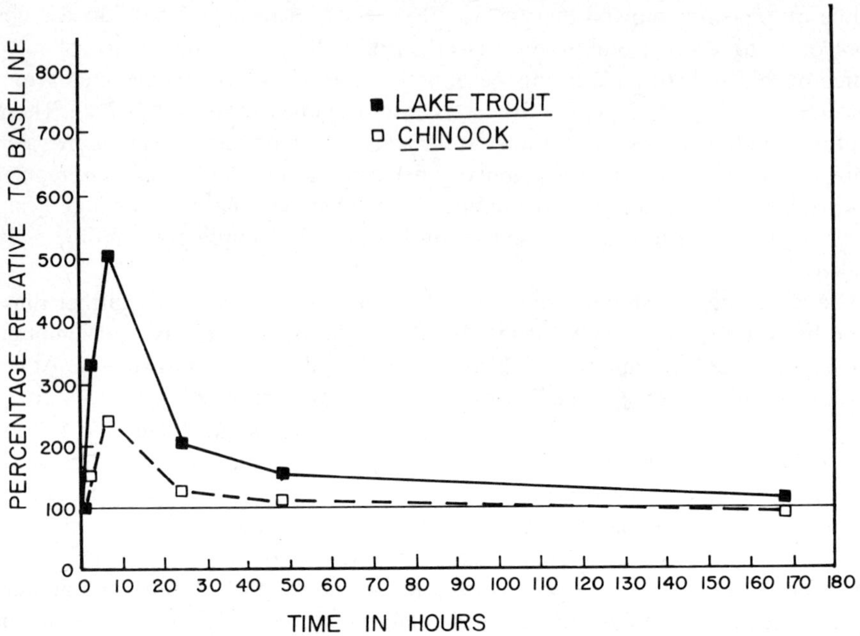


**Figure 7.1.** Median concentration of the chromatographic elution peaks of the PCBs found in human serum. (From Humphrey, 1987, with permission).

serum. Persons exposed to contaminated fish had higher concentrations for each of the peaks and had two peaks (58 and 528) that were not detected in the comparison group. Except for peak 47, the median concentrations for all peaks detected in fisheater serum were significantly (Mann-Whitney *U* test, $P < 0.001$) higher than those for the comparison group.

## 5. TOXICOLOGICAL IMPLICATIONS OF EXPOSURE

### 5.1. PCB Congener Toxicity

Each of the PCB elution peaks contain one or more PCB homologs. Some of the specific congeners that are major components in these elution peaks have been identified (Wolff et al., 1982; Webb and McCall, 1972). The toxicological properties for a number of the PCB congeners or structurally similar brominated congeners have been evaluated and reported (Safe et al., 1985; Safe, 1984; Aust et al., 1982). Seven of the elution peaks constitute the major proportion of the total PCBs quantified in human serum (Table 7.6). Each of these peaks has at least one identified congener component for which the toxicological properties have been evaluated. As shown, the more highly chlorinated homologs constitute the major proportion of the PCB mixture found in human serum and several of these are congeners that possess cytotoxic, enzyme-induction, or tumor-promotion toxico-

**Table 7.6 Toxicity of Several PCB Congeners Associated with Prominent Aroclor 1260 Peaks Found in Human Sera**

| Peak No. | Congener Assignment | Reported Toxicity[a] |
|---|---|---|
| 174 | 2 3 4 2′ 4′ 5′ | Cytotoxic |
| 146 | 2 4 5 2′ 4′ 5′ | Tumor promotor, PB inducer[b] |
| 125 | 2 4 5 3′ 4′ | Cytotoxic and mixed inducer[c] |
| 203 | 2 3 4 2′ 3′ 4′ | Mixed inducer[c] |
| 232, 244 | 2 3 4 5 3′ 4′ | Cytotoxic and mixed inducer[c] |

[a]From Safe (1984), Safe et al. (1985), Aust et al. (1982).
[b]Phenobarbital-type enzyme induction.
[c]3-Methylcholanthrene- and phenobarbital-type enzyme induction.

logical characteristics. The transfer of these toxic forms of PCB from the environment to humans is enhanced through consumption of contaminated fish.

## 5.2. Post Meal Concentrations

Regular consumption of contaminated fish not only increases the baseline body burden of PCB (Tables 7.2–7.4), but each meal also causes a temporary rise in the blood serum concentration (Humphrey, 1987). Following a meal, serum PCB levels immediately rise, peak within 10 hours and gradually decline to the premeal baseline level over the succeeding 7 days (Fig. 7.2). The serum concentration

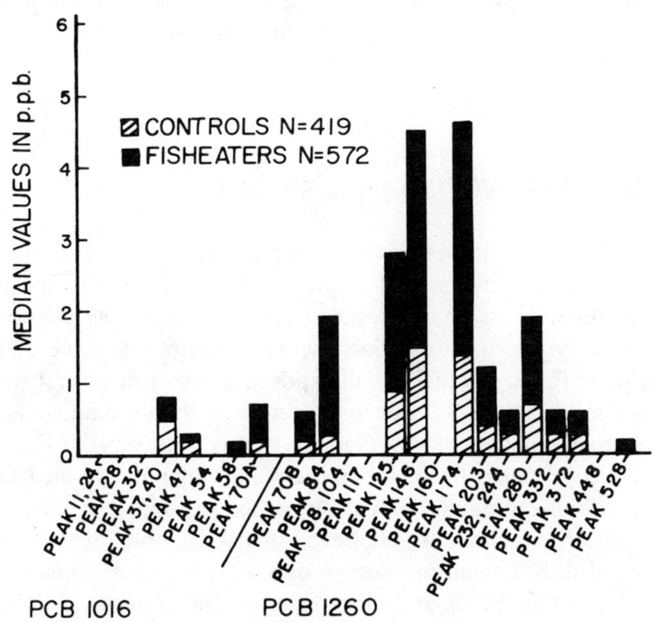


**Figure 7.2.** Percent change in baseline serum PCB levels following a fish meal. (From Humphrey 1987, with permission).

spikes, which are directly related to the concentration of the contaminant in the cooked meal (trout vs. salmon), represented a 230 to 500% increase in baseline values. When one considers that serum lipids that contain the lipophilic compounds already carry median concentrations of 4 ppm on a fat basis (Humphrey, 1983a), these spikes in concentration following a meal represent sizable temporary concentration increases. The long-term toxicological significance of this for humans is unknown.

### 5.3. Body Burden

Although exposure assessment has primarily been on the basis of serum concentration, related work (Humphrey, 1983b; Kreiss, 1985; Eyster et al., 1983) has shown that other body tissues, especially adipose, breast milk, and cord blood, contain the halogenated contaminants. Personal baseline PCB levels and concentration increases following a meal become especially important in females because the female body burden can become multigenerational as lipid-soluble contaminants such as PCBs are passed to the unborn and newborn offspring with unknown consequences. It should be noted that collaborative work intended to evaluate this aspect in the children of cohort mothers is underway (Jacobson and Jacobson, Chapter 18). The reported slow decrease in halogenated biphenyl concentrations in exposed humans (Kreiss et al., 1982) demonstrates the persistence of these contaminants in the human body. The accumulative effect of multiple doses (meals) creates a situation where the contaminant has a long period of time within the body to express its toxic effect alone or in combination with other contaminants that may be encountered. Persistence and latency are important factors because some of the PCB congeners believed to be associated with the dominant PCB homologs received from contaminated fish are enzyme inducers and/or tumor promoters (Table 7.6).

## 6. HEALTH IMPLICATIONS OF EXPOSURE

Whether or not human exposure to PCBs and other compounds at the concentrations found in the food they eat will result in disease conditions or latent disease such as cancer remains an open question. At the time of the cohort study, the exposed fisheater group did not report medical histories that were significantly different in disease frequency than the unexposed comparison participants from the same communities (Table 7.7). The persistence of these compounds allows the continuous internal exposure of target organs that is necessary for the disease process if it is to occur. The mix of other chemical contaminants in fish and other foods or the environment may provide the basis for an interaction with stored PCBs and resultant tumor induction. It is believed that the cohorts of humans identified and characterized in Michigan provide an opportunity to determine the long-term outcome of exposure to PCBs and other aquatic contaminants. Exposure assessment indicates that humans are final biological receptors for some of these chemical contaminants in the aquatic ecosystem.

**Table 7.7 Percentage of Medical Complaints Reported by Participants in the Cohort Study**

| Condition | Fisheaters | Comparisons |
|---|---|---|
| Diabetes | 4.6 | 5.9 |
| Heart attack | 6.8 | 3.9 |
| Hypertension | 21.1 | 18.4 |
| Kidney problems | 2.2 | 1.2 |
| Cancer | 5.5 | 4.7 |
| Liver problems | 2.4 | 2.0 |
| Other conditions | 11.5 | 14.3 |

*Source*. Humphrey (1983a).

## 7. DIRECTION OF FUTURE RESEARCH

The human population represents a final link in the aquatic ecosystem food chain. The data presented in this chapter clearly demonstrate that humans are receptors for chemical contaminants in this ecosystem. Speculation about the significance of such exposure, its duration, and the toxicological outcome of accumulated body burdens fuels public interest in the debate between risk estimation and risk reality. In the absence of clear-cut scientific evidence concerning human outcomes, perceived risk drives regulatory program options toward conservative assessments in order to provide some margin of safety for the public. These decisions and the state of scientific knowledge could greatly benefit from contemporary research on human exposure outcomes from contact with contaminants found in the Great Lakes aquatic ecosystem.

In Michigan, we have identified several population groups with unique exposure to chemical contaminants. The data set generated by the fisheater cohort study described in this chapter is the largest of its kind in the world and in combination with our other concurrent studies provides a unique opportunity to further evaluate the impact of a single toxin or combination of toxins in a human population. The importance of continuing and expanding this type of research has been underscored on several occasions by the International Joint Commission as well as by professionals in the fields of toxicology, epidemiology, and public health.

Existing human population studies, such as the fisheater cohort, provide the foundation for development of investigations that could test hypotheses in the following research areas:

1. *Body Burden*. Changes in accumulated body burden over time, partitioning of contaminants within human tissues and organs, release of contaminants from fat stores into the bloodsteam, and excretion and removal of contaminants from the body.
2. *Multigenerational Effects*. Evaluation of children born of or breastfed by exposed females; including cognitive development, behavioral and learning development, and sexual maturation.

3. *Metabolic Impact.* Application of new and sensitive techniques to measure metabolic damage, enzyme induction, and the significance of these on disease states and physiological function.
4. *Immunological Impact.* Application of new and sensitive techniques to measure altered immunocompetence; correlation of this with exposure and physiological function and development of disease.
5. *Disease Conditions.* Observation of life events, mortality, and the frequency of disease development and severity in exposed populations in comparison with general population rates. This includes hypertension; chronic disorders, such as arthritis or decrease in vital lung capacity; and development of diseases with long latency, such as cancer.
6. *Disease Predictors.* Application of emerging technology such as measurement of DNA adducts to correlate exposure with the development of DNA lesions, which are predictive of cancer, or other illness outcomes.

The existence of toxic substances in the Great Lakes, which are used by hundreds of thousands of persons for recreational fishing, creates an opportunity and an obligation to scientifically address questions concerning the health of this ecosystem including its human population. The bottom line in the minds of many citizens is—What happens to fisheaters as they grow old? The opportunity to answer that question is before us today.

## ACKNOWLEDGMENTS

Support for this work was provided by Contract 223-73-2209 from the Food and Drug Administration and by Cooperative Agreement Grant CR-807192 from the U.S. Environmental Protection Agency. The author acknowledges the efforts of colleagues Marvin Budd, Harold Price, and Caralee Roberts of the Michigan Department of Public Health and the professional staff at the EPA Large Lakes Research Station, Grosse Ile, Michigan.

## REFERENCES

Aust, S. D., Dannan G. A., Sleight, S. D., Fraker, P. J., Ringer, R. K., and Polin, D. (1982). Toxicology of polybrominated biphenyls. In M. A. Q. Khan (Ed.) *Toxicology of Halogenated Hydrocarbons,* Pergamon, Oxford, pp. 73–96.

Chou, S. F., Jacobs, L. W., Penner, D., and Tiedje, J. M. (1978). Absence of plant uptake and translocation of polybrominated biphenyls (PBBs). *Environ. Health Perspect.* **23,** 9–12.

Connor, M. S. (1984). Comparison of the carcinogenic risks from fish vs groundwater contamination by organic compounds. *Environ. Sci. Technol.* **18,** 628–631.

Eyster, J. T., Humphrey, H. E. B., and Kimbrough, R. D. (1983). Partitioning of polybrominated biphenyls in serum, adipose tissue, breast milk, placenta, cord blood, biliary fluid and feces. *Arch. Environ. Health* **38**(1), 47–53.

Food and Drug Administration (1983). *FDA/HOAC Residue Monitoring Data—Pesticide Residue Levels in Food in the USA July 1, 1969–June 30, 1976.* FDA, Washington, DC.

Fries, G. F., and Marrow, G. S. (1978). Uptake of chlorobiphenyls by the soybean plants. Abstr. Pop. 175th Am. Chem. Soc. Natl. Meet.

Hallett, D. J. (1988). Ecosystem management of persistent toxic chemicals—back to basics. In M. S. Evans (ed.) *Toxic Contaminants and Ecosystem Health: A Great Lakes Focus*. Wiley, New York.

Humphrey, H. E. B. (1976). *Evaluation of Changes of the Level of Polychlorinated Biphenyls in Human Tissues,* Final Report to the Food and Drug Administration, Contract No. 223-73-2209. FDA, Washington, DC.

Humphrey, H. E. B. (1983a). *Evaluation of Humans Exposed to Water Borne Chemicals in the Great Lakes,* Final Report to the Environmental Protection Agency, Cooperative Agreement CR-807192. USEPA, Washington, DC.

Humphrey, H. E. B. (1983b). Population studies of PCB's in Michigan residents. In F. M. D'Itri and M. A. Kamrin (Eds.), *PCB's Human and Environmental Hazards,* Ann Arbor Sci. Publ., Ann Arbor, MI, pp. 299–310.

Humphrey, H. E. B. (1987). The human population—An ultimate receptor for aquatic contaminants. *Hydrobiol.* **149,** 75–80.

International Joint Commission (1981). *Committee on the Assessment of Human Health Effects of Great Lakes Water Quality, 1981, Annu. Rep.* IJC, Windsor, Ontario.

International Joint Commission (1983). *Report to the Great Lakes Water Quality Board, An Inventory of Chemical Substances Identified in the Great Lakes Ecosystem,* Vols. 1–6. IJC, Windsor, Ontario.

Jacobson, J. L., and Jacobson, S. W. (1988). New methologies for assessing the effects of prenatal toxic exposure on cognitive functioning in humans. In M. S. Evans (ed.). *Toxic Contaminants and Ecosystem Health: A Great Lakes Focus*. Wiley, New York.

Kreiss, K. (1985). Studies on populations exposed to polychlorinated biphenyls. *Environ. Health Perspect.* **60,** 193–199.

Kreiss, K., Roberts, C., and Humphrey, H. E. B. (1982). Serial PBB levels, PCB levels and clinical chemistries in Michigan's PBB cohort. *Arch. Environ. Health* **37**(3), 141–146.

Needham, L. L., Burse V. W., and Price, H. A. (1981). Temperature-programmed gas chromatographic determination of polychlorinated and polybrominated biphenyls in serum. *J. Assoc. Off. Anal. Chem.* **64**(5), 1131–1137.

Safe, S. (1984). Polychlorinated biphenyls (PCB's) and polybrominated biphenyls (PBB's): Biochemistry, toxicology and mechanism of action. *CRC Crit. Rev. Toxicol.* **13**(4), 319–395.

Safe, S., Bandiera, S., Sawyer, T., Robertson, L., Safe, L., Parkinson, A., Thomas, P. E., Ryan, D., Reik, L. M., Levin, W., Denomme, M. A., and Fujita, T. (1985). PCB's: Structure-function relationships and mechanisms of action. *Environ. Health Perspect.* **60,** 47–56.

Swain, W. R. (1983). An overview of the scientific basis for concern with polychlorinated biphenyls in the Great Lakes. In F. M. D'Itri and M. A. Kamrin (Eds.), *PCB's: Human and Environmental Hazards.* Ann Arbor Sci. Publ., Ann Arbor, MI, pp. 11–48.

Webb, R. G., and McCall, A. C. (1972). Identities of polychlorinated biphenyl isomers in aroclor. *J. Assoc. Off. Anal. Chem.* **55,** 746–752.

Webb, R. G., and McCall, A. G. (1973). Quantitative PCB standards for electron capture gas chromatography. *J. Chromatogr. Sci.* **11,** 366–373.

Wolff, M. S., Thornton, J., Fischbein, A., Lillis, R., and Selikoff, I. J. (1982). Disposition of polychlorinated biphenyl congeners in occupationally exposed persons. *Toxicol. Appl. Pharmacol.* **62,** 294–306.

# PART TWO

# APPROACHES FOR INVESTIGATING TOXIC STRESS

# 8

# A REVIEW OF SELECTED CLINICAL INDICATORS OF STRESS-INDUCED CHANGES IN AQUATIC ORGANISMS

***John P. Giesy, Donald J. Versteeg, and Robert L. Graney***

*Pesticide Research Center, Department of Fisheries and Wildlife, Center for Environmental Toxicology, Michigan State University, East Lansing, Michigan 48824*

---

## 1. INTRODUCTION

The effects of chemical and physical alterations of the environment have traditionally been monitored by observing changes at the population or community levels of biotic organization (Gaufin, 1973; Lugo, 1978; Ward et al., 1978). This type of monitoring is limited because adverse effects in the environment have already occurred. While responses to stressors are observed at the ecosystem, community, and population levels of organization, the effects are manifested at the organismal level by causing death or impairing function. More specifically, chemical toxicants exert their effects by interacting with receptors, which are biomolecules. For predictive purposes, the tolerance of organisms to toxicants has been reported as acute lethality or effects on reproduction or behavior. More recently, as the discipline of aquatic toxicology has matured and environmental regulation has become more complex, responses to toxicants at the suborganismal levels of organization, such as the physiological, histological, and biochemical levels, which are referred to generally as "clinical" measures, have been considered as viable monitors of responses of organisms to stressors (Wedemeyer and Yasutake, 1977; Cairns et al., 1984; Barrett and Rosenberg, 1981; Pickering, 1981; Goldstein, 1975).

### 1.1. Stressors

The term stress is often used in relation to the effects of toxicants on individuals, populations, or ecosystems. Unfortunately, this term has been used in so many different toxicological contexts that it has lost any critical connotation. Stressors can be defined as any force that pushes the functioning of a critical subsystem beyond its ability to restore homeostasis through ordinary, nonemergency adjustment processes (Lugo, 1978). In our discussions we distinguish between stressors and stress in the manner of Selye (1956) and Fitch and Johnson (1977). A stressor is the causative agent, condition, or situation that causes strain or deviation from the normal homeostatic range and results in an organism mobilizing its resources and increasing its energy expenditure. Stress is the organism's response to the stressor.

Tolerance to a stressor is an organism's ability to operate normally within its homeostatic range without resisting the stressor in a demonstrable way. Resistance is an adaptive response to the stressor where energy is used to do work to resist the effects of the stressor. This could be the synthesis of replacement or alternative functional or structural proteins or substrates.

### 1.2. General Adaptation Syndrome

Selye (1976) described the response of an individual vertebrate organism to stressors in terms of a succession of physiological and biochemical reactions. These responses were observed to occur in a variety of organisms exposed to a variety of stressors, thus Selye referred to the responses as the "general adaptation syndrome" (GAS). The GAS is divided into three main parts: the alarm stage, the resistance stage, and the exhaustion stage.

The alarm stage is characterized by thc "flight or fight" response, during which the stressor elicits a response to mobilize the organism's defenses (Fig. 8.1). Epinephrine released into the blood system from the adrenal glands increases the activity of the sympathetic portion of the autonomic nervous system, which results in a wide range of physiological and biochemical changes (Pickering, 1981; Cairns et al., 1984; Fig. 8.1). The resistance stage offers general long-term protection against stressors. The final stage, exhaustion, occurs after long-term, continuous exposure to a stressor. In the exhaustion phase, energy stores become depleted, the ability of the organism to continue to resist the stressor is reduced, and death of the stressed organism follows rapidly.

A model of an individual's response to a stressor can be constructed based on the above discussion. An undisturbed individual operates within a normal homeostatic range (Fig. 8.2). Also, there is a wider range of conditions for any given parameter within which an organism can continue to function normally for some finite period of time. Organisms can adapt to even greater deviations from the normal range by alteration of function or activity. Such alterations will, in time, affect other systems. The dimensions of homeostatic ranges vary within and among species, both spatially and temporally. The response parameter can be either an enzyme activity or a substrate concentration and can increase or decrease in response to a stressor. The goal of clinical measures of toxicant-induced stress is to relate short-term changes in biochemical and physiological parameters to imminent adverse effects on individuals and thus populations and communities.

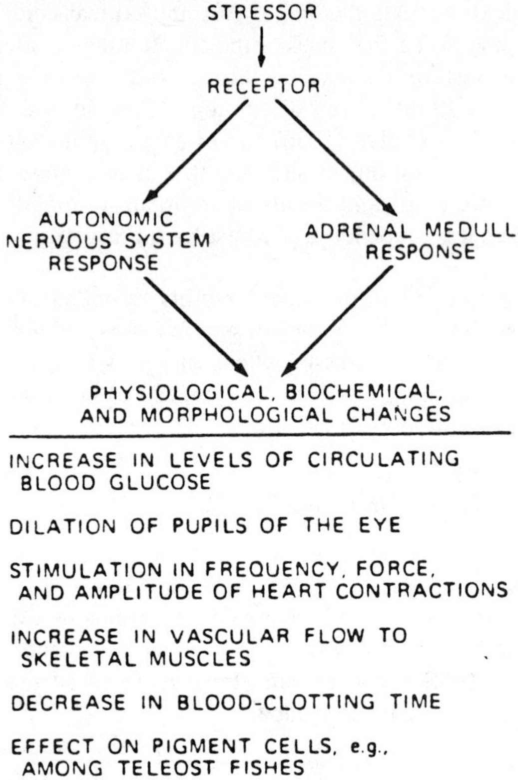


**Figure 8.1.** Alarm stage of the general adaptation syndrome. Reprinted from Esch and Hazen (1978), with permission.

The concept of a general monitor of stressor-induced responses in organisms is important because one limitation of clinical indicators is their specificity. That is, advanced knowledge of the potential stressor is necessary before an appropriate clinical measure can be devised. Thus, while individual clinical indicators may be sensitive and measurable antecedent to population-level effects, the probability of choosing the appropriate system to monitor may be small. It is for this reason that nonspecific indicators of stress have been advocated.

## 1.3. Advantages of Clinical Measures

Clinical indicators of the effects of chemical and physical stressors are attractive alternatives to more traditional measures for several reasons. Under field conditions, organisms are exposed to a multiplicity of both chemical and physical stressors, against a backdrop of naturally occurring seasonal fluctuations and ontogenetic processes, which in and of themselves are potentially stressful to the organism. One goal of biochemical ecotoxicology is to investigate general inte-

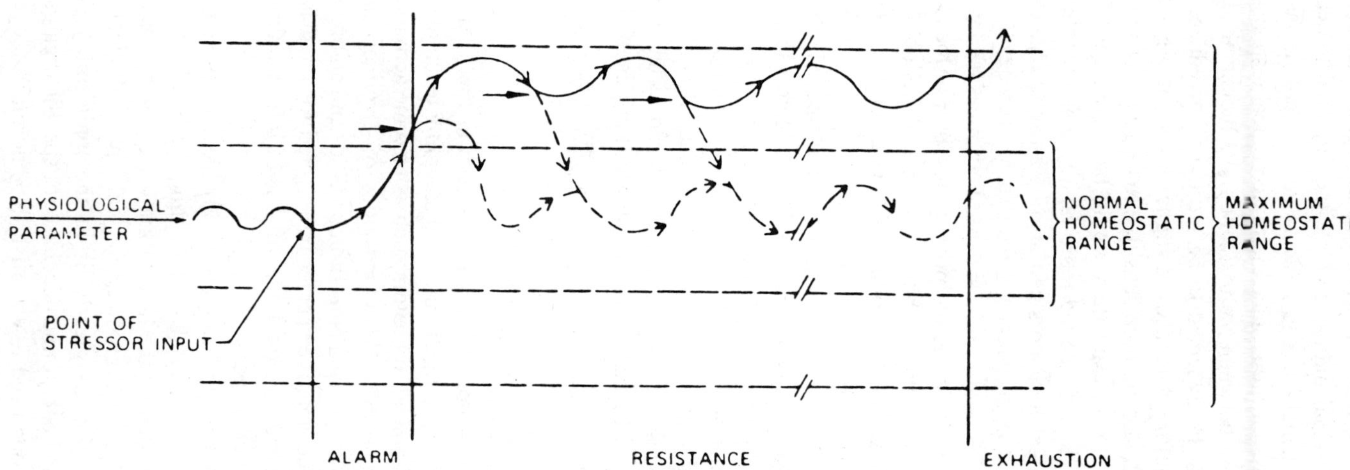


**Figure 8.2.** Consequences of stressor on an individual organism. Dashed line with arrows indicates withdrawal of stressor before point of no return. The organism is able to reestablish or readjust homeostatic equilibrium. Redrawn from Esch and Hazen (1978), with permission.

grative measures of stress, which will allow one to evaluate the subsequent strain elicited in individuals. Clinical indicators have the potential to act as integrative measures at the suborganismal level to indicate adverse conditions antecedent to population-level effects. In addition, since the toxicological response to a chemical is caused by the interaction between the toxicant and a biochemical receptor, biochemical responses would be expected to be the most immediate. That is, these responses would occur before responses at higher levels of organization. Therefore, clinical indicators should be more sensitive to toxicants. They should change at smaller concentrations of toxicant than would be required to elicit organism-level responses and they should respond more rapidly than the whole organism.

The specificity of clinical indicators can also be a useful characteristic. While nonspecific clinical indicators of adverse effects have been developed, most clinical indicators of toxic effects are more specific than measures at more complex levels of organization. In many cases the primary receptor or locus of toxic action can be identified. Examples include inhibition of acetylcholinesterase activity in brain tissue by organophosphate compounds (Coppage et al., 1975) and $\alpha$-aminolevulinic acid dehydratase activity by lead (Hodson et al., 1984).

Clinical measures of toxicant effects can be useful under both laboratory and field conditions. Previously, we noted that clinical indicators can be useful under field conditions because the responses are manifested at the biochemical level of organization before population-level effects occur. This is also an important attribute in laboratory testing schemes because of the costs of conducting bioassays. If shorter-term endpoints, rather than growth or reduced fecundity, can be developed, screening tests could be conducted for shorter periods of time and be less costly.

## 1.4. Limitations of Clinical Indicators

Within a single species a number of parameters such as diet, genetic strain, sex, age, reproductive state, water temperature, and sampling techniques can affect the "normal" range of biochemical indicators such as enzyme activities and substrate concentrations (Barnhart, 1969). Also, activities of enzymes can vary greatly among tissues within the same organism (Bouck, 1980) and among species. Therefore, we suggest that representative species be selected for intensive study in several different habitats. Because of the large volume of information presently available, rainbow trout are such a species.

Mayer (1983) and Mehrle and Mayer (1980) have discussed the paradox of clinical tests in aquatic toxicology. Understanding the biochemical toxicology of aquatic organisms is impeded by the lack of knowledge of the basic biochemistry of the organisms, including background "normal" physiological ranges (Mehrle and Mayer, 1980). This paradox greatly limits the applicability of many clinical tests. Developing a strong background data base for specific organisms is essential; however, not only does one need to establish the "normal" range of a particular indicator, but one also must establish the statistical confidence in determining that the value for a parameter is out of the normal range. Confidence, expressed as statistical power, is dependent on the variance of the parameter values, as well as the sample size. We advocate determining the standard deviation of the parameter values and power analysis to establish the required number of samples to demonstrate a statistically significant effect at a given power (Type I and Type II errors) (Giesy and Allred, 1985; Wedemeyer and Yasutake, 1977; Bayne et al., 1981). Few studies provide information on the range of values for a parameter, let alone allow one to calculate a standard deviation. However, Wedemeyer and Nelson (1975) provide the basis for calculating both Gaussian (parametric) and nonparametric tolerance intervals to estimate ''normal'' ranges for physiological and biochemical parameters in aquatic organisms.

It is extremely important that the response of clinical indicators be related to ecologically relevant responses of individuals and populations (Mehrle and Mayer, 1980). Clinical measures of pathological effects are successful in humans because a large body of correlative information exists. Because this type of information is lacking for aquatic organisms, biochemical indicators will never be as useful in assessing toxicant-induced effects in aquatic organisms as they are for mammalian species.

The remainder of this manuscript presents a cursory review of the rationale and potential applications of a variety of clinical indicators of stress. In preparing this chapter we were unable to cover all of the potential indicators but rather have selected a few general areas that we feel have been much studied in the past and have the greatest potential for future development. More detailed discussions of many of these indicators can be found in books by Bayne et al. (1985).

## 2. HORMONES

### 2.1. Rationale

To maintain homeostasis, aquatic organisms must compensate for the physiological and biochemical alterations incurred by exposure to a stressor. Coordination of the physiological and biochemical processes necessary to regain homeostasis involves the coordination of neuronal and hormonal elements. Thus, an assessment of the hormonal status of an organism can yield information on the intensity of stress impinging on an organism. This approach has been used extensively with a few well-understood hormones in fish. However, little research on the effects of stressors on the hormonal systems of invertebrates has been performed due to the much smaller body of information on invertebrate endocrinology.

Endocrine responses to stress may be (1) primary response, aimed to increase the survival of the organism, (2) a secondary or tertiary response, an attempt to maintain biochemical homeostasis within the organism, (3) an effect on plasma clearance of toxicant due to metabolism and tissue uptake, or (4) a toxic effect on the endocrine organ that causes altered hormone synthesis or release.

## 2.2. Fish

### *2.2.1. Hypothalamus–Pituitary Interrenal (HPI) Axis*

The HPI axis is stimulated by a wide variety of stressors, including physical stressors such as confinement and handling of toxicants (Donaldson, 1981; Wedemeyer and Yasutake, 1977). The HPI axis is rapidly stimulated by exposures to physical stressors, which leads to increased concentrations of cortisol in plasma.

Exposure of fish to toxicants causes alterations in plasma corticosteroids and interrenal ascorbic acid concentrations. The magnitude and duration of these effects are dependent on the type of toxicant, toxicant concentration, species, and acclimation temperature (Wedemeyer, 1971; Swift, 1981, 1982; Thomas et al., 1981a; Davis and Parker, 1983; Gluth and Hanke, 1984). In general, plasma cortisol concentrations in fish increase rapidly following onset of exposure and return to pre-exposure concentrations within 6 to 12 hours (Thomas et al., 1981b). In addition to stressors, numerous biotic variables affect the activity of the HPI axis in fish. These factors must be considered in utilizing an HPI axis assay in the laboratory and will make field studies on plasma corticosteroid concentrations difficult to interpret.

Quantification of the HPI axis and chromaffin tissue activity are useful indicators of stress in fish; however, it has not yet been demonstrated that toxicant exposures cause increased catecholamine concentrations in fish plasma. Indices of HPI axis and chromaffin cell activity are not expected to be especially useful long-term indicators of stress because (1) concentrations of corticosteroids and catecholamines in plasma occur for only a short time after the initial stress; (2) stressor-mediated effects on the HPI axis and chromaffin cell activity have not been directly related to ecologically relevant effects on growth, reproduction, or mortality. Thus, we feel their use in establishing safe levels of stress is currently limited and should be confined to experiments in which blood is sampled shortly after a stress such as handling, transportation, and stocking of fish. In fact, these acute stressors will probably mask any chronic toxicant-induced effects. We do not recommend these techniques be utilized to assess the effects of long-term, toxicant-induced effects in fish.

### *2.2.2. Catecholamines*

A variety of physical stressors and hypoxia alter concentrations of catecholamines in both plasma and chromaffin tissue. Unfortunately, the effects of chemical stressors on these hormones have not been investigated. Much of the work that has been conducted on the effects of stress on catecholamines in fish has been reviewed by Mazeaud and Mazeaud (1981). In teleosts subjected to a variety of stressors, concentrations of both adrenaline and noradrenaline in plasma increase. Forced swimming, hypoxia, cannulation, and handling all result in rapid and usually transient increases in plasma catecholamine concentrations (Nakano and Tomlinson, 1967; Butler et al., 1978; Wahlqvist and Nilsson, 1980; Mazeaud and Mazeaud, 1981). During physical stress, the time course of the increase of catecholamine concentrations in plasma and the specific hormone exhibiting the greatest increase is species-specific. There is little information on the effects of biotic factors on plasma catecholamine concentrations in fish.

### 2.2.3. *Other Hormones*

The effects of stressors on concentrations of hormones other than the corticosteroids and catecholamines in plasma have not been well studied. However, effects would be expected since hormones are involved in maintaining homeostasis and chemical stressors are known to perturb this homeostasis. Thyroid hormones are important in regulating intermediary metabolism in fish (Plisetskaya et al., 1983) and appear to play a role in the primary stress response; however, additional research into the function of and effects of biotic factors on these hormones is needed.

Prolactin is produced and released by the pituitary and is believed to be under inhibitory control by a hypothalamic hormone (Turner and Bagnara, 1976). It is generally believed that prolactin affects osmoregulation in fish and has been shown to affect branchial water permeability and ion flux at the gill and kidney (Folmar and Dickhoff, 1979; Flick et al., 1984; Wendelaar Bonga et al., 1984). Insulin is produced in the pancreas and secreted to promote cellular glucose uptake and utilization. Insulin causes hypoglycemia, decreased hepatic glycogen, decreased serum fatty acids, and decreased plasma amino acids (Thorpe and Ince, 1974; Fontaine, 1975; Wendelaar Bonga et al., 1984). Plasma insulin concentrations may prove useful in understanding the response of plasma glucose and hyperglycemic hormones during stress in fish.

Thyroid hormones, prolactin, insulin, and other hormones may be excellent biochemical tools for assessing a fish's response to stress. However, at present, insufficient information on the function of these hormones and their response to stress is available. To date no research has been performed to relate ecologically relevant effects on growth, reproduction, or survival to changes in hormones. We recommend further research be conducted on the effects of stressors on these hormones and that quantification in blood be considered for future use to better understand the mode of toxic action of persistent environmental contaminants.

In contrast to the effects of stressors on other hormones, concentrations of circulating sex-related steroids, such as the androgens and estrogens, are altered directly by toxicant effects on hormone synthesis or steroid metabolism. Altered concentrations of the sex steroids in plasma are believed to cause reproductive impairment. Determination of sex steroid concentrations in plasma and metabolism of these hormones may be of use in assessing the effects of stressors on reproduction in fish. Unfortunately, few studies have linked effects on steroid production, metabolism, and concentrations in blood with altered reproduction (Sangalang and Freeman, 1974). Thus, a determination of effects of stressors on reproduction has not been possible. We recommend research be conducted to link effects of stressors on the fertilization process, gamete production, and embryo survival with sex hormone dynamics in fish. At that time it will be possible to determine how stressor-induced effects on reproductive hormones should be interpreted. Sex hormone dynamics may prove useful for assessing reproduction effects of stressors; however, their general use as indicators of stress is not recommended or biologically supportable.

### 2.3. Invertebrates

Information on invertebrate endocrinology is available (Gaillard and Boer, 1978) and some studies have been conducted on the effects of chemical stressors on the endocrine system of aquatic invertebrates (Vernberg et al., 1982). However, the number of studies is much fewer than that conducted on the effects of chemical stressors on fishes. This fact, along with the limited understanding of the regulatory processes of these systems, severely limits the use of changes in the hormonal milieu of invertebrates as an in situ indicator of toxicant-induced stress. In addition, sensitive analytical techniques for measuring specific hormones on a routine basis generally are not available. Also, as with many of the fish hormones, the sensitivity of the response to physical (handling) and normal chemical stresses, such as salinity changes, may impede one's ability to identify chemical-induced effects. Given that many of these limitations can be overcome by basic research on invertebrate endocrinology and the development of a strong data base on natural variations in hormone titers, endocrine alterations may, in the future, have applicability as a monitoring tool.

## 3. ENERGETICS

### 3.1. Rationale

There are a multitude of responses to toxicant-induced stress that are related to the energetics of organisms. These responses can be directly related to the general adaptation syndrome (GAS) concept. The basis for the use of energetics is that energy is required to maintain homeostasis. For an organism to resist a stressor, it must perform work, which requires energy. This energy can be used for movement, to transport ions or biomolecules, to synthesize structural or functional proteins to replace those that have been damaged or biotransformed, and to metabolize toxicants. Acute responses, which are generally hormonally controlled, are considered to be the initial response to the stressor and often involve the increase of energy-related substrates in the plasma/hemolymph. These changes are transient, however, and will only occur under fairly severe stress (Mazeaud and Mazeaud, 1981). During gradual, sublethal stress, more subtle changes occur and initiate compensatory physiological adjustments such that changes in energy metabolism may be required to maintain homeostasis. Growth and maintenance metabolism demand the majority of an organism's energy expenditure, so any increased maintenance requirement will result in reduced growth. This concept has been developed into the scope for growth and scope for reproduction (Warren, 1971). The increased energy demand associated with stress may reduce particular energy stores such as glycogen and lipids and/or alter their distribution among tissues. Scope for growth is generally considered to be a nonspecific indicator responsive to most any kind of stress, natural or synthetic, that can be applied in both field-monitoring and laboratory hazard assessment schemes. Also, since the reproductive potential of many invertebrate species is directly related to the size of the organism, reduced growth can have direct effects on fecundity and potentially the population. Therefore, a comparative assessment of the energy status of an organism can be indicative of overall condition.

### 3.2. Glucose

Blood glucose concentrations of fish can be decreased by starvation or increased in response to acute and chronic stressors. Glucose concentrations are one of the most widely recognized and consistent responses to stressors. Acute exposure of fish to a variety of toxicants can alter the serum glucose profile (Leach and Taylor, 1982; Swift, 1982; Thomas, 1978). Short-term exposure, of as little as a few minutes to several hours, to a physical stressor causes an increase in concentrations of glucose in the plasma of fish (Hattingh, 1976; Morata et al., 1982; Carmichael et al., 1983; Nieminen et al., 1983).

Compared to the number of studies done on fish, few have been conducted on toxicant-induced effects on glucose in invertebrates. Hyperglycemic responses have been observed in freshwater crabs exposed to DDT (Fingerman et al., 1981), sumithion (Bhagyalakshmi et al., 1983b), and BHC (Sreenivasula Reddy et al., 1983), and in blue crabs and polychaete worms exposed to pentachlorophenol (Coglianese and Neff, 1982; Thomas et al., 1981b).

### 3.3. Glycogen

Glycogen is a branched polymer of glucose residues and represents the readily mobilizable storage form of glucose for most organisms. Both increases and decreases in glycogenolysis can occur due to toxicant-induced stress and can result in either depletion or accumulation of glycogen. In the majority of cases, the increased energy demand associated with stress causes a depletion of glycogen reserves (Bhagyalakshmi et al., 1984; Thomas et al., 1981b). Glycogen concentrations in liver, skeletal muscle, and heart tissue decrease concomitantly with increases of glucose concentrations in plasma (Nakano and Tomlinson, 1967; Miles et al., 1974; Nieminen et al., 1983). Increased concentrations of the catecholamine and corticosteroid hormones during response to stressors mobilize these energy stores (Nakano and Tomlinson, 1967; Pickering et al., 1982). As with blood glucose concentrations, concentrations of glycogen vary considerably among tissues and organisms (Prosser, 1973), and with diet (Bayne, 1973), reproductive conditions, and season (Newell and Bayne, 1980; Gabbott and Bayne, 1973). Changes are not as transitory or sensitive to nontoxicant stress as glucose concentrations in plasma and, therefore, are more promising as a monitor of stress.

### 3.4. Lipids

Alterations in the lipid and associated fatty acid composition of organisms exposed to a variety of stressors have been studied extensively in both vertebrates and invertebrates. Reasons for observed changes are not always apparent and a variety of mechanisms exist by which the lipid composition of an organism can be affected by toxicant-induced stress (Leatherland and Sonstegard, 1984). Increased energy demand may result in the mobilization of lipids from storage depots such that plasma transport lipoproteins and associated triglycerides are elevated. Direct toxicant interference with lipid synthesis and degradation and/or lipid transport processes could also alter the lipid profiles of the hemolymph. However, interpretations of the toxicological significance of changes in hemolymph lipids can be extremely difficult. Given the wide array of toxicants potentially present in a given ecosystem and the numerous potential target sites for interference with lipid metabolism, the monitoring of hemolymph lipid profiles in invertebrates would provide, at best, a general, nonspecific indicator of toxicant exposure.

The quantitative and qualitative lipid composition of specific tissues may also be altered in organisms under stress. Lipids provide an essential, readily available energy source for many organisms and have been shown to be mobilized in organisms exposed to toxicants (Lee et al., 1981; Rao et al., 1984). This energy pool may be mobilized for the purpose of maintaining homeostatis during toxicant-induced stress; however, it may also vary depending upon developmental stage, diet, reproductive condition, and molt status of the organism (Voogt, 1983; Clarke, 1977; O'Connor and Gilbert, 1969).

Monitoring changes in lipid content represents a nonspecific indicator of the relative health of an organism but only with respect to appropriate control organisms. Since lipid deposition in eggs can represent a major energy source for the developing embryo, any interference with lipid metabolism in the adult can directly influence the survival of the next generation. In controlled laboratory experiments, the lipid content of larvae can be indicative of the relative health of their parents. Also, lipid contents, monitored over time, can be used as an indicator of growth and potential reproductive performance.

### 3.5. Proteins

Serum and whole-body protein concentrations have been shown to be influenced by a number of environmental factors (Claybrook, 1983). Protein associated with the hemolymph can have a wide variety of functions, such as enzymatic, transport, or hormonal, and can therefore be influenced by toxicants via a variety of mechanisms. However, given the wide range of natural events that can influence hemolymph protein concentrations, it will be very hard for one to distinguish toxic effects from background variability.

The majority of the changes in structural protein that have been reported can be directly related to the oxidation of amino acids for energy. During starvation of invertebrates, both hemolymph and structural proteins are utilized as energy sources

(Claybrook, 1983; Florkin and Sheer, 1970; Gilles, 1970) and have been shown to be mobilized under stressful conditions (Bayne, 1973). During toxicant-induced stress, protein catabolism was increased in oysters exposed to naphthalene (Riley and Mix, 1981) and in freshwater crabs exposed to sumithion (Bhagyalakshmi et al., 1983a). Under laboratory conditions, McKee and Knowles (1986) have found protein to be a sensitive indicator of *Daphnia* growth and believe that total protein may be a quick, easy predictor of future reproductive potential.

### 3.6. Adenylate Energy Charge

The adenylate energy charge (AEC) (Eq. 3.6.1) has been suggested as a measure of the metabolic energy (anhydride-bound phosphate) available to an organism from the adenylate pool (Atkinson, 1977; Knowles, 1977; Haya and Waiwood, 1983). ATP is the organism's "energy currency." In general, the concentrations of adenylates and cations of all the adenylates (AEC) seem to serve as an overall regulator of many catabolic and anabolic processes. The advantage of an energy-based nonspecific measure of stress is that it should act as an integrative measure across several independent or synergistic stressors and should be a function of the physiological state of the organism. The AEC is a linear function, which theoretically can range between 0 (all AMP) and 1.0 (all ATP).

$$\frac{\text{ATP} + \frac{1}{2}\text{ADP}}{\text{ATP} + \text{ADP} + \text{AMP}} \qquad (3.6.1)$$

Values close to 1.0 indicate a greater metabolic energy potential, while smaller values reflect lesser energy potential. This ratio represents the internal energy status of the cell, from which the condition of the organism can be estimated (Chapman et al., 1971).

Energy charge is highly regulated (Atkinson, 1977) and seems to change only when organisms are moved out of their region of homeostasis. Ivanovici (1979; Ivanovici and Wiebe, 1981) in extensive reviews of stress and energy charge indicated the operational ranges of AEC. The results of experiments with microorganisms in synchronous culture indicate that an energy charge of 0.8 to 0.9 is indicative of viable organisms in a nonlimiting (no stress) environment, while those exhibiting energy charges of between 0.50 and 0.75 are somewhat limited (partially stressed) but viable and able to return to an unstressed condition when the stressors are removed. Animals exhibiting an energy charge of below 0.50 are severely stressed and will not regain viability when returned to optimal conditions.

The important advantage of energy charge measurements is that they consistently appear to correspond well to biochemical condition in many organisms studied (Chapman et al., 1971; Ivanovici, 1979; Montague and Dawes, 1974; Ching and Ching, 1972; Haya and Waiwood, 1983). Much of the early work on adenylate energy charge was conducted with microorganisms such as algae and bacteria (Walker-Simmons and Atkinson, 1977; Montague and Dawes, 1974; Ball and Atkinson, 1975; Atkinson, 1977; Knowles, 1977). An extensive review of the effects of anoxia on AEC in aquatic organisms is given by Haya and Waiwood (1983). The AEC is a monitor of anaerobic metabolism and a large body of literature on the effects of anaerobosis on aerobic organisms has been developed (Ridge, 1972; Miller and Wood, 1974; Derr and Zieve, 1972; Van den Thillar et al., 1976; Reinert and Hohreiter, 1984; Jorgensen and Mustafa, 1980). Investigations specifically designed to evaluate the AEC as a sublethal stress indicator in macroorganisms have only recently been initiated. The AEC has been shown to be altered in crayfish, lobster, clams, and fish exposed to a variety of organic and inorganic toxicants (Haya et al., 1980; Sklar and McKee, 1984; Ivanovici and Wiebe, 1981; Giesy et al., 1983; Heath, 1984).

When interpreted against a background of natural seasonal fluctuations, the dynamics of readily available organismal energy in the form of adenylates may serve as an early warning of stressful conditions for selected organisms. While this measure seems to be more sensitive than standard bioassays of lethality, the problems associated with the choice of target indicator species remain. Adenylate concentrations and the AEC have potential, but more work is needed to (1) determine the responses of a variety of organisms to a variety of stressors, and (2) relate changes in AEC to more traditional measures of chronic stress, such as growth and reproduction. While the measurement of stress-induced biochemical variation in populations may be useful, it is by no means a panacea to preclude population-, community-, or system-level studies. The AEC will eventually be most useful in shortening the time required to conduct screening toxicity tests.

## 4. ENZYMES

### 4.1. Rationale

The catalytic function of enzymes in biochemical reactions renders them essential for the normal functioning of all organisms. Enzyme activity is generally regulated such that specific substrates or entire pathways may be homeostatically adjusted to compensate for endogenous or exogenous changes. There are literally hundreds of enzymes that could be used as sublethal indicators of stress. If specific chemicals with known biochemical effects are suspected to be present, then the measurement of enzymes that are effected directly by these chemicals may be useful. However, when the chemical contaminant is unknown or extremely complex, which is often the case, then enzymes indicative of carbon flux through specific metabolic pathways or homeostatic regulation may be more useful.

Other factors that need to be considered when choosing an appropriate enzyme include the influence of season, sex, diet, location, reproductive stage, and molt status on the enzymes of interest (Barnhart, 1969). In addition, interpretation of the biological and/or ecological significance of sublethal alterations in enzyme activity can be extremely difficult and comparisons of biochemical indicators with population- and community-level effects have seldom been made (Gilfillan et al., 1979). In laboratory hazard assessment applications, one is primarily interested in the relative or comparative toxicity of specific chemicals. Therefore, application to biologically or ecologically important processes is not essential. However, when adapting biochemical indicators of stress to in situ monitoring programs, it is essential to relate any alteration in the parameter of interest to an adverse effect on the organism's growth, reproduction, or survival.

The rationale for using alterations in enzyme activity as an indicator toxicant-induced stress can be divided into four sections: (1) direct inhibitions, (2) enzyme induction, (3) serum enzymes, and (4) homeostatic adjustments.

## 4.2. Direct Inhibition

Alteration in the activity of enzymes directly inhibited by specific toxicants has been proposed as an indicator of exposure to that toxicant. Inhibition of enzyme activity by a variety of endogenous and exogenous chemical substances has been extensively investigated (Cairns et al., 1984; Stegeman and Heath, 1983). Examples of enzymes shown to be directly inhibited by specific toxicants include (1) ATPase inhibition by metals (Staurner et al., 1984; Yap, 1975) and organics (Poston, 1979), (2) acetylcholinesterase inhibition by organophosphate and carbamate insecticides (Weiss, 1961; Coppage, 1972), and (3) $\alpha$-aminolevulinate dehydratase inhibition by lead (Hodson et al., 1977). Inhibitions can occur by direct interaction with the active site such that activity is reduced; interaction with other portions of the enzyme such as sulfhydryl groups, causing conformational changes that affect activity; or interaction with enzyme substrates, which will subsequently affect activity or changes in membrane bound enzymes. The scenario proposed by most investigators entails the collection of ''monitoring'' organisms from an ecosystem of suspected contamination, measurement of activity of specific enzymes that are inhibited by the potential contaminants, and extrapolation of the degree of inhibition to the level of contamination. There are instances where application of the technique has been successful (Coppage, 1972). However, general applicability is limited by complications due to accessory factors, the difficulty of data interpretation due to lack of knowledge of intensity of the stressor (dose) and duration of exposure, and the requirement for a priori knowledge of the contaminant.

## 4.3. Inducible Enzymes

The activity or synthesis of specific enzymes can increase in organisms exposed to certain types of chemicals. For example, mixed function oxidase (MFO) activity has been shown to be increased in organisms exposed to polynuclear aromatic hydrocarbons. Monitoring the activity of these enzymes has been proposed as an in situ indicator of exposure to specific classes of compounds. This approach warrants continued investigation. Detailed discussions of MFO activity are given in chapters by both Lee and Melancon (Chapters 9 and 10).

## 4.4. Serum Enzymes

Elevation of the activity of tissue-specific enzymes in serum has been used by mammalian toxicologists as an indicator of tissue damage or dysfunction. Extensive knowledge of the enzyme and isozyme distribution within and among tissues enables mammalian toxicologists to isolate the source and potential cause of elevated serum enzymes. Mechanisms proposed for increased serum enzyme activity include (1) increased number of cells producing the enzyme, (2) increased leakage of enzyme from cells in which synthesis has been stimulated, (3) increased leakage of enzyme from cells due to alterations in membrane permeability, (4) release of enzyme from necrotic or damaged cells, and (5) impaired mechanism for removal of serum enzymes (Galen, 1975). Since many of the diagnostic enzymes utilized are fairly nonspecific, proper assessment of tissue injury requires the analysis of a large array of enzymes. Although analysis of enzyme alterations can lead to the identification of structurally and/or functionally damaged tissues, this technique is generally not toxicant-specific and, in terms of aquatic organisms, it is difficult to relate these changes to biologically significant effects. Likewise, the tissue distribution of specific enzymes and isozymes is extremely species-specific and interpretations of elevations of serum enzyme activities in one species using data generated in other species, such as extrapolation from mammalian to aquatic organisms, can be extremely misleading.

A large array of enzymes have been investigated with respect to toxicant-induced damage in aquatic organisms. Serum transaminases, specifically aspartate aminotransferase (ASAT; EC 2.6.1.2) and alanine aminotransferase (ALAT; EC 2.6.1.1), have been widely utilized in mammalian toxicology as indicators of specific organ dysfunction (Wroblewski and LaDue, 1956). Exposure of fish to a variety of metals and organic xenobiotics causes increases in the transaminases ASAT and ALAT (Versteeg and Giesy, 1985; Lockhart et al., 1975; Lane and Scura, 1970). Elevations in serum activity of the hydrolase enzymes, acid phosphatase (ACP; EC 3.1.3.1) and alkaline phosphatase (ALP; EC 3.1.3.2), have been used in clinical toxicology to assess cellular damage (Hayes, 1984) and have been proposed as a measure of poisoning by metals in fish (Jackim et al., 1970). Lactate dehydrogenase (LDH; EC 1.1.1.27) is increased in serum of fish exposed to a variety of organic and inorganic xenobiotics (Lockhart et al., 1975; Williams and Wooten, 1981). Identification of the organ releasing LDH to serum is possible due to the presence of organ-specific LDH isozymes. Stress has been demonstrated to cause liver damage in fish, which would cause a release of leucine aminonapthylamidase (LAN, 3.4.1.1) into the plasma (Bouck, 1980). Bouck (1980) studied the LAN activities of rainbow trout (*Salmo gairdneri*) and proposed the use of a "tissue equivalent" unit of LAN to provide a means of relating increased LAN in plasma to an amount of liver destruction.

## 4.5. Homeostatic Adjustment

Alterations in specific enzymes that regulate a variety of metabolic pathways can result from stress-related homeostatic adjustments induced by toxicant exposure.

The rationale for many of these changes can be found in the general adaptation syndrome, which as already discussed, results in the adjustment of many catabolic and/or anabolic processes required to maintain a constant internal environment. These changes may involve (1) shifts in catabolic pathways, such as glycogenolysis, glycolysis, and lipolysis, for maximization of energy production to meet demands created by the stressor; (2) concomitant adjustments in oxidative capacity; (3) changes in anabolic processes, such as protein synthesis, for the purpose of tissue repair; (4) maintenance of redox balance via reoxidation of glycolytically produced NADH; (5) enzymatic changes for the purpose of maintaining osmotic balance; and (6) anaerobically induced shifts in key branchpoint enzymes. Measurement of the rate-limiting or regulatory enzymes controlling these processes can be a useful indicator of the relative health of the organism.

Although the nonspecific nature of these enzyme alterations makes their correlation with specific toxicants impossible, this same quality makes them potentially useful in assessing the overall health status of organisms. Inherent variability among individuals due to season, geographic clines, genotype, and natural perturbations such as turbidity makes it difficult to demonstrate toxicant-induced alterations. However, by choosing the appropriate organism, tissue, and enzyme for a particular ecosystem, the effects of these problems can be minimized. We feel that measuring alterations in these enzymes is most applicable to a field-monitoring scheme and has limited application in laboratory hazard assessment.

There are literally hundreds of enzymes that could be used as this type of stress indicator. Spatial limitations of this chapter make it impossible to discuss the rationale and potential application of each of these enzymes. However, below is a list of enzymes that have been investigated. For further discussion of these enzymes, see review by Bayne et al. (1985).

1. Catabolic processes
   A. Glycogenolysis
      - Glycogen phosphorylase
      - Phosphoglucomutase
      - Glucose-6-phosphatase
   B. Glycolysis
      - Hexokinase
      - Phosphofructokinase
      - Pyruvate kinase
   C. TCA cycle
      - Citrate synthetase
      - Isocitrate dehydrogenase
      - $\alpha$-Ketoglutarate dehychogenase
2. Anabolic processes
   A. Pentose–phosphate shunt
      - Glucose-6-phosphate dehydrogenase
   B. Gluconeogenesis
      - Phosphoenolpyruvate carboxykinase
      - Fructose-1,6-diphosphatase

C. Glycogenesis
   Glycogen synthase
3. Protein metabolism
   Glutamate dehydrogenase
   Aspartate aminotransferase
   Alanine aminotransferase
4. Redox maintenance
   Lactate dehydrogenase
   Malate dehydrogenase
   Octopine dehydrogenase
   Glycerolphosphate dehydrogenase

In the context of a ''toxicant-specific'' monitoring program, inherent limitations associated with measuring inhibition of specific enzymes by specific toxicants presently limit the useful application of this technique in the aquatic environment. These include insufficient background information on (1) the role of specific enzymes in specific tissues of a large number of aquatic organisms; (2) normal ranges of activities and variation due to age, sex, season, reproductive condition, temperature, nutritional status, or developmental stage; (3) mechanisms of compensatory biochemical adaptation that allow an organism to survive during exposure to individual toxicants and combinations of toxicants; and (4) relationship of specific clinical measures to ecologically relevant population-level effects on survivorship, growth, and reproduction.

## 5. OSMOREGULATION

The maintenance of a constant internal milieu is essential for normal cellular functioning in aquatic organisms. In both fresh water and seawater, there is a large osmotic gradient between the extracellular fluid of an aquatic organism and its environment. In freshwater organisms, inorganic ions constitute the bulk of extracellular osmotic substances and, to insure normal cellular metabolism, they must be maintained within homeostatic limits. The primary sites for osmoregulation are the organism–water interface such as gills and the excretory organs. Freshwater organisms are hyperosmotic regulators. That is, they actively maintain an ionic composition that is greater than the surrounding water. Osmoregulatory ability can be inhibited by both naturally occurring chemical stressors, such as changes in pH or ionic strength, as well as synthetic toxicants, such as metals or organochlorine pesticides, or it may be altered as a result of a hormonally mediated generalized stress response (Mazeaud and Mazeaud, 1981). Therefore, measuring changes in the extracellular ionic composition should be directly related to an organism's ability to maintain an osmotic gradient (Wedemeyer and Yasutake, 1977). Since maintenance of a consistent internal environment is essential, changes in the ionic composition would be indicative of a breakdown in normal homeostatic process or direct inhibition of osmoregulatory mechanisms.

Effects of stressors on osmoregulation have not been extensively related to effects on survival, growth, and reproduction so the utility of osmoregulatory system monitoring for predicting chronic toxicity is currently low. Since so many factors alter osmoregulation, it may not be possible to utilize osmoregulatory status as a stress indicator. We recommend that research be conducted to ascertain the relationship between toxicity at the population level and effects on osmoregulation. In addition, the importance of proper osmoregulatory state for optimal growth, survival, and reproduction should be determined.

## 6. LYSOSOMAL ENZYME RELEASE ASSAY

The lysosomal enzyme release assay (LERA) is a measure of the stability (or lability) of the lysosomal membrane. Lysosomes are a morphologically heterogeneous group of membrane-bound subcellular organelles, which contain acid hydrolases (Karp, 1979) and function in the degradations of intracellular protein. In stable lysosomes, hydrolases are prevented from reacting with substrate by an intact membrane. Theoretically, membrane stability decreases and enzyme activity (release) increases as membrane permeability increases in response to stress. Currently, it is believed that any stressor that damages cell macromolecules will alter cellular metabolic requirements. Changes in the lysosomal membranes occur as a result of altered cell macromolecule requirements and in response to increases in autophagic activity. Lysosomal membrane stability has been demonstrated to be a useful measure of environmental stressors in aquatic organisms (Moore and Stebbing, 1976; Mensi et al., 1982; Versteeg and Giesy, 1985).

Organic and inorganic toxicants have been shown to alter lysosomal membrane stability in fish (Versteeg and Giesy, 1985) and invertebrates (Moore and Stebbing, 1976; Moore et al., 1982). Xenobiotically induced alterations in lysosomal membrane stability have been related to effects at the population level of organization for a few toxicants. All of the toxicants studied to date have caused lysosomal membrane destabilization; thus this assay may prove useful as a general indicator of xenobiotic stress in aquatic organisms. We recommend research be conducted on the environmental effects of stressors on Great Lakes mollusks. The success that has been obtained with marine mollusks and the benefits of mollusks as monitors of water toxicant concentrations combine to make lysosomal membrane destabilization a potentially useful indicator of environmental quality.

## 7. RNA, DNA, AND PROTEIN

The quantification of ribonucleic acid (RNA), deoxyribonucleic acid (DNA), and protein have been used to assess growth rates in aquatic organisms (Munro and Fleck, 1966; Buckley, 1979; Hartree, 1972). RNA content is directly related to the synthesis of structural protein. The ratio of RNA to DNA is generally believed to more accurately reflect growth since this ratio is not affected by cell number or size. RNA/DNA has been proposed as an integrative measure of stress effects, based on the assumption that exposure to stressors directly inhibits growth or that the energetic expenditure reduces scope for growth. Quantification of protein, on the other hand, is a measure of actual tissue (protein) accumulation and represents a "static" indicator of an organism's growth status and gives no information concerning its growth rate.

RNA/DNA ratios have been used successfully to assess short- and long-term growth rates in fish (Bulow, 1970; Bulow et al., 1981; Haines, 1973; Sower et al., 1983). This ratio varies rapidly and is directly related to nutrition, age, and season (Bulow et al., 1981). A number of toxicants have been shown to affect RNA/DNA under laboratory (Barron and Adelman, 1984; Verma et al., 1984), and field conditions (Kearns and Atchinson, 1979; Wilder and Stanley, 1983). To our knowledge only one study has been conducted on the effect of toxicant-induced stress on the nucleic acid content of aquatic invertebrates. McKee and Knowles (1986) used the nucleic acid and protein content of the freshwater cladoceran *Daphnia magna* as a short-term indicator of chronic reproductive effects. The use of such biochemical measurements as short-term endpoints in chronic reproduction studies routinely used in hazard assessment procedures has potential and should be further investigated.

## 8. FREE AMINO ACIDS

The free amino acids (FAA) of aquatic organisms represent a highly dynamic pool of biomolecules responsive to changes in biochemical and physiological processes induced by environmental perturbations (Ivanovici et al., 1981). The majority of the work on the biochemical and/or physiological regulation of intracellular FAA pools has been conducted with marine or euryhaline invertebrates. Investigations of mechanisms by which organisms regulate their FAA pool led to the realization that environmental stressors other than salinity, such as temperature or toxicants, can alter both the absolute and relative concentrations of intracellular FAA (Schafer, 1961; Jeffries, 1972; DuPaul and Webb, 1970). Although the mechanism has yet to be resolved, it has become clear that during exposure to toxicants, the intracellular FAA pool of certain marine organisms changes (Roesijadi et al., 1976; Kasschau et al., 1980).

The free amino acid composition of marine invertebrates has been shown to change under toxicant-induced stress (Schafer, 1961, 1963, 1968; Bayne et al., 1976; Sansome et al., 1978; Augenfeld et al., 1980; Schafer, 1961). However, few studies have been conducted with freshwater invertebrates (Bhagyalakshmi et al., 1983b; Gardner et al., 1981). The FAA pool fluctuates considerably and can be influenced by the temperature (Cook et al., 1972; DuPaul and Webb, 1970) and salinity (Ivanovici et al., 1981; Bishop et al., 1981) of the ambient environment as well as by the organism's diet (Riley, 1980), reproductive and/or developmental status (Ferguson, 1975; Kasschau and McCommas, 1982), sex (Ferguson, 1975), and molt status (McWhinnie et al., 1972; Tucker and Costlow, 1975).

A limited amount of information is available on the effect of toxicants (Mehrle and Mayer, 1975; Arillo et al., 1981) and accessory factors (Mommesen, 1980; Plakas et al., 1980) on the FAAs of fish.

Very few studies have been conducted on the effect of toxicants on the concentration of FAA in freshwater organisms. Differences in total FAA were reported for freshwater bivalves collected from polluted and unpolluted environments (Gardner et al., 1981). The mantle tissue of organisms collected from habitats contaminated by acid and metals had a significantly greater total concentration of FAA. Similarly, the bivalve *Corbicula fluminea* exposed under controlled laboratory conditions to sodium dodecyl sulfate had a significantly greater total FAA pool and a significantly altered FAA profile (Graney, 1986). In contrast, the total concentration of FAA in amphipods exposed to pentachlorophenol was significantly reduced in acute and chronic laboratory exposures (Graney and Giesy, 1986). Inconsistent responses of FAA pools makes interpretations of the significance of changes extremely difficult.

For many saltwater or euryhaline invertebrates, the data base exists to utilize alterations in specific amino acids as an indicator of toxicant induced stress. However, similar information on freshwater invertebrates is lacking and presently limits the use of FAA as a stress indicator. Additional research on the function and metabolism of amino acids in freshwater organisms and the effect of toxicants on these processes is required before judgment can be made on potential applications as stress monitors.

## 9. SUBSTRATES

The rationale for measuring concentrations of substrate biomolecules in assessing toxicant effects is that concentrations of these compounds may be altered because tissue concentrations will either be depleted due to an increased demand, such as with ascorbic acid, or increased or decreased due to blockage of a pathway such that products or reactants in these pathways are depleted or accumulated.

A number of substrates are used in biochemical processes. Some of these have direct toxicological significance, while others are indirectly involved in responses to toxicants. We have addressed some substrates in other sections (see Sections 3 and 4). Here we address several substrates that have not been discussed elsewhere.

### 9.1. Ascorbic Acid

Ascorbic acid (ASA) participates in a variety of biochemical processes, including catecholamine production and release (Lewin, 1976), collagen metabolism (Barnes, 1975), cytochrome P-450 mixed function oxygenase system (Zannoni and Lynch, 1973; Street and Chadwick, 1975), and immunocompetence (Lewin, 1976). Many of these processes are increased under toxicant-induced stress such that body reserves of ascorbic acid may be depleted (Wedemeyer, 1969). For organisms in which ascorbic acid is a dietary requirement, such as insects (Gupta et al., 1972), crustaceans (Chatterjee, 1973), and most fish (Chatterjee et al., 1975; Mahajan and Agrawal, 1979), increased utilization of this micronutrient may have severe biological consequences.

A large number of organic and inorganic compounds have been shown to affect ASA concentrations in fish. However, the effects of different compounds on vitamin C concentration is not constant (Mayer et al., 1978; Verma et al., 1981; Thomas et al., 1982a). Both increases and decreases in ASA have been reported. Toxicants have also been found to affect the ASA content of invertebrates. However, the paucity of knowledge on the metabolic role of ASA in these organisms makes interpretation of the differences extremely difficult.

## 9.2. Bone Collagen Formation

Collagen formation has been utilized as a measure of growth and in investigating toxicant-specific effects on collagen biochemistry. Although, collagen formation is dependent upon ascorbic acid, not all of the effects observed can be related to ASA.

A biochemical assessment of bone structure is indicative of biochemical effects on collagen and bone formation. Exposure to toxaphene altered bone development and quality in the fathead minnow and this was reflected in decreased collagen content (Mehrle and Mayer, 1975).

Whole-body collagen content has also been used as an indicator of growth potential. Exposure of brook trout fry to toxaphene caused significant decreases in whole-body and backbone hydroxyproline concentrations before significant decreases in growth could be measured.

## 9.3. Glutathione

The tripeptide glutathione (GSH) is a reactive nucleophile present in most organisms and primarily associated with the detoxification of environmental xenobiotics.

There are three primary mechanisms by which GSH can protect organisms from the toxic effects of environmental contaminants. The first involves the phase II conjugation of GSH to reactive compounds and their subsequent biliary excretion. Under normal conditions, a sufficient supply of GSH exists to metabolize background levels of xenobiotics. However, during continual, elevated exposure regimes, GSH can be depleted such that proper detoxification is impaired (Chasseaud, 1976). Glutathione also functions to detoxify hydrogen peroxide and thus prevent lipid peroxidation of cellular membranes (Flohe et al., 1976). Increases in chemically mediated lipid peroxidation can create a greater demand for GSH, thus depleting its body reserve. The third mechanism by which GSH protects organisms is via binding of heavy metals (Maines and Kappas, 1977). Although metallothioneins (MT) are considered the primary metal binding and detoxification proteins, the sulfhydryl group of GSH can also sequester and inactivate free metal ions.

Given the variety of interrelated functions of GSH in mammals, it is very difficult to predict whether concentrations will increase or decrease during toxicant exposure. The lack of information on the precise role of this tripeptide in aquatic organisms adds to the uncertainty. It is obvious that depletion of GSH can occur if demand for the substrate exceeds its synthesis. However, under what conditions and for which compounds this occurs and the biological significance of the observed depletion have yet to be resolved.

In mammalian species, glutathione concentrations have been shown to increase or decrease after exposure to environmental pollutants (Wong and Klaassen, 1981; Bayer et al., 1981). A limited amount of work has been conducted on nonmammalian species. Stress has been shown to cause increases in GSH in both invertebrates (Ilivicky and Agosin, 1967) and fish (Thomas et al., 1982b).

Accessory factors can also influence the concentration of tissue GSH and can thus be a potential limiting factor in utilizing GSH as an indicator of toxicant stress (Higashi et al., 1977; Thomas and Wofford, 1984).

## 10. CONCLUSIONS AND RECOMMENDATIONS

We are of the opinion that environmental regulation is a very complex science and that to be effective ecotoxicologists must make use of all available tools. Ultimately, the level of organization that we want to protect is the ecosystem and all of its components. Therefore, ecosystem-level monitoring will continue to be important, as will population- and community-level studies. Biochemical indicators will not replace measures of toxicant effects on individuals, populations, communities, and ecosystems but rather will add to and enrich our understanding of toxicant-induced effects. Specifically, an understanding of biochemical responses will elucidate the modes of action of toxicants, which will provide greater generality. This increased understanding will subsequently serve as the basis of improved predictability of toxic effects among toxicants and species and allow better extrapolation to different environmental conditions. Furthermore, understanding of biochemical toxicology will allow us to better predict the interactions of toxic mixtures.

Few clinical measures have been sufficiently well developed to be used in routine monitoring programs. We presently lack the knowledge to implement many of the clinical measures. Therefore, we endorse the continued development of these sensitive assays for use under field conditions.

## ACKNOWLEDGMENTS

Preparation of this manuscript was supported in part by the toxicology program of the Michigan Agricultural Experiment Station from which this is contribution 12003, the National Oceanographic and Atmospheric Administration Great Lakes Environmental Research Laboratory through Cooperative Agreement NA84-WC-H-06094, and the Michigan Sea Grant College Program Project R/TS-21 under

Grant NA080AA-D-00072 from the National Sea Grant College Program, National Oceanic and Atmospheric Administration, U.S. Dept. of Commerce, from which this is contribution MICHU-86-308.

## REFERENCES

Arillo, A., Margiocco, C., Melodia, F., Mensi, P., and Schenone, G. (1981). Ammonia toxicity mechansim in fish: Studies on rainbow trout (*Salmo gairdneri* Rich). *Ecotoxicol. Environ. Safety.* **5,** 316–328.

Atkinson, D. E. (1977). *Cellular Energy Metabolism and Its Regulation.* Academic Press, New York.

Augenfeld, J .M., Anderson, J. W., Woodruff, D. L., and Webster, J. L. (1980). Effects of Prudhoe Bay crude oil-contaminated sediments on *Protothaca staminea*: Hydrocarbon content, condition index, free amino acid level. *Mar. Environ. Res.* **4,** 135–143.

Ball, W. J., Jr., Atkinson, D. E. (1975). Adenylate energy charge in *Saccharomyces cerevisiae* during starvation. *J. Bacteriol.* **121,** 975–982.

Barnes, M. J. (1975). Function of ascorbic acid in collagen metabolism. *Ann. N.Y. Acad. Sci.* **258,** 264–277.

Barnhart, R. A. (1969). Effects of certain variables on hematological characteristics of rainbow trout. *Trans. Am. Fish. Soc.* **3,** 411–448.

Barrett, G. W., and Rosenberg, R. (1981). *Stress Effects on Natural Ecosystems.* Wiley, New York.

Barron, M. G., and Adelman, I. R. (1984). Nucleic acid, protein content and growth of larval fish sublethally exposed to various toxicants. *Can. J. Fish. Aquat. Sci.* **41,** 141–150.

Bayer, V., Younes, M., and Siegers, C. P. (1981). Enhancement of benzo (a) pyrene-induced sister chromatid exchanges as a consequence of glutathione depletion in-vivo. *Toxicol. Lett.* **9,** 339–343.

Bayne, B. L. (1973). Aspects of the metabolism of *Mytilus edulis* during starvation. *Neth. J. Sea Res.* **7,** 399–410.

Bayne, B. L., Livingstone, D. R., Moore, M. N., and Widdows, J. (1976). A cytochemical and a biochemical index of stress in *Mytilus edulis* L. *Mar. Pollut. Bull.* **7,** 221–224.

Bayne, B. L., Clarke, K. R., and Moore, M. N. (1981). Some practical considerations in the measurement of pollution effects on bivalve molluscs, and some possible ecological consequences. *Aquat. Toxicol.* **1,** 159–174.

Bayne, B. L., Brown, D. A., Burns, K., Dixon, D. R., Ivanovici, A., Livingstone, D. R., Lowe, D. M., Moore, M. N., Stebbing, A. R. D., and Widdows, J. (1985). *The Effects of Stress and Pollution on Marine Animals.* Praeger, New York.

Bhagyalakshmi, A., Sreenivasula Reddy, P., and Ramamurthi, R. (1983a). Muscle nitrogen metabolism of freshwater crab, *Oziotelphusa senex senex,* Fabricuis, during acute and chronic sumithion intoxication. *Toxicol. Lett.* **17,** 89–93.

Bhagyalakshmi, A., Sreenivasula Reddy, P., and Ramamurthi, R. (1983b). Changes in hemolymph glucose, hepatopancreas glycogen, total carbohydrates, phosphorylase and amino transferases of sumithion-stressed freshwater rice-field crab (*Oziotelphusa senex senex*). *Toxicol. Lett.* **18,** 277–284.

Bhagyalakshmi, A., Sreenivasula Reddy, P., and Ramamurthi, R. (1984). Subacute stress induced by sumithion on certain biochemical parameters in *Oziotelphusa senex senex,* the freshwater rice-field crab. *Toxicol. Lett.* **21,** 127–134.

Bishop, S. H., Greenwalt, D. E., and Burcham, J. M. (1981). Amino acid cycling in ribbed mussel tissues subjected to hyperosmotic shock. *J. Exp. Zool.* **213,** 277–287.

Bouck, G. R. (1980). Concentration of leucine aminonapthylamidase (LAN) and soluble protein in tissues of rainbow trout, *Salmo gairdneri. Can. J. Fish Aquat. Sci.* **37,** 116–120.

Buckley, L. J. (1979). Relationships between RNA-DNA ratio, prey density and growth rate in Atlantic cod (*Gadus morhua*) larvae. *J. Fish. Res. Board Can.* **36,** 1497–1502.

Bulow, F. J. (1970). RNA-DNA ratios as indicators of recent growth rates of a fish. *J. Fish. Res. Board Can.* **27,** 2343–2349.

Bulow, F. J., Zeman, M. E., Winningham, J. R., and Hudson, W. F. (1981). Seasonal variations in RNA-DNA ratios and in indicators of feeding, reproduction, energy storage, and condition in a population of bluefill sunfish, *Lepomis macrochirus* Rafinesque. *J. Fish Biol.* **18,** 237–244.

Butler, P. J., Taylor, E. W., Capra, M. F., and Davison, W. (1978). The effect of hypoxia on the levels of circulating catecholamines in the dogfish *Schyliohinus canicula. J. Comp. Physiol.* **727,** 325–330.

Cairns, V. W., Hodson, P. V., and Nriagu, J. O. (Eds.) (1984). *Contaminant Effects on Fisheries.* Wiley, New York.

Carmichael, G. J., Wedemeyer, A., McCraren, J. P., and Millard, J. L. (1983). Physiological effects of handling and hauling stress on smallmouth bass. *Prog. Fish. Cult.* **45,** 110–113.

Chapman, A. G., Fall, L., and Atkinson, D. E. (1971). Adenylate energy charge in *Escherichia coli* during growth and starvation. *J. Bacteriol.* **108,** 1072–1086.

Chasseaud, L. F. (1976). Conjugation with glutathione and mercapturic acid excretion. In I. M. Arias and W. B. Jakoby (Eds.), *Glutathione: Metabolism and Function.* Raven Press, New York, pp. 77–114.

Chatterjee, I. B. (1973). Vitamin C synthesis in animals: Evolutionary trends. *Sci. Cult.* **5,** 210–212.

Chatterjee, I. B., Majumder, A. K., Nandi, B. K., and Subramanian, N. (1975). Synthesis and some major functions of vitamin C in animals. *Ann. N.Y. Acad. Sci.* **258,** 24–47.

Ching, T. M., and Ching, K. K. (1972). Content of adenosine phosphate and adenylate energy charge in germinating ponderosa pine seeds. *Plant Physiol.* **50,** 536–540.

Clarke, A. (1977). Lipid class and fatty acid composition of *Chorismus antarcticus* (Pfeffer) (Crustacea: Decapoda) at south Georgia. *J. Exp. Mar. Biol. Ecol.* **28,** 297–314.

Claybrook, D. L. (1983). Nitrogen metabolism. In L. H. Mantel (Ed.), *The Biology of Crustacea.* Academic Press, New York, Vol. 5, pp. 163–213.

Coglianese, M. P., and Neff, J. M. (1982). Biochemical responses of the blue crab, *Callinectes sapidus* to pentachlorophenol. In W. B. Vernberg, A. Calabruse, F. P. Thurberg, and E. J. Vernberg (Eds.), *Physiological Mechanisms of Marine Pollution Toxicity.* Academic Press, New York, pp. 127–143.

Cook, P. A., Gabbott, P. A., and Youngson, A. (1972). Seasonal changes in the free amino acid composition of the adult barnacle, *Balanus balmordes. Comp. Biochem. Physiol. B* **42B,** 409–421.

Coppage, D. L. (1972). Organophosphate pesticides: Specific levels of brain AChE inhibition related to death in sheepshead minnows. *Trans. Am. Fish. Soc.* **101,** 534–536.

Coppage, D. L., Matthews, E., Cook, G. H., and Knight, J. (1975). Brain acetylcholinesterase inhibition in fish as a diagnosis of environmental poisoning by malathion, O,O,-dimethyl S-(1,2-dicarbethy oxyethyl) phosphorodithioute. *Pestic. Biochem. Physiol.* **5,** 536–542.

Davis, K. B. and Parker, N. C. (1983). Plasma corticosteroid and chloride dynamic in rainbow trout, Atlantic salmon, and lake trout during and after stress. *Aquacult.* **32,** 189–194.

Derr, R. F., and Zieve, L. (1972). Adenylate energy charge: Relation to guanylate energy charge and the adenylate kinase equilibrium constant. *Biochem. Biophys. Res. Commun.* **49,** 1385–1390.

Donaldson, E. M. (1981). The pituitary-interrenal axis as an indicator of stress in fish. In A. D. Pickering (Ed.), *Stress and Fish.* Academic Press, London, pp. 11–47.

DuPaul, W. D., and Webb, K. L. (1970). The effect of temperature on salinity-induced changes in the free amino acid pool of *Mya arenaria. Comp. Biochem. Physiol.* **32,** 785–801.

Esch, G. W., and Hazen, T. C. (1978). Thermal ecology and stress: a case history for red-sore disease in large mouth bass. In J. H. Thorp and J. W. Gibbons (Eds.), Energy and Environmental Stress in Aquatic Ecosystems. Conference 771114, Technical Information Center, U.S. Department of Energy, Springfield, VA, pp. 331-363.

Ferguson, J. C. (1975). The role of free amino acids in nitrogen storage during the annual cycle of a starfish. *Comp. Biochem. Physiol. A* **51A**, 341–350.

Fingerman, M., Hanumate, M. M., Deshpunde, V. P., and Nagabhushan, R. (1981). Increase in the total reducing substances in the hemolymph of the freshwater crab, *Barytelphusa guerini*, produced by a pesticide (DDT) and an indolealkylamine (serotonin). *Experientia* **37**, 178–179.

Fitch, K., and Johnson, P. (1977). *Human Life Science.* Holt, Rinehart & Winston, New York.

Flick, G., Wendelaar Bonga, S. E., and Fenwick, J. C. (1984). $Ca^{2+}$-dependent phosphatase and $Ca^{2+}$-dependent ATPase activities in plasma membranes of eel gill epithelium-III. Stimulation of branchial high-affinity $Ca^{2+}$-ATPase activity during prolactin-induced hypercalcemia in American eels. *Comp. Biochem. Physiol. B* **79B,** 521–524.

Flohe, L., Gunzler, W. A., and Ladenstein, P. (1976). Glutathione peroxidase. In I. M. Arias and W. B. Jakoby (Eds.), *Glutathione: Metabolism and Function.* Raven Press, New York, pp. 115–135.

Florkin, M., and Scheer, B. T. (Eds.) (1970). *Chemical Zoology, Vol. 6.* Academic Press, New York.

Folmar, L. C., and Dickhoff, W. W. (1979). Plasma thyroxine and gill $Na^+$-$K^+$ ATPase changes during seawater acclimation of coho salmon, *Oncorhynchus kisutch. Comp. Biochem. Physiol. A* **63A,** 329-332.

Fontaine, Y. A. (1975). Hormones in fishes. In D. C. Malins and J. R. Sargent (Eds.), *Biochemical and Biophysical Perspectives in Marine Biology.* Academic Press, New York, Vol. 2, pp. 139–212.

Gabbott, P. A., and Bayne, B. L. (1973). Biochemical effects of temperature and nutritive stress on *Mytilus edulis.* L. *J. Mar. Biol. Assoc. U.K.* **53,** 269–286.

Gaillard, P. J., and Boer, H. H. (Eds.) (1978). *Comparative Endocrinology.* Elsevier/North-Holland Biomedical Press, New York.

Galen, R. S. (1975). Multiphasic screening and biochemical profiles: State of the Art, 1975. *Prog. Clin. Pathol.* **6,** 83–110.

Gardner, W. S., Miller, W. H., and Imlay, M. J. (1981). Free amino acids in mantle tissues of the bivalve *Amblema plicata*: Possible relation to environmental stress. *Bull. Environ. Contam. Toxicol.* **26,** 157–162.

Gaufin, A. R. (1973). Use of aquatic invertebrates in the assessment of water quality. *ASTM Spec. Tech. Publ.* **STP 528,** 96–116.

Giesy, J. P., and Allred, P. M. (1985). Replicability of aquatic multispecies test systems. In J. Cairns (Ed.), *Multispecies Toxicity Testing.* Pergamon, New York, pp. 187–247.

Giesy, J. P., Duke, C. S., Bingham, R. D., and Dickson, G. W. (1983). Changes in phosphoadenylate concentrations and adenylate energy charge as an integrated biochemical measure of stress in invertebrates: The effects of cadmium on the freshwater clam *Corbicula fluminea. Toxicol. Environ. Chem.* **6,** 259–295.

Gilfillan, E. S., Page, D. S., Foster, J. C., Vallus, D., Gonzalez, L., Luckerman, A., Hotham, J. R., Pendergast, E., and Hebert, S. (1979). A comparison of stress indicators at the biochemical, organismal and community level of organism. *Mar. Environ. Res.* **14,** 503–504.

Gilles, R. (1970). Intermediate metabolism and energy production in some invertebrates. *Arch. Int. Physiol. Biochim.* **78,** 313–326.

Gluth, G. and Hanke, W. (1984). A comparison of physiology changes in carp, *Cyprinus carpio*, induced by several pollutants at sublethal concentrations II. The dependency on the temperature. *Comp. Biochem. Physiol.* **79C,** 39–45.

Goldstein, A. (1975). Biochemical indicators of environmental pollution. In W. A. Thomas (Ed.), *Indicators of Environmental Quality.* Plenum, New York, pp. 109–131.

Graney, R. L. (1986). Toxicological and comparative study of the concentration of free amino acids in benthic freshwater invertebrates. Ph.D. Dissertation, Michigan State University, East Lansing.

Graney, R. L., and Giesy, J. P. (1987). The effects of short term exposure to pentachlorophenol and osmotic stress on the free amino acid pool of the freshwater amphipod *Gammarus pseudolimnaeus* Bousfield. *Arch. Environ. Contam. Toxicol.* **16,** 167–176.

Gupta, S. D., Chaudhuri, C. R., and Chatterjee, I. B. (1972). Incapability of L-ascorbic acid synthesis by insects. *Arch. Biochem. Biophys.* **152,** 889–890.

Haines, T. A. (1973). An evaluation of RNA-DNA ratio as a measure of long-term growth in fish populations. *J. Fish. Res. Board Can.* **30,** 195–199.

Hartree, E. F. (1972). Determination of protein: A modification of the Lowry method that gives a linear photometric response. *Anal. Biochem.* **48,** 422–427.

Hattingh, J. (1976). Blood sugar as an indicator of stress in the freshwater fish, *Labeo capensis* (Smith). *J. Fish Biol.* **10,** 191–195.

Haya, K., and Waiwood, B. A. (1983). Adenylate energy charge and ATPase activity: Potential biochemical indicators of sublethal effects caused by pollutants in aquatic organisms. In J. O. Nriagu (Ed.), *Aquatic Toxicology.* Wiley, New York, pp. 307–333.

Haya, K., Johnston, C. E., and Waiwood, B. A. (1980). Adenylate energy charge and ATPase activity in American lobster (*Homarus americanus*) from Belledune Harbour. In J. F. Uthe and V. Zitco (Eds.), *Cadmium Pollution of Belledune Harbour,* Tech. Rep. Fish. Aquat. Sci. New Brunswick, pp. 85–91.

Hayes, A. W. (1984). *Principles and Methods of Toxicology.* Raven Press, New York.

Heath, A. G. (1984). Changes in tissue adenylates and water content of bluegill, *Lepomis macrochirus,* exposed to copper. *J. Fish Biol.* **24,** 299–309.

Higashi, T., Tateishi, N., Naruse, A., and Sakamoto, Y. (1977). A novel physiological role of liver glutathione as a reservoir of L-cysteine. *J. Biochem. (Tokyo)* **82,** 117–124.

Hodson, P. V., Blunt, B. R., Spry, D. J., and Austen, K. (1977). Evaluation of erythrocyte δ-amino levulinic acid dehydratase activity as a short-term indicator in fish of a harmful exposure to lead. *J. Fish. Res. Board Can.* **34,** 501–508.

Hodson, P. V., Blunt, B. R., and Whittle, D. M. (1984). Monitoring lead exposure of fish. In V. W. Cairns, P. V. Hodson, and J. O. Nriagu (Eds.), *Contaminant Effects on Fisheries.* Wiley, New York, pp. 87–98.

Ilivicky, J., and Agosin, M. (1967). Glutathione turnover in *Triatoma infestans* treated with 2,2,-bis-(p-chlorophenyl)-1,1,1-trichloro ethane. *Exp. Parasitol.* **20,** 345–356.

Ivanovici, A. M. (1979). Adenylate energy charge: Potential value as a tool for rapid determination of toxicity effects. Tech. Rep. 862 Fish. Mar. Serv. (Can.) pp. 241–255.

Ivanovici, A. M., and Wiebe, W. J. (1981). For a working definition of "stress": A review and critique. In G. W. Barrett and R. Rosenberg (Eds.), *Stress Effects on National Ecosystems.* Wiley, New York, pp. 13–27.

Ivanovici, A. M., Rainer, S. F., and Wadley, V. A. (1981). Free amino acids in three species of mollusc: Responses to factors associated with reduced salinity. *Comp. Biochem. Physiol. A* **70A,** 17–23.

Jackim, E., Hamlin, J. M., and Sonis, S. (1970). Effects of metal poisoning on five liver enzymes in the killifish (*Fundulus heteroclitus*). *J. Fish. Res. Board Can.* **27,** 383–390.

Jeffries, H. P. (1972). A stress syndrome in the hard clam, *Mercenaria mercenria. J. Invertebr. Pathol.* **20,** 242–251.

Jorgensen, J. B., and Mustafa, T. (1980). The effect of hypoxia on carbohydrate metabolism in flounder, *Platichthys flesus* (L.). II. High energy phosphate compounds and the role of glycolytic and gluconegenetic enzymes. *Comp. Biochem. Physiol. B* **67B,** 249–256.

Karp, G. (1979). *Cell Biology.* McGraw-Hill, New York.

Kasschau, M. R., and McCommas, S. A. (1982). Glycine concentration as a biochemical indicator of sex and maturation in the sea anemone *Bunodosoma cavernata*. *Comp. Biochem. Physiol. A* **72A,** 595–597.

Kasschau, M. R., Skaggs, M. M., and Chen, E. C. M. (1980). Accumulation of glutamate in sea anemones exposed to heavy metals and organic amines. *Bull. Environ. Contam. Toxicol.* **25,** 873–878.

Kearns, P. K., and Atchison, G. J. (1979). Effects of trace metals on growth of yellow perch (*Perca flavescens*) as measured by RNA-DNA ratios. *Environ. Biol. Fishes* **4,** 383–387.

Knowles, C. J. (1977). Microbial metabolic regulations by adenine nucleotide pools. *Symp. Soc. Gen. Microbiol.* **27,** 241–283.

Lane, C. E., and Scura, E. D. (1970). Effects of dieldrin on glutamic oxaloacetic transaminase in *Poecilia latipinna*. *J. Fish. Res. Board. Can.* **27,** 1869–1871.

Leach, G. J., and Taylor, M. H. (1982). The effects of cortisol treatment on carbohydrate and protein metabolism in *Fundulus heteroclitus*. *Gen. Comp. Endocrinol.* **48,** 76–83.

Leatherland, J. F., and Sonstegard, R. A. (1984). Pathobiological responses of feral teleosts to environmental stressors: Interlake studies of the physiology of Great Lakes salmon. In V. W. Cairns, P. V. Hodson, and S. O. Nriagu (Eds.), *Contaminant Effects on Fisheries,* Wiley, New York, pp. 116–139.

Lee, R. F. (1988). Possible linkages between mixed function oxygenase systems, steroid metabolism, reproduction, molting, and pollution in aquatic animals. In M. S. Evans (Ed.), *Toxic Contaminants and Ecosystem Health: A Great Lakes Focus*. Wiley, New York.

Lee, W. Y., Macko, S. A., and Nicol, J. A. C. (1981). Changes in nesting behavior and lipid content of a marine amphipod (*Amphithoe valida*) to the toxicity of a No. 2 fuel oil. *Water, Air, Soil Pollut.* **15,** 185–195.

Lewin, S. (1976). *Vitamin C: Its Molecular Biology and Medical Potential.* Academic Press, New York.

Lockhart, W. L., Wagemann, R., Clayton, J. W., Graham, B., and Murray, D. (1975). Chronic toxicity of a synthetic tri-aryl phosphate oil to fish. *Environ. Physiol. Biochem.* **5,** 361–369.

Lugo, A. E. (1978). Stress and ecosystems. In J. H. Thorp and J. W. Gibbons (Eds.), *Energy and Environmental Stress in Aquatic Systems.* Tech. Inf. Cent., U.S. Dept. of Energy, Washington, DC, pp. 62–101.

Mahajan, C. L., and Agrawal, N. K. (1979). Vitamin C deficiency in *Channa punctatus* Block. *J. Fish Biol.* **15,** 613–622.

Maines, M. D., and Kappas, A. (1977). Regulation of heme pathway enzymes and cellular glutathione content by metals that do not chelate with tetrapyroles; Blockade of metal effects by thiols. *Proc. Natl. Acad. Sci. U.S.A.* **74,** 1875–1878.

Mayer, F. L. (1983). Clinical tests in aquatic toxicology: A paradox. *Environ. Toxicol. Chem.* **2,** 139–140.

Mayer, F. L., Mehrle, P. M., and Crutcher, P. L. (1978). Interactions of toxaphene and vitamin C in channel catfish. *Trans. Am. Fish. Soc.* **107,** 326–333.

Mazeaud, M. M., and Mazeaud, F. (1981). Adrenergic responses to stress in fish. In A. D. Pickering (Ed.), *Stress and Fish.* Academic Press, London, pp. 49–76.

McKee, M. J., and Knowles, C. O. (1986). Protein nucleic acid and adenylate levels in *Daphnia magna* during chronic exposure to chlordecone. *Environ. Pollut., Ser. A* **42,** 335–351.

McWhinnie, M. A., Kirchenberg, R. J., Urbanski, R. J., and Schwarz, J. E. (1972). Crustecdysone mediated changes in crayfish. *Am. Zool.* **12,** 357–372.

Mehrle, P. M., and Mayer, F. L. (1975). Toxaphene effects on growth and bone composition of fathead minnow, *Pimephales promelas*. *J. Fish. Res. Board Can.* **32,** 593–598.

Mehrle, P. M., and Mayer, F. L. (1980). Clinical tests in aquatic toxicology: State of the Art. *Environ. Health Perspect.* **34,** 139–143.

Melacon, M. J., Binder, R. L., and Lech, J. J. (1988). Environmental induction of monoxygenase activity in fish. In M. S. Evans (Ed.), *Toxic Contaminants and Ecosystem Health: A Great Lakes Focus*. Wiley, New York.

Mensi, P., Arillo, A., Margiocco, C., and Schenone, G. (1982). Lysosomal damage under nitrate intoxication in rainbow trout (*Salmo gairdneri* Rich.). *Comp. Biochem. Physiol. C* **73C,** 161–165.

Miles, H. M., Loehner, S. M., Michard, D. T., and Salivar, S. L. (1974). Physiological responses of hatchery reared muskellunge (*Esox masquinongy*) to handling. *Trans. Am. Fish. Soc.* **103,** 336–342.

Miller, A. T., and Wood, J. J. (1974). Effects of acute carbon monoxide exposure on the energy metabolism of rat brain and liver. *Environ. Res.* **8,** 107–111.

Mommesen, T. P., French, C. J., and Hochachka, P. W. (1980). Sites and patterns of protein and amino acid utilization during the spawning migration of salmon. *Can. J. Zool.* **58,** 1785–1799.

Montague, M. D., and Dawes, E. A. (1974). The survival of *Peptococcus prevotii* in relation to the adenylate energy charge. *J. Gen. Microbiol.* **80,** 291–299.

Moore, M. N., and Stebbing, A. R. D. (1976). The quantitative cytochemical effects of three metal ions on a lysosomal hydrolase of a hydroid. *J. Mar. Biol. Assoc. U.K.* **56,** 995–1005.

Moore, M. N., Pipes, R. K., and Farrar, S. V. (1982). Lysosomal and microsomal responses to environmental factors in *Littoria littorea* from Sullom Voe. *Mar. Pollut. Bull.* **13,** 340–345.

Morata, P., Faus, M. J., Perez-Palomo, M., and Sanchez-Medina, F. (1982). Effect of stress on liver and muscle glycogen phosphorylase in rainbow trout (*Salmo gairdneri*). *Comp. Biochem. Physiol. B* **72B,** 421–425.

Munro, H. N., and Fleck, A. (1966). Recent developments in the measurement of nucleic acids in biological materials: A supplementary note. *Analyst* **91,** 78–88.

Nakano, T., and Tomlinson, N. (1967). Catecholamine and carbohydrate concentrations in rainbow trout (*Salmo gairdneri*) in relation to physical disturbance. *J. Fish. Res. Board Can.* **24,** 1701–1715.

Newell, R. I. E., and Bayne, B. L. (1980). Seasonal changes in the physiology, reproductive condition and carbohydrate content of the cockle *Cardium* (=Cerastoderma) *edule* (Bivalvia: Cardiidae). *Mar. Biol. (Berlin)* **56,** 11–19.

Nieminen, M., Pasanen, P., and Laitinen, M. (1983). Effects of formalin treatment on the blood composition of salmon (*Salmo salar*) and rainbow trout (*Salmo gairdneri*). *Comp. Biochem. Physiol. C* **76C,** 265–269.

O'Connor, J. D., and Gilbert, L. I. (1969). Alterations in lipid metabolism associated with premolt activity in a land crab and freshwater crayfish. *Comp. Biochem. Physiol.* **29,** 889–904.

Pickering, A. D. (Ed.) (1981). *Stress and Fish.* Academic Press, New York.

Pickering, A. D., Pickering, T. G., and Christie, P. (1982). Recovery of the brown trout, *Salmo trutta* L., from acute handling stress: A time course study. *J. Fish Biol.* **20,** 229–244.

Plakas, S. M., Katayama, T., Tanaka, Y., and Deshemaru, O. (1980). Changes in the levels of circulating plasma free amino acids of carp (*Cyprinus carpio*) after feeding a protein and an amino acid diet of similar composition. *Aquaculture* **21,** 307–322.

Plisetskaya, E., Woo, N. Y. S., and Murat, J. C. (1983). Thyroid hormones in cyclostomes and fish and their role in regulation of intermediate metabolism. *Comp. Biochem. Physiol. A* **74A,** 179–187.

Poston, T. M. (1979). Effects of phenol on ATPase activities in crude gill homogenates of rainbow trout (*Salmo gairdneri* Richardson). Battelle Pac. Northwest Lab. Rep. PNL-2697-UC48, 1–52.

Prosser, C. L. (1973). *Comparative Animal Physiology.* Saunders, Philadelphia, PA.

Rao, D. M., Murty, A. S., and Swarup, P. A. (1984). Relative toxicity of technical grade and formulated carbaryl and 1-naphthol to, and carbaryl-induced biochemical changes in the fish (*Cirrhinus mrigala.*) *Environ. Pollut., Ser. A* **34,** 47–54.

Reinert, R. E., and Hohreiter, D. W. (1984). Adenylate energy charge as a measure of stress in fish.

In V. W. Cairns, P. V. Hodson, and J. O. Nriagu (Eds.), *Contaminant Effects on Fisheries.* Wiley, New York, pp. 151–161.

Ridge, J. W. (1972). Hypoxia and the energy charge of the cerebral adenylate pool. *Biochem. J.* **127,** 351–355.

Riley, R. T. (1980). The effect of prolonged starvation on the relative free amino acid composition of the extracellular body fluids and protein bound amino acids in the oyster, *Crassostria virginica. Comp. Biochem. Physiol. A* **67A,** 279–281.

Riley, R. T., and Mix, M. C. (1981). The effects of naphthalene on glucose metabolism in the European flat oyster *Ostrea edulis. Comp. Biochem. Physiol. C* **70C,** 13–20.

Roesijadi, G., Anderson, J. W., and Giam, C. S. (1976). Osmoregulation of the grass shrimp *Palaememetes pugio* exposed to polychlorinated biphenyls (PCBs). II. Effect on free amino acids of muscle tissue. *Mar. Biol. (Berlin)* **38,** 357–363.

Sangalang, G. B., and Freeman, H. C. (1974). Effects of sublethal cadmium on maturation and testosterone and 11-hetotestosterone production in-vivo in Brook Trout. *Biol. Reprod.* **11,** 429–435.

Sansome, G., Biondi, A., and Noviello, L. (1978). Free amino acids in fluids and tissues of *Mytilus galloprovincialis* in relation to the environment. Their behavior as an index of normality of metabolism. *Comp. Biochem. Physiol. A* **61A,** 133–139.

Schafer, R. D. (1961). Effects of pollution on the free amino acid content of two marine invertebrates. *Pac. Sci.* **15,** 49–55.

Schafer, R. D. (1963). Effects of pollution on the amino acid content of *Mytilus edulis. Pac. Sci.* **17,** 246–250.

Schafer, R. D. (1968). Variations in free amino acid composition in some marine invertebrates subjected to specific types of pollutants. *Rev. Int. Oceanogr. Med.* **10,** 55–56.

Selye, H. (1956). *The Stress of Life.* McGraw-Hill, New York.

Selye, H. (1976). *Stress in Health and Disease.* Butterworth, Boston, MA.

Sklar, F. H., and McKee, K. L. (1984). Adenylate energy charge (AEC) response to stress and extraction technique in the Louisiana swamp crayfish, *Procambarus clarkii. Bull. Environ. Contam. Toxicol.* **33,** 584–591.

Sower, S. A., Schreck, C. B., and Evenson, M. (1983). Effects of steroids and steroid antagonists on growth, gonadal development, and RNA/DNA ratios in juvenile steelhead trout. *Aquaculture* **32,** 243–254.

Sreenivasula Reddy, P., Bhagyalakshmi, A., and Ramamurthi, R. (1983). In-vivo acute physiological stress induced by BHC in hemolymph biochemistry of *Oziotelphusa senex senex,* the Indian rice field crab. *Toxicol. Lett.* **18,** 35–38.

Staurnes, M., Sigholt, T., and Reite, O. B. (1984). Reduced carbonic anhydrase and Na-K-ATPase activity in gills of salmonids exposed to aluminum-containing acid water. *Experientia.* **40,** 226–227.

Stegeman, J. J., and Heath, G. W. (1983). *Response of Marine Organisms to Pollutants,* Vols. 1 and 2. Elsevier, London.

Street, J. C., and Chadwick, R. W. (1975). Ascorbic acid requirements and metabolism in relation to organochlorine pesticides. *Ann. N.Y. Acad. Sci.* **258,** 132–143.

Swift, D. J. (1981). Changes in selected blood component concentrations of rainbow trout, *Salmo gairdneri* Richardson, exposed to hypoxia or sublethal concentrations of phenol or ammonia. *J. Fish Biol.* **19,** 45–61.

Swift, D. J. (1982). Changes in selected blood component values of rainbow trout, *Salmo gairdneri* Richardson, following blocking of the cortisol stress response with betamethasone and subsequent exposure to phenol or hypoxia. *J. Fish Biol.* **21,** 269–277.

Thomas, P. (1978). Effect of cadmium exposure on plasma cortisol levels and carbohydrate metabolism in mullet (*Mugil cephalus*). *J. Endocrinol. Suppl.* **94,** 35 pp.

Thomas, P., and Wofford, H. W. (1984). Elevated acid-soluble thiol content in fish hepatic tissue: A response to pollutants. *Mar. Environ. Res.* **14,** 486–488.

Thomas, P., Wofford, H. W., and Neff, J. M. (1981a). Biochemical stress responses of striped mullet (*Mugil cephalus* L.) to fluorene analogs. *Aquat. Toxicol.* **1,** 329–342.

Thomas, P., Carr, R. S., and Neff, J. M. (1981b). Biochemical stress responses of mullett *Mugil cephalus* and polychaete worms *Neanthes virens* to pentachlorophenol. In J. Vernberg, A. Calabrese, F. P. Thurberg, and W. B. Vernberg (Eds.), *Biological Monitoring of Marine Pollutants.* Academic Press, New York, pp. 73–103.

Thomas, P., Bally, M., and Neff, J. M. (1982a). Ascorbic acid status of mullet, *Mugil cephalus* L., exposed to cadmium. *J. Fish Biol.* **20,** 183–196.

Thomas, P., Wofford, H. W., and Neff, J. M. (1982b). Effect of cadmium on glutathione content of mullet (*Mugil cephalus*) tissues. In W. B. Vernberg, A. Calabrese, F. P. Thurberg, and F. J. Vernberg (Eds.), *Physiological Mechanisms of Marine Pollutant Toxicity.* Academic Press, New York, pp. 73–103.

Thorpe, A., and Ince, B. W. (1974). The effects of pancreatic hormones, catecholamines, and glucose loading on blood metabolites in Northern pike (*Esox lucius* L.) *Gen. Comp. Endocrinol.* **23,** 29–44.

Tucker, R. K., and Costlow, J. D., Jr. (1975). Free amino acid changes in normal and eyestalkless megalopa larvae of the blue crab, *Callinectes sapidus,* during the course of the molt cycle. *Comp. Biochem. Physiol. A* **51A,** 75–78.

Turner, C. D., and Bagnara, J. T. (1976). *General Endocrinology.* Saunders, Philadelphia, PA.

Van den Thillar, T. G., Kesbede, F., and Van Waarde, A. (1976). Influence of anoxia on the energy metabolism of goldfish *Carasslus auratus* (L.). *Comp. Biochem. Physiol. A* **55A,** 329–336.

Verma, S. R., Tink, I. P., Gupta, A. K., and Dalela, R. C. (1981). Role of ascorbic acid in the toxicity of pesticides in a fresh water teleost. *Water, Air, Soil Pollut.* **16,** 107–114.

Verma, S. R., Saxena, M., and Tonk, I. P. (1984). The influence of Idet 20 on the biochemical composition and enzymes in the liver of *Clarias batrachus. Environ. Pollut. Ser. A* **33,** 245–255.

Vernberg, W. B., Calabrese, A., Thurberg, F. P., and Vernberg, F. J. (Eds.) (1982). *Physiological Mechanisms of Marine Pollutant Toxicity.* Academic Press, New York.

Versteeg, D. J., and Giesy, J. P. (1985). Lysosomal enzyme release in the bluegill sunfish (*Lepomis macrochirus* Rafinesque) exposed to cadmium. *Arch. Environ. Contam. Toxicol.* **14,** 631–640.

Voogt, P. A. (1983). Lipids: Their distribution and metabolism. In P. W. Hochachka (Ed.), *The Mollusca.* Vol. I. Academic Press, London, pp. 329–370.

Wahlqvist, I., and Nilsson, S. (1980). Adrenergic control of the cardiovascular system of the Atlantic cod, *Gadus morhua,* during "stress." *J. Comp. Physiol.* **137,** 145–150.

Walker-Simmons, M. and Atkinson, D. E. (1977). Functional capacities and the adenylate energy charge in *Escherichia coli* under conditions of nutritional stress. *J. Bacteriol.* **130,** 676–683.

Ward, J. V., Canton, S. P., and Gray, L. J. (1978). The stream environment and macroinvertebrate communities: Contrasting effects of mining in Colorado and the Eastern United States. *DOE Symp. Ser.* **48,** 176–187.

Warren, C. F. (1971). *Biology and Water Pollution Control.* Saunders, Philadelphia, PA.

Wedemeyer, G. A. (1969). Stress-induced ascorbic acid depletion and cortisol production in two salmonid fishes. *Comp. Biochem. Physiol.* **29,** 1247–1251.

Wedemeyer, G. A. (1971). The stress of formalin treatments in rainbow trout (*Salmo gairdneri*) and coho salmon (*Oncorhynchus kisutch*). *J. Fish. Res. Board Can.* **28,** 1899–1904.

Wedemeyer, G. A., and Nelson, N. C. (1975). Statistical methods for estimating normal blood chemistry ranges and variance in rainbow trout (*Salmo gairdneri*), shasta strain. *J. Fish. Res. Board Can.* **32,** 551–554.

Wedemeyer, G. A., and Yasutake, W. T. (1977). Clinical methods for the assessment of the effects of environmental stress on fish health. *Tech. Pap. U.S. Fish Wildl. Serv.* **89,** 1–18.

Weiss, C. M. (1961). Physiological effects of organic phosphorus insecticides on several species of fish. *Trans. Am. Fish. Soc.* **90,** 143–152.

Wendelaar Bonga, S. E., van der Meij, J. C. A., and Flik, G. (1984). Prolactin and acid stress in the teleost *Oreochromis* (formerly *Sarotherodon*) *mossambicus. Gen. Comp. Endocrinol.* **55,** 323–332.

Wilder, I. B., and Stanley, J. G. (1983). RNA-DNA ratio as an index to growth in salmonid fishes in the laboratory and in streams contaminated by carbaryl. *J. Fish Biol.* **22,** 165–172.

Williams, H. A., and Wooten, R. (1981). Some effects of therapeutic levels of formalin and copper sulphate on blood parameters in rainbow trout. *Aquaculture* **24,** 341–353.

Wong, K. L., and Klaassen, C. D. (1981). Relationship between liver and kidney levels of glutathione and metallothionein in rats. *Toxicology* **19,** 39–47.

Wroblewski, F., and LaDue, J. S. V. (1956). Glutamic pyruvic transaminase in cardiac and hepatic disease. *Proc. Soc. Exp. Biol. Med.* **91,** 569–571.

Yap, H. H., Desaiah, D. Cutkomp, L. K., and Koch, R. B. (1975). In vitro inhibition of fish Brain ATPase activity by cyclodiene insecticides and related compounds. *Bull. Environ. Contamn. Toxicol.* **14,** 163–167.

Zannoni, V. G., and Lynch, M. M. (1973). The role of ascorbic acid in drug metabolism. *Drug Metab. Rev.* **2,** 57–69.

# 9

# POSSIBLE LINKAGES BETWEEN MIXED FUNCTION OXYGENASE SYSTEMS, STEROID METABOLISM, REPRODUCTION, MOLTING, AND POLLUTION IN AQUATIC ANIMALS

***Richard F. Lee***

*Skidaway Institute of Oceanography, Savannah, Georgia 31416*



## 1. INTRODUCTION

In spite of the fact that many biochemical and physiological processes in aquatic animals are sensitive to pollutant-mediated alterations (Capuzzo, 1981; Perkins, 1979; Sastry and Miller, 1981; Thurberg, 1980; Waldichuk, 1979), many of these responses because they are not linked to population effects are of limited use in assessing pollution-related damage to aquatic populations and ecosystems. However, pollutant-induced biochemical changes associated or correlated with deleterious effects on reproductive processes can be used to diagnose early damage in aquatic populations. Even small decreases in reproductive success may cause population declines (Cushing, 1979; Verberg et al., 1978).

The mixed function oxygenase (MFO) systems play an important role in reproduction, invertebrate molting, and foreign compound metabolism. The MFO system is involved in the biosynthesis of steroids by steroidogenic tissues (Bollenbacher et al., 1977; Chang and O'Conner, 1978; Conney and Klutch, 1973; Feyereisen and Durst, 1978; Gower, 1975; Simpson and Boyd, 1967) and in the metabolism of foreign organic compounds, for example, certain pollutants, that convert lipophilic toxic organics to metabolites, which are more water soluble. I discuss in separate sections the role of the MFO system in metabolizing endogenous and exogenous substrates and linkages between the MFO system, reproduction, molting, and pollution. Exogenous substrates are often referred to as xenobiotics or foreign compounds; xenobiotics with toxic properties that are present in natural waters as a result of human activities are referred to as aquatic pollutants.

## 2. METABOLISM OF POLLUTANTS BY THE MFO SYSTEM

The MFO system of vertebrates and invertebrates is a multicomponent enzyme system composed of phospholipid, cytochrome P-450, and NADPH cytochrome P-450 reductase (Lee, 1981; Lu, 1976; Stegeman, 1981). The MFO system oxidizes foreign compounds by hydroxylation, *O*-dealkylation, *N*-dealkylation, or epoxidation. Highest MFO activity and P-450 content in fish is in their liver microsomes. Significant MFO activity and P-450 is also found in the intestine and kidney (Stegeman et al., 1979; Van Veld et al., 1985). The intestinal MFO system appears to metabolize foreign compounds carried in food, whereas the liver MFO system is involved in the metabolism of pollutants entering the fish via the water and food (Van Veld et al., 1985; Lech et al., 1982).

In aquatic invertebrates, the stomach microsomes of crabs and the intestinal microsomes of polychaetes have high MFO activity and P-450 content (Lee, 1981). Because MFO inhibitors are released during homogenization of crustacean hepatopancreas, measured MFO activity is low in this tissue, even though it has an important role in metabolizing pollutants in vivo (Pohl et al., 1974; Lee, 1981). As noted above, these oxidative reactions convert lipophilic foreign compounds to more water-soluble metabolites, thereby facilitating their elimination. However, in mammals, some foreign compounds, such as benzo[*a*]pyrene, are converted by the MFO system to reactive arene oxides that damage DNA and initiate carcinogenesis (Ames et al., 1972; Miller and Miller, 1971).

In mammals, exposure to certain inducers results in elevated MFO activity and P-450 content and often in the production of new forms of cytochrome P-450. Foreign compounds that induce new cytochromes P-450 are grouped into phenobarbital and polycyclic aromatic types. The phenobarbital class includes barbiturates, some insecticides, and noncoplanar derivatives of polychlorinated and polybrominated biphenyls (Goldstein, 1979). Compounds belonging to the polycyclic aromatic hydrocarbon class include benzo[*a*]pyrene, 5,6-benzoflavone, and coplanar polychlorinated and polybrominated biphenyls (Cantrell and Bresnick, 1971; Goldstein, 1979). Steroid hormones have been considered a third class of inducers, since the cytochromes P-450 induced by the synthetic compound, pregnenolone-16α-carbonitrile, differ from forms induced by phenobarbital and polycyclic aromatic hydrocarbons (Elshourbagy and Guzelian, 1980).

The MFO systems of fish and aquatic invertebrates do not respond to phenobarbital-type inducers but do respond to polycyclic aromatic hydrocarbon-type inducers (Lee, 1981; Stegeman, 1981). MFO activity and P-450 content in fish liver and kidneys increase after exposure to polycyclic aromatic hydrocarbon-type inducers (Addison et al., 1979; Bend et al., 1973; Carlson, 1978; Gruger et al., 1977a,b; Statham et al., 1978). There is good evidence that fish exposed to various pollutants in the field show an increase in MFO activity (Kurelec et al., 1977; Payne, 1976; Spies et al., 1982; Stegeman, 1978). Work by Kurelec et al. (1977) showed a large increase in the MFO activity of fish collected from an oil spill site (Fig. 9.1). Two months after the spill the MFO activity returned to near normal. Laboratory studies have shown the production of certain forms of liver cytochrome P-450 in fish after exposure to polycyclic aromatic hydrocarbons (Klotz et al., 1983, 1984; Williams and Buhler, 1984).

The polychaete, *Nereis virens,* exposed to a mixture of polychlorinated biphenyls had significantly higher MFO activity and P-450 content than control worms (Fries and Lee, 1984). Injection of polycyclic aromatic hydrocarbons into lobsters did not effect P-450 content of the hepatopancreas (James and Little, 1984).

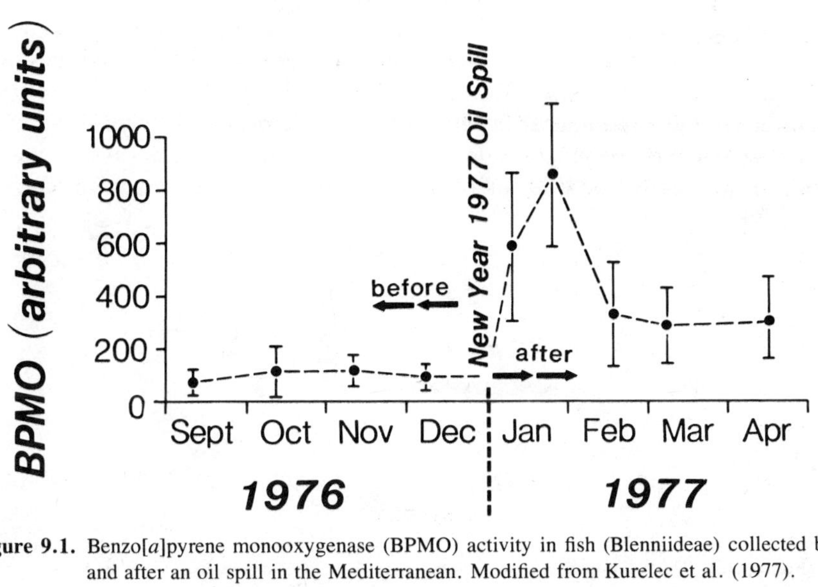


**Figure 9.1.** Benzo[*a*]pyrene monooxygenase (BPMO) activity in fish (Blenniideae) collected before and after an oil spill in the Mediterranean. Modified from Kurelec et al. (1977).

## 3. THE CYTOCHROME P-450 SYSTEM AND STEROID METABOLISM

The pituitary, liver, and gonads coordinate oogenesis in fish (Fig. 9.2). Fish gonadotropins stimulate the production of ovarian estrogens that control exogeneous vitellogenesis (Idler, 1982; Sundararaj et al., 1982). During gonad development female fish show a large increase in plasma estradiol levels (Lambert et al., 1978; Whitehead et al., 1978). Later stages of oocyte maturation and ovulation appear to be controlled mainly by $C_{21}$ steroids. In some fish 17$\alpha$-hydroxy-20$\beta$-dihydroprogesterone controls maturation (Campbell et al., 1980; Jalabert et al., 1978; Scott et al., 1982).

MFO systems play an important role in the biosynthesis of steroid hormones in vertebrates (Fig. 9.3). The hydroxylases and lyases involved in the synthesis of the various hormones are all cytochrome P-450 dependent enzyme systems. Different tissues and organelles are involved at various steps. For example, the ovary of the eel (*Anguilla anguilla*) hydroxylates progesterone to form 17$\alpha$-hydroxyprogesterone (Colombo and Belvedere, 1976). The 11$\beta$-hydroxylation of testosterone in fish gonads can play a role in the formation of 11-ketotestosterone, a potent fish androgen (Idler et al., 1971). The oxidation of steroids by the MFO system is important in both steroid activation and inactivation (Cooper et al., 1967; Simpson and Boyd, 1967).

There is good evidence that endocrine systems in crustaceans controls both reproduction and molting. Molt glands, also known as Y organs, secrete the crustacean molting hormone, 20-hydroxyecdysone (Chang and O'Conner, 1978). In many crustaceans mating occurs shortly after the final female molt. Thus, the molt cycle and ovarian development are closely linked and presumably many of the hormones controlling these processes are produced by cytochrome P-450-dependent MFO systems. Extracts from the testes of crabs can hydroxylate the Y-organ secretory product, ecdysone, to the active molting hormone, 20-hydroxyecdysone (Chang and O'Conner, 1978). Recent studies by James and Shiverick (1984) showed that P-450 from lobster hepatopancrease microsomes catalyzed the 16$\alpha$- and 6$\beta$-hydroxylation of progesterone and testosterone and the 21-hydroxylation of progesterone. Ecdysone was not a substrate for the hepatopancreas microsomal

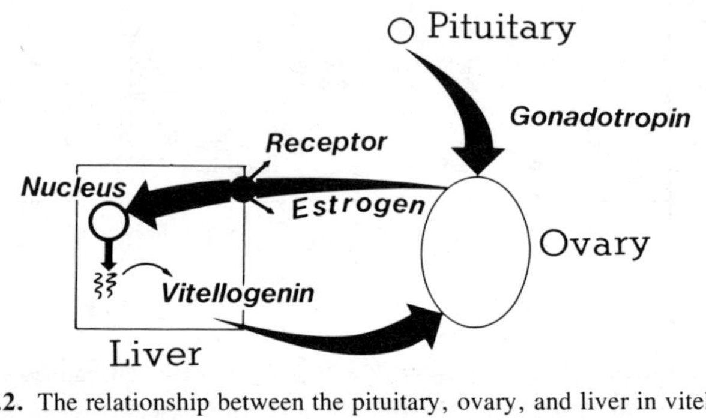


**Figure 9.2.** The relationship between the pituitary, ovary, and liver in vitellogenesis.

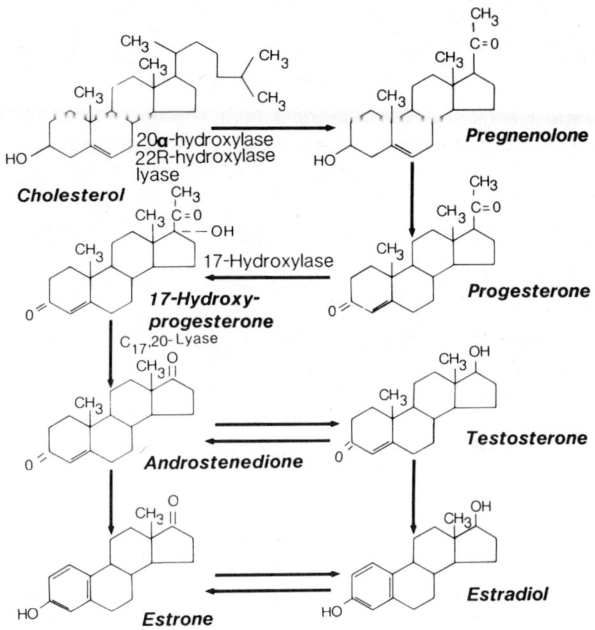


**Figure 9.3.** Biosynthetic pathways from cholesterol to testosterone and estrogens.

cytochrome P-450. High ecdysone monooxygenase activity was found in green gland mitochondria.

## 4. STEROID METABOLISM, XENOBIOTIC METABOLISM, REPRODUCTION, MOLTING, AND POLLUTION

As described above, steroids and foreign organic chemicals share common metabolic pathways that gives rise to important interactions (Brown, 1975; Conney and Klutch, 1963; Gessner, 1976). The similarity of the structure of steroids and some xenobiotics is illustrated in Fig. 9.4 where the benzo[*a*]pyrene and its diol metabolite are compared with estradiol and ecdysone. Such similarities in structure could account for the fact that some P-450 forms, isolated from fish or rat liver microsomes, can catalyze the hydroxylation of both benzo[*a*]pyrene and testosterone (Klotz et al., 1984; Lu, 1979).

Steroids affect xenobiotic metabolism and the injection of estradiol into fish resulted in a decrease of their MFO activity and P-450 content (Forlin et al., 1984; Stegeman et al., 1982). This effect appeared to be due to suppression of a constitutive form of P-450. In contrast, androgens increased fish P-450 content (Stegeman et al., 1982). Xenobiotics can affect steroid metabolism and exposure to P-450-inducing compounds can change the rate of hormone production or clearance (Kupfer and Bugler, 1976; Levin et al., 1974). The organochlorines stimulate

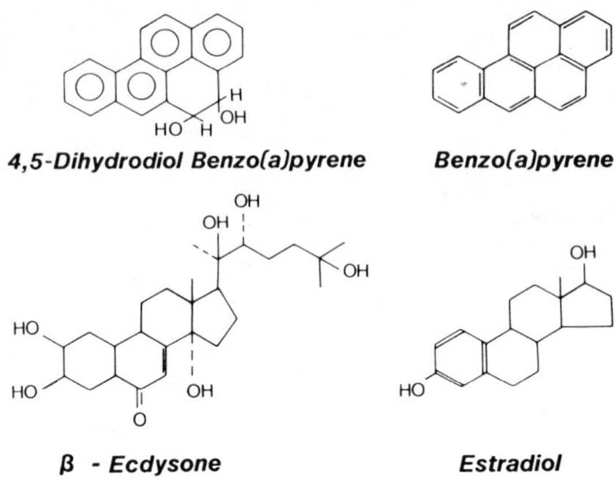


**Figure 9.4.** Structures of steroids, benzo[*a*]pyrene, and a benzo[*a*]pyrene metabolite.

steroid hydroxylase activity and alter the relative proportion of steroid metabolites in rats (Conney and Burns, 1972). Pentachlorobiphenyl enhanced the 7$\alpha$-hydroxylation of progesterone and testosterone in rats but suppressed the 16$\alpha$- and 6$\beta$-hydroxylation of both steroids (Yoshihara et al., 1982). Methycholanthrene increased the 7$\alpha$-hydroxylation of testosterone but did not affect the 16$\beta$-hydroxylation of the steroids.

Several environmental pollutants have been shown to have a direct stimulatory or inhibitory effect on adrenal and testicular steroidogenesis in fish (Freeman and Idler, 1975; Freeman and Sangalang, 1977). Exposure of trout to 3-methylcholanthrene increased the activity of liver microsomal 6$\beta$-hydroxylase, which acts on androstenedione (Hansson et al., 1980). When trout were exposed to water from a wastewater treatment plant both benzo[*a*]pyrene hydroxylase and androstenedione 6$\beta$-hydroxylase activity increased (Forlin and Hansson, 1982). Truscott et al. (1983) noted lowered levels of androgens in plasma of fish exposed to petroleum in the water. Injections of polychlorinated biphenyls into trout and carp caused a significant reduction of androgens and estrogens and an increase in cytochrome P-450 content (Sivarajah et al., 1978).

The reproduction of fish as discussed above involves signals from the pituitary gland to the gonads to produce estradiol (Fig. 9.2). The estradiol in turn induces the liver to synthesize vitellogenin, which is carried through the blood to the gonads where it becomes the major egg protein. Thus, organic pollutants that enter the fish can act at several points that control reproduction. Chen and Sonstegard (1984) found that production of vitellogenin, induced by estradiol, was reduced in fish exposed to polychlorinated biphenyls and that other MFO inducers, such as $\beta$-naphthoflavone, reduced synthesis of vitellogenin messenger RNA.

Perhaps the best evidence linking MFO activity with reproductive success is the work of Spies et al. (1984, 1985) with starry flounder (*Platicthys stellatus*) collected from different parts of San Francisco Bay. Fertilization success was inversely related to MFO activity (Fig. 9.5). High MFO activity was found in fish from areas of this bay that had the highest concentrations of chlorinated hydrocar-

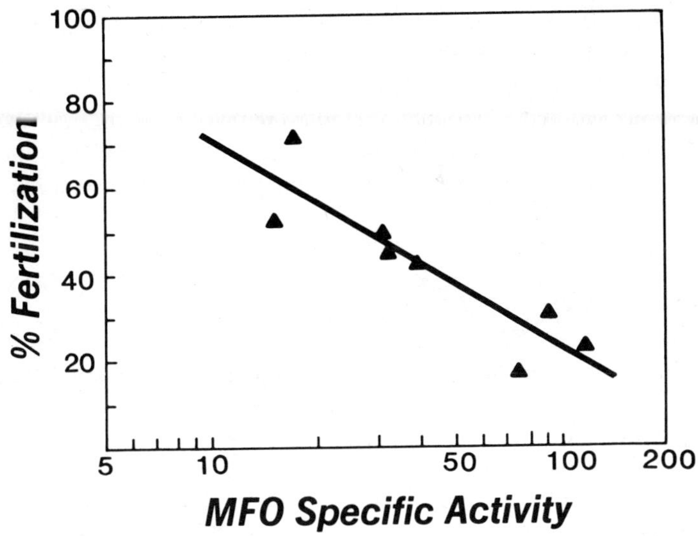


**Figure 9.5.** Mixed-function oxygenase (MFO) activity versus percent fertilization in starry flounder collected from different parts of San Francisco Bay. Modified from Spies et al. (1984).

bons. There was no correlation of MFO activity with sediment petroleum hydrocarbon concentrations. Thus, in this area, high MFO activity manifested effects at the population level. Stein et al. (1986) found that extracts of a sediment high in polycyclic aromatic hydrocarbons from Puget Sound in Washington impaired reproductive success of gravid English sole (*Parophrys uetulus*). An interesting related observation in mammals is the work of Mattison (1980) who found that a number of polycyclic aromatic hydrocarbons destroyed primordial oocytes in mouse ovary and the rate of oocyte destruction was proportional to the ovarian MFO activity.

As discussed earlier, MFO systems in crustaceans play an important role in the biosynthesis and metabolism of molting hormones. A variety of xenobiotics can act as a substrate for this MFO system and thus interfere with molting. A variety of pollutants are known to inhibit crustacean molting. The polychlorinated mixture, Aroclor 1254, inhibited fiddler crab molting (Fingerman and Fingerman, 1977), the pesticide methoxychlor prolonged larval development in several crab species (Armstrong et al., 1976; Bookhout et al., 1976), exposure to oil decreased molting success in Tanner crabs (Karinen and Rice, 1974), and Kepone-contaminated oysters fed to blue crabs resulted in less frequent molts by the crabs (Schimmel et al., 1979). We can only speculate that such exposure to pollutants interferes with MFO systems that control molting.

Polychaetes *Nereis virens* collected from a marine harbor containing sediments high in petroleum had six times the MFO activity as worms from a ''clean'' reference site (Fries and Lee, 1984). While many of the worms collected from the reference site had fully developed gametes, worms from the oiled area lacked or had undeveloped gametes in spring, summer, and fall. Failure of *Nereis virens* to form gametes at the petroleum-contaminated site suggests that these worms were unable to reproduce normally. Presumably, the population is maintained by larvae brought in by tidal currents. We speculate that ovulation in the worms was inhibited by oil in the sediment, possibly related to effects on steroid-regulating MFO systems. As noted above, Mattison (1980) found destruction of mouse ovary oocytes by aromatic hydrocarbons with destruction proportional to ovarian MFO activity.

## 5. SUMMARY AND CONCLUSIONS

The work described above shows that the reproduction in aquatic animals is affected by a variety of organic pollutants. These pollutants, as well as reproductive hormones, are metabolized by the MFO system. Since steroids and some organic pollutants show structural similarities and common metabolic pathways, the introduction of organic pollutants can change the rate of hormone production and/or clearance. Field studies have shown a linkage with MFO activity and the reproductive success of coastal fish and invertebrates. High MFO activity and impaired reproductive success was found in fish collected in polluted sites. I am not aware of field studies with freshwater fish or invertebrates relating MFO activity, reproductive activity, and pollution. There appears to be a need for such studies.

In addition to the MFO system, there are other enzyme systems involved with both reproductive hormones and xenobiotics. For example, UDP-glucuronyltransferases, a conjugating enzyme, can conjugate glucuronic acid to a xenobiotic or a steroid hormone (Rao et al., 1977). Thus, such conjugating enzyme systems, which control concentrations of steroid hormones, can be affected by xenobiotics. Possible linkage between conjugating enzyme systems, reproduction, and pollution in aquatic species remains to be demonstrated.

Finally, it should be noted that there are a number of field studies where both pollutants and viable reproducing populations were present. In two different field studies it was found that fish that lived in polluted areas for several years had elevated MFO activity but yet reproduction appeared to be "normal" (Kezic et al., 1983; Payne and May, 1979; Payne and Penrose, 1975; Payne et al., 1984). Trout exposed to crude oil, a good inducer of MFO activity, in the diet during sexual maturation showed no significant impairment of reproductive success (Hodgins et al., 1977). There may need to be an "overloading" of the detoxification system (e.g., MFO system) by some organic pollutants before effects are observed on the reproduction of aquatic species. Also, some organic pollutants may induce MFO activity in fish without a corresponding effect on reproduction.

## ACKNOWLEDGMENTS

The work reported here was supported by National Science Foundation Grant OCE-8017893.

## REFERENCES


Addison, R. F., Zinck, M. E., Willis, D. E., and Darrow, D. C. (1979). Induction of hepatic mixed function oxidases in trout by polychlorinated biphenyl and butylated monochlorodiphenyl ethers. *Toxicol. Appl. Pharmacol.* **49,** 245–248.

Ames, B. N., Sims, P., and Grover, P. L. (1972). Epoxides of carcinogenic polycyclic aromatic hydrocarbons are frameshift mutagens. *Science* **176,** 47–49.

Armstrong, D. A., Buchanan, D. V., Mallon, M. H., Caldwell, R. S., and Milleman, R. E. (1976). Toxicity of the insecticide methoxychlor to the dungeness crab *Cancer magister. Mar. Biol. (Berlin)* **38,** 239–252.

Bend, J. R., Pohl, R. J., and Fouts, J. R. (1973). Further studies of microsomal mixed function oxygenase systems of the little skate *Raja erinacea* including its response to some xenobiotics. *Bull. Mt. Desert Isl. Biol. Lab.* **13,** 9.

Bollenbacher, W. E., Smith, S. L., Wielgus, J. J., and Gilbert, L. I. (1977). Evidence for an α-ecdysone cytochrome P-450 mixed function oxidase in insect fat body mitochondria. *Nature (London)* **268,** 660–663.

Bookhout, C. G., Costlow, J. D., and Monroe, R. (1976). Effects of methoxychlor on larval development of mud crab and blue crab. *Water, Air, Soil Pollut.* **5,** 349–365.

Brown, S. S. (1975). Metabolic interactions involving drugs or foreign compounds. In D. E. Hathway (Ed.), *Foreign Compound Metabolism in Mammals.* Chemical Society, London, pp. 591–630.

Campbell, C. M., Fostier, A., Jalabert, B., and Truscott, B. (1980). Identification and quantification of steroids in the serum of rainbow trout during spermiation and oocyte maturation. *J. Endocrinol.* **85,** 371–378.

Cantrell, E., and Bresnick, E. (1971). Evidence for type II induction of microsomal enzymes by β-naphthoflavone in rat liver preparations. *Life Sci.* **10,** 1195–1200.

Capuzzo, J. M. (1981). Predicting pollution effects in the marine environment. *Oceanus* **24,** 25–33.

Carlson, G. P. (1978). Induction of cytochrome P-450 by halogenated benzenes. *Biochem. Pharmacol.* **27,** 361–363.

Chang, E. S., and O'Conner, J. D. (1978). *In vitro* secretion and hydroxylation of α-ecdysone as a function of the crustacean molt cycle. *Gen. Comp. Endocrinol.* **36,** 151–160.

Chen, T. T., and Sonstegard, R. A. (1984). Development of a rapid sensitive and quantitative test for the assessment of the effect of xenobiotics on reproduction in fish. *Mar. Environ. Res.* **14,** 429–430.

Colombo, L., and Belvedere, P. C. (1976). Steroid biosynthesis by the ovary of the European eel, *Anguilla anguilla* L., at the silver stage. *Gen. Comp. Endocrinol.* **28,** 371–385.

Conney, A. H., and Burns, J. J. (1972). Metabolic interactions among environmental chemicals and drugs. *Science* **178,** 576–586.

Conney, A. H., and Klutch, A. (1963). Increased activity of androgen hydroxylases in liver microsomes of rats pretreated with phenobarbital and other drugs. *J. Biol. Chem.* **238,** 1611–1617.

Cooper, D. Y., Novack, B., Foroff, O., Slade, A., Saunders, E., Narasimhulu, S., and Rosenthal, O. (1967). Photochemical action spectrum of reconstituted steroid 11β-hydroxylase of bovine adrenocortical mitochondria. *Fed. Proc., Fed. Am. Soc. Exp. Biol.* **26,** 341.

Cushing, D. H. (1979). The monitoring of biological effects: The separation of natural changes from those induced by pollution. *Philos. Trans. R. Soc. London, Ser. B* **286,** 597–609.

Elshourbagy, N. A., and Guzelian, P. S. (1980). Separation, purification, and characterization of a novel form of hepatic cytochrome P-450 from rats treated with pregnenolone-16α-carbonitrile. *J. Biol. Chem.* **255,** 1279–1285.

Feyereisen, R., and Durst, F. (1978). Ecdysterone biosynthesis: A microsomal cytochrome P-450-linked ecdysone 20-monooxygenase from tissues of the African migratory locust. *Eur. J. Biochem.* **88,** 37–47.

Fingerman, S. W., and Fingerman, M. (1977). Effects of a polychlorinated biphenyl and a polychlorinated dibenzofuran on molting of the fiddler crab, *Uca pugilator*. *Bull. Environ. Contam. Toxicol.* **18,** 138–142.

Forlin, L., and Hansson, T. (1982). Effects of treated municipal wastewater on the hepatic, xenobiotic and steroid metabolism in trout. *Ecotoxicol. Environ. Saf.* **6,** 41–48.

Forlin, L., Andersson, T., Koivusaari, U., and Hansson, T. (1984). Influence of biological and environmental factors on hepatic steroid and xenobiotic metabolism in fish: Interaction with PCB and β-naphthoflavone. *Mar. Environ. Res.* **14,** 47–58.

Freeman, H. C., and Idler, D. R. (1975). The effect of polychlorinated biphenyl on steroidogenesis and reproduction in the brook trout (*Salvelinus fontinalis*). *Can. J. Biochem.* **53,** 666–670.

Freeman, H. C., and Sangalang, G. B. (1977). A study of the effects of methyl mercury, cadmium, arsenic, selenium, and a PCB (Aroclor 1254) on adrenal and testicular steroidogenesis *in vivo* by the gray seal *Halichocrus grypus*. *Arch. Environ. Contam. Toxicol.* **5,** 369–383.

Fries, C. R., and Lee, R. F. (1984). Pollutant effects on the mixed function oxygenase (MFO) and reproductive systems of the marine polychaete *Nereis virens*. *Mar. Biol. (Berlin)* **79,** 187–193.

Gessner, T. (1976). A comparison of steroid glucuronyltransferases of rat liver and hepatoma, and effect of phenobarbitone and 3-methylcholanthrene induction. *Biochem. Soc. Trans.* **4,** 522–523.

Goldstein, J. A. (1979). The structure-activity relationships of halogenated biphenyls as enzyme inducers. *Ann. N.Y. Acad. Sci.* **320,** 164–178.

Gower, D. B. (1975). Properties and subcellular location of enzymes involved in steroidogenesis (and the role of cytochrome P-450). In H. L. H. Makin (Ed.), *Biochemistry of Steroid Hormones*. Blackwell, London, pp. 105–125.

Gruger, E. H., Jr., Wekell, M. M., Numoto, P. T., and Craddock, D. R. (1977a). Induction of aryl hydrocarbon hydroxylase in salmon exposed to petroleum dissolved in seawater and to petroleum and polychlorinated biphenyls, separate and together, in food. *Bull. Environ. Contam. Toxicol.* **17,** 512–520.

Gruger, E. H., Jr., Wekell, M. M., and Robisch, P. A. (1977b). Effects of chlorinated biphenyls and petroleum hydrocarbons on the activity of hepatic aryl hydrocarbon hydroxylase of coho salmon (*Oncorhynchus kisutch*) and chinook salmon (*O. tshawytscha*). In D. A. Wolfe (Ed.), *Fate and Effects of Petroleum Hydrocarbons in Marine Organisms and Ecosystems*. Pergamon, New York, pp. 323–332.

Hansson, T., Rafter, J., and Gustafsson, J. -A. (1980). Effects of some common inducers on the hepatic microsomal metabolism of androstenedione in rainbow trout with special reference to cytochrome P-450-dependent enzymes. *Biochem. Pharmacol.* **29,** 583–587.

Hodgins, H. O., Gronlund, W. D., Mighell, J. L., Hawkes, J. W., and Robisch, P. A. (1977). Effect of crude oil on trout reproduction. In D. A. Wolfe (Ed.), *Fate and Effects of Petroleum Hydrocarbons in Marine Ecosystems and Organisms*. Pergamon, New York, pp. 143–150.

Idler, D. R. (1982). Some perspectives on fish gonadotropins. In C. J. J. Richter and H. J. Th. Goos (Eds.), *Reproductive Physiology of Fish*. Pudoc, Wageningen, Netherlands, pp. 4–13.

Idler, D. R., Horne, D. A., and Sangalang, G. B. (1971). Identification and quantification of the major androgens in testicular and peripheral plasma of Atlantic salmon (*Salmo salar*) during sexual maturation. *J. Comp. Endocrinol.* **16,** 257–267.

Jalabert, B., Goetz, F. W., Breton, B., Fostier, A., and Donaldson, E. M. (1978). Precocious induction of oocyte maturation and ovulation in coho salmon *Oncorhynchus kisutch*. *J. Fish. Res. Board Can.* **35,** 1423–1429.

James, M. O., and Little, P. J. (1984). 3-Methylcholanthrene does not induce *in vitro* xenobiotic metabolism in spiny lobster hepatopancreas, or affect *in vivo* disposition of benzo(a)pyrene. *Comp. Biochem. Physiol. C* **78C,** 241–245.

James, M. O., and Shiverick, K. T. (1984). Cytochrome P-450 dependent oxidation of progesterone, testosterone, and ecdysone in the spiny lobster, *Panulirus argus*. *Arch. Biochem. Biophys.* **233,** 1–9.

Karinen, J. F., and Rice, S. D. (1974). Effects of Prudhoe Bay crude oil on molting Tanner crabs, *Chionoecetes bairdi. Mar. Fish. Rev.* **36,** 31–37.

Kezic, N., Britvic, S., Protic, M., Simmons, J. E., Rijavec, M., Zahn, R. K., and Kurelec, B. (1983). Activity of benzo(a)pyrene monoxygenase in fish from the Sava River, Yugoslavia: Correlation with pollution. *Sci. Total Environ.* **21,** 59–69.

Klotz, A. V., Stegeman, J. J., and Walsh, C. (1983). An aryl hydrocarbon hydroxylating hepatic cytochrome P-450 from the marine fish *Stenotomus chrysops. Arch. Biochem. Biophys.* **226,** 578–592.

Klotz, A. V., Stegemen, J. J., and Walsh, C. (1984). Multiple isozymes of hepatic cytochrome P-450 from the marine teleost fish scup (*Stenotomus chrysops*). *Mar. Environ. Res.* **14,** 402–404.

Kupfer, D., and Bugler, W. H. (1976). Interactions of chlorinated hydrocarbons with steroid hormones. *Fed. Proc., Fed. Am. Soc. Exp. Biol.* **35,** 2603–2608.

Kurelec, B., Britvic, S., Rijavec, M., Muller, W. E. G., and Zahn, R. K. (1977). Benzo[a]pyrene monooxygenase induction in marine fish—molecular response to oil pollution. *Mar. Biol. (Berlin)* **44,** 211–216.

Lambert, J. G. D., Bosman, G. I. C. G., van den Hurk, R., and van Oordt, P. G. W. J. (1978). Annual cycle of plasma estradiol-17$\beta$ in the female trout *Salmo gairdneri. Ann. Biol. Anim., Biochim., Biophys.* **18,** 923–927.

Lech, J. J., Vodicnik, M. J., and Elcombe, C. R. (1982). Induction of monoxygenase activity in fish. In L. J. Weber (Ed.), *Aquatic Toxicology.* Raven Press, New York, pp. 107–148.

Lee, R. F. (1981). Mixed function oxygenase (MFO) in marine invertebrates. *Mar. Biol. Lett.* **2,** 87–105.

Levin, W., Welch, R. M., and Conney, A. H. (1974). Increased liver microsomal androgen metabolism by phenobarbital: Correlation with decreased androgen action on the seminal vesicles of the rat. *J. Pharmacol. Exp. Ther.* **188,** 287–292.

Lu, A. Y. H. (1976). Liver microsomal drug-metabolizing enzymes: Functional components and their properties. *Fed. Proc., Fed. Am. Soc. Exp. Biol.* **35,** 2460–2463.

Lu, A. Y. H. (1979). Multiplicity of liver drug metabolizing enzymes. *Drug Metab. Rev.* **10,** 187–208.

Mattison, D. R. (1980). Morphology of oocyte and follicle destruction by polycyclic aromatic hydrocarbons in mice. *Toxicol. Appl. Pharmacol.* **53,** 249–259.

Miller, J. A., and Miller, E. C. (1971). Chemical carcinogenesis: Mechanisms and approaches to its control. *J. Natl. Cancer Inst. (U.S.)* **47,** 5–14.

Payne, J. F. (1976). Field evaluation of benzopyrene hydroxylase induction as a monitor for marine petroleum pollution. *Science* **191,** 945–946.

Payne, J. F., and May, N. (1979). Further studies on the effect of petroleum hydrocarbons on mixed-function oxidases in marine organisms. In M. A. W. Khan, J. J. Lech, and J. J. Menn (Eds.), *Pesticide and Xenobiotic Metabolism in Aquatic Organisms.* Am. Chem. Soc., Washington, DC, pp. 339–347.

Payne, J. F., and Penrose, W. R. (1975). Induction of aryl hydrocarbon benzo[*a*]pyrene hydroxylase in fish by petroleum. *Bull. Environ. Contam. Toxicol.* **14,** 112–116.

Payne, J. F., Bauld, C., Dey, A. C., Kiceniuk, J. W., and Williams, U. (1984). Selectivity of mixed-function oxygenage enzyme inductions in flounder (*Pseudo-pleuronectes americanus*) collected at the site of the Baie Verte, Newfoundland, oil spill. *Comp. Biochem. Physiol. C* **79C,** 15–19.

Perkins, E. J. (1979). The need for sublethal studies. *Philos. Trans. R. Soc. London, Ser. B* **286,** 424–442.

Pohl, R. J., Bend, J. R., Guarino, A. M., and Fouts, J. R. (1974). Hepatic microsomal mixed-function oxidase activity of several marine species from coastal Maine. *Drug Metab. Dispos.* **2,** 545–555.

Rao, G. S., Haueter, G., Rao, M. L., and Breuer, H. (1977). Steroid glucuronyl transferases of rat liver. *Biochem. J.* **162,** 545–556.

Sastry, A. N., and Miller, D. C. (1981). Application of biochemical and physiological responses to water quality monitoring. In F. J. Vernberg, A. Calabrese, F. P. Thurberg, and W. B. Vernberg (Eds.), *Biological Monitoring of Marine Pollutants.* Academic Press, New York, pp. 265–294.

Schimmel, S. C., Patrick, J. M., Fass, L. F., Oglesby, J. L., and Wilson, A. J. (1979). Kepone: Toxicity to and bioaccumulation by blue crabs. *Estuaries* **2,** 9–15.

Scott, A. P., Sheldrick, E. L., and Flint, A. P. F. (1982). Measurement of 17, 20-dihydroxy-4-pregren-3-one in plasma of trout (*Salmon gairdneri* Richardson): Seasonal changes and response to salmon pituitary extract. *Gen. Comp. Endocrinol.* **46,** 444–451.

Simpson, E. R., and Boyd, G. S. (1967). The cholesterol side-chain cleavage system of bovine adrenal cortex. *Eur. J. Biochem.* **2,** 275–285.

Sivarajah, K., Franklin, C. S., and Williams, W. P. (1978). The effects of polychlorinated biphenyls on plasma steroid levels and hepatic microsomal enzymes in fish. *J. Fish Biol.* **13,** 401–409.

Spies, R. B., Felton, J. S., and Dillard, L. (1982). Hepatic mixed function oxidases in California flatfish are increased in contaminated environments and by oil and PCB ingestion. *Mar. Biol. (Berlin)* **70,** 117–127.

Spies, R. B., Rice, D. W., and Ireland, R. R. (1984). Preliminary studies of growth, reproduction and activity of hepatic mixed-function oxidase in *Platichthys stellatus. Mar. Environ. Res.* **14,** 426–428.

Spies, R. B., Rice, D. W., Montagna, P. A., and Ireland, R. R. (1985). Reproductive success, xenobiotic contaminants and hepatic mixed-function oxidase (MFO) activity in *Platichthys stellatus* populations from San Francisco Bay. *Mar. Environ. Res.* **17,** 117–121.

Statham, C. N., Elcombe, C. R., Szyjka, S. P., and Lech, J. J. (1978). Effect of polycyclic aromatic hydrocarbons on hepatic microsomal enzymes and disposition of methyl naphthalene in rainbow trout *in vivo. Xenobiotica* **8,** 65–71.

Stegeman, J. J. (1978). Influence of environmental contamination on cytochrome P-450 mixed function oxygenases in fish: Implications for recovery in the Wild Harbor Marsh. *J. Fish. Res. Board Can.* **35,** 668–674.

Stegeman, J. J. (1981). Polynuclear aromatic hydrocarbons and their metabolism in the marine environment. In H. V. Gelboin and P. O. P. Ts'o (Eds.), *Polycyclic Hydrocarbons and Cancer.* Academic Press, New York, Vol. 3, pp. 1–60.

Stegeman, J. J., Binder, R. L., and Orren, A. (1979). Hepatic and extrahepatic microsomal electron transport components and mixed-function oxygenases in the marine fish *Stenotomus versicolor. Biochem. Pharmacol.* **28,** 3431–3439.

Stegeman, J. J., Pajor, A. M., and Thomas, P. (1982). Influence of estradiol and testosterone on cytochrome P-450 and monooxygenase activity in immature brook trout (*Salvelinus fontinalis*). *Biochem. Pharmacol.* **31,** 3379–2989.

Stein, J. E., Sanborn, H. R., Collier, T. K., and Varanasi, U. (1986). Effects of contaminant exposure on gravid English sole (*Parophrys vetulus*). Pac. Northwest Assoc. Toxicol. Meet., Abstract.

Sundararaj, B. I., Nath, P., and Burzawa-Gerard, L. (1982). Synthesis of vitellogenin and its uptake by the ovary in catfish *Heteropneustes fossilis* (Bloch) in response to carp gonadotropin and its subunits. *Gen. Comp. Endocrinol.* **46,** 93–98.

Thurberg, F. P. (1980). The use of physiological techniques in monitoring pollution: A consideration of its problems and current research. *Rapp. P.-V. Reun., Cons. Int. Explor. Mer.* **179,** 82–87.

Truscott, B., Walsh, J. M., Burton, M. P., Payne, J. F., and Idler, D. R. (1983). Effect of acute exposure to crude petroleum on some reproductive hormones in salmon and flounder. *Comp. Biochem. Physiol. C* **75C,** 121–130.

Van Veld, P. A., Vetter, R. D., Lee, R. F., and Patton, J. S. (1985). Factors affecting the metabolism of lipid solubilized benzo(a)pyrene by the intestine of spot (*Leiostomus xanthurus*). *Mar. Environ. Res.* **17,** 154–155.

Vernberg, W. B., McKellar, H., Jr., and Vernberg, J. F. (1978). Toxicity studies and environmental impact assessment. *Environ. Manage.* **1,** 239–243.

Waldichuk, M. (1979). Review of the problems. *Philos. Trans. R. Soc. London, Ser. B* **286,** 399–424.

Whitehead, C., Bromage, N. R., and Foster, J. R. M. (1978). Seasonal changes in reproductive function of the rainbow trout (*Salmo gairdneri*). *J. Fish Biol.* **12,** 601–608.

Williams, D. E., and Buhler, D. R. (1984). Benzo(a)pyrene-hydroxylase catalyzed by purified isozymes of cytochrome P-450 from $\beta$-naphthoflavone fed rainbow trout. *Biochem. Pharmacol.* **33,** 3743–3753.

Yoshihara, S., Nagata, K., Wada, I., Yoshimura, H., Kuroki, H., and Masuda, Y. (1982). A unique change of steroid metabolism in rat liver microsomes induced with highly toxic polychlorinated biphenyl (PCB) and polychlorinated dibenzofuran (PCDF). *J. Pharm. Dyn.* **5,** 994–1004.

# 10

# ENVIRONMENTAL INDUCTION OF MONOOXYGENASE ACTIVITY IN FISH

***Mark J. Melancon, Robert L. Binder,* and John J. Lech***

*Department of Pharmacology and Toxicology, Medical College of Wisconsin, Milwaukee, Wisconsin 53201*



**Present address:* Human and Environmental Safety Division, The Procter and Gamble Company, Miami Valley Laboratories, Cincinnati, OH.

## 1. INDUCTION OF MONOOXYGENASE ACTIVITY IN FISH

### 1.1. Background

Much of the capability of organisms to oxidize xenobiotic compounds can be isolated from tissue homogenates in the microsome fraction. Although microsomal fractions can be prepared from most tissues, in organisms that possess a liver, the liver is the major site of xenobiotic metabolism. The enzyme system responsible for this metabolism contains cytochrome P-450, so-called because of the absorption maximum at approximately 450 nm, of the reduced carbon monoxide spectrum. There are actually a number of cytochromes P-450, some of which are present in a control organism (one that has not been exposed to an inducing agent) and others that appear in significant quantities only in response to inducing agents. In this context, an inducing agent, that is an inducer of cytochrome P-450, is any chemical that either will increase the amount of a cytochrome P-450 that is present in a control organism or will cause the formation of a new cytochrome P-450 that is not present in a control organism.

Each of these cytochromes P-450 is capable of metabolizing a wide variety of substrates. Among these metabolizable substrates may be endogenous compounds, PAHs, chlorinated hydrocarbons, and so on. To assay for the presence of cytochromes P-450 by these enzyme activities, the substrates are generally selected for the sensitivity of the assay and/or for its specificity for a particular type of cytochrome P-450. The enzyme activities of cytochromes P-450 may be described in a general way by the terms monooxygenase and mixed function oxidase, but individual enzyme activities also may be identified by the name of the substrate and the reaction catalyzed.

The various cytochromes P-450 may differ in absorbance maxima over a range of approximately 447.5 to 452 nm, in molecular weights, in electrophoretic mobilitics, and in cnzymc activities. In some cases these properties may be so similar that HPLC or antibodies are required to distinguish between various cytochromes P-450.

A wide variety of chemicals have the ability to induce the formation (increase the amounts) of one or more cytochrome P-450 in an organism. The early studies on cytochromes P-450 and their induction were done mainly in mammalian species, particularly the rat. Most of the known inducing agents fall into the historical categories of phenobarbital-type inducers and 3-methylcholanthrene-type inducers. These categories are based on the properties of the cytochromes P-450 induced rather than on the properties of the inducing agents. Some of the chemicals of environmental interest that are phenobarbital-type inducers are DDT, nonplanar PCBs, and nonplanar PBBs. Among the chemicals of environmental interest that are 3-methycholanthrene-type inducers are benzo[a]pyrene, 7,12-dimethylbenz[*a*]anthracene, 2,3,7,8-tetrachlorodibenzodioxin, coplanar PCBs, and coplanar PBBs.

Since the mid-1970s, there have been numerous reports on the induction of microsomal enzyme activity in fish. Generally, these reports have been concerned with hepatic microsomal monooxygenase activities. A summary of these reports is presented in Table 10.1. The important aspects of these data are that all of the studies demonstrated induction of monooxygenase activity by a variety of 3-methylcholanthrene-type inducers and that this induction occurred in numerous fish species. Basically 3-methylcholanthrene-type induction has been observed in every fish species studied.

The effectiveness of phenobarbital-type inducers in fish is summarized in Table 10.2. The lack of consistant response to a variety of these inducers in a number of fish species is obvious. The positive responses are generally slight and in one case (Burns, 1976) baseline activity was very low and the induced fish had activity comparable to usual controls. Overall, there is little if any response of monooxygenase activity in fish to phenobarbital-type inducing agents. Overall, it appears reasonable to conclude that under most circumstances phenobarbital-type inducers have no significant effect on cytochromes P-450 and related monooxygenase activities in fish.

For a review of induction of cytochromes P-450 and monooxygenase activity in fish see Lech et al. (1982).

### 1.2. Dose Response

Although many of the chemicals that are inducing agents in fish are found in the aquatic environment, most of the studies of induction of hepatic microsomal monooxygenase activity in fish have utilized doses of inducing agents larger than those generally seen in the environment. Of the pollutants that are inducers in fish there are probably more data available on the concentrations of PCBs in fish than for any other class of compounds. At various times fish in the Great Lakes, the Mississippi River, the Hudson River, and other locations have been found to contain PCBs at 40 μg/g or even higher concentrations. In general the PCB concentrations in fish in these locations are no longer that high, but the present action level for PCBs in fish for interstate commerce is 2 μg/g and it is not unusual for large trout, salmon, and carp from these locations to exceed this limit.

Several studies that utilized relatively low levels of inducing agents have been reported and are presented in Figures 10.1–10.4. The response of rainbow trout to a range of 1 to 1000 mg/kg of a commercial PCB mixture is shown in Figure 10.1. Although the lowest dose of PCB at which there was a significant increase in monooxygenase activity was 10 mg/kg, there appeared to be a trend toward increases in monooxygenase activity even at the lower doses. The lack of significant induction at the lower doses is probably due more to the small number of fish used rather than to a lack of responsiveness. A similar study with greater numbers of rainbow trout (Fig. 10.2) showed a significant increase in hepatic microsomal monooxygenase activity after a dose of 0.2 mg/kg of Aroclor 1254. When rainbow trout were injected with a range of doses of benzo[*a*]pyrene the lowest dose at which a significant elevation of hepatic microsomal monooxygenase activity was seen was 0.3 mg/kg (Fig. 10.3). A significant elevation in hepatic microsomal

**Table 10.1 Induction of Monoxygenase Activity in Fish by 3-Methylcholanthrene-type Inducers**

| Species | Tissue | Inducer[a] | References |
|---|---|---|---|
| Rainbow trout | Liver | PCBs | Lidman et al. (1976); Sivarajah et al. (1978); Elcombe and Lech (1979) |
| | | PBBs | Elcombe and Lech (1978) |
| | | 3-MC | Statham et al. (1978); Forlin (1980); Stegeman et al. (1981) |
| | | DMBA, BNF | Statham et al. (1978) |
| | | TCDD | Vodicnik et al. (1981) |
| Brook trout | Liver | PCB, 3-MC | Addison et al. (1978) |
| Brown trout | Liver | Petroleum | Payne and Penrose (1975) |
| Coho salmon | Liver | PCB | Gruger et al. (1976) |
| Carp | Liver | PCB | Sivarajah et al. (1978) |
| | Liver | BP, AFX, crude oil | Kurelec et al. (1979) |
| | Liver, kidney | PCB, BNF | Melancon et al. (1981) |
| Goldfish | Liver | BNF | Gooch and Matsumura (1983) |
| Northern pike | Liver | 3-MC | Balk et al. (1980) |
| Catfish | Liver | PCB | Hill et al. (1976) |
| Walleye | Liver | BNF | Melancon and Lech (unpublished observations) |
| Scup | Liver | 3-MC, BNF | Stegeman et al. (1981); Chevion et al. (1977) |
| Croaker | Liver | 3-MC | Stegeman et al. (1981) |
| Mullet | Liver | PCB | Narbonne and Gallis (1979) |
| Sculpin | Liver | Refinery effluent | Ridlington et al. (1982) |
| Killifish | Liver | PCBs, No. 2 fuel oil | Binder and Stegeman (1980) |
| Flounder | Liver | 3-MC | Pohl et al. (1974) |
| Sheepshead | Liver | 3-MC | Little et al. (1981) |
| | | PCBs | James and Little (1981) |

[a]PCBs, polychlorinated biphenyls; PBBs, polychlorinated biphenyls; 3-MC, 3-methylcholanthrene; DMBA, 7,12-dimethylbenzanthracene; BNF, $\beta$-naphthoflavone; TCDD, 2,3,7,8-tetrachlorodibenzodioxin; BP, benzo[*a*]pyrene; AFX, aflatoxin.

**Table 10.2 Effect of Phenobarbital-type Inducers on Monoxygenase Activity in Fish Hepatic Microsomes**

| Species | Inducer[a] | Effect | References |
|---|---|---|---|
| Brook trout | DDT | 0 | Addison et al. (1977) |
| | DDE | 0 | |
| | Mirex | 0 | Law and Addison (1981) |
| Carp | 2,4,2′,4′-TCB | 0 | Melancon et al. (1981) |
| | Phenobarbital | + | Gutman and Kidron (1971) |
| Mummichog | Phenylbutazone | + | Burns (1976) |
| | DDT | + | Pohl et al. (1974) |
| Northern pike | Phenobarbital | 0 | Balk et al. (1980) |
| Rainbow trout | DDT | 0 | Buhler and Rasmussen (1968) |
| | DDT | + | Dewaide (1971) |
| | Dieldrin | + | |
| | 2,4,5,2′,4′,5′-HCB | + | Förlin and Lidman (1978) |
| | | 0 | Vodicnik et al. (1981) |
| | 2,4,5,2′,4′,5′-HBB | 0 | Franklin et al. (1981) |
| | 2,4,5,3′,4′,5′-HBB | 0 | |
| | 2,3,4,5,2′,4′,5′-HPBB | 0 | |
| | Kepone | 0 | Vodicnik et al. (1981) |
| | Mirex | 0 | |
| | Phenobarbital | 0 | Buhler and Rasmussen (1968); Eisele et al. (1984); Elcombe and Lech (1979); Forlin (1980); Hansson et al. (1980) |
| | Phenylbutazone | 0 | Buhler and Rasmussen (1968) |
| | 2,4,2′,4′-TCB | 0 | Vodicnik and Lech (1982) |
| Sheepshead | Phenobarbital | 0 | James and Little (1981) |
| | 2,4,5,2′,4′,5′-HBB | 0 | |

[a]TCB, tetrachlorobiphenyl; HCB, hexachlorobiphenyl; HBB, hexabromobiphenyl; HPBB, heptabromobiphenyl.

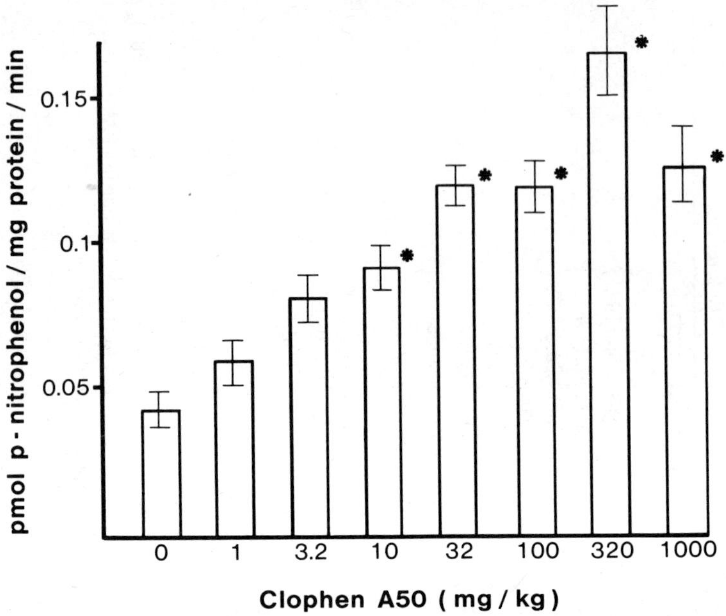


**Figure 10.1.** Effects of different doses of Clophen A50 on the *O*-demethylation of *p*-nitroanisole in rainbow trout hepatic microsomes. The columns correspond to mean values ± standard error. $n = 2$ or 3. * Denotes a significant difference for a particular dose as compared to the group receiving no PCBs ($P < 0.05$). Revised from Förlin and Lidman (1978).

monooxygenase activity was observed in carp in response to a single dose of 0.2 mg/kg of Aroclor 1254 (Fig. 10.4).

## 2. INDUCTION OF MICROSOMAL MONOOXYGENASE ACTIVITY IN FISH IN THE ENVIRONMENT

The fact that many aquatic pollutants such as PCBs and polycyclic aromatic hydrocarbons (PAHs) are capable of inducing monooxygenase activity in fish has led to a number of studies attempting to relate the induction of hepatic microsomal monooxygenase activity to environmental exposure to pollutants.

### 2.1. Approaches for Studying/Demonstrating Environmental Induction

Various research groups have used different approaches for studying factors relating to environmental induction in fish and/or demonstrating its occurrence. These approaches involve comparing microsomal monooxygenase activity (1) in fish from different areas, (2) with and without the addition of a specific inhibitor of mono-

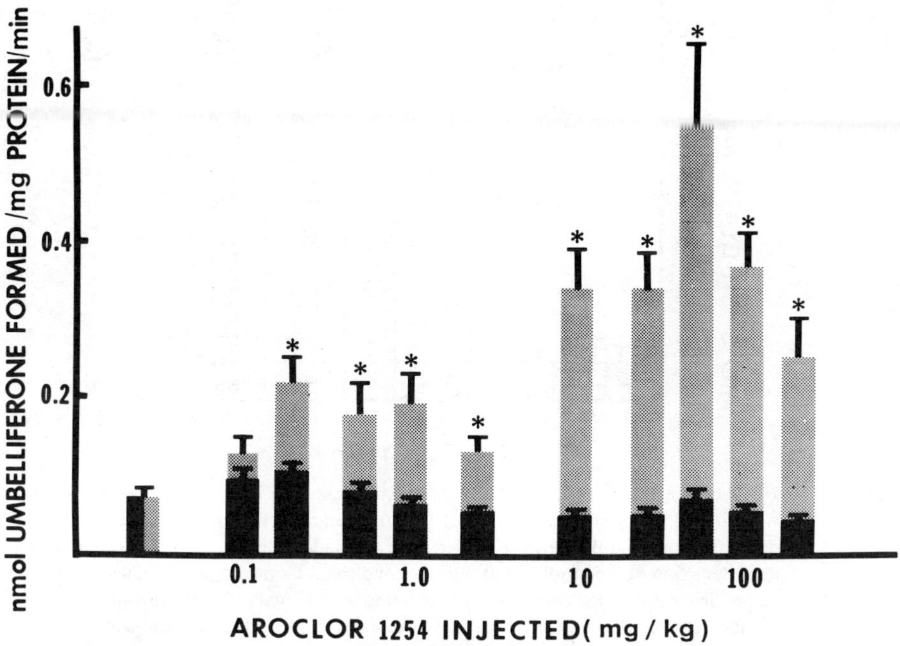


**Figure 10.2.** Hepatic microsomal ethoxyresorufin-*O*-deethylase activity in rainbow trout receiving graded doses of Aroclor 1254. The tops of the stippled bars represent the means for each dose of Aroclor 1254, the solid bars represent the means for corn oil-injected trout assayed on the same days as the dose of Aroclor 1254, the vertical lines represent standard error, and * denotes a significant difference between trout receiving a particular dose of PCBs and the indicated control group ($P < 0.05$). The $n$ values for the dose levels 0.025, 10, and 50 mg/kg are from 4 to 8 while at other doses they range from 13 to 27. From Melancon and Lech (1983) with permission.

oxygenase activity, (3) at different times in the same area, and (4) in embryos and fry from fish (eggs) from different areas.

### 2.1.1. *Comparing Microsomal Monooxygenase Activity in Fish from Different Locations*

A number of laboratories have reported studies comparing hepatic microsomal monooxygenase activity in fish from two or more locations, one a control, clean or nonpolluted location and the other(s) with known or suspected sources of pollution. In the published studies of this type, the fish from the polluted area generally have higher hepatic microsomal monooxygenase activity, but this is not always the case. Data from a number of studies of this type are summarized in Table 10.3. A problem with this type of study is the possibility that there might be some difference between the fish from the two locations other than their pollutant levels. Such differences might include water temperature, diet, or genetic

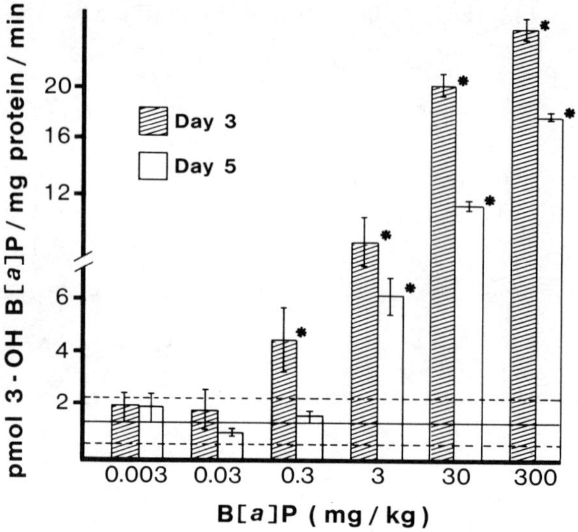


**Figure 10.3.** Hepatic microsomal benzo[*a*]pyrene hydroxylase activity in rainbow trout injected i.p. with graded doses of benzo[*a*]pyrene. Column heights represent the mean of duplicate determinations and the vertical lines represent standard deviation. The mean and standard deviation of 7 control fish samples are indicated by the horizontal lines. * Denotes a significant difference between trout receiving a particular dose of benzo[*a*]pyrene and the indicated control values ($P < 0.05$). Adapted from Gerhart and Carlson (1978).

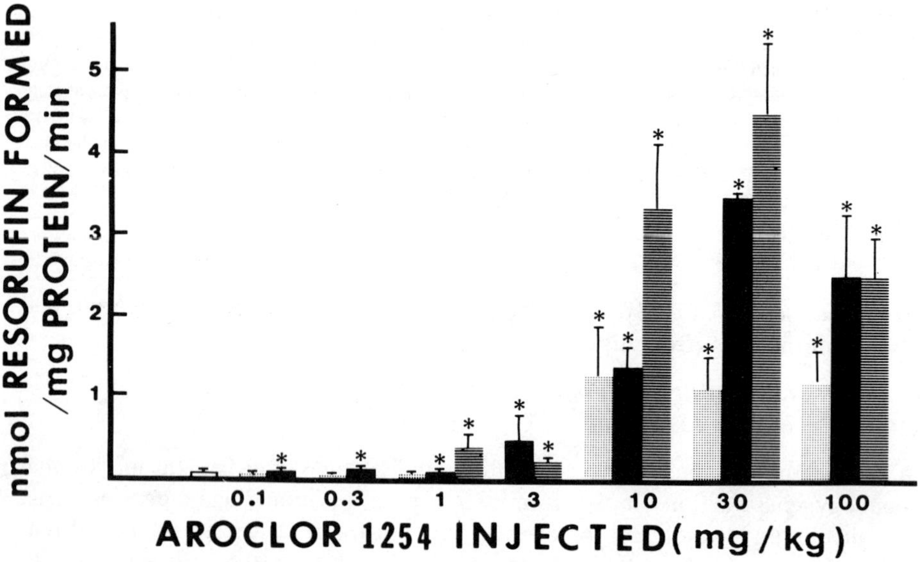


**Figure 10.4.** Hepatic microsomal ethoxyresorufin-*O*-deethylase activity in carp receiving graded doses of Aroclor 1254. The open bar represents the mean for all corn oil-injected carp, the stippled bars represent the means for 3 days after Aroclor 1254 injection, the solid bars represent the means for 6 days after Aroclor 1254 injection, the bars with horizontal lines represent the means for 10 days after Aroclor 1254 injection, vertical lines represent standard errors, and * indicates a significant increase ($P < 0.05$) compared to corn oil-injected carp. There were 35 corn oil-injected carp and 3 to 9 in the other daily groups. From Melancon and Lech (1983) with permission.

**Table 10.3 Hepatic Monoxygenase Activity in Fish from Different Areas**

| Species | Enzyme Activity Studied[a] | Source of Fish | Enzyme Activity[b] (Units/unit protein) | References |
|---|---|---|---|---|
| Brown trout | AHH | "Clean" lake | 26.5 ± 19.4 (8) | Payne and Penrose (1975) |
| | | "Polluted" lake | 362 ± 51 (3) | |
| Cunner | AHH | "Clean" area | 16.0 ± 7.6 (10) | Payne (1976) |
| | | "Clean" area | 19.9 ± 6.6 (9) | |
| | | Refinery area | 53.2 ± 25.2 (10) | |
| | | "Polluted" area | 46.6 ± 6.7 (8) | |
| Fundulus | AE | "Clean" areas | 0.34 ± 0.07 (3) | Burns (1976) |
| | | "Polluted" areas | 0.63 ± 0.08 (5) | |
| Fundulus | AHH | "Clean" area | 89 ± 28 (64) | Stegeman (1978) |
| | | "Clean" area | 74 ± 33 (49) | |
| | | "Polluted" area | 101 ± 19 (53) | |
| | | "Polluted" area | 109 ± 32 (149) | |
| Northern pike | AHH | "Clean" lake | 18 ± 12 (6) | Ahokas et al. (1976) |
| | | "Polluted" lake | 3 ± 2 (7) | |
| Carp | EROD | Hatchery | 0.21 ± 0.10 (15) | Melancon et al (1987) |
| | | "Polluted" area | 4.72 ± 2.99 (13) | |

[a]AHH, aryl hydrocarbon hydroxylase; AE, aldrin epoxidation; EROD, ethoxyresorufin-*O*-deethylase.
[b]Mean ± SD (*n*).

differences. None of these studies examined whether the elevated enzyme activity of the fish from the polluted areas would drop to the level of the fish from the clean areas if the fish from the polluted areas were held in clean water.

### 2.1.2. *Examining the Effect of an Inhibitor of Monooxygenase Activity on the Monooxygenase Activity of Captured Fish*

One approach to studying environmental induction takes advantage of the fact that α-naphthoflavone inhibits induced but not control monooxygenase activity in flounder hepatic microsomes (Little et al., 1981). In these studies numerous specimens of the same species, winter flounder, were collected from the same area (Foureman et al., 1983). Table 10.4 shows that low levels of hepatic microsomal aryl hydrocarbon hydroxylase are unaffected or slightly stimulated by α-naphthoflavone, while higher levels of the enzyme activity are inhibited. The levels of the two monooxygenase activities measured, aryl hydrocarbon hydroxylase and ethoxyresorufin-*O*-deethylase, are approximately proportionate. A summary of the data from 4 years of such studies is presented in Table 10.5. Based on their earlier studies on the specificity of α-naphthoflavone on the hepatic microsomal monooxygenase activity in this species, these researchers concluded that the lower activity samples, which were not inhibited by α-naphthoflavone, were noninduced, and the higher activity samples, which were inhibited by α-naphthoflavone, were induced. In addition they reported that when these microsomes were solubilized in sodium dodecyl sulfate and examined by polyacrylmide gel electrophoresis, the amount of a polypeptide with a molecular weight of approximately 57,000 paralleled the amount of monooxygenase activity found.

Although these researchers have reported on the specificity of α-naphthoflavone as an inhibitor of induced monooxygenase activity in hepatic microsomes of both

**Table 10.4 Monooxygenase Activity in Flounder Hepatic Microsomes and the Effect of α-Naphthoflavone (ANF)**

| | Arylhydrocarbon Hydroxylase Activity (units/min/mg protein) | | Ethoxyresorufin-*O*-deethylase Activity (pmol/min/mg protein) |
|---|---|---|---|
| No. of Fish | Without ANF | With ANF | |
| 1 | 0.55 | 3.36 | <2 |
| 2 | 1.65 | 3.21 | 175 |
| 3 | 2.86 | 2.38 | 219 |
| 4 | 3.06 | 2.72 | 305 |
| 5 | 4.51 | 3.40 | 450 |
| 6 | 8.36 | 3.56 | 219 |
| 7 | 9.59 | 2.91 | 610 |
| 8 | 23.10 | 5.84 | 947 |

*Source:* Revised from Foureman et al. (1983).

**Table 10.5 Annual Summaries of the Effect of α-Naphthoflavone (ANF) on Aryl Hydrocarbon Hydroxylase Activity and the Presence of Ethoxyresorufin-*O*-deethylase Activity in Flounder Hepatic Microsomes**

| | | Aryl Hydrocarbon Hydroxylase Activity | | |
|---|---|---|---|---|
| Year | No. of Fish | Units/min/mg Protein[a] | % Inhibition by ANF | Ethoxyresorufin-*O*-deethylase Activity (pmol/min/mg protein) |
| 78 | 172 | 1.8 ± 1.7 [0.04–8.8] | 49 | 131 ± 128 [<2–837] |
| 79 | 81 | 2.0 ± 1.9 [0.10–8.0] | 72 | 114 ± 143 [<2–796] |
| 80 | 86 | 1.8 ± 1.6 [0.08–6.1] | 84 | 237 ± 216 [<2–1165] |
| 81 | 74 | 1.3 ± 1.2 [0.06–6.4] | 68 | 62 ± 55 [<2–196] |

[a]Means ± SD [range].

*Source:* Revised from Foureman et al. (1983).

winter flounder and little skate, this specificity does not appear to extend to all fish species. With other saltwater species it was found that α-naphthoflavone discriminated well between induced and control (noninduced) monooxygenase activity in croaker hepatic microsomes (Stegeman et al., 1981), poorly in scup hepatic microsomes (Stegeman et al., 1981), and irregularly in sheepshead hepatic microsomes depending on the time of year (James and Little, 1981). In freshwater species the specificity of α-naphthoflavone for control (noninduced) and induced microsomal enzyme activity is highly variable. Similar inhibition of control and induced hepatic microsomal monooxygenase activity has been reported in rainbow trout (Elcombe and Lech, 1979) and carp (Melancon et al., 1981), while slightly greater inhibition of induced than control activity has been reported for lake trout (Binder and Lech, 1984), and a high degree of discrimination between control and induced activity in chub and carp has been reported by Kezić et al. (1983).

Clearly, it cannot be assumed that an inhibitor will discriminate between control and induced monooxygenase for every combination of species, tissue, and monooxygenase assay.

### *2.1.3. Monitoring Monooxygenase Activity in Fish from the Same Area at Different Times*

Another approach involved studying hepatic microsomal monooxygenase activity in fish from the same location over time. Studies of this type have utilized native free-swimming fish in a particular area, cages of clean fish suspended in a particular area at different times, and fish exposed in the laboratory to water pumped into the laboratory from an area of interest.

*2.1.3.1. Environmental Exposure of Free-Swimming Fish.* While doing laboratory studies on the induction of hepatic microsomal monooxygenase activity in free-swimming fish, Kurelec and co-workers (1977) were also measuring this activity in Blenniideae from a site in the Adriatic Sea. During the course of this study an oil spill occurred. From the data in Figure 10.5 we can see the significantly elevated monooxygenase activity at 1 and 3 weeks after the spill. At later times the monooxygenase activity appears to be about threefold the prespill activity, but it is not clear whether these are also significant elevations.

*2.1.3.2. Environmental Exposure of Caged Fish.* The same research group (Kezić et al., 1983) performed studies in which carp were suspended in cages in the Sava River. They were sampled at various times after exposure and the hepatic microsomal monooxygenase activity was assayed. The data (Table 10.6) show that the monooxygenase activity in carp exposed to river water increased in a time-dependent manner when compared to that of carp maintained in the laboratory.

*2.1.3.3. Exposure of Fish to Environmental Water Pumped Into the Laboratory.* In this study (Melancon et al., 1987) hatchery-raised carp were maintained at the University of Wisconsin–Great Lakes Research Facility (Milwaukee, Wisconsin) in flowing dechlorinated city water. One group of carp was maintained continuously in this water while another was exposed to Kinnikinnic River water that was pumped into the laboratory. Using head tanks and chillers, both types of water were kept at the same temperature and all fish were maintained with the same light/dark cycle and the same food. The data in Figure 10.6 show the results of one such experiment. It can be seen that the hepatic microsomal monooxygenase activity increased rapidly and substantially in the river water exposed fish. When some of these river water-exposed fish were transferred to clean water the monooxygenase activity rapidly dropped to control levels. In a single experiment done

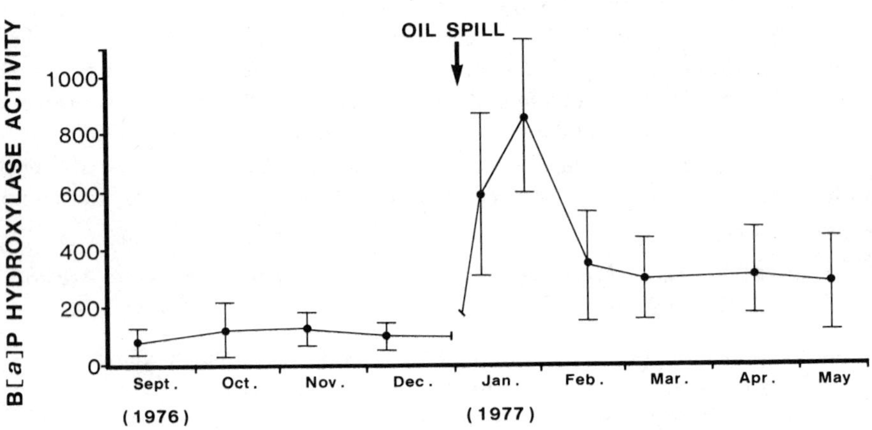


**Figure 10.5.** Benzo[*a*]pyrene monooxygenase activity in postmitochondrial liver homogenate fractions from Blenniideae collected at a site in the Adriatic near Rovinj before and after an oil spill incident. Redrawn from Kurelec et al. (1977).

**Table 10.6 Hepatic Microsomal Benzo[*a*]pyrene Hydroxylase Activity in Carp Exposed in Cages to Sava River Water**

| Days of Exposure | No. of Fish | Benzo[*a*]pyrene Hydroxylase Activity (units/min/mg protein) |
|---|---|---|
| | *River Water* | |
| 0 | 12 | 12.4 ± 5.4[a] |
| 5 | 5 | 46.0 ± 29.8 |
| 10 | 5 | 65.4 ± 34.4 |
| 140 | 4 | 270.2 ± 89.8 |
| | *Controls* | |
| 10 | 5 | 12.9 ± 6.7 |
| 20 | 5 | 13.2 ± 5.7 |

[a]Means ± SD.
*Source:* Revised from Kezić et al. (1983).

with bullheads there was a significant but lesser increase in monooxygenase activity.

### *2.1.4. Monitoring Monooxygenase Activity in Embryos and Fry from Eggs Collected from Fish in Different Locations*

This approach arose from the observation that hepatic monooxygenase activity was inducible in *Fundulus heteroclitus* embryos by PCBs (Binder and Stegeman, 1983). It was known that lake trout in Lake Michigan contain substantial body burdens of PCBs, and that the eggs of these fish contain PCBs at the level of several micrograms per gram. Studies were done with eggs from Lake Michigan, from Green Bay, and from a state fish hatchery to determine whether this contamination of gametes was adequate to induce monooxygenase activity during early development. The developing young from these three groups of fertilized eggs were laboratory cultured under the same conditions.

The majority of lake trout currently present in Lake Michigan were raised at the state hatchery and were stocked in the lake as fingerlings. Therefore, embryos from the three sources were likely to be from similar genetic stocks.

High levels of hepatic monooxygenase activity were found in the embryos and fry from Green Bay and Lake Michigan as compared to the hatchery group (Table 10.7). This difference was still present in swim up fry and in 55-day-old fry, but was no longer apparent when the fish had reached approximately 7 months of age. The data also show that experimental treatment of the hatchery offspring with PCBs could induce monooxygenase activity to levels like those in the Lake Michigan- and Green Bay-derived young.

In addition, when sodium dodecyl sulfate-solubilized microsomes were examined by polyacrylamide gel electrophoresis, it was found that a protein with a molecular weight of approximately 58,000 was present in Lake Michigan fry and

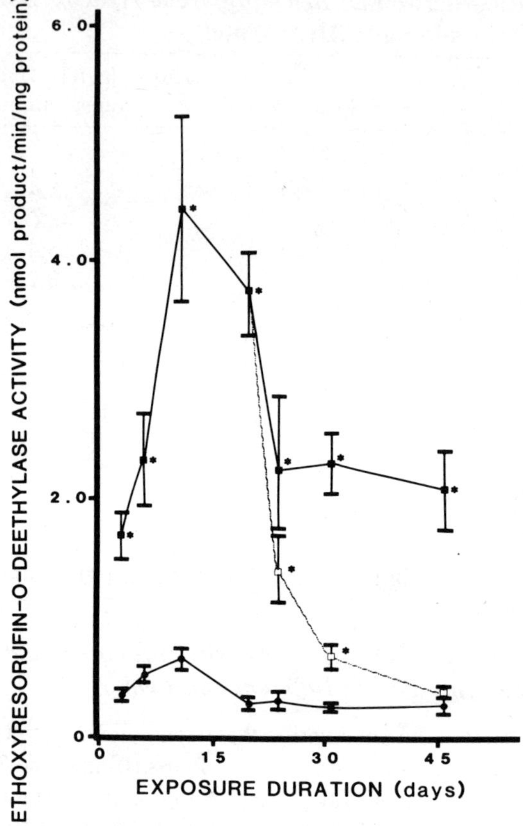


**Figure 10.6.** Effect of exposure to river water on hepatic microsomal ethoxyresorufin-*O*-deethylase activity in carp. Solid circles represent carp maintained in dechlorinated city water, solid squares represent carp exposed to river water, and open squares represent carp exposed to river water for 22 days after which they were transferred to dechlorinated city water. Vertical bars indicate standard error and * represents significant difference ($P < 0.05$) from values for carp maintained in dechlorinated city water. Redrawn from Melancon et al., 1987.

Aroclor 1254-treated hatchery fry, but not in untreated hatchery fry. This band was faint or absent in 210-day-old fry from the hatchery and from Lake Michigan. However, in both groups of fry the protein band was present following treatment with Aroclor 1254.

The concentration of PCBs in the young fish is given in Table 10.8. The correlation of higher monooxygenase activity with higher PCB levels is clear. The fact that the young fish from the hatchery source could have their monooxygenase activity raised by PCBs to the same level as that in the wild fish, and that the monooxygenase activity in the wild fish was reduced to hatchery levels after PCBs

**Table 10.7 Hepatic Monooxygenase Activity in Laboratory-Cultured Lake Trout**

| Group | Arylhydrocarbon Hydroxylase Activity[a] (units/mg protein) | | |
|---|---|---|---|
| | Hatchery | Lake Michigan | Green Bay |
| Embryos[b] | 16 ± 3 | 99, 106[c] | 66[c] |
| Swim-up fry[b] | 44 ± 3 (269 ± 8) | 156 ± 15 | 195 ± 8 |
| Swim-up fry[d] (55 days post hatch) | 108 ± 2 (670 ± 26) | 929 ± 67 | 657 ± 49 |
| Juveniles[d] (210 days post hatch) | 131 ± 15 (641 ± 91) | 131 ± 13 (608 ± 10) | |

[a]Means ± SD. Values in parentheses represent PCB-induced activity.
[b]Data were obtained with liver whole homogenates.
[c]Individual values for pooled liver homogenates.
[d]Data were obtained with microsomal preparations.
*Source:* Revised from Binder and Lech (1984).

**Table 10.8 Total PCB Concentration (μg/g) in Laboratory-Cultured Lake Trout**

| Location | Group | |
|---|---|---|
| | Swim-up Fry (47 days post hatch) | Juveniles (210 days post hatch) |
| Hatchery | 0.18 ± 0.02[a] | 0.22 ± 0.02[a] |
| Lake Michigan | 4.30 ± 0.30 | 0.40 ± 0.02 |
| Green Bay | 2.19 ± 0.08 | |

[a]Means of two samples ± range.
*Source:* Revised from Binder and Lech (1984).

levels dropped by elimination or by dilution by growth, demonstrate clearly that this was induction from environmental exposure.

## 2.2. Induction of Monooxygenase Activity in Extrahepatic Tissues

As indicated in Section 1.1, induction of renal microsomal monooxygenase activity has been observed in fish. Interestingly, the responses in liver and kidney to certain inducing agents are not always comparable. Data from Payne et al. (1984) are presented that show benzo[*a*]pyrene hydroxylase activity in preparations from liver and kidney of flounder from a clean site and a polluted site (Fig. 10.7). Of the two tissues, only kidney showed a difference in monooxygenase activity between the two sites. Studies of carp by Melancon et al. (1981) show that at the

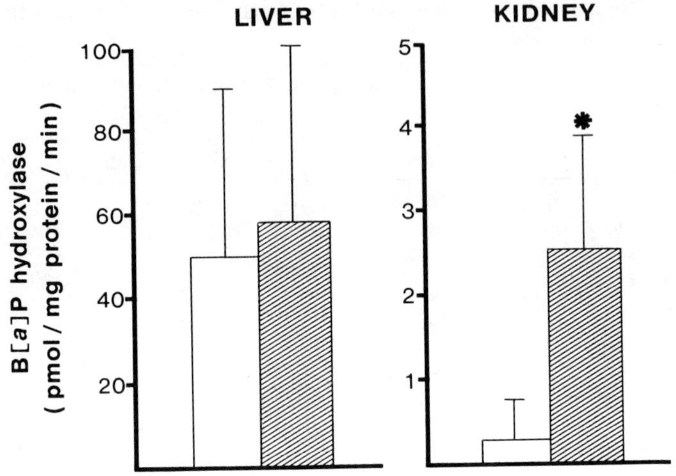


**Figure 10.7.** Benzo[*a*]pyrene hydroxylase activity in renal and hepatic microsomes from winter flounder from two sites in Newfoundland. The open bars represent fish from a control site in Coachman's Cove and the cross-hatched bars represent fish from a contaminated (oil spill) site at Baie Verte. * Represents a significant difference ($P < 0.05$). Redrawn from Payne et al. (1984).

doses utilized PCBs elevated monooxygenase activity in both liver and kidney while β-naphthoflavone elevated monooxygenase activity only in the liver (Fig. 10.8).

## 3. IMPLICATIONS OF ENVIRONMENTAL INDUCTION

Interest in environmental induction has revolved around two general areas. One concerns the utility of monooxygenase induction as a monitoring tool, and the other concerns whether induction of monooxygenase activity could lead to significant changes in the survival or health of the affected fish.

### 3.1. Monitoring for Pollutants That Are Inducing Agents

It is clear that environmental induction is occurring and can be used as an indicator for pollutants that are inducers in fish. It is not clear whether any information beyond a simple indication of such exposure can be generated. Some information that might be obtained in such studies would be the nature of the inducing agent and the tissue levels of the inducing agent.

It is not likely that we will ever be able to identify a particular chemical to which a fish was exposed by a monooxygenase profile (species and tissue responsiveness for induction of monooxygenase activity, responsiveness of particular

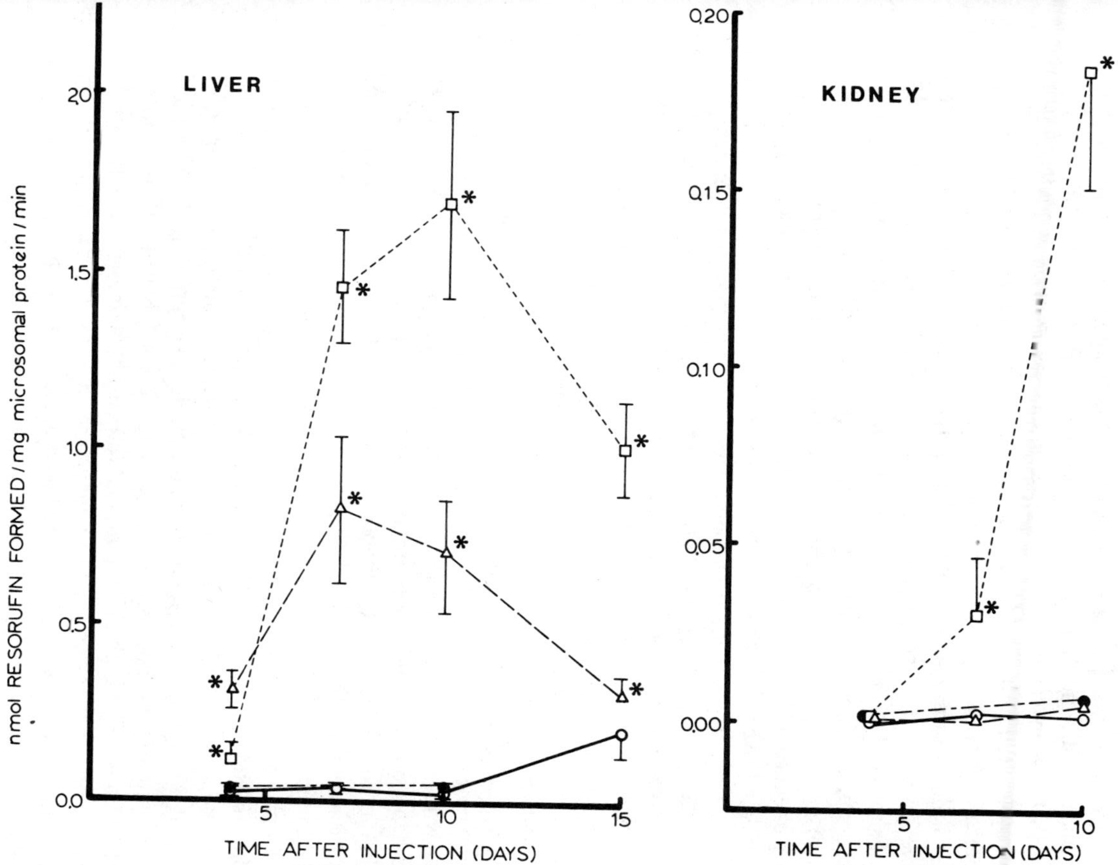


**Figure 10.8.** Ethoxyresorufin-*O*-deethylase activity in renal and hepatic microsomes from carp injected with corn oil (CO), 200 mg/kg Aroclor 1254(A1254), 100 mg/kg $\beta$-naphthoflavone (BNF), or 200 mg/kg 2,4,2′,4′-tetrachlorobiphenyl (TCB). Carp were sacrificed the indicated number of days following injection. Open circles represent CO-treated, open squares represent A1254-treated, open triangles represent BNF-treated, and solid circles represent TCB-treated carp. The symbols and vertical lines represent the means and standard errors for individual preparations from 3 fish. * Indicates significant increase ($P < 0.05$) compared to CO-injected. From Melancon et al. (1981), with permission.

monooxygenase activities, effects of inhibitors, polyacrylamide gel electrophoresis of cytochromes P-450, and immunological techniques). Perhaps the limited ability to distinguish between categories of chemicals such as PCBs and PAHs might become possible after further development of these techniques, particularly the immunological techniques.

There is little information available relating tissue levels of inducing chemical(s) to the amounts of monooxygenase activity observed. The data presented in Section 1.2 showed that rainbow trout and carp are able to respond to single doses of PCBs at about 0.2 ppm. This is well within the range of concentrations of PCBs found in fish in the environment, but it is not possible to equate a single dose of an inducing chemical to a body burden at the same concentration, which may have been accumulated over days or years via the water or the diet.

There are a few studies that have related tissue levels of inducing chemicals to hepatic microsomal monooxygenase activity. Addison et al. (1981) administered Aroclor 1254 to brook trout. Although many of these doses of Aroclor 1254 were high, up to 200 mg/kg, results from some of the lower doses showed a direct relationship between filet PCB concentrations and microsomal ethoxycoumarin-*O*-deethylase activity. They calculated from their results that filet levels of 4 ppm PCBs would correspond to a doubling of ethoxycoumarin-*O*-deethylase activity over background levels. Gerhart and Carlson (1978) exposed rainbow trout to benzo[*a*]pyrene by injection or by water exposure in a continuous flow delivery system. They measured benzo[*a*]pyrene concentrations in the gutted fish and AHH activity in hepatic microsomes. Regardless of the route of exposure, they found a direct relation of induction of monooxygenase activity to concentrations of benzo[*a*]pyrene in the fish samples. Induction of monooxygenase activity was not found with gutted fish concentrations of 140 ppb benzo[*a*]pyrene or less, was found in some samples at 249 to 350 ppb, and was found in all samples at 368 ppb or greater. James and Little (1981) determined tissue levels of PBBs and hepatic microsomal monooxygenase activity in sheepshead after administration of Firemaster FF1 at 15 mg/kg. They reported that muscle concentrations of PBBs averaged 0.15 ppm and did not correlate to enzyme activity, but that liver concentrations of PBBs were as high as 16 ppm and did correlate to enzyme activity.

To increase the monitoring value of induction of monooxygenase activity in fish, more data are needed comparing monooxygenase activity to tissue levels of the inducing agents. Muscle is one of the tissues in which such determinations should be made because this is the tissue of concern for regulatory purposes. In particular, data are needed relating monooxygenase activity to tissue levels of inducing chemicals to which the fish have been exposed by dietary and water routes, which are the ways in which environmental exposures would occur, over various periods of time.

### 3.2. Possible Effects of Induction on Fish

The possible effects of induction on fish would have to arise from differences in the metabolism of endogenous or of xenobiotic chemicals in fish. Although effects of induction of monooxygenase activity on steroid metabolism have been demonstrated in isolated microsomes (Hansson et al., 1980; Förlin and Hansson, 1982) and in intact fish (Sivarajah et al., 1978), it has not been demonstrated that this has any significant effect on the health or survival of the fish. Perhaps developing young that have been induced environmentally, such as the lake trout described in Section 2.1.4 that are not surviving in the environment, would be good subjects for such studies.

In the environment a fish would rarely be exposed to only a single pollutant. Therefore, in regard to other xenobiotic chemicals, a pollutant that causes induction of monooxygenase activity could lead to increased formation of toxic metabolites or to increased detoxication of toxic chemicals. Hendricks and co-workers have examined the formation of tumors caused by a variety of chemicals in fish treated with inducing amounts of PCBs. When PCBs were fed to rainbow trout at the same time as diethylnitrosamine there was increased incidence of tumors (Bailey et al., 1984), but when PCBs were fed to trout at the same time as aflatoxin $B_1$ there was an inhibition of tumor formation (Hendricks et al., 1977). On the other hand when females were treated with PCBs and the eggs from these females were later exposed to aflatoxin $B_1$ there was an increased incidence of tumor formation as compared to aflatoxin exposure of eggs from untreated females (Hendricks et al., 1980).

## 4. CONCLUSIONS

It is clear that induction of fish is occurring in the environment. It is also clear that care must be taken in performing studies relating to environmental induction to ensure that the presence of pollutants that are inducing agents is the only variable that could cause any observed elevations of monooxygenase activity. Additional studies are required (1) to determine how much information on the nature and quantity of pollutants to which fish have been exposed can be acquired from cytochrome P-450/monooxygenase-related data, and (2) to evaluate the effect of environmental induction on the well-being of the fish.

## ACKNOWLEDGMENTS

This work was supported in part by EPA Grant R810610-01-0, by the Wisconsin Sea Grant College Program, and by NIEHS Aquatic Biomedical Research Center Grant ES01985.

## REFERENCES

Addison, R. F., Zinck, M. E., and Willis, D. E. (1977). Mixed function oxidase enzymes in trout (*Salvelinus fontinalis*) liver: absence of induction following feeding of p,p′-DDT or p,p′-DDE. *Comp. Biochem. Physiol. C* **57C,** 39–43.

Addison, R. F., Zinck, M. E., and Willis, D. E. (1977). Induction of hepatic mixed-function oxidase (MFO) enzymes in trout *Salvelinus fontinalis* by feeding Aroclor 1254 or 3-methylcholanthrene. *Comp. Biochem. Physiol. C* **61C,** 323–325.

Addison, R. F., Zinck, M. E., and Willis, D. E. (1981). Time-and-dose dependence of hepatic mixed function oxidase activity in brook trout *Salvelinus fontinalis* on polychlorinated residues: Implications for biological effects monitoring. *Environ. Pollut., Ser. A* **25,** 211–218.

Ahokas, J. T., Karki, N. T., Oikari, A., and Soivio, A. (1976). Mixed function monooxygenase of fish as an indicator of pollution of aquatic environment by industrial effluent. *Bull. Environ. Contam. Toxicol.* **16,** 270–274.

Bailey, G. S., Hendricks, J. D., Nixon, J. E., and Pawlowski, N. E. (1984). The sensitivity of rainbow trout and other fish to carcinogens. *Drug Metab. Rev.* **15,** 725–752.

Balk, L., Meijer, J., Seidegård, J., Morganstein, J., and Depierre, J. W. (1980). Initial characterization of drug-metabolizing systems in the liver of the northern pike, *Esox lucius. Drug Metab. Dispos.* **8,** 98–103.

Binder, R. L., and Lech, J. J. (1984). Xenobiotics in gametes of Lake Michigan lake trout *Salvelinus namaycush* induce hepatic monooxygenase activity in their offspring. *Fundam. Appl. Toxicol.* **4,** 1042–1054.

Binder, R. L., and Stegeman, J. J. (1980). Induction of aryl hydrocarbon hydroxylase activity in embryos of an estuarine fish. *Biochem. Pharmacol.* **29,** 949–951.

Binder, R. L., and Stegeman, J. J. (1983). Basal levels and induction of hepatic aryl hydrocarbon hydroxylase activity during embryonic period of development in brook trout. *Biochem. Pharmacol.* **32,** 1324–1327.

Buhler, D. R., and Rasmussen, M. E. (1968). The oxidation of drugs by fishes. *Comp. Biochem. Physiol.* **25,** 223–239.

Burns, K. A. (1976). Microsomal mixed function oxidases in an estuarine fish, *Fundulus heteroclitus*, and their induction as a result of environmental contamination. *Comp. Biochem. Physiol. B* **53B,** 443–446.

Chevion, M., Stegeman, J. J., Peisach, J., and Blumberg, W. E. (1977). Electron paramagnetic resonance studies on hepatic microsomal cytochrome P-450 from a marine teleost fish. *Life Sci.* **20,** 895–900.

DeWaide, J. H. (1971). Metabolism of xenobiotics. Ph.D. Thesis, University of Nijmegem, The Netherlands.

Eisele, T. A., Coulombe, R. A., Pawlowski, N. E., and Nixon, J. E. (1984). The effects of route of exposure and combined exposure of mixed function oxidase inducers and suppressors on hepatic parameters in rainbow trout (*Salmo gairdneri*). *Aquat. Toxicol.* **5,** 211–226.

Elcombe, C. R., and Lech, J. J. (1978). Induction of monooxygenation in rainbow trout by polybrominated biphenyls: A comparative study. *Environ. Health Perspect.* **23,** 309–314.

Elcombe, C. R., and Lech, J. J. (1979). Induction and characterization of hemoprotein(s) P-450 and monooxygenation in rainbow trout (*Salmo gairdneri*). *Toxicol. Appl. Pharmacol.* **49,** 437–450.

Förlin, L. (1980). Effects of Clophen A50, 3-methylcholanthrene, pregnenolone-16-carbonitrile, and phenobarbital on the hepatic microsomal cytochrome P-450 dependent monooxygenase system in rainbow trout, *Salmo gairdneri*, of different age and sex. *Toxicol. Appl. Pharmacol.* **54,** 420–430.

Förlin, L., and Hansson, T. (1982). Effects of treated municipal wastewater on the hepatic xenobiotic, and steroid metabolism in trout. *Ecotoxicol. Environ. Saf.* **6,** 41–48.

Förlin, L., and Lidman, V. (1978). Effects of Clophen A50, 2,5,2′,5′-tetra- and 2,4,5,2′,4′,5′-hexachlorobiphenyl on the mixed-function oxidase system of rainbow trout, *Salmo gairdneri rich.* liver. *Comp. Biochem. Physiol. C* **60C,** 193–197.

Foureman, G. L., White, N. B., Jr., and Bend, J. R. (1983). Biochemical evidence that winter flounder, *Pseudopleuronectes americanus*, have induced hepatic cytochrome P-450-dependent monooxygenase activities. *Can. J. Fish. Aquat. Sci.* **40,** 854–865.

Franklin, R. B., Vodicnik, M. J., Elcombe, C. R., and Lech, J. J. (1981). Alterations in hepatic mixed-function oxidase activity of rainbow trout after acute treatment with polybrominated biphenyl isomers and Firemaster BP-6. *J. Toxicol. Environm. Health* **7,** 817–827.

Gerhart, E. H., and Carlson, R. M. (1978). Hepatic mixed-function oxidase activity in rainbow trout exposed to several polycyclic aromatic compounds. *Environ. Res.* **17,** 284–295.

Gooch, J. W., and Matsumura, F. (1983). Characteristics of the hepatic monooxygenase system of the goldfish, *Carassius auratus*, and its induction with β-naphthoflavone. *Toxicol. Appl. Pharmacol.* **68,** 380–391.

Gruger, E. H., Jr., Hruby, T., and Karrick, N. L. (1976). Sublethal effects of structurally related tetrachloro-, pentachloro-, and hexachlorobiphenyl on juvenile coho salmon. *Environ. Sci. Technol.* **10,** 1033–1037.

Gutman, Y., and Kidron, M. (1971). Liver N-demethylating activity-temperature effect and phenobarbital induction in different species. *Biochem. Pharmacol.* **20,** 3547–3550.

Hansson, T., Rafter, J., and Gustafsson, J.-A. (1980). Effects of some common inducers on the hepatic microsomal metabolism of androstenedione in rainbow trout with special reference to cytochrome P-450-dependent enzymes. *Biochem. Pharmacol.* **29,** 583–587.

Hendricks, J. D., Putman, T. P., Bills, D. D., and Sinnhuber, R. O. (1977). The inhibitory effects of a polychlorinated biphenyl (Aroclor 1254) on aflatoxin B1 carcinogenesis in rainbow trout, *Salmo gairdneri. J. Natl. Cancer Inst. (U.S.)* **59,** 1545–1551.

Hendricks, J. D., Stott, W. T., Putman, T. P., and Sinnhuber, R. O. (1980). Enhancement of aflatoxin B1 hepatocarcinogenesis in rainbow trout, *Salmo gairdneri*, embryos by prior exposure of gravid females to dietary Aroclor 1254. *ASTM Spec. Tech. Publ.* **STP 737,** 203–214.

Hill, D. W., Hejtmancik, E., and Camp, B. J. (1976). Induction of hepatic microsomal enzymes by Aroclor 1254 in *Ictalurus punctatus* (channel catfish). *Bull. Environ. Contam. Toxicol.* **16,** 495–502.

James, M. O., and Little, P. J. (1981). Polyhalogenated biphenyls and phenobarbital: Evaluation as inducers of drug metabolizing enzymes in the sheepshead, *Archosargus probatocephalus. Chem.-Biol. Interact.* **36,** 229–248.

Kezić, N., Britvić, S., Protić, M., Simmons, J. E., Rijavec, M., Zahn, R. K., and Kurelec, B. (1983). Activity of benzo(a)pyrene monooxygenase in fish from the Sava River, Yugoslavia: Correlation with pollution. *Sci. Total Environ.* **27,** 59–69.

Kurelec, B., Britvić, S., Rijavec, M., Müller, W. E. G., and Zahn, R. K. (1977). Benzo(a)pyrene monooxygenase induction in marine fish-molecular response to oil pollution. *Mar. Biol. (Berlin)* **44,** 211–216.

Kurelec, B., Matijasevic, Z., Rijavec, M., Alacevic, M., Britvić, S., Müller, W. E. G., and Zahn, R. K. (1979). Induction of benzo(a)pyrene monooxygenase in fish and the *Salmonella* test as a tool for detecting mutagenic/carcinogenic xenobiotics in the aquatic environment. *Bull. Environ. Contam. Toxicol.* **21,** 799–807.

Law, F. C. P., and Addison, R. F. (1981). Response of trout hepatic mixed-function oxidases to experimental feeding of ten known or possible chlorinated environmental contaminants. *Bull. Environ. Contam. Toxicol.* **27,** 605–609.

Lech, J. J., Vodicnik, M. J., and Elcombe, C. R. (1982). Induction of monooxygenase activity. In L. J. Weber (Ed.), *Aquatic Toxicology.* Raven Press, New York, pp. 107–148.

Lidman, U., Förlin, L., Molander, O., and Axelson, G. (1976). Induction of the drug metabolizing system in rainbow trout (*Salmo gairdneri*) liver by polychlorinated biphenyls (PCBs). *Acta Pharmacol. Toxicol.* **39,** 262–272.

Little, P. J., James, M. O., Bend, J. R., and Ryan, A. J. (1981). Imidazole derivatives as inhibitors of cytochrome P-450-dependent oxidation and activators of epoxide hydrolase in hepatic microsomes from a marine fish. *Biochem. Pharmacol.* **30,** 2876–2880.

Melancon, M. J., and Lech, J. J. (1983). Dose-effect relationship for induction of hepatic monooxygenase activity in rainbow trout and carp by Aroclor 1254. *Aquat. Toxicol.* **4,** 51–61.

Melancon, M. J., Elcombe, C. R., Vodicnik, M. J., and Lech, J. J. (1981). Induction of cytochromes P-450 and mixed-function oxidase activity by polychlorinated biphenyls and β-naphthoflavone in carp (*Cyprinus carpio*). *Comp. Biochem. Physiol. C* **69C,** 219–226.

Melancon, M. J., Yeo, S. E., and Lech, J. J. (1987). Induction of hepatic monooxygenase activity in fish by exposure to river water. *Environ. Toxicol. Chem.* **6,** 127–135.

Narbonne, J. F., and Gallis, J. L. (1979). In vivo and in vitro effects of Phenoclor DP6 on drug metabolizing activity in mullet liver. *Bull. Environ. Toxicol.* **23,** 338–343.

Payne, J. F. (1976). Field evaluation of benzopyrene hydroxylase induction as a monitor for marine pollution. *Science* **191,** 945–946.

Payne, J. F., and Penrose, W. R. (1975). Induction of aryl hydrocarbon (benzo[a]pyrene) hydroxylase in fish by petroleum. *Bull. Environ. Contam. Toxicol.* **14,** 112–116.

Payne, J. F., Bauld, C., Dey, A. C., Kiceniuk, J. W., and Williams, V. (1984). Selectivity of mixed-function oxygenase enzyme induction in flounder (*Pseudopleuronectes americanus*) collected at the site of the Baje Verte, Newfoundland oil spill. *Comp. Biochem. Physiol. C* **79C,** 15–19.

Pohl, R. J., Bend, J. R., Guarino, A. M., and Fouts, J. R. (1974). Hepatic microsomal mixed-function oxidase activity of several marine species from coastal Maine. *Drug Metab. Dispos.* **2,** 545–555.

Ridlington, J. W., Chapman, P. E., Boese, B. L., and Johnson, V. G. (1982). Petroleum refinery wastewater induction of the hepatic mixed-function oxidase system in Pacific staghorn sculpin. *Arch. Environ. Contam. Toxicol.* **11,** 123–127.

Sivarajah, K., Franklin, C. S., and Williams, W. P. (1978). The effects of polychlorinated biphenyls on plasma steroid levels and hepatic microsomal enzymes in fish. *J. Fish Biol.* **13,** 401–409.

Statham, C. M., Elcombe, C. R., Szyjka, S. P., and Lech, J. J. (1978). Effect of polycyclic aromatic hydrocarbons on hepatic microsomal enzymes and disposition of methylnaphthalene in rainbow trout in vivo. *Xenobiotica* **8,** 65–71.

Stegeman, J. J. (1978). Influence of environmental contamination on cytochrome P-450 mixed function oxygenases in fish: Implications for recovery in the Wild Harbor Marsh. *J. Fish. Res. Board Can.* **35,** 668–674.

Stegeman, J. J., Klotz, A. V., Woodin, B. R., and Pajor, A. M. (1981). Induction of hepatic cytochrome P-450 in fish and the indication of environmental induction in scup (*Stenotomus chrysops*). *Aquat. Toxicol.* **1,** 197–212.

Vodicnik, M. J., and Lech, J. J. (1982). Characterization of the induction response of the hepatic microsomal monooxygenase (MO) system of fishes. In E. Heitanen, M. Laitinen, and O. Hanninen (Eds.), *Cytochrome P-450, Biochemistry, Biophysics and Environmental Implications.* Elsevier/North-Holland Biomedical Press, Amsterdam, pp. 225–231.

Vodicnik, M. J., Elcombe, C. R., and Lech, J. J. (1981). The effect of various types of inducing agents on hepatic microsomal monooxygenase activity in rainbow trout. *Toxicol. Appl. Pharmacol.* **59,** 364–374.

# 11

# AMBIENT TOXICITY TO ASSESS BIOLOGICAL IMPACT

***Donald I. Mount***

*U.S. Environmental Protection Agency, Environmental Research Laboratory, Duluth, Minnesota 55804*



## 1. INTRODUCTION

Two of the most important reasons for water pollution control programs are to protect human health and to protect biological communities. While human health might be considered the more important, there is not a direct connection between water quality and health effects because of the removal of contaminants during drinking water treatment and the limited exposure to surface water. Many of the past health concerns have been because of bioaccumulation. Protection of biological communities has probably received as much or more attention than human health problems, perhaps because there is a more obvious connection between water quality and effects on aquatic organisms.

For decades, biologists have measured characteristics of aquatic communities to assess their well-being. Much of the effort has been an attempt to find "the all-encompassing measure" that will define the damage and point out the cause. With today's emphasis on chemical contaminants and knowledge of the large number of them that are to be found in surface water, it is easy to blame "toxics" for all undesirable impacts found and to assume that the next discharge upstream is the source. The presence of chemicals in a water body does not mean they are having adverse impacts. The impact on biological systems from toxic chemicals is a function of time and concentration. Rarely do our monitoring measurements tell us much about the duration of the measured concentrations.

Perhaps we need first to examine the generic types of factors that influence the health of aquatic communities, and then decide if the chemicals present are causing a toxic effect. Biological cycles, over-fishing, weather, flow alteration, habitat changes, temperatures, excessive nutrients, dissolved oxygen, disease, predator–prey relationships, energy flow changes, and many other causes may be equally or more important than toxic chemicals in causing changes. As difficult as analytical measurements may be, they are often easier than measuring the biological effect of these other factors and perhaps we are jumping to conclusions because chemicals are present. I suggest we first determine whether toxicity is present, not just assume toxicity occurs because chemicals are present. Providing upgraded waste treatment plants to remove chemicals will not help if the problem is caused by weather or nutrients.

Aquatic communities are too complex and too poorly understood for us to study them directly to discern if toxicity is causing impairment, with any practical amount of effort. Furthermore, many of the needed answers must be made available before substances are released into the environment and before environmental alterations occur. Usually such answers must come from laboratory tests.

## 2. A TOOL

### 2.1. Type of Test Needed

Fortunately, detecting toxicity is probably easier and quicker than detecting effects of other stresses mentioned above. Toxicity can be measured by following many of the same principles used to measure chemicals, by conventional analytical procedures. Toxicity should be viewed as a potential characteristic of water and as such can be measured by sampling and laboratory analyses. This is not true of over-fishing, climatic changes, or habitat changes. Toxicity then should be one of the first factors measured when attempting to identify the cause of community impairment.

Most surface-water samples do not show acute toxicity, so to determine if toxicity occurs in water samples, chronic toxicity must be measured. Until recently, available chronic tests on aquatic species were months to years in length. Such tests are too expensive to use for purposes discussed here both because of the length of time required and the volume of sample needed. For this type of monitoring (hereafter called ambient testing) one needs to measure toxicity on samples from various sampling sites that may be far separated geographically. Obviously then, samples must be shipped to a central point for testing, which makes conventional flow-through testing impractical.

Fortunately, some new tests have been developed that allow chronic toxicity to be measured in one week (Mount and Norberg, 1984; Norberg and Mount, 1985). Any test that measures chronic toxicity can be used. The advantage of these tests is that they take the shortest time to run and require the least volume of sample (Steve Schimmel of EPA, Narragansett, RI, is completing similar tests for marine organisms). The particular species used have no other unusual inherent qualities and certainly the fathead minnow growth test method (Norberg and Mount, 1985) could be modified and used for other species of larval fish. *Daphnia magna* chronic tests (Mount and Norberg, 1984), used so widely for toxicity testing, should also be amenable to scaling down volumes and shortening the test period.

The specific details of animal culture, handling, replicates, and data analyses can be found in the published papers. Here the samples to be tested, using the published methods, are discussed.

In usual toxicity tests, various dilutions of the sample are made to determine the threshold effect concentration. Ambient samples of surface water are rarely toxic enough to require dilution. Therefore, samples from designated locations are tested for toxicity as they are received after temperature adjustment. (*Note:* Temperature adjustment that can change toxicants is necessary since only at appropriate temperatures will adequate reproduction and growth occur. McNaught and Mount (1986) found that *Ceriodaphnia* can be tested over a wide temperature range, but at cooler temperatures much longer test periods are needed.) Each sample can be viewed as a separate toxicant or treatment and by comparing the response in samples from various stations, a relative toxicity value can be obtained for each station.

## 2.2. Station Selection and Sampling

Selection of stations is a site-specific decision depending on geography and hydrology and the degree of resolution desired. In flowing streams a series of upstream–downstream stations are usually selected after proper consideration is given to discharges, tributaries, reservoirs, and especially important niches or segments. If long reaches (e.g., 100 miles) are to be sampled, a "screening" phase in which a station every 10 miles and/or a station below each major tributary may need to be done first to locate where the most toxic conditions occur. This phase can be followed by a more intensive one in which the segments found to be toxic are sampled throughout relative to potential inputs to pinpoint more exactly sources of toxicity.

The eight site studies referenced in Section 3 reveal some cautions about sample station location. Never assume that headwater stations will be nontoxic because often they will contain old mines, dumpsites, springs contributing contaminated groundwater, pesticides, or naturally occurring toxicants such as heavy metals. Even if some sought-after species such as rainbow trout are found, toxicity may exist but not of sufficient strength to obliterate the trout population. Second, when a tributary or a discharge enters, even if toxic, the toxicity of both may disappear in the mixture. ''Antagonism'' has occurred frequently during these studies. For lake or reservoir situations, sampling stations can be located in reference to tributaries, harbors, discharges, and stratification. In water subject to tides or currents, these factors must be considered.

Once the stations have been chosen, sampling protocols become important. Contaminants are not usually present continuously or at constant concentrations. A grab sample may give a totally different test result than a composite sample. While a grab sample has the potential to show maximum toxicity it also has the maximum chance of missing maximum toxicity completely. A composite sample is a compromise for both concerns.

Composite samples may be difficult to collect. Since the tests are renewal tests, or can be, a good plan is to obtain a new sample on each of the 7 days and expose the test organisms to it for 24 hours. In this way, the exposure occurring at the station is best simulated in the toxicity test. Especially for tidal locations, the use of daily composite samples provides the best practical exposure simulation.

The results of the tests can be affected drastically by rainfall or flow changes due to flow regulation. If discharges from treatment plants, seepage, or springs are of concern, the concentration of these in the receiving water will be greatest at low-flow conditions. On the other hand, insecticides and other runoff-related contaminants may be at the highest concentration during high flows. The hydrologic conditions for proper sampling time must be selected based on such considerations. The test results will be no better than the representation of the samples tested. For variable conditions, sampling may have more bearing on the results than choice of test species or test type and may well be the most expensive of all. But expense of sampling is well justified in the validity of the results.

## 2.3. Methods Considerations

Finding a good control for ambient samples is difficult. There are not yet specified foods or standardized test waters that will mimic test organism response in a given natural water. Such characteristics as nutrient content, suspended solids, and other unknown factors will change the organism performance. Therefore, the best approach now available is to use as a quality assurance requirement a treatment in which a good quality water is used and for which organism performance is known. This will reveal if the lot of test organisms was of good quality and if test conditions were acceptable. These data should not be used for statistical comparison for data analyses, however, for reasons described above.

Since there is no predetermined way to know which ambient samples are not contaminated, the station sample in which the organisms performance is best is used for statistical comparison. Thus, the results give a relative toxicity value rather than an absolute one. If the organisms' performance at all stations is very low, for example, 5 young/female when the QA treatment gave 20, one should assume that all samples have toxicity. However, if one or more stations had 15 young, such numbers may not be statistically different from the QA sample and therefore may not display toxicity. In such cases, judgment must be used. The concern is that if only relative toxicity is used, one may get relative differences among the stations but none will be without toxicity and so no station would be considered acceptable.

Ambient tests can be used to determine if toxicity is a contributing cause to community impairment and should be viewed as an analytical measurement; in this case, however, the "detector" is an organism rather than an electronic "black box." Like every analytical measurement, the "detector" has a limit of sensitivity. The limit can be lowered by using more species for testing. The test will measure every substance but the detection level will vary greatly depending on the substance. Ironically, the sensitivity is greatest to the most toxic substances, and those are the ones we most want to identify.

## 3. AN EXAMPLE

A series of eight reports (Mount and Norberg-King, 1985, 1986; Mount et al., 1984, 1985, 1986a,b,c; Norberg-King and Mount, 1986) have described a large scale effort to determine if such tests on ambient water samples correlate with the species richness of the sampling stations from which the tested water was taken. Eight sites were evaluated. An example of a correlation is shown in Figure 11.1. The data are from the Kanawha River in West Virginia (Mount and Norberg-King, 1986). Nearly 100 miles of river were sampled for zooplankton and macroinvertebrate species richness. This river historically has been one of the more seriously polluted rivers in the United States, mostly from a heavy concentration of chemical industry. The study reach was all within three navigation pools and since there are no major tributaries, the effect of habitat over the study reach was lessened. The toxic effects as measured by the toxicity tests and the changes in species richness of zooplankton were not large, yet a correlation analysis was significant at $P < 0.001$. The macroinvertebrates were sampled using artificial substrates and they showed a decreasing species richness from upstream to downstream. There was no correlation between the toxicity results and the macroinvertebrates.

These results illustrate several important observations that can be drawn from the eight site studies in general and in conjunction with what is known about the sensitivity of various aquatic species. [The reader is referred to Blanck (1984) for a more thorough review of the matter of inferring community impact using single-species tests.] From the toxicity data base now in the literature, differences in the sensitivity of organism within a major group, for example, aquatic insects, are frequently three or four orders of magnitude. Differences in the sensitivity of a given organism to a variety of chemicals are known to span seven orders of magnitude (Blanck, 1984). Suter et al. (1985) showed that the more different the species are taxonomically, the greater the differences in sensitivity. These toxicity

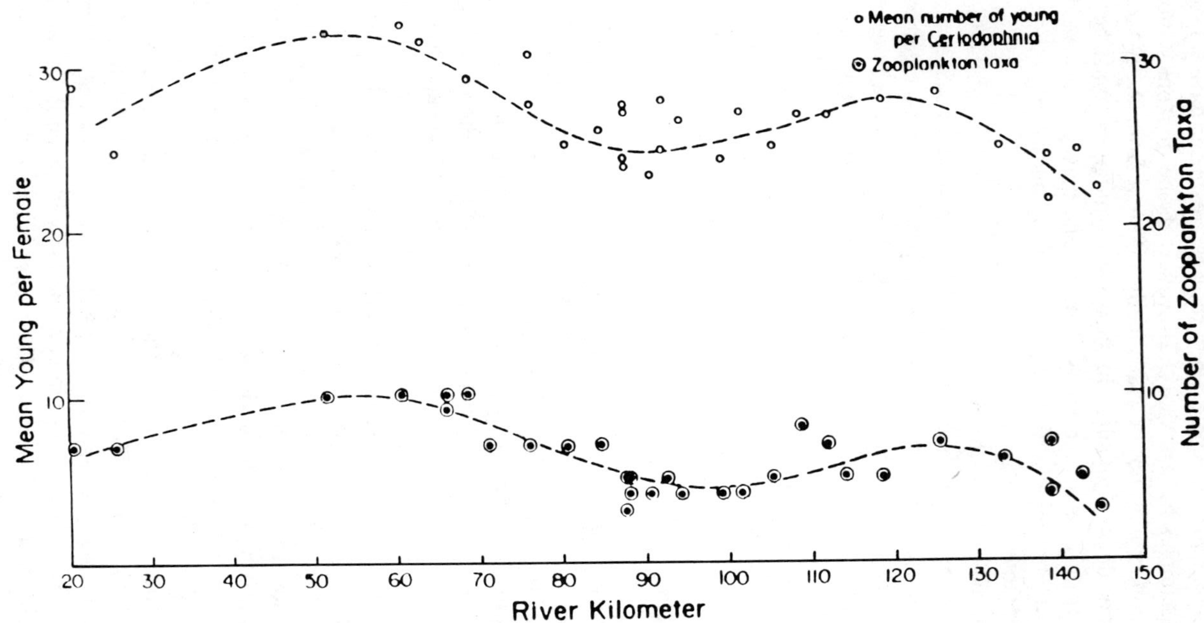


**Figure 11.1.** Young per female and zooplankton species richness at ambient stations on the Kanawha River. From Mount and Norberg-King (1986).

data then imply that within a community we should expect various groups (benthos, zooplankton, fishes, phytoplankton) to respond quite differently to a toxic chemical or mixture of chemicals. The implication one gets when reading many biological survey reports on contaminated reaches is that all groups of the community should decline and recover in harmony. If not, why would so many surveys examine only benthos or fishes? However, the toxicity data base clearly indicates that we should expect various groups to respond differently. The eight site studies also clearly show that to be true. Finally, if various groups do respond differently, then obviously toxicity data from a single species cannot correlate with all groups of organisms in the community. The more severe the impact, the better and the more correlations there should be.

Thus, the Kanawha data fit the expected. The benthos and zooplankton responded differently and so naturally the toxicity data could correlate with only one aquatic group. Also, the species richness was high enough to infer that the impacts are not massive.

## 4. TOXICITY PROFILE

If sampling stations are located over a reach of stream or segment of a lake and the toxicity measured in samples from those locations, a toxicity profile can be constructed from the data. The profile will show where toxicity occurs and where it is absent. One such profile is shown in Figure 11.2 (Mount et al., 1984) in which the reproductive rate of *Ceriodaphnia* is shown on the vertical scale and river stations are shown from upstream to downstream, over approximately 27 miles of stream, on the horizontal scale. Increased toxicity occurs below Station 2 as shown by the reduced number of young produced. Two discharges were located downstream of Stations 2 and 3. These effluents had chronically unsafe concentrations less than the waste concentrations occurring at those stations (Mount et al., 1984). Through fate processes causing decay of effluent components, and by dilution, the toxicity decreases at downstream Stations 7 to 9 where reproduction is as good or better than at Stations 2 and 3. The greater young production is attributed to enrichment by nutrients from agricultural runoff and domestic wastewater from small downstream plants.

A different type of plot would be needed for a lake situation where there is not a clear upstream–downstream relationship. Isopleths of toxicity could be used, for example, to describe data from a lake or harbor to give a comparable profile.

The eight site studies have shown that such profiles are predictive of community impact. As discussed above, the profiles did not, nor were they expected to, predict the impact for all community groups. More predictions could be made by gathering toxicity data on more species. The paper by DiToro, Holden, and Plafkin contained in this volume (Chapter 20), illustrates how such data can be used in modeling toxicity as well.

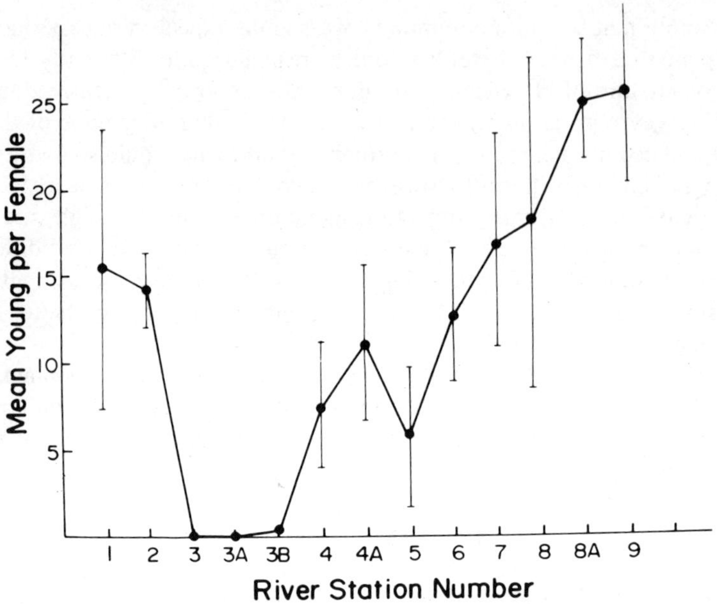


**Figure 11.2.** Ambient toxicity profile at 13 ambient stations on the Ottawa River. Vertical bars are standard deviations. Redrawn from Mount et al. (1984).

## 5. SUMMARY

The use of toxicity tests is a cost-effective tool for conducting biological community assessments, especially where the cause of impact is to be identified. The ambient toxicity test is especially good for this purpose and in these eight site studies it revealed many unsuspected sources of toxicity. If toxicity is not the cause of impact or if the wrong effluents are suspected of being the toxicity source, no expenditure for waste treatment plants to remove contaminants will alleviate the impact. Presence of chemicals does not mean adverse impact, but presence of toxicity does. Only organism responses can measure toxicity.

## REFERENCES

Blanck, H. (1984). Species dependent variation among aquatic organisms in their sensitivity to chemicals. *Ecol. Bull.* (Stockholm) **36**(19), 13 pp.

Di Toro, D. M., Hallden, J. A., and Plafkin, J. L. (1988). Modeling *Ceriodaphnia* toxicity in the Naugatuck River using additivity and independent action. In M. S. Evans (Ed.). *Toxic Contaminants and Ecosystem Health: A Great Lakes Focus*. Wiley, New York.

McNaught, D. C., and Mount, D. I. (1986). Appropriate duration and measures for *Ceriodaphnia* toxicity tests. *ASTM Spec. Tech. Publ.* **STP 891,** pp. 375–381.

Mount, D. I., and Norberg, T. J. (1984). A seven day life-cycle cladoceran toxicity test. *Environ. Toxicol. Chem.* **3,** 425–434.

Mount, D. I., and Norberg-King, T. J. (Eds.) (1985). Validity of effluent and ambient toxicity tests for predicting biological impact, Scippo Creek, Ohio. U.S. Environ. Prot. Agency, Washington, DC. EPA-600/3-85-044.

Mount, D. I., and Norberg-King, T. J. (1986). Validity of effluent and ambient toxicity tests for predicting biological impact, Kanawha River, Charleston, West Virginia. U.S. Environ. Prot. Agency, Off. Res. Dev. Rep. EPA/600/3-86-006.

Mount, D. J., Thomas, N. A., Norberg, T. J., Barbour, M. T., Roush, T. H., and Brandes, W. F. (1984). Effluent and ambient toxicity testing and instream community response on the Ottawa River, Lima, Ohio. U.S. Environ. Prot. Agency, Off. Res. Dev. Rep. EPA/600/3-84-080.

Mount, D. I., Steen, A. E., and Norberg-King, T. J. (Eds.) (1985). Validity of effluent and ambient toxicity tests for predicting biological impact, Five Mile Creek, Birmingham, Alabama. U.S. Environ. Prot. Agency, Washington, DC. EPA/600-8-85-015.

Mount, D. I., Steen, A. E., and Norberg-King, T. J. (Eds.) (1986a). Validity of effluent and ambient toxicity tests for predicting biological impact, Ohio River, Wheeling, West Virginia. U.S. Environ. Prot. Agency, Washington, DC. EPA/600/3-85-071.

Mount, D. I., Steen, A. E., and Norberg-King, T. J. (Eds.) (1986b). Validity of effluent and ambient toxicity tests for predicting biological impact, Back River, Baltimore Harbor, Maryland.U.S. Environ. Prot. Agency, Washington, DC. EPA/600/8-86-001.

Mount, D. I., Norberg-King, T. J., and Steen, A. E. (Eds.) (1986c). Validity of effluent and ambient toxicity tests for predicting biological impact, Naugatuck River, Waterbury, Connecticut. U.S. Environ. Prot. Agency, Washington, DC. EPA/600/8-86-005.

Norberg, T. J., and Mount, D. I. (1985). A new fathead minnow (*Pimephales promelas*) subchronic toxicity test. *Environ. Toxicol. Chem.* **4,** 711–718.

Norberg-King, T. J., and Mount, D. I. (1986). Validity of effluent and ambient toxicity tests for predicting biological impact, Skeleton Creek, Enid, Oklahoma. U.S. Environ. Prot. Agency, Off. Res. Dev. Rep. EPA/600/8-86-002.

Suter, G. W., II, Bainthouse, L. W., Breck, J. E., Gardner, R. H., and O'Neill, R. V. (1985). Extrapolating from the laboratory to the field: How uncertain are you? *ASTM Spec. Tech. Publ.* **STP 854,** pp. 400–413.

# 12

# FISH TESTS THAT GIVE USEFUL INFORMATION

*J. B. Sprague*

*Department of Zoology, University of Guelph, Guelph, Ontario, Canada N1G 2W1*



*Ask . . . the fishes of the sea and they shall declare unto thee.*

*Job 12: 7, 8*

## 1. INTRODUCTION

The above advice from an ancient scholar deserves a qualification: the fishes will declare useful information unto thee if you ask them the right questions. Perhaps my definition of useful should be stated. In the context of this chapter, useful information is knowledge of what concentrations or conditions in a waterbody are harmful to aquatic organisms and what conditions have no effect on overall well-being of the organisms. More or less, I am talking about numbers that are useful in producing water quality criteria or objectives.

During the last year I have gained some perspective on the need to channel our scientific resources toward obtaining this kind of practical information, and applying it with vigor in the Great Lakes basin. I have lived most of my life in that region and perhaps like most people, have been lulled into accepting second- or third-rate quality in some lakes and rivers. My first boyhood memory of Lake Erie is a holiday beach with a distasteful layer of black sawdust washing ashore. By contrast I have just spent a sabbatical leave on the edge of some British Columbia ecosystems that have not yet been completely remanufactured by humans. In fact some of the systems look suspiciously similar to what they must have been in the old days. Deer stand bemused on the roadside and gangs of sea lions play around in the bay. Water birds loaf and eagles soar. There are clean oysters on the beach, and the streams and bays teem with fish, at least they teem by Ontario and Ohio standards.

The Great Lakes area must have been a parallel paradise; remember that Lake Ontario was once full of Atlantic salmon (Hartman, Chapter 5). There is no reason that a semblance of the paradise cannot be restored, even in the face of the epidemic of humans in the region. Part of the restoration involves the selection of tests that really tell us what conditions make for happy fish living in supportive ecosystems.

## 2. TESTS OF ACUTE LETHALITY

The first category of useful tests includes the old standard lethal tests, such as determination of a 96-hour LC50 (median lethal concentration). Such tests are merely tools for preliminary exploration, but they continue to be useful for three purposes: (1) they can establish an initial benchmark for degree of toxicity of a substance; (2) they can monitor effluents in the real world; and (3) they can provide a start in sorting out complex topics in toxicology.

It is a pretty obvious first step to use a rapid and cheap lethal test to decide whether an unstudied pollutant is relatively innocuous, or potentially dangerous and deserving of further study. Having established the benchmark, there is no need for the process to be repeated ad nauseum. I am currently reviewing the effects of lead, zinc, and cadmium on marine mollusks and crustaceans, and was astonished to find in the primary literature, 47 determinations of the LC50 of cadmium for various shellfish. As usual, the between-species variation in LC50 was not much greater than the within-species variation from work of different investigators. This repetitive and unproductive use of research money is a little ridiculous, especially considering that the same review revealed only seven investigations that had tackled a much more important and challenging topic, that of joint toxicity of any of the three metals with any other pollutant.

A second use of lethal tests, that of monitoring the toxicity of effluents, is a monotonous task but most beneficial to the environment. A dead fish is understandable and convincing to a person in the street, to an industrialist, or to a judge. End-of-the-pipe lethality standards have been built into Canadian regulations for over a decade. Although completely illogical, they work, and have been instrumental in establishing satisfactory refinery effluents across the country, and in greatly improving pulp mill effluents. I was pleased to hear that they are being built into some discharge permits in the United States.

As a final use, lethal tests provide a fast and easily interpreted way to make initial explorations of complex topics. An example is the toxicity of mixtures, already mentioned. Enough exploration has been done to know that the topic is indeed very complex and that we cannot expect any simple algorithm to predict the joint action of all combinations of pollutants (Marking, 1985). The reaction to the complexity has been curious. Researchers have shied away from the topic (see above), the International Joint Commission (IJC) has never adopted an approach suggested by one of its scientific teams (IJC, 1980), and in the late 1970s, suggestions for dealing with pollutant mixtures quietly disappeared from compendia of water quality criteria in the United States and Ontario. Neither IJC nor the "red book" (U.S. Environmental Protection Agency, 1976) gives any guidance on how to deal with two toxicants occurring simultaneously. The subject is of great importance since mixed pollution is the rule rather than the exception, and Wong et al. (1978) have shown that a mixture of metals at concentrations of the IJC objectives is strongly inhibitory to algae. Much more research effort should be concentrated on asking the fishes about this important question.

A more encouraging example of using lethal tests to explore a complex and important topic is current research on chemical structure as related to biological activity. If we are to get ahead of the game, and prevent tomorrow's PCBs and dioxins from spreading through the environment before we are aware of them, we must be able to predict, from chemical structure, which materials are dangerous. There is at least one massive structure/activity program underway, under the auspices of the Duluth laboratory of the EPA, and it uses lethal tests for the initial sorting out. Bless the people who are doing such research.

## 3. SUBLETHAL TESTS

There are all kinds of sublethal responses that may be measured but I will divide them into whole-organism responses and within-organism ones which are often physiological or biochemical. It is the whole-organism studies that have given us most of the numbers that are useful for water quality criteria or otherwise dealing with loadings of pollutants in a practical way. This was true a decade ago, in an examination (Sprague, 1976) of the bases for criteria recommended for aquatic organisms in the "blue book" (National Academy of Science/National Academy of Engineering, 1974). Of 35 recommendations that had reasonable scientific bases, about 12 were derived from reproductive success of organisms, 7 from tissue residues, 6 from field observations, at least 6 from whole-organism responses such as growth, swimming, or respiratory rates, but charitably, only about 4 from within-organism effects. Things have not changed much, judging by the above-mentioned review of metals and shellfish and another one that I am completing on freshwater fish. In both cases, whole-organism responses such as growth and reproduction provided meaningful ranges of both harmful and no-observed-effect levels (NOEL). Equally large piles of reprints showed that within-organism changes were found in the same range of concentrations (no lower), and sometimes only at absurdly high lethal concentrations. Even then, a major question was whether a detectable within-organism change was a deleterious effect or merely a normal response within the adaptive range of the organism. A long time ago, the question was succinctly addressed by Mount and Stephan (1967): "An exposure causing death is obviously significant, but even the best fish physiologist would have difficulty establishing that a 10 percent reduction in hematocrit would result in an undesirable effect on a population." And indeed, in my shellfish review I found seven cases in which there was a decrease in tissue respiration or respiration of an organism exposed to sublethal levels of metal, a response that should perhaps be interpreted as a deleterious effect. But there were six other cases of respiratory increase in response to similar levels of metal; in all logic, should these then be interpreted as a beneficial effect?

The real reason for within-organism toxicology using biochemical and physiological techniques should be to explain the mechanism of poisoning that has been manifested at higher levels of integration, *if* we wish or need to understand the mechanism. Indeed, there is a strong reason for categorizing the modes of action of toxicants in order to advance the structure/activity research mentioned at the end of Section 2. On reviewing the papers, I was disappointed that many of the researchers did not express such a goal, and in fact few of the papers started with any hypothesis concerning mechanism of effect. Often it seemed that a battery of tests was run on ATPase, mixed-function oxidases, and so forth, in the hope that something would turn up. Funds for such research would buy many corks to plug effluent pipes.

## 4. CHRONIC EXPOSURES AND SHORTCUTS

It was only a couple of decades ago that we were attempting to predict sublethal no-effect levels by means of application factors. That is, one would multiply the LC50 for a sensitive fish by one-tenth or one-twentieth and use the results as a tentative objective for water quality. The 1960s saw the emergence and blooming of a technique long utilized in pharmacology—lifetime exposure of one generation, continued through the reproductive period and into the beginning of the next generation. The opportunity to assess *any* kind of deleterious effect in the course of such exposures leads to direct estimates of the no-observed-effect level.

Pioneered for fish by Mount and Stephan (1967), these life-cycle tests required almost a year for fathead minnows and about 2.5 years for trout. The tedious exposures were soon shortened to 3 or 4 months, once the skeptics were convinced that tests with a toy tropical fish gave results that were relevant to game fish in northern waters. McKim (1977) documented that, and concluded that "life-cycle toxicity tests have provided the most reliable information on the long-term effects of toxic chemicals on fish and the basis for many of the established water-quality criteria for aquatic life."

Time and effort were further reduced to about a month when it was shown that the most sensitive responses in a life-cycle exposure were almost always displayed by the early life-stages (embryonic, larval, and early juvenile development). Early-life-stage results generally predicted findings of the full life-cycle test within a factor of two, and in fact were usually much closer than that (McKim, 1977, 1985).

Sublethal exposures have recently been reduced to 4 days by relating biochemical measurements to whole-organism effects. Quantitative assessments of RNA, DNA, and protein in exposed larval fish are highly predictive of the effects on early life stages, the "safe" ranges being within or very near to those of the longer test (Barron and Adelman, 1984). The advantage lies with the speed, and at last there is hope of exploring complex toxicological topics with a meaningful sublethal test instead of a lethal one. The increased sophistication and cost of exploring with a sublethal technique is warranted by the closer approach to answers that can be used in protecting real-world environments.

In parallel with these sublethal tests on fish we have seen the development of 1- or 2-week chronic tests with other organisms such as daphnids (Biesinger and Christensen, 1972). These are equally useful or perhaps more useful, but fish tests will probably always be with us, if only because most humans attach a special importance to fish among all aquatic organisms.

## 5. RUBBER-BOOT TOXICOLOGY

All the above laboratory work utilizes, for reasons of good scientific procedure, steady conditions of exposure and favorable conditions for test organisms aside from the presence of the pollutant. There is adequate documentation that these lab tests will indeed predict what would happen in a real lake or river, if there were a neat situation with a single kind of pollutant at relatively constant concentration, with no binding or detoxifying agents present, and if there were sufficient understanding of modifying factors such as temperature and pH. We seldom find these ideal pollution situations, and so there is a need for a new wave of aquatic toxicology to go into the field, or at least into an artificial stream or other decent-sized mesocosm. Validation has been stressed by several authors in this volume, particularly by Cairns (Chapter 1) and Mount (Chapter 11).

As well as the real-world variability, there are also conceptual problems in utilizing laboratory results. Just as one cannot predict upward in levels of integration from hematocrit to fish populations, one cannot really predict from fish reproduction in the lab to populations in the wild, and it would be foolish to extrapolate upward by several levels of integration to the functioning of a community. For all these reasons it is, and always has been, essential to validate our laboratory findings by checking them out under field conditions.

Mount (Chapter 11) has outlined a major program in such field validation. The research is a curious blend of new and old, and I for one am delighted and hope the fever spreads. On the one hand, we have the latest in shortcut chronic exposures carried to the river bank. On the other hand we have a return to biological surveys, similar in approach to the early work of Richardson (1928) in the Illinois River, but more sophisticated. Biological effects, or lack of effects, are associated with physicochemical measurements of pollutants and other conditions, an approach that dates back half a century to Ellis (1937), but an approach that pays off. Indeed the existing criteria for suspended solids are almost entirely derived from field observations.

Another approach to rubber-boot toxicology is to experiment with controlled communities such as large enclosures in lakes, or in artificial ponds or streams. The plastic-enclosed mesocosms in lakes seldom include fish, or more than a few token individuals, but artificial streams have successfully evaluated growth, reproductive success, and survival of small populations of fish. The work of Zischke et al. (1983), as an example, shows that there are many advantages of artificial streams for validation of laboratory findings. The size can be modest (1.4 m wide in riffle areas), and the constant flow of new water can be continually dosed to a steady level of toxic substance for interpretational ease. The fish must earn a living under seminatural conditions, and the other components of the community may also be evaluated by standard survey techniques. Several concentrations may be run, and replicated in similar stream channels, in order to approach an estimate of the NOEL. Although it is expensive to conduct an experiment in artificial streams for several months, the cost is much less than that required to manipulate a real stream or creek, with all the variation in flow rate, physical problems with debris, and fluctuation in water quality (Geckler et al., 1976).

Attempts at field validation can produce some real surprises, as described by Mount (this volume), and that is why the shuttling back and forth between field and lab is such an effective way to make progress. A similar surprise occurred in the field experiment with copper some years ago in Ohio (Geckler et al., 1976). Lab tests predicted with accuracy the copper concentrations suitable for spawning of fish, but it turned out that a more sensitive avoidance response of wild fish was more important. Proposed biochemical indices should also be field tested for surprises, as was bravely done by some Canadian workers. Fish exposed to elevated metals in an Ontario lake showed the expected increase in metallothioneins (MT) and increased tolerance of lethal levels of the metal (Duncan and Klaverkamp, 1983). However, fish in a metal-polluted lake in British Columbia showed only the MT increase, apparently without the predicted change in tolerance (Roch and

McCarter, 1984). These kinds of surprises make us think, and that may lead us more quickly to ultimate truths.

## 6. EPIDEMIOLOGY

A somewhat parallel and meritorious combination of field and laboratory approach is occurring for the topic of pollutant-caused disease in fish, particularly the presence of tumors. In a thought-provoking critique, Gilbertson (1984) makes the case that toxicological research should *start* with field work to document anomalies in wild populations of fish. Subsequent lab work would establish the cause (Black, 1983), in a fashion analogous to the classical approach for showing that a certain bacterium is the causative organism for a human disease. The approach is also being used in other locations, notably Puget Sound (McCain et al., 1986), where there are surprising incidences of neoplasms in bottom-living fish from some inshore locations.

We should get a lot of useful information if enough people follow the lead of these researchers, and put on rubber boots to ask the fishes.

## 7. SUMMARY AND CONCLUSIONS

Classical lethal tests are merely a first step in sorting out dangerous pollutants or exploring problems, such as developing predictive relationships between chemical structure and toxicity.

At sublethal levels, simple whole-organism responses have proven much more useful for pollution control than within-organism ones (e.g., biochemical). Short whole-organism tests with early life stages have been of greatest benefit; they have provided cost-effective and readily used information for producing water quality criteria and they make it possible to tackle important but neglected topics such as joint sublethal action in mixtures.

The new wave of fish toxicology takes the short sublethal tests into the field to assess real polluted waters and test the accuracy of the laboratory predictions. This also involves validation by old-wave biological surveys in affected communities. The *combination* of whole-organism bioassay with community surveys, although low technology, is one of the cutting edges of applied aquatic toxicology, insofar as making valid decisions about safe and unsafe levels of water pollutants. There is similar promise in combining lab tests on fish abnormalities and neoplasms with epidemiology in wild populations.

## ACKNOWLEDGMENT

The biblical scholarship of Dr. Jerry F. Paine, Fisheries and Oceans Canada, provided the initial quotation.

## REFERENCES

Barron, M. G., and Adelman, I. R. (1984). Nucleic acid, protein content, and growth of larval fish sublethally exposed to various toxicants. *Can. J. Fish. Aquat. Sci.* **41,** 141–150.

Biesinger, K. E., and Christensen, G. M. (1972). Effects of various metals on survival, growth, reproduction, and metabolism of *Daphnia magna. J. Fish. Res. Board Can.* **29,** 1691–1700.

Black, J. J. (1983). Field and laboratory studies of environmental carcinogenesis in Niagara River fish. *J. Great Lakes Res.* **9,** 326–334.

Cairns, J. (1988). Politics, economics, science: going beyond disciplinary boundaries to protect aquatic ecosystems. In M. S. Evans (Ed.). *Toxic Contaminants and Ecosystem Health: A Great Lakes Focus*. John Wiley, New York.

Duncan, D. A., and Klaverkamp, J. F. (1983). Tolerance and resistance to cadmium in white suckers (*Catostomus commersoni*) previously exposed to cadmium, mercury, zinc, or selenium. *Can. J. Fish. Aquat. Sci.* **40,** 128–138.

Ellis, M. M. (1937). Detection and measurement of stream pollution. Bull. U.S. Bur. Sport Fish. Wildl. **48,** 365–437.

Geckler, J. R., Horning, W. B., Neiheisel, T. M., Pickering, Q. H., Robinson, E. L., and Stephan, C. E. (1976). Validity of laboratory tests for predicting copper toxicity in streams. U.S. Environ. Prot. Agency, Off. Res. Dev. Rep. EPA-600/3-76-116, pp. 1–192.

Gilbertson, M. (1984). Need for development of epidemiology for chemically induced diseases of fish in Canada. *Can. J. Fish. Aquat. Sci.* **41,** 1534–1540.

Hartman, W. (1988). Historical changes in the major fish resources of the Great Lakes. In M. S. Evans (Ed.). *Toxic Contaminants and Ecosystem Health: A Great Lakes Focus*. John Wiley, New York.

International Joint Commission (1980). *Report of the Aquatic Ecosystem Objectives Committee*. Great Lakes Science Advisory Board, Windsor, Ontario.

Marking, L. L. (1985). Toxicity of chemical mixtures. In G. M. Rand and S. R. Petrocelli (Eds.), *Fundamentals of Aquatic Toxicology: Methods and Applications*. Hemisphere, Washington, DC, pp. 164–176.

McCain, B. B., Myers, M. S., Brown, D. W., Rhodes, L. D., Landahl, J. T., Chan, S.-L., and Malins, D. C. (1986). Diseases of marine fishes in Puget Sound and relationships to pollution. Canadian Technical Report of Fisheries and Aquatic Sciences No. 1480, p.126.

McKim, J. M. (1977). Evaluation of tests with early life stages of fish for predicting long-term toxicity. *J. Fish. Res. Board. Can.* **34;** 1148–1154.

McKim, J. M. (1985). Early life stage toxicity tests. In G. M. Rand and S. R. Petrocelli (Eds.), *Fundamentals of Aquatic Toxicology: Methods and Applications*. Hemisphere, Washington DC, pp. 58–95.

Mount, D. I. (1988). Ambient toxicity to assess biological impact. In M. S. Evans (Ed.). *Toxic Contaminants and Ecosystem Health: A Great Lakes Focus*. John Wiley, New York.

Mount, D. I., and Stephan, C. E. (1967). A method for establishing acceptable limits for fish—Malathion and the butoxyethanol ester of 2, 4-D. *Trans. Am. Fish. Soc.* **96,** 185–193.

National Academy of Science/National Academy of Engineering (1974). Water Quality Criteria 1972. U.S. Envir. Prot. Agen., Ecol. Res. Ser. EPA-R3-73-033. Washington, D.C.

Richardson, R. E. (1928). The bottom fauna of the middle Illinois River, 1913–1925. *Bull. Ill. Nat. Hist. Surv.* **17,** 387–485.

Roch, M., and McCarter, J. A. (1984). Hepatic metallothionein production and resistance to heavy metals by rainbow trout (*Salmo gairdneri*). II. Held in a series in contaminated lakes. *Comp. Biochem. Physiol. C* **77C,** 77–82.

Sprague, J. B. (1976). Current status of sublethal tests of pollutants on aquatic organisms. *J. Fish. Res. Board Can.* **33,** 1988–1992.

U.S. Environmental Protection Agency (1976). *Quality Criteria for Water.* U.S. Envir. Prot. Agen., Washington, DC.

Wong, P. T. S., Chau, Y. K., and Luxon, P. O. (1978). Toxicity of a mixture of metals on freshwater algae. *J. Fish. Res. Board Can.* **35,** 479–481.

Zischke, J. A., Arthur, J. W., Nordlie, K. J., Hermanutz, R. O., Standen, D. A., and Henry, T. P. (1983). Acidification effects on macroinvertebrates and fathead minnows (*Pimephales promelas*) in outdoor experimental channels. *Water Res.* **17,** 47–63.

13

# USE OF BIOASSAYS IN DETERMINING THE TOXICITY OF SEDIMENT TO BENTHIC ORGANISMS

*J. O. Lamberson and R. C. Swartz*

*U.S. Environmental Protection Agency, Mark O. Hatfield Marine Science Center, Newport, Oregon 97365*

**1. Introduction**

**2. Overview of sediment bioassay methods**

2.1. Marine

2.2. Fresh water

2.3. Great Lakes

**3. Comparison of freshwater sediment bioassay methods**

3.1. Species

3.2. Response criteria

3.3. Duration

3.4. Sediment phase

3.5. Conclusions

**4. The *Rhepoxynius* sediment bioassay**

4.1. Method

4.2. Field validation

4.3. Laboratory applications

**5. Problems with sediment bioassays**

**6. Applications of sediment bioassays**

**7. Recommendations for research**

**8. Conclusions**

**References**

## 1. INTRODUCTION

Sediments have often been called a cumulative sink or a major repository for persistent toxic substances in the aquatic environment (Chapman, 1986; Oschwald, 1972; Swartz and Lee, 1980). Contaminated sediments may contain depauperate benthic community assemblages even in water systems with "healthy" overlying water quality (Laskowski-Hoke and Prater, 1984). Through the processes of diffusion, physical disruption of the sediments, or bioturbation and other biological processes, toxicants may be transferred from sediments to the overlying water column (Lee and Swartz, 1980; Nalepa et al., 1980; Swartz and Lee, 1980). Toxic substances sorbed on sediment particulates may be ingested by sediment-feeding benthic organisms or may be dissolved in interstitial water and assimilated through respiration, ingestion, or other absorptive processes. Toxicants may thus directly effect benthic organisms or enter the food chain as these organisms are consumed.

Methods employed to determine the degree of contamination of sediments frequently involve chemical analyses of the sediments or studies on benthic community structure. However, with currently available chemical methods, it is costly and difficult to analyze for the hundreds of potentially toxic chemical compounds that may be present in sediments (Malins et al., 1984). Chemical analyses give little information on bioavailability of compounds, biological effects on benthic organisms, or effects of interactions among chemicals on the behavior of the chemicals in the sediment or in living organisms. Structural analyses of benthic communities are time-consuming and thereby limited in the number of samples that can be conveniently analyzed, and may fail to take into account the inherent spatial and temporal variability of benthic ecosystems due to seasonal, biological, or physical processes.

Sediment bioassays are comparatively inexpensive and simple to conduct, and can provide rapid information on the degree of toxicity of potentially contaminated sediments to benthic organisms. Acute and chronic toxicity bioassays have been widely used since the 1950s to evaluate the effects of wastes on organisms in receiving waters (Buikema et al., 1982) and to develop water quality criteria, but comparatively few toxicity bioassays have been developed for sediments. Because sediment bioassays can provide direct evidence of adverse biological effects occurring from exposure of organisms to toxicants in sediments, their use has been recommended in combination with other methodologies to assess the toxicity of contaminated sediment (Chapman, 1986; Chapman and Long, 1983; Long, 1983; Long and Chapman, 1985).

In Puget Sound, Washington, a variety of histopathological conditions in demersal fishes and crustaceans were found to be correlated with elevated contaminant concentrations in sediments (Krahn et al., 1986; Long, 1982; Malins et al., 1982, 1984; Quinlan et al., 1986). Hargis et al. (1984) reported skin lesions, pancreatic and liver alterations, and mortality in fish experimentally exposed to field-collected contaminated sediments in Chesapeake Bay. The occurrence of tumors in demersal fish and deformities in sediment-dwelling invertebrates in the Great Lakes (Black, 1984; Black et al., 1980; Hare and Carter, 1976; Warwick,

1980) suggests that sediment toxicity research is also highly relevant for the Great Lakes area.

## 2. OVERVIEW OF SEDIMENT BIOASSAY METHODS

### 2.1. Marine

Since the EPA/COE implementation manual (Environmental Protection Agency/Corps of Engineers, 1977) provided guidance for development of sediment bioassays to test the toxicity of dredged materials, a variety of procedures have been developed to evaluate sediment contamination in the marine environment (reviewed by Chapman, 1986; Swartz, 1987). These range in complexity from short-term acute tests of the effects of individual contaminants on single species, to long-term determinations of the effects of chemical mixtures on the function of benthic microcosms. Sediments tested are usually solid phase, although sometimes suspended solids, elutriates, or sediment extracts are examined. Amounts of sediments tested range from a few grams to over 800 L, and systems may be static or flow-through. Test procedures involve organisms such as fish and a variety of benthic, epibenthic, and pelagic invertebrates including nematodes, oligochaetes, polychaetes, pelecypods, crustaceans, echinoderms, and tunicates. Response criteria, ranging from acute mortality through changes in mesocosm structure and function, include sublethal physiological responses such as respiration, cellular function, and behavioral and developmental variations. Experimental duration extends from 10 minutes to over a year. Bioassay test procedures have been applied to materials collected from potentially contaminated field sites, or less frequently, to uncontaminated sediments "spiked" with specific chemicals. In our laboratory, acute, static bioassay procedures for marine sediments have been developed using the infaunal amphipod *Rhepoxynius abronius* (Swartz et al., 1985a) and have been extensively applied to both field-collected and laboratory-spiked sediments (see Section 4).

### 2.2. Fresh Water

Methods for determining toxicity of contaminated freshwater sediments have been reviewed by Chapman (1986) and Nebeker et al. (1984), and a variety of methods are summarized in Table 13.1. Most research involving freshwater sediment bioassays has been concerned with methods development, and field and laboratory studies on heavy metal contamination. Several bioassay methods have been employed to rank field sediments on the basis of toxicity to test organisms (LeBlanc and Surprenant, 1985; Malueg et al., 1984b; Prater and Anderson, 1977a; Prater and Hoke, 1978; Samoiloff et al., 1983). Malueg et al. (1984a,b) used acute sediment bioassays to test sediments from metal-contaminated areas in Wisconsin, Michigan, and California. Bioassay methods have also been used to determine toxicity

**Table 13.1 Summary of Methods for Determining the Toxicity of Freshwater Sediments to Aquatic Organisms**

| Reference | Organism | Response Criterion | Duration | Sediment Phase | Toxicant or Sediment Source |
|---|---|---|---|---|---|
| Adams et al. (1985) | *Chironomus* | Mortality, growth | 14 d | Solid | Kepone |
| Bahnick et al. (1980) | *Hexagenia*<br>*Daphnia*<br>*Pontoporeia*<br>*Lepomis* | Mortality, cough frequency | 96 hr | Solid, interstitial water | Duluth-Superior Harbor |
| Brinkhurst et al. (1983)<br>Chapman and Brinkhurst (1984)<br>Chapman et al. (1982d) | Oligochaetes | Respiration rate | 1–2 hr | Extract | Various toxicants |
| Buikema et al. (1980) | *Daphnia*<br>*Hexagenia* | Acute mortality | 48 hr<br>96 hr | Solid | Superior-Duluth Harbor |
| | Enzymes | Enzyme inhibition | 1.5 hr | Extract | |
| Cairns et al. (1984)<br>Nebeker et al. (1984)<br>Schuytema et al. (1984) | *Daphnia*<br>*Hyalella*<br>*Chironomus*<br>*Gammarus*<br>*Hexagenia* | Acute mortality | 48 hr<br>10 d | Solid | Test sediment, cadmium, copper |
| Chapman et al. (1982a–d) | Oligochaetes | Acute mortality | 96 hr | Solid | Various toxicants |

| | | | | | |
|---|---|---|---|---|---|
| Chapman et al. (1982d)<br>Chapman (1986)<br>Landolt and Kocan (1984)<br>Landolt et al. (1984) | Trout cells | Mitotic abnormalities | 72 hr | Organic extract | Puget Sound |
| Gannon and Beeton (1969, 1971) | *Pontoporeia*<br>*Gammarus*<br>*Chironomus* | Acute mortality, sediment selection | 24–48 hr<br>48–60 hr | Solid | Field sediment, Great Lakes harbors |
| Jones and Lee (1978)<br>Lee et al. (1978) | *Palaemonetes*<br>*Daphnia* | Acute mortality | 96 hr | Settled elutriate | 19 U.S. waterways |
| LeBlanc and Surprenant (1985) | *Pimephales*<br>*Paratanytarsus*<br>*Daphnia*<br>Microorganisms | Survival, growth, reproduction, population changes | 21 d | Elutriate, settled elutriate | Dredged sediments |
| Malueg et al. (1983, 1984a,b)<br>Nebeker et al. (1984)<br>Schuytema et al. (1984) | *Daphnia*<br>*Hexagenia* | Acute mortality | 48 hr<br>5–10 d | Solid | Cadmium |
| Nebeker et al. (1984) | (a) *Daphnia*<br>(b) See Cairns et al. (1984)<br>(c) See Malueg et al. (1983, 1984a,b) | Acute mortality | 48 hr | Elutriate | Test sediment |

**Table 13.1** (*Continued*)

| Reference | Organism | Response Criterion | Duration | Sediment Phase | Toxicant or Sediment Source |
|---|---|---|---|---|---|
| Nebeker et al. (1984) (*Continued*) | (d) *Chironomus* | Larval survival and growth | 15 d | Solid | Test sediment |
| | (e) *Chironomus* | Adult emergence | 25 d | Solid | Test sediment |
| | (f) *Hyalella* | Adult and young survival | 28 d | Solid | Test sediment |
| | (g) *Daphnia* | Adult and young | 10 d | Solid | Test sediment |
| | (h) *Hyalella* *Gammarus* *Hexagenia* *Chironomus* Fish | Acute mortality | 96 hr or longer | Undisturbed | Field (in situ) |
| Prater and Anderson (1977a,b) Prater and Hoke (1978, 1980) | *Daphnia* *Hexagenia* *Asellus* *Pimephales* *Lirceus* | Acute mortality | 96 hr | Solid | Otter Creek, Duluth Harbor, Ashtabula Harbor, etc. |
| Samoiloff et al. (1983) | *Salmonella* *Panagrellus* | Mutagenesis, developmental inhibition, mortality | 96 hr | Extract | Tobin Lake, Saskatchewan |
| Wentsel et al. (1977, 1978) | *Chironomus* larvae | Growth and survival, emergence | 17–20 d | Solid | Palestine Lake, Indiana |

in freshwater sediments experimentally spiked with copper (Cairns et al., 1984) and cadmium (Schuytema et al., 1984).

A commonly used freshwater sediment bioassay method was developed by Prater and Anderson (Prater and Anderson, 1977a,b; Prater and Hoke, 1978), and modified by others (LeBlanc and Surprenant, 1985; Malueg et al., 1983, 1984a,b; Schuytema et al., 1984). In this system burrowing or epibenthic organisms (usually the insect larvae *Chironomus* or *Hexagenia,* the amphipods *Hyalella* or *Gammarus,* or the isopod *Asellus*) are exposed to test sediment, while swimming organisms (the cladoceran *Daphnia* or the fathead minnow *Pimephales*) are exposed to water recirculating above the sediment. Malueg et al. (1983) reported that the presence of *Hexagenia* in the sediment in this system increased the toxicity of the water to *Daphnia,* as particulates released from the sediments by the physical activity of the *Hexagenia* may have resulted in greater release of toxic materials into solution.

In solid phase static bioassays, test organisms are exposed to sediment with overlying water in a beaker. Organisms and contaminants may be present in the sediment or water or both. Cairns et al. (1984) employed acute static bioassays to test toxicity of copper-spiked sediments to *Daphnia, Chironomus, Hyalella,* and *Gammarus.* Wentsel et al. (1977, 1978) used *Chironomus* larval growth and survival and adult emergence as response criteria in solid phase static bioassays with field-contaminated sediments.

A variety of procedures have been developed to test biological effects of sediment extracts or elutriates. Sediment elutriate tests are designed to assess the effects of chemical contaminants released from sediments to the water column during disposal of dredged sediments in open water. Chemicals dissolved in interstitial water or adsorbed to suspended particulates and settleable solids are released to the water column as the dredged material settles. Bioassays exposing epibenthic or water column organisms to elutriate water and settled sediments have generally indicated little toxicity associated with the discharged material (Jones and Lee, 1978; Lee et al., 1978). However, these tests do not reveal biological effects of toxic sediment in situ or chronic adverse effects on infaunal organisms at the disposal site.

Bioassay methods developed to test the toxicity of chemicals dissolved in water have been adapted and applied to sediment extracts. Chapman and Brinkhurst (1984) and Chapman et al. (1982a,b,c) described test procedures using acute mortality and respiration response of marine and freshwater oligochaetes to toxicants in water, and have applied these methods to marine sediment extracts (Chapman et al., 1982d). Presumably these methods could be applied to freshwater sediment extracts as well. Another method, utilizing mitotic abnormalities in cultured rainbow trout cells in response to marine sediment extracts (Landolt and Kocan, 1984; Landolt et al., 1984), has recently proven to be sensitive with freshwater sediment extracts (Chapman, 1986). Samoiloff et al. (1983) described bioassay methods using bacterial mutagenicity and inhibition and nematode survival, growth, and mutagenicity in response to sediment extract fractions from a lake in Saskatchewan.

Nebeker et al. (1984) described a series of three acute and four chronic sediment bioassay procedures. Acute procedures tested the toxicity of sediment elutriates to *Daphnia* or of sediment with overlying water to *Daphnia, Hyalella, Chironomus,* or *Gammarus* in beakers. Results using these methods compared favorably with those using the Prater-Anderson system with *Daphnia* and *Hexagenia.* Chronic methods used *Daphnia* and *Hyalella* life cycle data and *Chironomus* larval survival and growth and adult emergence as response criteria. Chapman (1986) reported that *Daphnia* is a very sensitive indicator of toxicity in both the acute and life cycle tests, but that the other organisms used by Nebeker et al. (1984) were less sensitive.

In a series of partial life cycle tests using static and flow-through systems, Adams et al. (1985) demonstrated that the key route of exposure of a persistent toxic chemical (Kepone) in sediment to the midge, *Chironomus tentans,* was via the interstitial water. The results provide insight into the mechanism of action of toxicants in sediments to benthic organisms.

## 2.3. Great Lakes

In the Great Lakes, sediment toxicity testing has been primarily applied to assessing the impact of dredged sediment disposal. The Prater-Anderson apparatus has frequently been used for testing sediments with recirculating water and the organisms *Daphnia, Hexagenia, Asellus,* and *Pimephales.* Although some positive correlations were found between mortality in bioassays using this system and individual chemical parameters examined in sediments from Duluth and Superior Harbors (Prater and Anderson, 1977b), and from five Lake Michigan harbors (Hoke and Prater, 1980; Laskowski-Hoke and Prater, 1984; Prater and Hoke, 1979, 1980), the significance of these results is difficult to interpret because of possible interactions between contaminants. Using a modification of the Prater-Anderson procedure with sediment from the Keweenaw Waterway, Lake Superior, Malueg et al. (1984a) did find positive correlations among sediment bioassay results, copper content in the sediments and in the overlying water in the bioassay chambers, and major disruptions in the benthic macroinvertebrate community structure.

Buikema et al. (1980) developed an enzyme-inhibition bioassay method that tested the inhibitory effects of extracts of sediments from several sites in Superior-Duluth Harbor. Results were compared with 24- to 48- and 96-hour mortality data from *Daphnia* and *Hexagenia* exposed to the same sediments.

Gannon and Beeton (1969, 1971) surveyed nine Great Lakes harbor sediments for toxicity using a 48- to 60-hour acute static bioassay system and a 24- to 48-hour sediment-selectivity procedure. Acute mortality and avoidance of polluted sediments were observed using both test procedures with the amphipod *Pontoporeia* and other invertebrates. Oxberry et al. (1978), using long-term (up to one year) static tests with *Pontoporeia* and *Mysis* concluded that taconite tailings in Lake Superior were not toxic to benthic organisms. Bahnick et al. (1980) performed extensive series of 96-hour bioassays using *Hexagenia, Daphnia,* and *Pontoporeia* to test sediments from Duluth-Superior Harbor and Lake Superior. They tested solid sediment with overlying water in beakers, interstitial water, and sediment elutriate water and demonstrated good survival with all animals in the test systems in controls and in most harbor sediments tested.

It is apparent that the development of sediment toxicity bioassays for the Great Lakes and other freshwater areas is still in preliminary stages. More work needs to be done in development and field testing of methods, and application of these methods to sediment toxicity surveys and laboratory research.

## 3. COMPARISON OF FRESHWATER SEDIMENT BIOASSAY METHODS

A variety of sediment toxicity bioassays that have been developed for the Great Lakes and other freshwater areas are summarized in Table 13.1. Factors to be considered in the selection of a bioassay method include the duration of the test, the complexity of the system, the species of test organism, response criteria, and sediment phase to be tested.

### 3.1. Species

The use of a native species that is considered to be a natural indicator of pollution facilitates a field validation for a bioassay method when the bioassay results are compared with the distribution of the species in the field. Native species such as *Pontoporeia, Hexagenia, Mysis,* and *Chironomus* in the Great Lakes area are generally ecologically important, easily collected and available year round, and not subject to laboratory artifacts that may affect cultured organisms. They are, however, affected by natural population variation and ecological factors such as seasonality, weather, and pollution. Cultured species are readily available in large quantities, are subject to less population variation, and in the case of species such as *Daphnia,* often have a large amount of comparative information available in the literature. Multispecies bioassays give interesting information on the effects of sediment contamination on more than one organism, but the bioassay apparatus is relatively complex and more expensive, and thus allows for fewer replicates. Small-scale bioassay systems, such as trout cell cultures, enzyme systems, and microorganisms, allow for ease of handling and greater replication and are relatively inexpensive.

### 3.2. Response Criteria

Mortality is usually the most easily measured and most readily understood response criterion, and data may be compared among species and among various chemicals or other parameters within a species. Results must be compared to control data. Bioassays using mortality data with some species such as *Daphnia* can be very sensitive to contamination; however, many species are insensitive or highly variable. Bioassays using developmental criteria such as growth, insect larval emergence, or maturation rate are complex, results are also often highly variable, and the bioassays take too long to be useful for large-scale surveys or monitoring work. An exception is the life cycle test with *Daphnia* (Nebeker et al., 1984), which has a short generation time. Behavioral response criteria such as cough frequency and sediment selectivity are highly variable, and may be due to factors other than sediment contamination. Physiological response criteria, such as respiration rate, enzyme activity, and mitotic abnormalities, give consistent results and are very sensitive, but test only one aspect of an organism's total metabolism.

### 3.3. Duration

The duration of various bioassay methods ranges from very short term (<1 hour) to 1 year or longer for survival studies. Very short term (1–2 hours for respiration rate and enzyme activity) and short-term (72 hours for mitotic abnormalities) bioassays produce fast results and large numbers of samples can be tested in a short time. Short-term (48- to 96-hour) acute mortality bioassays also give relatively fast results and a large amount of comparative data are available, but these tests may be insensitive in that they are too brief to detect the more subtle, sublethal effects that would produce mortality with a longer exposure time. If possible, concurrent sublethal response criteria should be monitored. Longer term (such as 10-day) mortality tests may be more sensitive and the results may be less variable (Swartz et al., 1985a). Growth and development tests (10–30 days) are useful in that they test the effects of toxicants on various stages of a life cycle, but the longer time scale makes them less useful for routine testing.

### 3.4. Sediment Phase

The sediment to be tested may be in a solid phase (intact or settled), or a liquid (an elutriate, an extract, or interstitial water). Elutriate bioassays test the effects of dissolved substances from sediments on water column organisms. Toxicants dissolved in interstitial water from the sediment are diluted by elutriate water to usually nontoxic levels. The test is relatively insensitive, does not test the effects of contaminated sediment on benthic organisms at the disposal site, and does not test the effects of contaminants bound to sediment particles. When settled sediment is tested after elutriation, dissolved contaminants may have been washed away. Bioassays of sediment extracts are often very sensitive, but may not test the specific toxic factor or the effects of interactions among contaminants in intact sediments. The toxicity of dissolved contaminants can be tested in interstitial water bioassays, but if benthic species are used as test organisms they may be stressed by the absence of sediment, and the toxicity of particle-bound contaminants is not tested. Toxicants bound to sediments may be especially important to benthic organisms that ingest sediment particles. Bioassays that test intact, solid sediments with sediment-dwelling organisms examine the effects of both bound and dissolved contaminants on species that are most likely to be affected by toxicants in sediments.

### 3.5. Conclusions

A single bioassay method cannot answer all possible questions regarding sediment toxicity, and a variety of methods must be used together to answer basic questions. Very short term tests should be compared with longer term tests to determine effects of chronic exposure, acute mortality tests should be run with sublethal tests, and tests with cultured or water column species should be compared with tests using field-collected, benthic species. A bioassay method selected for routine field surveys should be compared with other bioassay methods to determine its sensitivity, and should initially be used in conjunction with other techniques of investigating sediment contamination.

## 4. THE *Rhepoxynius* SEDIMENT BIOASSAY

### 4.1. Method

The *Rhepoxynius* sediment bioassay is a revision of the multispecies marine sediment bioassay method developed in our laboratory for dredged material evaluation (Swartz et al., 1979). The present method is a 10-day acute static test with solid phase sediment and overlying water in 1-L aerated beakers (see Swartz et al., 1985a, for a detailed description).

The test animal, *Rhepoxynius abronius,* is a 3- to 5-mm burrowing amphipod that does not make tubes and does not normally extend body parts from the sediment. The amphipod is thus in intimate contact with the sediment and is directly exposed to sediment contaminants, which may be taken up by direct ingestion of small particles or prey, or absorption through the gills. *R. abronius* inhabits primarily the upper 2 cm of sediment and leaves the sediment infrequently, although emergence and swimming behavior have been observed under some conditions, as in the presence of infaunal predators (Ambrose, 1984). *R. abronius* has been found from the lower intertidal to depths of 274 m offshore, and is present year round at our collection site at densities of 150–750/$m^2$ (Barnard and Barnard, 1982; Kemp et al., 1985). *R. abronius* is readily collected from the field in large quantities, is easily handled in the laboratory, and is sensitive to sediment contamination.

Amphipods in general are among the first species to disappear from benthic communities in contaminated areas (Swartz et al., 1982, 1985b), and thus may be considered to be indicators of sediment pollution. Amphipods are also ecologically important as a food source for fish and benthic invertebrates.

The bioassay chamber is a standard 1-L beaker containing 2 cm of test sediment and filtered 25–28 ppt (parts per thousand) seawater up to the 950-ml mark on the beakers. Twenty amphipods are distributed to each beaker and allowed to bury. Beakers are covered with watch glasses, aerated, and held at 15°C under constant illumination for 10 days. Emergence of amphipods from the sediment is often related to degree of sediment toxicity and is monitored daily during the course of the bioassay.

After 10 days the sediment in the beakers is sieved and the numbers of live and dead amphipods are counted. Live amphipods are transferred to clean sediment and the numbers that are able to rebury within 1 hour are counted. Failure of surviving amphipods to rebury in clean sediment is directly related to degree of sediment contamination and amphipod mortality within a treatment, and is a sublethal indicator of sediment toxicity.

Survival and reburial data for test sediment are statistically compared with control data. Control survival almost always exceeds 95%. Survival of amphipods and reburial ability may be compared with results of chemical analyses of test sediments. Experimentally introduced contaminants may be used to calculate $LC_{50}$ and $EC_{50}$ values for individual chemicals in sediments.

The *Rhepoxynius* bioassay has been applied to both field-collected marine sediments and to uncontaminated sediments spiked in the laboratory with specific contaminants. Field sediments are collected with a core or grab to minimize disruption of the sediments, and generally only the upper 2 cm of sediment are used in a bioassay. Test sediments are collected within a few days of the initiation of a bioassay if possible and are stored at 4°C. Freezing or longer storage has been shown to change the toxicity of the sediment (Malueg et al., 1986; R. C. Swartz, unpublished). Sediments are thoroughly stirred before they are put into bioassay beakers but field sediments are not sieved. Uncontaminated sediment from the amphipod collection site can be spiked with individual or combined contaminants. After spiking, sediment is mixed in a rolling mill to evenly distribute the contaminant (G. R. Ditsworth, unpublished).

The static bioassay system represents a "worst-case" situation in which contaminants remain within the system and are available to the test organism. The amphipods have been shown to be tolerant of a wide range of physical and chemical sediment parameters such as grain size and organic enrichment, but are appropriately sensitive to toxicants (Swartz et al., 1984, 1985a).

## 4.2. Field Validation

It is important to test the ecological relevance of a laboratory bioassay technique in terms of correspondence between bioassay results and pollution effects in real ecosystems. A bioassay method can be field validated by comparing results of laboratory bioassays, sediment chemical measurements, and analyses of benthic community structure from field sediment collected simultaneously (Long and Chapman, 1985). The *Rhepoxynius* bioassay has been used to relate sediment toxicity with chemical measurements and community structure in sediments from Puget Sound, Washington, and the Palos Verdes Shelf in Southern California (Swartz et al., 1982, 1985b, 1986b). In Commencement Bay (Puget Sound), amphipods were abundant and diverse in deeper parts of the bay, in areas of low toxicity to *Rhepoxynius*, but amphipods were rare and phoxocephalids were absent in the highly industrialized waterways adjacent to the bay, where sediments were often highly toxic in bioassays (Table 13.2, Figs. 13.1–13.3). On the Palos Verdes

**Table 13.2 Distribution of Amphipods in Commencement Bay and Adjacent Waterways[a]**

| | All Amphipods | | Phoxocephalid Amphipods | | |
|---|---|---|---|---|---|
| Location (Station) | *S* | *N* | *S* | *N* | No. of Samples |
| Commencement Bay transect | | | | | |
| Deep disposal site (DSI) | 6.4 | 15.4 | 1.2 | 3.0 | 5 |
| Transect station (DSII) | 5.4 | 12.0 | 2.4 | 6.2 | 5 |
| Transect station (DSIII) | 5.6 | 18.6 | 1.6 | 9.6 | 5 |
| Transect station (DSIV) | 2.8 | 6.2 | 0.8 | 2.2 | 4 |
| Shallow disposal site (DSV) | 2.2 | 3.0 | 0.5 | 0.8 | 4 |
| Browns Point (B) | 5.0 | 11.4 | 0 | 0 | 5 |
| Hylebos Waterway (HI) | 1.4 | 2.6 | 0 | 0 | 5 |
| Sitcum Waterway (SI) | 1.2 | 1.6 | 0 | 0 | 5 |
| Blair Waterway (LI) | 1.4 | 1.4 | 0 | 0 | 5 |
| City Waterway (CII) | 0 | 0 | 0 | 0 | 5 |

[a]Density data are mean number of individuals (*N*) and species (*S*) per 0.1 $m^2$ grab. There were significant differences between the 10 stations in mean $N$ ($F = 19.35$, $P < 0.01$), and mean $S$ ($F = 20.58$, $P < 0.01$) for all amphipods. The results of Student–Newman–Keuls' multiple-range test are shown below, where any two means not underscored by the same line are significantly different ($P < 0.01$).

Number of species

| CII | SI | HI | LI | DSV | DSIV | B | DSII | DSIII | DSI |
|---|---|---|---|---|---|---|---|---|---|
| 0 | 1.2 | 1.4 | 1.4 | 2.2 | 2.8 | 5.0 | 5.4 | 5.6 | 6.4 |

Number of individuals

| CII | LI | SI | HI | DSV | DSIV | B | DSII | DSI | DSIII |
|---|---|---|---|---|---|---|---|---|---|
| 0 | 1.4 | 1.6 | 2.6 | 3.0 | 6.2 | 11.4 | 12.0 | 15.4 | 18.6 |

*Source:* Swartz et al. (1982), with permission.

Shelf, transect stations closer to a large sewage outfall exhibited decreased survival of *Rhepoxynius* in bioassays compared with more distant stations. This was related to decreased biomass, density, species richness, and dominance in the macrobenthic community close to the outfall, and increased concentrations of chemical contaminants relative to stations farther from the outfall where *Rhepoxynius* survival was high (Table 13.3). Other authors have also used the *Rhepoxynius* bioassay and reported similar sensitivity to contaminated sediments and correspondence between the *Rhepoxynius* and other sediment bioassays, sediment chemistry, and histopathological abnormalities of benthic fish (Chapman, 1986; Chapman et al., 1982d, 1985; Ott, 1986; Quinlan et al., 1986; Williams et al., 1986). Mearns et al. (1986) reported a very low coefficient of variation among cadmium $LC_{50}$ esti-

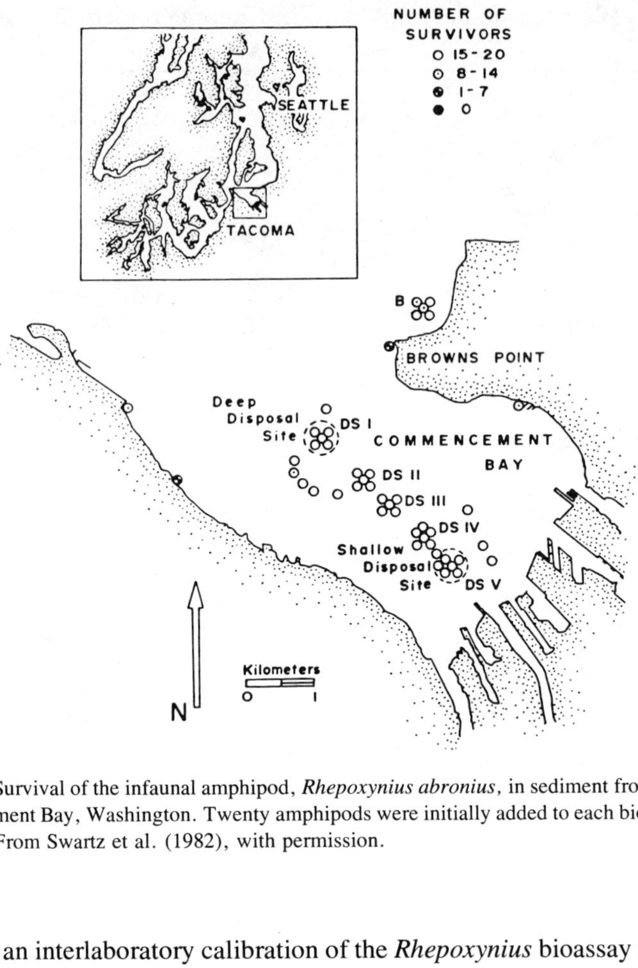


**Figure 13.1.** Survival of the infaunal amphipod, *Rhepoxynius abronius*, in sediment from Commencement Bay, Washington. Twenty amphipods were initially added to each bioassay sample. From Swartz et al. (1982), with permission.

mates when an interlaboratory calibration of the *Rhepoxynius* bioassay method was conducted.

## 4.3. Laboratory Applications

The bioassay can be applied to test the toxicity of individual chemicals added to unpolluted sediments or to determine the toxicological effects of interactions among chemicals in artificially spiked sediments. Swartz et al. (1986a) determined the $LC_{50}$ of cadmium in sediments and water in 96-hour and 10-day bioassays with *Rhepoxynius*. They found that sewage sludge or the fine fraction of natural sediment particulates added to cadmium-spiked sediment reduced the toxicity, because the added components apparently adsorbed some of the cadmium. Studies on individual chemicals, however, provide only a part of the information necessary to evaluate the biological effects of complex mixtures of chemicals deposited in sediments. Toxicological interactions among chemicals may be additive, synergistic,

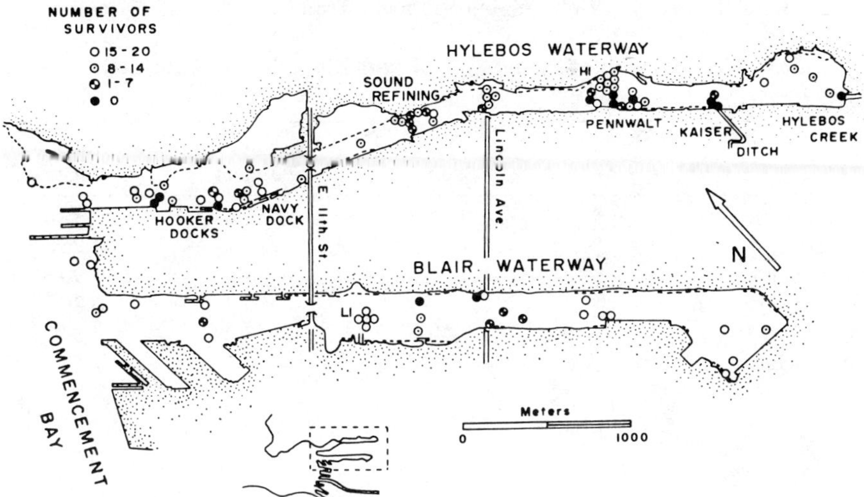


**Figure 13.2.** Survival of the infaunal amphipod, *Rhepoxynius abronius*, in sediment from the Hylebos and Blair Waterways adjacent to Commencement Bay, Washington. Twenty amphipods were initially added to each bioassay sample. From Swartz et al. (1982), with permission.

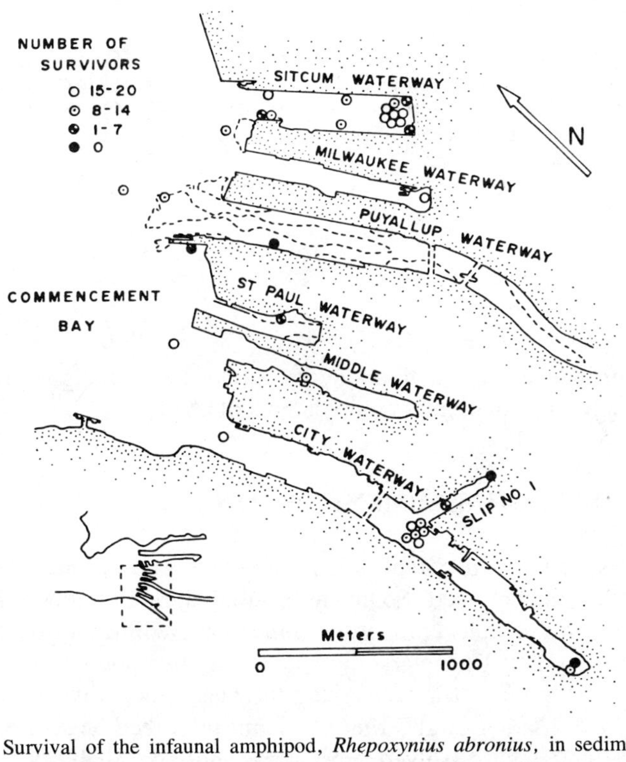


**Figure 13.3.** Survival of the infaunal amphipod, *Rhepoxynius abronius*, in sediment from several waterways adjacent to Commencement Bay, Washington. Twenty amphipods were initially added to each bioassay sample. From Swartz et al. (1982), with permission.

**Table 13.3 Comparison of Geochemical and Biological Conditions in Toxic and "Nontoxic" Sediment[a]**

| | Toxic Sediment | "Nontoxic" Sediment | Statistical Comparison, *t* |
|---|---|---|---|
| Station numbers | 1, 2, 3 | 4, 5, 6 | — |
| Survival | 15.8 | 18.1 | 3.25[b] |
| Infaunal index | 8.6 | 50.6 | 10.90[b] |
| Biomass (g/0.1 $m^2$) | 9.4 | 50.3 | 6.26[b] |
| Density ($N$/0.1 $m^2$) | 446.7 | 1963.2 | 3.53[b] |
| Species richness ($S$/0.1 $m^2$) | 20.9 | 70.1 | 7.58[b] |
| Dominance ($S$/75% $N$) | 2.2 | 5.4 | 2.67[c] |
| Dominant species | *Capitella* | *Parvilucina* | |
| Amphipod density ($N$/0.1 $m^2$) | 0.8 | 33.3 | 4.94[b] |
| Phoxocephalid density ($N$/0.1 $m^2$) | 0.0 | 23.8 | [d] |
| Eh | −54.3 | 223.7 | 5.59[b] |
| Total volatile solids (%) | 16.2 | 13.0 | 4.20[b] |
| BOD (g/kg) | 12.8 | 4.4 | 10.00[b] |
| Total oil/grease (g/kg) | 15.3 | 3.6 | 8.68[b] |
| Cadmium (mg/kg) | 28.7 | 11.1 | 6.94[b] |
| Chromium (mg/kg) | 669.3 | 375.8 | 3.21[b] |
| Bis(2-ethylhexyl)phthalate (mg/kg) | 15.6 | 4.4 | 2.64[c] |
| 4,4′-DDE (mg/kg) | 5.2 | 3.4 | 0.89 NS[e] |

[a]Toxic sediment was collected from stations on the Palos Verdes Shelf, CA, where survival in sediment bioassays was significantly less than in control sediment. "Nontoxic" sediment was collected at stations where survival was not significantly different from control survival.
[b]$P < 0.01$.
[c]$P < 0.05$.
[d]Statistical comparison not possible.
[e]NS, not significant.
*Source:* Swartz et al. (1985b), with permission.

antagonistic, or independent, and laboratory studies may be applied to determine or predict the effects of particular combinations of chemicals.

## 5. PROBLEMS WITH SEDIMENT BIOASSAYS

Sediment bioassays are more complex than water bioassays because of the variable properties of the sediment itself. Sediment handling during sampling and bioassay setup may change the chemical properties of the intact sediment through oxidation, mixing of layers, removal of interstitial water, and disruption of sediment structure. Storage of sediments has been found to change their toxicity and freezing should be avoided (Malueg et al., 1986). Chemicals sorbed on sediment particles are less bioavailable than dissolved substances, and bioavailability changes with factors such as organic content and oxidation state.

Toxicity of sediments to some organisms may depend on sediment properties other than chemical contamination, such as grain size and, in the case of marine species, interstitial water salinity. Monitoring of toxicant effects on infaunal test organisms is difficult because the animals are not normally seen until the bioassay is terminated, and the presence of predators or other organisms in the test sediment could affect results. When toxicity is found, it is difficult to determine whether the observed effects are due to toxicants associated with the sediment particles ingested by the test organism or to chemicals dissolved in interstitial water.

Finally, the significance of positive correlations between bioassay results and concentrations of individual chemicals in field sediments is difficult to interpret, since toxicity is often due to interactions among several or many contaminants. All of these "problem" areas offer opportunities for research leading to a greater understanding of the mechanisms of toxic effects of contaminants in sediments to aquatic organisms.

## 6. APPLICATIONS OF SEDIMENT BIOASSAYS

Sediment toxicity bioassays may be applied in various ways in field and laboratory research as well as in regulatory programs. The spatial or temporal distribution of sediment toxicity can be investigated or monitored in relation to pollution gradients or potential sources. Sediment bioassays reveal unsuspected areas of toxicity that should be analyzed in more detail using sediment chemical analyses or other ecological monitoring techniques. Bioassays can be used to rank sediments in terms of toxicity to benthic organisms, to establish priorities, and to determine the limits of small-scale contaminated problem areas for possible cleanup.

Sediment bioassays can be a part of the assessment of toxicity in sediments subject to regulatory decisions, and are being used successfully by the Corps of Engineers in its dredged material permit program. Bioassays could be used to establish sediment quality criteria, regulatory limits, and guidelines for disposal permit programs for complex wastes such as sewage sludge and industrial effluents.

In the laboratory, chemicals can be added to uncontaminated sediments to determine toxicity to benthic organisms, and to identify which chemicals or groups of chemicals may be causing biological effects observed in the field (Long, 1983).

## 7. RECOMMENDATIONS FOR RESEARCH

The following recommendations are made for research in the use of sediment toxicity bioassays in the Great Lakes and other areas of concern:

1. Develop biological testing methods for assessment of contamination in benthic ecosystems. Bioassays using the amphipod *Pontoporeia* have been employed by Bahnick et al. (1980), Gannon and Beeton (1969, 1971), and Oxberry et al. (1978). *Pontoporeia* is an abundant indigenous resident of the Great Lakes,

ecologically important as a food source for fishes, easy to maintain in laboratory and test systems, and tolerant of a wide range of natural sediment characteristics, but is sensitive to sediment contaminants (Alley, 1968; Kraft, 1979; Marzolf, 1965; Mozley and Howmiller, 1977; Nalepa and Thomas, 1976; Smith, 1972). These are important criteria in selection of species for bioassays (Buikema et al., 1982), and it is suggested that bioassay methods may be developed using this amphipod. The genus *Pontoporeia* includes estuarine as well as freshwater species. Possibly a bioassay system such as that used in our laboratory for *Rhepoxynius* could be developed for *Pontoporeia* for use in low-salinity and freshwater areas. Field validation should be an important part of the development of any new bioassay method.

2. Compare existing sediment bioassay methods to determine which are most suitable for specific purposes. No single technique can meet all requirements for sediment testing, and results of various methods should be compared and correlated with other indicators of sediment contamination. In Puget Sound (Chapman, 1986; Chapman et al., 1985) and in the Great Lakes area (Nebeker et al., 1984), comparisons of several bioassay methods indicated that sublethal and chronic techniques may indicate toxicity undetected in acute toxicity bioassays. In addition to comparisons of results using various bioassay methods, interlaboratory comparisons of potentially useful methods should be undertaken to determine their general applicability and the reproducibility of results (Mearns et al., 1986).

3. Other ecological assessment techniques should be used in combination with bioassays to define geographical areas of concern and to determine the current extent of pollution problem areas. A stepwise multidisciplinary approach for the assessment of sediment pollution in coastal areas has been devised and successfully applied in Puget Sound (Chapman and Long, 1983; Long, 1983; Long and Chapman, 1985), and a similar approach is recommended for the Great Lakes and other areas of concern.

4. Perform chemical analyses on contaminated sediments revealed as toxic by bioassay results. Correlation of sediment toxicity with single contaminants is rare, since toxicity is most often due to a complex mixture of chemicals, but where causative agents can be identified, it may be possible to locate and eliminate sources of toxicity. Statistical correlation may be increased and costs reduced if replicate samples for bioassays and chemical analyses or benthic community structure are taken simultaneously from the same grab or at the same station.

5. Determine effects of specific toxic substances as deposited in sediments on resident biota. Sedimentological properties such as organic content and particle size can change the bioavailability of sediment toxicants (Swartz et al., 1986a), and much more research is needed in determining the biological and ecological effects of individual and groups of chemicals deposited in sediments (Swartz, 1986). Laboratory studies may be employed to assess effects of individual chemicals in sediments, but high-priority research needs are to assess the interactions of toxic substances in combination, and the effects of sediment partitioning and organic content on availability of chemicals to test organisms. Initial research may

be accomplished with sediment bioassays using experimentally spiked sediments. In the long term, results may be applied to research involving benthic–pelagic trophic dynamics or functional changes in benthic ecosystems in microcosms, mesocosms, or natural environments.

## 8. CONCLUSIONS

Sediment bioassays alone do not provide all of the answers in assessment of sediment pollution effects on benthic ecosystems, but they can be used as a cost-effective tool in initial sediment toxicity surveys. They can be used in combination with other ecological assessment techniques in more detailed evaluations of sediment contamination or in laboratory studies of biological effects of toxic substances. Sediment bioassays are being used successfully in regulatory programs. Field validation is an important part of the development of any new bioassay method, and more research is needed in the Great Lakes and in other areas on the ecological consequences of persistent toxic substances in sediments.

## ACKNOWLEDGMENTS

Appreciation is extended to Dr. Paul Kemp, Dr. Alan Nebeker, and to the IJC reviewers for helpful comments in reviewing the manuscript.

## REFERENCES

Adams, W. J., Kimerle, R. A., and Mosher, R. G. (1985). Aquatic safety assessment of chemicals sorbed to sediments. *ASTM Spec. Tech. Publ.* **STP 854,** 429–453.

Alley, W. P. (1968). Ecology of the Burrowing Amphipod *Pontoporeia affinis* in Lake Michigan, Spec. Rep. No. 36. Great Lakes Res. Div., University of Michigan, Ann Arbor.

Ambrose, W. G. (1984). Increased emigration of the amphipod *Rhepoxynius abronius* (Barnard) and the polychaete *Nephtys caeca* (Fabricius) in the presence of invertebrate predators. *J. Exp. Mar. Biol. Ecol.* **80,** 67–75.

Bahnick, D. A., Swenson, W. A., Markee, T. P., Call, D. J., Anderson, C. A., and Morris, R. T. (1980). *Development of Bioassay Procedures for Defining Pollution of Harbor Sediments,* Final Report of U.S. EPA Proj. No. R804918-01. Cent. Lake Superior Environ. Stud., University of Wisconsin, Superior.

Barnard, J. L., and Barnard, C. M. (1982). The genus *Rhepoxynius* (Crustacea: Amphipoda: Phoxocephalidae) in American Seas. *Smithson. Contrib. Zool.* **357,** 26–28.

Black, J. J. (1984). Environmental implications of neoplasia in Great Lakes fish. *Mar. Environ. Res.* **14,** 529–534.

Black, J. J., Holmes, M., Dymerski, P. P., and Zapisek, W. F. (1980). Fish tumor pathology and aromatic hydrocarbon pollution in a Great Lakes estuary. In B. K. Afghan and D. Mackay (Eds.), *Hydrocarbons and Halogenated Hydrocarbons in the Aquatic Environment.* Plenum, New York, pp. 559–565.

Brinkhurst, R. O., Chapman, P. M., and Farrell, M. A. (1983). A comparative study of respiration rates of some aquatic oligochaetes in relation to sublethal stress. *Int. Rev. Gesamten Hydrobiol.* **68,** 683–699.

Buikema, A. L., Jr., Rutherford, C. L., and Cairns, J., Jr. (1980). Screening sediments for potential toxicity by in vitro enzyme inhibition. In R. A. Baker (Ed.), *Contaminants and Sediments.* Ann Arbor Sci. Publ., Ann Arbor, MI, Vol. 1, pp. 463–476.

Buikema, A. L., Jr., Niederlehner, B. R., and Cairns, J., Jr. (1982). Biological monitoring. Part IV. Toxicity testing. *Water Res.* **16,** 239–262.

Cairns, M. A., Nebeker, A. V., Gakstatter, J. H., and Griffis, W. L. (1984). Toxicity of copper-spiked sediments to freshwater invertebrates. *Environ. Toxicol. Chem.* **3,** 435–445.

Chapman, P. M. (1986). Sediment bioassay tests provide toxicity data necessary for assessment and regulation. In G. H. Geen and K. L. Woodward (Eds.), *Proceedings of the Eleventh Annual Aquatic Toxicology Workshop*, Nov. 13–15, 1984, Vancouver, Canada. Tech. Rpt. 1480 Fish. Aquat. Sci., pp. 178–197.

Chapman, P. M., and Brinkhurst, R. O. (1984). Lethal and sublethal tolerances of aquatic oligochaetes with reference to their use as a biotic index of pollution. *Hydrobiologia* **115,** 139–144.

Chapman, P. M., and Long, E. R. (1983). The use of bioassays as part of a comprehensive approach to marine pollution assessment. *Mar. Pollut. Bull.* **14,** 81–84.

Chapman, P. M., Farrell, M. A., and Brinkhurst, R. O. (1982a). Relative tolerances of selected aquatic oligochaetes to individual pollutants and environmental factors. *Aquat. Toxicol.* **2,** 47–67.

Chapman, P. M., Farrell, M. A., and Brinkhurst, R. O. (1982b). Relative tolerances of selected aquatic oligochaetes to combinations of pollutants and environmental factors. *Aquat. Toxicol.* **2,** 69–78.

Chapman, P. M., Farrell, M. A., and Brinkhurst, R. O. (1982c). Effects of species interactions on the survival and respiration of *Limnodrilus hoffmeisteri* and *Tubifex tubifex* (Oligochaeta, Tubificidae) exposed to various pollutants and environmental factors. *Water Res.* **16,** 1405–1408.

Chapman, P. M., Vigers, G. A., Farrell, M. A., Dexter, R. N., Quinlan, E. A., Kocan, R. M., and Landolt, M. L. (1982d). Survey of biological effects of toxicants upon Puget Sound biota. I. Broad-scale toxicity survey. NOAA Tech. Memo. OMPA-25, 1–98.

Chapman, P. M., Dexter, R. N., Kocan, R. M., and Long, E. R. (1985). An overview of biological effects testing in Puget Sound, Washington: Methods, results and implications. *ASTM Spec. Tech. Publ.* **STP 854,** 344–363.

Environmental Protection Agency/Corps of Engineers (1977). *Ecological Evaluation of Proposed Discharge of Dredged Material into Ocean Waters.* Environ. Effects Lab., U.S. Army Engineer Waterways Experiment Station, Vicksburg, MI.

Gannon, J. E., and Beeton, A. M. (1969). *Studies on the Effects of Dredged Materials from Selected Great Lakes Harbors on Plankton and Benthos,* Spec. Rep. No. 8. Cent. Great Lakes, University of Wisconsin, Milwaukee.

Gannon, J. E., and Beeton, A. M. (1971). Procedures for determining the effects of dredged sediments on biota-benthos viability and sediment selectivity tests. *J. Water Pollut. Control Fed.* **43,** 392–398.

Hare, L., and Carter, J. C. H. (1976). The distribution of *Chironomus* (*s.s.*)? *cucini* (*salinarius* group) larvae (Diptera: Chironomidae) in Parry Sound, Georgian Bay, with particular reference to structural deformities. *Can. J. Zool.* **54,** 2129–2134.

Hargis, W. J., Jr., Roberts, M. H., Jr., and Zwerner, D. E. (1984). Effects of contaminated sediments and sediment-exposed effluent water on an estuarine fish: Acute toxicity. *Mar. Environ. Res.* **14,** 337–354.

Hoke, R. A., and Prater, B. L. (1980). Relationship of percent mortality of four species of aquatic biota from 96-hour sediment bioassays of five Lake Michigan harbors and elutriate chemistry of the sediments. *Bull. Environ. Contam. Toxicol.* **25,** 394–399.

Jones, R. A., and Lee, G. F. (1978). *Evaluation of the Elutriate Test as a Method of Predicting Contaminant Release During Open Water Disposal of Dredged Sediment and Environmental Impact of Open Water Dredged Material Disposal,* Vol. 1, Tech. Rep. D78-45. U.S. Army Engineer Waterways Experiment Station, Vicksburg, MI.

Kemp, P. F., Cole, F. A., and Swartz, R. C. (1983). Life history and productivity of the phoxocephalid amphipod *Rhepoxynius abronius* (Barnard). *J. Crustacean Biol.* **5,** 449–464.

Kraft, K. J. (1979). *Pontoporeia* distribution along the Keweenaw shore of Lake Superior affected by copper tailings. *J. Great Lakes Res.* **5,** 28–35.

Krahn, M. M., Rhodes, L. D., Myers, M. S., Moore, L. K., MacLeod, W. D., and Malins, D. C. (1986). Associations between metabolites of aromatic compounds in bile and the occurrence of hepatic lesions in English sole (*Parophrys vetulus*) from Puget Sound, Washington. *Arch. Environ. Contam. Toxicol.* **15,** 61–67.

Landolt, M. L., and Kocan, R. M. (1984). Lethal and sublethal effects of marine sediment extracts on fish cells and chromosomes. *Helgol. Meersunters.* **38,** 125–139.

Landolt, M. L., Kocan, R. M., and Dexter, R. N. (1984). Anaphase aberrations in cultured fish cells as a bioassay of marine sediments. *Mar. Environ. Res.* **14,** 497–498.

Laskowski-Hoke, R. A., and Prater, B. L. (1984). Multivariate statistical analyses of 96-hour sediment bioassay and chemistry data. *Bull. Environ. Contam. Toxicol.* **33,** 400–409.

LeBlanc, G. A., and Surprenant, D. C. (1985). A method of assessing the toxicity of contaminated freshwater sediments. *ASTM, Spec. Tech. Publ.* **STP 854,** 269–283.

Lee, G. F., Jones, R. A., Saleh, F. A., Mariani, G. M., Homer, D. H., Butler, J. S., and Bandyopadhyay, P. (1978). *Evaluation of the Elutriate Test as a Method of Predicting Contaminant Release During Open Water Disposal of Dredged Sediment and Environmental Impact of Open Water Dredged Material Disposal,* Vol. 2, Tech. Rep. D78-45. U.S. Army Engineer Waterways Experiment Station, Vicksburg, MI.

Lee, H., II, and Swartz, R. C. (1980). Biological processes affecting the distribution of pollutants in marine sediments. Part II. Biodeposition and bioturbation. In R. A. Baker (Ed.), *Contaminants and Sediments.* Ann Arbor Sci. Publ., Ann Arbor, MI, Vol. 2, pp. 555–606.

Long, E. R. (1982). An assessment of marine pollution in Puget Sound. *Mar. Pollut. Bull.* **13,** 380–383.

Long, E. R. (1983). A multidisciplinary approach to assessing pollution in coastal waters. In Q. T. Magoon and H. Converse (Eds.), *Proceedings of the Third Symposium on Coastal and Ocean Management.* Am. Soc. Civ. Eng., San Diego, CA, pp. 163–178.

Long, E. R., and Chapman, P. M. (1985). A sediment quality triad: Measures of sediment contamination, toxicity and infaunal community composition in Puget Sound. *Mar. Pollut. Bull.* **16,** 405–415.

Malins, D. C., McCain, B. B., Brown, D. W., Sparks, A. K., Hodgins, H. O., and Chan, S.-L. (1982). Chemical contaminants and abnormalities in fish and invertebrates from Puget Sound. NOAA Tech. Memo. OMPA-19, 1–168.

Malins, D. C., McCain, B. B., Brown, D. W., Chan, S.-L., Myers, M. S., Landahl, J. T., Prohaska, P. G., Friedman, A. J., Rhodes, L. D., Burrows, D. G., Gronlund, W. D., and Hodgins, H. O. (1984). Chemical pollutants in sediments and diseases of bottom dwelling fish in Puget Sound, Washington. *Environ. Sci. Technol.* **18,** 705–713.

Malueg, K. W., Schuytema, G. S., Gakstatter, J. H., and Krawczyk, D. F. (1983). Effect of *Hexagenia* on *Daphnia* response in sediment toxicity tests. *Environ. Toxicol. Chem.* **2,** 73–82.

Malueg, K. W., Schuytema, G. S., Krawczyk, D. F., and Gakstatter, J. H. (1984a). Laboratory sediment toxicity tests, sediment chemistry and distribution of benthic macroinvertebrates in sediments from the Keweenaw Waterway, Michigan. *Environ. Toxicol. Chem.* **3,** 233–242.

Malueg, K. W., Schuytema, G. S., Gakstatter, J. H., and Krawczyk, D. F. (1984b). Toxicity of sediments from three metal-contaminated areas. *Environ. Toxicol. Chem.* **3,** 279–291.

Malueg, K. W., Schuytema, G. S., and Krawczyk, D. F. (1986). Effects of sample storage on a copper-spiked freshwater sediment. *Environ. Toxicol. Chem.* **5,** 245–253.

Marzolf, G. R. (1965). Substrate relations of the burrowing amphipod *Pontoporeia affinis* in Lake Michigan. *Ecology* **46,** 579–592.

Mearns, A. J., Swartz, R. C., Cummins, J. M., Dinnel, P. A., Plesha, P., and Chapman, P. M. (1986). Inter-laboratory comparison of a sediment toxicity test using the marine amphipod, *Rhepoxynius abronius. Mar. Environ. Res.* **19,** 13–37.

Mozley, S. C., and Howmiller, R. P. (1977). *Environmental Status of the Lake Michigan Region,* Vol. 6, ANL/ES-40. Environ. Control Technol. Earth Sci. (UC-11), Argonne Natl. Lab., Argonne, IL.

Nalepa, T. F., and Thomas, N. A. (1976). Distribution of macrobenthic species in Lake Ontario in relation to sources of pollution and sediment parameters. *J. Great Lakes Res.* **2,** 150–163.

Nalepa, T. F., White, D. S., Pringle, C. M., and Quigley, M. A. (1980). *The Biological Component of Phosphorus Exchange and Cycling in Lake Sediments,* Spec. Rep. No. 83. Great Lakes Res. Div., University of Michigan, Ann Arbor, pp. 93–109.

Nebeker, A. V., Cairns, M. A., Gakstatter, J. H., Malueg, K. W., Schuytema, G. S., and Krawczyk, D. F. (1984). Biological methods for determining toxicity of contaminated freshwater sediments to invertebrates. *Environ. Toxicol. Chem.* **3,** 617–630.

Oschwald, W. (1972). Sediment-water interactions. *J. Environ. Qual.* **1,** 360–366.

Ott, F. S. (1986). Amphipod sediment bioassays: Effects of grain size, cadmium, methodology, and variations in animal sensitivity on interpretations of experimental data. Ph.D. Dissertation, University of Washington, Seattle.

Oxberry, J. R., Doudoroff, P., and Anderson, D. W. (1978). Potential toxicity of taconite tailings to aquatic life in Lake Superior. *J. Water Pollut. Control Fed.* **50,** 240–251.

Prater, B. L., and Anderson, M. A. (1977a). A 96-hour bioassay of Otter Creek, Ohio. *J. Water Pollut. Control Fed.* **49,** 2099–2106.

Prater, B. L., and Anderson, M. A. (1977b). A 96-hour bioassay of Duluth and Superior harbor basins (Minnesota) using *Hexagenia limbata, Asellus communis, Daphnia magna,* and *Pimephales promelas* as test organisms. *Bull. Environ. Contam. Toxicol.* **18,** 159–168.

Prater, B. L., and Hoke, R. A. (1978). *Ninety-six Hour Toxicity Bioassay Tests of Ashtabula Harbor (Ohio),* Tech. Rep. DACW49-78-C-0083. Buffalo District, U.S. Army Corps of Engineers, Buffalo, NY.

Prater, B. L., and Hoke, R. A. (1979). *A Sediment Quality Evaluation of Five Harbors of the Great Lakes Using a 96-hour Sediment Bioassay and Bulk Chemistry,* Final Rep., Grant No. R-005338-010. U.S. Environ. Prot. Agency, Washington, DC.

Prater, B. L., and Hoke, R. A. (1980). A method for the biological and chemical evaluation of sediment toxicity. In R. A. Baker (Ed.), *Contaminants and Sediments.* Ann Arbor Sci. Publ., Ann Arbor, MI, Vol. 1, pp. 483–499.

Quinlan, E. A., Chapman, P. M., Dexter, R. N., Konasewich, D. E., Ebbesmeyer, C. C., Erickson, G. A., Kowalski, B. R., and Silver, T. A. (1985). Toxic chemicals and biological effects in Puget Sound: Status and scenarios for the future. NOAA Tech. Memo. NOS OMS-10, 1–334.

Samoiloff, M. R., Bell, J., Birkholz, D. A., Webster, G. R. B., Arnott, E. G., Pulak, R., and Madrid, A. (1983). Combined bioassay-chemical fractionation scheme for the determination and ranking of toxic chemicals in sediments. *Environ. Sci. Technol.* **17,** 329–334.

Schuytema, G. S., Nelson, P. O., Malueg, K. W., Nebeker, A. V., Krawczyk, D. F., Ratcliff, A. K., and Gakstatter, J. H. (1984). Toxicity of cadmium in water and sediment slurries to *Daphnia magna. Environ. Toxicol. Chem.* **3,** 293–308.

Smith, W. E. (1972). Culture, reproduction and temperature tolerance of *Pontoporeia affinis* in the laboratory. *Trans. Am. Fish. Soc.* **101,** 253–256.

Swartz, R. C. (1988). Toxicological methods for determining the effects of contaminated sediment on marine organisms. *Proc. 1984 Pelleston Conf.* (in press).

Swartz, R. C., and Lee, H., II (1980). Biological processes affecting the distribution of pollutants in marine sediments. Part I. Accumulation, trophic transfer, biodegradation and migration. In R. A. Baker (Ed.), *Contaminants and Sediments.* Ann Arbor Sci. Publ., Ann Arbor, MI, Vol. 2, pp. 533–553.

Swartz, R. C., DeBen, W. A., and Cole, F. A. (1979). A bioassay for the toxicity of sediment to marine macrobenthos. *J. Water Pollut. Control Fed.* **51,** 944–950.

Swartz, R. C., DeBen, W. A., Sercu, K. A., and Lamberson, J. O. (1982). Sediment toxicity and the distribution of amphipods in Commencement Bay, Washington, U.S.A. *Mar. Pollut. Bull.* **13,** 359–364.

Swartz, R. C., Schults, D. W., Ditsworth, G. R., and DeBen, W. A. (1984). Toxicity of sewage sludge to *Rhepoxynius abronius,* a marine benthic amphipod. *Arch. Environ. Contam. Toxicol.* **13,** 207–216.

Swartz, R. C., DeBen, W. A., Jones, J. K. P., Lamberson, J. O., and Cole, F. A. (1985a). Phoxocephalid amphipod bioassay for marine sediment toxicity. *ASTM Spec. Tech. Publ.* **STP 854,** 284–307.

Swartz, R. C., Schults, D. W., Ditsworth, G. R., DeBen, W. A., and Cole, F. A. (1985b). Sediment toxicity, contamination, and macrobenthic communities near a large sewage outfall. *ASTM Spec. Tech. Publ.* **STP 865,** 152–175.

Swartz, R. C., Ditsworth, G. R., Schults, D. W., and Lamberson, J. O. (1986a). Sediment toxicity to a marine infaunal amphipod: Cadmium and its interaction with sewage sludge. *Mar. Environ. Res.* **18,** 133–153.

Swartz, R. C., Cole, F. A., Schults, D. W., and DeBen, W. A. (1986b). Ecological changes in the Southern California Bight near a large sewage outfall: Benthic conditions in 1980 and 1983. *Mar. Ecol.: Prog. Ser.* **31,** 1–13.

Warwick, W. F. (1980). Palaeolimnology of the Bay of Quinte, Lake Ontario: 2800 years of cultural influence. *Can. Bull. Fish. Aquat. Sci.* **206,** 1–117.

Wentsel, R. A., McIntosh, A., and Atchison, G. (1977). Sublethal effects of heavy metal contaminated sediment on midge larvae (*Chironomus tentans*). *Hydrobiologia* **56,** 153–156.

Wentsel, R. A., McIntosh, A., and McCafferty, P. C. (1978). Emergence of the midge *Chironomus tentans* when exposed to heavy metal contaminated sediments. *Hydrobiologia* **57,** 195–196.

Williams, L. G., Chapman, P. M., and Ginn, T. C. (1986). A comparative evaluation of marine sediment toxicity using bacterial luminescence, oyster embryo and amphipod sediment bioassays. *Mar. Environ. Res.* **19,** 225–249.

# 14

# MORPHOLOGICAL DEFORMITIES IN CHIRONOMIDAE (DIPTERA) LARVAE AS BIOLOGICAL INDICATORS OF TOXIC STRESS

***W. F. Warwick****

*National Water Research Institute, 501 University Crescent, Winnipeg, Manitoba, Canada R3T 2N6*



**Present address:* National Hydrology Research Institute, Saskatoon, Saskatchewan, Canada S7N 3H5

## 1. INTRODUCTION

### 1.1. Overview

Reliable methods for detecting and assessing the significance of contaminants in water supplies are urgently needed in view of increasing domestic, industrial, and agricultural demands for water. Because all synthetic chemicals tend eventually to be transferred in one form or other to water, contamination of these resources has become one of the most serious threats not only to humans, but to the global community.

The number of chemicals entering the environment today does not make analysis for specific contaminants cost-effective. Individual chemicals rapidly change their characteristics in the aquatic environment and interaction among pollutants is more often present than absent. Individual chemicals are modified and integrated through physical, chemical, and biological processes to the extent that the total effect of a large number of minor pollutants may be as great as, or greater than, that of a single major pollutant. The more serious aspects of environmental contamination therefore derive from a multitude of chemicals that interact simultaneously and synergistically, whose effects are long term and often irreversible, and that have deleterious effects even at low levels of contamination (Goldsmith et al., 1972). The effect of the total contaminant burden may be impossible to estimate except by direct observation of its overall effect on the aquatic ecosystem.

Biological indicators provide the potential for direct observation of these effects by virtue of their role in aquatic ecology. Of the many choices available for a suitable biological indicator, morphological deformities in chironomid larvae (Chironomidae: Diptera) have considerable potential as a biological screening tool for detecting and assessing the effects of contaminants. Although the exact relationship between the incidence and severity of deformities and toxic substances remains to be established, the objectives of this paper are (1) to provide an update on the status of research on morphological deformities in chironomid larvae, (2) to identify the need for a conceptual framework as a prerequisite for using these deformities as a biological screening tool, (3) to standardize preparation and mounting techniques, and (4) to identify future needs and direction for developing a functional assessment technique.

## 1.2. Chironomid Larvae as Biological Indicators

Chironomid larvae have a number of advantages as environmental indicators. The family Chironomidae is ubiquitous and represents the single largest family of aquatic insects. Its species number in the thousands and these occupy virtually every type and condition of aquatic habitat (Coffman and Ferrington, 1984). Their extraordinary ecological range and environmental sensitivity make them particularly useful for assessing and interpreting changes in aquatic habitats with a high degree of resolution (Warwick, 1980a).

The life cycle of chironomids is short compared to other biological indicators such as fish or birds. In temperate latitudes, many species are uni- or bivoltine, but the occurrence of three or more annual generations is not uncommon (Pinder, 1986). Species inhabiting the profundal of deep lakes may require more than a year to complete larval development, whereas in the far north, larvae may require from 2 to 7 years to develop. The intermediate pupal stage lasts only a few days while adults only live a few days to a few weeks at most (Oliver, 1971). Breeding takes place during the adult stage and the female lays her eggs, usually in close proximity to her original emergence site.

Chironomids in lakes are exposed to contaminants during the most critical stage of their life cycle. Not only is the larval stage the longest stage in the cycle, but generally it is the only stage during which feeding and energy storage in the form of body tissue takes place. (Pupae and adults do not feed as a rule.) Chironomid larvae can be divided into four types based on feeding behavior (filter-feeders, scrapers, predators, and mixed feeders), but few species appear to be restricted to a single mode of feeding (Pinder, 1986). Because many toxic substances deposit into the sediments bound to detrital organics, many filter-feeding species such as *Chironomus plumosus* and *Rheotanytarsus* are directly exposed through their diet. Most lake species also live in the easily sampled surface layers of soft sediments where they are directly exposed to water/soluble contaminants or contaminants mobilized by biochemical processes near or at the sediment/water interface.

The concentration of chironomid communities in the near-surface sediments and their comparatively short life cycle also mean that these communities reflect conditions in the near term. Unlike fish or birds which may take years to develop overt signs of physical malformation, morphological responses in chironomid larvae probably are a reasonably accurate reflection of conditions prevailing in the sediments at the time of sampling.

This immediacy of response is important from another point of view. The remains of chironomid larvae, particularly the head capsule, preserve very well in most sediments. These remains will include the remains of morphologically deformed larvae in the yearly increment to the sediments. Palaeoanalysis of the frequency and severity of deformities in these remains not only can provide a temporal measure of deteriorating conditions, but may also prove useful in identifying sources of contaminants in the historical context (Warwick, 1980a).

The usefulness of chironomid communities in environmental assessment has already been demonstrated in the lake classification system. The development of this system began with the early work of Thienemann (1913, 1922, 1931) and Naumann (1917, 1929) and evolved (Brundin, 1949; Saether, 1975, 1979; Wiederholm, 1976, 1980; Warwick, 1980a) into a relatively sophisticated system for classifying trophic conditions in lakes. The system is based primarily on the response of chironomid communities to changes in oxygen supply, food availability, temperature, sedimentation patterns, and other environmental parameters that effect changes in the composition and relative abundance of species in the community.

Unlike the lake classification system, a system for detecting contaminants based on morphological deformities would focus more on the response of individual larvae than on the chironomid community as a whole. Although comparatively little is known about how individual organisms react to contaminants, Petersen and Petersen (1983) suggested that changes at the level of the individual organism might be more useful than changes at the community level because (1) individual responses occur before community responses and can thus provide an earlier warning of environmental contamination and (2) the toxicant can be determined by bioassay on the responding species and specific criteria established for the receiving system. A biological screening system based on deformed chironomid larvae ultimately would take into account (1) the incidence of deformities, (2) the severity of the deformities, (3) the type of deformities, and (4) the structure and composition of the contemporary benthic community.

## 2. DESCRIPTIONS OF MORPHOLOGICAL DEFORMITIES

The morphology of the type genus *Chironomus* is presented in part in Figures 14.1 and 14.2. The illustrations are from Oliver et al. (1978) while the nomenclature follows Saether (1980) and Cranston and Reiss (1983). These authors should be consulted for a fuller treatment of larval morphology.

It is evident even from these simplified drawings that the potential for abnormal variation in an animal as complex as this is immense and the number of possible

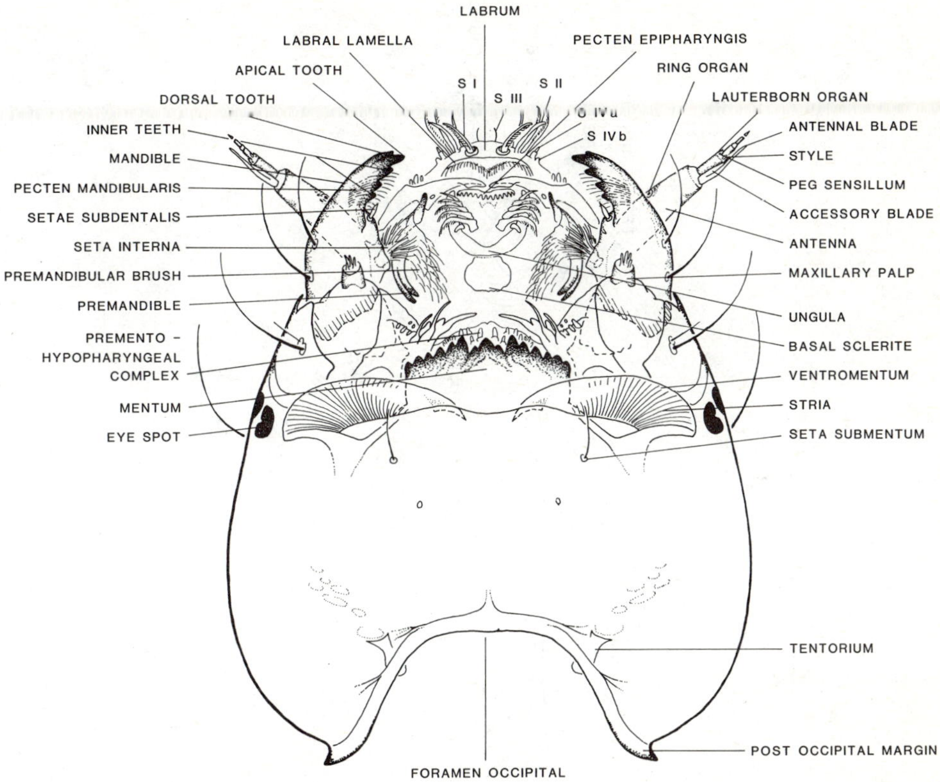


**Figure 14.1.** Ventral view of *Chironomus* head capsule. Adapted from Oliver et al. (1978). Nomenclature follows Saether (1980) and Cranston and Reiss (1983).

expressions of response correspondingly high. For the present, I regard any morphological feature that departs from the normal configuration as a deformity. I have deliberately kept the definition of a deformity simple until such time as we are able to distinguish between normal and abnormal variation with a higher degree of precision and certainty.

Morphological deformation may range from mildly abnormal mouthparts to the grotesque thickening and fusing of all body structures (Fig. 14.3). Deformities in the mentum range from mild asymmetry in the numbers of lateral teeth, to overlapping in the outer lateral teeth, or to overlapping of lateral and median teeth (Fig. 14.4c). More extensive deformation may include the addition of countless numbers of teeth to the lateral or median areas of the mentum (Fig. 14.4d) or the loss of many of its teeth (Fig. 14.4b). In Tanypodinae, the ligula (analogous to the mentum in all other subfamilies; see Figs. 14.3b,d) may also display missing teeth, twisted or misshapen teeth or numerous additional teeth. Deformed mandi-

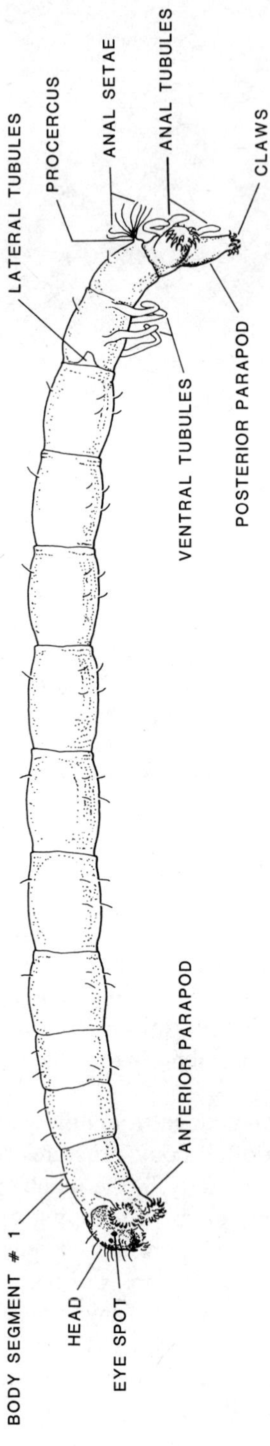
LATERAL TUBULES
PROCERCUS
ANAL SETAE
ANAL TUBULES
CLAWS
POSTERIOR PARAPOD
VENTRAL TUBULES
ANTERIOR PARAPOD
BODY SEGMENT # 1
HEAD
EYE SPOT

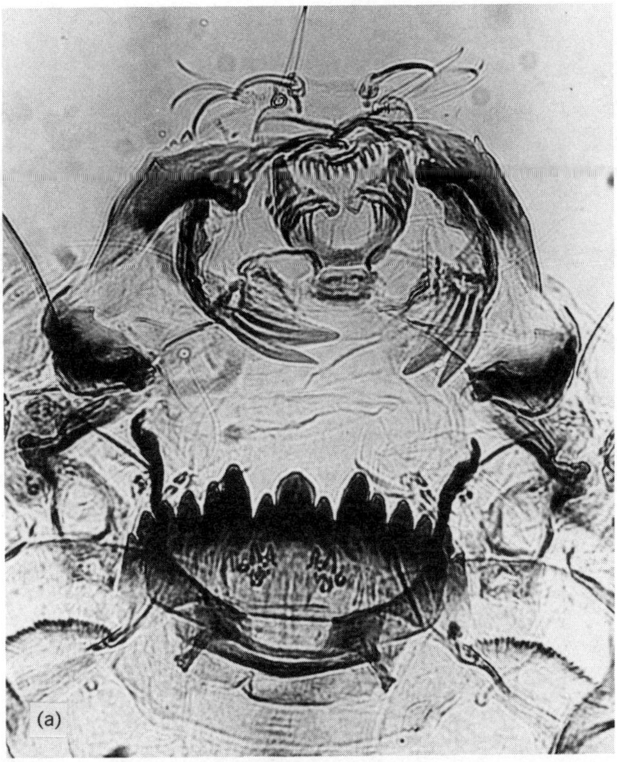


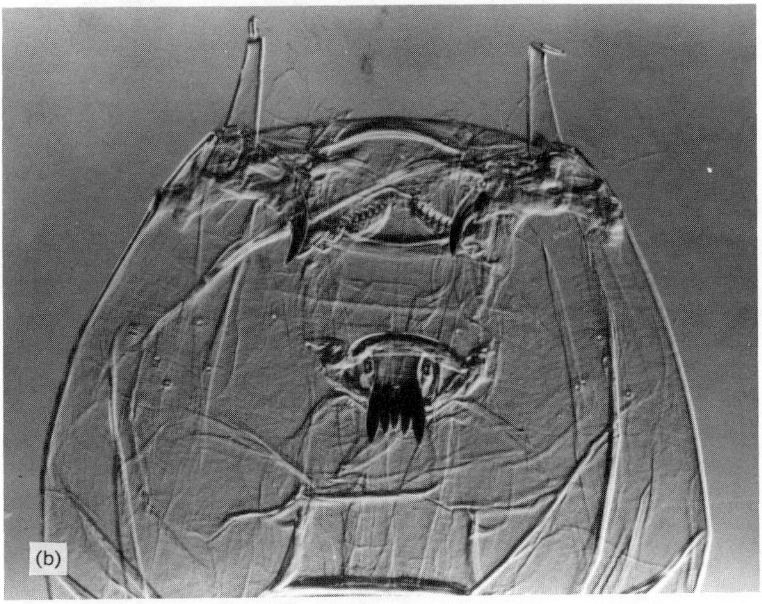


**Figure 14.3.** Thickening and fusing in the head capsule of massively deformed larvae of *Chironomus* Meigen and *Procladius* Skuse: buccal area of (a) normal and (c) massively deformed *Chironomus* larvae and head capsules of (b) normal and (d) massively deformed *Procladius* larvae.

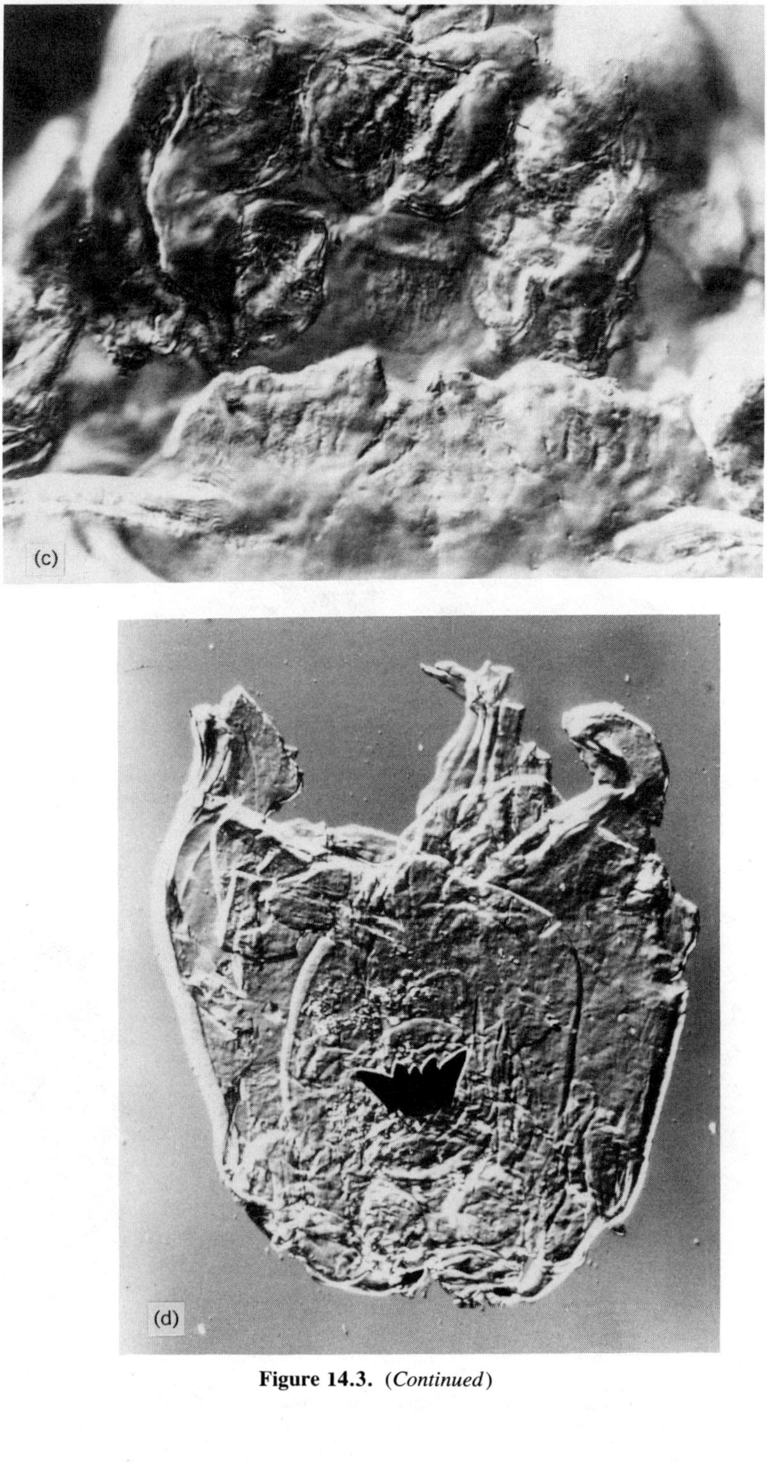

Figure 14.3. (*Continued*)

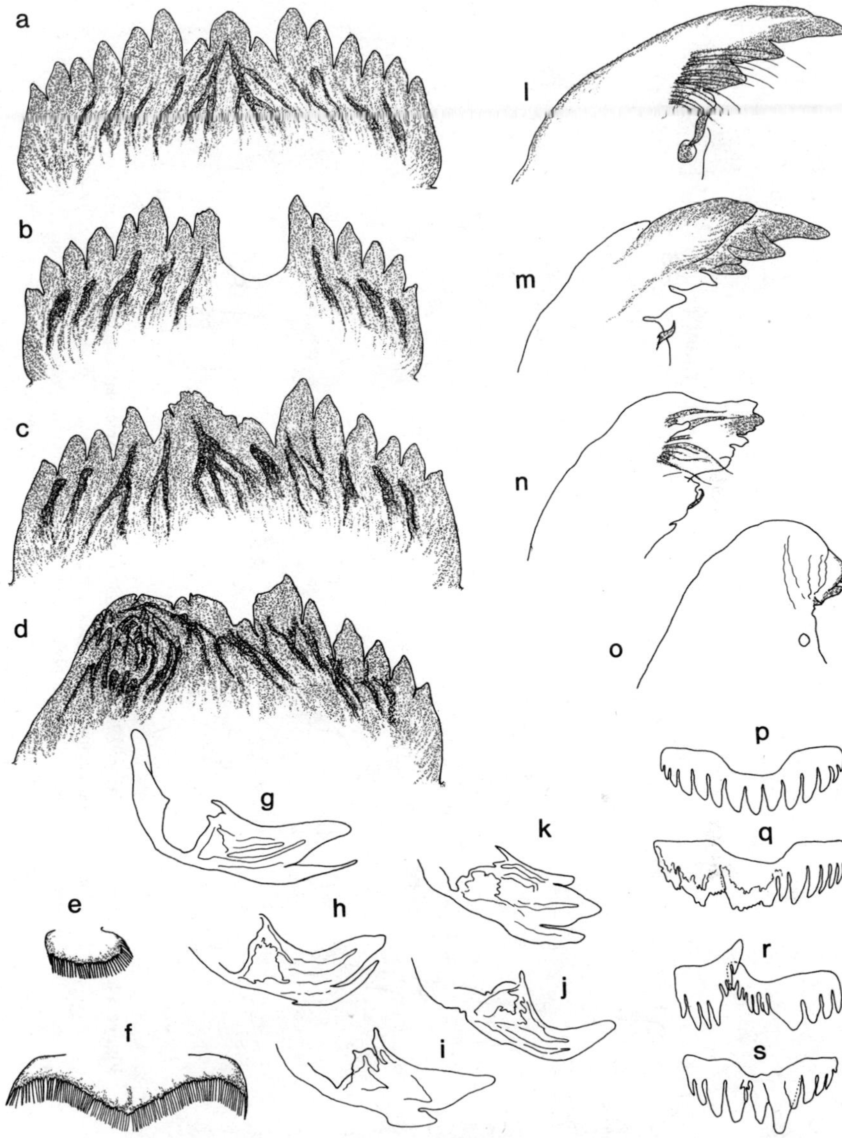


**Figure 14.4.** Morphological deformities in the mouthparts of *Chironomus* Meigen: (a) normal mentum; deformed menta showing (b) Koehn gap, (c) central/lateral overlap and (d) extensive lateral disorganization; (e) deformed and (f) normal labral lamellae; (g) normal and (h–k) deformed premandibles; (l) normal and (m–o) deformed mandibles; and (p) normal and (q–s) deformed epipharyngeal pecten. Specimen (b) is from the Teltowkanal; specimen (c) is from Last Mountain Lake; all other specimens are from Tobin Lake.

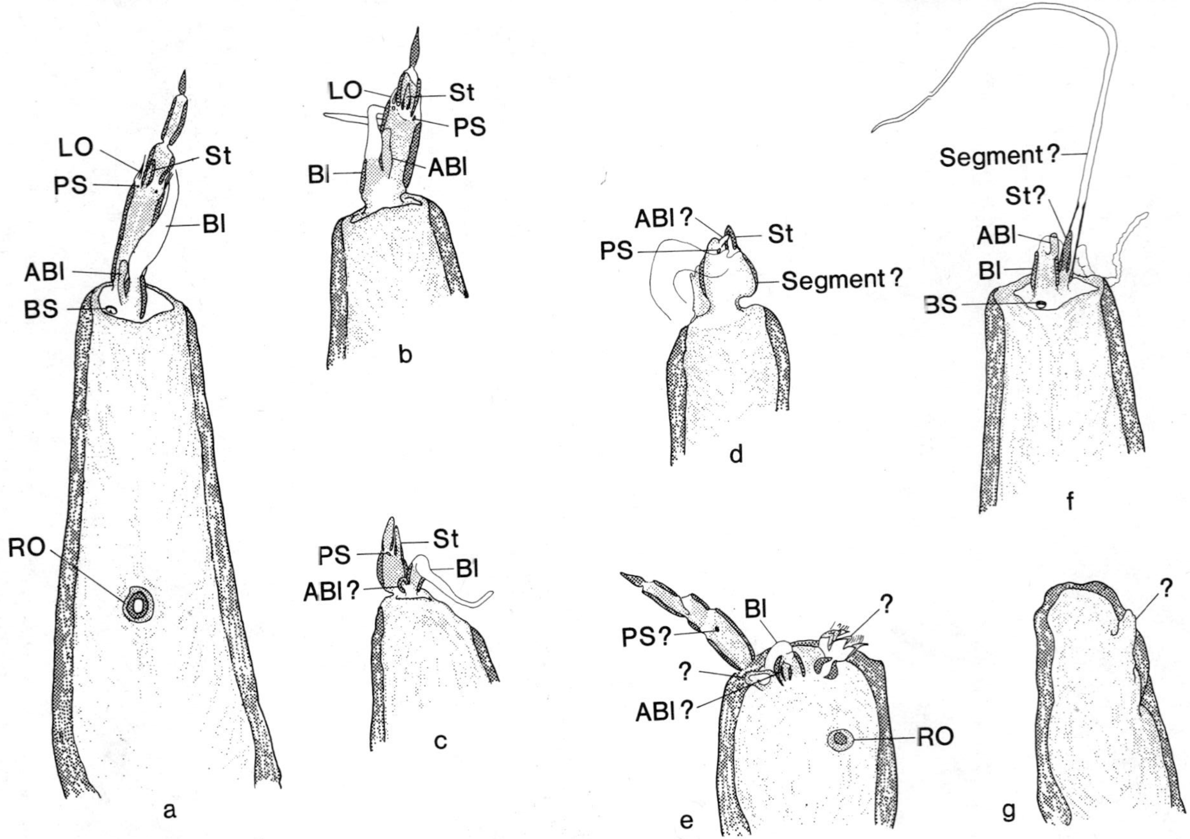


**Figure 14.5.** Deformed antennae of *Chironomus* Meigen: (a) normal antenna; deformed antennae showing (b,c) loss of distal segments, (d) questionable second segment, (e) displacement of blade (Bl) and accessory blade (ABl) complex, structures of unknown homology and reduction in length, (f) fusing of distal segments and displacement of the peg sensillum (PS) and (g) fusing of apex of basal segment, loss of ring organ (RO) and structure of unknown homology. Other structures indicated include the organs of Lauterborn (LO), style (St), and the basiconic sensillum (BS). Nomenclature after Warwick (1985).

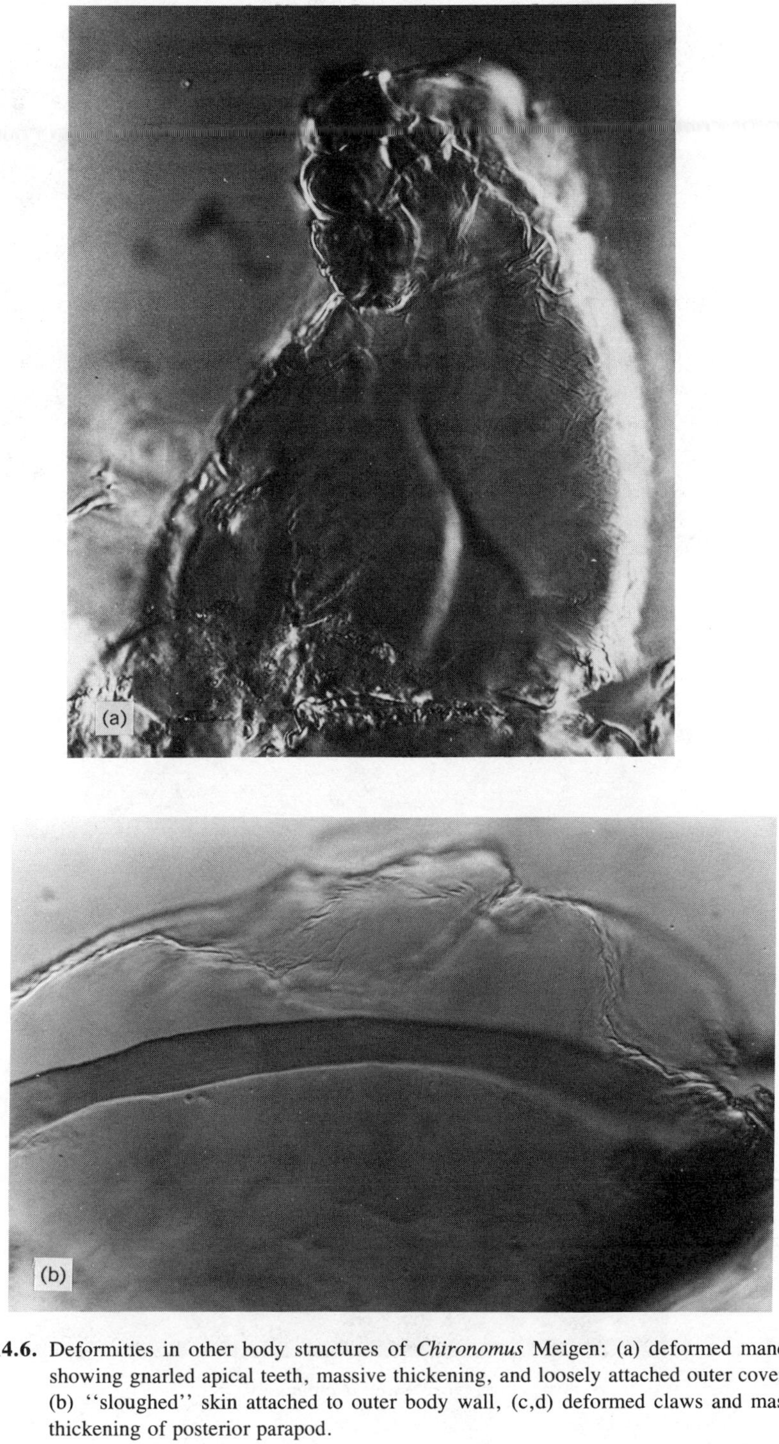


**Figure 14.6.** Deformities in other body structures of *Chironomus* Meigen: (a) deformed mandible showing gnarled apical teeth, massive thickening, and loosely attached outer covering, (b) ''sloughed'' skin attached to outer body wall, (c,d) deformed claws and massive thickening of posterior parapod.

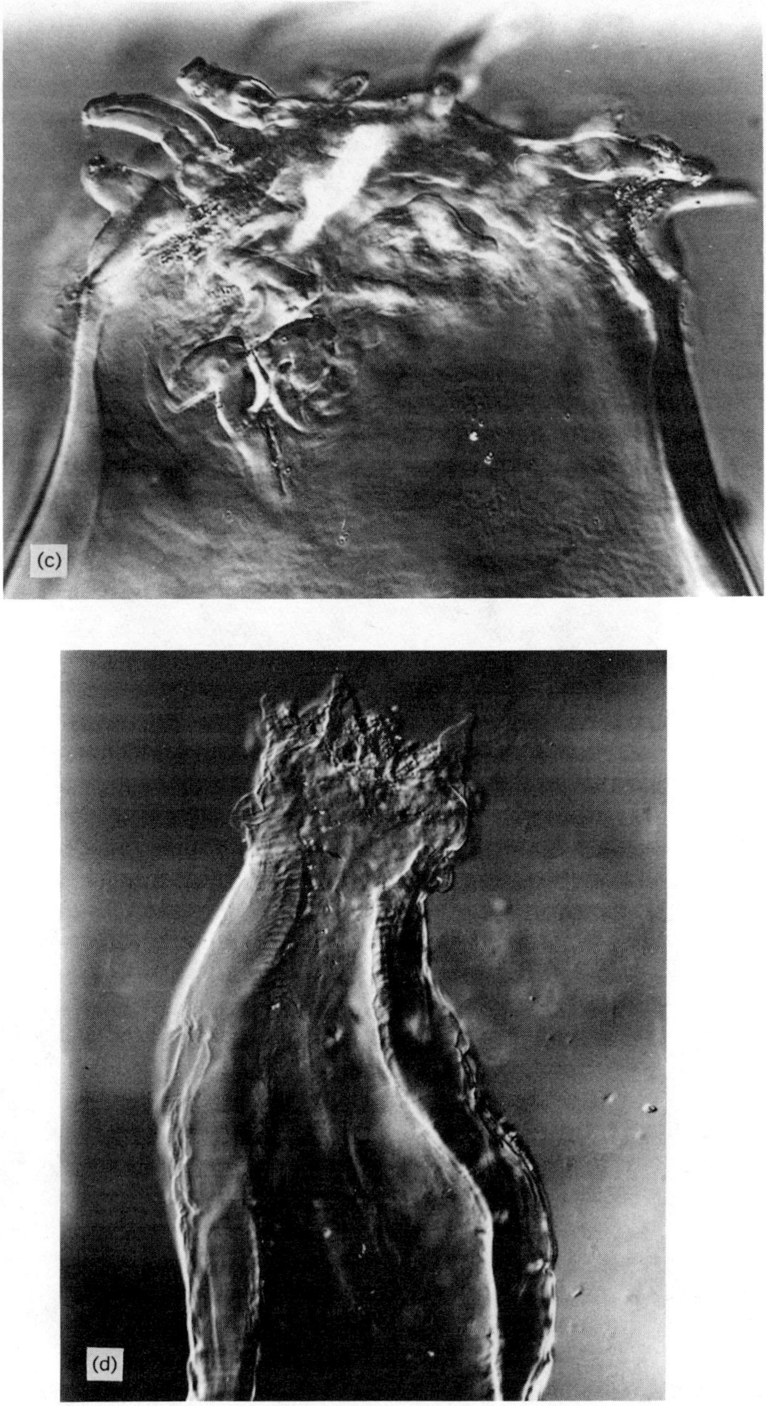


**Figure 14.6.** (*Continued*)

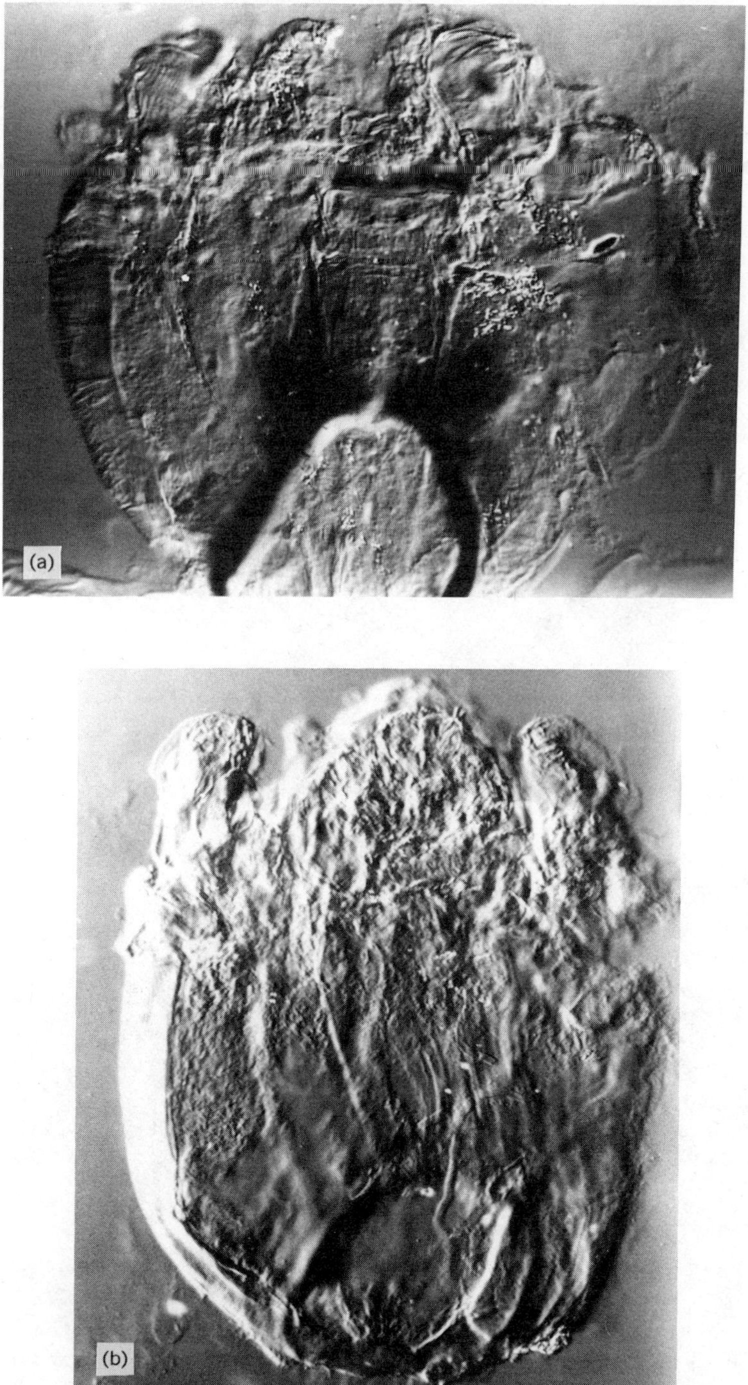


**Figure 14.7.** Head capsules of massively deformed larvae of *Chironomus* Meigen from Pasqua Lake, southeastern Saskatchewan.

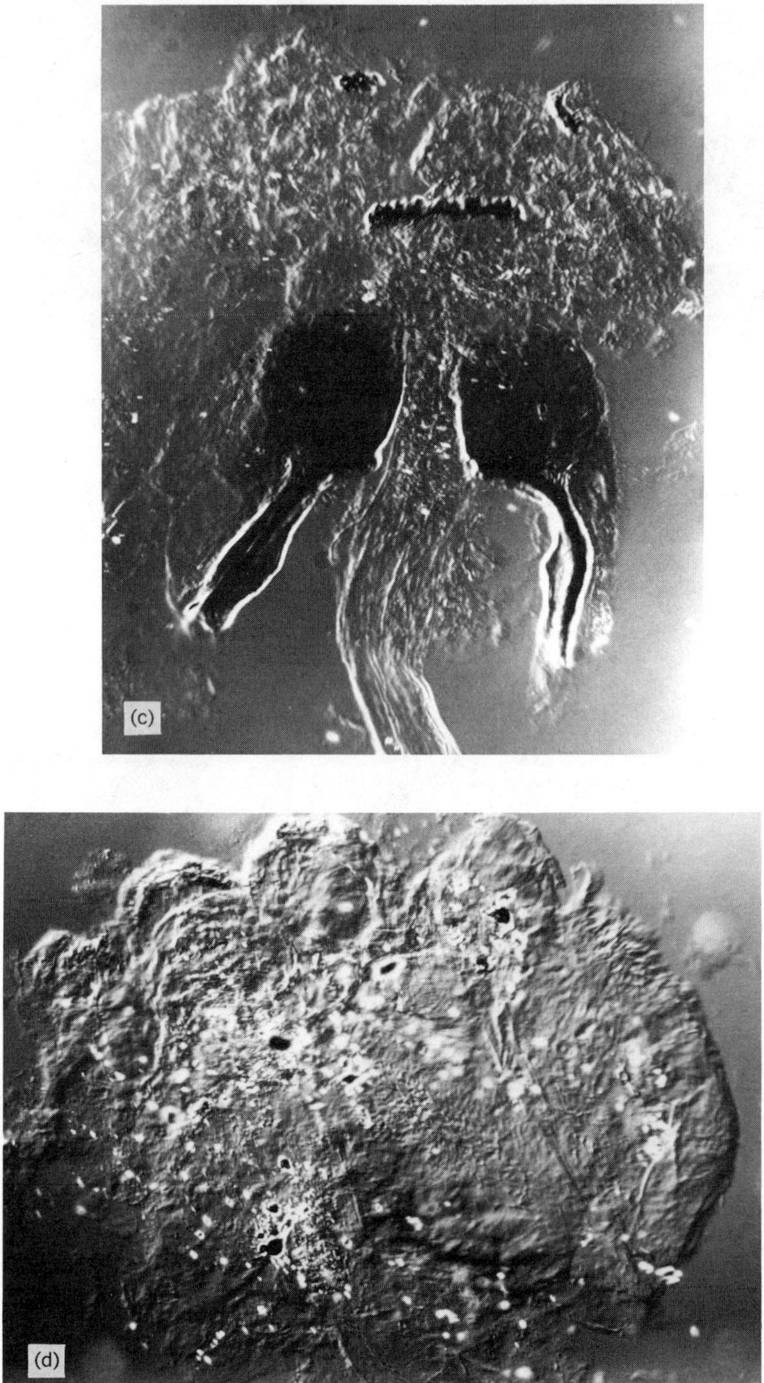


**Figure 14.7.** *(Continued)*

bles often have apical teeth added or deleted (Figs. 14.4l,m), fused (Fig. 14.4n), or stunted (Fig. 14.4o). The subapical teeth of premandibles are sometimes shorter than the primary tooth (Fig. 14.4h), but occasionally they are much reduced in length (Fig. 14.4i), absent (Fig. 14.4j), or accompanied by additional teeth (Fig. 14.4k). The teeth of the epipharyngeal pecten often are fused (Fig. 14.4q), separated and disorganized (Fig. 14.4r), or stunted and cramped (Fig. 14.4s). Deformation of the labral lamellae (Fig. 14.4c), ventromentum, and other less obvious structures have also been observed, but only infrequently; at this time, there is not enough information available to provide an adequate description of the range of deformation possible in these structures.

Antennal deformities range from mild shortening of the basal segment to the more severe reduction in length of one or all antennal segments (Fig. 14.5e). The more obvious deformities involve the loss of individual segments (Figs. 14.5b,c), the fusion of one or more adjacent segments (Fig. 14.5f), or the presence of segments of questionable equivalence (Fig. 14.5d). In severe cases, only the fused remnant of the basal segment may remain (Fig. 14.5g). In some cases, sensory organs such as the blade/accessory blade complex or the peg sensillum may be displaced from their normal positions at the apices of the first (basal) or second segments, respectively. The ring organ on the basal segment may also be displaced or entirely lacking (Fig. 14.5g). In some cases, structures are present for which no homology is known. These may range from simple lumps and bumps to more complex trident- or palmate-shaped structures (Fig. 14.5e).

More severely deformed larvae display massive thickening of the body and head capsule walls and disorganization of the structures of the head capsule. These may range from thickening and fusing of individual structures (Fig. 14.6) to the complete loss of structural integrity in the head (Fig. 14.7). The thickened structures of massively deformed larvae characteristically are laminated (Figs. 14.6b,d) and the outer body wall is often sheathed with a loosely attached layer of thin integument or ''sloughed'' skin (Figs. 14.6a,b). The chitinous claws of the posterior parapods often are reduced to pudgy, fingerlike protuberances that appear to lack any physical strength (Figs. 14.6c,d). It is difficult to comprehend how the larvae from Pasqua Lake, southeastern Saskatchewan, were able to survive to reach the state of ''development'' shown in Fig. 14.7.

## 3. MORPHOLOGICAL DEFORMITIES AND CONTAMINANTS

The evidence for a cause and effect relationship between morphological deformities and environmental contaminants remains largely circumstantial. Hamilton and Saether (1971) formulated the hypothesis that deformities were caused by contaminants from their knowledge of the distribution of deformed larvae. They noted that deformed larvae usually were found in areas receiving industrial or agricultural contaminants, but not in areas polluted by domestic sewage. Little experimental evidence exists to support the Hamilton/Saether hypothesis, but results from field studies reported by subsequent workers tend to substantiate the relationship.

## 3.1. Spatial Gradients

### *3.1.1. The St. Lawrence Great Lakes*

Deformed chironomids have been reported from a number of contaminated sites in the St. Lawrence Great Lakes. Brinkhurst et al. (1968) found three specimens of badly deformed larvae of *Chironomus s.s.* in samples taken from the western basin of Lake Erie off the mouth of Maumee River. The head capsules were thickened and heavily pigmented and the mouthparts exhibited a variety of aberrations. The most conspicuous feature of these larvae, however, was their exceedingly thick exoskeletons. These ranged in thickness from 99 to 131 $\mu$m, values 7 to 17 times greater than normal. Although unable to pinpoint the specific cause, they noted that the deformed larvae all came from an area exposed to effluents from the industrial complex around Toledo, Ohio. They also noted that dredge samples collected along the Maumee–Detroit River line in the area of heaviest contamination contained no larvae whereas more than 1700 specimens, all of which appeared normal, were collected from stations throughout the remainder of the lake.

Cook and Veal (1968) reported that most of the *Chironomus* spp. collected in Port Hope Harbour, Lake Ontario, near the cooling water discharge from the uranium refining plant of Eldorado Nuclear Limited, had deformed mouthparts similar to those described in Brinkhurst et al. (1968). They attributed the high incidence of deformities observed in fauna from the inner harbor to high levels of gross alpha-(1400 pCi), beta-(2200 pCi), and radium-226 (110 pCi) radioactivity introduced into the harbor in refinery wastes.

Hare and Carter (1976) reported that about 77% of *Chironomus (s.s.)? cucini* (*salinarius* group) larvae* collected in Parry Sound Harbour, Lake Huron, had deformed mouthparts. Of 73 *C. (s.s.)? cucini* larvae examined from outside the harbor, only one specimen (1.4%) from Deep Harbour was slightly deformed. Single specimens of *Procladius* sp. and *Micropsectra* sp. (*praecox* group) were also found in samples in or near Parry Sound Harbour. The deformed larvae were collected in areas where the intensity of industrial and commercial activity was greatest. Hare and Carter (1976) suggested that the deformed larvae likely were near the limit of their "industrial effluent tolerance range."

Krieger (1984) found three deformed larvae in a study of environmental degradation in the southern near-shore zone of the central basin of Lake Erie. These included a specimen of *Chironomus* near the mouth of the Grand River at Fairport Harbor and two specimens of *Procladius* near the mouths of the Black River at Lorain and the Cuyahoga River at Cleveland, respectively. With only single specimens from three widely separated sites, Krieger declined to ascribe the deformities to anthropogenic causes.

Crowther and Luoma (1984) found deformed larvae of *Chironomus salinarius* in an area of Lake Superior that received wastes from a pulp and paper mill near

*The proportion of deformed *cucini* larvae in Table 1 is contradicted in Hare and Carter's (1976) Discussion, which states that 83 of 85 *Chironomus* larvae in Parry Sound were deformed.

Thunder Bay, Ontario. They interpreted the presence of the deformed larvae as an indication of habitat impairment.

Warwick et al. (1987) linked the incidence of deformities in *Chironomus* larvae to the degree of sediment contamination in a preliminary note on Port Hope Harbour, Lake Ontario. The harbor is contaminated by uranium-238 decay chain radionuclides, several trace metals, and cooling water from radium and/or uranium refining operations. Contamination in the inner harbor generally exceeds that in the outer harbor by an order of magnitude. In the inner harbor, thorium-230 activities in the sediments were 6.78 Bq/g, whereas in the outer harbor they were 0.42 Bq/g. Nickel concentrations similarly ranged from 58 to 5 $\mu$g/g between the inner and outer harbors, respectively. The incidence of deformities in the inner harbor was 83% compared to 14% in the outer harbor. Although the small number of specimens available for analysis precluded any rigorous statistical treatment, the separation between the inner and outer harbors was well defined. The effects of elevated water temperatures are not known, but these may also have played a role in inducing the deformities observed.

### *3.1.2. The Canadian Prairies*

Deformed larvae have been reported from two sites on the Canadian prairies. In a preliminary report on Pasqua Lake, southeastern Saskatchewan, Warwick (1980b) identified seven deformed larvae of *Chironomus* spp. and one of *Procladius* sp. in the western deep-water basin of the lake and only one deformed larva of *Chironomus* among the remnant fauna of the eastern basin. The deformities in these larvae were confined to the mouthparts; there was no abnormal thickening of the body walls or unusual pigmentation. Since this publication, however, analysis of the data set from the western basin has been completed and a set of massively deformed larvae has been found (see Fig. 14.7). These larvae are characterized by grossly thickened body walls and head capsules that lack the organization of normal larvae entirely. Of the 1813 larvae examined, 2.26% were massively deformed. In the western basin as a whole, the incidence of deformation is estimated at 8.2%—up from the 2.3% reported initially. In the eastern basin, the incidence of deformation remains at 3.1% for the 32 larvae found. Although the incidence of deformation is lower in the eastern basin, conditions there probably are much harsher, since only a remnant population of *Chironomus* larvae survives to represent the benthic community; all other forms, including oligochaetes, have been eliminated from the eastern basin. Pasqua Lake receives waste effluents primarily from the cities of Regina and Moose Jaw, Saskatchewan.

Birkholz et al. (1980) found deformed larvae in Tobin Lake, Saskatchewan, a reservoir on the Saskatchewan River that acts as a sink for contaminants introduced into the system from across the greater Canadian prairie region. The incidence of deformities in preliminary samples collected in October 1979 was 19.6% in a sample of 1178 larvae. The majority of the deformed larvae were *Chironomus anthracinus* type, but a few *C. plumosus* type were also found.

### *3.1.3. The British Columbia Interior*

Saether (1970) found eight* deformed larvae belonging to a number of different species in the Okanagan lakes of British Columbia. In Okanagan Lake itself, he found deformed specimens of *Protanypus* cf. *morio* Zett. (2), *Stictochironomus* sp. (1), and single specimens of unidentified Chironomini (1) and Tanytarsini (1) larvae. In downstream Skaha Lake, he found deformed *Chironomus semireductus* type (1), *Procladius* (*Psilotanypus*) sp. (1), and *Stictochironomus* sp. (1) larvae. Although a total of 703 larvae were examined from 37 sites, all of the deformed larvae were found at sites located off tributaries feeding into the two lakes. The majority of the most seriously deformed larvae in Okanagan Lake were collected off the mouths of Vernon Creek below Vernon (28.6%) and, in Skaha Lake, the Okanagan River below Penticton (21.4%). Seriously deformed larvae were also collected off the mouths of creeks near Kelowna (11.1%) and West Bank (1.8%). Saether speculated on the cause of the deformities and suggested that these might occur in individuals living at the borderline of their ''ecological range.'' He discounted the theory, however, because the deformities, which were so conspicuous, had been reported only once previously in the literature by Brinkhurst et al. (1968). Since no major industries were located in the area, Saether suspected that insecticides used in the numerous orchards surrounding the lakes were responsible for the deformed larvae.

Saether and McLean (1972) resampled the September 1969 survey sites again in the spring of 1971, but found only one deformed larvae (a specimen belonging to a genus near *Polypedilum*) despite the fact that almost three times as many samples were taken. To explain the difference in the numbers of deformed larvae in the two sample sets, Saether suggested that deformed larvae may not be able to survive for any length of time and, since the second survey was done in the second week in May, very few larvae were brood of the year and deformed larvae were not likely to be present.

### *3.1.4. Europe and Scandinavia*

Köhn and Frank (1980) found numerous deformed larvae of *Chironomus thummi* Kieffer in urban sections of the Teltowkanal (Berlin, GDR) polluted by thermal and industrial effluents. The incidence of deformities in *Chironomus thummi* and *C. plumosus* larvae in 5 sites along the canal ranged between 25.4 and 37.6%. Many of the larvae displayed a very characteristic gap in the mentum. No correlations were found between the higher water temperatures and the incidence of deformities and few mermithids were found to suggest mermithid infestation. Köhn and Frank attributed the deformities to industrial effluents, particularly heavy metal contamination. In related studies, Frank (1981) found that the activities of glyceraldehyde-3-phosphate dehydrogenase and lactate dehydrogenase were 2.1× lower in deformed larvae of *Chironomus plumosus* and *C. riparius* exposed to heavy

*This number is misquoted in Hare and Carter (1976) from a typographical error in Table 6 in Saether (1970).

metals in the Teltowkanal than in larvae from uncontaminated sites. However, he stated that the effects of heat and high levels of cadmium, copper, lead, and zinc in the sediments could not alone explain the decrease in enzymatic activities and the deformites in the chironomid populations. Frank (1983) also reported that the incidence of deformities in offspring from deformed larvae from the Teltowkanal declined from about 23% in contaminated canal water to between 1 and 2% when offspring were reared in clean, uncontaminated water.

Wiederholm (1984) found that the incidence of deformed teeth in the mentum of several chironomid genera was significantly higher in contaminated Swedish lakes than in uncontaminated lakes. In uncontaminated areas of lakes Vättern and Mälaren, only one deformed larva (0.3%) was found in 287 specimens of *Micropsectra* and *Tanytarsus*. In contrast, in two bays in Lake Vänern contaminated by mercury, zinc, cadmium, and pulp mill waste, the incidence of deformities increased significantly. In Kattfjorden Bay, 14 out of 370 larvae (3.8%) were deformed while in Åsfjorden Bay, 11 out of 63 larvae (17.5%) were deformed. The incidence of deformities in these bays was about 10 and 50 times greater than that of the reference areas, respectively. Wiederholm also cited incidence of deformities of 8.3 and 25% in *Chironomus* larvae from lakes Norra Barken and Stora Aspen, where several heavy metals and oil are among the known pollutants. He noted that the material available from these lakes was limited because environmental conditions were such that few living chironomids could be found.

Klink (1984) found that the incidence of deformities was high in contemporary *Chironomus* communities of the IJssel River in The Netherlands; these were 43.9 and 30.7% for samples collected in 1981 and 1982, respectively. Klink cautioned that the number of specimens available (41 and 75, respectively) was not large enough for good statistical analysis. He used considerably larger samples to establish the incidences of deformities in *Chironomus* communities from the sandpits at Huizen and Harderwijk (458) and the Oude Gooren nature reserve (378); these were 2.4 and 2.6%, respectively.

Urk et al. (1985) found a high percentage (25–50%) of deformed larvae in the remnant fauna of the IJssel River in The Netherlands. So few living larvae remained that no statistical trends were evident. They also found deformed larvae in the Hollandsch Diep and the Haringvliet. The incidence of deformities in the delta of the Hollandsch Diep was about twice as great (43–46%) in water deeper than 2 m than along the shoreline (20%). In the Haringvliet, however, the incidence of deformities was roughly equivalent (21 vs. 23%) in fauna from the shoreline and deeper-water zones.

### *3.1.5. The Continental United States*

Tennessen and Gottfried (1983) reported deformities in the ligula of six out of nine Tanypodinae genera examined from two artificial lakes and three coal stripmine ponds in northern Alabama. Deformed ligulae were found in *Ablabesmyia annulata* (Say) (4.0%), *Clinotanypus pinquis* (Loew) (3.2%), *Djalmabatista pulcher* (Johannsen) (3.2%), *Larsia* sp. (0.9%), *Procladius* (*Procladius*) *bellus* (Loew)

(4.5%), *P. (Psilotanypus) subletti* Roback (1.7%), and *Tanypus punctipennis* Meigen (8.3%). The authors were unable to draw any definitive conclusions about what might be causing the deformities.

## 3.2. Temporal Gradients

### *3.2.1. The St. Lawrence Great Lakes*

Warwick (1980a) found that the incidence of deformities in the mentum and ligula of subfossil chironomid remains in the Bay of Quinte, Lake Ontario, was a recent event that coincided with the beginning of the "chemical age." He found that the incidence of deformities increased from 0.09% in sediments several thousand years old to 1.99% in contemporary (1972) fauna. The 20-fold increase began in sediments dated at 1951. This date closely approximates the date (1949) that Bakelite Thermal Sets Ltd., a division of Union Carbide, established facilities in Belleville, Ontario (Mika and Mika, 1973), and began releasing chemical waste products into the Bay of Quinte (D. R. Oliver, Biosystematics Research Institute, Ottawa, personal communication).

### *3.2.2. Europe and Scandinavia*

Wiederholm (1984) also reported that the incidence of deformed mouthparts was greater in contemporary chironomid communities than in subfossil communities from sediments pre-dating the "chemical age" in Swedish Lake Mälaren. In unpolluted Ekoln Bay, only one deformed mentum (0.8%) was found in 126 *Chironomus* specimens recovered from pre-1950 sediments (20–75 cm). In slightly polluted Granfjärden Bay, where elevated concentrations of zinc and probably mercury, cadmium, nickel, and lead are found, the incidence of deformities in contemporary communities was 1.4% or about twice that in Ekoln Bay. In Västeråsfjärden Bay, subfossil material from 50–60 cm of sediment contained no deformed *Chironomus* specimens whereas contemporary communities (1969) contained as many as 10.7% deformed *Chironomus* larvae. Wiederholm cautioned, however, that the number of subfossil and contemporary specimens available for examination from this bay was limited.

Klink (1984) similarly found that the incidence of deformed menta was greater in contemporary *Chironomus* communities than in subfossil communities from sediment deposits of the Meuse River in the Netherlands. No deformed specimens were found in specimens deposited in 1830 or 1880, but in sediments deposited in 1984, the incidence of deformities was 13.9%.

## 3.3. Dose–Response Experiments

Hamilton and Saether (1971) performed a series of preliminary experiments in an effort to produce deformities similar to those found in Lake Erie and the Okanagan lakes. Cultures of *Chironomus tentans* Fab. were exposed to single doses of several different concentrations of 1,1-dichloro-2,2-bis(*p*-chlorophenyl)ethane (DDE), nitrilotriacetic acid (NTA), polychlorobiphenyls (PCBs), Aldrin, and 50–50 mixtures of 2,4-D and 2,4,5-T. Two egg masses were hatched in each solution and the larvae reared to the fourth instar before harvesting. No attempt was made to monitor chemical concentrations over the course of the exposure period. Hamilton and Saether reported largely negative results, although they did note that a small proportion (<10%) of the larvae exposed to DDE were deformed in a manner somewhat similar to those observed in the field.

Cushman (1984) correlated the incidence of deformities in the mentum of larvae of *Chironomus decorus* Johannsen exposed to varying concentrations of synthetic (coal-derived) crude oil. He found that deformities in the medial area of the mentum increased significantly at higher concentrations, but was unable to discern any response at lower dosages. Cushman was unable to correlate the overall incidence of deformities with crude oil concentrations and concluded that the traditional measures of toxicity (abundance, biomass, and diversity) were more effective (Cushman and Goyert, 1984) than the occurrence of deformities as an indicator of coal-liquid contamination.

## 4. EVALUATION OF MORPHOLOGICAL DEFORMITIES

### 4.1. Requirements for a Biological Screening Tool

Morphological deformities in chironomid larvae have been increasingly reported in the literature as evidence of contamination. Until recently, however, little attention has been devoted to devising a systematic approach to deal with these phenomenon. To provide a sound basis for an effective biological screening tool, both the quantitative and qualitative aspects of morphological deformities must be evaluated.

Most previous studies have been hampered by inappropriate or inadequate preparation and mounting techniques. Hamilton and Saether (1971) were unable to appreciate the significance of their dose–response experiments because the methods used to prepare and mount their material were not suitable for examining fine morphological detail. Their specimens were mounted in the traditional manner in which the natural shape of the head capsule and body is maintained. Either the resulting mounts were too thick to examine under high-power objective lenses or, even when higher magnifications were possible, interpretations of fine morphological structures like the antennae were seriously hindered by depth of field problems.

Cushman's (1984) experimental study on the effects of a synthetic, coal-derived oil product (PDU 9) on *Chironomus decorus* populations probably suffered in the same way. His comment that the mentum was used in his analysis because it was clearly visible in "almost all the mounted larvae" indicates that his mounting technique may not have been sufficiently exacting for a detailed analysis of his material. Also, he examined his specimens only "at 100–400× magnification." Both facts suggest Cushman may have missed the more subtle responses of the fauna, particularly the results of the low-dose exposures, for technical reasons.

The inability of these authors to examine their specimens consistently and accurately at high-power magnifications may have led to inaccurate or improper conclusions. Certainly, Hamilton and Saether (1971), whose material I have been able to reexamine, missed the significance of their pioneering work. The same probably applies to many of the other studies. Unfortunately, their methods all too frequently are not described in sufficient detail to allow direct comparison. One of the purposes of this paper is to propose a standard mounting technique that will permit direct comparison between studies in the future.

Previous studies have also been limited by the absence of a conceptual framework within which the broader range of deformities can be evaluated and documented and even by the lack of a good definition of what does or does not constitute a deformity. Klink (1984) made several comparisons with Wiederholm (1984), but it is clear that their definitions of mentum deformities differed considerably. Klink's alphanumeric system for cataloging deformities in the mentum is very detailed and exacting. On the other hand, Wiederholm acknowledges that the incidence of deformities he reported probably would have been higher "if the numbers of specimens with an aberrant number of teeth in an otherwise symmetrical mentum had been included in (his) calculations." For the present, I define a deformity as any morphological structure that departs from the normal configuration. The definition should be kept as simple as possible until we are better able to discriminate between normal and abnormal variation.

## 4.2. Standard Preparation Techniques

Warwick (1985) proposed standardized preparation and mounting techniques based on those described by Warwick (1980a,b) and Warwick and Casey (1982). These are essentially the same as the methods described by Schlee (1966) and Saether (1969) except that the head capsule and body of the larvae are fully flattened in the mounting process. Larvae are cleared initially by digesting in either warm or cold 8% KOH. The time required for clearing in warm KOH is largely dependant on the size of the individual specimen; clearing with cold KOH requires a longer period of time, but can be used effectively to control the rate at which specimens are processed. The gut and body contents are removed by piercing the body wall, removing the gut tract and flushing out solubilized fats and oils. Cleared larvae are then passed through successive baths of glacial acetic acid, anhydrous ethanol, ethanol layered over cedarwood oil, and cedarwood oil. The head and bodies of medium to large larvae are generally mounted separately; smaller larvae are mounted intact. Dissection of the head capsule normally is carried out on the glass slide in Canada balsam rather than at the cedarwood oil stage to prevent loss of parts during manipulation. The head capsule is placed ventral side up and the mandibles snapped back to prevent them from obscuring the antennae. A coverslip is put in place and the head capsule flattened by exerting increasing pressure to the coverslip with a pair of forceps. A rotary motion can be applied to the coverslip during the flattening process to spread or shift mouthparts that threaten to obscure other diagnostic features. The body is similarly flattened.

The technique differs only slightly from traditional methods in that the structures of the head capsule and body are fully flattened and spread out. This is done to permit the examination of fine morphological detail and measurement of individual structures using high-power, oil-immersion lenses. Some traditional measurements may be precluded by flattening, but these can be replaced by other measurements. For example, head capsule width could be replaced by mentum width. I prefer Canada balsam as a mounting medium because it is easily softened in xylene vapor if remounting is necessary. Other mounting media may be used after suitable adjustments to the clearing technique.

Specimens with worn teeth on the mentum or mandibles are easily distinguished from those with deformed teeth. The margins of worn teeth appear crinkly or pebbled at high power magnifications because the dorsal and ventral plates of the mentum wear or break asynchronously. In contrast, the margins of unworn, normally shaped teeth or deformed teeth appear smooth because the two plates are properly fused together.

## 4.3. Indexing Morphological Responses

### *4.3.1. Antennal Deformities*

Warwick (1985) presented an indexing scheme for classifying the severity of antennal deformities in the genus *Chironomus* as a preliminary step in developing a biological screening tool. The antennae of *Chironomus* were chosen over other morphological features as a starting point for several reasons: (1) intuitively, as a receptor organ, antennae may be expected to be more sensitive to contaminants; (2) the effects of contaminants would be more readily quantifiable because antennae possess a quantifiable number of segments; (3) *Chironomus* is one of the most common, widely distributed and well-documented chironomid genera; (4) recognition of abnormal specimens within a particular genus requires a high degree of familiarity with the genus to be able to distinguish between normal and abnormal variation; and (5) deformities are known to occur in other genera, but they appear to be more common in detritivores like *Chironomus*.

The degree of severity of deformation in individual antenna is assessed according to the ranking system summarized in Table 14.1. Ranking of severity is carried out according to a sequence of steps, each of which is assigned a separate scale according to the type of deformation involved. In Step 1, antennae are assigned to basic categories depending on the number of genuine and/or questionable segments present. Deformed antennae with five segments are assigned a value of 1; antennae lacking segments are assigned four additional points for each missing segment or 2 points where the presence of a segment is questionable. Thus, a deformed antenna having two genuine and one questionable segment (i.e., 3? segments) would be assigned a value of 1 plus 8 (two missing segments) plus 2 (one questionable

**Table 14.1 Summary of the Values Assigned to Individual Types of Deformities in Antennae of *Chironomus* spp. Larvae**

| Step | Feature | Indices of Morphological Response (IMR) |
|---|---|---|
| 1 | Basic classification categories | |
| | (a) Loss of genuine segments | 1, 5, 9, 13, 17, and 21 |
| | (b) Presence of questionable segments | 3, 7, 11, 15, and 19 |
| 2 | Reduction in antennal length | 1–9 |
| 3 | Displacement or loss of ring organ | 1–2 |
| 4 | Apical fusing of basal segments | 2 |
| 5 | Displacement of sensory organs | 2 |
| 6 | Structures of unknown homology | |
| | (a) Major structures | 2 |
| | (b) Minor structures | 1 |
| Σ | IMR | |

*Source.* Adapted from Warwick (1985).

segment) for a value of 11. In Step 2, the first 10% variation in antennal length is regarded within normal or "zero" variation, but a value of 1 is assigned for each additional 10% reduction in length. In Step 3, values of 1 and 2 are assigned, respectively, if the ring organ has been either displaced or lost. In Steps 4 and 5, values of 2 are assigned if the apex of the basal segment has been completely fused over or if any of the accessary sensory organs have been displaced. In Step 6, values of 1 or 2 are assigned, respectively, for the presence of simple (minor) or more complex (major) structures for which no homology is known. The values assigned at each of these steps are then summed to provide IMR, the Index of Morphological Response, for the antennae of individual larvae. Individual IMR values are then summed to provide a measure of the morphological response of a chironomid community to toxic stress. The Index of Severity of Antennal Deformation (ISAD) is then defined as

$$\text{ISAD} = \sum \text{IMR}/n$$

where IMR represents the total of the indices of morphological response describing the degree of severity of deformation in individual antenna and $n$ is the total number of specimens examined. The procedures and more complex interpretive aspects of the scheme are detailed in Warwick (1985).

Although developed originally for *Chironomus* species, ISAD can be easily adapted to other chironomid species because the indexing system is open ended.

Chironomid species with different numbers of segments in their antennae can be accommodated simply by adjusting the range of basic classification categories.

### *4.3.2. Application of ISAD*

The severity of antennal deformations in *Chironomus tentans* larvae exposed to DDE by Hamilton and Saether (1971) was analyzed by ISAD after the original specimens were remounted using our techniques. Unexpectedly, the results showed an inverse relationship between ISAD and the concentration of DDE (Fig. 14.8). Although based on small sample populations, regression analysis for the line described by

$$y = 1.29 - 0.05x$$

indicated that the inverse relationship was highly significant ($r = 0.971$ where $r_{0.01} = 0.959$ for 3 df).

The incidence of antennal deformities for the same data set also decreased as the concentration of DDE increased (Fig. 14.8). The inverse relationship described by

$$y = 24.01 - 1.05x$$

was significant between 95 and 99% probability ($r = 0.946$ where $r_{0.05} = 0.878$).

The reason for the inverse relationship between antennal deformities and the

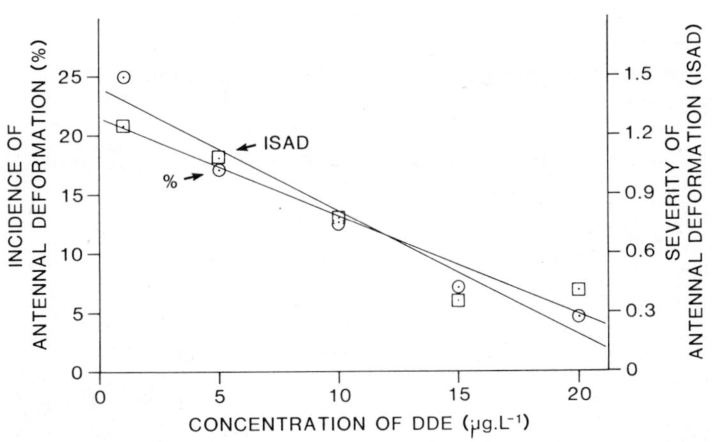


**Figure 14.8.** Dose–response relationships between the incidence (%) and severity (ISAD) of deformities in *Chironomus tentans* Fab. larvae exposed to DDE (μg/L). Highly significant regression equation for severity of deformation (ISAD) is $y = 1.29 - 0.05x$; significant regression equation for incidence of deformation (%) is $y = 24.01 - 1.05x$. From Warwick (1985; Fig. 22). With permission.

concentration of DDE is not immediately clear. Warwick (1985) speculated that the relationship represents an isolated portion of the total response of antennae to DDE (Fig. 14.9). Quantal dose–response curves may be regarded as a graded response with the familiar "bell-shaped" distribution if, as in ISAD, the entire population of antennae is regarded as an individual (Chan et al., 1984). Individual members of such a population respond differently to a particular stimulus such as a contaminant. Some members will be highly sensitive while others will be very resistant. In these experiments, the range of DDE concentrations used probably span a range of antennal response in which resistance to DDE is increasing. If this hypothesis is correct, the range of greatest antennal sensitivity probably is less than 1 $\mu$g DDE/L or its chemical equivalent.

The hypothesis is supported by evidence from several field sites. In Tobin Lake, Saskatchewan, both the incidence (%) and severity (ISAD) of antennal deformation declined along the hydraulic gradient established between the inflow of the Saskatchewan River and the Squaw Rapids dam (Fig. 14.10). Comparison of data sets from Sites V and VI show that ISAD declined (Table 14.2) from 1.64 to 1.19 between the two sites, respectively. The incidence of antennal deformities similarly decreased from 8.43 to 6.97% between the two sites. According to the relationship shown in the DDE experiments, Site VI, located 5.6 km farther downstream, should be the more severely contaminated of the two sites.

Three observations suggest that Site VI is the more severely contaminated site. Reservoir hydraulics are such that, as hydraulic energy dissipates, waterborne par-

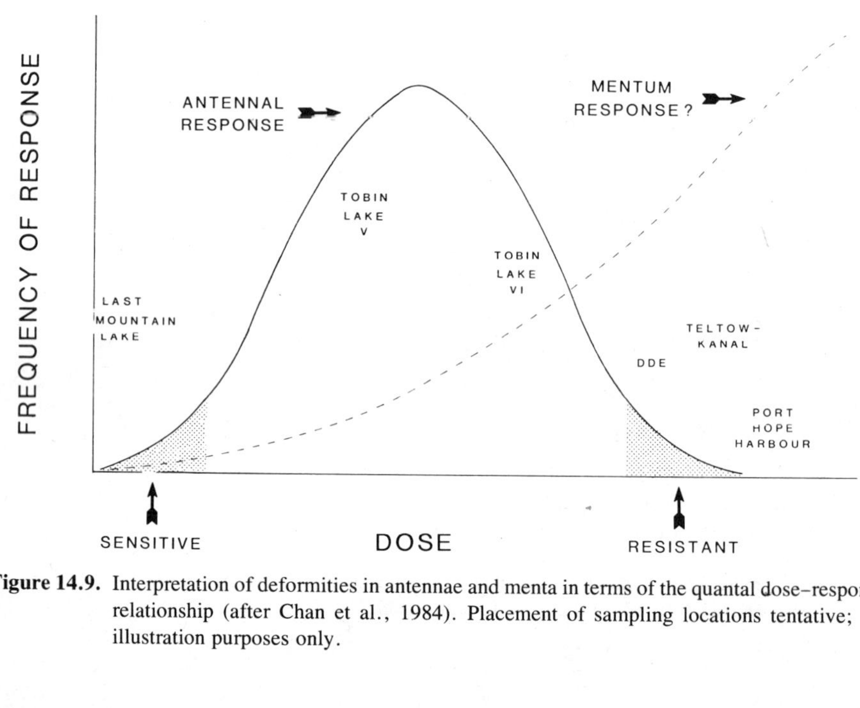


**Figure 14.9.** Interpretation of deformities in antennae and menta in terms of the quantal dose–response relationship (after Chan et al., 1984). Placement of sampling locations tentative; for illustration purposes only.

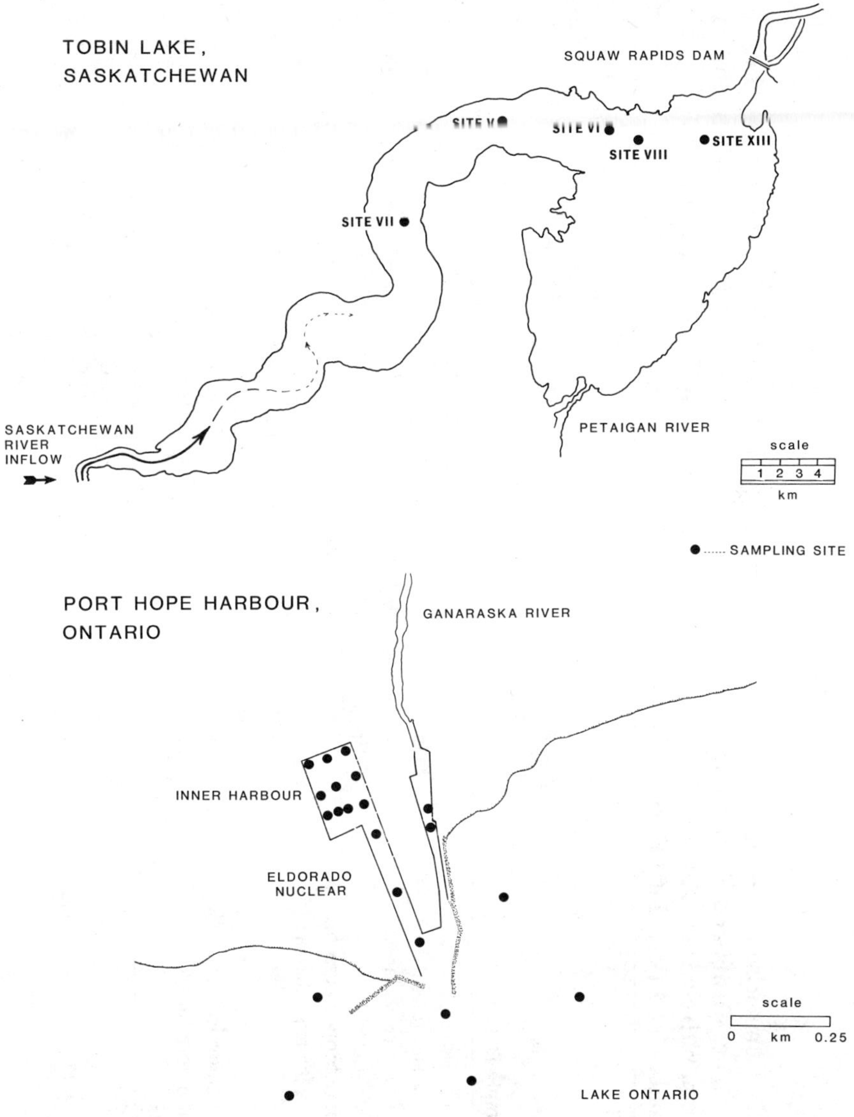


**Figure 14.10.** Sampling locations in Tobin Lake and Port Hope Harbour, Lake Ontario. Adapted from Birkholz et al. (1980) and Hart et al. (1986), respectively.

ticulates of decreasing density are deposited into the sediments. Because different chemical species tend to bind selectively to different waterborne fractions, the hydraulic sorting process should establish a gradient of contaminants between the inflow area and the dam (Fig. 14.10). Samoiloff et al. (1983) established that a chemical gradient does exist along this line. They found that, out of 52 potentially toxic compounds isolated from sediments from Sites VII and VIII (located one on

**Table 14.2 Comparison of the Incidence (%) and Severity (ISAD) of Antennal Deformation and the Incidence (%) of Menta and Total (%) Deformation in *Chironomus* spp. Larvae from Last Mountain Lake (Comparatively Uncontaminated), Tobin Lake (Contaminated), the Teltowkanal (Highly Contaminated), and Port Hope Harbour (Very Highly Contaminated) Sites**

| Sample Designation | Last Mountain Lake II | Tobin Lake V | Tobin Lake VI | Teltowkanal 3 (incl.) | Port Hope Harbour All Sites (incl.) |
|---|---|---|---|---|---|
| Depth of water (m) | | 14.3 | 23.5 | | |
| Number of larvae examined | 935 | 690 | 488 | 83 | 251 |
| Number of *Chironomus* larvae | 837 | 688 | 488 | 83 | 48 |
| Number of deformed antennae | 10 | 58 | 34 | 2 | 0 |
| Incidence of deformed antennae (%) | 1.2 | 8.4 | 7.0 | 2.4 | 0 |
| Number of evaluated antennae (d) | 9 | 58 | 34 | 2 | 0 |
| Total IMR | 99 | 1126 | 582 | 38 | 0 |
| ISAD | 0.12 | 1.64 | 1.19 | 0.46 | 0 |
| Incidence of deformed menta (%) | 1.0 | 4.8 | 13.1 | 28.9 | 47.9 |
| Incidence of deformities (%) (all types) | 7.2 | 15.1 | 26.0 | 34.9 | 70.8 |

either side of Sites V and VII), only five were common to both sites. Evidence for a concentration gradient was found even farther downstream. In recent work, massively deformed larvae of *Chironomus anthracinus* type were found at Site 13 (W. F. Warwick, unpublished data). At the present time, the presence of massively deformed larvae is believed to indicate a greater degree of contamination. The sediments in the deeper-water areas also contain large quantities of light, fibrous, organic materials believed to be pulp-fiber wastes from pulp mills located along the Saskatchewan River. The presence of these materials is important not only because it indicates hydraulic energy is negligible near the dam, but also because many chemicals selectively bind with organic materials (Karickhoff 1981; Carter and Suffet, 1982). Although not confirmatory, these field data agree with the pattern suggested by the experimental data.

The hypothesis that larval antennae become increasingly resistant to deformation at higher concentrations of contaminants is also supported by data from the Teltowkanal (Table 14.2). Only two larvae out of 83 *Chironomus thummi* specimens from this highly contaminated canal had deformed antennae. The ISAD value for these antennae was only 0.46. This compares more closely with the value for comparatively uncontaminated Last Mountain Lake (0.12) than with those for either of the Tobin Lake sites and yet the canal is acknowledged to be highly contaminated.

This trend is even more apparent in the data from Port Hope Harbour where no deformed antennae were found among the 48 *Chironomus* spp. examined (Table 14.2). The harbor is very highly contaminated with radionuclide wastes, heavy metals, and elevated water temperatures (Warwick et al., 1987) and yet there is little indication of deformities in the larval antennae. The only observation to this effect, but as yet not quantified, is that the physical dimensions of the antennae seem unusual. This may be a function of species differences, but it may also reflect growth problems in the *Chironomus* populations (Wentsel et al., 1977).

The relationships between the field sites are shown in Fig. 14.9 in relation to the quantal dose–response curve. Their positioning is only tentative at this time, but the figure serves to show graphically the line of reasoning being followed.

#### *4.3.3. Mentum Deformities*

Reanalysis of the *Chironomus tentans* larvae exposed to DDE by Hamilton and Saether (1971) also suggests that, as concentrations increase, response may shift from deformation of the more sensitive antennae to deformation of the less sensitive, harder mouthparts such as the mentum and mandibles. In Figure 14.11, the amount of splitting of the median tooth of the mentum increases as the concentration of DDE increases. The evidence is only subjective, however, because the teeth of the larvae were badly worn. The experimental populations were reared on a sand substrate which badly abraded the teeth of the mentum. Certain information in this data set also indicates that chemicals like DDE weaken the physical structure of chitin (W. F. Warwick, unpublished data). This would make the teeth of these larvae even more suceptible to abrasion. Although wearing effectively pre-

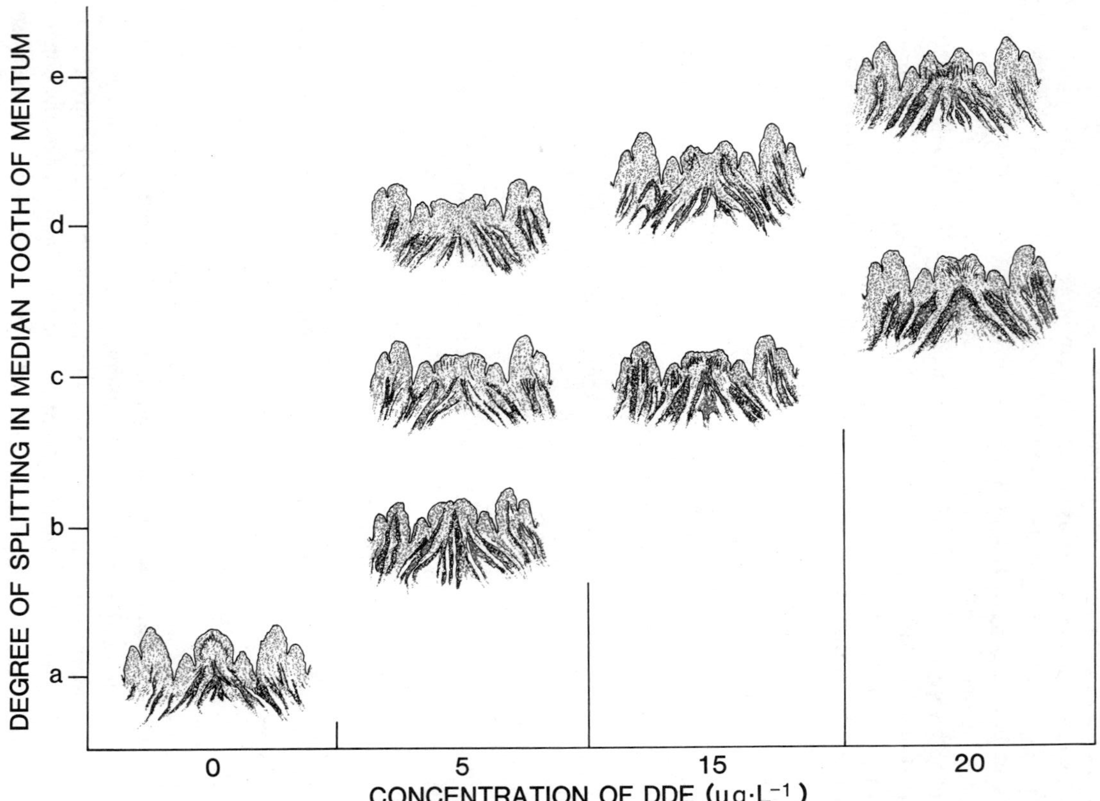


**Figure 14.11.** Dose–response relationship between splitting in the median tooth of the mentum of *Chironomus* and the concentration of DDE.

cludes any direct quantification at this time, splitting in the median teeth is still detectable in the patterns of stria remaining even in badly worn teeth.

The shift from deformation of the antennae to deformation of the more heavily sclerotized mouthparts as concentrations of contaminants increase is also evident in the field data. The incidence of deformities in the mentum (Table 14.2) increases steadily from 1.0% in comparatively uncontaminated Last Mountain Lake to 47.9% in very highly contaminated Port Hope Harbour. The same trend is evident when the incidence of deformities of all types is considered. The incidence increases from 7.2% in Last Mountain Lake to 70.8% in Port Hope Harbour. Some of these numbers may be revised as our ability to define deformities in other structures improves, but the general trend is evident.

The percentage for the incidence of deformities of all types in Port Hope Harbour (70.8%) is also deceptive in that it relates only to the number of deformed larvae and says nothing about the severity of the deformities. These larvae display the greatest number of multiple deformities (that is, more than one type of morphological deformity in a single head capsule) that I have so far encountered. Out of the 34 deformed specimens, 22 larvae, or 64.7%, had more than one type of deformity. In Last Mountain Lake, multiple deformities only occurred in 13.3% of the deformed larvae. Even these figures mask considerable information because the degree of severity of the deformities in the Port Hope Harbour fauna is much greater than that of those in Last Mountain Lake. These inadequacies highlight the requirement for methods of indexing the severity of deformities in the other structures of the larval head capsule including the ventromentum, mandibles, premandibles, epipharyngeal pecten, parapod claws, and labral lamellae.

Klink (1984) has gone a long way in this respect by developing an alphanumeric system for classifying deformities in teeth of the larval mentum. The system is set up in six steps, each separated by a /. Step 1 identifies the larval instar; Step 2, the normal number of teeth; and Step 3, the actual number of teeth. Step 4 identifies the difference between Steps 2 and 3 alphabetically as a deformity (D) or as wearing (W?) followed by the numerical identification of the tooth (or teeth) involved numbering on the mentum from left to right. Step 5 identifies teeth that are absent because of gaps (G), those that are regularly (symmetrically) absent (R), or those that are irregularly (asymmetrically) absent (IR), followed by the numerical identification of the missing tooth (or teeth) on the mentum. Step 6 identifies the presence of additional teeth as those in line with the normal teeth (IN), as those superimposed over the normal teeth (SUP), or those below the normal teeth (SUB). The position of the extra tooth is identified by the numerical designation of the two normal teeth on either side. The system basically is a cataloging system and does not try to quantify the severity of mentum deformities.

### *4.3.4. Massive Deformities*

Massively deformed larvae represent another category in which deformation is less specific, but much more serious in that the life cycle almost certainly could not be completed. Hamilton and Saether (1971) suggested that the thickened exoskeleton may have been secreted as a protective layer against some external irritant or toxic substance. The characteristic laminations in the body and head capsule walls of these larvae (Figs. 14.6b and 14.7a) alternatively suggest that something may have interfered with the molting process. Perhaps instead of progressing to the next stage in the life cycle, the larvae continues, like a broken record, to repeat the current stage of development. Slàma and Williams (1965) found that larvae of the bug *Pyrrhocoris apterous* L. continued to develop through sixth and seventh instar stages into "monster" larvae instead of molting into normal adults after the fifth larval instar. They found that extracts from several North American pulpwood species interfered with the juvenile hormone to prevent normal metamorphosis. The presence of massively deformed larvae in Pasqua Lake, Saskatchewan, in which the structural integrity of the head capsule has been almost completely lost (Fig. 14.7), indicates that other contaminants can induce the same type of deformation. The mechanism or mechanisms by which these changes are affected and their interpretive importance remain to be resolved.

Massive deformities do not seem to be confined to any particular stage of larval development. The massively deformed specimens of *Chironomus* sp. found in Lake Erie by Brinkhurst et al. (1968) were all fourth instar larvae. The massively deformed specimens of *Chironomus* sp. from Pasqua Lake also were fourth instars, but those from Tobin Lake were all second instar larvae (W. F. Warwick, unpublished data). Although the instar stage of the specimen of *Protanypus morio* figured in Hamilton and Saether (1971) was unspecified, it probably was a second instar larvae as well. No massively deformed third instar larvae have been reported yet, but these will undoubtedly be found.

## 5. FACTORS AFFECTING MORPHOLOGICAL RESPONSES

### 5.1. Life History Factors

Little information is available on the occurrence of deformities in different instar stages, but observation suggests that the greater majority occur in fourth instar larvae. This may be the result of sampling techniques, which may be biased toward collecting larger larvae. Larger mesh sieve nets may retain larger third and fourth instar larvae, but lose a high proportion of earlier instars. Sampling brood classes that have developed in synchrony would also bias the proportion of other instars collected in a sample.

Emergence time may influence the incidence of deformities another way. If a population is sampled after the greater proportion have emerged as adults, estimates of the larval population may be unrealistically low. Those larvae remaining may also represent physiologically weaker members of the population who were slower or unable to conclude metamorphosis. Urk et al. (1985) reared larvae collected from contaminated Ketelmeer and uncontaminated Oostvaardersplassen in clean water aquaria. The adults of the larvae from the contaminated site were both slower to emerge and were smaller than those from the uncontaminated site. When the experiment was repeated again 3 weeks later, the same pattern persisted, but emergence success was reduced in both cases. The authors questioned the reproducibility of their sampling, but it may also be that they were sampling an older population of physiologically weaker animals. These individuals may represent a segment of the population that was already predisposed to malformation. Both factors reduced population estimates and older larval populations would tend to inflate an estimate of the incidence of deformities.

Certain chironomid genera appear to be more susceptible to morphological deformities than other genera. W. F. Warwick (unpublished data) found large numbers of *Cryptochironomus* sp., *Cryptotendipes* sp., *Procladius*, and *Chironomus plumosus* type larvae coexisting in the southern (contaminated) end of Last Mountain Lake, Saskatchewan. Although the larvae of most of these taxa were apparently normal, about 18% of the *Chironomus* larvae were deformed. In Lake Mälaren, Wiederholm (1984) found deformed *Chironomus* larvae in a predominantly *Microchironomus* fauna in Västeråsfjärden Bay. Although none of the *Microchironomus* larvae were deformed, about 10% of the *Chironomus* larvae were. In Åsfjarden Bay, he found apparently normal larvae of *Heterotrissocladius* coexisting with *Micropsectra* larvae, many of which (17.5%) were deformed. Urk et al. (1985) found few deformed *Procladius* larvae in areas where numerous deformed *Chironomus* coexisted. They wondered whether predaceous *Procladius* were less sensitive to pollutants or whether the lack of response was due to different food requirements. Deformities have been reported in a number of other species, but there appears to be no common bond, either taxonomic or in life style, to link morphologically susceptible taxa. Hare and Carter (1976) suggested that taxa which do not display any physical response to contaminants may respond in other, more subtle ways, perhaps physiological or behavioral in nature.

### 5.2. External Factors

A number of biotic and abiotic factors could potentially cause enough stress on chironomid communities to induce morphological deformities. The effects of mermithid parasitism on the morphology of chironomid larvae is not known, but, in the adult fly, infestation can induce morphological and anatomical abnormalities, intersexes, and gynandromorphs (Rempel, 1940; Thienemann, 1954; Wülker, 1964, 1975; Aagard, 1974; Saether and Galloway, 1980; LeSage and Harrison, 1980). No mermithids have been associated with deformed chironomid larvae in our work to date and Köhn and Frank (1980) reported mermithid parasites were scarce in chironomid populations from the Teltowkanal. The effects of mermithid infestation on larval morphology should be investigated, however, because the rates of infestation can be quite high (Golini and Corliss, 1981). Infestation can be serious and frequently leads to the death of the host larvae (Golini and Sherry, 1979). Such an investigation should include other parasites and/or symbionts such as water mites, viruses and microsporidia, and ciliate protozoans.

Some circumstantial evidence also suggests that water temperature may be a factor in the development of morphological deformities. The Teltowkanal studied by Köhn and Frank (1980) received cooling water from two thermal generating plants. The highest incidences of deformities—32.2 and 37.6%—came from sampling sites immediately opposite the Steglitz and Lichterfelde generating stations where water temperatures reached 27°C during the summer months. The incidence of deformities was lower below both stations where water temperatures were cooler. No correlation was found between water temperature and the incidence of deformities, but any relationship, if it existed, would be greatly complicated by the presence of industrial contaminants, heavy metals, and sewage polluting the canal.

Port Hope Harbour studied by Cook and Veal (1968) was similarly influenced by industrial cooling water. No temperature data were reported, but the greater majority of deformed larvae came from Station B located 30 m from a cooling water discharge and 60 m from an aqueous waste discharge. There were no live animals found at Station A located 15 m from the cooling water outlet and only a few oligochaetes at Station C located 15 m from aqueous waste outlet. Although the importance of radionuclides in the sediments cannot be discounted, the areal distribution of deformed larvae suggest the effects of temperature should be investigated as well. It may be that the effects of radionuclides and other toxicants are increased at higher water temperatures.

W. F. Warwick (unpublished data) examined specimens of *Cryptochironomus* sp. larvae from two Saskatchewan sites. The southern end of Last Mountain Lake is contaminated by spring flood waters from the Qu'Appelle River which are temporarily diverted into the lake to reduce flooding farther downstream in the Qu'Appelle Valley. Less than 1% of the *Cryptochironomus* larvae from this site were deformed. In contrast, about 19% of the *Cryptochironomus* larvae from Maskwa Lake, northwestern Saskatchewan, were mildly deformed. The latter lake is spring fed and there are no known sources of contamination. The only apparent differences between the two sites were the water depths and temperature. The larvae from Last Mountain Lake were collected in 4 m of water whereas those from Maskwa Lake were collected in only 10–12 cm of water. The beach in Maskwa Lake shelves very gradually and water temperatures in the shallows become very warm (25–28°C) during the summer months. Although temperature is the obvious difference, UV radiation and salinity may also be factors.

Under certain circumstances, factors such as salinity, oxygen depletion, changes in water depth, or the erosion and deposition of mineral sediments could possibly cause sufficient stress on chironomid communities to induce deformities. Changes in these factors normally would initiate successional changes in community structure. However, if these changes took place at such a rate that the normal processes of succession were rendered ineffectual, the stress of rapid change could possibly induce morphological responses in individual larvae.

The trophic state of receiving waters may also influence the effect contaminants have on chironomid communities. In highly eutrophic lakes where concentrations of dissolved oxygen may be limited, the impact of even small amounts of toxic chemicals may have an effect out of all proportion to their actual concentration. This may be particularly true in higher latitude lakes where long periods of ice cover occur. Contaminant burdens may be diluted by the abundant supplies of organic materials in eutrophic lakes where production rates are high, but in oligotrophic lakes where productivity is lower, contaminant burdens probably are concentrated in what little organic materials are available. The rate of burial probably is more rapid in eutrophic lakes, but the effects of burial probably would be offset to a large degree by the more efficient remobilization processes in highly organic sediments. In systems receiving large inputs of mineral sediments, remobilization processes probably would be less efficient and the removal of contaminants through burial more effective.

## 6. REQUIREMENTS FOR FUTURE DEVELOPMENT

Morphological deformities in chironomid larvae show considerable potential as a biological screening tool, but research in this field is still much in its infancy. Considerable work remains to be done before deformities can be used in an operational sense for detecting and assessing the nature, extent, and significance of toxic materials in aquatic ecosystems.

Preliminary research has shown that the response of chironomid larvae to contaminants probably is not simple, but involves a moving, changing spectrum of response in many different morphological structures. An early requirement, therefore, is for a conceptual framework in which the range and severity of these morphological responses can be indexed for inclusion in a "total morphological index." The frequency or incidence of deformities is relatively easy to derive, but provides only limited information about total response. On the other hand, the severity and multiplicity of deformities in individual larvae probably are more important than frequency for this purpose. I have initiated a scheme for assessing the severity of deformities in the antennae of *Chironomus* spp. (Warwick, 1985), but quantifying the range of variation in other morphological structures is going to take time to accomplish. I have urged the adoption of standard preparation and mounting techniques so valid comparisons may be made with the results of other workers. I would also encourage individual workers to provide detailed illustrations of each deformity or series of deformities with their descriptions so that the full extent to which individual structures can be deformed can be established more quickly.

The preliminary pilot studies initiated by Hamilton and Saether also need to be expanded upon to confirm the linkages between contaminants and morphological deformities. Simple "black-box" experiments like these could be used to screen a number of suspect chemicals for further study into cause and effect.

The theory that deformation in the larval head capsule shifts from the more sensitive structures like the antennae to harder structures like the mentum and mandibles as the concentration of contaminant increases should be examined under rigorously controlled conditions. It might be logical to start with DDE since some documentation already exists on its effects. The original single-dose experiments should also be expanded to include continuous concentration exposures and multiple generation exposures to answer such question as bioaccumulation effects and the importance of exposure time in culture testing. Such experiments may also determine the significance of massive deformities. Currently, these are thought to signify a response to higher concentrations of contaminants, but it may be that they can occur at any point on a concentration gradient. These experiments could also be modified to determine the effect on contaminant impact of such abiotic factors as temperature, oxygen regime, food type and availability, and substrate conditions.

Longer term laboratory research should be directed toward improving the measuring and detection capabilities of the technique for operational use. For example, morphological responses could be calibrated against a number of "reference" contaminants to provide a yardstick for assessing the severity of contamination in field situations. It is becoming increasingly unlikely in our complex society, with the long-range transport of atmospheric pollutants (LRTAP), increased use of pesticides and herbicides, and increasing industrialization, that the degradation of aquatic environments will be the result of a single contaminant. More likely, degradation will be due to a mixture of chemicals whose net effect may bear little or no resemblance to its component parts. A yardstick based on a set of "reference" contaminants would be very useful for determining the net effect of such mixtures without having to identify their component parts. And it would be even more useful if the yardstick was based on a number of responsive species, each characteristic of a different type of aquatic ecosystem.

On the other hand, examination of the factors that interact to increase, modify, decrease, or alleviate toxic effects may prove very useful in predicting the impact of new chemicals before they are introduced into the environment. An example might be to study morphological responses to interaction between metal concentrations and acid rain-induced pH changes.

The mechanisms or processes through which chemicals exert their effects should also be explored experimentally. A number of mechanisms have been proposed which vary from phenotypic (Hare and Carter, 1976) to enzymatic (Frank, 1981) and hormonal (Slàma and Williams, 1965). From the evidence available so far, it seems likely that larvae may respond to different chemicals in different ways. If this is the case, such specificity of response may be important in identifying chemicals or classes of chemicals in contaminated environments. Such information would add immeasureably to the diagnostic value of the tool. The effects of chemicals on chromosomal structure and histochemical makeup should also be investigated, particularly in view of their emerging use in taxonomic work.

Research can be usefully continued on field sites of differing degrees of contaminant stress to test the hypotheses put forward here and to expand the data base on the range of deformation possible in each of the different morphological structures. Such studies would best be accompanied by some measure of the stress being imposed on the communities, but often this is difficult, if not impossible, to determine without great cost. The rates of deformation should also be determined in uncontaminated sites to determine background rates of deformation in unstressed populations. Such studies should include sites of varying productivity to determine the effects of trophic state. Uncontaminated sites are becoming increasingly difficult to find, however, and we may have to resort to palaeotechniques to obtain reliable estimates of background rates.

Field studies can also be usefully employed to determine the effect of life history factors such as emergence patterns, generation differences, instar susceptibility, physiological state, and inter- and intraspecific differences. All field studies should include the more traditional measures of abundance, biomass, and species diversity in accompanying benthic communities. In the long run, the inclusion of this information in a comprehensive index would greatly improve the usefulness and accuracy of biological monitoring.

## 7. SUMMARY AND CONCLUSIONS

Morphological deformities in chironomid larvae show considerable potential as a biological screening tool for detecting and assessing the nature, extent, and significance of toxic chemicals in freshwater ecosystems. Morphological deformities occur in unstressed populations of chironomid larvae, but the incidence and severity of deformities increase significantly in the presence of environmental contaminants. Field observations show that deformities increase in populations exposed to industrial wastes and/or agricultural residues, but not in populations exposed to domestic wastes. Although the linkage between deformities and contaminants is based on circumstantial evidence, there is some experimental evidence to support the hypothesis of cause and effect. Reanalysis of preliminary dose–response experiments carried out by Hamilton and Saether (1971) show that aldrin, dieldrin, 2,4-D/2,4,5-T, and DDE can induce deformities in cultures of *Chironomus tentans* similar to those observed in larvae from contaminated ecosystems. The experiments with DDE indicate that the response of larvae to contaminants is not a simple process, but may involve graded responses that shift to different morphological structures as concentration increases. Warwick (1985) suggested that responses in individual structures may follow independent quantal dose–response relationships. Field data from sites ranging from comparatively uncontaminated to very highly contaminated show that response shifts from the highly sensitive antennae to harder structures like the mentum and mandibles as the severity of contamination increases. Some field evidence also suggests that certain chemicals or classes of chemicals may induce characteristic deformities. If this were to prove correct, such information would add immeasurably to the usefulness of deformities as a diagnostic tool.

## ACKNOWLEDGMENTS

I thank Dr. A. L. Hamilton for his continued support and dialogue. Dr. J. Klaverkamp also gave valuable suggestions and advice. Dr. E. Scherer and D. R. Kimmett critically reviewed the manuscript. C. A. Casey provide skilled technical assistance through all phases of the project from sample preparation to drafting the final manuscript. Dr. T. Köhn provided the samples from the Teltowkanal and Dr. J. Fitchko the samples from Port Hope Harbour. E. Van der Veen provided Dutch translations and V. Schranks assisted in sampling the northwestern Saskatchewan lakes. Illustrations were by N. A. Heaton, photographs by R. K. Beach. Funding was provided by under the Toxic Chemicals Management Program and by the Inland Water Directorate, Western and Northern Region.

## REFERENCES

Aagard, K. (1974). Morphological changes caused by nematode parasitism in Tanypodinae (Diptera, Chironomidae). *Nord. Entomol. Tidsskr.* **21,** 11–14.

Birkholz, D. A., Samoiloff, M. R., Warwick, W. F., Webster, G. R. B., and Witteman, J. (1980). *The Tobin Lake Project,* Project Proposal and Progress Report, November 1980. Off. Ind. Res., University of Manitoba, Winnipeg.

Brinkhurst, R. O., Hamilton, A. L., and Herrington, H. B. (1968). *Components of the Bottom Fauna of the St. Lawrence, Great Lakes* (No. PR33). Great Lakes Inst., University of Toronto, Toronto.

Brundin, L. (1949). Chironomiden und andere Bodentiere der südschwedischen Urgebirgsseen. *Rep. Inst. Freshwater Res. Drottningholm* **30,** 1–914.

Carter, C. W., and Suffet, I. H. (1982). Binding of DDT to dissolved humic materials. *Environ. Sci. Technol.* **16**(11), 735–740.

Chan, P. K., O'Hara, G. P., and Hayes, A. W. (1984). Principles and methods for acute and subchronic toxicity. In A. W. Hayes (Ed.), *Principles and Methods of Toxicology.* Raven Press, New York, Student Edition, pp. 1–51.

Coffman, W. P., and Ferrington, L. C. (1984). Chironomidae. In R. W. Merritt and K. W. Cummins (Eds.), *An Introduction to the Aquatic Insects of North America.* Kendall/Hunt, Dubuque, IO, 2nd ed., pp. 551–652.

Cook, W., and Veal, D. (1968). *A Preliminary Biological Survey of Port Hope Harbour.* Report to the Ministry of the Environment for Ontario.

Cranston, P. S., and Reiss, F. (1983). The larvae of Chironomidae (Diptera) of the Holarctic region—Key to subfamilies. *Entomol. scand., Suppl.* **19,** 11–15.

Crowther, R. A., and Luoma, M. E. (1984). Pattern recognition techniques to determine benthic invertebrate trophic and community responses to industrial input. *Verh. Int. Ver. Limnol.* **22**(4), 3981–3989.

Cushman, R. M. (1984). Chironomid deformities as indicators of pollution from a synthetic, coal-derived oil. *Freshwater Biol.* **14,** 179–182.

Cushman, R. M., and Goyert, J. C. (1984). Effects of a synthetic crude oil on pond benthic insects. *Environ. Pollut., Ser. A* **33,** 163–186.

Frank, C. (1981). Glycolytic capacity of chironomid larvae from polluted and unpolluted waters. *Verh. Int. Ver. Limnol.* **21,** 1627–1630.

Frank, C. (1983). Beeinflussung von Chironomidenlarven durch Umweltchemikalien und ihre Eignung als Belastungs- und Trophieindikatoren. *Verh. Dtsch. Zool. Ges.*, pp. 143–146.

Goldsmith, E., Allen, R., Allaby, M., Davoll, J., and Laurence, S. (1972). A blueprint for survival. *Ecologist* **2**(1), 1–43.

Golini, V. I., and Corliss, J. O. (1981). A note on the occurrence of the hymenostome ciliate *Tetrahymena* in chironomid larvae (Diptera: Chironomidae) from the Laurentian Great Lakes. *Trans. Am. Microsc. Soc.* **100**(1), 89–93.

Golini, V. I., and Sherry, J. P. (1979). *Chironomus plumosus* (Diptera: Chironomidae) from Lake Ontario parasitized by a mermithid nematode with subsequent colonization by a saprolegniaceous fungus. *Trans. Am. Microsc. Soc.* **98**(4), 572–576.

Hamilton, A. L., and Saether, O. A. (1971). The occurrence of characteristic deformities in the chironomid larvae of several Canadian lakes. *Can. Entomol.* **103**(3), 363–368.

Hare, L., and Carter, J. C. H. (1976). The distribution of *Chironomus (s.s.)? cucini* (*salinarius* group) larvae (Diptera: Chironomidae) in Parry Sound, Georgian Bay, with particular reference to structural deformities. *Can. J. Zool.* **54,** 2129–2134.

Hart, D. R., McKee, P. M., Burt, A. J. and Goffin, M. J. (1986). Benthic community and sediment quality assessment of Port Hope Harbour, Lake Ontario. *J. Great Lakes Res.* **12,** 206–220.

Karickhoff, S. W. (1981). Semi-empirical estimation of sorption of hydrophobic pollutants on natural sediments and soils. *Chemospere* **10**(8), 833–846.

Klink, A. (1984). Kopafwijkingen bij Chironomidae-larven onder invloed van chemische verontreiniging. *Meded. Hydrobiol. Adviesburo K. Wageningen Rapp.* **12,** 1–31.

Köhn, T., and Frank, W. (1980). Effect of thermal pollution on the chironomid fauna in an urban channel. In D. A. Murray (Ed.), *Chironomidae: Ecology, Systematics, Cytology and Physiology.* Pergamon, Oxford, pp. 187–194.

Krieger, K. A. (1984). Benthic macroinvertebrates as indicators of environmental degradation in the southern nearshore zone of the central basin of Lake Erie. *J. Great Lakes Res.* **10**(2), 197–209.

LeSage, L., and Harrison, A. D. (1980). The biology of *Cricotopus* (Chironomidae: Orthocladiinae) in an algal-enriched stream. Part III. Effects of parasitism. *Arch. Hydrobiol., Suppl.* **58**(1/2), 1–25.

Mika, N., and Mika, H. (1973). *Belleville: Friendly City.* Mika Publ. Ltd., Belleville, Ontario.

Naumann, E. (1917). Undersökningar öfver fytoplankton och under den pelagiska regionen försiggående yttje-och dybildningsr inom vissa sud-och mellansvenska erbergsvatten. *K. Sven. Vetenskapskad. Handl.* **56**(6), 1–615.

Naumann, E. (1929). Einige neue Gesichtspunkte zur Systematik der Gewässertypen. Mit besonderer Berucksichtigung der Seetypen. *Arch. Hydrobiol.* **20,** 191–198.

Oliver, D. R. (1971). Life history of the Chironomidae. *Annu. Rev. Entomol.* **16,** 211–230.

Oliver, D. R., McClymont, D., and Roussel, M. E. (1978). A key to some larvae of Chironomidae (Diptera) from the Mackenzie and Porcupine River watersheds. Tech. Rep. 791 Fish. Mar. Serv. (Can.), pp. 1–73.

Petersen, L. B-M., and Petersen, R. C. (1983). Anomalies in hydropsychid capture nets from polluted streams. *Freshwater Biol.* **13,** 185–191.

Pinder, L. C. V. (1986). Biology of freshwater Chironomidae. *Annu. Rev. Entomol.* **31,** 1–23.

Rempel, J. G. (1940). Intersexuality in Chironomidae induced by nematode parasitism. *J. Exp. Zool.* **84,** 261–289.

Saether, O. A. (1969). Some Nearctic Podonominae, Diamesinae, and Orthocladiinae (Diptera: Chironomidae). *Bull. Fish. Res. Board Can.* **170,** 1–154.

Saether, O. A. (1970). A survey of the bottom fauna of the Okanagan Valley, British Columbia. Tech. Rep. 196 Fish. Mar. Serv. (Can.) pp. 1–29.

Saether, O. A. (1975). Nearctic chironomids as indicators of lake typology. *Verh. Int. Ver. Limnol.* **19,** 3127–3133.

Saether, O. A. (1979). Chironomid communities as water quality indicators. *Holarctic Ecol.* **2,** 65–74.

Saether, O. A. (1980). Glossary of chironomid morphology terminology (Diptera: Chironomidae). *Entomol. scand., Suppl.* **14,** 1–51.

Saether, O. A., and Galloway, T. D. (1980). Sexual anomalies in Chironomini (Chironomidae: Diptera) from Lake Winnipeg, Manitoba, with observations on mermithid (Nematoda) parasites. *Acta Univ. Carol., Biol., 1978,* pp. 193–211.

Saether, O. A., and McLean, M. P. (1972). A survey of the bottom fauna in Wood, Kalamalka and Skaha lakes in the Okanagan Valley, British Columbia. Tech. Rep. 342, Fish. Res. Board Can. pp. 1–27.

Samoiloff, M. R., Bell, J., Birkholz, D. A., Webster, G. R. B., Arnott, E. A., Pulak, R., and Madrid, A. (1983). Combined bioassay-chemical fractionation scheme for the determination and ranking of toxic chemicals in sediments. *Environ. Sci. Technol.* **17**(6), 329–334.

Schlee, C. (1966). Präparation und Ermittlung von Messwerten an Chironomidae (Diptera). *Gewässer Abwässer* **41/42,** 169–193.

Slàma, K., and Williams, C. M. (1965). Juvenile hormone activity for the bug *Pyrrhocoris apterus. Proc. Natl. Acad. Sci. U.S.A.* **54,** 411–414.

Tennessen, K. J., and Gottfried, P. K. (1983). Variation in structure of ligula of Tanypodinae larvae (Diptera: Chironomidae). *Entomol. News* **94**(4), 109–116.

Thienemann, A. (1913). Der Zusammenhang zwischen dem Sauerstoffgehalt des Tiefenwassers und der Zusammensetzung det Tiefenfauna unserer Seen. *Int. Rev. gesamten Hydrobiol. Hydrogr.* **6,** 243–249.

Thienemann, A. (1922). Die beiden Chironomusarten der Tiefenfauna der norddeutschen Seen. *Arch. Hydrobiol.* **13,** 609–646.

Thienemann, A. (1931). Tropische Seen und Seetypenlehre. *Arch. Hydrobiol., Suppl.* **9,** 205–231.

Thienemann, A. (1954). Chironomus. Leben, Verbreitung und wirtschaftliche Bedeutung der Chironomiden. *Binnengewasser* **20,** 1–834.

Urk, G. van, Kerkum, F. C. M., and Wiersma, S. M. (1985). Bodemfauna in verontreinigde onderwaterbodems. *$H_2O$* **24,** 509–513.

Warwick, W. F. (1980a). Palaeolimnology of the Bay of Quinte, Lake Ontario: 2800 years of cultural influence. *Can. Bull. Fish. Aquat. Sci.* **206,** 1–117.

Warwick, W. F. (1980b). Pasqua Lake, southeastern Saskatchewan: A preliminary assessment of trophic status and contamination based on the Chironomidae (Diptera). In D. A. Murray (Ed.), *Chironomidae: Ecology, Systematics, Cytology, and Physiology.* Pergamon, Oxford, pp. 255–267.

Warwick, W. F. (1985). Morphological abnormalities in Chironomidae (Diptera) larvae as measures of toxic stress in freshwater ecosystems: Indexing antennal deformities in *Chironomus* Meigen. *Can. J. Fish. Aquat. Sci.* **42**(12), 1881–1914.

Warwick, W. F., and Casey, C. A. (1982). Sampling chironomid communities in lakes. Can. Dep. Environ., Inland Waters Dir., Natl. Water Res. Inst., Tech. Rep. W.N.R.-PR-82-02, pp. 1–43.

Warwick, W. F., Fitchko, J., McKee, P. M., Hart, D. R., and Burt, A. J. (1987). The incidence of deformities in *Chironomus* spp. from Port Hope Harbour, Lake Ontario. *J. Great Lakes Res.* **13,** 88–92.

Wentsel, R., McIntosh, A., and Atchison, G. (1977). Sublethal effects of heavy metal contaminated sediments on midge larvae (*Chironomus tentans*). *Hydrobiologia* **56,** 153–156.

Wiederholm, T. (1976). Chironomids as indicators of water quality in Swedish lakes. *Natl. Swed. Environ. Prot. Board,* NLU Information **10,** 1–17.

Wiederholm, T. (1980). Use of benthos in lake monitoring. *J. Water Pollut. Cont. Fed.* **52**(3), 537–547.

Wiederholm, T. (1984). Incidence of deformed chironomid larvae (Diptera: Chironomidae) in Swedish lakes. *Hydrobiologia* **109,** 243–249.

Wülker, W. (1964). Parasite-induced changes of internal and external sex characters in insects. *Exp. Parasitol.* **15,** 561–597.

Wülker, W. (1975). Parasite-induced castration and intersexuality in insects. In R. Reinboth (Ed.), *Intersexuality in the Animal Kingdom.* Springer-Verlag, Berlin, pp. 121–134.

# 15

# MESOCOSMS: A TOOL FOR STUDYING MICROBES AND EFFECTS AND BEHAVIOR OF TOXIC MATERIALS

***John E. Hobbie and Stuart G. Wakeham***

*Marine Biological Laboratory and Woods Hole Oceanographic Institution, Woods Hole, Massachusetts 02543*



## 1. INTRODUCTION

In recent years, aquatic ecology has moved from a purely observational science to one that also incorporates elements of experimental science. Banse (1982) points out that ecologists can never expect to attain the rigor of a pure laboratory experiment but that the use of mesocosms is an important step in that direction. Mesocosms are defined as enclosures of ecosystems larger than 1 $m^3$. These can be as simple as small plastic tubes hanging in the water column and as complicated as large, rigid tanks on land that contain both water and sediment and that are equipped for heating and stirring. Much of the information on types of mesocosms and on the results of various studies is summarized in a book by Grice and Reeve (1982).

There are two main advantages of research with mesocosms. First, a part of the aquatic environment can be contained in a mesocosm and sampled in replicate over long periods. This is extremely important for plankton research because the patchiness of plankton in nature often makes it very hard to gain information from a series of measurements over time. Containment is often the only way to sample the same parcel of water again and again. Second, mesocosms allow ecosystems to be manipulated as a tool for understanding how ecosystems work, how organisms respond to change or to toxic materials, and how toxic materials behave in the aquatic environment. For example, the top predator in an ecosystem may be removed to determine the effect on the community structure. Another possibility is to add nutrients or toxic chemicals to investigate processes, survival of individual species, or the fate of the nutrient or toxic chemical.

There are also disadvantages to mesocosm research. One problem, a biological one, is that their size is always smaller than the real world and so some parts of the food chain, usually the top predators, may be excluded. The general rules of mesocosm size and ecosystem studies have been laid out by Parsons (1982). He points out that a mesocosm as big as a small lake or as one end of a fjord must be used if effects of large fish are to be studied (Fig. 15.1). Obviously some aspects of size or species structure in an ecosystem will be changed when a top predator is missing and these effects may propagate downward to the phytoplankton. Another problem is the physical and chemical artifacts that occur when an ecosystem is enclosed. The enclosure walls, in effect, become a hard-bottom for colonization by microbes, algae, and animals. Also, the enclosure and temperature regime may

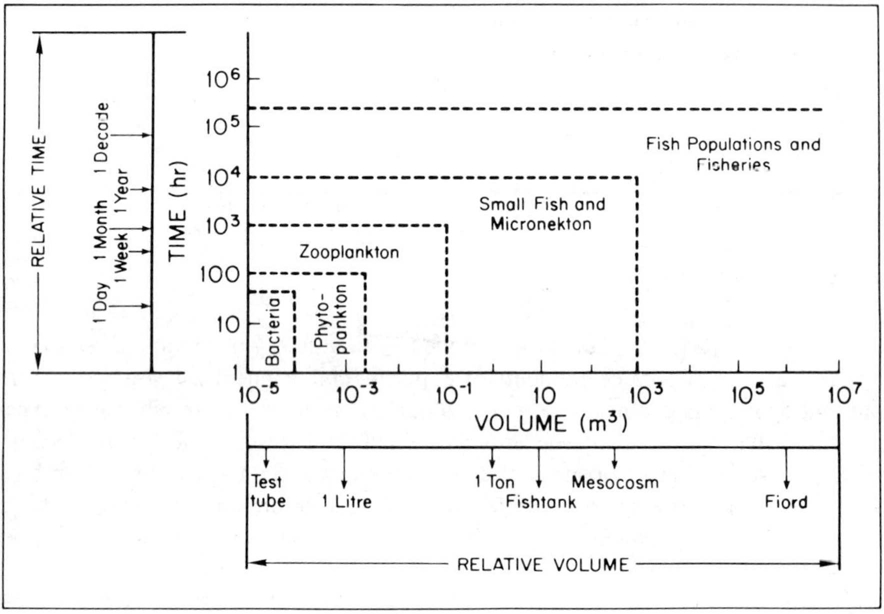


**Figure 15.1.** Mesocosm size related to organism life cycle. Redrawn from Parsons (1982).

reduce the currents, especially in the horizontal direction. In an undisturbed aquatic system, currents often transport oxygen to the sediments, alter the exchange of materials across the sediment–water interface, and influence the transport or deposition of particulate material.

Given the problems of mesocosm research, interpretations of data must be cautious. Mesocosms are models of nature and, like mathematical models, are useful but do not duplicate the real world. However, they are the best world we have since they are biotic systems and contain most of the organisms, interactions of organisms, and interactions between organisms and physical and chemical factors of their environment. Our knowledge of biotic interactions and processes and of their controls can be greatly improved by mesocosm research. Indeed, isn't the successful construction of a mathematical model of a mesocosm a necessary step to construction of a model of an entire natural ecosystem?

Mesocosms are probably most successful as imitators of the environment of the smallest, rather than the largest, organism of an ecosystem. In the case of microbes, for instance, there are at least $10^{12}$ organisms in a 1-$m^3$ mesocosm, enough for a variety of species and processes. In contrast, microcosms of a few tens or thousands of milliliters will not include enough of the predators who control the microbial abundance. It is also true that microbial processes usually appear to be unchanged in mesocosms and these processes are often the focus of study rather than the individual or the species.

## 2. MESOCOSMS IN MICROBIAL RESEARCH

Microbes, defined here as bacteria, have been an important part of every mesocosm project whether recognized or not. In one of the first experiments (McAllister et al., 1961; Antia et al., 1963) in a 120-$m^3$ plastic sphere in the ocean off Nanaimo, British Columbia, there was an immediate peak of oxygen consumption after filling and a later peak after a phytoplankton bloom. It is very likely these peaks of respiration were caused by peaks of bacteria. In a later experiment, Harrison et al. (1977) and Azam and Hodson (1977) demonstrated that most of the heterotrophic uptake of [$^3$H]glucose was due to organisms that passed through a 1-$\mu$m-diameter (pore size) filter. When better techniques became available for counting bacteria (e.g., fluorescent DNA dyes plus epifluorescence), it became evident that bacteria were abundant and showed cycles of abundance similar to classic predator–prey relationships (King et al., 1980). Respiration within marine microcosms was directly investigated by Williams (1982, 1984) in measurements of oxygen use by different size fractions of planktonic particles less than 250 $\mu$m. Particles greater than 10 $\mu$m were responsible for 30% of the respiration, while those less than 1 $\mu$m were responsible for 50%. These small forms are usually entirely bacteria (except for small cyanobacteria); thus, through these and other studies in mesocosms it has now been demonstrated that bacteria are important in carbon cycling in marine foodchains.

Bacteria are a part of a larger food web which produces and decomposes organic compounds. It is only recently that it has been possible to estimate how important bacteria are in the total flow of carbon within a system. Mesocosms have proven a useful tool here, and Davies (1984) found that some 63% of the carbon fixed in photosynthesis each day reached the bacteria through a solubilization pathway in a marine mesocosm. Recent research, which uses the incorporation of [$^3$H]thymidine to measure the production of bacteria (Hobbie and Cole, 1984), has demonstrated the tight coupling between algal production and bacterial production; in eight mesocosms with different rates of primary production, bacterial carbon cycling was always close to 30% of algal primary production.

Bacteria also take part in nutrient cycling in the plankton. Their exact role in phosphate uptake and release is still uncertain, but Harrison et al. (1977) found that most of the $^{32}PO_4^{3-}$ added to a mesocosm was rapidly taken up into particles less than 1 $\mu$m. The nitrogen cycle has been better studied in part because of the limiting role of nitrogen in phytoplankton production. Mesocosms have been useful for nitrogen cycle research because isotopes of $^{15}$N could be added to an entire system and their fate investigated (Harrison and Davies, 1977). This and other studies summarized by Davies (1984) showed that the ammonia produced by the microbial mineralization of seston and of primary amines was 45 to 81% of the total production. Thus, the picture of nutrient cycling that emerges from mesocosm studies is different from the classical view of the dominance of zooplankton feeding and their excretion of ammonia.

Mesocosm research in the last few years has helped to establish beyond any doubt that microflagellate predation controls the numbers of bacteria in the plankton. Much of the research has been carried out in the sea (Andersen and Fenchel, 1985); a recent study by J. E. Hobbie and J. V. K. Helfrich (unpublished) in a lake indicates that the same processes occur in fresh water. The data in Figure 15.2 were collected in a 60-m$^3$ mesocosm placed in an arctic lake and fertilized

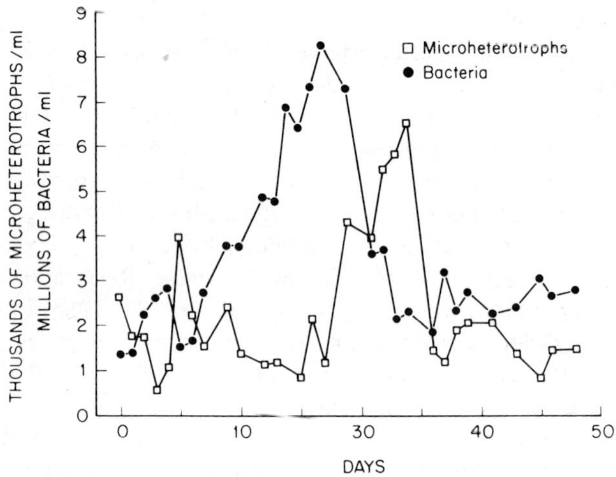


**Figure 15.2.** Bacterial (millions/mL) and protozoan (thousands/mL) numbers in a fertilized mesocosm of an arctic lake.

with N and P daily. Bacterial and microflagellate numbers, determined by direct counts, rose and fell in a classic predator–prey relationship. Although this is not proof, the relationship may reasonably be attributed to the predation by microflagellates.

There is one type of manipulation that has not been successfully carried out for microbes with mesocosms. That is to add or remove other species in the microbial food chain to test theories about the structure of the microbial community. Removal of organisms larger than bacteria has been carried out by filtration in microcosms and the result is always rapid growth and a rather dramatic increase in size of the bacteria (e.g., Ferguson et al., 1984). Given the results, a mesocosm experiment that removed only protozoans would be predictable and not worthwhile. However, it would be interesting to examine the bacteria in a mesocosm where the predators were manipulated but at several trophic levels above the bacteria. For example, does the size structure of the bacterial community change when *Daphnia* are abundant and can the change due to a reduction in predation be separated from a change due to increased levels of substrate from the excretion of the *Daphnia?*

## 3. MESOCOSMS IN TOXIC CHEMICAL RESEARCH

There have been only a few studies that combine mesocosms, microbes, and toxic materials. Microcosm research has been more popular, yet there are several reasons why microcosms may not be adequate models for microbial response to toxic materials. First, the physiological response of the microbes themselves (adaptation, growth effects) may be realistic but the overall response of the community may be unrealistic because of indirect effects on microbes. For example, oil added to small ponds appeared to cause a change in the dominant alga from a rhodomonad to a chrysophyte species (Barsdate et al., 1980). However, later tests revealed that a large zooplankton, *Daphnia middendorffiana,* was killed by the oil and its absence allowed the growth of the chrysophyte. It is possible that the predator food chain of bacteria could be similarly disrupted in a microcosm that excluded certain higher predators. Second, toxic materials may be modified, stored, or concentrated in large ecosystems but not in microcosms. For example, it is known that certain hydrocarbons can be preserved for years in sediments where they were buried by the activities (bioturbation) of benthic organisms (e.g., Hinga et al., 1980; Wade and Quinn, 1980; Gearing et al., 1980). An experiment in a microcosm would show the direct effect of the hydrocarbon on planktonic animals, but miss the potential long-term effect.

There is a universal response of a microbial community to the addition of toxic materials to ecosystems: first a decrease in activity and then a recovery. For example, in seawater mesocosms in British Columbia the addition of as little as 10 $\mu$g Cu/L (Vaccaro et al., 1977) and 5 $\mu$g Hg/L (Azam et al., 1977) reduced the uptake of labeled organic compounds by bacteria, but only for a short time. The bacteria may have been able to adapt as individuals or a new population may have grown up that was resistant to the metals. It is also well known (Wood, 1974) that bacteria can transform some heavy metals into nontoxic forms by such mechanisms as methylation and reduction to volatile forms.

Some members of a bacterial community will quickly become resistant to toxic materials. As Westman (1985, p. 364) stated, "The result of this differential sensitivity will be the usurpation of relinquished resources by the more pollution-tolerant species and, as a result, increased dominance by the tolerant species." A good example of this comes from a study of a microbial community from Lake Michigan by Szczepanik-van Leeuwen and Penrose (1983). They found that the metabolic functions of the microbes were initially inhibited by 81 $\mu$g/L of 2,4-dichlorophenol (DCP) but recovered after 6 days. At first, less than 5% of the organisms were able to grow in the presence of DCP; after 23 days some 80% were resistant.

The effects of added oil and its various derivatives was an important part of the two large mesocosm studies in North America, the MERL (Marine Ecosystems Research Laboratory) and the CEPEX (Controlled Ecosystem Pollution Experiment) studies. In both, there was a lag of one to several weeks before a population of microbes developed that could break down hydrocarbons (Santschi, 1982). Some compounds were degraded very slowly yet were removed from the water by other processes such as volatilization and sedimentation.

A MERL study of benz[$\alpha$]anthracene (BA) and pentachlorophenol illustrates the advantages of the mass-balance approach to studies of the fate of toxic materials that are possible in mesocosms (Lee et al., 1978, 1982). When BA was added to a mesocosm (1.8 m diameter, 5.5 m deep, 13 $m^3$), its concentration (Fig. 15.3) decreased rapidly with a half-life of 24 hours (BA plus its metabolites had a half-life of 50 hours). The radioactivity in $CO_2$ reached a maximum by day 12. The interpretation of these (and other) experiments was the following: (1) BA concentration in the water column first decreased because of sorption onto particles and sedimentation to the bottom (62% of the radioactivity); (2) photochemical degradation occurred in the water column and microbes broke down the products; (3) BA was mixed into the sediments by benthic animals where it was broken down slowly by microbes over the next 2 years.

Pentachlorophenol degradation followed quite a different pathway in a MERL tank (Lee et al., 1978, 1982) (Fig. 15.4). This compound is less susceptible to sorption than BA and so remained in the water column with a half-life of 22 days. A great deal of photooxidation occurred during this period and eventually the products were degraded by microbes to $CO_2$. Less than 1% of the radioactivity entered the sediment. Thus, both types of compounds were eventually degraded by microbes, one rapidly and one slowly. However, the initial removal from the water column was very different (sorption–sedimentation vs. photodegradation–biotic breakdown).

Mesocosms also permit the study of volatile organic compounds as exemplified by toluene (Wakeham et al., 1985) and octadecane/decane (Wakeham et al., 1986).

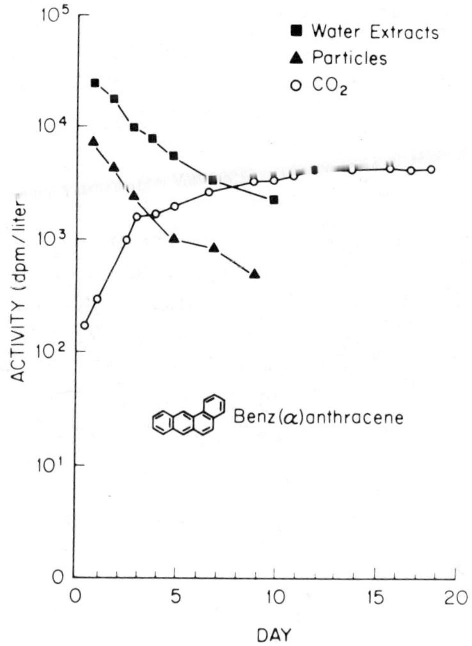


**Figure 15.3.** Radioactivity in MERL tanks in dissolved (BA), particles (DMBA), and $CO_2$ forms after addition of [$^{14}$C]benz[α]anthracene (BA). Redrawn from Lee et al. (1982).

The fate of these compounds was studied by adding toluene (2 μg/L), radiolabeled toluene ([$^{14}$C]toluene), and Freon 12 (difluorodichloromethane) to a 13-$m^3$ MERL tank. Similarly, 0.5 μg/L of cold decane or octadecane plus the $^{14}$C-labeled compounds and Freon were added to other tanks. The Freon acted as a tracer of the loss of these volatile compounds to the atmosphere.

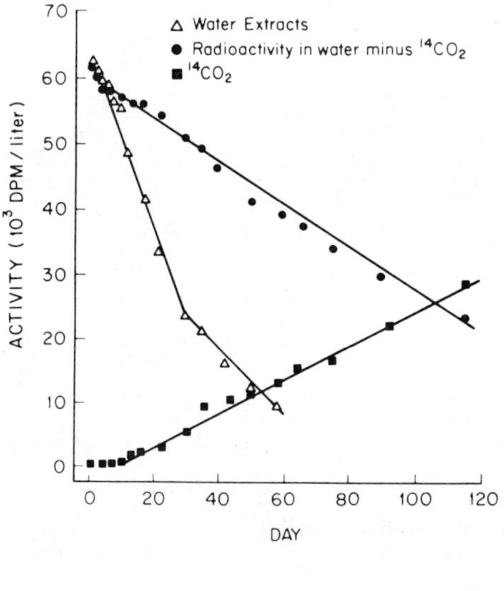


**Figure 15.4.** Radioactivity in MERL tanks in dissolved, particulate, and $CO_2$ forms after addition of [$^{14}$C]pentachlorophenol. Redrawn from Lee et al. (1982).

Toluene is a good example of a compound that does not sorb onto particles but remains dissolved (Wakeham et al., 1985). In the winter experiment (Table 15.1), the volatile/dissolved toluene slowly decreased (to 32% of total activity by day 47) while the amount that had been oxidized to $CO_2$ increased to 48% (of activity in the tank on day 47). Decomposition products (nonvolatile) reached 10% of the total. However, there was volatilization from the surface as seen by the loss of 80% of the [$^{14}$C]toluene initially introduced into the tank. In contrast, during summer the $^{14}$C remained in the tank and most ended up as $H^{14}CO_3^-$. The explanation for these differences is given in Figure 15.5 where the two experiments are compared on day 21. During the winter, biotic activity was slow (little $CO_2$ formed) and there was enough time for most of the toluene to volatilize and enter the atmosphere. During the summer, there was a sharp peak of biotic degradation shortly after day 5 (seen as a sharp peak of bacterial growth in Fig. 15.6) and most of the toluene was transformed to $CO_2$, which did not readily enter the atmosphere. Be-

**Table 15.1 Microcuries of $^{14}$C in Four Fractions and As a Total Count of Activity in the Water in Mesocosms[a]**

| | Day 5 | | Day 13 | | Day 47 | |
|---|---|---|---|---|---|---|
| Form | μCi | % | μCi | % | μCi | % |
| | | *Winter* | | | | |
| Volatile | 210 | (86)[b] | 150 | 80 | 10 | 32 |
| Nonvolatile | 9 | (4)[b] | 3 | 2 | 3 | 10 |
| Particulate | 1 | (0.4)[b] | 1 | 0.7 | 3 | 10 |
| $^{14}CO_2$ | (25)[c] | (10)[b] | 35 | 17 | 15 | 48 |
| Sum | (245)[b] | — | 190 | — | 31 | — |
| Total in water | 245 | — | 170 | — | 52 | — |

*Summer*

| | Day 2 | | Day 4 | | Day 10 | | Day 18 | |
|---|---|---|---|---|---|---|---|---|
| Form | μCi | % | μCi | % | μCi | % | μCi | % |
| Volatile | 280 | 95 | 155 | 70 | 1 | 0.4 | 1 | 0.5 |
| Nonvolatile | 2 | 0.7 | 7 | 3 | 10 | 4 | 9 | 4 |
| Particulate | 2 | 0.7 | 8 | 4 | 30 | 12 | 10 | 5 |
| $^{14}CO_2$ | 12 | 4 | 50 | 23 | 210 | 84 | 200 | 91 |
| Sum | 296 | — | 220 | — | 251 | — | 220 | — |
| Total in water | 330 | — | 250 | — | 240 | — | 230 | — |

[a]The budgets are for individual days. The isotope was added as 250 μCi [methyl-$^{14}$C]toluene in winter and 350 μCi in summer.
[b]Values calculated assuming 25μCi $^{14}CO_2$.
[c]$^{14}CO_2$ data are not available for the first 13 days of this experiment, but a value of 25μCi is possible.
*Source*. Wakeham et al. (1985), with permission.

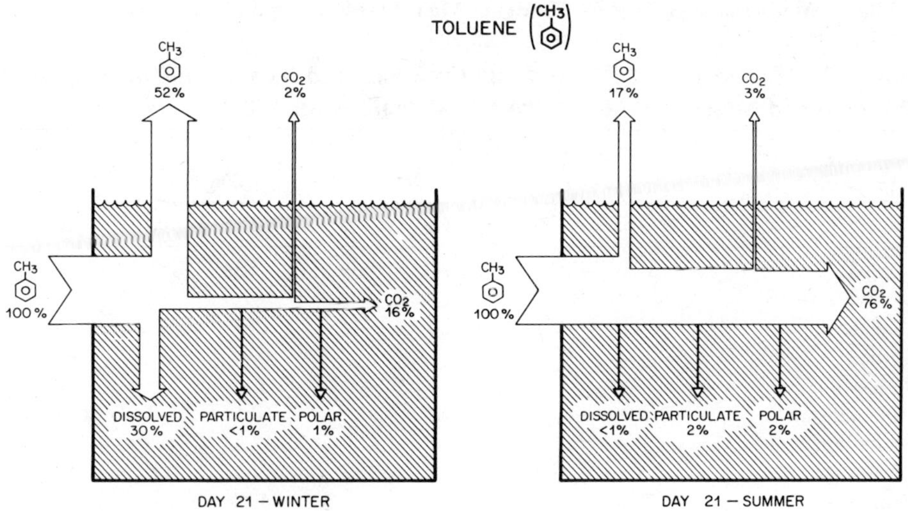


**Figure 15.5.** Mass balances in MERL tanks 21 days after addition of [$^{14}$C]toluene. The "polar" category is decomposition products. From Wakeham et al. (1985), with permission.

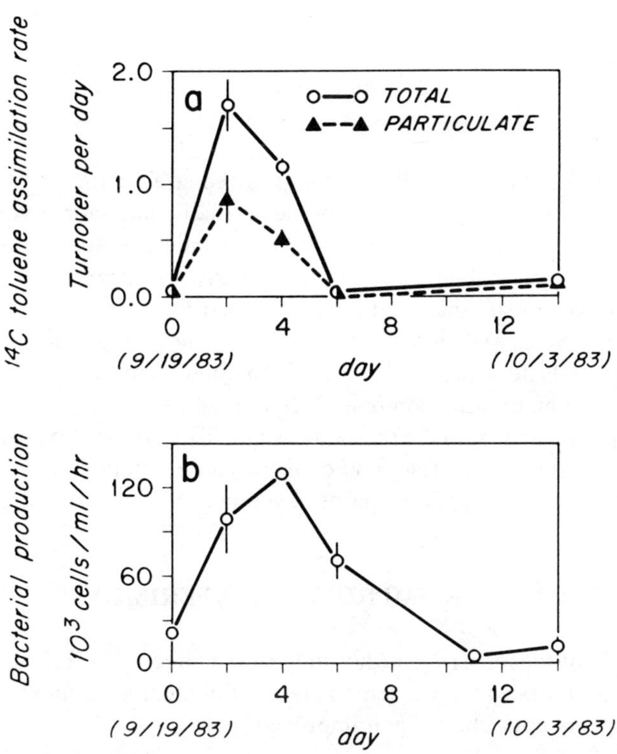


**Figure 15.6.** (a) Rates of particulate uptake and total (particulate + $CO_2$) turnover of [$^{14}$C]toluene in a MERL tank. (b) Bacterial production in the same tank measured by [$^{3}$H]thymidine incorporation. From Wakeham et al. (1985), with permission.

**Table 15.2 Microcuries of $^{14}C^a$ in Four Fractions and As a Total Count of Activity in the Water in Mesocosms. (data from Wakeham et al. 1986).**

| | Day 0 (2 hr) | | Day 3 | | Day 8 | | Day 17 | |
|---|---|---|---|---|---|---|---|---|
| | μCi | % | μCi | % | μCi | % | μCi | % |
| Dissolved (Volatile) | 374 | 53 | 192 | 39 | 60 | 14 | 10 | 2 |
| Particulate (Sorbed + biomass) | 300 | 43 | 220 | 45 | 100 | 23 | 24 | 5 |
| Intermediate metabolites | 10 | 1 | 45 | 9 | 15 | 3 | 22 | 5 |
| $CO_2$ | 16 | 3 | 32 | 7 | 260 | 60 | 430 | 88 |
| Sum | 700 | — | 490 | — | 435 | — | 486 | — |
| Total in water | 670 | — | 650 | — | 450 | — | 484 | — |

[a]The budgets are for individual days. The isotope was added as 670 μCi octadecane.

*Source*. Wakeham et al. (1986).

cause of the higher temperature in summer compared to the winter experiment (18–20°C vs. 2–8°C), the toluene was oxidized so quickly that only 17% entered the atmosphere.

Octadecane behavior in a MERL tank followed still another pattern (Wakeham et al., 1986). Octadecane is considerably less volatile and more hydrophobic than toluene. In this experiment (Table 15.2), the octadecane sorbed to particles extremely rapidly (43% in 2 hours) and then quickly was degraded to $CO_2$. By day 17, 65% of the original spike had been degraded by the biota, while only about 15% had been exchanged into the atmosphere. These experiments with volatile organic compounds demonstrate how rapidly a population of microbes can respond to the addition of an organic compound. Initial lag periods following introduction of the tracer compounds into the mesocosms quickly gave way to periods of intense biodegradation. Substantial fractions of the volatile compounds introduced were degraded within a few days, especially in summer.

## 4. SUGGESTIONS FOR IMPROVING EXPERIMENTS

The usefulness of mesocosms for the study of the effect of pollutants on microbes, and vice versa, has been amply demonstrated. But there are a number of ways that these experiments could have been improved.

1. The microbial events that occur when a mesocosm is filled should be separated from those that occur when a pollutant is added. There should be a period

of at least a week after a mesocosm is set up before an experiment is begun, because it is possible that some larger organisms are killed and that the microbes respond to the release of organic matter into the water. In addition, there may be a release of (uncharacterized) material (possibly toxic) from sediments during filling of the tank. When the pollutant has transitory effects, as in the case of toluene or octadecane, then the experimental addition should be made again in the same tank to confirm the results through quasi replication.

2. A variety of techniques should be tested for measuring the effect of various materials on microbes. There are millions of living bacteria per milliliter of water yet only a few hundred or thousand may be characterized by growth on agar plates. Furthermore, only a small proportion of the total bacterial biomass may be responsible for its activity with respect to degradation of a pollutant chemical. Before we can understand the possible effects of toxics on bacteria community structure, the types of bacteria involved must be characterized. One possible method for this is immunofluorescence techniques (Bohlool and Schmidt, 1980). Another problem is to find ways to distinguish what portion of the microbial community is involved in a response to added pollutants. This might be done by autoradiography (Hoppe, 1976).

3. The possibility of indirect effects of added pollutants should be considered and investigated. Thus, the integrity of the aquatic ecosystem, such as survival and health of algae, the source of much of the substrate of microbial growth, or of microflagellates, the main microbial predator, must be examined during experiments. Otherwise, the effects of a pollutant on the producers or predators might be confused with direct effects on the microbial community.

## ACKNOWLEDGMENTS

The collection of some of the data and the writing of this paper have been supported by NSF Grants OCE-8415687 and DPP-8320544 and by Grant NA81RAD00015 from the Office of Marine Pollutant Assessment, National Oceanic and Atmospheric Administration.

## REFERENCES

Andersen, P., and Fenchel, T. (1985). Bacterivory by microheterotrophic flagellates in seawater samples. *Limnol. Oceanogr.* **30,** 198–202.

Antia, N. J., McAllister, C. D., Parsons, T. R., Stephens, K., and Strickland, J. D. H. (1963). Further measurements of primary production using a large volume plastic sphere. *Limnol. Oceanogr.* **8,** 166–183.

Azam, F., and Hodson, R. E. (1977). Size distribution and activity of marine microheterotrophs. *Limnol. Oceanogr.* **22,** 492–501.

Azam, F., Vaccaro, R. F., Gillespie, P. A., Mousalli, E. I., and Hodson, R. E. (1977). Controlled ecosystems pollution experiment: Effect of mercury on enclosed water columns. II. Marine bacterioplankton. *Mar. Sci. Commun.* **3,** 313–329.

Banse, K. (1982). Experimental marine ecosystem enclosures in a historical perspective. In G. D. Grice and M. R. Reeve (Eds.), *Marine Mesocosms: Biological and Chemical Research in Experimental Ecosystems.* Springer-Verlag, New York, pp. 11–24.

Barsdate, B. J., Miller, M. C., Alexander, V., Vestal, J. R., and Hobbie, J. E. (1980). Oil spill effects. In J. E. Hobbie (Ed.), *Limnology of Tundra Ponds: Barrow, Alaska.* Dowden, Hutchinson & Ross, Stroudsburg, PA, pp. 388–406.

Bohlool, B. B., and Schmidt, E. L. (1980). The Immunofluorescence approach to microbial ecology. In M. Alexander (Ed.), *Advances in Microbial Ecology*, Vol. 4. Plenum, New York, pp. 203–241.

Davies, J. M. (1984). The use of large enclosures in marine microbial research. In J. E. Hobbie and P. J. leB. Williams (Eds.), *Heterotrophic Activity in the Sea.* Plenum, New York, pp. 465–479.

Ferguson, R. L., Buckley, E. N., and Palumbo, A. V. (1984). Response of marine bacterioplankton to differential filtration and confinement. *Appl. Environ. Microbiol.* **47,** 49–55.

Gearing, P. J., Gearing, J. N., Pruell, R. J., Wade, T. L., and Quinn, J. G. (1980). Partitioning of No. 2 fuel oil in controlled estuarine ecosystems. Sediments and suspended particulate matter. *Environ. Sci. Technol.* **14,** 1129–1136.

Grice, G. D., and Reeve, M. R. (Eds.) (1982). *Marine Mesocosms: Biological and Chemical Research in Experimental Ecosystems.* Springer-Verlag, New York.

Harrison, W. G., and Davies, J. M. (1977). Nitrogen cycling in a marine planktonic food chain: Nitrogen fluxes through the principal components and the effects of adding copper. *Mar. Biol. (Berlin)* **43,** 299–306.

Harrison, W. G., Azam, F., Renger, E. H., and Eppley, R. W. (1977). Some experiments on phosphate assimilation by coastal marine plankton. *Mar. Biol. (Berlin)* **40,** 9–18.

Hinga, K. R., Pilson, M. E. Q., Lee, R. G., Farrington, J. W., Tiezsen, K., and Davis, A. C. (1980). Biogeochemistry of benzanthracene in an enclosed marine ecosystem. *Environ. Sci. Technol.* **14,** 1136–1143.

Hobbie, J. E., and Cole, J. J. (1984). Response of a detrital food web to eutrophication. *Bull. Mar. Res.* **35,** 357–363.

Hoppe, H.-G. (1976). Determination and properties of actively metabolizing heterotrophic bacteria in the sea, investigated by means of micro-autoradiography. *Mar. Biol. (Berlin)* **36,** 291–302.

King, K. R., Hollibaugh, J. T., and Azam, F. (1980). Predator-prey interactions between the larvacean *Oikopleura dioica* and bacterioplankton in enclosed water columns. *Mar. Biol. (Berlin)* **56,** 49–57.

Lee, R. F., Gardner, W. S., Anderson, J. W., Blaylock, J. W., and Barwell-Clarke, J. (1978). Fate of polycyclic aromatic hydrocarbons in controlled ecosystem enclosures. *Environ. Sci. Technol.* **12,** 832–838.

Lee, R. F., Hinga, K., and Almquist, G. (1982). Fate of radiolabeled polycyclic aromatic hydrocarbons and pentachlorophenol in enclosed marine ecosystems. In G. D. Grice and M. R. Reeve (Eds.), *Marine Mesocosms: Biological and Chemical Research in Experimental Ecosystems.* Springer-Verlag, New York, pp. 123–135.

McAllister, C. D., Parsons, T. R., Stephens, K., and Strickland, J. D. H. (1961). Measurements of primary production in coastal sea water using a large volume plastic sphere. *Limnol. Oceanogr.* **6,** 237–258.

Parsons, T. R. (1982). The future of controlled ecosystem enclosure experiments. In G. D. Grice and M. R. Reeve (Eds.), *Marine Mesocosms: Biological and Chemical Research in Experimental Ecosystems.* Springer-Verlag, New York, pp. 411–418.

Santschi, P. H. (1982). Application of enclosures to the study of ocean chemistry. In G. D. Grice and M. R. Reeve (Eds.), *Marine Mesocosms: Biological and Chemical Research in Experimental Ecosystems.* Springer-Verlag, New York, pp. 63–80.

Szczepanik-van Leeuwen, P. A., and Penrose, W. R. (1983). Functional properties of a microcosm of the freshwater benthic zone and the effect of 2,4-dichlorophenol. *Arch. Environ. Contam. Toxicol.* **12,** 427–437.

Vaccaro, R. F., Azam, F., and Hodson, R. E. (1977). Response of natural marine bacterial populations to copper: Controlled ecosystem pollution experiment. *Bull. Mar. Sci.* **27,** 17–22.

Wade, T. L., and Quinn, J. G. (1980). Incorporation, distribution and fate of saturated petroleum hydrocarbons in sediments from a controlled marine ecosystem. *Mar. Environ. Res.* **3,** 15–33.

Wakeham, S. G., Canuel, E. A., Doering, P. H., Hobbie, J. E., Helfrich, J. V. K., and Lough, G. G. (1985). The biogeochemistry of toluene in coastal seawater: Radiotracer experiments in controlled ecosystems. *Biogeochemistry* **1,** 307–328.

Wakeham, S. G., Canuel, E. A., and Doering, P. H. (1986). Behavior of aliphatic hydrocarbons in coastal seawater: Mesocosm experiments with [$^{14}$C]Octadecane and [$^{14}$C]Decane. *Environ. Sci. Technol.* **20**(6), 574–580.

Westman, W. E. (1985). *Ecology, Impact Assessment, and Environmental Planning.* Wiley, New York.

Williams, P. J. leB. (1982). Microbial contribution to overall plankton community respiration—studies in enclosures. In G. D. Grice and M. R. Reeve (Eds.), *Marine Mesocosms. Biological and Chemical Research in Experimental Ecosystems.* Springer-Verlag, New York, pp. 305–321.

Williams, P. J. leB. (1984). A review of measurements of respiration rates of marine plankton populations. In J. E. Hobbie and P. J. leB. Williams (Eds.), *Heterotrophic Activity in the Sea.* Plenum, New York, pp. 357–389.

Wood, J. M. (1974). The biological cycles for toxic elements in the environment. *Science* **183,** 1049–1052.

# 16

# EFFECTS OF NUTRIENTS AND COPPER ON COPEPOD POPULATION DYNAMICS: A MESOCOSM STUDY

**_Barbara K. Sullivan and P. J. Ritacco_**

*Marine Ecosystems Research Laboratory, University of Rhode Island, Narragansett, Rhode Island 02882-1197*

## 1. INTRODUCTION

Studies of toxic substances in enclosed experimental ecosystems are appealing because they permit observation of behavior and effects in systems that may more closely resemble nature than do laboratory experiments. There is a growing history of use of enclosed experimental ecosystems for study of fates and effects of pollutants in aquatic ecosystems that has recently been reviewed by Cairns and Buikema (1984), Grice (1984), and Santschi et al. (1984). In general, the use of mesocosms has permitted more accurate measurement of pollutant fate in complex systems; as a result, pollutant effects can be more precisely related to exposure concentration in the water column or sediments.

Because of the very complex nature of model ecosystems there are processes that as yet are inadequately understood. Grice (1984) emphasizes that effects at the population level (e.g., reduced numbers or changes in community structure) have sometimes been observed at contaminant levels below those that cause mortality in the laboratory. This was also seen in a freshwater mesocosm study (Gachter, 1979). Because of results like these it can be important to determine the reasons underlying the difference between contaminant behavior in single- and multispecies systems. It should be recognized, however, that because of the inherent complexity of large experimental systems, mechanisms controlling the population dynamics of various species are difficult to determine. There are many examples of investigations of toxics in enclosures in which it has been very difficult to distinguish between direct toxicity to the organisms (the effect measured in laboratory bioassays) and secondary effects caused by complex interactions among the biota (Gamble et al., 1977; Gibson and Grice, 1977; Reeve et al., 1977a; Gachter, 1979; Urech, 1979; Marshall and Mellinger, 1980; Kuiper, 1982). Nevertheless, the use of an appropriate combination of laboratory tests run concurrently with mesocosm studies does permit some analysis of factors leading to the observed population response in the enclosed ecosystems. It then becomes possible to infer the relationship between contaminant behavior in the laboratory and in the mesocosm.

Experimental manipulation of the mesocosms to produce systems with properties that mimic important variables in natural systems (such as nutrient input) is another important application of mesocosms to studies of toxic compounds. A range of natural environmental variables that may effect toxicity and behavior of toxic compounds may then be examined. In the Marine Ecosystems Research Laboratory (MERL) mesocosm study we describe here we have used an experimental design that we hoped would permit both (1) examination of the mechanisms of toxicity of the introduced compound (copper), and (2) manipulation of the mesocosms to create a range of trophic states (nutrient enrichment) across which to observe the behavior and toxicity of copper.

Results of other mesocosm studies that have investigated the effects of copper on marine ecosystems, the Loch Ewe experiments (Gamble et al., 1977) and Controlled Ecosystem Pollution Experiments or CEPEX (Menzel, 1977), have shown a number of common responses by phytoplankton but have been more difficult to interpret with respect to zooplankton responses. Copper (10 μg/L) was added to two Loch Ewe bags, one of which was nutrient enriched. Concentrations of chlorophyll *a* increased following the addition, and the phytoplankton community structure was altered so that small flagellated species dominated over diatoms. This effect persisted longest in the unenriched bag. Zooplankton numbers decreased in all treatments, with and without copper. Determination of effects of copper on zooplankton were complicated by the fact that ctenophores were exerting strong predation pressure on the copepods and because no studies to measure copepod mortality rates independent of predation were conducted. Laboratory studies did suggest that copper concentrations were high enough to have caused sublethal reduction of copepod feeding and reproductive rates.

Copper additions of 5, 10, and 50 μg/L were studied in CEPEX bags at control nutrient levels only (Gibson and Grice, 1977; Reeve et al., 1977a). As in the Loch Ewe bags, blooms of small phytoplankton species were produced following copper additions. Interpretation of the zooplankton response was again complicated by the large numbers of predators affecting copepod populations. However, decreases in copepod numbers were greatest in mesocosms dosed with 50 μg/L copper. Reduced feeding and egg production by copepods withdrawn from the mesocosms and fed both naturally available and standard food indicated that there were sublethal effects at 10–50 μg/L (Reeve et al., 1977a). Thus, a direct though sublethal effect of the copper at all treatment levels was documented.

Results of these studies would have avoided the ambiguity concerning the degree to which copper affected mortality rates of the copepods if there had been direct measurement of mortality. Because sublethal effects were shown to occur, however, it appears that copper should have an effect on copepod population dynamics via reduced reproductive potential even at concentrations of 5–10 μg/L.

There is also evidence that effects of copper could vary according to the trophic state or phytoplankton productivity of the system. For *Acartia tonsa,* the dominant copepod species in MERL experiments, 72-hour $LC_{50}$ values had a very wide range, from 9 to 78 μg/L, for adults collected at different times from Narragansett Bay (Sosnowski and Gentile, 1978; Sosnowski et al., 1979). Reeve et al. (1976) also reported a great range in the 96-hour $LC_{50}$ values of field-collected *Acartia.* Sosnowski et al. (1979) attributed this variability to changes in the phytoplankton food resources available to the copepods at different collection times. In the MERL mesocosm experiment described here food availability has been included as a controlled variable. In addition, the chemistry of copper may be altered in systems with high dissolved organic concentrations such that it is less toxic to phytoplankton and other biota (Sunda and Lewis, 1978).

The MERL experiment provides several improvements over previous mesocosm experiments. There were 9 tanks available for the experiment, enough to permit a wider range of nutrient conditions to be simulated, and a benthos was included, so that contaminant behavior in nearshore or estuarine environments could be modeled. The response to nutrient addition was followed at 7 different levels of nutrient enrichment in which chlorophyll levels were positively correlated with nutrient treatment. The response to copper additions was followed in 4 of

**Table 16.1 Experimental Design and Data Collection for Nutrient and Copper Studies in MERL Mesocosms**

| Nutrient Treatments | Copper | Phytoplankton + Zooplankton Analysis | Survival of Nauplii | Copepod Reproduction |
|---|---|---|---|---|
| Control | 0 | + | + | + |
| Control | 0 | + | 0 | 0 |
| Control | + | + | + | + |
| 1× | 0 | 0 | 0 | 0 |
| 2× | + | + | 0 | 0 |
| 4× | 0 | + | 0 | 0 |
| 8× | 0 | 0 | 0 | 0 |
| 16× | + | + | + | + |
| 32× | + | + | 0 | 0 |

these nutrient treatments (Table 16.1). To help interpret population data we have measured mortality rates of juvenile copepods in the mesocosm by in situ incubations, and followed in situ reproductive rates for 1 month following the additions. Our results indicate that copper had its impact via direct toxicity to sensitive larval stages at approximately those concentrations predicted from single-species bioassays. These effects were clearly correlated with population numbers immediately following the copper addition. However, following the return to normal survival rates, mechanisms controlling subsequent changes in population size during the following weeks were difficult to explain. Population recovery was fastest in moderately enriched tanks but much slower in highly eutrophic mesocosms. Our results may indicate that populations in highly eutrophic systems have a lower chance of recovery after being exposed to toxic substances.

## 2. DESIGN AND SAMPLING OF THE MERL MESOCOSMS

### 2.1. Mesocosm Description

The MERL ecosystems are scaled to the natural system, Narragansett Bay, Rhode Island, in terms of sunlight, temperature, tidal mixing, and flushing (3.7% volume exchange of unfiltered bay water per day). The 13-$m^3$ tanks (1.83 m diam, 5.49 m height) contain both sediment and water column communities, and while they have a slow exchange of seawater, any ecosystem changes resulting from controlled perturbations of the tanks can be observed and quantified (e.g., Oviatt et al., 1982). It has been demonstrated that in many respects the MERL systems mimic natural systems, maintaining similar concentrations of nutrients, phytoplankton, and zooplankton biomass throughout the annual cycle (Pilson et al., 1979, 1980).

A nutrient experiment was begun in June 1981 and run continuously until October 1983 to determine ecosystem response to a wide range of nutrient loading rates. Copper was added to selected mesocosms in the final phase of the 28-month experiment, on August 23, 1983, because it necessitated an end to the nutrient study. Nutrients were added daily to six tanks to provide input rates that were multiples (1×, 2×, 4×, 8×, 16×, 32×) of the average annual nutrient loading to Narragansett Bay calculated on a per square meter basis. These treatments were unreplicated. No nutrients were added to three replicate control tanks. Solutions of $NH_4Cl$, $KH_2PO_4$, and $Na_2SiO_37H_2O$ were added in a molar ratio of 12.8N:1.0P:0.9Si simulating ratios in sewage (Nixon et al. 1984). Additions to each treatment were multiples of the 1× level which was 7.57, 0.59, 0.54 mmol/tank/day of N, P, and Si. For the copper experiment, a single dose of 40 $\mu$g/L $CuSO_4$ was added to one of the three control tanks and to the 2×, 16×, and 32× treatments. Treatments and measurements presented in this paper are summarized in Table 16.1. Concentrations of particulate and dissolved copper were monitored for 34 days following the addition (Hunt and Donaghay, 1984).

## 2.2. Population Analyses

Measurements on phytoplankton included weekly samples for chlorophyll *a* (Lorenzen, 1966), cell counts, and species identification from live and preserved samples. Primary production was estimated weekly from consecutive dawn-dusk-dawn oxygen measurements (Odum and Hoskins, 1958). Production by the carbon-14 $CO_2$ method was also determined but the oxygen method was considered more accurate. A detailed discussion of these methods is described elsewhere (Oviatt et al., 1986b).

Zooplankton abundances and species composition of each mesocosm were estimated from weekly net tows (two replicates) sampling from bottom to top of each tank. Nets were 30 cm diameter, 64-$\mu$m mesh, and sampled 0.25 $m^3$ per tow. Subsampling was done with a Stempel pipette to produce a count of 200 or more animals. Zooplankton biomass was determined weekly from a single net tow using a 150-$\mu$m mesh net to reduce contamination from phytoplankton. Samples were rinsed with deionized water, frozen, freeze-dried in a Virtis automatic freeze dryer, and weighed on an electrobalance. All of the above measures were taken from June 1981 to October 1983 with indicated frequencies except just prior to and following the copper addition when sampling frequency in selected tanks was increased to every 2 days.

## 2.3. Measuring Survival of Juvenile Copepods

Rates of survival were determined for larval stages (nauplius I–III) of the dominant summer copepod, *Acartia tonsa,* prior to and on successive 2-day periods following the copper addition in three MERL mesocosms, a control with no copper addition, a second control with copper added, and 16× nutrient with copper added. Eggs were obtained from females collected from the appropriate tank; they were then incubated in MERL tanks for 48 hours in cages of clear polycarbonate tubing. Cages enclosed a volume of 500 ml, were capped at each end with plankton netting

(64-$\mu$m mesh), and suspended at 1.0 m depth. Flushing rate of these cages was quite rapid but, to ensure that the internal and external environment of the cages remained similar, the water was manually exchanged every 8 hours. Three replicate cages, containing 40 eggs each, were used per tank. Eggs hatched during the first 4 hours of incubation. Numbers of hatched and surviving nauplii were counted at the end of the 48-hour incubation. Some eggs were also held in control tank water to determine the hatch success in the absence of copper. Juvenile stages were chosen for study because they are considered to be more sensitive to toxics than adults and because laboratory bioassays with nauplii of another estuarine species (*Eurytemora affinis*) had indicated $LC_{50}$ values of 30 $\mu$g/L (Sullivan et al., 1983) i.e., near the range of concentrations expected soon after the copper spike.

The relationship between percent survival and copper concentration in the water column was examined using least-squares linear regression of the arcsine transformed percentages (Sokal and Rohlf, 1981). A $t$ test (Snedecor and Cochran, 1967) was used to determine significant differences between slopes of regressions in different treatment tanks.

### 2.4. Copepod Reproduction

Reproductive rates of female *Acartia tonsa* were monitored in the same three MERL tanks as described above, before and after the copper addition. Females were removed from the tanks to 1.0-L glass jars filled with tank water filtered through a 64-$\mu$m mesh and incubated for 16 hours. There were 10 females per jar, a number that did not exceed normal densities in the tanks, and 3 replicates per tank. The jars were placed on a rotating plankton wheel in a controlled temperature room with dim fluorescent lighting (30 lux) set to the ambient temperature and day length. Eggs and any hatched nauplii were collected and counted or retained for use in mortality experiments.

## 3. PHYTOPLANKTON AND ZOOPLANKTON RESPONSE TO NUTRIENT AND COPPER ADDITIONS

### 3.1. Phytoplankton

Both primary production and chlorophyll standing stocks increased linearly with increasing nutrient additions to the mesocosms (Oviatt et al., 1986b). When calculated on an annual basis net ecosystem production increased from 1.3 mol $O_2$/$m^2$/year in controls to 10.6 mol $O_2$/$m^2$/year in the 32× treatment (Oviatt et al., 1986a). Variability in production and chlorophyll over shorter sampling intervals was quite high, however (Table 16.2, Fig. 16.6). For example, during the period of the copper experiment, production and chlorophyll were highest in the 32× treatment, but in the 16× treatment, production was often lower than at the 2× level. Phytoplankton species composition was quite different across the nutrient gradient, with numbers of diatoms being highest in the 2× treatment while flagellates and monads were most abundant in the 16 and 32× treatments (Table 16.2).

**Table 16.2 Phytoplankton Numbers, Chlorophyll *a*, and Production in Control and Nutrient Enriched (2, 4, 16, 32×) Mesocosms during Copper Experiment**[a]

| Day | Diatoms (No/L) | Microflagellates (No./L) | Monads (No./L) | Chlorophyll *a* (μg/L) | Production (mmol $O_2/m^2$/day ) |
|---|---|---|---|---|---|
| | | | *Control 1 (no copper)* | | |
| 0 | 245 | 1250 | 1562 | 2.3 | 109.4 |
| 1 | 180 | 1719 | 2500 | — | — |
| 2 | 180 | 938 | 5781 | — | — |
| 3 | 146 | 2031 | 8594 | — | — |
| 6 | 27 | 2031 | 11250 | 2.9 | 20.6 |
| 14 | 2 | 2188 | 6094 | 3.2 | 123.1 |
| 20 | 1 | 2188 | 6875 | 2.9 | 78.4 |
| | | | *Control 2 + Cu* | | |
| 0 | 300 | 1406 | 3438 | 7.6 | 111.9 |
| 1 | 264 | 1719 | 3672 | — | 67.8 |
| 2 | 322 | 2656 | 6406 | — | 72.2 |
| 3 | 334 | 3125 | 10938 | — | — |
| 6 | 833 | 3906 | 10312 | 11.3 | 52.8 |
| 14 | 177 | 3906 | 10718 | 4.3 | 86.9 |
| 20 | 9 | 1562 | 16094 | 9.5 | 78.8 |
| | | | *2× + Cu* | | |
| 0 | 3072 | 2812 | 5625 | 16.2 | 141.9 |
| 1 | 236 | 3281 | 3906 | — | 78.8 |
| 2 | 2886 | 2656 | 4688 | — | 81.3 |
| 3 | 2110 | 2812 | 18906 | — | 133.4 |
| 6 | 2363 | 4375 | 12344 | 20.6 | 30.9 |
| 14 | 24 | 4062 | 9062 | 4.4 | 148.8 |
| 20 | 8 | 2812 | 12187 | 6.2 | 92.8 |
| | | | *4× (no copper)* | | |
| 0 | 1960 | 5781 | 21719 | 27.3 | 176.6 |
| 1 | — | — | — | — | — |
| 2 | — | — | — | — | — |
| 3 | — | — | — | — | — |
| 6 | 99 | — | — | 14.9 | −59.1 |
| 14 | 2 | 12500 | 12188 | 9.9 | 117.5 |
| 20 | 1 | — | — | 14.8 | 162.8 |
| | | | *16× + Cu* | | |
| 0 | 220 | 2031 | 5625 | 15.7 | 91.9 |
| 1 | 270 | 6230 | 5625 | 0 | 106.8 |
| 2 | 263 | 1094 | 7969 | — | 84.2 |
| 3 | 263 | 15000 | 58281 | — | — |
| 6 | 448 | 5625 | 36719 | 24.9 | 19.7 |
| 14 | 182 | 5312 | 35625 | 22.1 | −2.8 |
| 20 | 73 | 3281 | 20625 | 5.3 | 80.6 |

**Table 16.2** (*Continued*)

| Day | Diatoms (No/L) | Microflagellates (No./L) | Monads (No./L) | Chlorophyll *a* ($\mu$g/L) | Production (mmol $O_2$/m$^2$/day) |
|---|---|---|---|---|---|
| | | | *32× + Cu* | | |
| 0 | 318 | 469 | 35156 | 43.0 | 467.5 |
| 1 | 216 | 3750 | 577500 | — | 408.8 |
| 2 | 148 | 6406 | 977500 | — | 358.4 |
| 3 | 148 | 17812 | 735000 | — | — |
| 6 | 416 | 5938 | 52656 | 6.9 | −26.8 |
| 14 | 2 | 2500 | 428750 | 33.1 | 163.8 |
| 20 | 3 | 5928 | 41250 | 18.5 | 190.0 |

[a]Only those treatments with copper added are shown, except for one control and 4× with no copper added. Copper dose (Aug. 23, 1983) occurred between days 0 and 1. Data courtesy of Drs. P. Lane and C.A. Oviatt.

With the exception of a large but very short-lived (1 day) reduction of diatom abundance in only one mesocosm (2× nutrients), phytoplankton cell numbers were not initially altered by the copper addition (Fig. 16.1). However, reductions in primary production and growth rate of cell cultures were observed on the first 2 days following the copper addition, with the largest effect in the 2× treatment (Table 16.2; P. L. Donaghay, personal communication). By 1 week following the addition there is some indication of increases in monads and flagellates (Table 16.2).

Concentrations of dissolved and particulate copper following introduction to the MERL mesocosms have been described by Hunt and Donaghay (1984) and data for selected mesocosms are reported here in Table 16.3. Concentrations of total copper in the water column fell rapidly from 39–41 $\mu$g/L to 13–15 $\mu$g/L during the week following the addition. This rate was faster than can be explained by the exchange rate of inflowing water to the mesocosms and was due to other processes (C. Hunt, personal communication). There were small differences among nutrient treatments in rate of removal from the water column and in the fractionation of the copper between particulate and dissolved phases (Table 16.3). The rate of removal from the water column was most rapid in the unenriched mesocosm and slowest in 32×; more of the copper was in the dissolved (and presumably toxic) form in the 16×, or more nutrient-enriched treatment, than in the control. The differences across nutrient treatment were small, however, and after 10 days all tanks had identical concentrations and distributions of copper.

## 3.2. Survival of Zooplankton and Population Numbers

While increases in zooplankton production and numbers were observed in nutrient-enriched treatments (Sullivan and Ritacco, 1985a) there was a great deal of variability over short time scales that did not follow any apparent pattern across nutrient treatments. For example, during the copper-spike experiment, copepods were

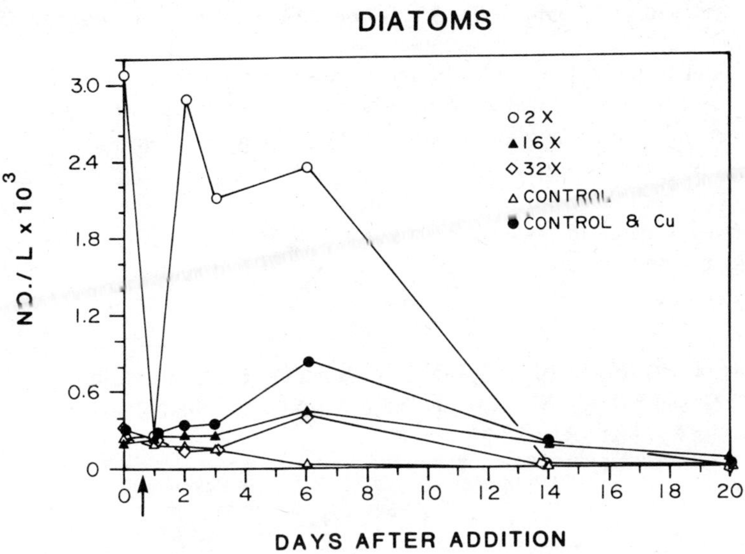


**Figure 16.1.** Diatom abundances (No./L) just prior to and following the copper addition in control and nutrient enriched MERL mesocosms. Data provided by P. Lane.

**Table 16.3 Survival (% of total) of Nauplii during 48-hr Incubations in MERL Tanks prior to and up to 19 Days Following Copper Addition**[a,b]

| | | | Average[c] Cu | | |
|---|---|---|---|---|---|
| | | | Total | Dissolved | |
| Date | Treatment | Mean ± S.D. | (μg/L) | (μg/L) | (%) |
| Aug. 21–23 | 0× | 97 ± 2.5 | <2 | <2 | — |
| (day −2 to 0) | 0× + Cu | 93.3 ± 3.8 | <2 | <2 | — |
| | 16× + Cu | 96.8 ± 2.8 | <2 | <2 | — |
| Aug. 23–25 | 0× | 100 | <2 | <2 | |
| (day 0 to 2) | 0× + Cu | 36.7 ± 5.2 | 40.98 | 32.4 | 79 |
| | 16× + Cu | 15.8 ± 11.2 | 39.5 | 34.4 | 82 |
| Aug. 25–27 | 0× | 100 | <2 | <2 | 78 |
| (day 2 to 4) | 16× + Cu | 21.3 ± 5.3 | 28.5 | 23.9 | 84 |
| Aug. 31–Sept. 2 | 0× | 100 | <2 | <2 | |
| (day 8 to 10) | 0× + Cu | 73.3 ± 6.3 | 13.2 | 12.2 | 92 |
| | 16× | 83.3 + 8.0 | 15.2 | 13.6 | 89 |
| Sept. 9–12 | 0× | 100 | <2 | <2 | |
| (day 17 to 19) | 0× + Cu | 100 | 7.62 | 4.7 | 62 |
| | 16× + Cu | 52.9 ± 9.0 | 6.35 | 5 | 79 |

[a] Copper was added on Day 0. Data on copper concentrations provided by C. Hunt.

[b] 48-hr $LC_{50}$ (95% C.I) values based on data in table:

0× + Cu 32.21 (13.2–38.8)
16× + Cu 21.41 (15.2–28.5)

[c] Mean value during 48-hr incubation, calculated from 2 replicates.

as abundant in the copper spiked-control mesocosm as in the 2× and 4× treatments, even prior to adding copper, while the other controls (no copper added) had lower numbers throughout (Table 16.4). Apparently, variability among controls was as great as across nutrient treatments during this particular period. As can be seen from the controls without copper, the late summer period is normally a time for increases in copepod population size (see day 0–14, Table 16.4). Increasing numbers were also evident in the low-nutrient (control and 2×) tanks to

**Table 16.4 Numbers of Zooplankton (No/L ± S.D.) in Control and Copper Dosed MERL Mesocosms from August 23, 1983 (Day 0) to September 6, 1983 (Day 14)[a]**

| | Pre copper | Post Copper | | |
|---|---|---|---|---|
| | Day 0 | Day 3 | Day 8 | Day 14 |
| | | *Control (no copper)* | | |
| Total copepods | 31.5 ± 1.5 | 35.9 ± 1.3 | 61.2 ± 14.5 | 77.8 ± 4.6 |
| Polychaete larvae | 9.5 ± 0.5 | 13.0 ± 1.7 | 12.4 ± 8.1 | 22.5 ± 1.4 |
| | | *Control (no copper)* | | |
| Total copepods | 35.9 ± 3.3 | — | — | 68.0 ± 3.5 |
| Polychaete larvae | 3.1 ± 3.6 | — | — | 18.3 ± 4.6 |
| | | *Control + Cu* | | |
| Total copepods | 104.0 ± 2.1 | 37.1 ± 4.1 | 106.8 ± 12.4 | 316.0 ± 3.1 |
| Polychaete larvae | 5.0 ± 2.7 | 4.9 ± 2.7 | 11.0 ± 6.7 | 17.0 ± 1.4 |
| | | *2× + Cu* | | |
| Total copepods | 59.4 ± 7.4 | 54.3 ± 7.4 | 116.5 ± 13.4 | 155.5 ± 57.3 |
| Polychaete larvae | 31.0 ± 4.2 | 60.5 ± 14.1 | 135.0 ± 3.5 | 210.0 ± 21.1 |
| | | *4× (no copper)* | | |
| Total copepods | 96.8 | 0 | 0 | 92.3 |
| Polychaete larvae | 111.3 | — | — | 45.8 |
| | | *16× + Cu* | | |
| Total copepods | 260.0 ± 5.6 | 18.8 ± 1.1 | 19.5 ± 6.4 | 60.5 ± 3.5 |
| Polychaete larvae | 66.5 ± 6.4 | 67.0 ± 16.6 | 87.0 ± 4.6 | 219.0 ± 11.3 |
| | | *32× + Cu* | | |
| Total copepods | 79.5 ± 33.2 | 6.1 ± 1.9 | 2.5 ± 0.84 | 4.3 ± 1.3 |
| Polychaete larvae | 234.0 ± 41.0 | 332.0 ± 63.0 | 215.0 ± 21.9 | 350.0 ± 49.5 |

[a] Level of nutrient enrichment of the mesocosm is indicted. See text for further explanations.

which copper was added, but not in any of the higher nutrient tanks, with or without copper.

An initial reduction in total zooplankton biomass (Fig. 16.2) and copepod numbers (Fig. 16.3, Table 16.4) was observed in all mesocosms to which copper was added, regardless of the level of nutrient enrichment. Reductions were 40–50% in low-nutrient mesocosms and higher (80–90%) in nutrient-rich mesocosms. In contrast, biomass and numbers in control tanks to which copper was not added remained stable. Direct measures of copepod survival rates during this initial period corroborated the implication of copper lethality obtained from biomass and numbers. There was good agreement between trends in survival of caged nauplii and changing numbers of nauplii in the tanks. Survival of copepod nauplii incubated in cages in copper dosed treatments was very low (16–37% surviving) during the first 24–48 hours following the copper addition, whereas in a control tank without copper, survival exceeded 97% (Fig. 16.4, Table 16.3). If the survival data from the first 48-hour incubation are used to predict numbers of nauplii that would remain in the tanks by day 3 (initial survival rate in the cages multiplied by starting numbers of nauplii) and assuming no successful recruitment during this period, the result matched the actual decline very closely ("E" in Fig. 16.4).

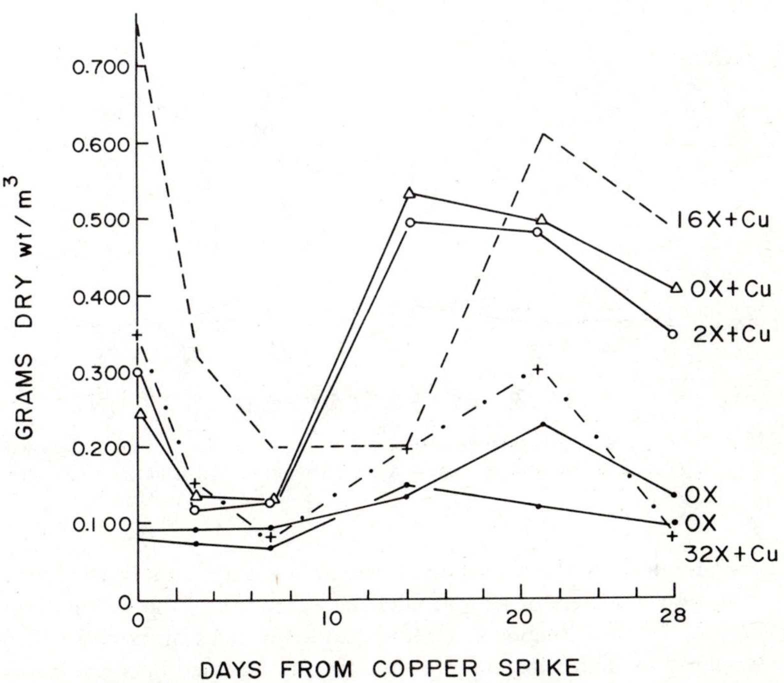


**Figure 16.2.** Zooplankton biomass (g/m$^3$) in 2 control and 4 copper-treated MERL mesocosms prior to and for 28 days following the copper addition. The level of nutrient enrichment of the mesocosm is indicated.

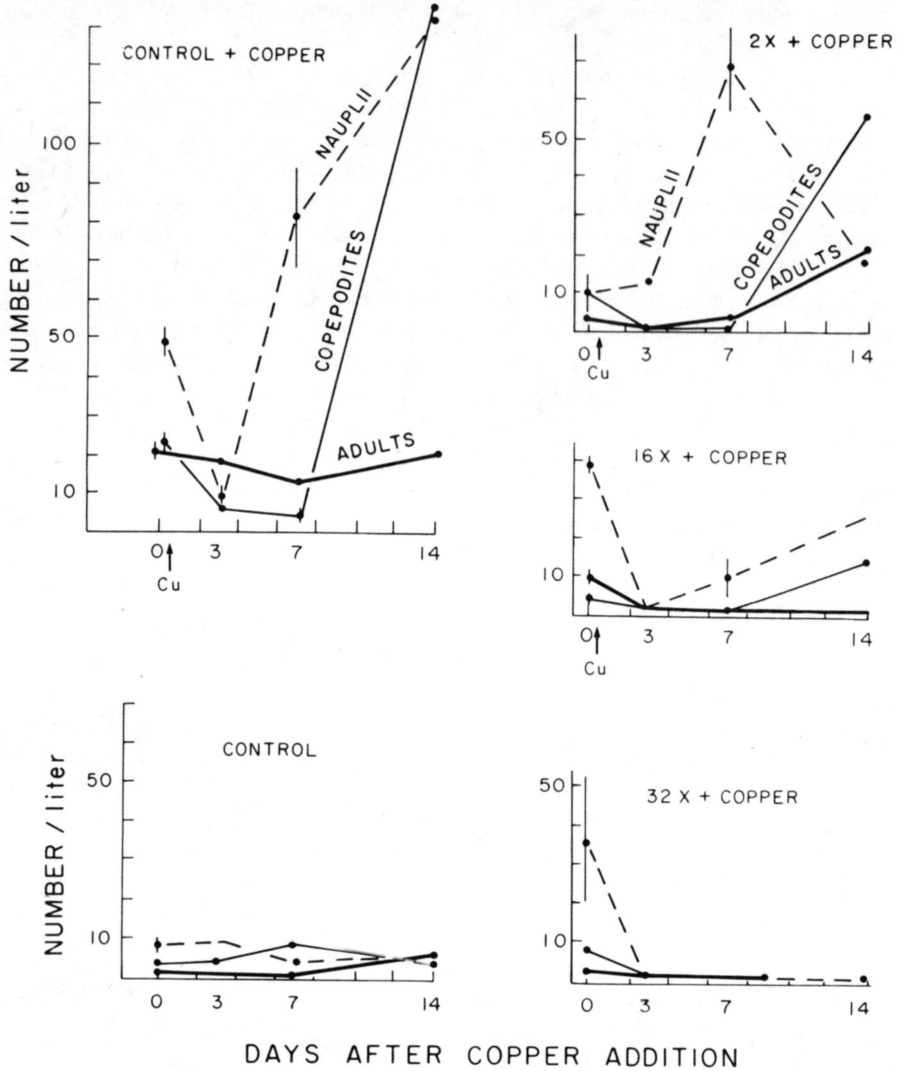


**Figure 16.3.** Numbers of *Acartia tonsa* in control and copper-treated mesocosms prior to and following the addition of copper. Standard deviation of replicate net tows is indicated.

In general, numbers of adult copepods were not as reduced following the copper addition as were numbers of nauplii with the exception of the 2× nutrient treatment (Fig. 16.3). This implies that adults had lower rates of mortality than juveniles. Mortality of adult copepods was not directly assessed in cage experiments.

By 72–96 hours after the copper spike, much of the copper had already been lost from the water column and the effective copper dose was much reduced; rates of survival in the cages were also higher, especially in the copper control tank

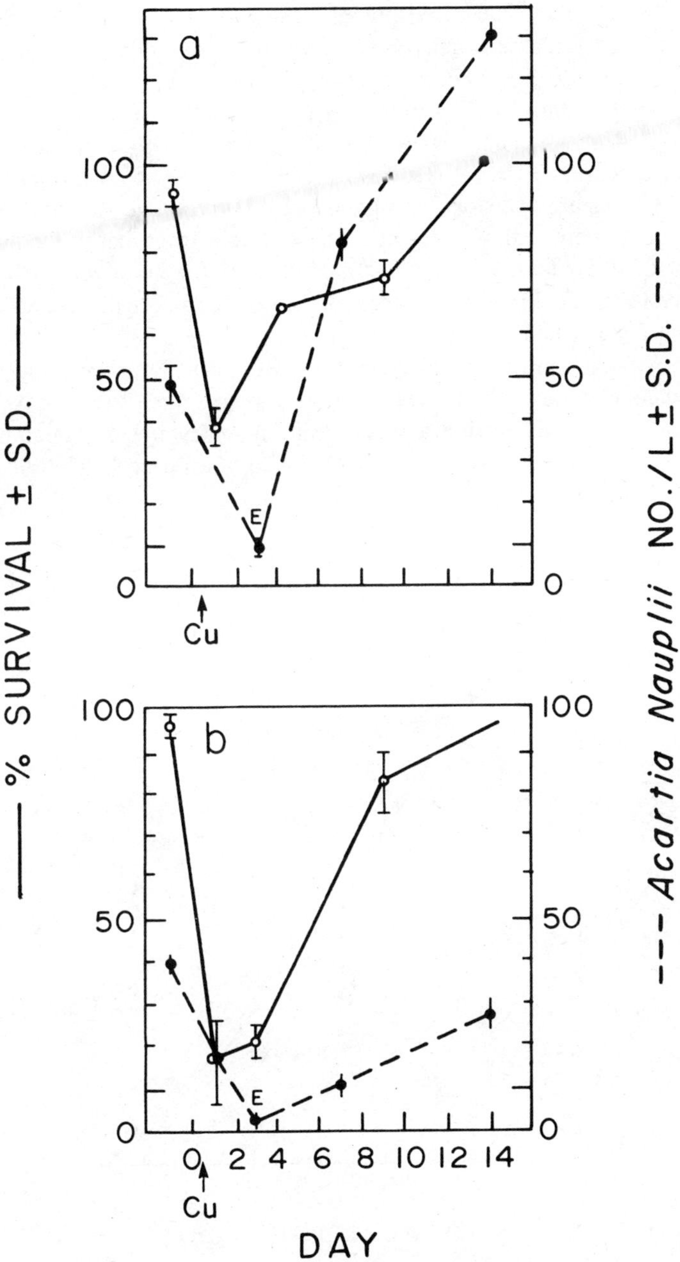


**Figure 16.4.** Percent survival of *Acartia tonsa* nauplii (solid line) in cages incubated in MERL mesocosms compared to numbers of nauplii per liter present prior to and following copper addition (dashed line): (a) Control + Cu, (b) 16× + Cu. "E" indicates the expected number of nauplii if survival estimates from the cages are multiplied by the number of nauplii present at the beginning of the incubation period (further explanation in text).

(Table 16.3, Fig. 16.4). An approximate 48-hour $LC_{50}$ for each tank in which nauplii were incubated was calculated using the temporally changing mortality measured with the decline in concentration of total copper (Fig. 16.5). The $LC_{50}$ was higher in the control + Cu treatment (32.2. μg/L) than in the enriched 16× treatment (21.4 μg/L) (Table 16.3). For a given value of either total, dissolved, or particulate copper, survival was lower in the 16× + copper treatment than in the control + copper. The slope of regressions of survival versus total or dissolved copper in the nutrient tank was significantly different from that in the control ($t$ = 3.62, $P \leq 0.01$ for total Cu; $t$ = 2.86, $P \leq 0.02$ for dissolved Cu). However, within a treatment there were no significant differences between regressions using total or dissolved copper.

*Acartia* populations recovered quickly as evidenced by increasing numbers of nauplii and copepodites only 1 week after the copper spike (Table 16.4, Fig. 16.3) in both control and 2× nutrient tanks. While nauplii were very abundant by day 7 in both these treatments, recovery was delayed until after day 7 in the 16× and 32× nutrient levels.

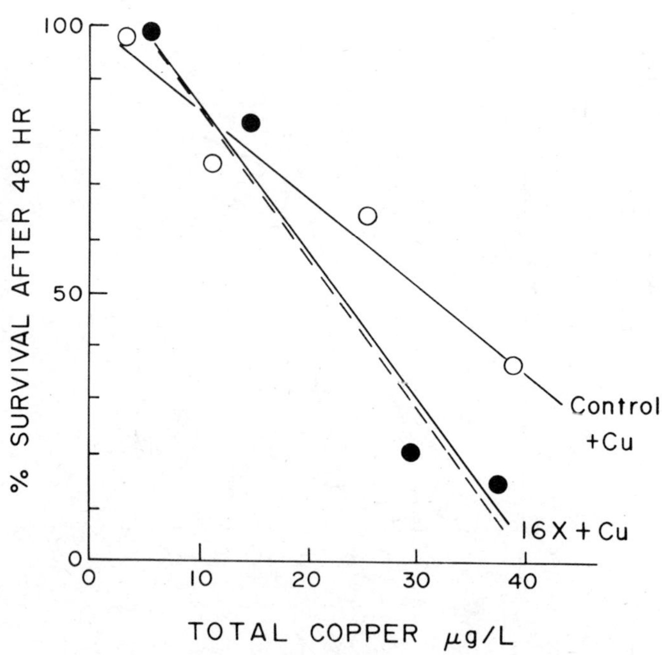


**Figure 16.5.** The relationship between total copper in the water column and percent survival of nauplii in cages incubated in MERL mesocosms. Data on copper concentration provided by C. Hunt. Mean values from 2–3 replicates per experiment are plotted. Control tank: $S$ = 75.1 − 0.94TC, $r^2$ = 92.8; 16× Nutrient Tank: $S$ = 86.6 − 1.70TC, $r^2$ = 90.0 ($S$ = arcsin transformed percent survival, TC = total copper).

Other species of copepods, including *Pseudodiaptomus coronatus* and *Oithona colcarva,* responded in a more or less similar manner to *A. tonsa* (Table 16.4). Their numbers decreased immediately after the copper addition and recovery was slow in highly enriched tanks. Polychaete larvae, on the other hand, appeared unaffected by copper and their numbers increased in all tanks.

### 3.3. Copepod Reproduction

There was considerable variability in rates of egg production by *Acartia tonsa* in all MERL tanks even prior to the addition of copper (Fig. 16.6). Throughout the nutrient experiment short-term fluctuations in egg production were large and while stepwise regression analysis selected chlorophyll *a* as a significant predictor of egg production, no more than 50% of the variability was explained (Sullivan and Ritacco, 1983). Maximum rates of egg production were higher in nutrient enriched tanks (see Fig. 16.6, 16× treatment) but were not sustained for more than a few days. Such variability, caused by natural factors, makes it difficult to use egg production as a measure of effects caused by toxic substances. Rates of egg pro-

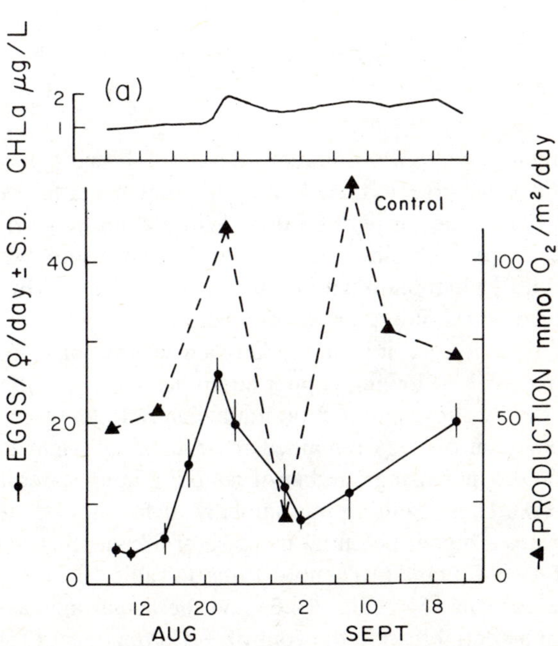


**Figure 16.6.** Egg production by female *Acartia tonsa* and phytoplankton production (mmol $O_2/m^2/$day) prior to and following the copper spike in control, control + Cu, and 16× + Cu mesocosms. Chlorophyll *a* (µg/L) is shown in the upper segment of the figure for each mesocosm.

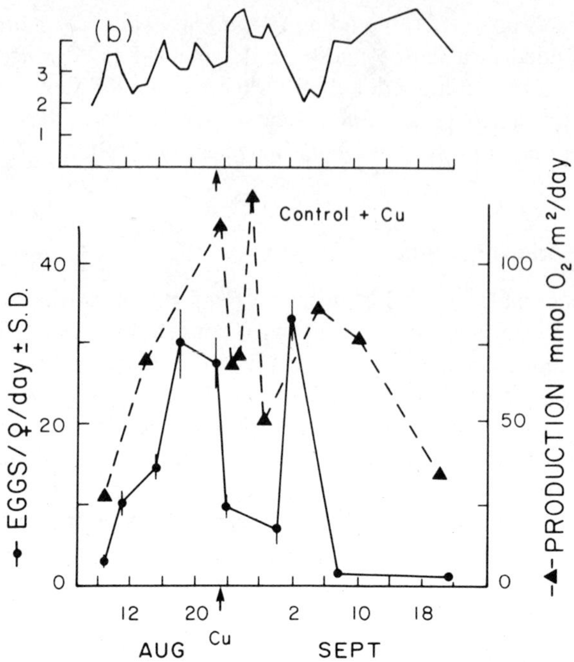


**Figure 16.6.** (*Continued*)

duction of surviving females did drop considerably following the addition of copper, relative to rates just prior to the addition, but they never fell below lower rates commonly seen before the copper addition. While chlorophyll *a* levels were not reduced by the copper (possibly because grazing pressure was reduced after copepod mortalities), phytoplankton production was lowered (Fig. 16.6) for 2 or more days. Reduced rates of egg production could therefore have been a response to this lower productivity, but it is more likely a direct response to copper toxicity, at least in the first week following copper additions.

Despite lower rates of egg production immediately following copper additions, the residual production of eggs remained more than sufficient to repopulate the tanks and explain the increasing number of nauplii that appeared in the water column 1 and 2 weeks later. Multiplying numbers of females per liter by their egg production indicates a higher potential increase of 88 nauplii/L/day in the control + Cu than in 16× + Cu tank (40 nauplii/L/day) (Table 16.5). In fact, population recovery was delayed in 16× (Fig. 16.3) and the actual increase an even lower percentage of the potential than in the control + Cu treatment (Table 16.5). In the no-copper control, where food resources were very low compared to other tanks (Fig. 16.6), the number of eggs produced was apparently sufficient only to maintain a small population of *Acartia* with constant density over time, and no increases were observed (Fig. 16.3).

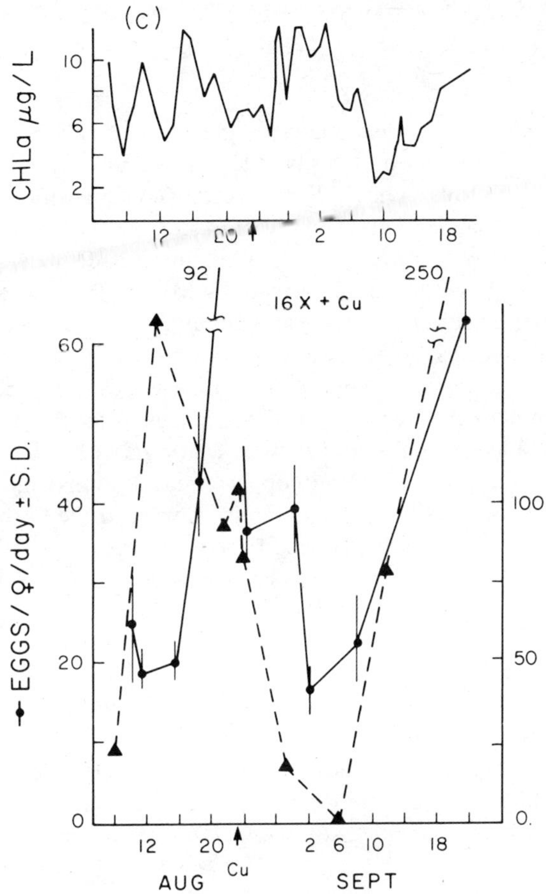


**Figure 16.6.** (*Continued*)

## Table 16.5 Potential and Actual Increases in Numbers of *Acartia tonsa* Nauplii during First and Second Weeks Following Copper Addition in Control, Control + Cu, and 16× + Cu Mesocosms[a]

| Mesocosm Treatment | Week after Copper Added | Potential No./L | Actual No./L | % |
|---|---|---|---|---|
| Control | 1 | 19.5 | 0 | 0 |
| | 2 | 8.6 | 0 | 0 |
| Control + Cu | 1 | 88.0 | 17.3 | 19.7 |
| | 2 | 210.5 | 7.0 | 3.3 |
| 16× + Cu | 1 | 40.0 | 1.9 | 4.8 |
| | 2 | 55.0 | 2.1 | 3.8 |

[a] The percent of the potential production observed as actual increases in numbers of nauplii in the water column is indicated. Potential increase = eggs/female/day multiplied by number of females/L. Egg development time was constant during this period and was assumed to be 24-hr. Actual increase based on counts from net tows.

## 4. DISCUSSION

The addition of 40 $\mu$g/L of copper to control and nutrient-enriched MERL tanks clearly had a direct toxic effect on the juvenile stages and, to a lesser extent, on adults of dominant copepods. These effects were persistent while water column concentrations of Cu remained high. The use of in situ cages for direct observation of naupliar survival was very useful in identifying this fact. In several respects the effects of this relatively high dose of copper to the MERL enclosures were similar to those that could have been predicted from laboratory studies. Rates of mortality observed in the multispecies MERL systems were relatively close to those measured in single species bioassays (Table 16.6). The in situ cage experiments clearly demonstrated that mortality of nauplii was the mechanism for the initial reduction in their numbers. While we could not calculate a true $LD_{50}$ for the nauplii, especially considering the rapidly changing copper concentrations, our approximate value of 20–30 $\mu$g/L for a 48-hour $LC_{50}$ is quite similar to the 96-hour $LD_{50}$ determined for nauplii of the estuarine copepod species *Eurytemora affinis* (Sullivan et al., 1983). Mortality of adults in the MERL mesocosms was not as great as for nauplii, implying an $LC_{50}$ of greater than 40 $\mu$g/L total Cu and confirming the conclusion of laboratory studies that juvenile stages are more sensitive than adults. $LC_{50}$ values reported from bioassays with adult *Acartia tonsa* from Narragansett Bay ranged as high as 78 $\mu$g/L (Sosnowski et al., 1979) indicating agreement with the mesocosm results.

While in this one instance, laboratory bioassays predicted field effects, there may be cases in which field conditions modify toxic effects. In another of our mesocosm studies, which concerned toxic effects of ammonia (Sullivan and Ri-

**Table 16.6 Toxicity of Copper and Ammonia to Copepods in Single Species Bioassays Compared to That Measured in Multispecies MERL Mesocosms**

| Species and Life Stage | $LC_{50}$ | Reference |
|---|---|---|
| | *Copper ($\mu$g/L)* | |
| *Eurytemora affinis*, nauplii | 96h : 31 ± 2.5 | Sullivan et al. (1983), laboratory |
| *Acartia tonsa*, nauplii | 48h : 32.2 | This study, control + Cu mesocosm |
| | 21.4 | This study, 16× + Cu mesocosm |
| *Acartia tonsa*, adult | 72h : 17 − 78 | Sosnowski et al. (1979), laboratory |
| | *Un-ionized Ammonia ($\mu M$/L)* | |
| *Acartia tonsa*, nauplii | 48h : 5.9 − 11.7 | Sullivan and Ritacco (1985b), laboratory |
| *Acartia tonsa*, nauplii | 48h : 5.0 | Sullivan and Ritacco (1985b), mesocosms |

tacco, 1985b), mortality in the mesocosms was significantly higher than predicted (Table 16.6). Nevertheless, the use of laboratory assays on the same copepod species and stages as found in the mesocosms was even in that instance, an extremely useful way of identifying at least one source of copepod mortality (un-ionized ammonia) in eutrophic mesocosms. The ammonia study is also relevant to the current study in that we were concurrently measuring unionized ammonia and copper concentrations. During the period of copper additions un-ionized ammonia was well below $LD_{50}$ concentrations. Nevertheless, we cannot exclude an additive effect of copper and ammonia decreasing the $LD_{50}$ of either substance in the 16× and 32× treatments and causing the greater apparent effect of copper in these highly enriched systems.

A second prediction from the laboratory studies of Sosnowski et al. (1979) was only partially in agreement with our results. They concluded that increased food availability should make copepods less sensitive to copper. Nutrient enrichment of the model ecosystems did alter the response to copper so that toxicity of copper was slightly lower and the recovery of copepods was faster in moderately nutrient-enriched tanks (2×) but slower in highly enriched mesocosms (16× and 32×). In fact, regression analysis indicated a significantly higher toxicity of copper to nauplii in the 16× treatment. Copepods in systems with very high levels of enrichment, even though accompanied by high standing stocks of chlorophyll *a,* seem to have been more affected by the copper. In addition, the higher organic content of eutrophic waters would be expected to bind more of the free ionic (and more toxic) form of copper (Sunda and Lewis, 1978; Pesch et al., 1979). While our methods did not include measurement of the organics in the water it is likely that they were higher in the more eutrophic treatments. Our results were therefore in the opposite direction from those expected from either food availability models or copper complexation models. Possibly other factors influenced the toxicity of copper in the 16× and 32× treatments, such as food quality (phytoplankton species composition) or the amount of particulate versus dissolved copper. While dissolved copper was slightly higher in the 16× treatment, regression analysis indicated that treatment differences in dissolved versus particulate phases did not explain different survival rates.

Laboratory tests (Reeve et al., 1977b) also indicated that egg production should be reduced at copper levels experienced throughout the period of observation (20 days). Unfortunately, while egg production by adult *Acartia* females may have been reduced by exposure to copper, we could not rule out other natural factors that may have caused this reduction because of the normally large variability in egg-laying rates and the confounding effects of changes in rates of primary production following the copper addition. Although Reeve et al. (1977b) showed in the laboratory that *Acartia* reproduction was sensitive to copper concentrations as low as 10 $\mu$g/L it was more difficult to assess the magnitude and mechanisms of effect of copper on reproductive rates for MERL mesocosm populations. Reproductive rates of copepods including *Acartia* vary in natural systems and in the MERL tank populations from high to low values over periods of only several days in response to changes in their food supply (Dagg, 1977; Sullivan and Ritacco,

1985a). Food quality may also be involved (Cahoon, 1981). In this experiment there was a short-term change of 48 hours in rates of primary production, although no change in standing stock (chlorophyll *a*) and usually no changes in cell number, while egg production was reduced for a period of at least 1 week. Therefore, we cannot determine conclusively whether reductions in egg production were a direct effect of copper toxicity or a response to changes in phytoplankton production. Exposure of the copepods to a standard food as well as to the natural food in the mesocosms, such as was done by Reeve et al. (1977a) might have helped resolve this issue.

This experiment also gave us the opportunity to observe recovery of copepod populations after exposure to copper and differences in this recovery in treatments with different levels of nutrient enrichment. The slow recovery in the 16× treatment was surprising because there were adequate numbers of females with sufficiently high rates of egg production to permit rapid repopulation. Why, then, were the numbers of *Acartia* (and other species) slower returning to precopper levels in the 16× and 32× nutrient treatments? In addition to the apparently greater toxicity of copper in these systems there may be other mechanisms responsible for failures of copepod populations that involve water quality and possibly food quality, which intermittently limit success of juvenile copepods. Sullivan and Ritacco (1985b) describe increases in naupliar mortality associated with high levels of un-ionized ammonia in the 16× and 32× treatments. These factors may reduce the probability of copepod repopulation after any perturbation of highly eutrophic systems. Episodes of poor egg hatch success ($<51\%$ versus a normal rate of $>90\%$) have also been frequently observed (B. K. Sullivan, unpublished) prior to the copper addition, and could not be associated with copper toxicity. Poor food quality (high numbers of flagellates and monads) in these treatments (Donaghay, 1985) may be responsible. Certainly, species composition of the phytoplankton was markedly different from less enriched mesocosms during the period of the copper study. Growth rates of copepods in the 16× treatment were also slower than in the control tank long after copper had returned to background levels. This occurred despite the presence of higher phytoplankton biomass in the 16× treatment (A. G. Durbin and E. G. Durbin, personal communication), a fact consistent with the interpretation of a food quality effect.

## 5. SUMMARY

1. Copper had its major impact on copepod populations via direct toxicity to the more sensitive juvenile stages. This implies that single-species bioassays are relevant to predicting effects of copper on marine communities in that they accurately predicted toxic levels and age-specific responses seen in the MERL tanks. However, they were not adequate to predict the differences in copper toxicity to copepods in systems of differing nutrient enrichment. Mortality was significantly higher at the same copper concentration in the highly enriched system. Only actual

experimentation in systems like MERL is likely to give correct predictions for behavior of contaminants in complex systems.

2. The effects of copper on toxicity to copepods and recovery of zooplankton populations are likely to be more pronounced in systems where either food availability is low (MERL control) or food quality is poor (highly nutrient enriched) than in moderately nutrient-rich systems. It was not possible to distinguish conclusively, however, whether the slower recovery of eutrophic systems relative to moderately enriched systems was due to food quality or differences in copper chemistry that affected copper availability and toxicity to the copepods.

3. Egg production was probably also affected by copper but it was difficult to determine the mechanism (direct or indirect effect of copper). Evaluation of egg production by females allowed some prediction of copepod population response, but other factors, including numbers of adult females in the population, survival and growth rates of young, and food quality are apparently more predictive of population success than are reproductive rates alone.

## ACKNOWLEDGMENTS

Data on phytoplankton cell counts and chlorophyll were provided by P. Lane, and on primary production by C. A. Oviatt. We also thank C. Hunt, C. A. Oviatt, P. Donaghay, A. Durbin, and E. Durbin who have helped in providing both data and discussions leading to the conclusions of this paper. This study was supported by NOAA Grant NA-80-RA-0-80046.

## REFERENCES

Cahoon, L. B. (1981). Reproductive response of *Acartia tonsa* to variations in food ration and quality. *Deep-Sea Res.* **28A,** 1215–1221.

Cairns, J., Jr., and Buikema, A. L., Jr. (1984). Verifying predictions of environmental safety and harm. In H. White (Ed.), *Concepts in Marine Pollution Measurements.* Maryland Sea Grant Program, University of Maryland, College Park, pp. 81–111.

Dagg, M. (1977). Some effects of patchy food environments on copepods. *Limnol. Oceanogr.* **22,** 99–107.

Donaghay, P. L. (1985). An experimental test of the relative significance of food quality and past feeding history to limitation of egg production of the estuarine copepod, *Acartia tonsa. Arch. Hydrobiol., Beih. Ergeb. Limnol.* **21,** 235–245.

Gachter, R. (1979). MELIMEX, an experimental heavy metal pollution study: Goals, experimental design and major findings. *Schweiz. Z. Hydrol.* **41,** 169–176.

Gamble, J. C., Davies, J. M., and Steele, J. H. (1977). Loch Ewe bag experiment, 1974. *Bull. Mar. Sci.* **27,** 145–175.

Gibson, V. R., and Grice, G. D. (1977). Response of macro-zooplankton populations to copper: Controlled Ecosystem Pollution Experiment. *Bull. Mar. Sci.* **27,** 85–91.

Grice, G. D. (1984). Use of enclosures in studying stress on plankton communities. In H. White (Ed.),

*Concepts in Marine Pollution Measurements.* Maryland Sea Grant Program, University of Maryland, College Park, pp. 563–573.

Hunt, C. D., and Donaghay, P. L. (1984). Alteration of metal depositional pathways by eutrophication. *EOS, Trans. Am. Geophys. Union* **65,** 938.

Kuiper, J. (1982). Ecotoxicological experiments with marine plankton communities in plastic bags. In G. D. Grice and M. R. Reeve (Eds.), *Marine Mesocosms: Biological and Chemical Research in Experimental Ecosystems.* Springer-Verlag, New York, pp. 182–193.

Lorenzen, C. J. (1966). A method for continuous measurement of *in vivo* chlorophyll concentration. *Deep-Sea Res.* **13,** 223–227.

Marshall, J. S., and Mellinger, D. L. (1980). Dynamics of cadmium-stressed plankton communities. *Can. J. Fish. Aquat. Sci.* **37,** 403–414.

Menzel, D. W. (1977). Summary of experimental results: Controlled Ecosystem Pollution Experiment. *Bull. Mar. Sci.* **27,** 142–145.

Nixon, S. W., Pilson, M. E. Q., Oviatt, C. A., Donaghay, P., Sullivan, B., Seitzinger, S., Rudnick, D., and Frithsen, J. (1984). Eutrophication of a coastal marine ecosystem—an experimental study using the MERL microcosms. In M. J. R. Fasham (Ed.), *Flows of Energy and Materials in Marine Ecosystems: Theory and Practice.* Plenum, New York, pp. 105–135.

Odum, H. T., and Hoskins, C. M. (1958). Comparative studies on the metabolism of marine waters. *Publ. Inst. Mar. Sci., Univ. Tex.* **5,** 16–46.

Oviatt, C. A., Frithsen, J., Gearing, J., and Gearing, P. (1982). Low chronic addition of No. 2 fuel oil: Chemical behavior, biological impact and recovery in a simulated estuarine environment. *Mar. Ecol.: Prog. Ser.* **9,** 121–136.

Oviatt, C. A., Rudnick, D. T., Keller, A., Sampou, P., and Almquist, G. (1986a). A comparison of system ($O_2$ and $CO_2$) and C-14 measurements of metabolism in estuarine mesocosms. *Mar. Ecol.: Prog. Ser.* **28,** 57–67.

Oviatt, C. A., Keller, A. A., Sampou, P. A., and Beatty, L. L. (1986b). Patterns of productivity during eutrophication: A mesocosm experiment. *Mar. Ecol.: Prog. Ser.* **28,** 69–80.

Pesch, G., Stewart N., and Pesch, C. (1979). Copper toxicity to the bay scallop (*Argopecten uradians*). *Bull. Environ. Contam. Toxicol.* **23,** 759–765.

Pilson, M. E. Q., Oviatt, C. A., Vargo, G. A., and Vargo, S. L. (1979). Replicability of MERL microcosms: Initial observations. In F. S. Jacoff (Ed.), *Advances in Marine Environmental Research* (EPA-600/9-79-035). Environ. Prot. Agency, Narragansett, RI, p. 409.

Pilson, M. E. Q., Oviatt, C. A., and Nixon, S. W. (1980). Annual nutrient cycles in a marine microcosm. In J. P. Giesy (Ed.), *Department of Energy Symposium Series* (CONF 781101). Natl. Tech. Inf. Serv., Augusta, GA, pp. 753–778.

Reeve, M. R., Grice, G. D., Gibson, V. R., Walter, M. A., Darcy, K., and Ikeda, T. (1976). A controlled environmental pollution experiment (CEPEX) and its usefulness in the study of larger marine zooplankton under toxic stress. In A. P. Lockwood (Ed.), *Effects of Pollutants in Aquatic Organisms.* Cambridge Univ. Press, London and New York, pp. 145–162.

Reeve, M. R., Gamble, J. C., and Walter, M. A. (1977a). Experimental observations on the effects of copper on copepods and other zooplankton: Controlled ecosystem pollution experiment. *Bull. Mar. Sci.* **27,** 92–104.

Reeve, M. R., Walter, M. A., Darcy, K., and Ikeda, T. (1977b). Evaluation of potential indicators of sublethal toxic stress on marine zooplankton (feeding, fecundity, respiration and excretion): Controlled ecosystem pollution experiment. *Bull. Mar. Sci.* **27,** 105–113.

Santschi, P. H., Nyffeler, U., Anderson, R., and Schiff, S. (1984). The enclosure as a tool for the assessment of transport and effects of pollutants in lakes. In H. White (Ed.), *Concepts in Marine Pollution Measurement.* Maryland Sea Grant Program, University of Maryland, College Park, pp. 549–562.

Snedecor, G. W., and Cochran, W. G. (1967). *Statistical Methods.* Iowa State Univ. Pres, Ames.

Sokal, R. R., and Rohlf, F. J. (1981). *Biometry.* Freeman, New York.

Sosnowski, S. L., and Gentile, J. H. (1978). Toxicological comparison of natural and cultured populations of *Acartia tonsa* to cadmium, copper and mercury. *J. Fish. Res. Board Can.* **35,** 1366–1369.

Sosnowski, S. L., Germond, D. J., and Gentile, J. H (1979). The effect of nutrition on the response of field populations of the calanoid copepod *Acartia tonsa* to copper. *Water Res.* **13,** 449–452.

Sullivan, B. K., and Ritacco, P. J. (1983). Temporal patterns of reproduction of copepod populations in model marine ecosystems and implications for the study of copepod population dynamics. *EOS, Trans. Am. Geophys. Union* **64,** 1103.

Sullivan, B. K., and Ritacco, P. J. (1985a). The response of dominant copepods species to food limitation in a coastal marine ecosystem. *Arch. Hydrobiol., Beih. Ergeb. Limnol.* **21,** 407–418.

Sullivan, B. K., and Ritacco, P. J. (1985b). Ammonia toxicity to larval copepods in eutrophic marine ecosystems: A comparison of results from bioassays and enclosed experimental ecosystems. *Aquat. Toxicol.* **7,** 205–217.

Sullivan, B. K., Buskey, E., Miller, D. C., and Ritacco, P. J. (1983). Effects of copper and cadmium on growth, swimming and predator avoidance in *Eurytemora affinis* (Copepoda). *Mar. Biol. (Berlin)* **77,** 299–306.

Sunda, W. G., and Lewis, J. M. (1978). Effect of complexation by natural organic ligands on the toxicity of copper to a unicellular alga, *Monochrysis lutheri. Limnol. Oceanogr.* **23,** 870–876.

Urech, J. (1979). MELIMEX, an experimental metal pollution study: Effects of increased heavy metal load on crustacea plankton. *Schweiz. Z. Hydrol.* **42**(1), 247–260.

# 17

# BIOPERTURBATIONS IN MESOCOSM AND NUMERICAL SIMULATION EXPERIMENTS

*T. R. Parsons*

*Department of Oceanography, University of British Columbia, Vancouver, British Columbia, Canada V6T 1W5*



## 1. INTRODUCTION

In the following discussion a toxic contaminant has been interpreted as any allochthonous substance that causes a bioperturbation to the natural ecosystem. Examples of such materials include inorganic nutrients and metabolizable substrates as well as such pollutants as metals, hydrocarbons, and silt. Using mesocosm ecosystem experiments to assess the effect of these pollutants has led to several different interpretations of results. The first of these is that the concept of a ''controlled'' experiment when dealing with manipulations involving 50 to 1000 tons of water may be misleading. Unlike laboratory experiments in flasks or chemostats, ''control'' of very large containers is not physically easy. For example, the initial step of adding a pollutant to a large volume of water may be done in several different ways, so as in fact to constitute different treatments. Second, the type of water that is selected either by geographic area (e.g., arctic vs. tropic) or season (e.g., summer vs. winter) may be expected to yield different results. A third source of variation in results pertains to the different starting conditions of an experiment. Several authors have recognized that if replicate mesocosms are not filled at exactly the same time, small scale differences in plankton populations and nutrients may become exaggerated over time to produce very different results, without any apparent perturbation having been made to part of the system.

In addition to these experimental problems, the actual effects of pollutants may vary with the amount of pollutant, the rate of addition, and the state of the ecosystem being examined. A combination of all these effects at one time may be difficult to diagnose and may give rise to different interpretations. In this chapter illustrations of different effects have been taken from (1) published reports of petroleum hydrocarbon and eutrophication mesocosm experiments, and (2) computer model simulations which show how changes in phasing between different levels of production may help to explain some of the results.

## 2. RESULTS

### 2.1. The Effects of Different Pollutants on Marine Pelagic Ecosystems

#### *2.1.1. Petroleum Hydrocarbons*

From laboratory scale experiments, Gordon and Prouse (1973) reported that the general effect of petroleum hydrocarbons on phytoplankton was to cause decreased photosynthesis. This occurred with crude oil and No. 2 fuel oil at water-soluble concentrates above 100 ppb; there was, however, a suggestion in their paper that photosynthesis was increased by ca. 120% at concentrations of crude oil below 50 ppb. In experiments with ca. 60-ton controlled ecosystems Parsons et al. (1976) showed that the Haptophyte, *Chrysochromulina kappa,* increased in the presence of ca. 75 to 220 ppb of aromatic hydrocarbons. This result could also be demonstrated experimentally; it was shown that below 200 ppb of aromatic hydrocarbons from No. 2 fuel oil, there was an increase in photosynthesis of up to 130% (the maximum occurring at 50 ppb of aromatic hydrocarbons). Vargo et al. (1982) have reported that the presence of 90- to 190-ppb hydrocarbons from No. 2 fuel oil caused an increase in the populations of diatoms (dominated by *Chaetoceros* spp.) and nanoflagellates. This was attributed to a reduced grazing pressure by herbivorous zooplankton combined with the nonlethal concentrations of No. 2 fuel oil. Dahl et al. (1983) found that Ekofisk crude oil in controlled ecosystems at 470 ppb caused a decrease in primary productivity and chlorophyll *a*, which was largely attributed to inhibition of diatom growth.

Herbivorous metazoan zooplankton and benthic fauna appear to be inhibited by petroleum hydrocarbons at concentrations of 90–190 ppb of No. 2 fuel oil (Oviatt et al., 1982). Herbivorous protozoans, on the other hand, grow well in the presence of petroleum hydrocarbons. Both Lee et al. (1977) and Dahl et al. (1983) have reported large numbers of ciliates that fed off a bacterial/zooflagellate food chain.

The presence of a dispersant, such as the biodegradable Corexit 9527, generally causes the oil/dispersant mixture to be more toxic toward zooplankton (e.g., Hartwick et al., 1982; Parsons et al., 1984b). However, bacteria and choanoflagellates (or unspecified zooflagellates) are generally reported to grow well in the dispersant/oil mixtures and this would enhance their decomposition (Traxler and Bhattacharya, 1978; Throndsen, 1982; Parsons et al., 1984b).

While the above references to hydrocarbon effects as observed in controlled ecosystems are by no means complete, they serve as examples, which are summarized in Figure 17.1. This figure has some diagnostic value in illustrating why apparent differences may occur in reactions of the plankton community to petroleum hydrocarbons. These include increased or decreased autotrophic production, increased protozoan but decreased metazoan production, and generally increased bacterial/zooflagellate production.

### *2.1.2. Nutrient Enrichment*

Nutrient enrichment can be interpreted very widely to include many inorganic and organic compounds. For the purposes of this discussion, nutrient enrichment will be interpreted to be either an inorganic source of nitrogen or an organic source of metabolizable substrate. The former has been chosen for illustrative reasons because nitrogen is often considered to be the limiting nutrient in seawater (e.g., Ryther and Dunstan, 1971); the latter substrate is included here because it provides an alternative source of energy to autotrophic production and may in fact represent a more ancient pathway in the aquatic food chain.

The effect of repeated additions of nitrate to seawater is to enhance diatom production providing all other nutrients (e.g., phosphate, silicate) remain abundant (Takahashi et al., 1982). Under conditions of silicate limitation, flagellates dominate (Thomas and Dodson, 1975). Flagellates and pennate diatoms appear to be more abundant at low nutrient levels; this appears to be particularly true in the case of flagellates, if ammonium is the only source of new inorganic nitrogen, or if there is a low nutrient flux (Harrison and Turpin, 1982). Excessive nutrients generally lead to the production of phytodetritus, which represents ungrazed phytoplankton available to the benthic community; at low levels of nutrient input, phytoplankton are usually grazed efficiently by the zooplankton, which results in a smaller input to the benthic community in the form of zooplankton fecal pellets (cf. Gamble et al., 1977; Parsons et al., 1977).

Henderson and Smith (1980) showed that benthic algal communities enriched

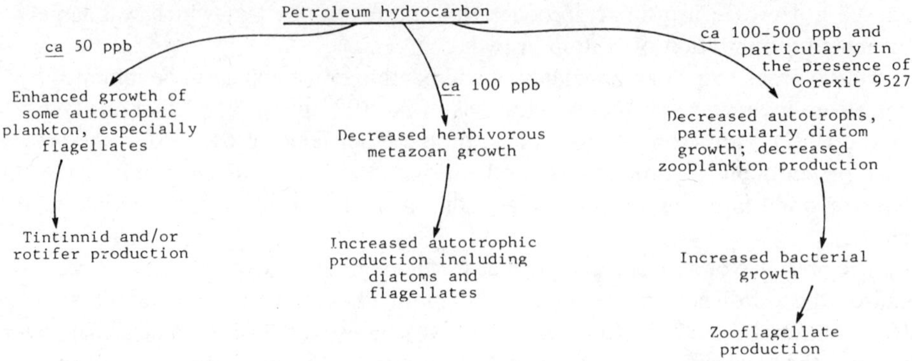


**Figure 17.1.** Different responses of the plankton community to petroleum hydrocarbons.

with nitrogen and phosphorus retained the phosphorus enrichment for longer time periods (over 2 months). Schiewer and Baader (1982) found that sporadic releases of ammonium from blue-green algal mat communities could be explained by periodic changes in the phosphorus flux. Thus the N:P ratio of initial enrichment conditions together with the different metabolic pathways of N and P in the plant communities could yield very different results, including the potential for ammonia toxicity if there was any shift in pH to alkaline conditions.

The addition of an organic substrate (sucrose) to river water was shown to increase benthic production and the yield of fish (Warren et al., 1964). In mesocosm experiments, the addition of glucose (1 and 5 ppm) to seawater caused a depression in autotrophic production and an increase in heterotrophic production that enhanced the growth of meroplanktonic benthic larvae (Parsons et al., 1981). In this experiment the bacteria outcompeted the phytoplankton for a limited supply of nitrate, which resulted in a ca. 70% depression in photosynthesis in the presence of glucose.

From these results it is possible to draw a second diagnostic chart (Fig. 17.2) which illustrates the various effects of nutrient enrichment (nitrate or glucose) on an aquatic ecosystem. This figure indicates that a variety of results can be obtained in controlled ecosystems ranging from increased secondary holoplanktonic production, increased secondary meroplanktonic production, decreased diatom production and changes in the amount and quality of food available to the benthos.

## 2.2. Numerical Model Simulations of the Effects of Different Pollutants

### *2.2.1. The Model and Its Application*

Some of the results of experimental simulations of food chains, as seen above inside controlled ecosystems, can also be simulated in a computer model. The model has been described in detail previously (Parsons and Kessler, 1986) and an outline of the interacting components is given in Figure 17.3 and Table 17.1. The

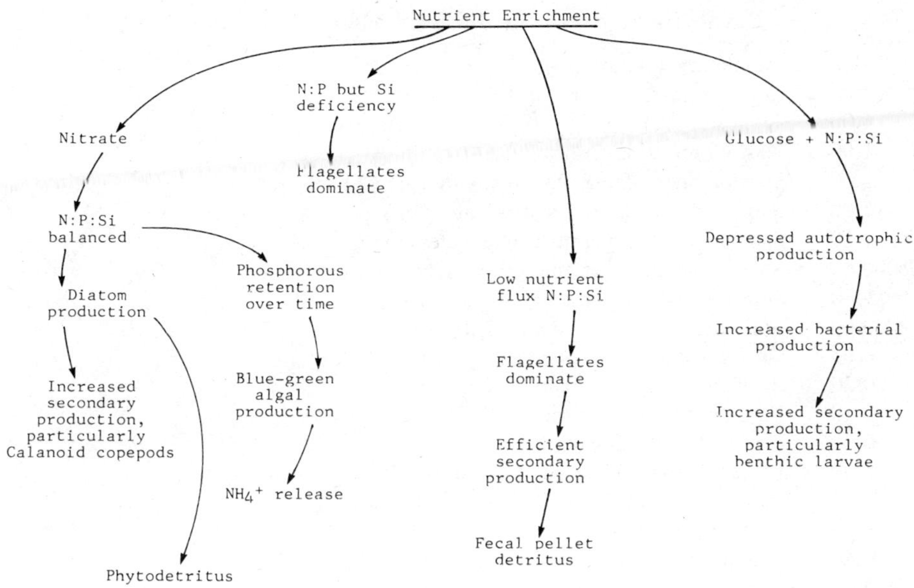


**Figure 17.2.** Different responses of the algal community to nutrient (inorganic and organic) enrichment.

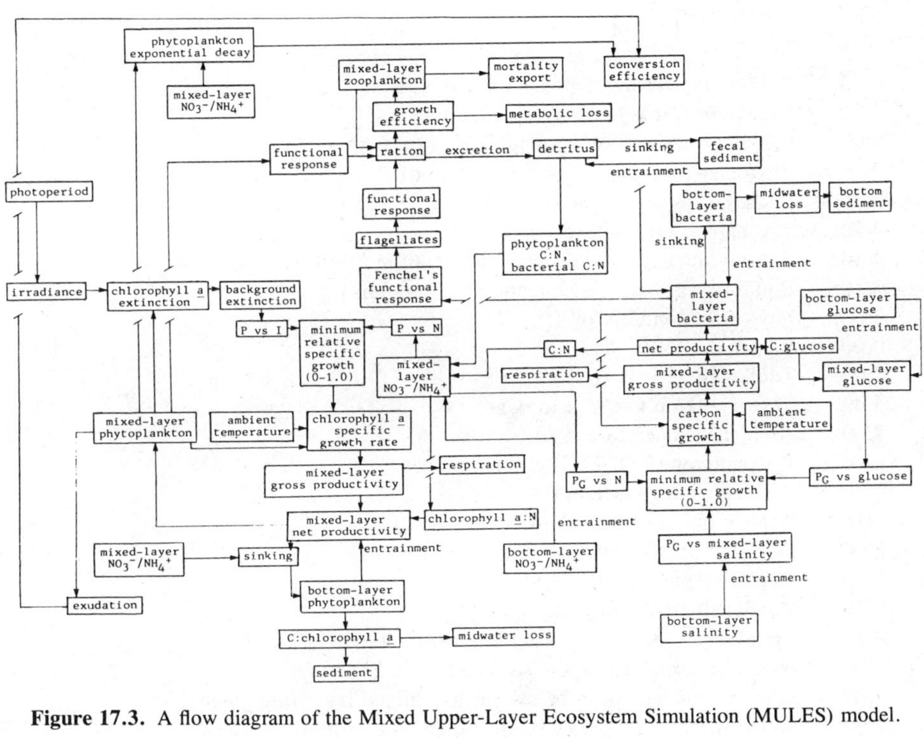


**Figure 17.3.** A flow diagram of the Mixed Upper-Layer Ecosystem Simulation (MULES) model.

**Table 17.1 Parameter Used in the MULES Model**

| | |
|---|---|
| 288.00 | Simulation duration (hr) |
| 5.00 | Depth of mixed-layer (m) |
| 1.00 | Time step (hr) |
| 1.00 | Starting mixed-layer nitrate ($\mu M$) |
| 2.00 | Starting mixed-layer phytoplankton (mg Chl $a$/m$^3$) |
| 2.00 | Starting mixed-layer bacteria (mg C/m$^3$) |
| 0.10 | Starting mixed-layer microflagellates (mg C/m$^3$) |
| 5000.00 | Starting mixed-layer glucose (mg glucose/m$^3$) |
| 5.00 | Starting mixed-layer zooplankton (mg C/m$^3$) |
| 2000.00 | Maximum surface irradiance ($\mu$E/m$^2$/sec) |
| 0.60 | Daylight fraction of 24-hr day (n.d.) |
| 0.50 | Background light extinction coefficient (/m) |
| 0.10 | Entrainment velocity (m/day) |
| 0.025 | Chlorophyll-specific extinction (m$^2$/mg Chl$a$) |
| 1.00 | 1/2 saturation for glucose-limited growth (mg glucose/m$^3$) |
| 0.10 | Bacterial 1/2 saturation for N-limited growth ($\mu$M) |
| 18.00 | 1/2 saturation for salt-inhibited bacterial growth (ppt) |
| 0.60 | Phytoplankton 1/2 saturation for N-limited growth ($\mu M$) |
| 200.00 | 1/2 saturation for L-limited growth ($\mu$E/m$^2$/sec) |
| 15.00 | 1/2 saturation for grazing on phytoplankton function (mg C/m$^3$) |
| 20.00 | Fenchel's 1/2-saturation bacterial grazing function (mgC/m$^3$) |
| 15.00 | 1/2 saturation for grazing on flagellate function (mg C/m$^3$) |
| 0.05 | Maximum grazing response on phytoplankton (mg C/mg C/hr) |
| 0.23 | Fenchel's maximum grazing response on bacteria (mg C/mg C/hr) |
| 0.15 | Maximum grazing response on zooflagellates (mg C/mg C/hr) |
| 5.00 | Grazing threshold on phytoplankton (mg C/m$^3$) |
| 5.00 | Grazing threshold on zooflagellates (mg C/m$^3$) |
| 0.10 | Phytoplankton respiration as fraction of $P_{max}$ (n.d.) |
| 0.10 | Bacterial respiration as fraction of $B_{max}$ (n.d.) |
| 0.20 | Zooplankton ration excreted (n.d.) |
| 0.10 | Bottom-layer phytoplankton concentration (mg Chl $a$/m$^3$) |
| 0.10 | Bottom-layer bacteria concentration (mg cm$^3$) |
| 6.00 | Phytoplankton C:N (n.d.) |
| 10.00 | Bacterial C:N (n.d.) |
| 6.00 | Zooflagellate C:N (n.d.) |
| 25.00 | Phytoplankton C:Chl $a$ (n.d.) |
| 18.00 | Ambient temperature (°C) |
| 0.60 | Zooplankton (G/T + G) efficiency (n.d.) |
| 0.80 | Zooflagellate growth efficiency (n.d.) |
| 0.10 | Replete phytoplankton sinking rate (m/day) |
| 2.00 | Depleted phytoplankton sinking rate (m/day) |
| 0.50 | Deplete/replete switch ($\mu M$) |
| 0.10 | Zooflagellate sinking rate (m/day) |
| 0.50 | Zooplankton detritus sinking rate (m/day) |
| 20.00 | $NO_3^-$ concentration below surface mixed layer ($\mu M$) |
| 1.00 | Glucose concentration below surface mixed-layer (mg glucose/m$^3$) |
| 31.00 | Salt concentration below surface mixed-layer (ppt) |

**Table 17.1** (*Continued*)

| | |
|---|---|
| 0.00 | Zooplankton detritus below surface mixed-layer (mg C/m$^3$) |
| 0.10 | Zooflagellate concentration below surface mixed layer (mg C/m$^3$) |
| 0.0275 | Slope for phytoplankton Arrhenius regulation |
| -0.07 | *Y* intercept for phytoplankton Arrhenius regulation |
| 0.10 | Slope for bacterial Arrhenius regulation |
| -1.50 | *Y* intercept for bacterial Arrhenius regulation |
| 48.00 | Hours between simulation snapshots (n.d.) |
| 0.15 | Zooplankton mortality rate (%/day) |
| 0.00 | Simulation start time after sunrise (hr) |
| 20.00 | Depth of bottom layer (m) |
| 0.00 | Constant phytoplankton leakage (fraction/day) |
| 0.10 | Efficiency of bacterial conversion of phytoplankton exudate (n.d.) |

Mixed Upper-Layer Ecosystem Simulation (MULES) model is designed to have four basic components that interact as shown in Figure 17.3. These are autotrophic, heterotrophic, and zooplankton production, and sedimented material to the benthic community.

Relationships between trophic levels within the model generally utilize equations to be found in Parsons et al. (1984a). Light and nutrients are forcing functions for autotrophic production, while metabolizable organic substrate (expressed as ''glucose'') is the forcing function for heterotrophic production. The model uses nitrate as the rate-limiting nutrient but recycling of nitrogen through the production of ammonium occurs as a result of zooplankton grazing. Bacterial production is a function of phytoplankton decay, phytoplankton exudation, or allochthonous ''glucose'' introduced by entrainment from below the mixed layer. Bacteria are consumed by zooflagellates according to the work of Fenchel (1982) and the zooflagellates themselves are eaten by the zooplankton. Benthic productivity is a function of production within the model and assignment of sinking rates to various components (e.g., phytoplankton, depending on the amount of nitrate remaining in the water column).

The model contains a large number of parameters, which would appear to make it very sensitive to starting conditions and choice of coefficients. In the case of the starting conditions, the model has been started at a low-plankton, high-nutrient level such as might occur in time during the spring bloom, or in space in an upwelling zone. Since these are the most important temporal/spatial events for plankton production, it was felt that they were the most logical on which to carry out perturbations. Thus the ratios of phytoplankton to bacteria to zooplankton to zooflagellates, and so forth, are approximately those for the initiation of a bloom. In the case of coefficients for grazing, photosynthesis, respiration, and so on, it was a matter of choice to enter values that were reasonably consistent with the literature for temperate plankton. Thus the effect of perturbations was considered to have simulated natural growth conditions, rather than randomizing events such as could be simulated by combining unreasonably conflicting values (e.g., an entrainment velocity to the mixed layer that flushed phytoplankton out of the system faster than they grew).

Coefficients for the equations and starting values for each variable are shown in Table 17.1. Changes in these parameters have been made to simulate various perturbations as follows.

For Figure 17.4, 5 ppm of glucose was added and production of phytoplankton, zooplankton, zooflagellates and bacteria were compared with the same system in the absence of glucose. This has the effect of simulating the addition of a metabolizable organic pollutant to a marine ecosystem.

For Figure 17.5, glucose was returned to 0 ppm and a control (i.e., batch) simulation using 5 $\mu M$ nitrate with no further additions (entrainment = 0) was compared with a starting nitrate level of 1 $\mu M$ nitrate with a daily entrainment (0.3 m/day) amounting to 1.2 $\mu M$ of nitrate per day (i.e., pulsed). The requirement in this experiment was to simulate eutrophication under conditions of continuous or single large loads of nutrients.

In Figure 17.6, the nitrate was returned to 5 $\mu M$ and the extinction coefficient was changed over the range 0.2, 0.7, and 2.0 $m^{-1}$ to simulate the effect of reduced light levels. Such conditions might be encountered during dredging operations or from the release of mine tailings into an estuary.

### *2.2.2. The Effect of Organic Loading*

The addition of 5 ppm of glucose in Figure 17.4 results in a suppression of autotrophic production (ca. 5 vs 11 $mg/m^3$ chlorophyll *a* maxima in the treated and control simulations, respectively). In spite of the suppressed autotrophic produc-

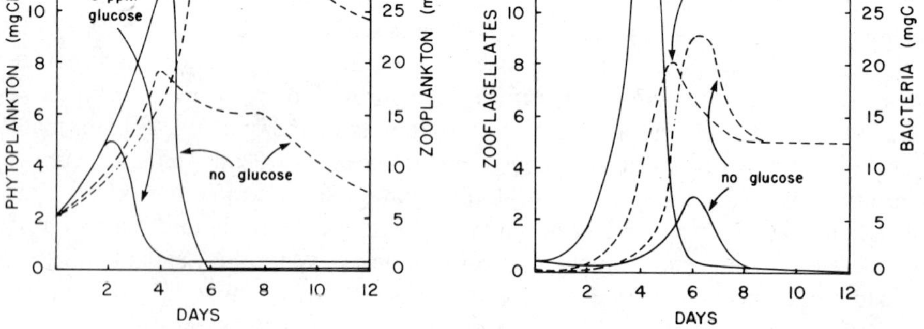


**Figure 17.4.** (A) Changes in phytoplankton (—) and zooplankton (---) in the presence (5 ppm) and absence of glucose. (B) Changes in bacteria (—) and zooflagellates (---) in the presence (5 ppm) and absence of glucose.

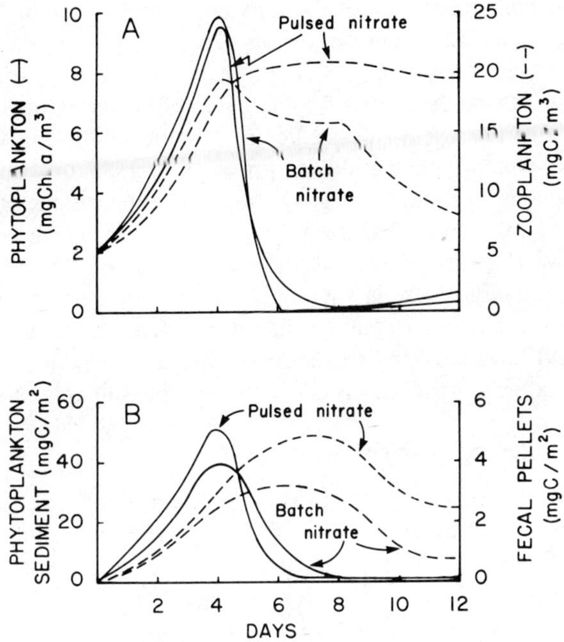


**Figure 17.5** (A) Response of phytoplankton (—) and zooplankton (---) to pulsed and batch nitrate additions. (B) Response of phytodetritus (—) and zooplankton fecal pellets (---) to pulsed and batch nitrate additions.

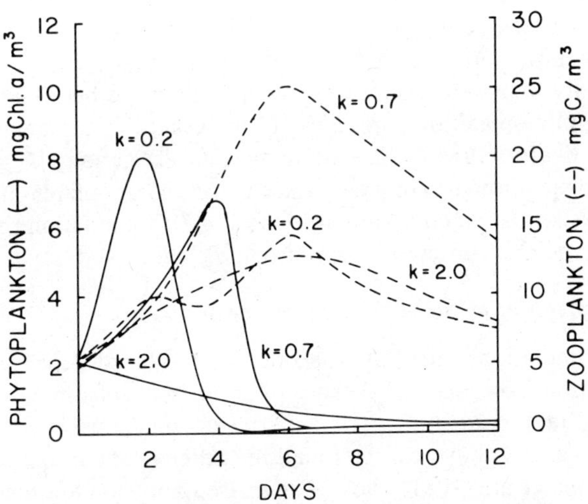


**Figure 17.6.** Phytoplankton (—) and zooplankton (---) production at extinction coefficients of 0.2, 0.7, and 2.0 $m^{-1}$.

tion, the zooplankton production is about twice as great in the presence of glucose than in its absence. This sequence of events is caused by bacteria taking up nitrate in competition with autotrophic production and in transferring much of the ''primary production'' through a heterotrophic cycle of bacteria/zooflagellates to zooplankton. Although this cycle is less efficient than the autotrophic cycle, it is nonetheless better *in phase* with the slower growing zooplankton. In the absence of glucose, phytoplankton production grows rapidly, exhausts the supply of nitrate, and then either sinks out or causes a pulse of bacteria and zooflagellates (Fig. 17.4B) that show up on day 8 as food for zooplankton (Fig. 17.4A). The presence of glucose causes a large pulse in bacterial biomass that is grazed by zooflagellates to provide food for zooplankton at the time when phytoplankton is declining (i.e., by day 4, Figs. 17.4A,B). This synchronization of zooplankton food supply enables the zooplankton to grow much better than on phytoplankton alone.

### 2.2.3. *The Effect of Nutrient Additions*

In Figure 17.5A the effect of pulsed additions of nitrate is to maintain the zooplankton standing stock much better than if a single (batch) addition of nitrate is made at the beginning of the experiment. Figure 17.5B. shows changes in sedimented material, with phytodetritus dominating when nitrate is still present, but zooplankton fecal detritus dominating (after day 8) at low levels, or in the absence of new nitrate.

### 2.2.4. *The Effect of Silt*

Figure 17.6 shows the effect of increasing extinction coefficients on phytoplankton and zooplankton production. At low values (i.e., clear water) phytoplankton grows rapidly, exhausts the nitrate supply, and sinks out or decays. The zooplankton cannot grow fast enough to benefit greatly from this growth pattern but they obtain some extra growth after day 6 from a bacteria/zooflagellate growth based on decaying phytoplankton. However, when light is reduced ($k = 0.7\ m^{-1}$) zooplankton growth is actually increased because the zooplankton are better able to track the slower growing phytoplankton (ca. 25 vs. 15 mg C/m$^3$ of zooplankton for $k = 0.7$ and 0.2, respectively). Finally, at extremely high extinctions ($k = 2.0$), phytoplankton growth is strongly suppressed such that the zooplankton production is also curtailed back to a quantity similar to the $k = 0.2$ simulation, but fed entirely by autotrophs instead of through a heterotrophic cycle.

### 2.2.5. *Verification of Simulations from Experiments*

In these three simulations, alternative results are produced depending on the rate of nutrient additions, presence or absence of an organic substrate, and the level of light allowed into the system. Experimental results that support some of the computer-simulated results above can be found in earlier work. In the case of glucose additions, Parsons et al. (1981) showed increased zooplankton production in the presence of 1 and 5 ppm of glucose. The pulsed addition of nitrate has been used extensively by persons operating controlled ecosystems. However, the interesting point is perhaps the fact that a single bloom of phytoplankton (e.g., a spring bloom) tends to cause a large amount of phytodetritus to sediment out, such as has been found to occur in the nutrient-rich waters of the Pacific (Stephens et al., 1967). By contrast, where nutrients are limiting in Atlantic coastal waters, Steele and Baird (1972) found that most of the material collected in sediment jars was zooplankton fecal pellets. Finally, the results of increased zooplankton production with a small increase in the extinction coefficient has also been found to occur under experimental conditions when mine tailings were added to controlled ecosystems (Parsons et al., 1986). In this reference it was shown that single additions of 40 and 300 ppm of mine tailings caused a phase lag in primary productivity that eventually resulted in a much larger production of both larvaceans and copepods as compared with the control mesocosm.

In conclusion it appears that the effect of various perturbations on aquatic ecosystems can result in a variety of consequences, depending on how the perturbation is applied. This seems to be important in a diagnostic sense in that it helps to explain otherwise apparently conflicting results. While the main pathways of organic production in the aquatic environment are reasonably well understood (e.g. the effects of light, temperature, and nutrients), it is the effects of minor perturbations on these ''forcing'' functions that cause very marked additional change, largely through small phase shifts. These ''phasing'' functions need to be investigated and a diagnostic manual should be developed for environmental managers which describes the alternative effects to be expected when certain treatments are applied to aquatic ecosystems.

## ACKNOWLEDGMENTS

The author acknowledges the financial support of the Natural Sciences Engineering and Research Council and the Max Bell Foundation. The programming for the computer model, MULES, was done by Dr. T. A. Kessler, Department of Oceanography, University of British Columbia.

## REFERENCES

Dahl, E., Laake, M., Tjessem, K., Eberlein, K., and Bohle, B. (1983). Effects of Ecofisk crude oil on an enclosed planktonic ecosystem. *Mar. Ecol.: Prog. Ser.* **14,** 81–91.

Fenchel, T. (1982). Ecology of heterotrophic microflagellates. II. Bioenergetics and growth. *Mar. Ecol.: Prog. Ser.* **9,** 35–42.

Gamble, J. C., Davies, J. M., and Steele, J. H. (1977). Loch Ewe bag experiment, 1974. *Bull. Mar. Sci.* **27,** 146–175.

Gordon, D. C., and Prouse, N. J. (1973). The effects of three oils on marine phytoplankton photosynthesis. *Mar. Biol. (Berlin)* **22,** 329–333.

Harrison, P. J., and Turpin, D. H. (1982). The manipulation of physical, chemical, and biological factors to select species from natural phytoplankton communities. In G. D. Grice and M. R. Reeve (Eds.), *Marine Mesocosms.* Springer-Verlag, New York, pp. 275–289.

Hartwick, E. B., Wu, R. S. S., and Parker, D. B. (1982). Effect of crude oil and an oil dispersant (Corexit 9527) on populations of the littleneck clam (*Protothaca staminea*). *Mar. Environ. Res.* **6,** 291–306.

Henderson, R. S., and Smith, S. V. (1980). Semitropical marine microcosms: Facility design and elevated nutrient effects experiment. In J. P. Giesy (Ed.), *Microcosms in Ecological Research.* Publ. Tech. Inf. Cent., U.S. Dept. of Energy, Washington, DC, pp. 869–910.

Lee, R. F., Takahashi, M., Beers, J. R., Thomas, W. H., Seibert, D. L., Koeller, P., and Green, D. R. (1977). Controlled ecosystems: Their use in the study of petroleum hydrocarbons on plankton. In F. J. Vernberg, A. Calabrese, F. P. Thurberg, and W. B. Vernberg (Eds.), *Physiological Responses of Marine Biota to Pollutants.* Academic Press, New York, pp. 323–341.

Oviatt, C., Frithsen, J., Gearing, J., and Gearing, P. (1982). Low chronic additions of No. 2 fuel oil: Chemical behaviour, biological impact and recovery in a simulated estuarine environment. *Mar. Ecol.: Prog. Ser.* **9,** 121–136.

Parsons, T. R., and Kessler, T. A. (1986). Computer model analysis of a pelagic ecosystem in estuarine waters. In S. Skreslet (Ed.), *The Role of Freshwater Outflow in Coastal Marine Systems.* NATO ASI Series, Vol. G7, Springer-Verlag, New York, pp 161–181.

Parsons, T. R., Li, W. K. W., and Water, R. (1976). Some preliminary observations on the enhancement of phytoplankton growth by low levels of mineral hydrocarbons. *Hydrobiologia* **51,** 85–89.

Parsons, T. R., von Bröckel, K., Koeller, P., Takahashi, M., Reeve, M. R. and Holm-Hansen, O. (1977). The distribution of organic carbon in a marine planktonic food web following nutrient enrichment. *J. Exp. Mar. Biol. Ecol.* **26,** 235–247.

Parsons, T. R., Albright, L. J., Whitney, F., Wong, C. S., and B. Williams, P. J. (1981). The effect of glucose on the productivity of seawater: An experimental approach using controlled aquatic ecosystems. *Mar. Environ. Res.* **4,** 229–242.

Parsons, T. R., Takahashi, M., and Hargrave, B. (1984a). *Biological Oceanographic Processes.* Pergamon, Oxford.

Parsons, T. R., Harrison, P. J., Acreman, J. C., Dovey, H. M., Thompson, P. A., Lalli, C. M., Lee, K., Li, G. and Chen, X. L. (1984b). An experimental marine ecosystem response to crude oil and Corexit 9527. Part 2. Biological effects. *Mar. Environ. Res.* **13,** 265–275.

Parsons, T. R., Lalli, C. M., Thompson, P., Wu, Y., Hou, S., and Xu, H.-S. (1986). The effect of mine tailings on the production of plankton. *Acta Oceanol. Sin.* **5,** 417–423.

Ryther, J. H., and Dunstan, W. M. (1971). Nitrogen, phosphorous and eutrophication in the coastal marine environment. *Science* **171,** 1008–1012.

Schiewer, J., and Baader, G. (1982). Pelagic $NH_4$ rhythm independent of the sediment in eutrophic shallow waters. *Int. Rev. Gesamten Hydrobiol.* **67,** 845–850.

Steele, J. H., and Baird, I. E. (1972). Sedimentation of organic matter in a Scottish sea loch. *Mem. Ist. Ital. Idrobiol. Dott., Marco de Marchi* **29** (Suppl.), 73–88.

Stephens, K., Sheldon, R. W., and Parsons, T. R. (1967). Seasonal variations in the availability of food for benthos in a coastal environment. *Ecology* **48,** 852–885.

Takahashi, M., Koike, I., Iseki, K., Beinfang, P. K., and Hattori, A. (1982). Phytoplankton species responses to nutrient changes in experimental enclosures and coastal waters. In G. D. Grice and M. R. Reeve (Eds.), *Marine Mesocosms.* Springer-Verlag, New York, pp. 333–351.

Thomas, W. H., and Dodson, A. M. (1975). On silicic acid limitation of diatoms in near surface waters of the eastern tropical Pacific Ocean. *Deep-Sea Res.* **22,** 671–677.

Throndsen, J. (1982). Oil pollution and plankton dynamics. III. Effects on flagellate communities in controlled ecosystem experiments in Lindaspollene, Norway, June 1980. *Sarsia* **67,** 164–169.

Traxler, R. W., and Bhattacharya, L. S. (1978). Effects of a chemical dispersant on microbial utilization of petroleum hydrocarbons. *ASTM Spec. Tech. Publ.* **659,** 180–187.

Vargo, G. A., Hutchins, M., and Almquist, G. (1982). The effect of low, chronic levels of No. 2 fuel oil on natural phytoplankton assemblages in microcosms. 1. Species composition and seasonal succession. *Mar. Environ. Res.* **6,** 245–264.

Warren, C. E., Wales, J. H., Davis, G. E., and Doudoroff, P. (1964). Trout production in an experimental stream enriched with sucrose. *J. Wildl. Manage.* **28,** 617–660.

# 18

# NEW METHODOLOGIES FOR ASSESSING THE EFFECTS OF PRENATAL TOXIC EXPOSURE ON COGNITIVE FUNCTIONING IN HUMANS

***Joseph L. Jacobson and Sandra W. Jacobson***

*Wayne State University, Detroit, Michigan 48202*

## 1. INTRODUCTION

The earliest research on prenatal exposure to toxic substances in humans focused on gross structural abnormalities or birth defects associated with intrauterine exposure to such substances as alcohol and thalidomide (see Abel, 1980; Tuchmann-Duplessis, 1975). The term teratogen was defined as any agent that raises the incidence of deviations or malformations in the embryo or fetus (see Fein et al., 1983). In recent years, there has been increasing interest in subtler functional deficits associated with exposure to teratogenic agents, including disturbances in arousal or sleep regulation, specific learning disabilities, and impaired motor coordination (Vorhees and Butcher, 1982). The term deficit is used in this context, rather than defect, to refer to a developmental lag or disturbance that may or may not persist through subsequent stages of development. Both animal and human studies have shown that in cases where high-level exposure to a teratogenic agent leads to gross physical anomalies, lower level exposure is frequently associated with subtler behavioral or functional deficits (Coyle et al., 1976; Fein et al., 1983; Weiss and Spyker, 1974). For example, children of alcoholic mothers are at risk for mental retardation and the distinctive facial dysmorphology associated with fetal alcohol syndrome, whereas children exposed prenatally to lower levels of alcohol exhibit subtler disturbances, such as slower response times and poorer attention, in the absence of physical deficits (Shaywitz et al., 1980; Streissguth et al., 1984).

This line of research has several implications for the assessment of risks to human health from exposure to persistent toxic substances found in the Great Lakes. First, given the levels at which most humans residing in the Great Lakes region have been exposed to these substances, subtle behavioral deficits appear more likely than gross physical anomalies (Fein et al., 1983). Research in behavioral teratology also suggests that subtle functional deficits may be used to determine the threshold for central nervous system (CNS) damage following developmental exposure to a particular substance. Since functional deficits are typically found at lower levels of exposure than mental retardation and other organic anomalies, where no behavioral effects are detected, the risk of CNS damage can probably be considered minimal.

Our research has focused on the effects of intrauterine exposure to polychlorinated biphenyls (PCBs). PCBs are a family of synthetic hydrocarbons used in a wide range of industrial products, particularly as insulators in electrical transformers and capacitors. Although the manufacture of PCBs was banned during the 1970s, they are highly resistant to biodegradation and are found world-wide (Carey and Gowen, 1976; International Agency for Research on Cancer, 1978; Veith and Lee, 1971; Young et al., 1976). Because they are lipophilic, PCBs tend to bioaccumulate particularly in the fatty tissue of fish and animals subsisting on contaminated fish. Concern with this source of environmental exposure led the U.S. Food and Drug Administration to impose an embargo of commercial fish catches exceeding specified tolerance levels in 1973 (Swain, 1983) and subsequently to lower the permissible levels in the early 1980s. However, catches of hundreds of thousands of sports fishermen were not covered by these restrictions. In a study of sports fisherman residing in western Michigan Humphrey (1983) found a direct correlation between the respondents' serum PCB levels and the amount of Lake Michigan fish they had consumed. Such data have led public health departments in Michigan and elsewhere to issue advisories recommending limitation of the amount of sports fish consumed, especially by women who expect to become pregnant and children.

The first phase of our research was conducted in collaboration with Greta G. Fein, Principal Investigator, who is now at the University of Maryland, and Pamela M. Schwartz. Since exposure early in development seems likely to pose the greatest risk, our study focused on the effects of intrauterine PCB exposure on the offspring of women who had consumed relatively large quantities of Lake Michigan fish. We also assessed exposure to polybrominated biphenyls (PBBs), a fire retardant that was accidentally mixed with cattle feed in Michigan in 1973–74, leading to widespread contamination of local dairy herds (see Barr, 1978; Brilliant et al., 1978; Wickizer et al., 1981).

## 2. STUDY DESIGN

The model shown in Figure 18.1 illustrates alternative routes of potential PCB exposure. Starting on the left, consumption of contaminated fish leads to the bioaccumulation of PCBs in the mother. Since PCBs are difficult to eliminate, the mother's body burden presumably reflects her lifetime exposure and can be as-

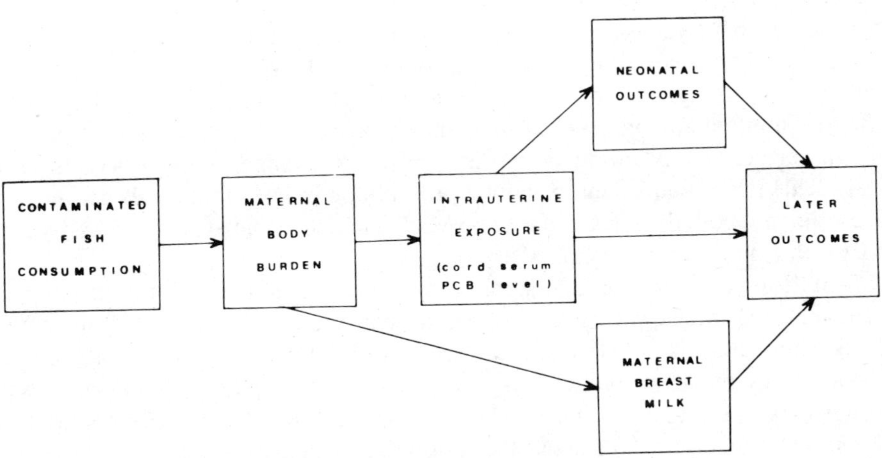


**Figure 18.1.** Alternative sources of pre- and postnatal exposure to PCBs. Modified from Jacobson et al. (1983a).

sessed from the PCB level in her serum. Similarly, umbilical cord serum samples can be obtained during delivery and analyzed for PCB content to provide a biological record of the infant's exposure during the prenatal period. While any newborn effects can be attributed to intrauterine exposure, later infant development can be influenced in several ways. For one, neonatal deficits may persist over time and interfere with subsequent development. On the other hand, intrauterine exposure may affect prenatal central nervous system development in ways that do not become evident until after the newborn period when the relevant cognitive or behavioral system matures or when compensatory mechanisms that initially masked the impairment cease to function. Contaminated breast milk provides a third source of exposure for infants who nurse.

To select the sample, 8482 mothers delivering infants in four western Michigan hospitals were screened between July 1980 and September 1981 (see Jacobson et al., 1983a, for a more detailed description of the sample selection). Each mother was interviewed on the day following delivery regarding the amount of each species of Lake Michigan fish she had consumed during the prior year. If she had eaten greater amounts at some time in the past, she was asked about her consumption rates for that period as well. The questions were broken down by species so that more highly contaminated species could receive greater weight in our estimate of toxic exposure. Wayland Swain of the EPA Large Lakes Research Station assisted us by weighting the various species according to their relative PCB content based on assays of fish performed by various investigators over a 10-year period (Table 18.1). "Contaminated fish consumption" was defined as the weighted sum of annual Lake Michigan fish consumption in the present or past, whichever was greater.

The final sample of 313 women included 242 who ate Lake Michigan fish on a regular basis and 71 who did not eat these fish. Since the distributions of serum and breast milk PCB levels in this sample ranged from negligible or nondetectible to relatively high values, exposure level was treated as a continuous variable in the statistical analyses. Variables that were highly skewed were subjected to logarithmic transformations prior to inclusion in correlational analyses. Contaminated fish consumption among exposed mothers averaged 6.7 kg or about 14.7 pounds of fish per year (standard deviation (SD) = 5.8 kg, range = 1.2 to 41.7 kg per year). This rate is equivalent to about 2 or 3 salmon or lake trout meals per month. The women reported having eaten Lake Michigan fish for an average of 16.1 years (SD = 9.0, range = 1.0 to 40.0 years).

Umbilical cord and maternal serum samples and maternal breast milk samples were analyzed for the presence of polychlorinated and polybrominated biphenyls (PCBs and PBBs) by the Michigan Department of Public Health. The analysis was performed by means of packed column gas chromatography. Quantification was provided by adapting the Webb-McCall method (Sawyer, 1978) to a computer data system with Aroclor 1260 used as the calibration standard for PCBs (see Jacobson et al., 1983a). PCB levels in the cord serum samples were low, with approximately

**Table 18.1 Weights of Lake Michigan Fish Species Used to Derive Summary Measure of Contaminated Fish Consumption**

| Type of Fish | Weights Indicating Relative PCB Content |
|---|---|
| Chinook salmon | 1.0 |
| Coho salmon | 1.0 |
| Lake trout | 1.0 |
| Carp | 1.0 |
| Catfish | 1.0 |
| Bullhead | 0.6 |
| Pike and pickerel | 0.6 |
| German brown trout | 0.4 |
| Perch | 0.4 |
| Walleye | 0.4 |
| White sucker | 0.4 |
| Alewife | 0.3 |
| Burbot (lawyer) | 0.2 |
| Steelhead and rainbow trout | 0.2 |
| Brook trout | 0.2 |
| Menominee (round whitefish) | 0.1 |
| Smelt | 0.1 |
| Whitefish and chub (herring) | 0.1 |

*Source.* Derived from data supplied by the Large Lakes Research Station, U.S. Environmental Protection Agency, Grosse Ile, MI.

two-thirds below the detection limit of 3 ng/mL or parts per billion set by the analytical laboratory (Table 18.2). This finding is not entirely surprising given that PCBs are lipophilic and cord blood is lean. From a statistical standpoint, the poor reliability of measurement at these low levels will tend to depress the magnitude of any effects that might be found in the analyses of the data.

**Table 18.2 PCB Levels in Umbilical Cord and Maternal Sera and in Human Breast Milk[a]**

| Source | *N* | Mean | SD | Range | |
|---|---|---|---|---|---|
| | | | | Low | High |
| Cord serum | 286 | 2.5 | 1.9 | 0.0 | 12.6 |
| Maternal serum | 208 | 5.5 | 3.7 | 0.2 | 23.1 |
| Breast milk (fat basis) | 138 | 812.7 | 379.4 | 185.7 | 2600.0 |

[a]Values are expressed in terms of ng/mL or parts per billion (ppb).

## 3. NEWBORN AND INFANT FINDINGS

### 3.1. Placental Passage

One Japanese study provided evidence of placental passage of PCBs (Kodama and Ota, 1977), but this was not confirmed in two other studies, possibly because the sample sizes were small and the contaminant levels low (Akiyama et al., 1975; Masuda et al., 1978). The data from our study (Table 18.3) show a significant relationship between the amount of contaminated fish consumption reported by the mother and the level of PCBs in her serum (Schwartz et al., 1983). Placental passage is indicated by the significant correlation between maternal and cord serum PCB levels. There also were significant correlations for polybrominated biphenyls (PBBs), the fire retardant that was accidentally mixed with cattle feed in Michigan in 1973–74 (Table 18.3). Because PCBs and PBBs are chemically very similar, the substantially higher PBB correlations were not expected.

The differences between the PCB and PBB correlations in Table 18.3 are probably due to differences in reliability of measurement (see Jacobson et al., 1984b). PBBs are easier to measure accurately because few other compounds appear in their region on the gas chromatograms. By contrast, several common chlorinated hydrocarbon pesticides appear in the same region as PCBs, making them difficult to discriminate in laboratory tests. In addition, only one mixture of PBB isomers entered the Michigan food chain, whereas a variety of PCB mixtures are found in the environment (Jacobson et al., 1984b). Assuming that certain isomers in each mixture cross the placenta more readily than others, those PBB isomers that cross readily are a relatively constant proportion of any maternal PBB serum sample. Total PBB level should, therefore, estimate the transfer of these isomers relatively accurately. By contrast, the proportion of PCB isomers that cross readily to the fetus may vary depending on the particular PCB mixtures to which the mother has been exposed. Overall PCB level may, therefore, provide a less accurate estimate of fetal exposure. This comparison with the PBB data suggests that, if PCB levels could be measured more reliably, the placental passage correlation would be substantially higher.

**Table 18.3 Transfer of PCBs and PBBs to Serum and Maternal Milk[a]**

| | PCBs | PBBs |
|---|---|---|
| Maternal fish consumption–maternal serum | 0.38* (196) | —[b] |
| Maternal serum–cord serum | 0.42* (148) | 0.81* (153) |
| Maternal serum–breast milk (fat basis) | 0.35* (91) | 0.71* (92) |

[a]Values are Pearson correlation coefficients. $N$ is indicated in parentheses.
[b]Not tested.
$^*P < .001$.

### 3.2. Potential Confounding Variables

Laboratory animal research in behavioral teratology utilizes experimental designs in which pregnant animals are deliberately exposed to discrete levels of potentially toxic substances. Since exposure level is determined by random assignment, group differences in outcome are directly attributable to differences in exposure. A unique issue in human teratological research is how to deal with other variables that may be confounded with exposure level due to lack of experimental control when random assignment is not possible. In this study, we started by compiling a list of more than 70 control variables that could affect infant outcome, including demographic background, infant health history, pregnancy and delivery complications, maternal smoking, and consumption of alcohol and caffeine. Dichotomous variables that did not occur in at least 15% of the subjects were considered too rare to evaluate statistically, thereby reducing the list to 36 variables (see Jacobson et al., 1986). In general, fish consumption did not vary systematically with socioeconomic status, education, or other background factors; some people just seem to like fish, while others do not. Most of the Pearson correlations of the control variables with the PCB exposure measures were below 0.10, and none exceeded 0.25. To be conservative, however, all control variables that were even weakly related to PCB exposure ($P < 0.10$) were treated as potential confounds.

In the analyses reported in Sections 3.3, 3.4, and 3.5 below, the influence of potential confounding variables was controlled statistically by means of stepwise multiple regression analyses, in which the order of entry of the variables was determined a priori. For example, in the analysis of the effects of cord serum PCB level on the newborn, four variables listed in Table 18.4 were considered potential confounds because they were significantly correlated with cord serum PCB level. These potential confounds were entered in the first four steps of each regression evaluating cord PCB effects on infant outcome, and cord serum PCB level was entered in the final step. A PCB effect was inferred only if the entry of PCB exposure significantly increased the variance explained in the outcome measure. This procedure is conservative in that it attributes to PCB exposure only the variance that cannot also be explained by the potential confounds entered in the earlier steps of the analysis. Separate regressions were performed to assess the effects of each

**Table 18.4 Correlations of Cord Serum PCB Level with Its Four Potential Confounding Variables**

| Potential Confound | $r$ |
|---|---|
| Maternal age | 0.12* |
| Maternal weight gain during pregnancy | −0.14* |
| Maternal alcohol consumption during pregnancy | 0.14* |
| Maternal smoking during pregnancy | 0.12* |

*$P < .10$.

exposure measure (e.g., cord serum PCB level, maternal contaminated fish consumption) on each outcome variable (birth size, neonatal behavior, etc.). In addition to the potential confounds, three suppressor variables, maternal height and weight and sex of infant, were used in the analyses of cord PCB effects on birth size to compensate for a tendency for cord serum PCB levels to be higher in the infants of larger mothers and in male infants.

### 3.3. Size Effects

Shortly after birth, the infants were assessed for size, gestational age, and behavioral functioning. The infant and child examiners in all phases of this research were blind with regard to both maternal report and biological measures of toxic exposure. This was accomplished by maintaining separate staffs for laboratory analyses, maternal interviews, and child assessments. Higher cord serum PCB levels were significantly related to smaller birth size and shorter gestation after controlling for potential confounding variables as indicated in Section 3.2 (Fein et al., 1984). Higher levels of maternal consumption of contaminated Lake Michigan fish were also significantly correlated with these outcomes, supporting our assumption that contaminated fish were a major source of PCB exposure for these infants. The effects of contaminated fish consumption on birth weight, head circumference, and gestational age were all dose dependent (see figure in Fein et al., 1984). Length of gestation was assessed on the basis of both the mother's report of her last menstrual period and the Ballard examination, which estimates gestational age from the infant's physical and neuromuscular maturity at birth. Interestingly, it was the mother's *lifetime* fish consumption level rather than her level of consumption during pregnancy that predicted the size and gestational age deficits (Fein et al., 1984). This finding suggests that the risk to the fetus is based more on the gradual accumulation of PCBs in the mother over time than on exposure during pregnancy per se.

Although these effects were statistically significant, their absolute magnitude was small. None of the infants of mothers who had consumed large quantities of fish weighed less than 1500 g, the medical criterion for "very low birth weight," and few were actually preterm, that is, less than 38 weeks gestation. As shown in Table 18.5, newborns with detectible levels of PCBs averaged about 160 g less at

**Table 18.5 Weight at Birth and 5 Months by Cord Serum PCB Level[a]**

| Weight (g) | Cord Serum PCB Level | | |
|---|---|---|---|
| | < 3 ppb | ≥ 3 ppb | Difference |
| At birth | $3575 \pm 536$ (166) | $3412 \pm 538$ (75) | 163 |
| 5 months | $7375 \pm 791$ (88) | $7112 \pm 791$ (45) | 263 |

[a]Values are means ± standard deviation. *N* is indicated in parentheses.

birth than infants with PCB levels under 3 parts per billion, a size deficit which is comparable to that associated with smoking during pregnancy (U.S. Public Health Service, 1979). However, whereas the offspring of women who smoke during pregnancy tend to catch up during the early postpartum months (Proulx, 1985; Russell et al., 1968), the size deficit associated with prenatal PCB exposure persisted. Higher cord serum PCB level continued to be correlated with smaller size at 5 months.

### 3.4. Newborn Behavior

The Brazelton (1973) Neonatal Behavioral Assessment Scale was administered to each infant on the third day following delivery. The Brazelton Scale is a 30-minute procedure designed to assess a broad range of newborn behaviors, including motor coordination, muscle tone, state changes (that is, arousal and self-quieting), and infant reflexes. (These items were then reduced to seven clusters, described in Jacobson et al. (1984a).) Higher levels of maternal contaminated fish consumption were associated with several neonatal behavioral deficits, including more jerky, unbalanced movement and startles, fewer state changes in response to external manipulation, and a greater number of abnormally weak reflexes (Jacobson et al., 1984c). Although smaller birth size and shorter gestation were significantly related to both contaminated fish consumption and cord serum PCB level, the behavioral effects were not correlated with the cord serum measure. How can we explain this discrepancy? As mentioned earlier, the reliability of the cord serum PCB laboratory analyses was limited. The finding that a greater number of outcomes were predicted by contaminated fish consumption suggests that the maternal report may actually provide a more reliable measure at these levels of exposure than cord serum. Alternatively, since the behavioral deficits were correlated with fish consumption but not with serum PCB level, it is possible that these deficits were due to other toxins from the same contaminated fish that were not measured by the analytical laboratory.

### 3.5. Infant Cognitive Functioning

The principal infant cognitive measures were the Bayley Scales of Infant Development (Bayley, 1969) administered at 5 months and a visual recognition memory task (Fagan and Singer, 1983) administered at 7 months. The Bayley Scales are currently the most popular standardized infant test and are designed to assess neurobehavioral development across a broad range of competences. While short-term test–retest reliability is good, providing evidence of concurrent validity, the Bayley has been criticized for not adequately assessing visual and auditory processing and for utilizing tasks requiring motoric proficiency to evaluate cognitive competence (Zelazo, 1979).

In the Visual Recognition Memory Test (Fagan and Singer, 1983), the infant is seated on the mother's lap in front of an observation chamber containing two stimulus plaques. An observer watches the infant through a small peephole in back of the stage and records the infant's gaze direction on an electronic event recorder. The infant is initially shown two identical target photos, which appear simultaneously in the right and left positions until he or she has looked at them for a total of 20 seconds. The familiar target is then paired with a novel target for two 5-second recognition periods, reversing left–right positions from one period to the next. Because the normative response at this age is to spend more time looking at a novel stimulus, infants who recognize the original target will spend more time looking at the new one. A propensity to look longer at the unfamiliar stimulus, therefore, indicates the ability to recall the original target and to discriminate it from the new one. Although this task requires relatively complex information processing involving stimulus discrimination and memory storage and retrieval, the motoric response—a shift in gaze direction—is minimal. In contrast to the Bayley Scales, which have poor predictive validity during the first year of life, this recognition memory test has been found to predict IQ based on a picture vocabulary test at 4 and 7 years of age (Fagan and McGrath, 1981). In a recent study, this test also predicted IQ on the Stanford-Binet Intelligence Scale at age 6 in an at-risk population (Rose and Wallace, 1985).

As can be seen in Figure 18.2, preference for the novel stimulus was related to cord serum PCB level in a dose-dependent fashion (Jacobson et al., 1985). The most highly exposed infants spent only about half the time looking at the novel stimulus, which suggests that they either failed to recognize the original target or were unable to discriminate it from the new one. The data in Table 18.6 indicate that the infants with detectible cord serum PCB levels were almost four times more likely to score greater than 1 standard deviation below the sample mean. Multiple regression analysis showed that the recognition memory effect was not mediated

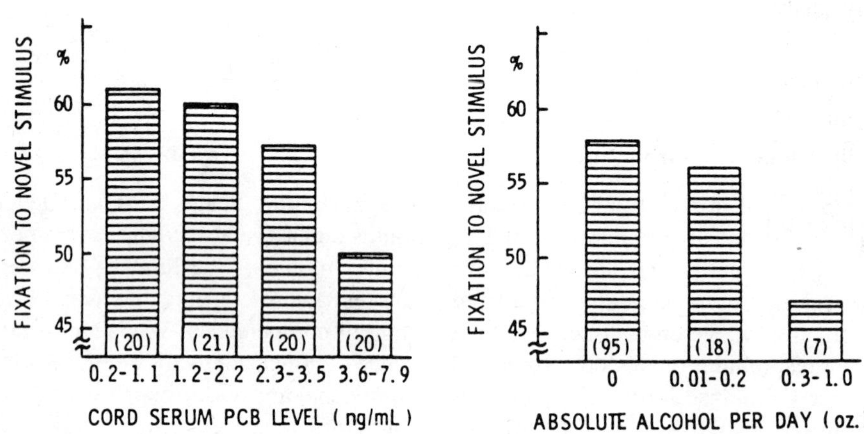


**Figure 18.2.** Dose–response relationships of visual recognition memory with cord serum PCB level and with absolute alcohol per day during pregnancy, after adjusting for relevant potential confounding variables. Modified from Jacobson et al. (1985).

**Table 18.6 Visual Recognition Memory Performance by Cord Serum PCB Level**

| | Cord Serum PCB Level | |
|---|---|---|
| Infants' Score | $<3$ ppb | $\geq 3$ ppb |
| More than 1 SD above mean | 15 (28.8%) | 1 (3.4%) |
| Within 1 SD of mean | 33 (63.5%) | 20 (69.0%) |
| More than 1 SD below mean | 4 (7.7%) | 8 (27.6%) |

by either the physical size or neonatal behavioral performance deficits (Jacobson et al., 1985). Looking back at the model illustrating alternative routes of exposure (Fig. 18.1), this effect seems to be linked directly to intrauterine exposure, even though it may not be evident at birth.

In contrast to prenatal exposure, PCB exposure from nursing was not related to any of the 5- and 7-month outcomes (Jacobson et al., 1985). This finding contradicts predictions by several investigators that postnatal exposure would pose the greater threat (Kodama and Ota, 1977; Masuda et al., 1978). Although contaminant levels in cord serum are substantially lower than in breast milk, our data underscore the greater vulnerability of the organism during the prenatal period. Relative to fetal body weight, even low-level prenatal exposure can be substantial. Intrauterine exposure is continuous and occurs before the development of protective barriers, such as the blood–brain barrier, and drug-metabolizing capacities that are found postnatally (Coyle et al., 1976).

Our experience with these infant tests suggests two conclusions. First, at contemporary levels in the Great Lakes region, the effects of intrauterine PCB exposure appear to be subtle; these deficits would not necessarily be evident in a routine clinical examination. As noted earlier, none of the affected infants had a very low birth weight according to standard medical criteria and, despite their poorer recognition memory performance, the exposed infants were not notably deficient outside the testing situation. Second, as predicted from research on other toxic substances, impairment is seen most clearly in the domain of cognitive processing, although it is not clear which specific aspects of cognition are most vulnerable. Poor performance on the recognition memory task may be due to slower speed in processing information, difficulties in encoding and retrieval, poor visual discrimination, or poorer attention to the target photo during the initial familiarization period. An important challenge to the behavioral teratologist is the identification and development of assessment procedures that are sensitive to subtle deficits and focus on the specific aspects of processing likely to be affected by teratogenic exposure.

## 4. FOUR-YEAR FOLLOW-UP ASSESSMENT PROCEDURES

We are currently conducting a 4-year follow-up of these infants under a grant from the National Institute of Environmental Health Sciences. This phase of the research is being performed in collaboration with Harold Humphrey of the Michigan Department of Public Health. The original cohort of children exposed to PCBs in utero has been expanded to include 150 additional children from PCB- and PBB-contaminated farms, whose residents Dr. Humphrey has been following for several years. A blood sample obtained from each child is being analyzed for levels of PCBs, PBBs, and nine chlorinated hydrocarbon pesticides by means of packed column gas chromatography, and also for the presence of particular toxic PCB congeners by means of capillary column gas chromatography (Mullin et al., 1984).

Each child is assessed at home on three standardized psychological tests (the McCarthy Scales of Children's Abilities, the Beery Test of Visual–Motor Integration, and the Peabody Picture Vocabulary Test) and four new tasks developed by researchers in the field of human information processing. The newer tasks are designed to focus on specific aspects of cognitive functioning, including reaction time, short-term memory processing efficiency, visual discrimination, and sustained attention. To illustrate how these new tasks differ from conventional psychological tests, we will describe a vigilance paradigm we are using to assess the child's ability to maintain attention in a context that is not cognitively demanding. This task has been adapted from a procedure used by Streissguth and her associates, who found poorer sustained attention among children whose mothers consumed moderate amounts of alcohol or smoked during pregnancy (Streissguth et al., 1984).

The child is seated in front of a video monitor, which displays a picture of a house. Any of three stimuli—an apple, a butterfly, or a cat—can appear in any of the three windows of the house, but the child is told to press the red button only when the cat comes to the window. The cat appears 42 times over a 12-minute period at random intervals ranging from 4.5 to 52.0 seconds. The number of times the child presses the button in response to the cat's appearance at the window provides a continuous measure of attention, which should be more sensitive to subtle deficits than the dichotomous pass/fail criteria used in most standardized IQ tests. Speed of processing is indicated by how long it takes the child to press the button once the cat has appeared. Speed of response during the last 4 minutes indicates the child's ability to sustain attention over the 12-minute period.

One issue that is of particular importance in research on young children involves controlling for variation in the intellectual environment in which the child is reared. Traditionally, socioeconomic status has been used as a surrogate measure of home environment, but it is probably not sufficiently sensitive. Two variables have been included in the 4-year assessment in an attempt to deal with this problem. Each mother is given the Peabody Picture Vocabulary Test, a brief vocabulary recognition measure, which is highly correlated with overall IQ test performance. This variable can be used to control for aspects of both genetic endowment and parental stimulation that are likely to influence intellectual functioning independently of teratogenic exposure. We are also administering the Home Observation for Measurement of the Environment (HOME) (Caldwell and Bradley, 1978), a procedure that combines observation of the child's home with a semistructured parental interview to provide a direct measure of intellectual stimulation and parental involvement. The validity and sensitivity of the HOME are indicated by recent evidence that it explains more of the variance in school performance than traditional measures of socioeconomic status (Barnard et al., 1984; Gottfried and Gottfried, 1984).

## 5. CONCLUSION

The effects of low-level teratogenic exposure may not be specific to particular toxic agents. Certain aspects of cognitive functioning, such as attention or speed of processing, may be sensitive to effects from exposure to a broad range of substances. Our own research has shown, for example, that performance on Fagan's recognition memory task is affected by intrauterine exposure to alcohol as well as PCBs (see Fig. 18.2; Jacobson et al., 1983b), and Streissguth (Streissguth et al., 1984) found effects on sustained attention from prenatal exposure to both alcohol and smoking. Until recently, many researchers in this area have relied primarily on psychological tests designed originally to assess mental retardation or developmental delay rather than subtle deficits in processing efficiency. Our experience suggests the need to develop a battery of tasks focusing on those aspects of attention and cognitive processing that seem to be affected most readily by teratogenic exposure. Since recent research suggests that cognitive and behavioral deficits often appear at the lowest levels of toxic exposure, these paradigms may be particularly useful in identifying thresholds of teratogenic effects on central nervous system development.

## ACKNOWLEDGMENTS

The infant phase of this research was funded under Grant CR808520010 to Greta G. Fein from the Environmental Protection Agency; the 4-year follow-up, under Grant ES03256 to Joseph L. Jacobson from the National Institute of Environmental Health Sciences. We thank our collaborators on the newborn and infant phase: Greta G. Fein, who served as Principal Investigator, and Pamela M. Schwartz, who directed the field operations; Wayland R. Swain, Nelson A. Thomas, and Harold E. B. Humphrey, for their assistance in developing this project; Joseph F. Fagan III, for helping us in the implementation and interpretation of his visual recognition memory task; and Ann P. Streissguth, for her suggestions regarding the adaptation of her vigilance task. We also thank Jeffrey K. Dowler, Gail A. Brumitt, and Karen F. Kopera for their contributions to this research and Susan L. Schantz for her editorial suggestions.

## REFERENCES


Abel, E. L. (1980). Fetal alcohol syndrome: Behavioral teratology. *Psychol. Bull.* **87,** 29–50.

Akiyama, K., Ohi, G., Fujitani, K., and Yagyu, H. (1975). Polychlorinated biphenyl residues in maternal and cord blood in Tokyo metropolitan area. *Bull. Environ. Contam. Toxicol.* **14,** 588–592.

Barnard, K. E., Bee, H. L., and Hammond, M. A. (1984). Home environment and cognitive development in a healthy, low-risk sample: The Seattle Study. In A. W. Gottfried (Ed.), *Home Environment and Early Cognitive Development: Longitudinal Research.* Academic Press, New York, pp. 117–149.

Barr, M., Jr. (1978). Pediatric health aspects of PBBs. *Environ. Health Perspect.* **23,** 291–294.

Bayley, N. (1969). *Bayley Scales of Infant Development.* Psychological Corporation, New York.

Brazelton, T. B. (1973). *Neonatal Behavioral Assessment Scale.* Lippincott, Philadelphia, PA, pp. 1–66.

Brilliant, L. B., Wilcox, K., Van Amburg, G., Eyster, J., Isbister, J., Bloomer, A. W., Humphrey, H., and Price, H. (1978). Breast-milk monitoring to measure Michigan's contamination with polybrominated biphenyls. *Lancet* **2,** 643–646.

Caldwell, B. M., and Bradley, R. H. (1978). *Manual for the Home Observation for Measurement of the Environment.* University of Arkansas, Little Rock.

Carey, A. E., and Gowen, J. A. (1976). PCBs in agricultural and urban soil. In *Proceedings of the National Conference on Polychlorinated Biphenyls, Chicago, 1975* (EPA-560/6-75-004). U.S. Environ. Prot. Agency, Washington, DC, pp. 195–198.

Coyle, I., Wayner, M. J., and Singer, G. (1976). Behavioral teratogenesis: A critical evaluation. *Pharmacol. Biochem. Behav.* **4,** 191–200.

Fagan, J. F., and McGrath, S. K. (1981). Infant recognition memory and later intelligence. *Intelligence* **5,** 121–130.

Fagan, J. F., and Singer, L. T. (1983). Infant recognition memory as a measure of intelligence. In L. P. Lipsitt (Ed.), *Advances in Infancy Research.* Ablex, Norwood, NJ, Vol. 2, pp. 31–78.

Fein, G. G., Schwartz, P. M., Jacobson, S. W., and Jacobson, J. L. (1983). Environmental toxins and behavioral development: A new role for psychological research. *Am. Psychol.* **38,** 1188–1197.

Fein, G. G., Jacobson, J. L., Jacobson, S. W., Schwartz, P. M., and Dowler, J. K. (1984). Prenatal exposure to polychlorinated biphenyls: Effects on birth size and gestational age. *J. Pediatr.* **105,** 315–320.

Gottfried, A. W., and Gottfried, A. E. (1984). Home environment and cognitive development in young children of middle-socioeconomic-status families. In A. W. Gottfried (Ed.), *Home Environment and Early Cognitive Development: Longitudinal Research.* Academic Press, New York, pp. 57–115.

Humphrey, H. E. B. (1983). Population studies of PCBs in Michigan residents. In F. M. D'Itri and M. A. Kamrin (Eds.), *PCBs: Human and Environmental Hazards.* Butterworth, Boston, MA, pp. 299–310.

International Agency for Research on Cancer (1978). Polychlorinated biphenyls and polybrominated biphenyls. *IARC Monogr. Eval. Carcinog. Risk Chem. Hum.* **18.**

Jacobson, S. W., Jacobson, J. L., Schwartz, P. M., and Fein, G. G. (1983a). Intrauterine exposure of human newborns to PCBs: Measures of exposure. In F. M. D'Itri and M. Kamrin (Eds.), *PCBs: Human and Environmental Hazards.* Butterworth, Boston, MA, pp. 311–343.

Jacobson, S. W., Jacobson, J. L., Dowler, J. K., Fein, G. G., and Schwartz, P. M. (1983b). Sensitivity of Fagan's recognition memory test to subtle intrauterine risk. Paper presented at the annual meeting of the American Psychological Association, Anaheim, CA.

Jacobson, J. L., Fein, G. G., Jacobson, S. W., and Schwartz, P. M. (1984a). Factors and clusters for the Brazelton Scale: An investigation of the dimensions of neonatal behavior. *Dev. Psychol.* **20,** 339–353.

Jacobson, J. L., Fein, G. G., Jacobson, S. W., Schwartz, P. M., and Dowler, J. K. (1984b). The transfer of polychlorinated biphenyls (PCBs) and polybrominated biphenyls (PBBs) across the human placenta and into maternal milk. *Am. J. Public Health* **74,** 378–379.

Jacobson, J. L., Fein, G. G., Jacobson, S. W., Schwartz, P. M., and Dowler, J. K., (1984c). Prenatal exposure to an environmental toxin: A test of the multiple effects model. *Dev. Psychol.* **20,** 523–532.

Jacobson, S. W., Fein, G. G., Jacobson, J. L., Schwartz, P. M., and Dowler, J. K. (1985). The effect of intrauterine PCB exposure on visual recognition memory. *Child Dev.* **56,** 853–860.

Jacobson, S. W., Jacobson, J. L., and Fein, G. G. (1986). Environmental toxins and infant development. In H. E. Fitzgerald, B. M. Lester, and W. Yogman (Eds.), *Theory and Research in Behavioral Pediatrics.* Plenum, New York, Vol. 3, pp. 96–146.

Kodama, H., and Ota, H. (1977). Studies on the transfer of PCB to infants from their mothers. *Japn. J. Hyg.* **32,** 567–573.

Masuda, Y., Kagawa, R., Kuroki, H., Kuratsune, M., Yoshimura, T., Taki, I., Kusuda, M., Yamashita, F., and Hayashi, M. (1978). Transfer of polychlorinated biphenyls from mothers to foetuses and infants. *Bull. Environ. Contam. Toxicol.* **16,** 543–546.

Mullin, M. E., Pochini, C. M., McCrindle, S., Romkes, M., Safe, S. H., and Safe, L. M. (1984). High resolution PCB analysis: Synthesis and chromatographic properties of all 209 PCB congeners. *Environ. Sci. Technol.* **18,** 468–476.

Proulx, G. A. (1985). The effects of prenatal exposure to smoking and postnatal feeding method on infant growth. Unpublished Master's Thesis, Wayne State University, Detroit, MI.

Rose, S. A., and Wallace, I. F. (1985). Visual recognition memory: A predictor of later cognitive functioning in preterms. *Child Dev.* **56,** 843–852.

Russell, I. S., Taylor, R., and Law, C. E. (1968). Smoking in pregnancy, maternal blood pressure, pregnancy outcome, baby weight and growth, and other related factors: A prospective study. *Br. J. Prev. Soc. Med.* **22,** 119–126.

Sawyer, L. D. (1978). Quantitation of polychlorinated biphenyl residues by electron capture gas-liquid chromatography. *J. Assoc. Off. Anal. Chem.* **61,** 272–291.

Schwartz, P. M., Jacobson, S. W., Fein, G. G., Jacobson, J. L., and Price, H. A. (1983). Lake Michigan fish consumption as a source of polychlorinated biphenyls in human cord serum, maternal serum, and milk. *Am. J. Public Health* **73,** 293–296.

Shaywitz, S., Cohen, D., and Shaywitz, B. (1980). Behavior and learning difficulties in children of normal intelligence born to alcoholic mothers. *J. Pediatr.* **96,** 978–982.

Streissguth, A. P., Martin, D. C., Barr, H. M., Sandman, B. M., Kirchner, G. L., and Darby, B. L. (1984). Intrauterine alcohol and nicotine exposure: Attention and reaction time in 4-year-old children. *Dev. Psychol.* **20,** 533–541.

Swain, W. R. (1983). The scientific basis for concern with polychlorinated biphenyls in the Great Lakes. In F. M. D'It and M. A. Kamrin (Eds.), *PCBs: Human and Environmental Hazards.* Butterworth, Boston, MA, pp. 11–48.

Tuchmann-Duplessis, H. (1975). Drug effects on the fetus. *Monographs on Drugs,* Vol. 2. ADIS Press, Sydney, Australia.

U.S. Public Health Service (1979). *Smoking and Health: A Report of the Surgeon General* (U.S. Dep. of Health, Education, and Welfare No. PHS 79-50060). U.S. Gov. Printing Office, Washington, DC.

Veith, G. D., and Lee, G. F. (1971). Polychlorinated biphenyls (PCBs) in the Milwaukee River. *Water Res.* **5,** 1107–1115.

Vorhees, C. V., and Butcher, R. E. (1982). Behavioral teratogenicity. In K. Snell (Ed.), *Developmental Toxicology.* Croom Helm, London, pp. 247–298.

Weiss, B., and Spyker, J. M. (1974). Behavioral implications of prenatal and early postnatal exposure to chemical pollutants. *Pediatrics* **53,** 851–856.

Wickizer, T. M., Brilliant, L. B., Copeland, R., and Tilden, R. (1981). Polychlorinated biphenyl contamination of nursing mothers' milk in Michigan. *Am. J. Public Health* **71,** 132–137.

Young, D. R., McDermott, D. J., and Heesen, T. C. (1976). Marine inputs of polychlorinated biphenyls off Southern California. In *Proceedings of the National Conference on Polychlorinated Biphenyls, Chicago, 1975* (EPA-560/6-75-004). U.S. Environ. Prot. Agency, Washington, DC, pp. 199–208.

Zelazo, P. R. (1979). Reactivity to perceptual-cognitive events: Application for infant assessment. In R. B. Kearsley and I. E. Sigel (Eds.), *Infants at Risk: Assessment of Cognitive Functioning.* Erlbaum, New York, pp. 49–83.

# 19

# TOXIC SUBSTANCES AND SURVIVAL OF LAKE MICHIGAN SALMONIDS: FIELD AND LABORATORY APPROACHES

***Michael J. Mac***

*U.S. Fish and Wildlife Service, National Fisheries Center—Great Lakes, Ann Arbor, Michigan 48105*

## 1. INTRODUCTION

Natural reproduction is more than a ''numbers'' game, in which enough of the young must survive to maintain the adult population. It is also a vehicle through which natural selection, adaptation, and genetic variation operate to produce individuals more capable than their predecessors of surviving in a changing or hostile environment. Certain fish populations are maintained without successful natural reproduction but these populations rely on the planting of fish of hatchery origin. As is seen in several Great Lakes fish populations, planting can provide adequate numbers of spawning adults in the population. Without natural reproduction, however, the selection processes are removed and genetic variability is decreased. The results can be a population of less fit individuals (Kincaid, 1983) and the potential for abnormal embryologic development (Leary et al., 1985). When successful reproduction is reduced or lost in a population, much more is lost than the potential biomass.

Persistent toxic substances such as those found in the Great Lakes can affect successful reproduction in feral fish populations. Mortality induced by contaminants in developing eggs or in fish during their first year of life reduces the capability of a population to produce the numbers needed to maintain adult stocks. Mortality of young-of-the-year fish need not be complete to result in the extinction of a population; even increases in mortality as small as 5% can result in observable decreases in adult stocks (Jensen, 1971).

## 2. ROUTES OF CONTAMINANT EXPOSURE

In the reproduction and early development of fish, offspring are exposed to contaminants via three routes: the female parent, water, and food. Exposure from the female parent is made possible by the deposition of contaminants in the egg yolk during vitellogenesis. This process consists of the synthesis of vitellogenin in the liver, its passage into circulation, and its incorporation into developing gametes (de Vlaming et al., 1980). This process is particularly sensitive to persistent contaminants because (1) the site of synthesis is the liver where many organic contaminants accumulate and are commonly metabolized, and (2) vitellogenin contains a high proportion of lipid, which provides substrate for the transfer of lipophilic contaminants to gonadal tissue. Newly fertilized eggs from contaminated females thus may contain significant concentrations of contaminants, which are sometimes higher than those in the rest of the fish (Niimi, 1983). The second route, exposure to waterborne contaminants, is critical to young fish. McKim (1977) found that in most life cycle toxicity tests, the embryo–larval and early juvenile life stages were the most sensitive. The third route consists of contaminants present in food, which may be accumulated when fish begin exogenous feeding.

## 3. HISTORICAL EVIDENCE

The potential effect of contaminants on reproduction in fish from the Great Lakes was recognized nearly 20 years ago. High mortality observed in the offspring of adult steelhead trout (*Salmo gairdneri*) and coho salmon (*Oncorhynchus kisutch*) that had been collected in Lake Michigan was attributed to high levels of DDT in the eggs (Johnson and Pecor, 1969). Other studies during the mid to late 1960s reported the deleterious effects of DDT on fish reproduction, as evidenced in both laboratory exposures and in fish captured from the wild (Allison et al., 1964; Burdick et al., 1964, 1972).

In light of the possible effects of DDT on reproduction, researchers began to examine the role that contaminants may play in the lack of successful reproduction by lake trout (*Salvelinus namaycush*) in Lake Michigan. The lake trout probably became extinct in Lake Michigan in the mid 1950s for a number of reasons including sea lamprey predation, overexploitation, and habitat deterioration (Willford et al., 1981). Beginning in 1965, a great effort to restore the fishery started with control of the sea lamprey and an extensive lake trout restocking program. This program has resulted in the planting of an average of over two million hatchery-reared fish annually. But despite these efforts, no naturally produced fish, one year old or older, have been found in the main part of the lake.

Through the early and mid 1970s, several studies demonstrated some differences in the capability of offspring of lake-reared adults when compared with relatively uncontaminated hatchery trout offspring; however, no observations of severe mortality similar to those reported by Burdick et al. (1964) and Johnson and Pecor (1969) were made. Stauffer (1979) reported good survival (50%) of Lake Michigan lake trout to 6 weeks after swim-up for eggs collected in 1974 and concluded that the burden of contaminants present in the eggs and fry were not deleterious to survival. Mac et al. (1981) reported similar survival rates for fry to yolk absorption for eggs collected in 1972 and Berlin et al. (1981) documented that fry hatched from eggs collected in 1975 suffered only 22% mortality through 176 days after hatch.

## 4. RECENT INFORMATION

### 4.1. Lake Trout Mortality

Beginning in 1978, an alarming change in the survival of young lake trout collected from southeastern Lake Michigan was observed (Fig. 19.1). The problem was characterized by extremely high mortality to the fry, beginning shortly after swim-up, and continuing for several weeks (Fig. 19.2). The mortality was severe, with less than 3% of the fry surviving to 6 months after hatching. Afflicted fry displayed characteristic behavior: swimming was erratic, with the fish swimming in circles

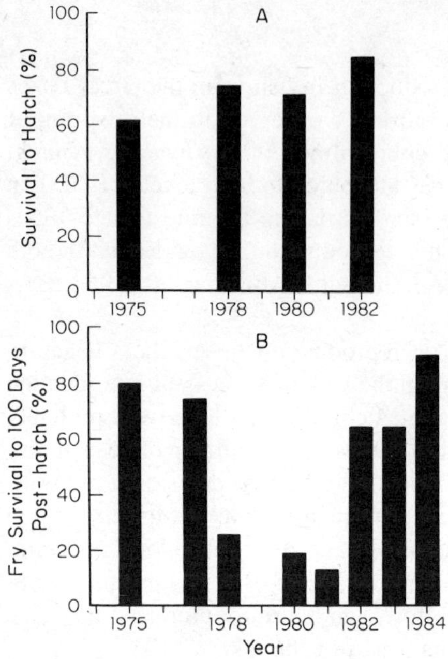


**Figure 19.1.** (A) Percent survival from fertilization to hatch for eggs taken from adult lake trout captured off Saugatuck between 1975 and 1982. (B) Percent survival of fry from Saugatuck eggs to 100 days post hatch.

or in a corkscrew fashion, and a loss of equilibrium was evident as these fish swam upside down or on their sides in the water column. The fish lay on their sides, rather than the normal upright position, on the bottom for several days before dying. Dead fish usually had vestiges of yolk material left, empty stomachs, and little or no air in their swim bladders. These observations are identical to those

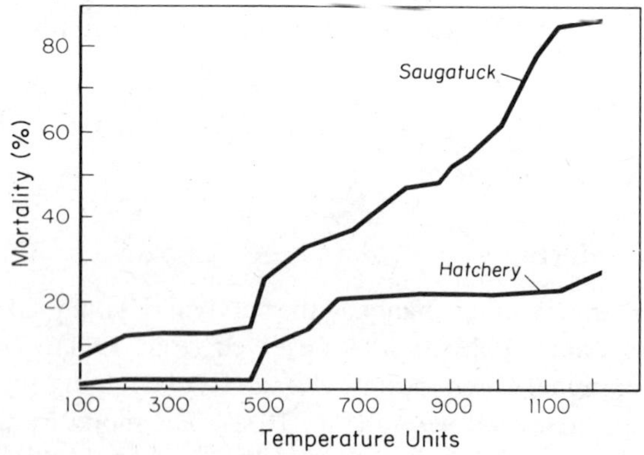


**Figure 19.2.** Percent mortality for eggs and fry that were offspring of adult trout from either Lake Michigan at Saugatuck or from hatchery broodstock and collected in 1980. Development is shown in temperature units (degree days).

**Table 19.1 Hatching Success of Lake Trout Eggs, 1980 (% of Fertilized Eggs)**

| | Egg Source | | | |
|---|---|---|---|---|
| Incubation site | Hatchery | Lake Superior | Lake Huron | Lake Michigan |
| Laboratory | 89 | 98 | 84 | 71 |
| Lake Superior | 84 | 86 | 87 | 72 |
| Lake Huron | 89 | 98 | 88 | 81 |
| Lake Michigan | 80 | 97 | 89 | 65 |

*Source.* From Mac et al. (1985). With permission of *The Journal of Great Lakes Research.*

reported by Johnson and Pecor (1969) but differed slightly from those reported by Burdick et al. (1964). Burdick did report finding distended air bladders in dying fry. Both groups of researchers found no evidence of disease in affected fish and this was also true for Lake Michigan lake trout fry.

At the time this high mortality first occurred in Lake Michigan trout (1978), it was unknown whether a problem inherent in the eggs was responsible or whether technical errors were made in obtaining eggs or in the care of fry. A study using eggs collected in the fall of 1980 (Mac et al., 1985) verified that something in the eggs and/or milt of adult trout from Lake Michigan was responsible for the high mortality. Mac et al. (1985) obtained fertilized eggs from lake trout captured in Lake Michigan at Saugatuck, Lake Superior at Marquette, Lake Huron in Hammond Bay, and from state hatchery broodstock. Eggs from each source were divided into four groups and incubated under laboratory conditions at four sites supplied with water from either Lake Michigan, Huron, or Superior, or with well water (Table 19.1). Fry hatched from the eggs were reared for 139 days after hatching, past the critical swim-up to yolk absorption period (Table 19.2). This enabled testing of lake-specific water quality as well as quality of offspring of adult trout from three different lakes. Adult trout that provided eggs and milt were all

**Table 19.2 Survival of Lake Trout Fry (%) to 139 Days Post Hatch, 1980**

| | Egg Source | | | |
|---|---|---|---|---|
| Incubation Site | Hatchery | Lake Superior | Lake Huron | Lake Michigan |
| Laboratory | 45 | 46 | 50 | 8 |
| Lake Superior | 42 | 21 | 14 | 3 |
| Lake Huron | 74 | 81 | 58 | 1 |
| Lake Michigan | 66 | 79 | 58 | 3 |

*Source.* From Mac et al. (1985). With permission of *The Journal of Great Lakes Research.*

planted hatchery fish, with similar genetic background. Thus, differences in the survivability of offspring from these trout would represent differences in environmental history.

The results of this study were dramatic. Average survival to 139 days after hatch at all incubation sites was only 4% for fry of Lake Michigan origin (Table 19.2). Meanwhile, mean survival for fry from the other sources was Lake Superior, 59%; Lake Huron, 48%; and hatchery, 52%. Again, most of the mortality to Lake Michigan fry occurred over several weeks, beginning shortly after swim-up (Fig. 19.2) and the same behavioral signs were observed in affected fry. During this study, only minor differences in survival from fertilization to hatching were observed (Table 19.1) (Mac et al., 1985).

The mortality graph from the 1980 study (Fig. 19.2) shows mortality for eggs and fry through development, as represented in temperature units (degree days), starting at fertilization and continuing for 139 days after hatch. Skea et al. (1985) used temperature units (TU) to separate the swim-up mortality from that due to starvation. Using the original data from DDT-poisoned lake trout from Lake George published by Burdick et al. (1964), Skea et al. (1985) determined swim-up at 750 TU, syndrome mortality occurred between 850 and 950 TU, and starvation losses occurred after 1300 TU. On the mortality graph for Lake Michigan lake trout fry from the 1980 collection (Fig. 19.2), hatch occurred around 530–600 TU, swim-up began at around 770 TU, and the mortality occurred between 900 and 1100 TU. Although some variability exists, the agreement between these two sets of observations shows that the mortality in lake trout from Lake Michigan occurred at a similar time in development as the DDT induced mortality in Lake George lake trout during the mid 1960s. From the work in New York (Skea et al., 1985), it is also apparent that this fry mortality is not due to starvation.

Annual monitoring of the survival of fry that were offspring of Lake Michigan adult lake trout continued from 1981 through 1984. In Fig. 19.1, survival of lake trout fry is normalized to 100 days after hatching so that success can be compared for the years 1975 through 1984. Although some temperature differences exist among these years, using 100 days assures that the critical mortality period would have been completed. Through this 10-year period, it is apparent that the poorest survival occurred in the years 1979–1981, with 1981 as the lowest. Fry hatched from eggs collected in the fall of 1982 gave the first indication that the problem was lessening. Survival to 100 days after hatch for that year was 65% (Fig. 19.1). Although there was a great improvement in survival, most of the fry that died did so shortly after swim-up and displayed the characteristic behavior. Eggs collected in the fall of 1983 displayed nearly identical survival rates as did eggs collected in 1982 and again the characteristic mortality was seen. In 1984, fry hatched from the fall collection of eggs showed no swim-up mortality and displayed 90% survival to 100 days after hatching. This survival rate was identical to that of fry of hatchery origin that were concurrently reared under identical conditions. Whatever factor or factors were responsible for the high mortality of lake trout fry from southeastern Lake Michigan between 1978 and 1983 were gone.

At the onset of this syndrome in lake trout fry, the viability of eggs and sperm from the parent trout, as measured by survival from fertilization to hatching, was good. In fact, during the period 1975–1982, when trout mortality at swim-up was increasing, a general trend of increasing survival to hatch was observed (Fig. 19.1). Because the swim-up mortality appeared to be unrelated to the hatchability of the eggs, survival to hatch was not monitored after 1982.

## 4.2. Chinook Salmon Mortality

The successful fishery for chinook salmon (*Oncorhynchus tshawytschaw*) in the Great Lakes is largely a product of hatchery-planted fish, although some natural reproduction occurs (Carl, 1982). The chinook hatchery system in Michigan obtains eggs from adult salmon that return to the Manistee River, a tributary that flows into Lake Michigan some 200 km north of Saugatuck. This hatchery operation has been so successful that the state has often been able to provide fertilized eggs to other states interested in planting chinook salmon. Beginning in 1979, hatchery personnel responsible for rearing these eggs and fry began to observe increased mortality in young fish shortly after swim-up (Fig. 19.3). Characteristic behavior that accompanied the mortality included erratic swimming behavior, which was referred to as ''spinning mortality'' (T. Amundson, Wisconsin DNR, personal communication). As in the case of lake trout, pathological examination indicated no evidence of disease problems in the salmon. In previous years, survival of chinook fry from swim-up through active feeding, was 98% (Fig. 19.3). In 1979, this survival was 96.6%, decreased to 92.5% in 1980, and was lowest in 1981 when only 80.5% of the fish survived the several weeks from swim-up to active feeding (Giesy et al., 1986). While these survival percentages are not as low as those seen for lake trout, they represent only a short period (2 weeks) in the early development of the chinook salmon rather than the 100 days after hatch as reported for lake trout. From 1982 on, survival from swim-up to active feeding was back to normal (pre-1979) levels (Giesy et al., 1986).

The similarities between the observed problems in lake trout and those in chinook salmon from Lake Michigan are more than coincidental. The onset of abnormal mortality occurred at the same developmental stage in both species: its

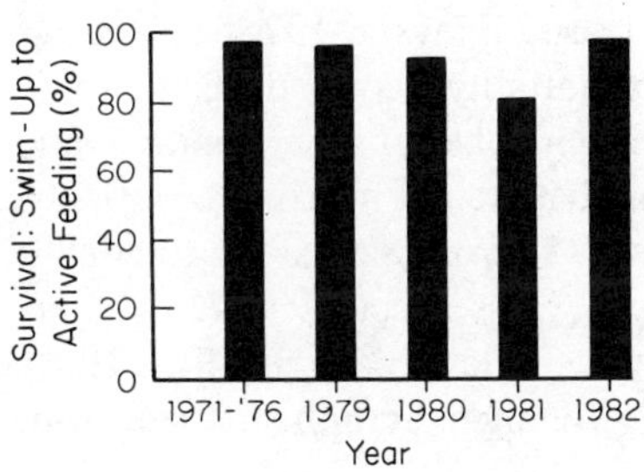


**Figure 19.3.** Percent survival from swim-up to active feeding for chinook salmon fry that were offspring of adult salmon from Lake Michigan for the years 1971–1982.

occurrence was first observed in lake trout in 1978 and in salmon in 1979, and the peak mortality took place in the same year (1981). These observations were made in the offspring of apparently healthy adult fish that share the same habitat and forage base. No evidence of a disease-related pathological condition was observed, and the problem disappeared in several years after its onset (1982 for chinook, 1984 for lake trout). The severity of the mortality differed between the two species and that would be expected if the cause of this problem was related to environmental exposure. The lake trout takes longer to reach sexual maturity and thus would accumulate contaminants over a longer period of time. This would be critical if temporal changes in the level or presence of a causative agent occurred in this area of the lake.

### 4.3. Evidence of Contaminant Involvement

Several observations indicated that the problems seen in lake trout fry were caused by contaminants:

1. The excessive mortality was restricted to lake trout from southern Lake Michigan, which have higher concentrations of organochlorine contaminants than trout from any of the other upper Great Lakes.
2. The swim-up stage, when the mortality began, has been previously reported to be sensitive to contaminants (Burdick et al., 1964; Johnson and Pecor, 1969) and is the stage at which contaminant concentrations in young lake trout are highest (Mac and Seelye, 1981).
3. The syndrome was observed over a period of 6 years, with mortality peaking in 1981. Similar problems were observed in chinook salmon collected in Lake Michigan 120 miles north of Saugatuck. Although the mortality was not as high in chinook salmon it also peaked in 1981 (Giesy et al., 1986).

Although this circumstantial evidence implicates environmental contaminants as the cause of the observed fish mortality, it was difficult to suspect any of the well-known organochlorines present in Lake Michigan. Chemical analyses of lake trout eggs and newly hatched fry used in these studies showed that levels of total PCBs, DDT, chlordane, and toxaphene had decreased between 1975, before observation of the mortality, and 1980, when high mortality occurred (Table 19.3; Willford et al., 1981; Mac et al., 1985). However, a number of other contaminants are present in adult lake trout from Lake Michigan that are not measured routinely. Hesselberg and Seelye (1982) found lake trout in 1977 from Saugatuck to contain 167 organic compounds using gas chromatography/mass spectroscopy. By comparison only 8 compounds were identified in trout of hatchery origin (Fig. 19.4). The variety and complexity of chemicals in Lake Michigan complicate measuring

**Table 19.3 Organochlorine Contaminants ($\mu g/g$) in Eggs or Fry from Southern Lake Michigan Lake Trout**

| Contaminant | Year 1975[a] | 1980[b] |
|---|---|---|
| PCB (Aroclor 1254) | 7.6 | 2.7 |
| DDT | 4.8 | 1.3 |
| Dieldrin | 0.06[c] | 0.05[c] |
| Chlordane | 0.12[c] | 0.02[c] |
| Toxaphene | 5.7[c] | 2.6[c] |

[a]From Willford et al. (1981).
[b]PCB and DDT values from Mac et al. (1985). With permission of *The Journal of Great Lakes Research*.
[c]Newly hatched fry.

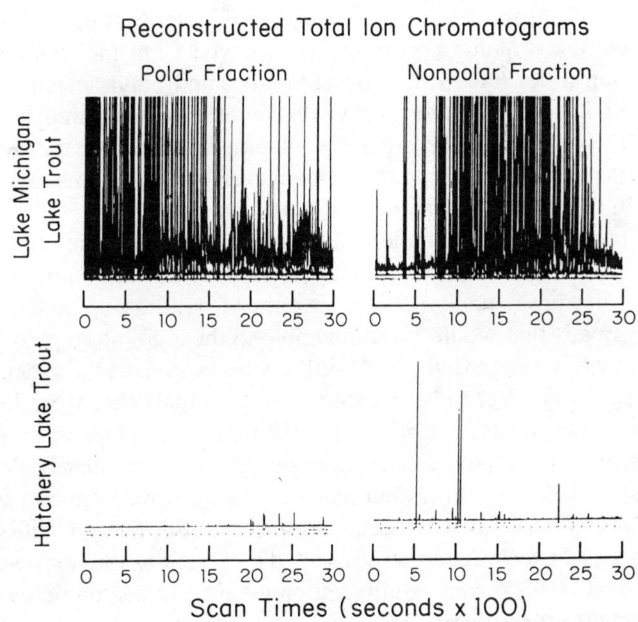


**Figure 19.4.** Reconstructed total ion chromatograms showing comparative number of contaminants in adult lake trout of either Lake Michigan or hatchery origin. From Hesselberg and Seelye (1982).

contaminant effects in the field. But, all these compounds need to be considered to assess the potential influence of contaminants on Great Lakes biota.

## 5. CURRENT RESEARCH

A determination of whether contaminants are responsible for the observed mortality of swim-up fry will require that high burdens of contaminants be built up in adult hatchery trout and the survival of their offspring be observed. Because of the many organic contaminants present in lake trout from southern Lake Michigan (Hesselberg and Seelye, 1982), and the indication that the more common organochlorine contaminants are not directly responsible, a laboratory exposure in which individually added chemicals are used seems fruitless. Rather, the fish must be exposed to the chemical array present in Lake Michigan fish. Experiments conducted in which frozen Great Lakes forage fish were fed to adult lake trout for more than a year resulted in contaminant concentrations in adult trout of about twice that in the forage fish (Dave Nortrup, Great Lakes Fishery Laboratory, personal communication). However, chemical analyses indicated that this concentration was not as high as that in adult Lake Michigan lake trout during the 1978–1983 period when fry mortality was high. As an alternative adult lake trout are currently being exposed to a multiple contaminant mixture that is obtained by chemical extraction of fish tissue. Large amounts of Lake Michigan adult lake trout are homogenized and organic contaminants removed from the tissue by using a column extraction procedure with petroleum ether and ethyl acetate (Hesselberg and Johnson, 1972). The extract is then evaporated to a desired concentration leaving a mixture of most of the organic contaminants accumulated by adult lake trout. This extract is being used to spike commercial food fed to adult trout in the laboratory and is also used in waterborne exposures.

The use of tissue extracts has alleviated some of the difficulties encountered in conducting environmental exposure studies; however, in examining the effects on reproduction, other problems exist. The process of building up contaminant concentrations in gonads that would be comparable to those found in wild fish from a polluted system takes a long time. In addition, studies such as these depend on the simultaneous ripening of gonads in experimental animals that were held for extended periods under artificial laboratory conditions. Although these restrictions are not insurmountable, a new technique is being developed that could eliminate the need for adult exposures in testing parentally transferred contaminants on egg development and fry survival. This technique involves the direct injection of toxins into fish embryos. The technique was developed by Metcalfe and Sonstegard (1984) and by Black et al. (1985) as a carcinogen bioassay, and has numerous other applications in aquatic toxicology. An initial attempt was made to use the embryo injection technique along with the tissue extract as a source of contaminants to determine whether the mortality in swim-up fry was indeed contaminant induced.

Although the results of this first effort made in cooperation with Dr. Black were not completely successful, the technique nevertheless shows promise and utility.

## 6. SUMMARY

1. Reproductive problems have recently been identified in several Great Lakes fishes. Circumstantial evidence indicates that contaminants are involved, although those routinely measured do not appear to be the cause.

2. Egg hatching success and fry survival may be good indicators of population health. It is critical in this type of monitoring of these life stages that studies must extend beyond the swim-up to yolk absorption period, because this period is particularly sensitive to parentally transferred contaminants. Proper handling of eggs and measurements of fertilization rates are essential to the demonstration of environmental effects.

3. The large array of chemical contaminants present in the Great Lakes must be considered when one is evaluating the effects of toxicants on reproduction and early development in fish. Fish tissue extracts may provide a source for complex toxicant mixtures for use in various laboratory exposures.

4. Embryo injection techniques may enable screening for teratogenicity, carcinogenicity, and toxicity of chemicals maternally inherited in yolk. Refinement of this technique for use with multiple contaminant mixtures, such as those found in tissue extracts, could provide an excellent methodology for determining the effects of contaminants on reproduction and development in Great Lakes fishes.

5. Sound assessment of the effects of contaminants on reproduction and development, require that all three routes of exposure (water, food, and parents) be examined. Culturing fish in situ or in lake water efficiently exposes fish to the numerous waterborne contaminants present at any one location. Similarily, natural food or food spiked with extracts of fish tissue may provide the best material for dietary exposures. Maternally inherited contaminants may be the most difficult to test; however, they also appear to have the potential for the greatest impact. Further development and refinement of the techniques for evaluating contaminants transmitted through yolk are needed.

## ACKNOWLEDGMENTS

Contribution 677 of the National Fisheries Center—Great Lakes, U.S. Fish and Wildlife Service, Ann Arbor, MI 48105. This article was written by an employee of the U.S. Government as part of his official duties and therefore cannot be copyrighted.

## REFERENCES

Allison, D., Kallman, B. J., Cope, O. B., and Van Valin, C. (1964). Some chronic effects of DDT on cutthroat trout. Res. Rep. 64, U.S. Fish Wildl. Serv., pp. 1–30.

Berlin, W. H., Hesselberg, R. J., and Mac, M. J. (1981). Growth and mortality of fry of Lake Michigan lake trout during chronic exposure to PCBs and DDE. In Chlorinated hydrocarbons as a factor in the reproduction and survival of lake trout (*Salvelinus namaycush*) in Lake Michigan. Tech. Pap. 105, U.S. Fish Wildl. Serv., pp. 11–22.

Black, J. J., Maccubbin, A. E., and Schiffert, M. (1985). A reliable, efficient, microinjection apparatus and methodology for the in vivo exposure of rainbow trout (*Salmo gairdneri*) and salmon (*Oncorhynchus kisutch*) embryos to chemical carcinogens. *J. Natl. Cancer Inst.* **75,** 1123–1128.

Burdick, G. E., Harris, E. J., Dean, H. J., Miller, T. M., Skea, J., and Colby, D. (1964). The accumulation of DDT in lake trout and the effect on reproduction. *Trans. Am. Fish. Soc.* **93,** 127–136.

Burdick, G. E., Dean, H. J., Harris, E. J., Skea, J., Karcher, R., and Frisa, C. (1972). Effect of rate and duration of feeding DDT on the reproduction of salmonid fishes reared and held under controlled conditions. *N.Y. Fish Game J.* **19,** 97–115.

Carl, L. M. (1982). Natural reproduction of coho salmon and chinook salmon in some Michigan streams. *North Am. J. Fish. Manage.* **4,** 375–380.

de Vlaming, V. L., Wiley, H. S., Delahunty, G., and Wallace, R. A. (1980). Goldfish (*Carassius auratus*) vitellogenin: Induction, isolation, properties and relationship to yolk proteins. *Comp. Biochem. Physiol. B* **67B,** 613–623.

Giesy, J. P., Newstead, J., and Garling, D. L. (1986). Relationships between chlorinated hydrocarbon concentrations and rearing mortality of chinook salmon (*Oncorhyncus tschawytscha*) eggs from Lake Michigan. *J. Great Lakes Res.* **12,** 82–98.

Hesselberg, R. J., and Johnson, J. L. (1972). Column extraction of pesticides from fish, fish food, and mud. *Bull. Environ. Contam. Toxicol.* **7,** 115–120.

Hesselberg, R. J., and Seelye, J. G. (1982). *Identification of Organic Compounds in Great Lakes Fishes by Gas Chromatography/Mass Spectrometry: 1977,* Admin. Rep. No. 82-1. Great Lakes Fish. Lab., Ann Arbor, MI.

Jensen, A. L. (1971). The effect of increased mortality on the young in a population of brook trout, a theoretical analysis *Trans. Am. Fish. Soc.* **100,** 456–459.

Johnson, H. E., and Pecor, C. (1969). Coho salmon mortality and DDT in Lake Michigan. *Trans. North Am. Wildl. Nat. Resour. Conf.* **34,** 159–166.

Kincaid, H. L. (1983). Inbreeding in fish populations used for aquaculture. *Aquaculture* **33,** 215–227.

Leary, R. F., Allendorf, F. W., and Knudsen, K. L. (1985). Development instability as an indication of reduced genetic variation in hatchery trout. *Trans. Am. Fish. Soc.* **114,** 230–235.

Mac, M. J., and Seelye, J. G. (1981). Patterns of PCB accumulation by fry of lake trout. *Bull. Environ. Contam. Toxicol.* **27,** 368–375.

Mac, M. J., Berlin, W. H., and Rottiers, D. V. (1981). Comparative hatchability of lake trout eggs differing in contaminant burden and incubation conditions. In Chlorinated hydrocarbons as a factor in the reproduction and survival of lake trout (*Salvelinus namaycush*) in Lake Michigan. Tech. Pap. 105, U.S. Fish Wildl. Serv., pp. 8–10.

Mac, M. J., Edsall, C. C., and Seelye, J. G. (1985). Survival of lake trout eggs and fry reared in water from the upper Great Lakes. *J. Great Lakes Res.* **11,** 520–529.

McKim, J. M. (1977). Evaluation of tests with early life stages of fish for predicting long-term toxicity. *J. Fish. Res. Board Can.* **34,** 1148–1154.

Metcalfe, C. D., and Sonstegard, R. A. (1984). Microinjection of carcinogens into rainbow trout embryos: An in vivo carcinogenesis assay. *J. Natl. Cancer Inst.* **73,** 1125–1132.

Niimi, A. J. (1983). Biological and toxicological effects of environmental contaminants in fish and their eggs. *Can. J. Fish. Aquat. Sci.* **40,** 306–312.

Skea, J. C., Symula, J., and Miccoli, J. (1985). Separating starvation losses from other early feeding fry mortality in steelhead trout *Salmo gairdneri,* chinook salmon *Oncorhyncus tshawytscha,* and lake trout *Salvelinus namaycush. Bull. Environ. Contam. Toxicol.* **35,** 82–91.

Stauffer, T. M. (1979). Effects of DDT and PCBs on survival of lake trout eggs and fry in a hatchery and in Lake Michigan, 1973–1976. *Trans. Am. Fish. Soc.* **108,** 178–186.

Willford, W. A., Bergstedt, R. A., Berlin, W. H., Foster, N. R., Hesselberg, R. J., Mac, M. J., Passino, D. R. M., Reinert, R. E., and Rottiers, D. V. (1981). Introduction and Summary. In chlorinated hydrocarbons as a factor in the reproduction and survival of lake trout (*Salvelinus namaycush*) in Lake Michigan. Tech. Pap. 105, U.S. Fish Wildl. Serv., pp. 1–7.

# 20

# MODELING *Ceriodaphnia* TOXICITY IN THE NAUGATUCK RIVER USING ADDITIVITY AND INDEPENDENT ACTION

***Dominic M. Di Toro and James A. Hallden***

*HydroQual, Inc., Mahwah, New Jersey 07430*

***James L. Plafkin***

*U.S. Environmental Protection Agency, Monitoring and Data Support Division, Office of Water, Washington, DC 20460*

## 1. INTRODUCTION

The development of techniques to assess and control toxicants in wastewater discharges has traditionally focused on individual chemicals. With the issuance of its "Policy for the Development of Water Quality-Based Permit Limitations for Toxic Pollutants" (EPA, 1984a), the U.S. Environmental Protection Agency (USEPA) has acknowledged the utility of effluent toxicity testing as a complementary technique for controlling toxic discharges. In its implementation guidance, (EPA, 1984b) the USEPA has recommended that aggregate receiving water toxicity from multiple discharges should be considered to be additive when wasteload allocations are performed. A study to test this basic assumption was conducted on the Naugatuck River, Connecticut, in 1983. An analysis of the extent to which observed instream toxicity can be quantitatively related to point-source toxicity is presented below.

The utility of toxicity, as determined by a specific bioassay, as a water quality-based control parameter for regulating wastewater discharges depends on three factors. The most fundamental of these is the necessity to establish a relationship between measured instream toxicity and stream biological degradation. This relationship depends on the degree to which the test organism response correlates with observed degradation of the natural biota. These correlations are currently being developed (Mount et al., 1984, 1986; EPA, 1984b).

The second requirement, which follows from the first, is that toxicity can be measured in the receiving water as well as in effluents. This requires a sensitive bioassay that can be used in a field setting. In addition, it should be practically possible to assay many samples. Certain of the mini-chronic bioassays being developed have these characteristics (Mount and Norberg, 1984).

The third practical requirement is that the fate of toxicity can be modeled in natural waters. In the present study, toxicity loads are determined from point-source toxicity tests. The stream toxicity concentrations are calculated using simple dilution methods. These computations are compared to instream toxicity test results. This is basically the standard water quality modeling approach of constructing a mathematical model and validating its results. The novelty is that the variable of concern is generic toxicity rather than a specific toxic substance.

The development of a bioassay technique to evaluate water quality impacts in receiving waters is not without precedent. The concept of biochemical oxygen demand (BOD) is similar in many ways to the concept of aggregate toxicity. The purpose of the BOD test is to determine the potential for oxygen consumption of a wastewater or stream sample without regard to the specific chemical identity of the oxidizable substances. This test provides a uniform basis of comparison between wastewater effluents of widely differing composition, and it has been shown that mathematical models can successfully reproduce observed stream BOD concentrations as a function of the known BOD sources and sinks in the water body. Further, BOD has proven useful in calculating the expected degree of oxygen depletion in the receiving water. It is hoped that measurements of aggregate toxicity will be of similar utility in addressing the impacts of toxic substances on the receiving water biota.

## 2. TOXICITY TESTS

The Mount-Norberg (1984) chronic toxicity assay used in this study measures the fecundity of *Ceriodaphnia* that are exposed to various sample dilutions. A single juvenile female *Ceriodaphnia* is placed in each of a series of vessels containing a sequence of dilutions (0, 1, 3, 10, 30, and 100% sample). Ten replicates are used for each dilution. Each animal is examined for a 7-day (3-brood) period, and its status is recorded at days 2, 5, and 7. At these three times a fresh dilution is made from the originally collected sample and the female is transferred. All samples are spiked with yeast to provide food. The fecundity (number of young produced) of each *Ceriodaphnia* is recorded over the 7-day period.

Interpreting the results of these experiments is complicated by the impact that mortality has on reproduction. Obviously, the total number of young produced during the test period is dependent on the survival of the *Ceriodaphnia.* Therefore, the mean number of young produced per organism, which is the test endpoint of interest, can be substantially affected by mortality. A method of interpreting these data has been developed by Hamilton (1984). This method incorporates data from both survivors and nonsurvivors by calculating the mean young produced at each time interval and summing the means to obtain an overall mean *Ceriodaphnia* fecundity. This method of data analysis is used to compute the mean fecundity at each dilution.

The choice of the source of dilution water for the toxicity tests depends on the use to be made of the results. The point of view adopted in this study is that the toxicity measurement is an assay, much like the BOD test. The question to be addressed is: Can the result of instream toxicity tests be quantitatively related to point-source toxicity? Thus, the assay is not an attempt to reproduce the toxicity downstream of a point source that is the result of mixing upstream and effluent flows (Mount et al., 1984). If it were, dilution water from immediately upstream of that point source, regardless of its quality, would be used. Rather, for this study, the purpose of the tests is to place toxicity measurements on a common basis. Therefore, the dilution water is the same for all tests. For the Naugatuck River, the upstream station farthest above all known point sources of toxicity (station N1, Fig. 20.1) was chosen.

## 3. NAUGATUCK RIVER STUDY AREA

The study area of the Naugatuck River, located in central Connecticut (Fig. 20.1), encompasses the lower 65 km of the river from above Torrington to the confluence with the Housatonic River. Annual average flows range from less than 3 $m^3$/second

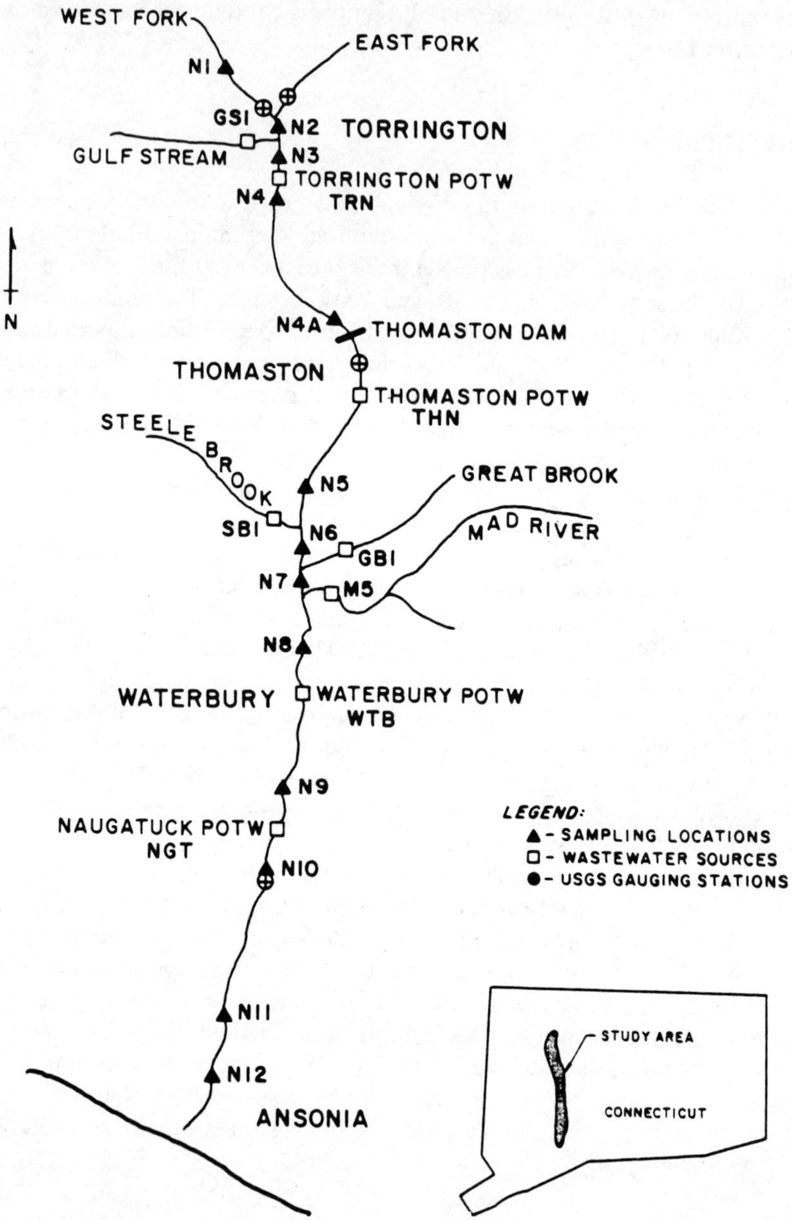


**Figure 20.1.** Map of the Naugatuck River, Connecticut, study area.

near the east and west fork confluence to approximately 15 $m^3$/second at Beacon Falls. The lower 5 km of the Naugatuck River exhibit the tidal influence of the Housatonic River. The four significant tributaries to the river are Gulf Stream, Great Brook, Steele Brook, and Mad River. All of these tributaries receive significant municipal and industrial discharges.

The four significant direct effluent discharges to the Naugatuck River are the Torrington, Thomaston, Waterbury, and Naugatuck Publically Owned Treatment Works (POTWs). The area is heavily industrial and the influent to the Waterbury POTW, in particular, contains some industrial effluent. These eight direct and tributary discharges comprise a variety of complex effluents. Thus, the river is an ideal location to test the additivity of toxicity. Both the direct and tributary sources are treated as point sources to the river.

## 4. DATA COLLECTION

The field-sampling program was designed to measure the flow and toxicity of the eight point sources and the resulting toxicity in the stream. An upstream control (N1) and stations below each point source (N2–N10) were established (Fig. 20.1). Table 20.1 lists the locations for both the stream stations and point sources. Significant effluent and stream toxicity variability was expected (Di Toro, 1986) and it was decided to collect sufficient data so that both the mean and variability of

**Table 20.1 Location of Stream Sampling Stations and Tributary and Point Discharges**

| Stream | Tributary and Point Discharge | River km |
|---|---|---|
| Station N2 | | 63.5 |
| | Gulf Stream (GS1) | 62.8 |
| Station N3 | | 62.4 |
| | Torrington POTW (TRN) | 61.2 |
| Station N4 | | 59.1 |
| | Thomaston POTW (THM) | 46.7 |
| Station N5 | | 46.5 |
| | Steele Brook (SB1) | 33.5 |
| Station N6 | | 33.0 |
| | Great Brook (GB1) | 32.5 |
| Station N7 | | 30.4 |
| | Mad River (M5) | 29.9 |
| Station N8 | | 28.3 |
| | Waterbury POTW (WTB) | 27.2 |
| Station N9 | | 21.9 |
| | Naugatuck POTW (NGT) | 19.5 |
| Station N10 | | 16.5 |

toxicity could be established. A balance between statistical and practical requirements dictated the sampling frequency. During the first week of field sampling (August 22, 1983) seven daily 24-hour composite samples were collected at each point source and stream station from N2 to N8 inclusive, the stations upstream of the Waterbury POTW. For the second week the point sources and stream stations below and including N8 were similarly sampled.

All the point-source samples (8 sources × 7 days) were collected, air shipped, and tested in the laboratory using the full dilution sequence (6 dilutions × 10 replicates) comprising over 3000 individual assays (EA Engineering Science and Technology (EAES&T), 1984). For the stream samples only 100% sample was assayed for each station (N2–N10) in the field (Mount et al., 1986). Comparisons between lab and field analyzed split samples revealed no significant differences (Di Toro and Hallden, 1985). In addition, N8, N9, and N10 were analyzed in the laboratory using the full dilution sequence since the river at these locations was expected to exhibit significant toxicity. The N9 and N10 samples were synthesized from the N8, Waterbury, and Naugatuck POTW samples, using measured stream and effluent flow rates for each day. Further details are available (Mount et al., 1986; EAES&T, 1984).

## 5. TOXICITY DATA ANALYSIS

The 7-day survival data can be analyzed using conventional probit methods (Finney, 1971) to establish the $LC_{50}$ of the sample. However, the reproduction data—mean young produced/female versus sample dilution—are more suited to a continuous dose–response analysis. The log-logistic dose–response function (Finney, 1971; Kooijman, 1981) was chosen:

$$R = R_o/\{1 + \exp[(\ln F - \ln \alpha)/\beta]\} \qquad (1)$$

where $R$ is the response (mean young/female), $R_o$ is the control response, $F$ is the fraction of sample in the assay vessel ($F$ = dilution %/100%), $\alpha$ is the fraction of the sample that reduces fecundity to one-half the control fecundity, and $\beta$ is the population sensitivity parameter. A small $\beta$ implies a rapid change of response with respect to sample fraction. Although the use of the term is imprecise, we refer to $\alpha$ as the $EC_{50}$ by analogy to the $LC_{50}$ computed from the survivor data.

The dose–response function, Equation 1, is fit to each set of fecundity data using a nonlinear least-squares computer program. Examples of this procedure are shown in Figure 20.2. The data and the fitted dose–response function are shown. The units of the $EC_{50}$ values estimated using this procedure are the sample fractions that reduced the fecundity rate to one-half of the control rate as illustrated by the dashed lines. However, sample fractions are not convenient units for dilution calculations. A more convenient and entirely equivalent parameter, a toxic

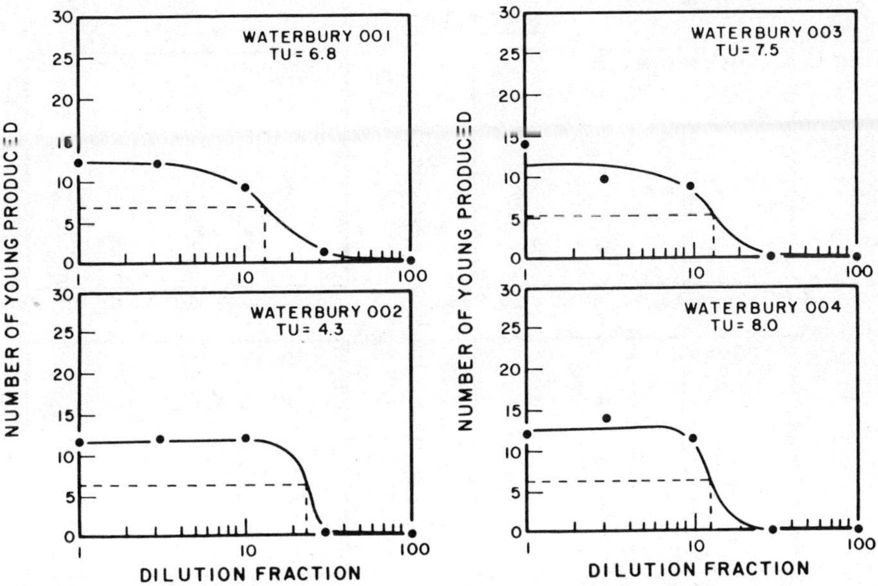


**Figure 20.2.** Dose–response data for the Waterbury POTW effluent. Mean number of young produced/female versus dilution percent. Solid line is fitted log-logistic Equation 1. $EC_{50}$ dilution fraction and toxic units as indicated.

unit introduced by Sprague and Ramsay (1965), is defined as the reciprocal of the sample fraction causing a 50% change in response:

$$TU = 1/(EC_{50} \text{ or } LC_{50} \text{ sample fraction}) \tag{2}$$

For example, if the $EC_{50}$ of a sample is at 20% sample dilution, the $EC_{50}$ of this sample is 100%/20% = 1/0.20 = 5 toxic units. Appendix A illustrates dilution calculations using toxic units. All dilution assay data (point sources and stations N8–N10) are analyzed in this fashion. For the sake of consistency the analysis of the survivor data is similar with survivor number replacing fecundity in Equation 1.

## 6. FIELD DATA ANALYSIS

Probability methods are employed in the analysis of the toxicity data because the data are quite variable and estimates of the variability are required. The use of log-probability plots greatly aids the analysis (Aitchison and Brown, 1957). Figure 20.3 presents log-probability plots of the point source fecundity data and Figure 20.4 presents the stream station data at which the dilution sequence were per-

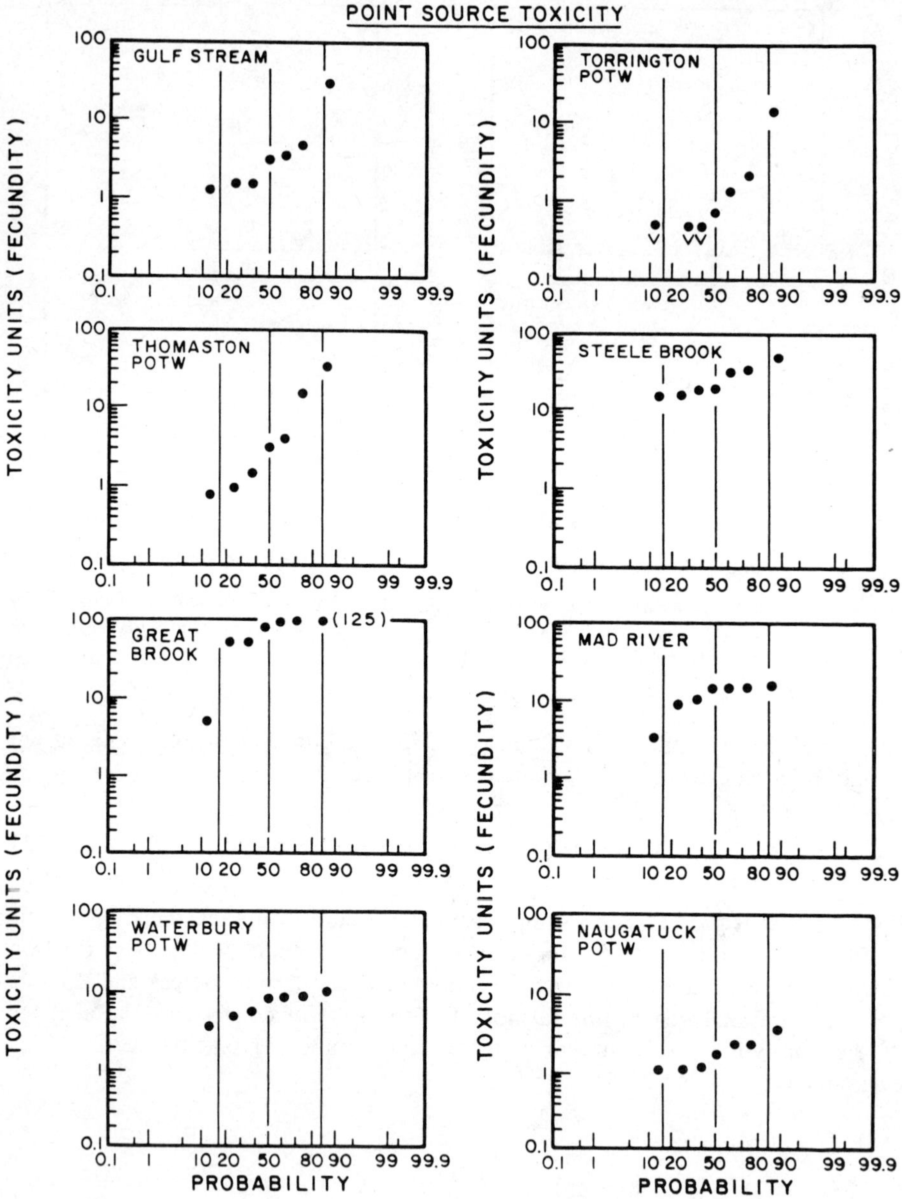


**Figure 20.3.** Log probability plots of point-source toxicity ($EC_{50}$ in toxic units).

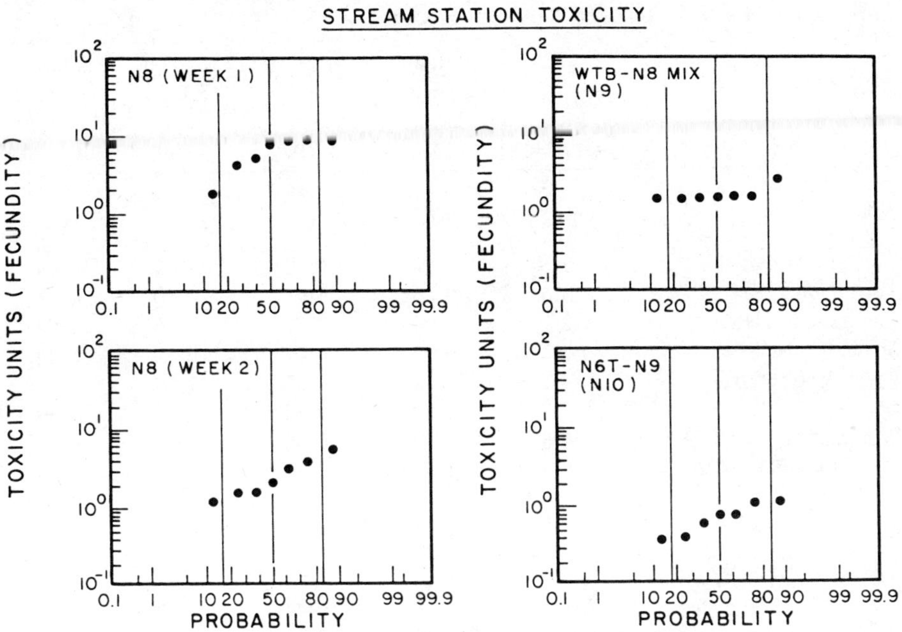


**Figure 20.4.** Log probability plots of stream stations N8–N10 toxicity ($EC_{50}$ in toxic units).

formed. The plots present the fecundity $EC_{50}$ in toxic units on a logarithmic scale (vertical axis) ordered from the lowest to the highest versus the cumulative probability of observing that ordered value or less. For the $i$th value in the ordering, that probability is $i/(N + 1)$ for $N$ values in the list. For example, with seven samples the probabilities are $1/(7 + 1) = 1/8 = 12.5\%$, 25%, and so forth. Thus, the seven $EC_{50}$ values estimated for each day's effluent are plotted versus these percentages in Figure 20.3. Samples that are estimated to have less than detectable toxicity are ranked first, and are indicated by the downward arrow (e.g., Torrington POTW). The horizontal axis scale (labeled probability) is constructed so that if the data plot on a straight line, the data are characterized by a log-normal probability distribution. The vertical lines are at the 16th, 50th, and 84th percentiles corresponding to the log mean ± the log standard deviation. With only seven samples it is difficult to be definitive but log-normal behavior is certainly not ruled out by the data in most cases. The 50th percentile (i.e., median) concentration will be used in the analyses that follow. The full probabilistic analysis (Di Toro and Hallden, 1985) employs all the data. Table 20.2 lists the medians and coefficients of variation for each of these point sources and stream stations.

The stream stations upstream of N8 were assayed at only 100% of sample so that dose–response analysis and, therefore, $EC_{50}$ values could not be computed because no dilution series bioassays were available. However, probability plots of these data (Fig. 20.5) are instructive. Station N8 exhibited no fecundity, which is

**Table 20.2 Flow Rates and Median Toxicity**

| Source | Flow ($ft^3$/sec) | Median Fecundity $EC_{50}$ (TU) | Coefficient of Variation of $EC_{50}$ | Median 7-Day $LC_{50}$ (TU) | Coefficient of Variation of $LG_{50}$ |
|---|---|---|---|---|---|
| | | *Tributaries and Point Sources* | | | |
| Gulf Stream (GS1) | 2.8 | 3.18 | 0.72 | 2.68 | 0.74 |
| Torrington (TRN) | 3.5 | 0.86 | 1.07 | 0.50 | 1.63 |
| Thomaston (THM) | 2.1 | 3.08 | 2.32 | 2.36 | 1.64 |
| Steele Brook (SB1) | 4.7 | 16.02 | 0.41 | 16.32 | 0.26 |
| Great Brook (GB1) | 1.7 | 78.41 | 1.52 | 77.24 | 1.40 |
| Mad River (M5) | 11.8 | 14.87 | 0.55 | 14.82 | 0.69 |
| Waterbury POTW[a] (WTB) | 26.6 | 7.72 | 0.38 | 9.43 | 0.24 |
| Naugatuck POTW[a] (NGT) | 6.4 | 1.89 | 0.50 | 0.92 | 1.39 |
| | | *Stream Stations* | | | |
| Upstream boundary | 7.0 | [b] | — | [b] | — |
| N8 | — | 7.75 | 0.63 | 5.29 | 0.54 |
| N8[a] | — | 2.03 | 0.57 | 2.52 | 0.51 |
| N9[a] | — | 1.60 | 0.21 | 2.40 | 0.24 |
| N10[a] | — | 0.73 | 0.19 | 0.79 | 0.49 |

[a] Week 2.
[b] Below detectable limit.

consistent with the results in Figure 20.4 indicating that the toxicity ranged from 2 to 10 TU in week 1.

A more speculative interpretation of the remaining data in Figure 20.5 is based on the following idea. Suppose, for any station, we knew that the reproduction rate for the equivalent control population at that station was, say, 20 young/adult. Then each of the samples from that station that assayed at 10 or more young/adult must be at a toxicity of less than the $EC_{50}$ since the fecundity at the $EC_{50}$ is exactly one-half of the control. Hence, if more than half of the seven samples had 10 or greater young/adult, then more than half of the samples had toxicity of less than 1 TU. Therefore, the median $EC_{50}$ toxicity at that station would be less than 1 TU. The problem, of course, is that the control reproduction is unknown. If it is assumed that the control reproduction is the maximum observed at each stream station, then the median toxicity at stations N3–N6 are all less than 1 TU since the median fecundity for these stations is greater than one-half of the maximum fecundity at that station, as shown in Figure 20.5. Station N7 exceeds 1 TU as does N9, because the median fecundity is less than one-half the observed maximum fecundity. The lab assay results for N9 (Table 20.2) yield a median toxicity

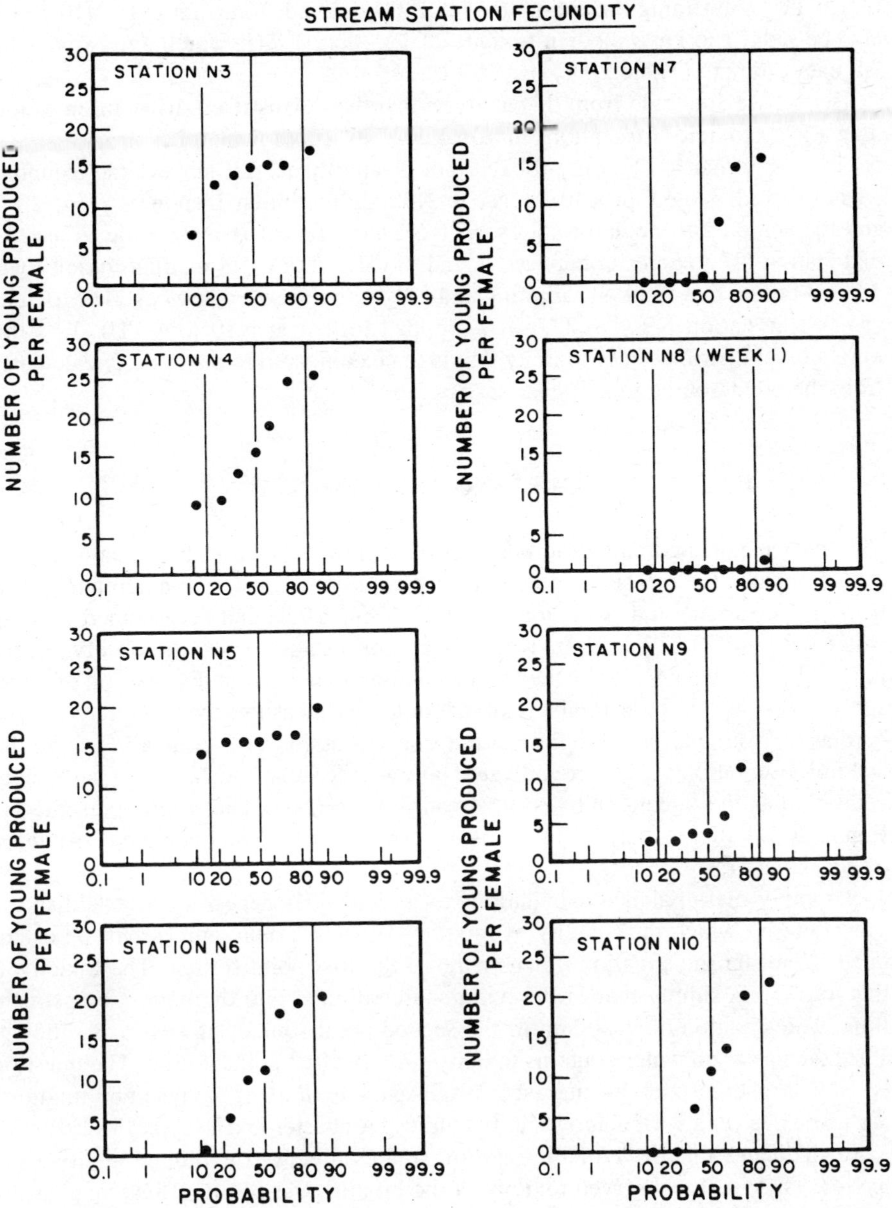


**Figure 20.5.** Log probability plots of stream stations N3–N10 fecundity data at 100% dilution.

of 1.6 TU, confirming the estimate of >1 TU derived from Figure 5. N10 would also be judged to have median toxicity of less than 1 TU, which agrees with the lab data estimate (Table 20.2) of 0.4 TU.

Hence, we conclude from this analysis that the downstream distribution of median stream toxicity for stations up to station N6 is less than 1 TU. It increases at N7 to greater than 1 TU, but probably not greater than 2 TU since some fecundity was observed. Note that stations N7 to N9 exhibit similar responses (Fig. 20.5) and the station N9 median toxicity = 1.6 TU so that it is reasonable to assume that station N7 toxicity is between 1 and 2 TU. Station N8 median toxicity was 7.7 TU for the first week of sampling and 2.0 TU for the second week. The toxicity dropped at station N9 (1.6 TU) and dropped further at N10 (0.4 TU). It is this longitudinal distribution of toxicity that is to be compared to the toxicity calculated from the point-source toxicities.

## 7. MASS BALANCE MODELING OF TOXICITY

The mass balance computation begins with a flow balance in the stream. Figure 20.6 compares the measured flow rates at various locations with a computed flow using the upstream and point-source flows (Table 20.2) and an assumed groundwater inflow of 0.017 $m^3$/km, which is chosen to agree with the observed flow rates. The flow balance is checked by calculating the instream distribution of tracer substances—in this case total dissolved solids as measured by conductivity and hardness—from measured point-source concentrations and comparing it to instream observations. The groundwater inflow is assumed to have the same concentration as the uncontaminated upstream station (N1). The results presented in Figure 20.6 indicate that the flow balance can be used to calculate the distribution of conservative substances.

To apply mass balance calculations to toxicity it is necessary to establish the procedure by which the toxicity of mixtures of toxic point sources will be calculated. Consider the situation downstream of the first point source. The toxicity at this location is simply that of the point source diluted into the nontoxic upstream flow. The situation downstream of the second point source, however, is different since the upstream water contains toxicity from the first point source. The question is: How do these toxicities interact to produce the toxicity of the resultant mixture? An extensive review of acute toxicity studies (Alabaster and Lloyd, 1982; Lloyd, 1986) indicates that 90% of the mixture studies reviewed exhibited additive behavior. That is, the observed toxicity of the mixture was 0.4 to 1.5 as toxic as the prediction using additivity of diluted toxic units. Hence, to within this accuracy mixtures of toxic components appear to exhibit additive toxicity. It is interesting to note that the toxic unit parameter was first introduced within the context of mixture toxicity investigations (Sprague and Ramsay, 1965).

The computation of instream toxicity concentration assuming additivity is exactly parallel to the procedure employed for the concentration of conservative sub-

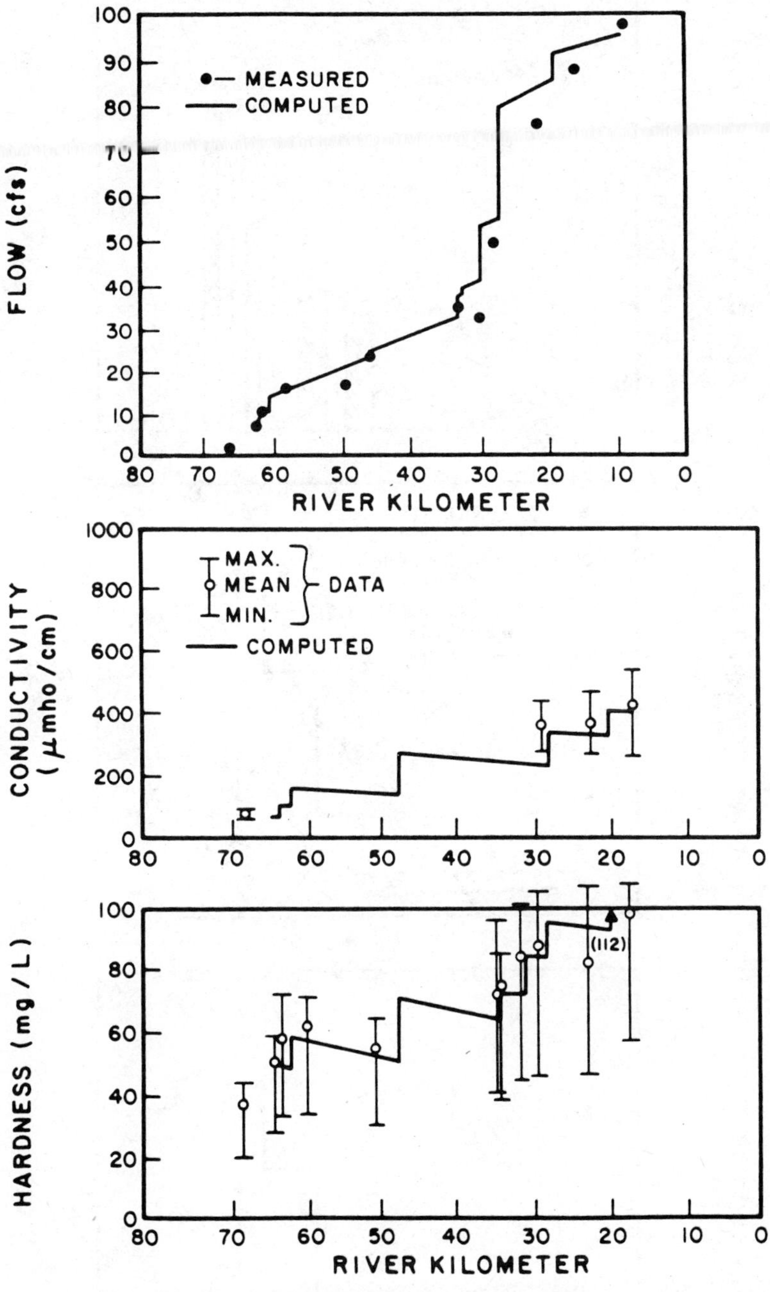


**Figure 20.6.** Naugatuck River flow rate versus distance downstream (top). Comparison of observed and computed conductivity (middle) and hardness (bottom).

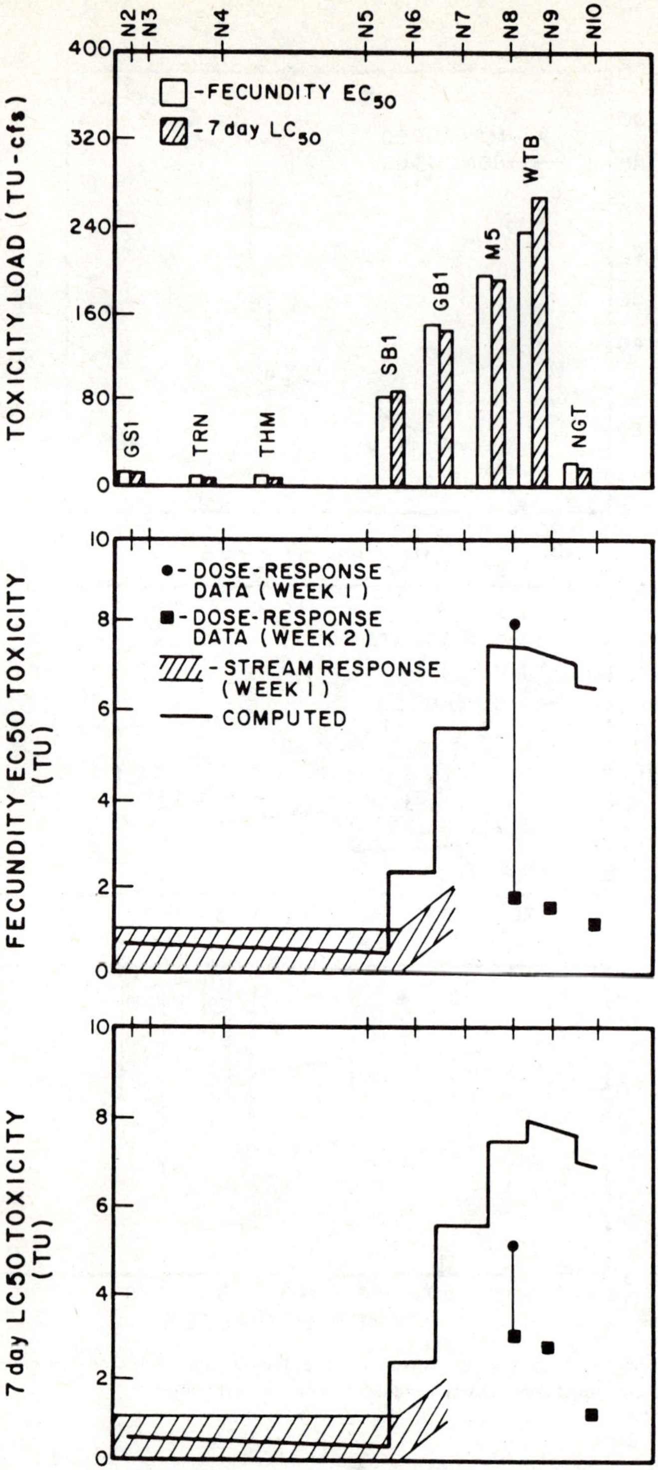
TOXICITY LOAD (TU-cfs)
N2 N3 N4 N5 N6 N7 N8 N9 N10
0 80 160 240 320 400
□ - FECUNDITY EC50
▨ - 7 day LC50
GS1
TRN
THM
SB1
GB1
M5
WTB
NGT
FECUNDITY EC50 TOXICITY (TU)
0 2 4 6 8 10
● - DOSE-RESPONSE DATA (WEEK 1)
■ - DOSE-RESPONSE DATA (WEEK 2)
▨ - STREAM RESPONSE (WEEK 1)
— COMPUTED
7 day LC50 TOXICITY (TU)
0 2 4 6 8 10

stances, assuming that toxicity is also conservative. The mass balance equation downstream from each point source is

$$TU_m = \frac{TU_s\, Q_s + TU_w\, Q_w}{Q_s + Q_w} \tag{3}$$

where $TU_m$ is the toxic unit concentration of the mixture, $TU_s$ and $Q_s$ are the toxic unit concentration and flow rate of the stream immediately upstream of the discharge, respectively; $TU_w$ and $Q_w$ are the toxic unit concentration and flow rate of the wastewater discharge, respectively. The median effluent toxicity and flows are used (Table 20.2), and the upstream boundary and incremental flows are assumed to have negligible toxicity. Thus, for additive toxicity the toxic unit concentration at any location in the stream is the flow-weighted sum of all upstream toxicity sources. The results of these computations are shown in Figure 20.7, together with the instream toxicity data.

The influence of a point source of toxicity on the river depends on the product of the effluent toxicity $TU_w$ and the flow rate $Q_w$ (see Eq. 3). In the terminology of water quality modeling this is a ''loading'' or mass flux rate of toxicity. The magnitudes of these loads are also presented in Figure 20.7 as an aid to interpreting the calculated $LC_{50}$ and $EC_{50}$ toxicity distributions. The major loads commence with Steele Brook (SB1) and the computed toxicity at N6 increases in response to this loading. Great Brook (GB1) and the Mad River (M5) loadings further increase the computed toxicity at N7 and N8. The largest loading is from the Waterbury POTW (WTB) but the instream toxicity does not increase appreciably since the effluent toxicity is approximately equal to the computed upstream toxicity. The Naugatuck POTW (NGT) effluent is less toxic but its flow rate is so small that it has only a small effect on the computed toxicity distribution.

By contrast, the observed instream toxicity remains below 2 TU up to N7 where the computed toxicity is 6 TU. The observed N8 toxicity range was 2.0 to 7.7 fecundity toxic units and 2.5 to 5.5 lethality toxic units during the 2 weeks of sampling by comparison to the 8 TU computed from additivity. The contrast between observed and computed toxicity below the Waterbury POTW is even more dramatic. The observed fecundity and lethality toxicity concentrations at N9 are 1.6 and 2.4 TU, respectively, whereas the computation remains above 7 TU. The situation at N10 is similar. Hence, the conclusion based on the median dilution calculation is that toxicity does not appear to be additive. Detailed probabilistic dilution calculations (Di Toro and Hallden, 1985) using bootstrap methods (Efron, 1982) arrive at the same conclusion.

**Figure 20.7.** Toxicity mass emission rate (''load'') for the point sources as indicated (top). Comparison of median observed and computed (lines) toxicity: $EC_{50}$ toxic units (middle) and $LC_{50}$ toxic units (bottom). Toxicity is computed by assuming it to be additive and conservative. Observed (symbols) median dose–response stream data from Figure 20.4. Stream data (shading) inferred from Figure 20.5 (see text). To increase legibility, the horizontal axis has not been drawn exactly to spatial scale.

## 8. ALTERNATIVE HYPOTHESES

A number of hypotheses can be offered to explain the failure of the additive mixture model of toxicity in the downstream region of the study area. It is possible that toxicity is not conservative and that it naturally decays with time via various physical (settling, volatilization) and/or chemical (hydrolysis, photolysis) reactions as it flows downstream. Although this hypothesis could be applied to the upper reaches of the river, it does not account for the results observed at N9 and N10. Two types of data are available from the lower reaches that contradict this hypothesis:

1. Mixtures of N8 plus Waterbury POTW and N9 plus Naugatuck POTW (in the correct flow proportions) were freshly prepared in the laboratory each day and would not be affected by decay.
2. Instream N9 and N10 samples were independently collected and analyzed. These data are in agreement with the laboratory mixtures. Perhaps as important, the travel time from the Waterbury POTW to N9 is brief, less than 1 day. Thus, decay of toxicity, even if it is occurring, is not the entire explanation.

A second possible contribution to the failure of additivity in the downstream region might be inherent in the data collection procedure. Due to logistic restrictions, sites downstream of station N8 were sampled during the second week of the survey, whereas sites upstream of station N8 were sampled during the first week. Data collected at station N8 during both weeks of the survey indicate a difference in toxicity response between the two periods. However, even if the lower toxicity concentrations observed at station N8 during week 2 were used as upstream boundary conditions, the additive model would still overestimate instream toxicity at station N10.

Another possibility is antagonistic behavior of toxic mixtures. That is, the presence of the second mixture, with its toxic components, aggressively destroys the toxicity of the first mixture by chemical reactions and, in turn, is itself reduced in toxicity by the first mixture. One example would be a strong acid mixing with a strong base. A more subtle example would be the presence of different complexing agents in two mixtures that selectively reduce the toxicity of the toxic components in the other toxic mixture. The difficulty with testing this hypothesis is that there appears to be no independent way of estimating the quantities and properties of these complexing agents in each of the point sources. Thus, although such behavior is clearly possible, it is not testable within the framework of the Naugatuck River study.

## 9. INDEPENDENT ACTION

An alternative to the additivity model is that the toxicity of mixtures is governed by independent action, that is, the magnitude of the toxicity of a mixture is determined only by its most toxic component and is independent of the other less toxic components (Hewlett and Plackett, 1959). To see why this hypothesis is useful for explaining the suppression of toxicity, consider what happens if toxic mixtures are themselves mixed. The toxicity of one mixture is diluted by the additional volume of the second mixture. If it is assumed that the aggregate toxicity is solely attributable to the most toxic component, it is possible for the toxicity of the final mixture to be less toxic than either was originally.

Symbolically, if $C_1$, $Q_1$ and $C_2$, $Q_2$ represent the toxicity concentrations (in toxic units) and flows of two point sources that mix with the upstream flow $Q_s$, then the mixture downstream of the second effluent is made up of $C_1Q_1/Q_T$ of the first effluent and $C_2Q_2/Q_T$ of the second effluent, where $Q_T = Q_s + Q_1 + Q_2$ so that each effluent is diluted not only by the upstream flow, but also by the flow of the other effluents. If toxicity were additive then the downstream toxicity $C_m$ would be

$$\text{Additive:} \quad C_m = \frac{C_1Q_1}{Q_T} + \frac{C_2Q_2}{Q_T} \tag{4}$$

But, if toxicity were acting independently, then the downstream toxicity would be the toxicity of the most toxic diluted component, that is,

$$\text{Independent:} \quad C_m = \left( \frac{C_1Q_1}{Q_T}, \frac{C_2Q_2}{Q_T} \right) \tag{5}$$

For multiple point sources, the generalization is

$$\text{Independent:} \quad C_m = \max \left( \frac{C_1Q_1}{Q_T}, \frac{C_2Q_2}{Q_T}, \ldots, \frac{C_nQ_n}{Q_T} \right) \tag{6}$$

where max $(x_1, x_2, \cdots, x_n)$ is equal to the maximum of the arguments, $x_1$, $x_2$ $\cdots$, $x_n$, and $Q_T = Q_1 + Q_2 + \cdots + Q_n$. Thus, where the additive mode of action accumulates toxicity as one proceeds downstream, independent action predicts that only the most toxic component, suitably diluted, is responsible for the observed toxicity at any point. This hypothesis is attractive because it is directly testable within the framework of the Naugatuck River study. It requires no further assumptions.

The median dilution analysis assuming independent action is performed as follows. For each point source a complete spatial profile of toxicity is computed using all the flows and only the toxicity of that point source. All other point sources are assumed to have zero toxicity concentration. Thus, at stream station $k$, the toxicity concentration due to point source $j$ is computed yielding $C_{jk}$. This computation is repeated for all point sources, $j = 1, \cdots, n$, upstream of station $k$. The independent action prediction for station $k$ is the maximum $C_{jk}$, $j = 1, \cdots, n$.

The resulting computed toxicity concentration profiles are presented in Figure 20.8. The dashed lines represent the toxicity profiles computed previously (Fig.

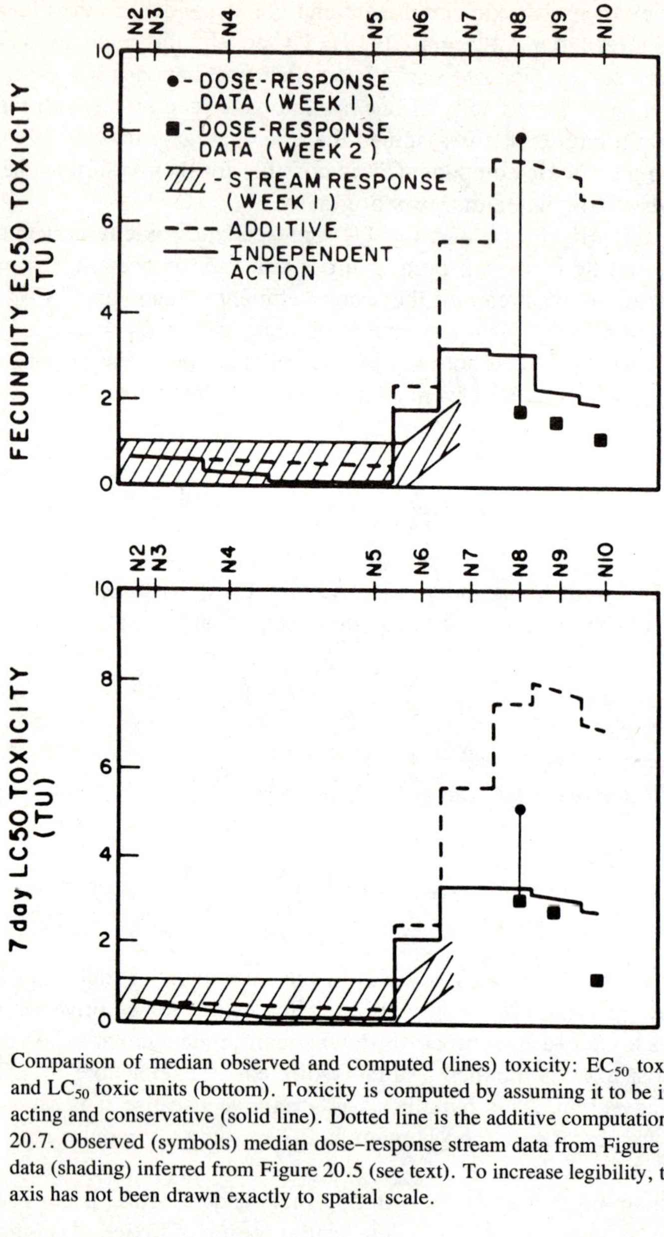


**Figure 20.8.** Comparison of median observed and computed (lines) toxicity: $EC_{50}$ toxic units (top) and $LC_{50}$ toxic units (bottom). Toxicity is computed by assuming it to be independently acting and conservative (solid line). Dotted line is the additive computation from Figure 20.7. Observed (symbols) median dose–response stream data from Figure 20.4. Stream data (shading) inferred from Figure 20.5 (see text). To increase legibility, the horizontal axis has not been drawn exactly to spatial scale.

20.7) assuming additivity of toxicity. Differences in the upstream concentration profiles are not significant since the point-source discharges are small in this area. However, the independent action results more closely resemble the observed downstream data. Computed median toxicity concentrations of 2 to 3 TU below station N8 are more representative of the data than toxicity levels computed assuming additivity. Although both the computed fecundity and 7-day lethality toxicity profiles are in reasonable conformity with observed concentrations, the median dilution model still predicts toxicity levels of about 2.5 toxic units at station N10 while the data indicate less than 1 TU.

In general, the assumption of independent action significantly improves the agreement between calculated and observed toxicity. The additive analysis predicts a strongly increasing toxicity to station N9 in complete contradiction to the observed dramatic drop of toxicity at N9 after the Waterbury POTW. The independent action analysis correctly predicts the gradually increasing toxicity profile from N5 to N8 and the decline at N9. However, the further decline at N10 is not reproduced. As previously noted, a somewhat stronger case could be made for additivity in the upper reaches if only the week 1 data from N8 are considered and week 2 computations are performed using N8 data as a boundary condition.

## 10. DAILY DATA COMPARISONS

The mixture experiments using N8–Waterbury POTW and N9–Naugatuck POTW samples can be analyzed individually to examine the question of whether additive or independent action more adequately depicts instream conditions. The results of these analyses are presented for fecundity and lethality (Table 20.3) in Figure 20.9 for both additive and independent action. For the additive assumption, the calculated toxicity is always greater than the observed toxicity. The toxicities computed using independent action are closer to the observed data in many instances for the simulated N9 mixture. In all cases, the computed toxicities for the simulated N10 mixture are greater than the observed values, with no substantial difference between the additive and independent action models. These analyses indicate that station N10 has lower toxicity levels than can be explained by independent action alone. Therefore, the possibility of an antagonistic toxicity interaction is suggested.

## 11. CONCLUSIONS

The point sources of toxicity in the Naugatuck River do not accumulate in an additive manner. The toxicities computed from this mixing assumption are much larger than the toxicities observed in the river. The assumption of independently acting point sources can be used to compute instream toxicities that are in closer agreement with observations. The principle utility of this mixing model is that it can be applied to the set of toxicity data without any additional data or arbitrary parameters. Whether, in fact, the agreement is fortuitous or the result of actually independently acting sources of toxicity remains to be determined.

An analysis of heavy metal data (Carlson et al., 1986) indicates that a significant portion of the observed toxicity can be attributed to copper. Additional work along these lines (Di Toro and Hallden, 1986) has confirmed the role of copper in the upstream stations and related its toxicity to hardness variations. However, the non-additivity of the toxicity at N8 and the Waterbury POTW effluent still is evident.

**Table 20.3 Comparison of Computed versus Observed $EC_{50}$ Toxicity August 31 to September 6, 1983[a]**

*Waterbury POTW—N8 Mixture*

| | Station N8 | | Waterbury POTW | | | Calculated Toxicity | | Observed Mixture Toxicity |
|---|---|---|---|---|---|---|---|---|
| Day | Flow | Toxicity | Flow | Toxicity | Total flow | Additive | Independent | (Station N9) |
| 1 | 1.58 | 3.63 | 0.81 | 4.59 | 2.39 | 3.96 | 2.40 | 1.55 |
| 2 | 1.46 | 5.05 | 0.77 | 9.21 | 2.23 | 6.49 | 3.41 | 1.68 |
| 3 | 1.43 | 1.20 | 0.74 | 7.90 | 2.17 | 3.48 | 2.70 | 1.57 |
| 4 | 1.23 | 3.04 | 0.69 | 7.93 | 1.92 | 4.80 | 2.85 | 2.77 |
| 5 | 1.15 | 2.03 | 0.65 | 7.72 | 1.80 | 4.08 | 2.79 | 1.60 |
| 6 | 1.19 | 1.54 | 0.67 | 5.19 | 1.86 | 2.85 | 1.87 | 1.68 |
| 7 | 1.19 | 1.52 | 0.67 | 3.41 | 1.86 | 2.20 | 1.23 | 1.55 |
| | | | | | | 3.98 ± 1.40 | 2.46 ± 0.72 | 1.77 ± 0.44 |

*Naugatuck POTW–N9 Mixture*

| | WTB–N8 Mixture | | Naugatuck POTW | | | Calculated Toxicity | | Observed Mixture Toxicity |
|---|---|---|---|---|---|---|---|---|
| Day | Flow | Toxicity | Flow | Toxicity | Total Flow | Additive | Independent | (Station N10) |
| 1 | 2.39 | 1.55 | 0.19 | 2.50 | 2.58 | 1.62 | 1.44 | <1.00 |
| 2 | 2.23 | 1.68 | 0.19 | 1.27 | 2.42 | 1.65 | 1.55 | <1.00 |
| 3 | 2.17 | 1.57 | 0.19 | 1.89 | 2.36 | 1.60 | 1.44 | .74 |
| 4 | 1.92 | 2.77 | 0.14 | 2.54 | 2.06 | 2.75 | 2.58 | .73 |
| 5 | 1.80 | 1.60 | 0.16 | 3.89 | 1.96 | 1.79 | 1.47 | 1.00 |
| 6 | 1.86 | 1.68 | 0.15 | 1.14 | 2.01 | 1.64 | 1.55 | 1.00 |
| 7 | 1.86 | 1.55 | 0.15 | 1.17 | 2.01 | 1.52 | 1.43 | <1.00 |

[a]Flows in cubic meters per second. Toxicity expressed as fecundity ($EC_{50}$) toxic units.

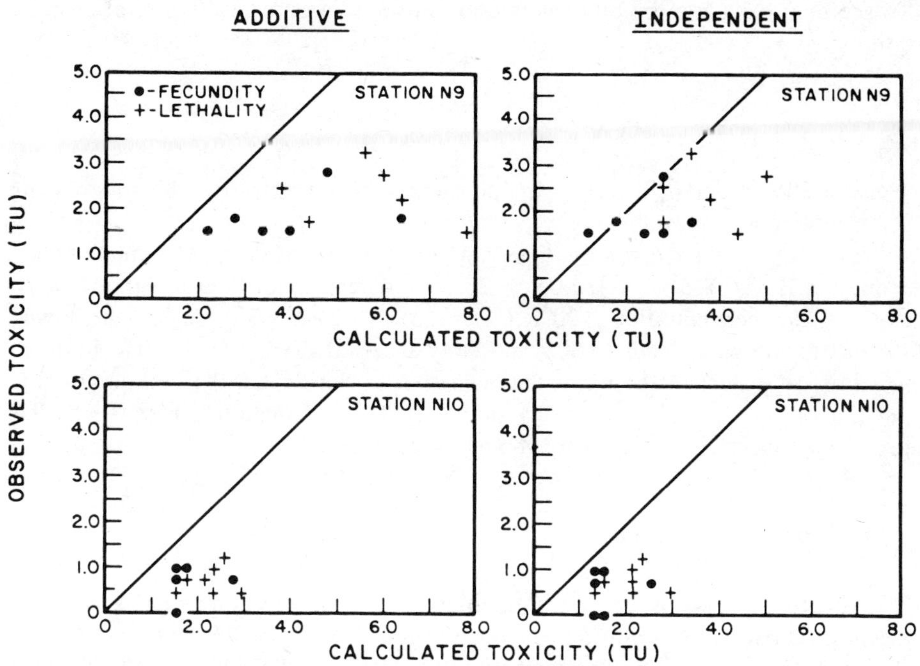


**Figure 20.9.** Comparison of observed and calculated toxicity (in toxic units) for $EC_{50}$ (circles) and $LC_{50}$ (pluses) at station N9 (top) and N10 (bottom). Calculations assuming additive toxicity (left side) and independently acting toxicity (right side). Line of perfect agreement is shown. Computations illustrated in Table 20.3 for station N9.

From a regulatory point of view, if toxicity from certain point sources in the Naugatuck River indeed act independently, the issue of appropriate abatement procedures would be greatly simplified. If each point source must comply with the requirement that its toxicity after mixing is less than 1 TU, then, since toxicity is acting independently, every location in the river will exhibit less than 1 TU. The issues of equitable allocation of assimilative capacity do not arise since if each point source contributes less than 1 TU after dilution, then for independent action, the mixture is always below 1 TU. The regulatory implications of independently acting toxicity loads are so important that the generality of this observation should be checked. In the absence of direct evidence for independent action, however, the reasonable course of action would appear to be the assumption of additive toxicity.

## APPENDIX A: TOXIC UNITS AND DILUTION CALCULATIONS

Consider a situation in which the upstream flow is 3 $m^3$/second and the point source flow is 1 $m^3$/second. If the effluent has an $LC_{50}$ of 5% dilution, then the downstream mixture, which is ¼ effluent, would have an $LC_{50}$ of 20% dilution.

To see this, calculate the effluent fraction in the downstream mixture after 20% dilution:

$$1 \text{ part effluent} \times (\tfrac{1}{4}) \times 20\% = 5\%$$

Thus, at 20% dilution, the downstream mixture is 5% effluent, which would indeed cause a 50% mortality.

Now consider the calculation from a toxic unit point of view. The effluent concentration is 100%/5% = 1/0.05 = 20 TU. After stream dilution, the toxicity concentration concentration is 20 × (¼) = 5.0 TU. That is, the $LC_{50}$ of the downstream mixture would cause a 50% mortality at 1/(5.0 TU) = 0.2 = 20% dilution, which is equivalent to the result obtained above. Notice that the toxic unit computation is exactly analogous to any simple dilution calculation, which is why the toxic unit concept is so useful for modeling purposes.

## ACKNOWLEDGMENTS

The preceding analysis was performed for the U.S. Environmental Protection Agency (USEPA), Monitoring and Data Support Division (MDSD), Office of Water, with support from Contracts 68-03-3131 and 68-01-6986. The field study was carried out as a cooperative effort among MDSD, the Permits Division of the Office of Water, and the Environmental Research Laboratory (ERL), Duluth, Minnesota. The conclusions presented represent the views of the authors and do not represent policies and recommendations of the USEPA. The contributions of Donald Mount, Nelson Thomas, and Ron Carlson of ERL; Lou Rossman of the USEPA, Water Engineering Research Laboratory, Cincinnati, Ohio; Mike Barbour, Wayne McCulloch, and Tony Maciorowski of EA Engineering, Science and Technology; and Lee Dunbar of the Connecticut Department of Environmental Protection, are greatly appreciated.

## REFERENCES

Aitchison, J., and Brown, J. A. C. (1957). *The Lognormal Distribution.* Cambridge Univ. Press, London and New York.

Alabaster, J. S., and Lloyd, R. (1982). *Water Quality Criteria for Freshwater Fish.* Butterworth, London.

Carlson, A. R., Nelson, H., and Hammermeister, D. (1986). *Development and Validation of Site Specific Criteria for Copper and Zinc: An Integration of Metal Addition Toxicity, Effluent and Receiving Water Toxicity, and Ecological Survey Data* (EPA/600/53-86/026). U.S. Environ. Prot. Agency, Environ. Res. Lab., Duluth, MN.

Di Toro, D. M. (1986). Exposure assessment for complex effluents: Principles and possibilities. In H. L. Bergman, R. A. Kimerle, and A. W. Maki (Eds.), *Environmental Hazard Assessment of Effluents.* Pergamon, New York, pp. 135–151.

Di Toro, D. M., and Hallden, J. A. (1985). *A Mass Balance of Analysis of Ceriodaphnia Toxicity in the Naugatuck River.* HydroQual Inc., Mahwah, *N.J.*

Di Toro, D. M., and Hallden, J. A. (1986). *Toxicity Modeling in the Naugatuck River Considering Specific Toxicants and Ambient Water Quality.* HydroQual, Inc., Mahwah, NJ.

EA Engineering Science and Technology (EAES&T) (1984). *Ceriodaphnia Reproductive Potential Test on Ambient Stations and Selected Effluents, Naugatuck River.* EAES&T, Sparks, MD.

Efron, B. (1982). *The Jackknife, the Bootstrap, and Other Sampling Plans.* Soc. Ind. Appl. Math., Philadelphia, PA.

Environmental Protection Agency (EPA) (1984a). Policy for the development of water quality based permit limitations for toxic pollutants. *Fed. Regist.* **49**(48), 9016–9019.

Environmental Protection Agency (EPA) (1984b). *Technical Support Document for Water Quality Based Toxics Control* (EPA-440/4-85-032). U.S. Environ. Prot. Agency, Washington, DC.

Finney, D. J. (1971). *Probit Analysis.* Cambridge Univ. Press, London and New York.

Hamilton, M. (1984). *A Statistical Analysis of the Seven Day Ceriodaphnia Reticulata Reproductivity Toxicity Test.* U.S. Environ. Prot. Agency, Environ. Res. Lab., Duluth, MN.

Hewlett, P. S., and Plackett, R. L. (1959). A unified theory for quantal response to mixtures of drugs: Noninteractive action. *Biometrics* **15,** 591–610.

Kooijman, S. A. L. M. (1981). Parametric Analysis of mortality rates in bioassays. *Water Res.* **15,** 107–119.

Lloyd, R. (1986). The toxicity of mixtures of chemicals to fish: An overview of European laboratory and field experience. In H. L. Bergman, R. A. Kimerle, and A. W. Maki (Eds.), *Environmental Hazard Assessment of Effluents.* Pergamon, New York, pp. 42–53.

Mount, D. I., and Norberg, T. J. (1984). A seven day life cycle cladoceran toxicity test. *Environ. Toxicol. Chem.* **3,** 425–434.

Mount, D. I., Thomas, N. A., Norberg, T. J., Barbour, M. T., Roush, T. H., and Brandes, W. F. (1984). *Effluent and Ambient Toxicity Testing and Instream Community Response on the Ottawa River, Lima, Ohio* (EPA-600/3-84-080). U.S. Environ. Prot. Agency, Environ. Res. Lab., Duluth, MN.

Mount, D. I., Norberg-King, T. J., and Steen, A. E. (1986). *Validity of Effluent and Ambient Toxicity Tests for Predicting Biological Impact, Naugatuck River, Waterbury, Conn.* (EPA-600/8-86/005). U.S. Environ. Prot. Agency, Environ. Res. Lab., Duluth, MN.

Sprague, J. B., and Ramsay, B. A. (1965). Lethal levels of mixed copper-zinc solutions for juvenile salmon. *J. Fish. Res. Board Can.* **22,** 425–432.

# 21

# APPROACHES TO MODELING THE FATE AND EFFECTS OF TOXICANTS IN PELAGIC SYSTEMS

***James E. Breck and Steven M. Bartell***

*Environmental Sciences Division, Oak Ridge National Laboratory, Oak Ridge, Tennessee 37831*

## 1. INTRODUCTION

The objectives of the symposium/workshop titled "Persistent Toxic Substances and the Health of Aquatic Communities" were (1) to synthesize information concerning the effects of persistent toxic substances on biota, (2) to explore innovative methods for assessing the effects of persistent toxic substances on Great Lakes biota, and (3) to identify gaps in our knowledge and research needs. Computer modeling of the fates and effects of toxic substances in pelagic systems can contribute to all three of these goals. Model formulation and development requires the integration of information (goal 1) that describes the physicochemical behavior of the contaminants, the functioning of aquatic communities, and the fates of contaminants and their effects on biota. The resultant model(s) can be thought of as a summary of hypotheses concerning the fates and effects of toxicants in aquatic systems. Models such as those we describe may contribute to the assessment of effects of toxicants on Great Lakes biota (goal 2). In the process of formulating models, gaps in our knowledge are often identified (goal 3), and the analysis of model behavior, as presented here, can be helpful in setting research priorities.

Our purpose is to describe three models that we have used to forecast the fates and effects of toxic chemicals in aquatic ecosystems: (1) a bioenergetics-based model of contaminant dynamics in fish, (2) a model of the fates of polycyclic aromatic hydrocarbons in aquatic systems, and (3) a model that simulates the toxic effects of chemicals in a pelagic food web. We briefly describe results of applying these models for selected chemicals and ecosystems. We also discuss the sensitivity of model forecasts to uncertainties associated with model development and parameter estimation.

## 2. BIOENERGETICS MODELS

Bioenergetics models have been used to estimate growth or consumption rates by accounting for the energy that enters, leaves, and is stored in an organism (e.g., Kitchell et al., 1974, 1977; Elliott, 1976; Ursin, 1967; Stewart et al., 1983; Rice et al., 1983; Rice and Cochran, 1984). These models provide frameworks for estimating the accumulation of contaminants (e.g., Fagerström et al., 1974; Norstrom et al., 1976; Weininger, 1978; Brown et al., 1982; Jensen et al., 1982; Thomann and Connolly, 1984; Mathers and Johansen, 1985) and for modeling the stresses on individuals, populations, and ecosystems (e.g., Bartell et al., 1984; Vaughan et al., 1984). We have used a bioenergetics approach to model the growth of aquatic organisms, the accumulation of contaminants by these organisms, and the effects of contaminants on growth. Adaptations of the bioenergetics models have been used to explore possible influences of aquatic biota on the transport and fate of toxicants in aquatic systems (Park et al., 1980; Bartell et al., 1981).

Contaminants introduced into an ecosystem might alter the growth rates of resident individual organisms. Changes in individual growth rates can result from the direct effects of toxicant accumulation by individuals (Buikema et al., 1979) or from the indirect effects of toxicants on other parts of the ecosystem, for example, on predators or prey (Schindler et al., 1985).

Changes in growth rate, whether caused by natural factors (Carpenter et al., 1985) or contaminants, can influence the contaminant concentration in individual organisms and the flux of contaminants through the biota. Changes in the growth rate of members of a population can alter energy flow, material cycling, and contaminant flux within the system. Changes or differences in growth rate may be correlated with the rate of contaminant accumulation (Mathers and Johansen, 1985), especially when the food chain represents the primary source of contaminants to successive trophic levels. Even when the major pathway of accumulation is directly from the water, altered growth rates can influence the toxicant concentration in the organism through rates of uptake and elimination that are in part determined by individual size (Murphy and Murphy, 1971; de Freitas et al., 1975; Buikema et al., 1979). Changes in growth might concentrate or dilute the toxicants present in individuals (Thomann, 1981; Niimi and Cho, 1981). Therefore, it is important to consider growth rates in models of contaminant fates and effects, and to understand the factors that determine rates of individual growth in aquatic systems.

Bioenergetics models simulate the physiological processes that determine individual growth, including consumption ($C$), respiration ($R$), egestion ($F$), excretion ($U$), and reproduction ($G$). A basic model can be formulated in terms of weight-specific rates:

$$dB/B\,dt = C - (R + F + U + G) \tag{1}$$

where $B$ is body weight and $dB/B\,dt$ is the weight-specific growth rate (see Kitchell et al., 1977; Stewart, 1980; Stewart et al., 1983; Rice et al., 1983). These processes are influenced by body weight, water temperature, activity level (swimming speed), and food quantity and quality. The growth rate also depends on the energy density (e.g., lipid level) of fish and their prey.

## 3. MODEL FOR CONTAMINANT ACCUMULATION (BIOCON)

### 3.1. BIOCON Description

The bioenergetics approach can be usefully applied to problems that involve the accumulation of chemical contaminants, as shown by Fagerström et al. (1974), Norstrom et al. (1976), Weininger (1978), Neely (1979), Jensen et al. (1982), Thomann and Connolly (1984), and Mathers and Johansen (1985). BIOCON is a modification of the models of Kitchell et al. (1977), Weininger (1978), Stewart (1980), and Hewett and Kitchell (1982). These models (including BIOCON) differ, in part, from some other bioenergetics approaches (Fagerström et al., 1974; Jensen et al., 1982; Thomann and Connolly, 1984) by accounting for the seasonal changes in water temperature that affect growth processes. The equations used in

BIOCON to model the bioenergetics of fish growth are expansions of Equation (1) to account for the influence of fish size, swimming speed, water temperature, and the changing energy density of fish and prey items; the equations are described in detail in Stewart et al. (1983). In BIOCON, equations were added to follow contaminant dynamics, and additional control options were added for running the model and fitting it to growth data. Because BIOCON permits multiple prey types in the diet, contaminant flux along different uptake routes can be compared, for example direct uptake from water versus uptake from benthic prey versus uptake from pelagic prey.

From seasonal data on water temperature and fish energy (or lipid) content and from initial and final (or annual) measurements of fish weight, the model calculates the energy intake rates (daily ration) and oxygen consumption necessary to produce the observed growth according to equation (1) (Kitchell et al., 1977; Stewart et al., 1983). Stewart et al. (1983) have shown the importance of considering the ontogenetic change in energy density in fish such as the lake trout; because older lake trout have about twice the energy density of younger ones but have the same diet, about twice as much food (and contaminants!) are ingested per gram of weight added.

The daily rate of contaminant uptake is computed from the rates of respiration and consumption and from information on the contaminant concentration in water and in each prey item. The weight-specific rate coefficient for contaminant uptake from water ($k_1$, mL/g/day) is computed as

$$k_1 = 1000 \cdot R \cdot (e_P/e_{O_2})/[O_2] \quad (2)$$

where $R$ is the respiration rate (mg $O_2$/g/day), $(e_P/e_{O_2})$ is the gill uptake efficiency of contaminant relative to that of oxygen, and $[O_2]$ is the concentration of oxygen in the water (mg $O_2$/L) (Norstrom et al., 1976; Neely, 1979) (Table 21.1).

The fraction of contaminants retained over the first few days after entering the body depends on several factors, including the assimilation efficiency for ingested contaminants, the rate of elimination of contaminants carried in the blood, and the rate at which contaminants carried in the blood are transferred to other tissues and made less available for elimination. Based on the experiments of Guiney and Peterson (1980) comparing the retention of 2,5,2′,5′-tetrachlorobiphenyl when force fed versus injected into fish, an initial assimilation efficiency of $e_G$ = 100% was assumed for PCBs in ingested food, with 30% of the PCBs entering a fish by any route assumed to be lost over the first few days, and the remaining 70% eliminated according to the long-term elimination rate (Table 21.1) (Guiney et al., 1977). All dieldrin entering a fish was assumed to be immediately subject to the long-term elimination rate.

Several hypotheses have been advanced concerning the factors that influence the rate of contaminant elimination from fish. The simplest hypothesis is that the elimination rate is a first-order process characterized by a constant loss-rate coefficient (e.g., Neely et al., 1974; Weininger, 1978), although loss rate appears to

**Table 21.1 Contaminant-Related Parameter Values Used in BIOCON**

| Symbol | Parameter Description | 2,5,2′,5′-Tetrachlorobiphenyl | Dieldrin |
|---|---|---|---|
| $(e_P/e_{O_2})$ | Relative gill uptake efficiency | 1.27[a] | 1.39[b] |
| $e_G$ | Gut uptake efficiency | 1.00[c] | 1.00[d] |
| $k_s$ | Short-term loss fraction | 0.30[c] | 0.00[d] |
| $k'$ | Long-term elim. coeff., independent of $R$ and $L$ | 0.0039[e] | 0.42[f] |

[a]Calculated from data in McKim and Heath (1983).
[b]Calculated from data for endrin in McKim and Goeden (1982).
[c]Assumption based on experiments of Guiney and Peterson (1980) and Guiney et al. (1977).
[d]Assumption based on experiments of Stewart and Stein (1974).
[e]Units: g/mg $O_2$, estimated from (i) data in Guiney et al. (1979), where $k_2$ = 0.0012/day is the geometric mean of estimates for males (0.00133/day) and females (0.00108/day); (ii) estimated lipid content of $L$ = 9% (Oliver and Niimi, 1983); and (iii) estimated respiration rate of $R$ = 2.80 mg $O_2$/g/day for rainbow trout weighing 260 g at 13°C (Stewart, 1980). Estimates of $k'$ for other chemicals (e.g., pentachlorobenzene) are based on $k' = 4800/K_{ow}$, estimated using log $K_{ow}$ = 6.09 for 2,5,2′,5′-tetrachlorobiphenyl (Rapaport and Eisenreich, 1984).
[f]Units: g/mg $O_2$; Estimated from (i) nonlinear fit to data in Reinert et al. (1974), simultaneously fitting data for water-only exposures and water-plus-food exposures; $k_2$ = 0.0133 ± 0.0014 (mean ± 1 asymptotic SEM, $n$ = 5); (ii) estimated lipid content of $L$ = 6% for lake trout of similar weight and age (Rottiers and Tucker, 1982); and (iii) estimated respiration rate of $R$ = 1.9 mg $O_2$/g/day for 200-g lake trout at 10°C.

be influenced by body size (de Freitas et al., 1975; Norstrom et al., 1976), lipid content (Roberts et al., 1977; Bruggeman et al., 1981), respiration rate (Fagerström et al., 1974; Bartell et al., 1981), and octanol:water partition coefficient ($K_{ow}$) (Bruggeman et al., 1981). In BIOCON, the contaminant elimination rate is assumed to be inversely proportional to fish lipid content and directly proportional to the weight-specific and temperature-dependent respiration rate:

$$k_2 = k' \cdot R/L \tag{3}$$

where $k_2$ is the elimination rate coefficient (per day), $k'$ (g/mg $O_2$) is a factor dependent on the chemical but independent of respiration rate ($R$, mg $O_2$/g/day) and lipid level ($L$, %) (Table 21.1).

For some classes of contaminants, for example, polycyclic aromatic hydrocarbons, rapid biotransformation acts as the primary mechanism for disappearance of the parent compound from an individual fish (Southworth et al., 1980; Binder et al., 1984; McCarthy and Jimenez, 1985). This contrasts with the elimination of many environmentally persistent chemicals, like polychlorinated biphenyls (PCBs), dieldrin, and DDT, where biotransformation rates can be comparatively much slower.

### 3.2. Application of BIOCON

BIOCON was used to predict and understand processes that lead to concentrations of persistent chemicals in fishes exposed to contaminated water and food in the laboratory. Values of the bioenergetics parameters were taken from Stewart (1980) for rainbow trout and Elliott (1976) for brown trout, except for the energy content (4520 cal/g) and the egestion rate (26% of the ingested energy) of the pellet diet, which were estimated from Windell et al. (1978). Using the reported values of fish lipid levels, food type, water temperature, dissolved contaminant concentrations, and contaminant concentrations in food, the daily consumption rates were adjusted by BIOCON so that the model matched the reported final fish weights from known initial weights. The energetic relationships expressed in BIOCON thus determined the food inputs that were required, given the growth and respiratory demands, and the consequent contaminant uptake and elimination.

In 11 comparisons with published laboratory data, all predicted concentrations from BIOCON were within a factor of 3 of the concentrations measured in the fish (Table 21.2). Of these reported studies, only Spigarelli et al., (1983) simultaneously exposed fish to contaminanted food and water; brown trout were exposed to contaminated Lake Michigan water and contaminated alewives collected from Lake Michigan. BIOCON predicted that 11% of the PCBs accumulated by brown trout came from contaminated water. Spigarelli et al. (1983) noted that the brown trout exposed only to dissolved PCBs accumulated ~10% of the PCB concentrations accumulated by trout exposed to both contaminated water and alewives. Jensen et al. (1982) modeled the accumulation of PCBs by brown trout and estimated that about 7.5% of their PCBs came from the water.

BIOCON was also used to simulate the accumulation of PCBs and dieldrin by different age classes of Lake Michigan alewives. Values of alewife energetics, seasonal water temperature, and seasonal diet composition were derived from Stewart (1980), Hewett and Kitchell (1982), and S. W. Hewett (Center for Limnology, University of Wisconsin, Madison, personal communication). Spawning losses of 10.3% of body mass were based on data in Norden (1967). Concentrations of dissolved toxicants were constant throughout the simulations at 1 ng/L for dieldrin and 5 ng/L for PCBs (Rice et al., 1982; Thomann and Connolly, 1984). Evans et al. (1982) provided estimates of dieldrin and PCB concentrations in the alewife prey represented in BIOCON: cladocerans, copepods, *Mysis relicta,* and the amphipod *Pontoporeia hoyi* (0.0046, 0.0072, 0.010, and 0.028 ppm dieldrin, respectively, and 0.029, 0.045, 0.099, and 0.26 ppm PCB, respectively, all estimated on a wet-weight basis).

Simulated direct uptake from the water was more important in age class 0 alewives than in age class III alewives (Fig. 21.1). The younger, smaller fish exhibit higher weight-specific rates of metabolism; therefore, per gram body weight, these fish move greater volumes of contaminated water across their gills than older, larger alewives. Younger alewives also have higher specific rates of consumption; however, they consume prey with lower average contaminant concentrations than

**Table 21.2 Comparison of Measured and Simulated Contaminant Concentrations in Laboratory Fish Exposed to Contaminants in Water or Food**[a]

| Fish Species | Duration of Exposure (days) | Final Weight (g) | Temperature (°C) | Contaminant | Concentration in Water (ng/L) | Concentration in Food (μg/g) | Concentration in fish Measured (μg/g) | Concentration in fish Model (μg/g) | Reference |
|---|---|---|---|---|---|---|---|---|---|
| Rainbow trout | 119 | 349 (67) | 15 (1) | Pentachlorobenzene | 0.34 (0.1) | 0.0 | 0.0052 (0.0005) | 0.0066 | Oliver and Niimi (1983) |
| Rainbow trout | 105 | 400 (130) | 15 (1) | Pentachlorobenzene | 9.3 (7.6) | 0.0 | 0.220 (0.025) | 0.190 | Oliver and Niimi (1983) |
| Brown trout | 57 | 478[b] | 12.9 (12–14)[c] | PCBs | 10.9 (1.3) | 2.50 (0.2) | 1.09 (0.65) | 1.17 | Spigarelli et al. (1983) |
| Rainbow trout | 168 | 118 (4) | 15 (0.8) | PCBs | 0.0 | 0.093 — | 0.083 (0.012) | 0.065 | Hilton et al. (1983) |
| Rainbow trout | 168 | 141 (9) | 15 (0.8) | PCBs | 0.0 | 1.20 — | 1.45 (0.10) | 0.84 | Hilton et al. (1983) |
| Rainbow trout | 168 | 124 (6) | 15 (0.8) | PCBs | 0.0 | 2.00 — | 2.73 (0.15) | 1.41 | Hilton et al. (1983) |
| Rainbow trout | 168 | 132 (6) | 15 (0.8) | PCBs | 0.0 | 3.39 — | 2.43 (0.27) | 2.51 | Hilton et al. (1983) |
| Rainbow trout | 168 | 118 (4) | 15 (0.8) | Dieldrin | 0.0 | 0.0093 — | 0.0031 (0.000) | 0.0029 | Hilton et al. (1983) |
| Rainbow trout | 168 | 141 (9) | 15 (0.8) | Dieldrin | 0.0 | 0.047 — | 0.033 (0.00) | 0.017 | Hilton et al. (1983) |
| Rainbow trout | 168 | 124 (6) | 15 (0.8) | Dieldrin | 0.0 | 0.066 — | 0.063 (0.013) | 0.023 | Hilton et al. (1983) |
| Rainbow trout | 168 | 132 (6) | 15 (0.8) | Dieldrin | 0.0 | 0.104 — | 0.073 (0.006) | 0.039 | Hilton et al. (1983) |

[a]Measured concentrations are mean values (SD) on a wet-weight basis.
[b]Final weight estimated from initial weight (375 g) and reported growth rate of 0.48%/day.
[c]Range.

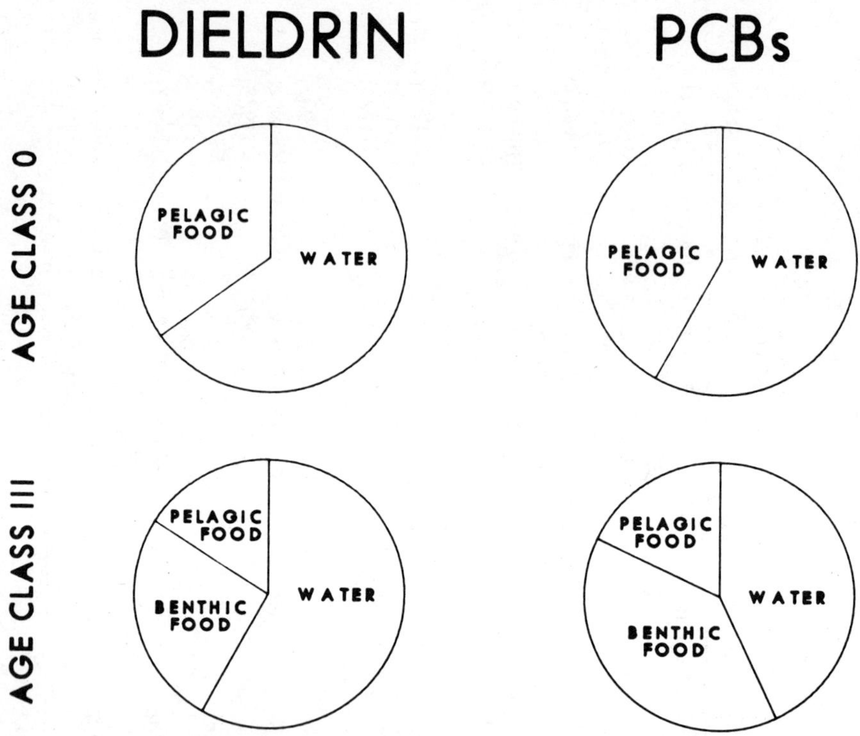


**Figure 21.1.** The relative importance of water, pelagic food, and benthic food as sources of PCBs and dieldrin for two age classes of Lake Michigan alewives. The estimates from the model BIOCON are based on the total mass of contaminant taken up during the year by an average individual of each age class. The major pelagic prey are copepods, cladocerans, and *Mysis relicta;* the main benthic prey, *Pontoporeia hoyi,* is not eaten by younger alewives.

older alewives. Relatively more dieldrin than PCBs was accumulated directly from the water: for age class III alewives, 58% of the accumulated dieldrin versus 43% of the PCBs came from the water; for age class 0 alewives, the values were 66% for dieldrin and 58% for PCBs.

The fraction of PCBs obtained from the water was much greater for alewives than the 11% for brown trout (Table 21.2), even though the PCB-related parameters in the model were the same for both fish species. This result demonstrates the importance of differences among species in physiology, body size, and diet in relation to contaminant accumulation, and shows the value of the bioenergetics modeling approach in accounting for these differences. Similar differences among five species of fish were noted by Jensen et al. (1982).

Additional simulations were performed to determine the influences of species differences in body size, diet, contaminant exposure, and other factors on the relative importance of pathways of PCB uptake. To examine the effects of fish size, the brown trout model was implemented for a trout of equivalent size to an age class I alewife. When simulated growth was from 12.5 to 15.9 g over 57 days, a growth rate (0.48%/day) equal to that of the larger trout used in the experiment, the smaller trout accumulated 15% of their PCB burden directly from the water, compared with the 11% value for the larger trout and the 46% value for age class I alewives in Lake Michigan. The 46% contribution is less than the 58% value for age class 0 alewives, but greater than the 43% value for age class III fish. In simulations where small trout were exposed to the prey and Lake Michigan PCB concentrations measured by Spigarelli et al. (1983), these fish accumulated 40% of their PCB burden directly from the water, similar to the 46% value for age class I alewives grown under the same conditions.

These comparisons suggest that differences in diet and exposure concentrations between the laboratory experiments and the alewives in Lake Michigan can account for differences in the relative importance of direct accumulation of PCBs from water. Similarly, Jensen et al. (1982) concluded that differences in PCB accumulation among fish were the result of differences in exposure, fish size, and rate of growth. Fish size also contributes to the differences in these BIOCON simulations, while differences in physiology between the species seems of lesser importance; this contrasts with the results of Jensen et al. (1982), who found important species differences due to metabolic parameters. Benthic prey, especially *Pontoporeia,* contributed more PCBs and dieldrin to the alewives than pelagic prey items (cladocerans, copepods, and *Mysis relicta*). For age class III alewives, 39% of the PCB accumulation came from benthic prey, and only 18% came from the pelagic prey. The corresponding contributions to accumulation of dieldrin were 26 and 16%.

Concentrations of hydrophobic contaminants tend to be higher in sediments and interstitial water than in bulk lakewater (Eisenreich et al., 1980), and the sediment concentrations are expected to decrease more slowly than concentrations in surface waters in relation to decreases in inputs. Thus, these results based on modeling alewives support the hypothesis (Weininger, 1978) that, even if contaminant concentrations in overlying surface waters were to decrease, a significant flux of contaminants to salmonid fishes in Lake Michigan could occur through inputs from benthic prey to alewives.

### 3.3. Sensitivity Analysis of BIOCON

BIOCON was analyzed to identify those parameters that exerted the greatest influence on the simulated accumulation of dieldrin by Lake Michigan alewives. The methods of error analysis described in Bartell et al. (1986) were applied. Of the 40 parameters that were varied by 2%, the parameter that specified lipid level was the most important. This parameter was a multiplier on the seasonally changing lipid level specified for the alewives, and consequently, influenced the amount of food (and contaminants) consumed as well as the contaminant elimination rate (Eq. 3). The relative influence changed seasonally but consistently outranked all other model parameters in importance. These results are consistent with observations that normalization of contaminant concentration to fish lipid content often reduces the variability among fish samples (Könemann and van Leeuwen, 1980; Brown et al., 1985). The second most important parameter was the rate coefficient of elimination, which in this model is also a function of the fish lipid level. These two parameters ranked ahead of uptake efficiency and the concentration of dieldrin in the water.

## 4. THE FATES OF AROMATICS MODEL (FOAM)

### 4.1. FOAM Description

The FOAM was developed to forecast the transport, degradation, and accumulation of PAHs in lotic ecosystems (Bartell et al., 1981, 1984). The hypothesis underlying the formulation of the FOAM was that the fundamental chemical characteristics of this homologous series of compounds carry relevant information concerning the behavior of PAHs in the environment. The FOAM parameters that determine the rates of photolytic degradation, sorption and desorption from suspended particulate matter and sediments, and the rates of uptake and depuration by the biota are estimated from molecular descriptors for the particular PAH of interest (Bartell, 1984). Structure–activity regressions are used to estimate the corresponding model parameters from the molecular weight, melting point, octanol:water partition coefficient, and the light absorption spectrum of the PAH. The effects of uncertainties associated with parameter values estimated from the regression models are directly considered in simulations by using residual variance about the regressions to assign the parameter values to statistical distributions. Repeated simulations using values sampled from these distributions produce expected distributions of PAH concentrations in water, sediments, and biota. Analyses of the simulated distributions have been used to summarize the sensitivity of FOAM forecasts to parameter uncertainties in specific model implementations (e.g., Bartell et al., 1983).

FOAM includes a dynamic food web model. Temporal differences in the biomass of primary producers and consumer populations are determined by a basic bioenergetics approach to growth (e.g., Park et al., 1974). The physiological process formulation of growth used in this model facilitated the incorporation of metabolically mediated rates of uptake and depuration of PAHs (Bartell et al., 1981).

### 4.2. Application of the FOAM

The FOAM was implemented to simulate the accumulation of phenanthrene by food web components of the Hersey River, downstream from Reed City, in west-central Michigan. Black et al. (1981) measured concentrations of phenanthrene in sediments, crayfish, lamprey, trout, and suckers in this river. Physical dimensions and flow rates for the Hersey River were used to define reach dimensions and flow rates in the FOAM. Initial biomass values for the stream biota were estimated from the literature. Structure–activity regressions were used to estimate rates of photolysis, volatilization, sorption, and rates of uptake and depuration by the biota (Bartell, 1984).

The apparent source of the PAHs entering the river was seepage from a former wood-processing plant that operated from 1902 to 1949. Because exact loading rates were not known, the application of the model was not conventional; simulated bioaccumulation of known system inputs could not be compared with measured bioaccumulation. Yet the data might still be useful for model evaluation. We hypothesized that if the internal dynamics of the FOAM were consistent with the pathways and processes that determine the accumulation of PAHs in aquatic systems, then a single loading rate for the system should be identifiable that would produce acceptable predictions of phenanthrene in all the food web components. The likelihood of quantifying a loading rate that would produce simultaneous agreement with all food web components by chance (e.g., a random walk model) becomes vanishingly small as the number of components increases.

Uncertainties in parameter values estimated from the structure–activity regressions were used to assign the parameters to distributions (Bartell et al., 1983). Two hundred Monte Carlo iterations were performed with each tested loading rate. Acceptability was defined in relation to the overlap between the ranges of predicted concentrations and the measured concentrations. Failure to identify a hypothetical loading rate that the FOAM translated into acceptable estimates of bioaccumulation by several food web components would argue that the internal dynamics of the model were inconsistent with the processes that determined the transport and fate of PAHs in the Hersey River. Given the model structure, an acceptable loading rate would also specify the concentration of dissolved phenanthrene expected in the Hersey River. This could be tested with additional sampling. Our strategy for model evaluation was analagous to fitting bioenergetics models of fish growth to measured growth data by varying the consumption rates (Kitchell et al., 1977; Stewart et al., 1983); the expected pattern of fish consumption rates (or phenanthrene loading rates) could, in principle, be verified with laboratory or field measurements (e.g., Rice and Cochran, 1984).

The loading rate that produced the best simultaneous fit to the observed phenanthrene concentrations in the Hersey River biota was 10.0 ng $m^{-2}$ $d^{-1}$. At this loading rate, the model consistently predicted the concentration of phenanthrene observed in Hersey River benthic insects. The range of model predictions, given uncertainties in the parameter estimates, included the observed concentration for most of the simulated growing season (Fig. 21.2). When the predictions were regressed against the observations, assuming steady conditions in the stream, the hypothesis that the model was within a factor of 2 of the data could not be easily rejected. A similar pattern obtained for comparisons of predicted and measured phenanthrene concentrations in Hersey River suckers (Fig. 21.3), although the model tended to overestimate accumulation of the PAH by these fish. Ranges of model predictions were less than the observed concentration by crayfish, while accumulation by trout was overestimated by the FOAM.

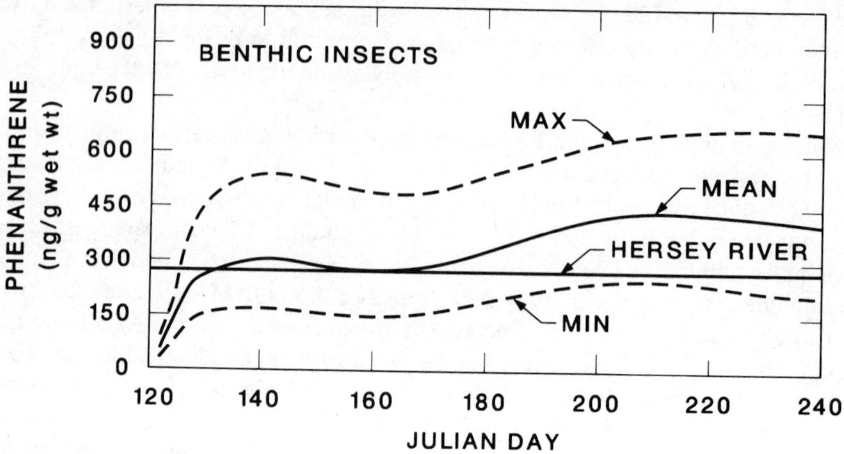


**Figure 21.2.** Mean, minimum, and maximum concentrations of phenanthrene predicted by the Fates of Aromatics Model compared with the measured concentration in benthic insects in the Hersey River, Michigan.

## 4.3. Sensitivity Analyses of the FOAM

The residual variance about the structure–activity regressions was used to define distributions for the FOAM parameters. In this sense, the FOAM can be no better than the data underlying the regressions. Given the parameter uncertainties, it might be possible for the model to generate acceptable results simply because of model imprecision. The worse the model, the greater the likelihood that ranges of predictions will include the observed concentrations. To determine if the acceptable

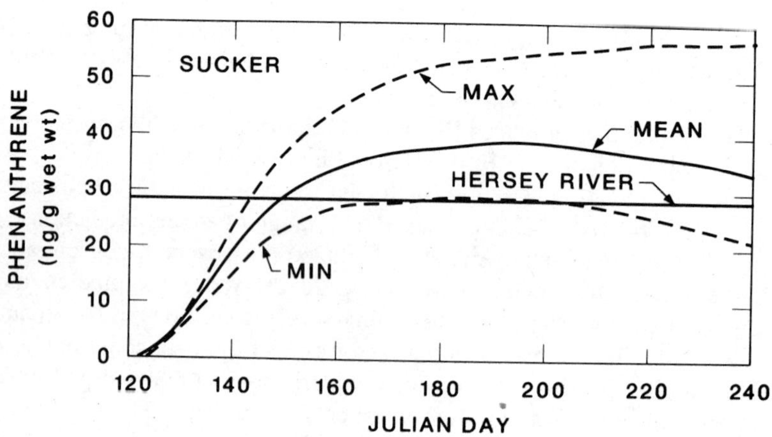


**Figure 21.3.** Mean, minimum, and maximum concentrations of phenanthrene predicted by the Fates of Aromatics Model compared with the measured concentration in suckers in the Hersey River. Michigan.

simulations were the result of model imprecision, we examined the parameter values that generated acceptable predictions. If the distributions that produced accurate results differed markedly from the initial distributions defined by the structure–activity regressions, successful predictions were likely generated by improbable combinations of extreme parameter values and the agreement between predictions and observations might be suspect.

The results of analyzing the parameter values demonstrated no shift in the mean or standard deviation in excess of 10%. Thus, the accurate predictions resulted from the central tendencies of the parameter values. When the initial parameter distributions were assigned coefficients of variation (CV) of 40%, which were greater than the CVs defined by the residual variance of the regression models, the FOAM produced fewer acceptable forecasts. The increase in potential parameter variability increased the model imprecision, which could increase the overlap between predicted and measured phenanthrene concentrations. However, the bias in model predictions also increased with the result that a smaller percentage of simulations overlapped the data. Examination of distributions created by parameter values that produced acceptable simulations showed shifts in mean values and standard deviations by as much as 30%. Minimum and maximum values changed by as much as 600%.

Results of the Monte Carlo simulations were used to identify the key parameters that explained the variance in the predicted concentration of phenanthrene in the Hersey River biota. Using the methods of Bartell et al. (1983), the relative partial sums of squares were used to rank order the sensitivity. The PAH flux along the five most sensitive pathways of accumulation by Hersey River suckers is indicated by the bold arrows in Figure 21.4. Predicted accumulation by suckers was sensi-

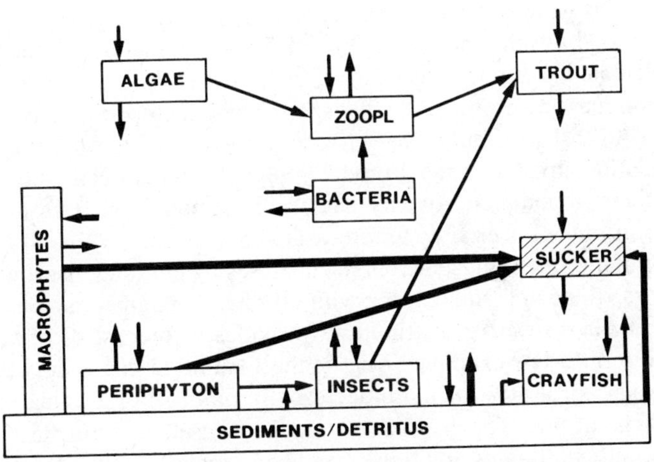


**Figure 21.4.** Illustration of the sensitivity of predicted accumulation of phenanthrene by suckers in the Hersey River. Bold arrows identify pathways that most contribute to variance in predicted concentration in suckers as a function of variation in model parameter values.

tive to the model parameters that determined the rates of feeding and assimilation of macrophytic vegetation, periphyton, and detritus. Consequently, the parameters that determined the accumulation of dissolved phenanthrene by these food sources were also important in determining phenanthrene accumulation by suckers. Importantly, the pattern of parameter sensitivities changed in examination of phenanthrene accumulation by benthic insects, sediments, trout, and crayfish.

## 5. THE STANDARD WATER COLUMN MODEL (SWACOM)

### 5.1. SWACOM Description

O'Neill et al. (1982) derived the SWACOM to examine the hypothesis that the results from single-species toxicity assays could be used to forecast the toxic effects of chemicals in pelagic systems. The SWACOM consists of 10 populations of phytoplankton, 5 zooplankton populations, 3 populations of forage fish, and a single piscivorous fish population. The biomasses of these populations change through time in relation to daily alteration of incident light intensity, water temperature, and nutrient availability. The growth equations are modifications of Equation 1. The model permits evaluation of how the predator–prey and competitive interactions among the SWACOM populations modify the expected direct effects of toxicants in a highly interconnected system (O'Neill et al., 1983). In the SWACOM, natural variability is modeled by assigning variance to the optimal temperature for growth of each population. Temperature optima are assumed not to be influenced by chemical exposure.

To extrapolate from bioassay data to potential effects on biomass production in a pelagic system, the SWACOM incorporates a bioassay submodel. Each of the 19 populations is grown "in isolation" under nonlimiting conditions (e.g., light, temperature, food or nutrient availability). Effects parameters for each population are then calculated as multipliers of the growth processes required to simulate the decrease in biomass measured in the bioassay. The effects parameters are then used with the SWACOM to simulate the effects of chemical exposure on all 19 interacting populations over an annual production cycle. Incomplete understanding of pelagic systems, incomplete toxicity data, and extrapolation from the laboratory to the field contribute uncertainty to forecasts of toxic effects (O'Neill et al., 1982). These uncertainties are represented in the analyses by defining distributions for the effects factors. Repeated simulations with all effects parameters chosen independently from their respective distributions provides a forecast of effects stated in probabilistic terms. For example, what is the likelihood that a doubling or more of noxious algae or a 50% or more decrease in biomass of a valuable game fish population will occur? The SWACOM has been used to estimate the expected effects of organic pollutants and trace metals on biomass production for hypothetical pelagic systems (Barnthouse et al., 1985).

## 5.2. Effects of Naphthalene

Figure 21.5 illustrates the likelihood of observing increased algal production or decreased game fish production in relation to the exposure concentration of naphthalene. Each point indicates the frequency in 300 simulations that the endpoint of interest was predicted. The results demonstrated that the likelihood of reaching or surpassing an endpoint increases in relation to the exposure concentration. For a single exposure concentration, the likelihood of observing greater effects is less than the likelihood of observing smaller effects. The nonzero probabilities of exceeding selected endpoints in the absence of chemical exposure underline the importance of considering natural system variability in estimating the effects of toxicants in interconnected systems.

## 5.3. Sensitivity of Effects Forecasts Using the SWACOM

Both direct toxicity on the SWACOM populations and indirect effects that propagate throughout the food web contribute to estimates of expected effects (O'Neill et al., 1982, 1983). Through analysis of the repeated simulations used to estimate the expected effects, it was possible to determine the relative contribution of direct

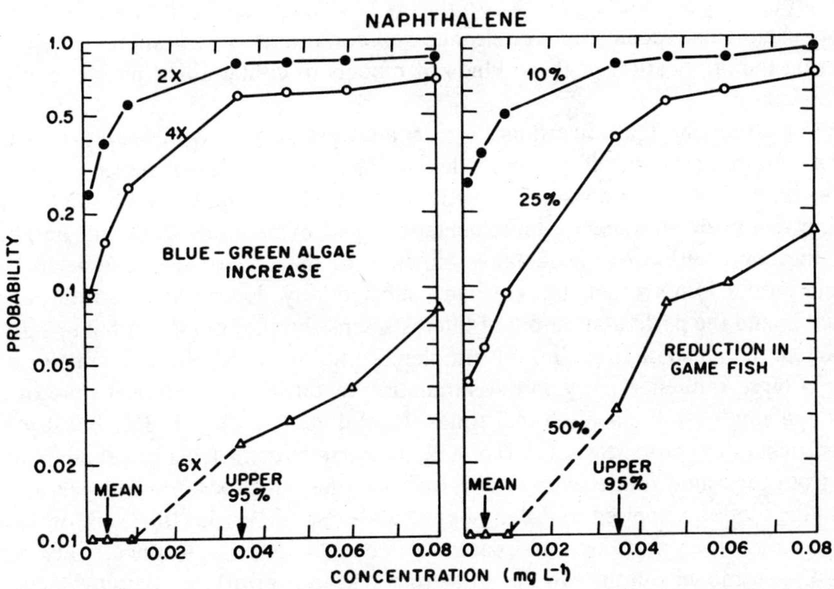


**Figure 21.5.** The probabilites of a greater than two-, four-, or sixfold increase in blue-green algal biomass and of a greater than 10, 25, or 50% reduction in game fish biomass as a function of naphthalene concentration, estimated using the Standard Water Column Model. The arrows indicate the mean and upper 95% confidence limit on the naphthalene concentration in one exposure scenario (Suter et al., 1985).

and indirect toxicities to expected effects over the range of exposure concentrations. At low concentrations, changes in biomass of other trophic levels contributed most to the variance in predicted biomass of algae and game fish. Competitive and predator–prey interactions contributed importantly to estimates of toxic effects. At higher exposure concentrations, direct chemical effects on rates of photosynthesis, feeding, and respiration at each trophic level dominated variance in the simulated effects.

## 6. DISCUSSION

The models outlined here share a common bioenergetics framework. This level of resolution in model structure allows the models to account for the influence of physiological processes on contaminant uptake and elimination, and for the modifying effects of contaminants on metabolic processes and, subsequently, on ecological interactions. When implemented at this level of resolution, the model parameters are measureable in laboratory experiments. These models separate the influence of biological (physiological), environmental, and contaminant-specific parameters. This permits more detailed comparisons to be made between laboratory experiments performed with different species or under different experimental conditions (e.g., Table 21.1). It also makes it easier to apply the models to additional chemicals, species, and environmental conditions, an advantage when considering the application of these kinds of models to contaminant problems in the Great Lakes.

We recommend that uncertainty (error) analysis of models be included as part of any future research efforts to model the effects of toxic substances on Great Lakes biota. Whenever possible, uncertainty in model output should be considered during model development, implementation, and evaluation. It should be noted that the contribution of particular parameters to the uncertainty associated with model output is not fixed, but can vary substantially depending upon the model structure and the particular output of interest. Sensitivity of model output to a given parameter can also change through simulated time; for example, a parameter may have a large influence early in the simulation or during the warmest time of the year and smaller influence at other times (Bartell et al., 1983, 1986). Because the physicochemical properties of different contaminants span such a broad range, output from the same model can be sensitive to different processes and parameters when the model is applied to different chemicals (Bartell et al., 1983). Information concerning which processes and parameters contribute most to uncertainty associated with model output can aid in setting research priorities, thereby focusing attention on research that will contribute most to reducing the uncertainty associated with the output of interest. Consideration of the model uncertainties also leads to consideration of natural variability in aquatic systems, and the potential difficulties in identifying the effects of low concentrations of toxic substances. Uncertainty analyses may be helpful in identifying those conditions under which experiments or environmental sampling are most likely to yield clear results.

The models we described serve as sets of operational hypotheses concerning the environmental behavior of xenobiotics. These models, and others, can be used to explore the implications of available information on forecasts of chemical fates and effects for different chemicals, species, and environmental conditions. When coupled with the methods of sensitivity and error analyses, the models can be used to examine the value of collecting additional information. The models should be viewed not as hypotheses to be wholly accepted or rejected, but as tools for continual improvement of understanding, for advancement of theory, and for making more accurate forecasts.

## ACKNOWLEDGMENTS

We thank D. L. DeAngelis, G. F. Cada, T. D. Fontaine, and an anonymous reviewer for comments on an earlier draft. This research was sponsored jointly by the National Oceanic and Atmospheric Administration, Great Lakes Environmental Research Laboratory, under Interagency Agreement 40-1302-82, the Office of Toxic Substances, U.S. Environmental Protection Agency, under Interagency Agreement DW89930690-01-0, and the Ecological Research Division of the Office of Health and Environmental Research, U.S. Department of Energy, under Contract DE-AC05-84OR21400 with Martin Marietta Energy Systems, Inc. Publication No. 2993, Environmental Sciences Division, Oak Ridge National Laboratory. Although the research described in this paper has been partially funded by the United States Environmental Protection Agency, it has not been subjected to the EPA's required peer and policy review and, therefore, does not necessarily reflect the views of the agency, and no official endorsement should be inferred.

## REFERENCES

Barnthouse, L. W., Suter, G. W., II, Baes, C. F., III, Bartell, S. M., Cavendish, M. G., Gardner, R. H., O'Neill, R. V., and Rosen, A. V. (1985). *Environmental Risk Analysis for Indirect Coal Liquefaction* (ORNL/TM-9120). Oak Ridge Natl. Lab., Oak Ridge, TN.

Bartell, S. M. (1984). Forecasting the fate and effects of aromatic hydrocarbons in aquatic systems. In K. E. Cowser (Ed.), *Synthetic Fossil Fuel Technologies: Results of Health and Environmental Studies*. Butterworth, Boston, MA, pp. 523–540.

Bartell, S. M., Landrum, P. F., Giesy, J. P., and Leversee, G. J. (1981). Simulated transport of polycyclic aromatic hydrocarbons in artificial streams. In W. J. Mitsch, R. W. Bosserman, and J. M. Klopatek (Eds.), *Energy and Ecological Modelling*. Elsevier, Amsterdam, pp. 133–144.

Bartell, S. M., Gardner, R. H., O'Neill, R. V., and Giddings, J. M. (1983). Error analysis of predicted fate of anthracene in a simulated pond. *Environ. Toxicol. Chem.* **2,** 19–28.

Bartell, S. M., Gardner, R. H., and O'Neill, R. V. (1984). The fates of aromatics model (FOAM): Description, application, and analysis. *Ecol. Modell.* **22,** 109–121.

Bartell, S. M., Breck, J. E., Gardner, R. H., and Brenkert, A. L. (1986). Individual parameter perturbation and error analysis of fish bioenergetics models. *Can. J. Fish. Aquat. Sci.* **43,** 160–168.

Binder, R. L., Melancon, M. J., and Lech, J. J. (1984). Factors influencing the persistence and metabolism of chemicals in fish. *Drug Metab. Rev.* **15,** 697–724.

Black, J. J., Hart, T. F., Jr., and Evans, E. (1981). HPLC studies of PAH pollution in a Michigan trout stream. In M. Cooke and A. J. Dennis (Eds.), *Polynuclear Aromatic Hydrocarbons: Chemical Analysis and Biological Fate.* Battelle, Columbus, OH, pp. 343–355.

Brown, M. P., McLaughlin, J. J. A., O'Connor, J. M., and Wyman, K. (1982). A mathematical model of PCB bioaccumulation in plankton. *Ecol. Modell.* **15,** 29–47.

Brown, M. P., Werner, M. B., Sloan, R. J., and Simpson, K. W. (1985). Polychlorinated biphenyls in the Hudson River. *Environ. Sci. Technol.* **19,** 656–661.

Bruggeman, W. A., Marton, L. B. J. M., Kooiman, D., and Hutzinger, O. (1981). Accumulation and elimination kinetics of di-, tri-, and tetra-chlorobiphenyls by goldfish after dietary and aqueous exposure. *Chemosphere* **10,** 811–832.

Buikema, A. L., Jr., McGinniss, M. J., and Cairns, J., Jr. (1979). Phenolics in aquatic ecosystems: A selected review of recent literature. *Mar. Environ. Res.* **2,** 87–181.

Carpenter, S. R., Kitchell, J. F., and Hodgson, J. (1985). Cascading trophic interactions and lake productivity. *BioScience* **35,** 634–639.

de Freitas, A. S. W., Gidney, M. A. J., McKinnon, A. E., and Norstrom, R. J. (1975). Factors affecting whole-body retention of methyl mercury in fish. *ERDA Symp. Ser.* **42,** 441–451.

Eisenreich, S. J., Hollod, G. J., Johnson, T. C., and Evans, J. (1980). Polychlorinated biphenyl and other microcontaminant-sediment interactions in Lake Superior. In R. A. Baker (Ed.), *Contaminants and Sediments.* Ann Arbor Sci. Publ., Ann Arbor, MI, Vol. 1, pp. 67–94.

Elliott, J. M. (1976). The energetics of feeding, metabolism and growth of brown trout (*Salmo trutta* L.) in relation to body weight, water temperature and ration size. *J. Anim. Ecol.* **45,** 923–948.

Evans, M. S., Bathelt, R. W., and Rice, C. P. (1982). Polychlorinated biphenyls and other toxicants in *Mysis relicta. Hydrobiologia* **93,** 205–215.

Fagerström, T., Åsell, B., and Jernelöv, A. (1974). Model for accumulation of methyl mercury in northern pike *Esox lucius. Oikos* **25,** 14–20.

Guiney, P. D., and Peterson, R. E. (1980). Distribution and elimination of a polychlorinated biphenyl after acute dietary exposure in yellow perch and rainbow trout. *Arch. Environ. Contam. Toxicol.* **9,** 667–674.

Guiney, P. D., Peterson, R. E., Melancon, M. J., Jr., and Lech, J. J. (1977). The distribution and elimination of 2,5,2',5'-[$^{14}$C] tetrachlorobiphenyl in rainbow trout (*Salmo gairdneri*). *Toxicol. Appl. Pharmacol.* **39,** 329–338.

Guiney, P. D., Melancon, M. J., Jr., Lech, J. J., and Peterson, R. E. (1979). Effects of egg and sperm maturation and spawning on the distribution and elimination of a polychlorinated biphenyl in rainbow trout (*Salmo gairdneri*). *Toxicol. Appl. Pharmacol.* **47,** 261–272.

Hewett, S. W., and Kitchell, J. F. (1982). *Implementation of Predator-prey Models for Great Lakes Fishes,* Final Rep. NOAA Grant 144P436. Lab. Limnol., University of Wisconsin, Madison.

Hilton, J. W., Hodson, P. V., Braun, H. E., Leatherland, J. L., and Slinger, S. J. (1983). Contaminant accumulation and physiological response in rainbow trout (*Salmo gairdneri*) reared on naturally contaminated diets. *Can. J. Fish. Aquat. Sci.* **40,** 1987–1994.

Jensen, A. L., Spigarelli, S. A., and Thommes, M. M. (1982). PCB uptake by five species of fish in Lake Michigan, Green Bay of Lake Michigan, and Cayuga Lake, New York. *Can. J. Fish. Aquat. Sci.* **39,** 700–709.

Kitchell, J. F., Koonce, J. F., O'Neill, R. V., Shugart, H. H., Jr., Magnuson, J. J., and Booth, R. S. (1974). Model of fish biomass dynamics. *Trans. Am. Fish. Soc.* **103,** 786–798.

Kitchell, J. F., Stewart, D. J., and Weininger, D. (1977). Applications of a bioenergetics model to yellow perch (*Perca flavescens*) and walleye (*Stizostedion vitreum vitreum*). *J. Fish. Res. Board Can.* **34,** 1922–1935.

Könemann, H., and van Leeuwen, K. (1980). Toxicokinetics in fish: Accumulation and elimination of six chlorobenzenes by guppies. *Chemosphere* **9,** 3–19.

Mathers, R. A., and Johansen, P. H. (1985). The effects of feeding ecology on mercury accumulation in walleye (*Stizostedion vitreum*) and pike (*Esox lucius*) in Lake Simcoe. *Can. J. Zool.* **63,** 2006–2012.

McCarthy, J. F., and Jimenez, B. D. (1985). Reduction in bioavailability to bluegills of polycyclic aromatic hydrocarbons bound to dissolved humic material. *Environ. Toxicol. Chem.* **4,** 511–521.

McKim, J. M., and Goeden, H. M. (1982). A direct measure of the uptake efficiency of a xenobiotic chemical across the gills of brook trout (*Salvelinus fontinalis*) under normoxic and hypoxic conditions. *Comp. Biochem. Physiol. C.* **72C,** 65–74.

McKim, J. M., and Heath, E. M. (1983). Dose determination for waterborne 2,5,2',5'-[$^{14}$C] tetrachlorobiphenyl and related pharmacokinetics in two species of trout (*Salmo gairdneri* and *Salvelinus fontinalis*): A mass-balance approach. *Toxicol. Appl. Pharmacol.* **68,** 177–187.

Murphy, P. G., and Murphy, J. V. (1971). Correlations between respiration and direct uptake of DDT in the mosquito fish *Gambusia affinis*. *Bull. Environ. Contam. Toxicol.* **6,** 581–588.

Neely, W. B. (1979). Estimating rate constants for the uptake and clearance of chemicals by fish. *Environ. Sci. Technol.* **13,** 1506–1510.

Neely, W. B., Branson, D. M., and Blau, G. E. (1974). Partition coefficient to measure bioconcentration potential of organic chemicals in fish. *Environ. Sci. Technol.* **8,** 1113–1115.

Niimi, A. J., and Cho, C. Y. (1981). Elimination of hexachlorobenzene (HCB) by rainbow trout (*Salmo gairdneri*), and an examination of its kinetics in Lake Ontario salmonids. *Can. J. Fish. Aquat. Sci.* **38,** 1350–1356.

Norden, C. R. (1967). Age, growth, and fecundity of the alewife, *Alosa pseudoharengus* (Wilson), in Lake Michigan. *Trans. Am. Fish. Soc.* **96,** 387–393.

Norstrom, R. J., McKinnon, A. E., and de Freitas, A. S. W. (1976). A bioenergetics-based model for pollutant accumulation by fish. Simulation of PCB and methylmercury residue levels in Ottawa River yellow perch (*Perca flavescens*). *J. Fish. Res. Board Can.* **33,** 248–267.

Oliver, B. G., and Niimi, A. J. (1983). Bioconcentration of chlorobenzenes from water by rainbow trout: Correlations with partition coefficients and environmental residues. *Environ. Sci. Technol.* **17,** 287–291.

O'Neill, R. V., Gardner, R. H., Barnthouse, L. W., Suter, G. W., Hildebrand, S. G., and Gehrs, C. W. (1982). Ecosystem risk analysis: A new methodology. *Environ. Toxicol. Chem.* **1,** 167–177.

O'Neill, R. V., Bartell, S. M., and Gardner, R. H. (1983). Patterns of toxicological effects in ecosystems: A modeling approach. *Environ. Toxicol. Chem.* **2,** 451–461.

Park, R. A. et al. (24 co-authors) (1974). A generalized model for simulating lake ecosystems. *Simulation* **23,** 33–50.

Park, R. A., Connolly, C. I., Albanese, J. R., Clesceri, L. S., Heitzman, G. W., Herbrandson, H. H., Indyke, B. H., Loehe, J. R., Ross, S., Sharma, D. D., and Shuster, W. W. (1980). *Modeling Transport and Behavior of Pesticides and Other Toxic Organic Materials in Aquatic Environments,* Rep. No. 7. Cent. Ecol. Model., Rensselaer Polytechnic Inst., Troy, NY.

Rapaport, R. A., and Eisenreich, S. J. (1984). Chromatographic determination of octanol-water partition coefficients ($K_{ow}$'s) for 58 polychlorinated biphenyl congeners. *Environ. Sci. Technol.* **18,** 163–170.

Reinert, R. E., Stone, L. J., and Bergman, H. L. (1974). Dieldrin and DDT: Accumulation from water and food by lake trout (*Salvelinus namaycush*) in the laboratory. Proc. 17th Conf. Great Lakes Res., pp. 52–58.

Rice, C. P., Eadie, B. J., and Erstfeld, K. M. (1982). Enrichment of PCBs in Lake Michigan surface films. *J. Great Lakes Res.* **8,** 265–270.

Rice, J. A., and Cochran, P. A. (1984). Independent evaluation of a bioenergetics model for largemouth bass. *Ecology* **65,** 732–739.

Rice, J. A., Breck, J. E., Bartell, S. M., and Kitchell, J. F. (1983). Evaluating the constraints of temperature, activity, and consumption on growth of largemouth bass. *Environ. Biol. Fishes* **9,** 262–275.

Roberts, J. R., de Freitas, A. S. W., and Gidney, M. A. J. (1977). Influence of lipid pool size on the bioaccumulation of the insecticide chlordane by northern redhorse suckers (*Moxostoma macrolepidotum*). *J. Fish. Res. Board Can.* **34,** 89–97.

Rottiers, D. V., and Tucker, R. M. (1982). Proximate composition and caloric content of eight Lake Michigan fishes. *Tech. Pap. U.S. Fish Wildl. Serv.* **108,** 1–8.

Schindler, D. W., Mills, K. H., Malley, D. F., Findlay, D. L., Shearer, J. A., Davies, I. J., Turner, M. A., Linsey, G. A., and Cruikshank, D. R. (1985). Long-term ecosystem stress: The effects of years of experimental acidification on a small lake. *Science* **228,** 1395–1401.

Southworth, G. R., Keffer, C. C., and Beauchamp, J. J. (1980). Potential and realized bioconcentration: A comparison of observed and predicted bioconcentration of azaarenes in the fathead minnow (*Pimephales promelas*). *Environ. Sci. Technol.* **14,** 1529–1531.

Spigarelli, S. A., Thommes, M. M., and Prepejchal, W. (1983). Thermal and metabolic factors affecting PCB uptake by adult brown trout. *Environ. Sci. Technol.* **17,** 88–94.

Stewart, D. J. (1980). Salmonid predators and their forage base in Lake Michigan: A bioenergetics-modeling synthesis. Ph.D. Thesis, University of Wisconsin, Madison.

Stewart, D. J., and Stein, R. A. (1974). Short-term fate of dietary dieldrin in the digestive tract of juvenile lake trout (*Salvelinus namaycush*). *Bull. Environ. Contam. Toxicol.* **11,** 563–566.

Stewart, D. J., Weininger, D., Rottiers, D. V., and Edsall, T. A. (1983). An energetics model for lake trout, *Salvelinus namaycush:* application to the Lake Michigan population. *Can. J. Fish. Aquat. Sci.* **40,** 681–698.

Suter, G. W., Barnthouse, L. W., Baes, C. F., III, Bartell, S. M., Cavendish, M. G., Gardner, R. H., O'Neill, R. V., and Rosen, A. E. (1985). *Environmental Risk Analysis for Direct Coal Liquifaction* (ORNL/TM-9074). Oak Ridge Natl. Lab., Oak Ridge, TN.

Thomann, R. V. (1981). Equilibrium model of fate of microcontaminants in diverse aquatic food chains. *Can. J. Fish. Aquat. Sci.* **38,** 280–296.

Thomann, R. V., and Connolly, J. P. (1984). Model of PCB in the Lake Michigan lake trout food food chain. *Environ. Sci. Technol.* **18,** 65–71.

Ursin, E. (1967). A mathematical model of some aspects of fish growth, respiration, and mortality. *J. Fish. Res. Board Can.* **24,** 2355–2453.

Vaughan, D. S., Yoshiyama, R. M., Breck, J. E., and DeAngelis, D. L. (1984). Modeling approaches for assessing the effects of stress on fish populations. In V. W. Cairns, P. V. Hodson, and J. O. Nriagu (Eds.), *Contaminant Effects on Fisheries.* Wiley, New York, pp. 259–278.

Weininger, D. (1978). Accumulation of PCBs in lake trout in Lake Michigan. Ph.D. Thesis (Water Chemistry), University of Wisconsin, Madison.

Windell, J. T., Foltz, J. W., and Sarokon, J. A. (1978). Effect of fish size, temperature, and amount fed on nutrient digestibility of a pellet diet by rainbow trout, *Salmo gairdneri. Trans. Am. Fish. Soc.* **107,** 613–616.

# 22

# STATISTICAL METHODS FOR STUDYING BIOTIC RESPONSES TO PERSISTENT TOXIC SUBSTANCES

***S.R. Esterby***

*Aquatic Ecology Division, National Water Research Institute, Burlington, Ontario, Canada*

## 1. INTRODUCTION

The scope for the application of statistical methods in the study of the effects of toxic substances on aquatic communities is very broad because statistical methodology is available for the design of studies and the analysis of results, and because both laboratory experiments and field observations are necessary for the elucidation of the effects of toxic substances. Further, structural and functional responses and many different types of organisms are being studied. There are, however, some important statistical concepts that are applicable to the many aspects of the present topic. These concepts will be discussed as they apply to laboratory experiments and to field studies.

Laboratory experiments and field studies are complementary, and each, on its own, is limited by different inadequacies. Examples of recommendations from the literature that acknowledge their complementary roles are the U.S. National Research Council (1981) and Sloof and de Zwart (1983). In the National Research Council report, field observations and experimental work are deemed necessary to the understanding of the structure and the function of baseline ecosystems. Sloof and de Zwart (1983) recommend that they be applied simultaneously since field studies have low specificity to toxic stress and low signal-to-noise ratio, while laboratory studies suffer from poor ecological significance. These distinctions are of consequence to the statistical methods that can be applied and also to the strength of the conclusions that can be drawn.

The statistical concept fundamental to the present topic involves the separation of variation due to different sources. For example, in a field study, it is the separation of the variation associated with natural fluctuations of the ecosystems or with other modifying factors from that associated with toxic substances that is important. The strength of the conclusions about cause and effect weaken as we go from laboratory experiments to field observations. There is a corresponding change in statistical methods and the nature of the statistical inferences that can be drawn. Thus the discussion will be developed using this correspondence.

The need to consider design and analysis as integral parts of a study will be documented. Study design will be emphasized because it is the critical step that determines how much information, related to the objectives of the study, can be extracted from the data produced. Several papers in which recommendations were made for use of statistical methods will be considered, and these will be related to the statistical terminology that will be defined. Methods of analysis will be discussed and examples from studies of benthic invertebrates will be given. In many cases, when a paper is cited, only one or a few aspects of the study will be discussed and thus the reader should not assume that the portion cited covers the entire contents.

## 2. STUDY DESIGN

The term study is used here to include both experiments and programs under which different types of field observations are collected. There are many texts on the design of experiments, but few on the design of environmental field studies. Elliott (1977) and Green (1979) consider aspects of both design and analysis, and a number of authors, for example, Miller (1984) and Hurlbert (1984), criticize other environmental scientists either for not using statistical methods or for using them improperly.

The incorporation of statistical principles in the design of field studies is perhaps the most important role for statistics. Miller (1984) observes that much biological research is conducted without addressing the question of how much data are required to estimate a quantity with a specified precision or to detect a change of a specified size. However, the points raised by Miller emphasize only one consideration in the design of a study, and it is this consideration that often results in a superficial use of statistics. Before discussing design techniques further, some definitions are required.

### 2.1. Experiment, Field Sampling Program, and Observational Study Defined

The distinction between experiments and planned surveys is that in the former the experimenter has control over the assignment of treatments to experimental units, whereas in the latter the observer has no control over the factors that determine the grouping of observational units (Cox, 1958). Thus, our major concern, the field study, would be classified as a planned survey. Neither Green (1979) nor Elliott (1977) have attempted to classify such field studies further than to call them sampling programs, and, as such, have given the standard statistical references for the theory of survey sampling (Yates, 1960; Stuart, 1962; Cochran, 1977). Hurlbert (1984) has called sampling programs mensurative experiments, and what statisticians call simply experiments, he has named manipulative experiments. In this chapter (Fig. 22.1), Cox's definition of an experiment will be used, and a field study in which randomization is used will be called a sampling program. When a

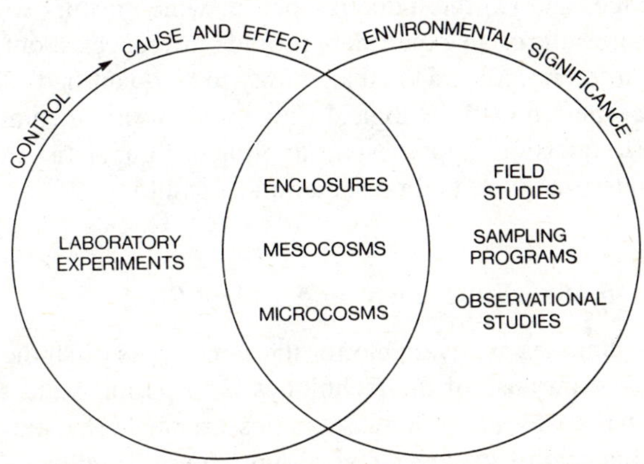


**Figure 22.1.** Relationships between types of studies.

study is subject to the limitation that randomization cannot be used, it will be called an observational study (Snedecor and Cochran, 1980).

Studies conducted in laboratory microcosms, mesocosms, and in situ enclosures have the potential for control of the assignment of treatments to experimental units and thus can be classified as experiments. Clearly, the ability to control sources of variation other than that due to treatments will be much less than in simpler laboratory experiments and larger error terms would be expected. The experimental unit is the entire microcosm, mesocosm, or enclosure and replication consists of applying the same treatment to two or more experimental units. Since the units are samples from natural ecosystems (necessarily simplified), it follows that the degree of heterogeneity between units will be much greater than that obtained in single species tests. Hurlbert (1984) cites coefficients of variation ranging from 11 to 172% for a number of ecosystem variables in studies where replicate microcosms were used to assess variability. Metcalf et al. (1971), using $^{14}C$-labeled DDT, found 10,000- and 13,000-fold magnification of DDT in fish over that in water from two replications of a model ecosystem test. The concentrations of $^{14}C$ in water were 0.002 and 0.004 ppm, respectively, in the two replicates that received the same rate of application.

## 2.2. The Steps in Planning a Study

The link between the design of an investigation and the analysis of the resulting data needs to be emphasized. Cochran and Cox (1957) state that there are three parts to the planning of an experiment: (1) a clear and specific statement of the objectives, including questions to be answered, hypotheses to be tested, and effects to be estimated; (2) description of the experiment covering all details related to treatments, experimental materials, measurements to be made, number of replicates, and mechanism for assigning treatments to experimental units; and (3) an outline of the methods of analysis. Box et al. (1978) take this further to position the experiment in the iterative process of scientific learning that includes (1) the deductive phase consisting of abstraction of specific and testable hypotheses from existing knowledge, and (2) the inductive phase, which occurs when conclusions obtained from the results of an experiment provide inferences about general knowledge. Statistical inference allows the uncertainty to be quantified. Similarly, Green (1979) states that there must be a logical flow in an environmental study between the steps: purpose, question, hypothesis, sampling design, statistical analysis, tests of hypotheses, interpretation, and presentation of results.

## 2.3. Comparison of an Experiment and a Field Study

Similar design techniques are available for the three types of studies (Table 22.1), and, in general, the purpose of the techniques is to permit valid statistical inferences to be drawn that are free from bias and possess the best achievable precision. The statistical concepts of experimental design, free of mathematical technicali-

**Table 22.1 Techniques for the Design of Studies and Their Consequences**

| Experiments | | Field Sampling Programs | | Observational Studies | |
|---|---|---|---|---|---|
| Design Technique | Consequence | Design Technique | Consequence | Design Technique | Consequence |
| Random assignment of treatments to units | Eliminate bias; valid tests of significance | Random selection of observational units | Eliminate bias; valid tests of significance | | |
| Replication | Error estimation | Replication | Error estimation | Replication | Error estimation |
| Blocking | Increase precision | Stratification | Increase precision | Stratification | Increase precision |
| Concomitant variable | Increase precision | Covariate | Increase precision | Covariate | Attempt to eliminate bias; increase precision |

ties, are given by Cox (1958). A clear distinction between the potential of experiments and surveys can be rendered by first reviewing the features of experiments.

Given that the objectives and hypotheses have been established and the treatments, experimental units, and the nature of the observations have been decided upon, Cox (1958) gives the requirements for a good experiment as freedom from systematic errors, sufficient precision, a wide range of applicability, simplicity, and provision for assessing the uncertainty in the conclusions. Techniques such as blocking (grouping of units into sets of like units) and the use of concomitant variables (adjustment of the main observation to account for different values of the concomitant variable) are used to increase precision by removing explainable variation from the error term (or unexplained variation). Random assignment of treatments to units protects against systematic errors by randomizing unknown patterns in the uncontrolled variation and provides a means of assessing uncertainty. Although exact significance tests and probability limits can be calculated from the permutation distribution provided by randomization, in practice, they are calculated by assuming that the errors follow a certain distribution (usually normal). The assumption that they constitute a random sample from the distribution of assumed form is effected by the physical procedure of randomization. In summary, the aim is to eliminate as much as possible of the uncontrolled variation, for example, by grouping of the units, and then to randomize the remainder.

The essential difference between an experiment and a field study is the control over the choice of levels of the factors and of the assignment of factors to units that is exercised by those planning an experiment but is unavailable in field studies. It is this control, together with the randomization of the uncontrolled variation, that permits the establishment of cause and effect. The lack of it permits conclusions about associations and possible causes only.

Differences in the degree of control over factors can be compared in the following two pairs of examples where the experiment and field study in each pair were designed to address similar questions. Only the features of the study that were needed for this comparison have been given. In an experiment to study chronic toxicity of lead to rainbow trout, Hodson et al. (1979) set the levels of exposure by controlling the concentration of lead in water at the nominal levels of 0, 10, 32, and 100 $\mu$g/L with four replicate tanks per level. This resulted in large differences in mean blood lead concentrations at the four levels of exposure (150, 420, 1700, and 1800 $\mu$g/L, respectively). However, in an observational study with a similar objective (Hodson et al., 1984), a very narrow range of blood lead concentrations in lake trout, carp, and white perch was found, since clearly no control over exposure levels was possible. In an experiment, the scientist can ensure that a reasonable range of treatment levels is used. If only a small range is observed for the explanatory variable in an observational study, then failure to detect an association may be due to this limited range.

Thyroid hyperplasia in Great Lakes coho and chinook salmon is well documented and appears to have an environmental etiology other than iodide deficiency, bioaccumulation of organochlorines being suspect agents (Leatherland and Sonstegard, 1981). In an experiment to determine if two organochlorines that are found in Lake Ontario coho salmon would alter thyroid activity, Leatherland and Sonstegard (1978) randomly assigned groups of 30 coho salmon to six aquaria, maintained under identical conditions, and fed fish one of the following diets: uncontaminated control diet, low PCB (50 $\mu$g/g), high PCB (500 $\mu$g/g), low Mirex (5 $\mu$g/g), high Mirex (50 $\mu$g/g) and mixed diet (50 $\mu$g PCB/g plus 5 $\mu$g Mirex/g), where concentrations are for dry feed. Both Mirex and PCBs were found to reduce serum thyroid hormone levels in yearling coho salmon, but no histological evidence of thyroid hyperplasia was found. In a complementary field study, Moccia et al. (1978) collected coho and chinook salmon from the Great Lakes and recorded the frequency of goiters at each site. Organochlorines were measured in pooled muscle tissue samples from at least 10 randomly selected fish, separated by sex, from each site. The authors concluded that the data provided no evidence of an association between goiter frequency and tissue organochlorine burden (their statement on p. 651 of no cause–effect relationship has been taken to mean no association). It was also noted by the authors that syngergistic action between organochlorines and factors not examined in the study may have led to the negative result.

Another limitation of the field study is the difficulty of performing physical randomization. Most of the well-known statistical methods, such as analysis of variance, *t* tests, and their nonparametric counterparts, are based on the assumption that the data are a random sample. The robustness to nonnormality of *t* tests and of the analysis of variance tests for comparing several means and the availability of transformations make the assumption of normality much less of a problem than failure of the assumption of randomness (Box et al., 1978). When the method of sampling is something other than random, bias may be introduced, perhaps due to periodic variation or a trend. Further, there is no reliable way of estimating the standard error of the primary quantities being estimated (Cox, 1958). For the dependence known to be of the form of serial correlation, Scheffé (1959) shows that, under the assumption of normality, the probability of a type I error, for a two-tail test of the mean, ranges from 0.00001 to 0.14 for serial correlations of $-0.4$ and 0.4, respectively, when the nominal significance level is 0.05. Thus, with negatively correlated errors the test would be conservative, that is, too few significant results would be obtained, and with positively correlated errors too many significant results would be obtained. In general, with correlated errors, the true value of the significance level is different from the nominal, but unknown.

Water quality measurements that are taken at short time intervals exhibit serial correlation. Lettenmaier (1976) analyzed 5 years of daily records on one or more of the parameters temperature, dissolved oxygen, suspended solids, and specific conductance in four United States rivers. After transformation to normality and estimation of two Fourier components corresponding to 6- and 12-month cycles, using Box and Jenkins model identification techniques, he found that a simple lag 1 autoregressive model was adequate for the residual series except in two of the seven series. For these two only a lag 2 model was required. The lag 1 correlations were all about 0.8. Similarly, chlorophyll measurements taken at close spatial intervals exhibit serial autocorrelation and Platt et al. (1970) give an example of spectral analysis of such data. From the analysis they drew conclusions about patch diameter. Another example of autocorrelated response is given by Green et al. (1983). Lengths at annual winter shell rings $L_t$ were measured for clams (*Macomba balthica*) collected near Fort Churchill, Manitoba, Canada, and these lengths were used to determine growth curves and ages for each tidal level. Positive serial correlation, which was expected since these successive $L_t$ values come from the same clam, was estimated to be 0.41. Unlike the previous two examples in which methods for dependent time series were used, Green used methods of analysis suitable for independent observations and then verified that the dependency was not large enough to alter the conclusion about statistical significance although it may have altered the observed significance level.

Methods for increasing precision, analogous to blocking and concomitant variables in experimental design, are also available for field study design (Table 22.1). Stratified random sampling consists of grouping units into homogeneous strata and of drawing a random sample from within each strata. Adjustment for sources of bias in field studies can be accomplished by the analysis of covariance if observations are made on a variable that will influence the response variable of interest. Again the principle is the removal of terms from the error component to the component of explainable variation. In some observational studies, it will be possible to use some of the design techniques (Table 22.1). Since randomization is not possible, considerably more effort must be expended in the statistical analysis to ascertain that the assumptions of the analysis are not violated. Despite this, considerably weaker conclusions are unavoidable. Not only are we limited to conclusions about associations, but also, in observational studies, bias and the validity of statistical inferences are a problem. Note that the analysis of covariance is used to remove bias in observational studies but only to increase precision when randomization is possible (Table 22.1). Recall from above that randomization protects against bias by randomizing unknown patterns in the uncontrolled variation.

In Section 4.3, a sampling program is cited (Green and Hobson, 1970) in which the analysis of variance was used to determine if there were differences in the density of *Gemma gemma* in two faunal zones. The authors also measured sediment characteristics and found that they differed little between faunal zones. However, if this parameter had been found to differ between zones and, for example, sediment particle size was known or expected to alter *G. gemma* density, an analysis of covariance could have been conducted to remove the bias in density estimates attributable to particle size. The purpose would be to remove bias even though randomization was performed here since the random selection was within each faunal zone. The analysis of covariance could be performed to improve precision if sites classified to each type of faunal zone were selected at random and sediment particle size varied between sites within each type of faunal zone.

The analysis of covariance is often used for another purpose in observational studies, namely, to study regressions in multiple classifications (Snedecor and Cochran, 1980). Examples can be found in Green et al. (1983) and Green (1968).

Precision is concerned with the amount of variability in repeated measurements and accuracy with the absence of systematic error (see Mandel, (1964) for more precise definitions). Clearly no number of repetitions of a measurement that contains bias or systematic error is large enough to provide an accurate estimate of the true value. Accuracy is determined by comparison with a reference value and adequate precision would generally be required to achieve agreement. An example of a reference value would be the concentration of lead in National Bureau of Standards reference material, bovine liver.

### 2.4. Further Considerations: Replication, Hypotheses, Efficiency

Estimation of the error variance from the experiment or field study requires replication. In an experiment it is replication of the treatments that is required; in field sampling it is replication at the level at which differences in response are being investigated. For example, in a quantitative study to estimate the number of species in a defined area, the replication must consist of sampling a number of sites within the area, not just replicate samples at one site.

Hurlbert (1984) discusses at length the misuse of statistical analyses in field experiments when improper or no replication was performed. He has drawn on many examples from the ecological literature. The major points he makes, related to replication, are that (1) methods of statistical inference should not be used if there is no replication in an experiment; (2) when factors such as spatial or temporal variation are important, treatments must be assigned within groups of units defined by spatial and temporal considerations; (3) subsampling within a treatment should not be confused with replication of the treatment; and (4) the term replicability, used, for example, in studies involving microcosms, to mean that experimental units are similar prior to application of treatments, should not be confused with replication, which means applying the same treatment to more than one of these similar experimental units. Point 2 is covered by the design method of blocking and thus Hurlbert's discussion of interspersion (the assignment of all treatments to each level of space and time occurring in the experiment) and randomization deals with the idea of separating experimental units into blocks to remove known sources of variation from the error term, followed by random assignment of treatments within the blocks.

Since many of the principles of design in experiments and field studies are similar, it follows that the similar considerations regarding replicates are appropriate. Green (1979) summarizes them succinctly in stating that replicate samples should be taken within each combination of time, location, and any other observable variable thought to influence response. Elliott (1977) gives some concrete advice on how to accomplish this objective in sampling programs for benthic invertebrates, but these methods are of much broader applicability.

Consider first the role of the hypothesis in a scientific study. In any particular investigation, the null hypothesis is formed and a test of significance provides the probability of obtaining a result at least as unfavorable to the null hypothesis as that observed in the current investigation. For example, in the two-tailed test of the null hypothesis that the mean is zero, the observed significance level is given by the probability of obtaining an estimate of the mean at least as far from zero as that observed. The observed significance level (*P* value) is then used together with other available information to draw conclusions (Cox, 1977). Formulation of an alternative hypothesis will help in the choice of a suitable statistic, one particularly sensitive to departures, from the null hypothesis, of the form specified by the alternative hypothesis. The alternative hypothesis will also determine whether a one- or two-tailed test of significance will be formed.

The theoretical question of suitable statistics for tests of hypotheses and estimation can be examined independently from particular scientific studies, although motivation may be derived from particular studies. The performance of a test statistic can be evaluated under a hypothetical null hypothesis $H_0$ relative to various alternative hypotheses $H_A$. It is here that type I and type II errors are used and are calculated as the probability under the null distribution of rejecting $H_0$ when it is true and the probability under the alternative distribution of accepting $H_0$ when it is false, respectively. The power of a test is given by the probability, calculated under the alternative hypothesis, of rejecting the null hypothesis when it is false. For example, the evaluation of two statistics proposed for trend detection might include the calculation of the power of each statistic under an alternative hypothesis of linear trend, with the statistic with higher power being recommended. Examples of power calculations can be found in Pettitt (1977), Gupta and Govindarajulu (1980), and O'Brien and Dyck (1985). Numerical power calculations are often performed using Monte Carlo methods (i.e., generation of random samples from a probability distribution) rather than evaluating an analytical expression.

One further aspect is the efficiency of the statistical procedure that is to be used. Efficient methods require fewer observations for estimation or hypothesis testing than nonefficient ones, and in hypothesis testing this will be balanced against the power of the test to detect departures from the null hypothesis. For the normal distribution, when the sample mean and median are calculated from large samples, the relative efficiency of the median to the mean is 0.637, because the variance of the median is larger than the variance of the mean (Snedecor and Cochran, 1980, p. 136). Thus, a sample 57% larger is required to estimate the population mean, with the same precision as obtained from the sample mean, when the sample median is used. Clearly, the results of such theoretical calculations of efficiency and power provide guidance in the choice of suitable statistics for a particular scientific study.

## 3. METHODS OF ANALYSIS

In the analysis of data from all types of studies, graphical examination of the data prior to more formal analysis is important. In fact, plots should be used throughout the analysis, the type depending upon the method of analysis. Plotting methods are particularly useful in the detection of departures from the assumptions made in the analysis. For observational studies, plots and descriptive statistics may be all that is possible, whereas, for well-designed studies (Green, 1979; Hurlbert, 1984), they may be all that is necessary.

## 3.1. Experiments

Methods for the analysis of experiments include analysis of variance, with the associated techniques of multiple comparisons, analysis of covariance and response surface methods, and the specific methods for bioassay experiments. There are many good texts on the analysis of variance and related topics, and these vary in mathematical difficulty (e.g., Cochran and Cox, 1957; Scheffé, 1959; Box et al., 1978). The scattered literature on multiple comparisons and simultaneous inference has been brought together by Miller (1981). General principles and techniques are illustrated in the introduction and a description of how to use each procedure is given as the first section in successive chapters. The classical references for bioassay are Finney's books (1964, 1971). More recent references, specifically for toxicity tests, can be found in Sládeček et al. (1984), including a review by Sprague (1969). Capizzi et al. (1985) propose improved methods of analysis of chronic aquatic toxicity tests that account for the dependence of response on concentration, the existence of censored data (i.e., assay terminated before exact lifetime or maturation time determined for some organisms), and the need for transformations.

Factorial experiments can be used to simultaneously assess the effects of more than one toxic substance. The techniques of confounding and partial replication can sometimes be used to obtain information about which factors and/or interactions are important using fewer experimental units than required for full replicates or a series of single-factor experiments. This could be useful in microcosms or in situ enclosures, where the number of experimental units that can be used is restricted. A clear explanation of the difficulties that can arise in the analysis of such designs is given by Cochran and Cox (1957).

By including the three additional combinations of low PCBs and high Mirex, high PCBs and low Mirex, and high PCBs and high Mirex, the design of the experiment of Leatherland and Sonstegard (1978), discussed in Section 2.3, becomes a complete 3 × 3 factorial experiment. The methods for the analysis of factorial experiments (e.g., Cochran and Cox, 1957; Sokal and Rohlf, 1969) could then be applied to study interactions more fully.

The functional responses of organisms to toxic substances have been studied mostly through laboratory experiments. Additional statistical methods that could be useful to survival studies are techniques for the study of failure–time data, that is, data for which the major endpoint is the time until a particular event occurs (Kalbfleisch and Prentice, 1980).

## 3.2. Field Studies

Since natural systems are under observation in field studies, before investigating the relationship between the response of the feature of interest to the level of another factor, it is necessary to determine how to characterize this feature. For example, if an organism exhibits a random spatial distribution, the number of organisms per unit area can be characterized by fitting a Poisson distribution. This leads to the consideration of three components: (1) the feature(s) of a population or organism that have been used to describe status and the statistical methods that have aided the description; (2) the ways in which these features might be expected to change due to toxic and other stresses; and (3) the modifications to the statistical methods or the adoption of new methods so that the relationship between the features of the population or organism and the factors representing stress can be examined.

Features that have been used to characterize populations include (1) properties of individual species, (2) composition of the population, (3) abundance of the components of the population, and (4) derived variables such as diversity indices, biotic indices, and linear combinations of abundances. Points 1 and 4 can be analyzed by univariate methods, whereas points 2 and 3 require multivariate methods with the distinction that point 2 consists of qualitative data (actually binary) and point 3 of quantitative data. For a univariate response, the mean level or other measure of location, measures of dispersion, and the form of frequency distribution (particularly with respect to random versus contagious spatial distribution) are of interest. This will usually be accomplished by fitting an appropriate probability distribution (Elliott, 1977; Snedecor and Cochran, 1980) to the data or by the use of procedures based on ranks. For the multivariate data sets, the aim is to condense the data so that relationships between variables can be seen and groups of similar objects can be identified and the methods generally used are the largely descriptive techniques of clustering and ordination (Pielou, 1969, 1984; Gordon, 1981).

The types of changes that might be expected as a result of toxic stress include (1) spatial differences in the properties of individuals or populations, (2) temporal changes in these properties, and (3) relationships between the properties of the individual or population and the level of the contaminant or other modifying factor, where level refers to both continuous and categorical variables. Because of the need to adjust for factors other than toxic stress, these other factors (e.g., temperature, pH) should be included in the model. Familiar methods for univariate responses are regression, analysis of covariance, estimation of the components of variance from nested designs, and fitting of frequency distributions (Draper and Smith, 1981; Snedecor and Cochran, 1980; Box et al., 1978). Alternatively, nonparametric procedures are available for comparing samples and for tests of randomness versus alternatives such as trend (Lehmann, 1975). Univariate procedures for changes in regression relationships (Esterby and El-Shaarawi, 1981), changes in distribution (Pettitt, 1979), and classification (El-Shaarawi et al., 1981) are available. In addition to the multivariate procedures of clustering and ordination, procedures such as canonical variables (Pielou, 1969) can be used to study the relationship between sets of biological and chemical and physical variables.

At the functional level, field studies can be used to characterize the occurrence of abnormalities such as tumors in fish (e.g., Black, 1983; Sonstegard and Leatherland, 1984). Here methods of categorical data analysis, such as contingency table analysis and methods for proportions, are appropriate (Snedecor and Cochran, 1980; Plackett, 1974).

## 3.3. Checking the Model

The specification of the design can be written in the form of a model that expresses the functional relationship between the response variables and the variables whose effect we want to study, and that contains an error term. In general, the form for one response variable is

$$y = \text{structural model} + \text{error}$$

or

$$y = f(\mathbf{x}_1) + \epsilon$$

When we speak of removing terms from the error component of the model to the component of explained variation this can be represented as moving from a model

$$y = f(\mathbf{x}_1) + \epsilon(\mathbf{x}_2)$$

to

$$y = f(\mathbf{x}'_1) + \epsilon(\mathbf{x}'_2)$$

where the dimension of $\mathbf{x}'_1$ is greater than that of $\mathbf{x}_1$ and the dimension of $\mathbf{x}'_2$ is smaller than that of $\mathbf{x}_2$, since, in the second case, fewer variables contribute to random error. Regression models are familiar, but all statistical procedures are based on models. When parametric procedures are used, tests of the validity of the assumptions of the statistical model might consist of (1) tests of fit of a particular distribution to a set of data (e.g., $\chi^2$ and Kolmogorov-Smirnov goodness-of-fit tests) or (2) residual analysis (Draper and Smith, 1981) after an analysis of variance or regression analysis. The residual

$$\epsilon = y - f(\mathbf{x}_1)$$

contains the information about failure of the assumptions of the model.

The assumption upon which most nonparametric tests are based is randomness and, in fact, many of the tests consist of testing a hypothesis of randomness using a statistic that is sensitive to departures from randomness of the form specified by the alternative hypothesis.

Several approaches may be taken for the use of methods of statistical inference in observational studies. A model can be postulated, the corresponding statistical analysis performed, and then the assumptions examined using diagnostic checking. This proceeds iteratively, redefining the model as indicated by the checking procedures in the previous step, until the diagnostic checks provide no evidence of serious departures from the assumptions. This is what must be done when regression methods are used on happenstance or historical data. The problems in the analysis of happenstance data and the remedies, in terms of experimental design methods, are given in Box et al. (1978).

## 4. EXAMPLES FROM BENTHIC STUDIES

The choice of the examples was largely based on the book by Elliott (1977) on statistical methods and sampling design for benthic invertebrates. A discussion of the methods for the design of sampling programs and the statistical analysis of the resulting data must be specific to the particular field of application. Elliott describes methods that have been shown to be useful in the analysis of benthic data and considers statistical as well as practical aspects of study design. In addition, a number of well-designed studies have been reported.

### 4.1. Quantitative Study Design

The objective of a quantitative study is to estimate the number per unit area of each species in some well-defined population, and may include further specifications related to the reasons for collecting these numbers. Elliott (1977) gives the major considerations in designing such a study as the determination of (1) the dimensions of the sampling unit, (2) the number of units to sample, and (3) the method of choosing the units to be sampled. It has been shown that a small sampling unit is most efficient when the spatial distribution is contagious (references given by Elliott are Beall (1939), Finney (1946), Taylor (1971)). The number of sampling units can be determined if the form of the frequency distribution of a species is known. The idea behind methods for choosing sampling units is to draw random samples from homogeneous populations. This leads to stratified random sampling if the substrate is not homogeneous. Optimum allocation of the total number of samples can be accomplished by choosing the number of samples for each stratum in proportion to the standard deviation of the stratum, the size of the stratum and the reciprocal of the square root of the cost of sampling the stratum (Cochran, 1977).

When the objective is the estimation of the population mean, the estimation techniques of survey sampling theory are appropriate (Cochran, 1977). Often, however, comparisons between strata are the primary concern, and then techniques such as the analysis of variance or covariance are required.

### 4.2. Design for Estimation of Population Mean

Three studies in which random sampling was used in the estimation of the population mean are Russell (1972), Cuff and Coleman (1979), and Resh (1979). Both Resh and Cuff and Coleman provide a clear explanation of how the random sampling was accomplished, the former being an example of simple random sampling of units and the latter of stratified random sampling.

Resh estimated the density of the larvae of the caddisfly *Chematopsyche pettitti* in a riffle of uniform depth (10–12 cm) with substrate particle size of 4–12 cm in diameter located in Rock Creek, Carroll Co., Indiana. The area was divided into a series of 30.5 × 61.0-cm rectangular quadrats and 26 quadrats were chosen at random. Two adjacent samples were taken at each location. The number of larvae of *C. pettitti* was counted in each sample, and the means, obtained from untransformed and logarithmically transformed counts, were reported.

A negative binomial distribution was fitted to the 52 samples and the conclusion was drawn that the spatial distribution of *C. pettitti* is contagious. This equal treatment of all 52 samples appears to have been made since Resh found large differences between pairs of samples at the same location. However, it does not separate the variation due to location from the variation between samples at the same location. To examine the evidence of nonrandomness at the distance between samples equal to that between locations, one solution would be to fit a distribution to the sum of the counts of the two samples at each location. The sum of the counts gives information about the mean at that location; the difference gives information about the within location variation. Inferences about the mean of a negative binomial distribution can be obtained from likelihood intervals using either the maximum likelihood function for the mean $\mu$, or the joint likelihood function for the two parameters $\mu$ and $k$ (Esterby and El-Shaarawi, 1984). To do this the likelihood function should be written with parameters $\mu$ and $k$ (see Elliott, 1977, p. 24) instead of $p$ and $k$, noting that $\mu = pk$.

Cuff and Coleman (1979) used stratified simple random sampling of station, with proportional allocation of the number of stations to the strata, to estimate the number of crustaceans, polychaetes, and mollusk individuals per 0.3 $m^2$ in Western Port (Victoria, Australia). The total number of stations that could be sampled was limited to 41 due to time and labor considerations. The random selection of stations in a stratum was achieved by subdividing a map into areas corresponding to 60 × 60-m areas and selecting the required number of stations for each stratum using a random number table. Proportional allocation means that the number of stations selected from a stratum is proportional to the total number of stations in each stratum, that is, to the total number of 60 × 60-m areas in this case. The location of the stations, marked on the map, was then determined using shipborne radar. The data were transformed using logarithms and 95% confidence limits for the population means were reported. No explanation for this choice of stratal boundaries was given. Comparison of within-stratum variances suggests that they did not obtain more homogeneous regions by stratification. This would explain why, in their evaluation of alternative sampling designs, stratified random sampling did not produce substantially more precise estimates than simple random sampling.

Russell (1972) also used stratified random sampling, with proportional allocation, and although he did not explain how the random sampling was accomplished, the objective of defining homogeneous regions by stratification seems to have been given more consideration. A preliminary survey was conducted to permit contours of equal abundance to be constructed, and from this, strata were determined.

### 4.3. Design for Comparison of Strata

Green (1968) used stratified random sampling of Crib Island, Queensland, for the filter-feeding bivalve *Notospisula parva* Petit. A preliminary random sampling showed high aggregation and a dependence of density $x$ on distance from mean low water. Thus a transect from mean low water to mean high water was divided into 10 strata and six 43-$cm^2$ core samples taken from each stratum on each sampling date. After transformation using log $(x + 1)$, an analysis of covariance was performed to assess changes in density with time. Again using the results of a preliminary study, Green and Hobson (1970) determined three faunal zones using association analysis on the presence or absence of 54 species of larger invertebrates in 400 samples in the region of Barnstable Harbor, Massachusetts. The center of two of these zones, which were at different tidal heights, were then randomly sampled monthly during one year for one of the more abundant species, *Gemma gemma.* An analysis of variance on the logarithmically transformed densities was conducted to determine if there were spatial and temporal differences in the density of *G. gemma.*

## 4.4. Separation of Natural and Contaminant-Induced Variability

Resh (1979) discussed why life history information should be used in designing benthic studies. For example, early instar larvae of *C. pettitti* may be more clumped than later instar larvae because individuals are dispersing from an egg mass. Quadrat size is known to influence the spatial pattern observed when the spatial distribution is nonrandom (Elliott, 1977). DeMarche (1976) demonstrated a relationship between the characteristics of a stream community and the temporal changes in the sediments of a small unpolluted river. Recognizable faunal assemblages were associated with sediments classified according to particle size and the characteristics of the sediments changed with season. Considerations such as these need to be made to design sampling programs that will allow the separation of the variability due to modifying factors other than toxic substances. Relative to the study designs described above, this means adding another level of subdivision, but the same principles are involved. Further, the types of changes expected due to exposure of an organism or community to a contaminant would be useful information. An example is the reduction in the coefficient of variation of the percentage of chironomids in samples in a section of a stream with higher copper concentrations relative to a section with lower concentrations (Winner et al., 1980). In Shayler Run, between the station just below the point of copper introduction, station 2, and the station farthest from the source of copper, station 6, the coefficient of variation increased from 26.4 to 132. The change in percentage of chironomids/sample was a decrease from 75 to 3.

## 4.5. Examples of Nonparametric and Cluster Analysis

Examples of nonparametric or cluster analysis of data from studies of benthic communities or organisms in the presence of toxic substances were not located. Methods appropriate for such studies will be briefly described for three observational studies and then similar methods will be applied to two observational studies of benthic organisms.

Univariate nonparametric methods have been used for assessing the change in a benthic community, as exhibited in an index derived from the species count vector, in relation to organic pollution (Bell et al., 1981), and for testing for differences between concentrations of a toxic substance in water at two locations (El-Shaarawi et al., 1985). The minimal assumptions of the nonparametric procedures were important since, in the first case, choice of an appropriate distribution is difficult and, in the second case, the presence of numerous samples below detection limits was handled by ranking observations. Ranks were also used for clustering stations on the basis of concentrations of many toxic substances measured in water and sediment when some samples were below detection limits (El-Shaarawi et al., 1985).

Bell et al. (1981) reduced the species-count vectors to a single number such as a diversity index or a biotic index and considered two situations: (1) samples were ordered with respect to the time since occurrence of the pollution event or the distance from the source of pollution; and (2) samples were collected from a number of regions where each region exhibited a different level of pollution. In both cases evidence of pollution would be in the form of evidence against the hypothesis of randomness. Permutation statistics were used, which, for the first case, are based on the regression coefficient and, for the second case, are an analog of the between-region sum of squares in the analysis of variance. For example, the statistic in the first case would be

$$h = \sum_{j=1}^{L} t_j f_j$$

where for a total of $L$ samples, $t_j$ is the time of the $j$th sample and $f_j$ is a biological descriptor. To test the null hypothesis of randomness (no relationship with time), the statistic $h$ is calculated for all possible permutations of the $L$ samples and the significance level is given by the proportion of samples at least as unfavorable to the hypothesis as the observed $h$. Note that randomness means that all permutation values are equally probable. For example, if a diversity index was used for $f_j$ in the form of the statistic $h$ which involves distance, high values of $h$ indicate a pollution effect since diversity would be expected to increase with the distance from the source of pollution. The rank tests that correspond to these two cases are the Spearman rank correlation coefficient and the Kruskal-Wallis statistic. Another possibility for the first case, which was not considered by Bell et al., is the detection of a change in a univariate biological variable at an unknown time due to a pollution event. This comes under the topic of change-point methods and a nonparametric method is given by Pettitt (1979). Parametric methods of detection (Brown et al., 1975) and estimation (Esterby and El-Shaarawi, 1981) are also available.

To test the hypothesis of no difference between the concentration of a toxic substance in water at a station in the lower and one in the upper Niagara River against the alternative of higher concentrations in the lower river, each pair of concentrations was ranked and the sign test was applied (El-Shaarawi et al., 1985). The two stations were sampled at about the same time (usually the same day) over

7 years and, by pairing concentrations by date, changes over time that affected both stations could be eliminated. Further, the analysis was made possible by ranking the data since the level of the toxic substance was reported either quantitatively, as below the detection limits, or as measurable.

The multivariate set of data consisted of the concentrations of 28 chlorinated organic substances in water and suspended sediments at five stations along the length of the Niagara River (El-Shaarawi et al., 1985). The stations were ranked separately for each substance and the standardized Euclidean distance calculated between the vectors of ranks for each pair of stations. A simple complete linkage clustering (Anderberg, 1973, p. 41) performed on these distances showed that the two lower river stations were similar and the three upper river stations were similar. Major effluent outfalls and dump sites occur along the length of the river but more occur in the lower river (Allen et al., 1983, Fig. 3).

### 4.6. Contaminants in Benthic Fauna and Sediments

Fox et al. (1983) reported the concentrations of 12 organochlorine contaminants in sediments, oligocheates and amphipods at five Lake Ontario stations sampled in April and July, 1981. Because of the limited nature of the data set, any analysis must be based on minimal assumptions and this is in agreement with the few analyses done by the authors. Only the relationship between contaminant concentration in the organisms and the sediment will be considered here. The important features can be determined from plots such as those shown for 1,3-DCB and HCB in Figure 22.2. From the plots for all 12 organochlorines, it was observed that, except for PCBs, there are usually only two groups rather than separate points corresponding to each station due largely to the separation of station 210 from the others. The stations were ranked separately for each contaminant and the Euclidean distance was calculated between the rank vectors of all pairs of stations for each combination of compartment and date. The complete linkage cluster analysis (Fig. 22.3) shows that, when the similarity of the stations is assessed using all 12 contaminants together, station 210 is different from the others only for sediments and oligocheates in April and that otherwise station 209 is similar to 210, as is 208 for amphipods in July. The Spearman rank correlation coefficient (Snedecor and Cochran, 1980) is seen to be more conservative than the usual product moment correlation coefficient (Table 22.2) but provides the same general conclusions for PCBs and HCB. It can also be calculated when some samples are below detection limits, as was the case for one sample when oligocheates collected in July were analyzed for 1,3,5-TCB (Table 22.2).

### 4.7. Major Benthic Taxa in Lake Ontario: Density versus Depth

Golini (1979) reported the abundance of the major benthic taxa obtained from a survey of surficial sediments of Lake Ontario in September 1977 in which sample locations were determined by a polygonal grid. He observed that the abundance

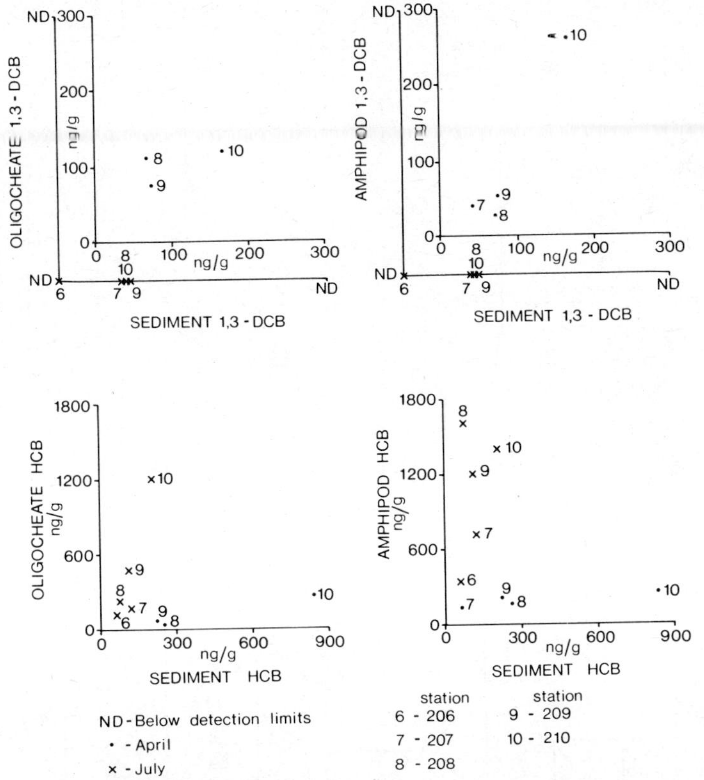


**Figure 22.2.** Examples of plots of contaminant level in organisms and sediments for the two sampling periods using the data of Fox et al. (1983). The outer axes in the first two plots show which stations had a 1,3-DCB level below detection limits for the organism or sediment or both.

varied inversely with depth. This was based upon the calculation of mean relative abundance of stations in depth zones as determined by contours on a lake map (Fig. 22.5). As a preliminary step in a further examination of this relationship, scatter plots of the density of the major taxa versus depth were prepared (examples are given in Fig. 22.4). For all major taxa, the relationship with depth, which is most pronounced, is the increase in the variability of the density at shallower depths. The increase in scatter began at about 100–130 for Nematoda, *Pontoporeia,* Sphaeriidae, and Chironomidae, whereas the scatter appeared to increase with decreasing depth over all depths for Oligochaeta. This suggests differences between stations at the same depth, for the shallower depths, that are not due to depth. A grouping of the stations on a basis other than depth was sought by applying to the densities of the seven major taxa Forgy's nonhierarchical, nearest-centroid clustering method, which uses the Euclidean distance and has the number of clusters fixed (Anderberg, 1973). The densities for each major taxa were stan-

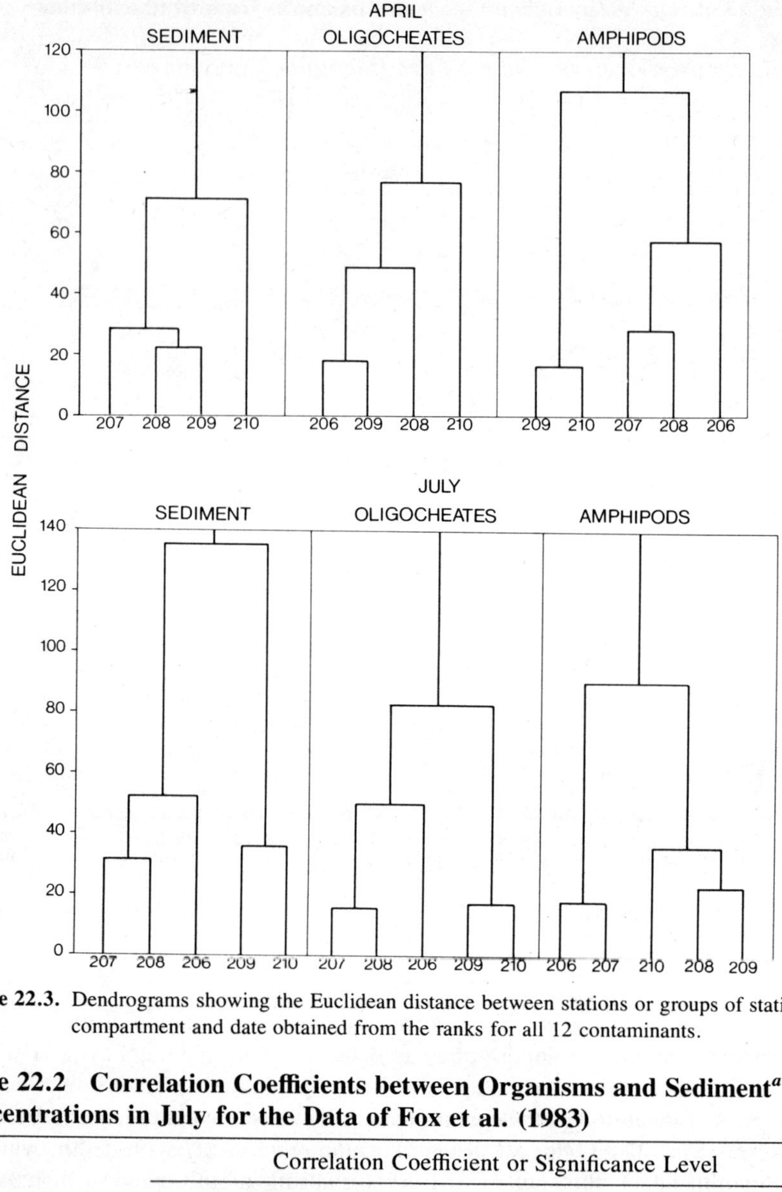


**Figure 22.3.** Dendrograms showing the Euclidean distance between stations or groups of stations by compartment and date obtained from the ranks for all 12 contaminants.

## Table 22.2 Correlation Coefficients between Organisms and Sediment[a] Concentrations in July for the Data of Fox et al. (1983)

| | Correlation Coefficient or Significance Level | | | | | |
|---|---|---|---|---|---|---|
| | Oligocheates | | | *Amphipods* | | |
| Contaminant | $r$ | $r_s$ | $\alpha_s$ | $r$ | $r_s$ | $\alpha_s$ |
| PCBs | 0.93 | 0.70 | 0.12 | 0.35 | 0.50 | 0.23 |
| HCB | 0.96 | 0.90 | 0.04 | 0.47 | 0.30 | 0.34 |
| HCBD | | 0.98 | 0.04 | | 0.90 | 0.04 |
| 1,3,5-TCB | | 1.00 | 0.008 | | 0.70 | 0.12 |

[a]Key: $r$ is the product moment correlation coefficient between concentrations (Fox et al., 1983). $r_s$ and $\alpha_s$ are the Spearman correlation coefficient and corresponding observed significance level. PCBs, polychlorinated biphenyls; HCB, hexachlorobenzene; HCBD, hexachlorobutadiene; 1,3,5-TCB, 1,3,5-trichlorobenzene.

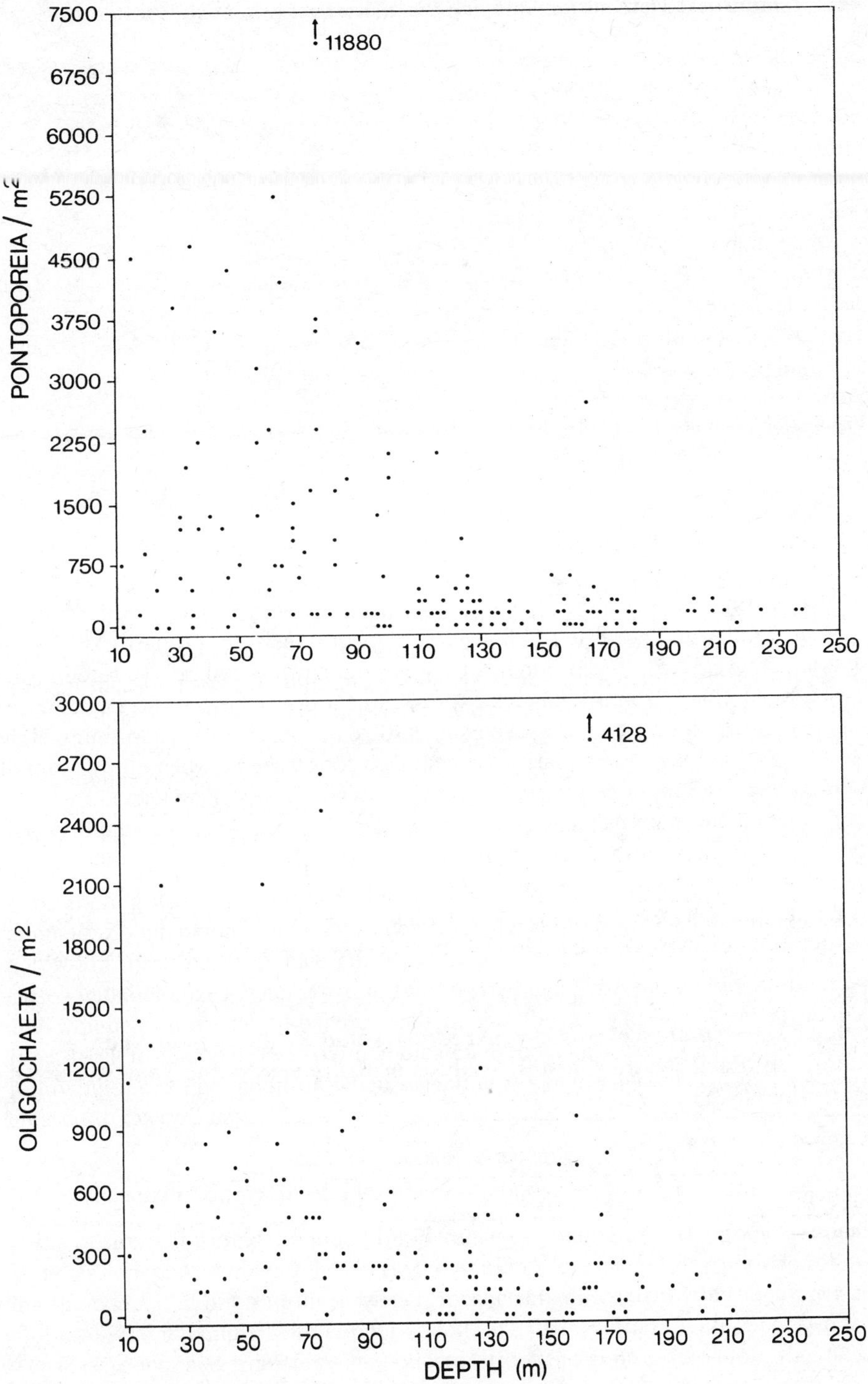


**Figure 22.4.** Scatter plots of the number/m² of *Pontoporeia* and Oligochaeta versus depth for 151 Lake Ontario stations as reported by Golini (1979).

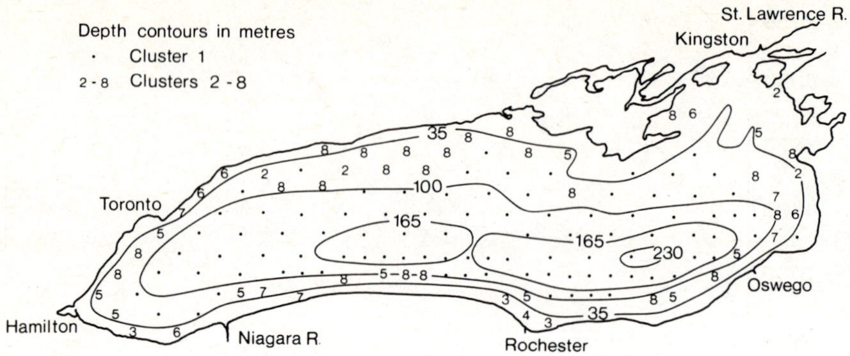


**Figure 22.5.** Location of the Lake Ontario stations in the eight clusters as determined using the standardized densities of the seven major benthic invertebrate taxa reported by Golini (1979).

dardized by subtracting the mean and dividing by the standard deviation of the taxa prior to clustering.

Dominating the distribution of stations among clusters for the total number of clusters *k* equal to 2 to 8 is the first large cluster consisting primarily of deeper stations (Table 22.3). The difference between the clusters can be measured by *R*, the between cluster sum of squares expressed as a proportion of the total sum of squares, both summed over all seven standardized densities. By determining eight clusters, 66% of the total variation is accounted for by the between-cluster sum of squares and the one large cluster is split into two. To better understand what is happening as the stations are divided into groups, various plots and summary statistics for the individual taxa were prepared. For eight clusters, a number of the stations in the 10- to 35-m and 36- to 99-m depth contours, which were previously in the first cluster, were assigned to the eighth cluster (Fig. 22.5). The effect on the densities of individual taxa in the clusters was examined and the example of

**Table 22.3 Number of Stations per Cluster and *R*, the Sum of Squares Ratio, Obtained for the Cluster Analysis of the Seven Major Taxa Reported by Golini (1979)**

| Number of Clusters | Number of Stations per Cluster | | | | | | | | |
|---|---|---|---|---|---|---|---|---|---|
| | 1 | 2 | 3 | 4 | 5 | 6 | 7 | 8 | *R* |
| 2 | 132 | 19 | | | | | | | 0.28 |
| 3 | 128 | 2 | 21 | | | | | | 0.42 |
| 4 | 127 | 2 | 15 | 7 | | | | | 0.53 |
| 5 | 125 | 3 | 1 | 13 | 9 | | | | 0.56 |
| 6 | 125 | 4 | 2 | 13 | 2 | 5 | | | 0.63 |
| 7 | 119 | 4 | 2 | 1 | 11 | 5 | 9 | | 0.65 |
| 8 | 100 | 4 | 3 | 1 | 10 | 5 | 5 | 23 | 0.66 |

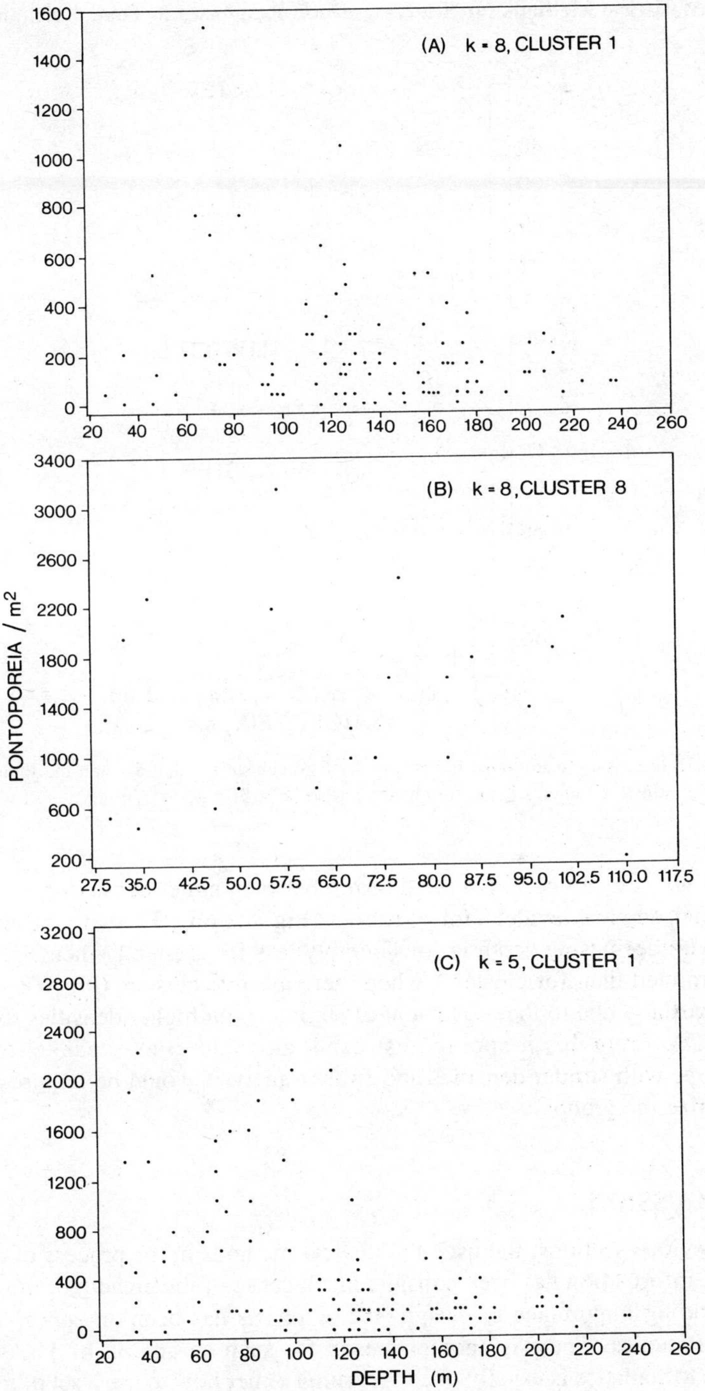


**Figure 22.6.** Scatter plots of the number/m$^2$ of *Pontoporeia* versus depth for stations in clusters 1 and 2 when a total of eight clusters, $k = 8$, were determined, and in cluster 1 when $k = 5$.

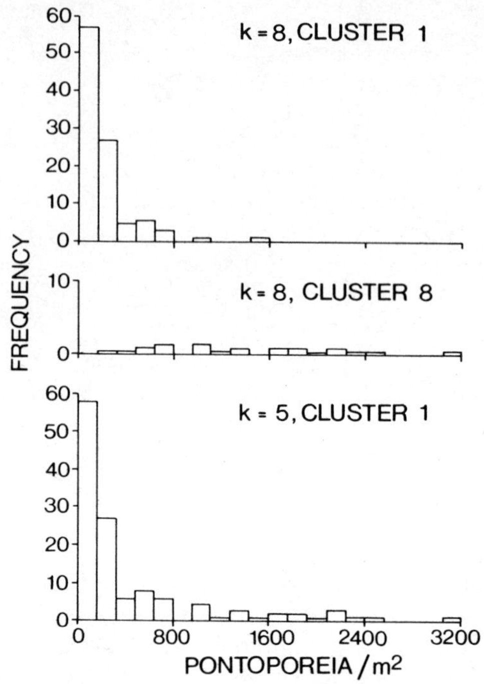


**Figure 22.7.** Frequency histograms for *Pontoporeia* for clusters 1 and 8 when the total number of clusters $k$ equals 8 and for cluster 1 when $k = 5$.

*Pontoporeia* is given here. No relationship between either density or the degree of scatter and depth is evident for cluster 8 (Fig. 22.6b). Further, the increase in scatter with decreasing depth is considerably less for cluster 1 when eight clusters are determined than for cluster 1 when there are five clusters (Fig. 22.6a and c). This is partially due to the movement of stations with higher densities to cluster 8 (Fig. 22.7). From this it appears that depth alone does not separate the stations into groups with similar densities and further analysis would be required to better characterize the groups.

## 5. DISCUSSION

In the previous sections, the use of statistical methods in the process of collecting scientific information has been considered. Because of the breadth of the topic and the overriding importance of design, the emphasis has been on concepts and the design of studies. There is another reason for such an approach. The scientist's exposure to statistics is usually one of learning either how to use a set of techniques or how to use a statistical computer package. This will be of little help in the critical design stage and often will discourage the individual from looking at the data carefully before using a readily available computer program. An understanding of concepts, such as the separation of components of variability and frequency distributions, together with the availability of techniques for graphical display and simple descriptive statistics will allow the individual to get a grasp on the information in the data without the confusion caused by improperly understood techniques. This will make the task of choosing a suitable formal or complicated procedure much easier. The initial examination of the data is appropriate even in a well-designed study where specification of the method of analysis was included in the design stage.

The major shortcoming in environmental studies is probably the failure to give enough thought to the design of studies. Having determined a well-defined objective, the investigator needs to attempt to account for as many complicating factors as possible so that, in the end, data that can be used to meet the objective are available. Preliminary small-scale studies will allow better planning of a major study when little is known about factors that will introduce bias or increase the variability. It is useful to think of the analysis of the data in terms of the simple model in Section 3.3, because this summarizes what has been discussed concerning explainable variation and random error terms.

Although much of the discussion has been very general, the usefulness for studies of toxic substances should be clear. Many of the examples that have been discussed involve studies on the characteristics of organisms or biological populations in uncontaminated environments. In Section 3.2, it was indicated that studies of toxic effects must look at the departures from these normal characteristics to assess toxic effects. Hence the methods must be at least as good as those used for the former studies, and must attempt to remove variability due to the normal characteristics of biological populations from the error term. For example, if an organism is more sensitive to a contaminant in one stage of the life cycle, the identity of the life cycle of each individual is needed so that the effect of the contaminant is not masked by the differing responses due to stage of life cycle. Another specific point is worth mentioning because it is a common practice. Information about reasons for variability is often lost by taking means over varying conditions or pooling samples for analysis. The identity of each unit on which measurements are made should be retained in the analysis until it is shown that a set of units do not vary. This is often the opposite of what is done. An example of retaining the identity of the unit is to do a paired $t$ test rather than the $t$ test for equality of means from two independent samples (Snedecor and Cochran, 1980, pp. 91–102).

In Section 2.4, the distinction was made between the scientist's objective of analyzing a set of data to extract information and the theoretical statistician's objective of evaluating statistical methods under well-defined hypothetical conditions (in terms of probability distributions). This is important because with the present accessibility of computers more scientists are conducting simulations. What the statistician often does in the evaluation of a statistic is to conduct a Monte Carlo experiment (i.e., generate samples from a probability distribution). Since the field of theoretical statistics does not always keep up with the needs in science, some of these studies, motivated by particular problems (e.g., in the analysis of water quality data), are being done by nonstatisticians. What is important to emphasize here is that these are methodological development techniques and most scientists do not need to use them. The results of the studies, which provide guidance on which statistical method to use in a particular study, is what is generally needed.

There are two other types of computer number-generation methods that may be encountered. The first is the purely deterministic generation of numbers that are intended to represent a response in a system for which well-defined conditions have been postulated. Sometimes, stochastic components are included. The objective is often to predict variables in real environmental systems, but, clearly, the predictions are for well-defined hypothetical systems and may or may not represent the real system. The second is the nonparametric bootstrap method (Efron and Gong, 1983), which is an attempt to provide a method for statistical estimation when the form of the probability distribution is unknown by randomly sampling the empirical cumulative distribution function. It is not a data-generation technique. All that is known about the real population for which an inference is sought is the sample, that is, the observed data set. As for all data analysis techniques, the assumption that the data are good must be made. Its strength is that of other nonparmetric techniques, namely, weaker distributional assumptions. Its shortcomings are also the same as those of other nonparametric techniques, in that it emphasizes uncertainty about the underlying distribution but ignores important assumptions such as statistical independence (Prentice, 1981).

There are several other points that are important because they are related to the motivation for many studies of toxic substances. Whether we are concerned with obtaining scientific information to make defensible predictions about the potential hazards of toxic substances or to develop water quality criteria for toxic substances, the variability of the system being assessed or monitored must be scientifically characterized and the characterization incorporated in the prediction or the statement of the criteria. It is clear that this cannot be given superficial treatment if the prediction or criterion is to be useful. In its simplest form this could consist of determining the form of the probability distribution that adequately fits the type of data collected under the conditions stated in a water quality regulation. Having established the critical level required to protect the system, the nature of the variability in the data can be used to set the exact specification of the regulation. The effect of the form of the probability distribution on the probability of noncompliance has been illustrated in the case of bacterial criteria (Esterby, 1982) and similar considerations apply to other types of water quality criteria.

A further consideration is the use of empirical models. Many environmental scientists equate models to systems-analysis-type models when predictions are required. It is important to recognize that to construct mechanistic models we must have a reasonable understanding of the system being modeled. The process of understanding a biological system will be aided by the construction of empirical models since this helps us to abstract the essential ideas from our experimental results and their consequences. Statistical methods are a means of constructing empirical models.

# REFERENCES


Allen, R. J., Mudroch, A., and Sudar, A. (1983). An introduction to the Niagara River/Lake Ontario pollution problem. *J. Great Lakes Res.* **9,** 111–117.

Anderberg, M. R. (1973). *Cluster Analysis for Applications.* Academic Press, New York.

Beall, G. (1939). Methods of estimating the population of insects in a field. *Biometrika* **30,** 422–439.

Bell, C. B., Conquest, L. L., Pyke, R., and Smith, E. P. (1981). Some nonparametric statistics for monitoring water quality using benthic species counts. *Envirometrics 81: Selected Papers.* SIAM, Philadelphia, PA, pp. 100–120.

Black, J. J. (1983). Field and laboratory studies of environmental carcinogens in Niagara River fish. *J. Great Lakes Res.* **9,** 326–334.

Box, G. E. P., Hunter, W. G., and Hunter, J. S. (1978). *Statistics for Experimenters.* Wiley, New York.

Brown, R. L., Durbin, J., and Evans, J. M. (1975). Techniques for testing the constancy of regression relationships over time (with discussion). *J. R. Stat. Soc. B* **37,** 149–192.

Capizzi, T., Oppenheimer, L., Mehta, H., and Naimie, H. (1985). Statistical considerations in the evaluation of chronic aquatic toxicity studies. *Environ. Sci. Technol.* **19,** 35–43.

Cochran, W. G. (1977). *Sampling Techniques,* 3rd ed. Wiley, New York.

Cochran, W. G., and Cox, G. M. (1957). *Experimental Designs,* 2nd ed. Wiley, New York.

Cox, D. R. (1958). *Planning of Experiments.* Wiley, New York.

Cox, D. R. (1977). The role of significance tests. *Scand. J. Stat.* **4,** 49–70.

Cuff, W., and Coleman, N. (1979). Optimal survey design: Lessons from a stratified random sample of macrobenthos. *J. Fish Res. Board Can.* **36,** 351–361.

DeMarche, B. G. E. (1976). Spatial and temporal patterns in macrobenthic stream diversity. *J. Fish Res. Board Can.* **33,** 1261–1270.

Draper, N. R., and Smith, H. (1981). *Applied Regression Analysis,* 2nd ed. Wiley, New York.

Efron, B., and Gong, G. (1983). A leisurely look at the bootstrap, the jackknife, and cross-validation. *Am. Statist.* **37,** 36–48.

Elliott, J. M. (1977). *Some Methods for the Statistical Analysis of Samples of Benthic Invertebrates,* 2nd ed., Freshwater Biol. Assoc. Sci. Publ. No. 25. Ferry House, U.K.

El-Shaarawi, A. H., Esterby, S. R., and Dutka, B. J. (1981). Bacterial density in water determined by Poisson or negative binomial distributions. *Appl. Environ. Microbiol.* **41,** 107–116.

El-Shaarawi, A. H., Esterby, S. R., Warry, N. D., and Kuntz, K. W. (1985). Evidence of contaminant loading to Lake Ontario from the Niagara River. *Can. J. Aquat. Sci.* **42,** 1278–1289.

Esterby, S. R. (1982). Fitting probability distributions to bacteriological data: Considerations for surveys and for water quality guidelines. *J. Fr. Hydrol.* **13,** 189–203.

Esterby, S. R., and El-Shaarawi, A. H. (1981). Inference about the point of change in a regression model. *Appl. Stat.* **30,** 277–285.

Esterby, S. R., and El-Shaarawi, A. H. (1984). Spatial and temporal distribution of total and fecal coliform concentrations, 1966–1970. In A. H. El-Shaarawi (Ed.), *Statistical Assessment of the Great Lakes Surveillance Program, 1966–1981, Lake Erie.* Environment Canada, Burlington, Ontario, Sci. Ser. No. 136, pp. 151–232.

Finney, D. J. (1946). Field sampling for the estimation of wireworm populations. *Biometrics* **2,** 1–17.

Finney, D. J. (1964). *Statistical Method in Biological Assay,* 2nd ed. Griffin, London.

Finney, D. J. (1971). *Probit Analysis,* 3rd ed. Cambridge Univer. Press, London and New York.

Fox, M. E., Carey, J. H., and Oliver, B. G. (1983). Compartmental distribution of organchlorine contaminants in the Niagara River and the western basin of Lake Ontario. *J. Great Lakes Res.* **9,** 287–294.

Golini, V. I. (1979). *Benthic Macro- and Meioinvertebrates of Lake Ontario 1977: Distribution and Relative Abundance,* Fish. Mar. Serv. Manuscript Rep. No. 1519. Department of Fisheries and Oceans, Burlington, Ontario.

Gordon, A. D. (1981). *Classification.* Chapman & Hall, London.

Green, R. H. (1968). Mortality and stability of a low diversity subtropical intertidal community. *Ecology* **49,** 848–854.

Green, R. H. (1979). *Sampling Design and Statistical Methods for Environmental Biologists.* Wiley, New York.

Green, R. H., and Hobson, K. D. (1970). Spatial and temporal structure in a temperate intertidal community, with special emphasis on *Gemma gemma* (Pelecypoda:Mollusca). *Ecology* **51,** 1000–1011.

Green, R. H., Singh, S. M., Hicks, B., and McCuaig, J. M. (1983). An Arctic intertidal population of *Macoma balthica* (Mollusca, Pelecypoda): Genotypic and phenotypic components of population structure. *Can. J. Fish. Aquat. Sci.* **40,** 1360–1371.

Gupta, G. P., and Govindarajulu, Z. (1980). Nonparametric tests of randomness against autocorrelated normal alternatives. *Biometrika* **67,** 375–379.

Hodson, P. V., Blunt, B. R., Jensen, D., and Morgan, S. (1979). Effect of fish age on predicted and observed chronic toxicity of lead to rainbow trout in Lake Ontario water. *J. Great Lakes Res.* **5,** 84–89.

Hodson, P. V., Blunt, B. R., and Whittle, D. M. (1984). Monitoring lead exposure of fish. In V. W. Cairns, P. V. Hodson, and J. O. Nriagu (Eds.), *Contaminant Effects on Fisheries.* Vol. 16, Advances in Environmental Science and Technology, Wiley, New York, pp. 87–98.

Hurlbert, S. H. (1984). Pseudoreplication and the design of ecological field experiments. *Ecol. Monogr.* **54,** 187–211.

Kalbfleisch, J. D., and Prentice, R. L. (1980). *The Statistical Analysis of Failure Time Data.* Wiley, New York.

Leatherland, J. F., and Sonstegard, R. A. (1978). Lowering of serum thyroxine and triiodothyronine levels in yearling coho salmon, *Oncorhynchus kisutch,* by dietary Mirex and PCBs. *J. Fish. Res. Board Can.* **35,** 1285–1289.

Leatherland, J. F., and Sonstegard, R. A. (1981). Thyroid dysfunction in Great Lakes coho salmon, *Oncorhynchus kisutch* (Walbaum): Seasonal and interlake differences in serum T3 uptake and serum total and free T4 and T3 levels. *J. Fish Dis.* **4,** 413–423.

Lehmann, E. L. (1975). *Nonparametrics, Statistical Methods Based on Ranks.* Holden-Day, Oakland, CA.

Lettenmaier, D. P. (1976). Detection of trends in water quality data from records with dependent observations. *Water Resour. Res.* **12,** 1037–1046.

Mandel, J. (1964). *The Statistical Analysis of Experimental Data.* Wiley Interscience, New York.

Metcalf, R. L., Sangha, G. K., and Kapoor, I. P. (1971). Model ecosystem for the evaluation of pesticide biodegradability and ecological magnification. *Environ. Sci. Technol.* **5,** 709–713.

Miller, D. R. (1984). Distinguishing ecotoxic effects. In P. J. Sheehan, D. R. Miller, G. C. Butler, and P. Bourdeau (Eds.), *Effects of Pollutants at the Ecosystem Level (Scope 22).* Wiley, New York, pp. 15–22.

Miller, R. G., Jr. (1981). *Simultaneous Statistical Inference,* 2nd ed. Springer-Verlag, New York.

Moccia, R. D., Leatherland, J. F., Sonstegard, R. A., and Holdrinet, M. V. H. (1978). Are goiter frequencies in Great Lakes salmon correlated with organochlorine residues? *Chemosphere* **8,** 649–652.

O'Brien, P. C., and Dyck, P. J. (1985). A runs test based on run length. *Biometrics* **41,** 237–244.

Pettitt, A. N. (1977). A Cramér-von Mises type of goodness of fit statistic related to $b_1$ and $b_2$. *J. R. Stat. Soc. B* **39,** 364–370.

Pettitt, A. N. (1979). A nonparametric approach to the change-point problem. *Appl. Stat.* **28,** 126–135.

Pielou, E. C. (1969). *An Introduction to Mathematical Ecology.* Wiley, New York.

Pielou, E. C. (1984). *The Interpretation of Ecological Data.* Wiley, New York.

Plackett, R. L. (1974). *The Analysis of Categorical Data* Griffin, London.

Platt, T., Dickie, L. M., and Trites, R. W. (1970). Spatial heterogeneity of phytoplankton in a near-shore environment. *J. Fish. Res. Board Can.* **27,** 1453–1473.

Prentice, R. L. (1981). Discussion of Efron, B. Nonparametric standard errors and confidence intervals. *Can. J. Stat.* **9,** 139–172.

Resh, V. H. (1979). Sampling variability and life history features: Basic considerations in the design of aquatic insect studies. *J. Fish Res. Board Can.* **36,** 290–311.

Russell, H. J. (1972). Use of a commercial dredge to estimate a hardshell clam population by stratified random sampling. *J. Fish Res. Board Can* **29,** 1731–1735.

Scheffé, H. (1959). *The Analysis of Variance.* Wiley, New York.

Sládeček, V., Hawkes, H. A., Alabaster, J. S., Daubner, I., Nöthlich, I., de L. G. Solbé, J. F., and Uhlmann, D. (1984). Biological examination. In M. J. Seuss (Ed.), *Examination of Water for Pollution Control,* Vol. 3. *Biological Bacteriological and Virological Examination.* Pergamon, Oxford, pp. 1-272.

Sloof, W., and de Zwart, D. (1983). Bio-indicators and chemical pollution of surface waters. *Environ. Monit. Assess.* **3,** 237–245.

Snedecor, G. W., and Cochran, W. G. (1980). *Statistical Methods,* 7th ed. Iowa State Univ. Press, Ames.

Sokal, R. R., and Rohlf, F. J. (1969). *Biometry.* Freeman, San Francisco, CA.

Sonstegard, R. A., and Leatherland, J. F. (1984). Comparative epidemiology: The use of fishes in assessing carcinogenic contaminants. In V. W. Cairns, P. V. Hodson, and J. O. Nriagu (Eds.), *Contaminant Effects on Fisheries.* Vol. 16, Advances in Environmental Science and Technology, Wiley, New York, pp. 223–231.

Sprague, J. B. (1969). Measurement of pollutant toxicity to fish. I. Bioassay methods for acute toxicity. *Water Res.* **3,** 793–821.

Stuart, A. (1962). *Basic Ideas of Scientific Sampling.* Griffin, London.

Taylor, L. R. (1971). Aggregation as a species characteristic. In G. P. Patil et al. (Eds.) *Statistical Ecology.* Pennsylvania State Univ. Press, University Park, Vol. 1, pp. 357–372.

U.S. National Research Council (1981). *Testing for Effects of Chemicals on Ecosystems.* Nat. Acad. Press, Washington, DC.

Winner, R. W., Boesel, M. W., and Farrell, M. P. (1980). Insect community structure as an index of heavy-metal pollution in lotic ecosystems. *Can. J. Aquat. Sci.* **37,** 647–655.

Yates, F. (1960). *Sampling Methods for Censuses and Surveys.* Griffin, London.

# 23

# DEVELOPMENT OF ECOSYSTEM OBJECTIVES FOR THE GREAT LAKES

***Andrew Robertson***

*National Marine Pollution Program Office, National Oceanic and Atmospheric Administration, Rockville, Maryland 20852*



## 1. INTRODUCTION

Canada and the United States share an invaluable resource in the St. Lawrence Great Lakes that stretch for more than 1000 miles (1600 km) along their border. Collectively these lakes contain almost one-fifth of the world's liquid surface fresh water. They are a major factor in the regional economy, being of importance for shipping, hydroelectric power generation, irrigation, waste disposal, and cooling and makeup water for industry, as well as serving as a drinking water source for many millions of people, the basis for a great deal of sport fishing and other recreational activities, and the site of a major commercial fishery.

Yet, as with all parts of our global ecosystem, these lakes are susceptible to degradation in environmental quality if misused by humans. By the late 1960s, it had become clear to Canada and the United States that this was happening due to uncontrolled use, especially indiscriminate shoreline development and waste disposal. To reverse this trend, the two countries negotiated and signed the Great Lakes Water Quality Agreement in 1972 (International Joint Commission, 1972) and then revised this agreement in 1978 (International Joint Commission, 1978). The Agreement affirms the commitment of the two countries to restore and enhance the water quality of the Great Lakes System. It provides the basis of a continuing binational pollution control program on the Great Lakes and has led to major cooperative efforts to control and reverse the deterioration in their environmental quality.

To help meet its overall goal of restoration and enhancement, the Agreement specifies both general and specific water quality objectives for the Great Lakes. The general objectives, while valuable in delineating the ambient condition desired for the waters of these lakes, are broad, general statements and too indefinite to be used directly as the basis for a regulatory program. The specific objectives, on the other hand, are much more definitive statements of the conditions desired for the Great Lakes. The definition of a specific objective in the Agreement states that it is "the concentration or quantity of a substance or level of effect that the Parties agree, after investigation, to recognize as a maximum or minimum desired limit for a defined body of water or portion thereof, taking into account the beneficial uses or level of environmental quality which the Parties desire to secure and protect" (International Joint Commission, 1978).

The development of these specific objectives is the first step in the four-step procedure by which Canada and the United States carry out their joint program to protect the Great Lakes. The second step is to make measurements throughout the lakes of the levels of substances and effects for which there are objectives. Locations where this monitoring shows that one or more of the objectives are not being met are identified as areas of concern. The third step is to determine the source or sources of any substances or effects that are out of compliance with the objectives, while the fourth and final step is to develop recommendations to the involved jurisdictions for remedial measures for solution of the identified problems.

It is clear that the specific objectives play a vital role in the pollution control efforts being conducted by Canada and the United States under the Water Quality Agreement. They provide the underpinning for the entire program and specify the conditions desired for the Great Lakes. Yet the philosophy of how these objectives should be developed and how they should be used has evolved substantially since 1972 when the Agreement was first signed.

These objectives provide the linkage whereby scientific information concerning toxic substances and their effects is brought to bear on the regulatory problems of restoring and enhancing the quality of the Great Lakes. Thus, it is important that this book, which is directed toward identifying research needs and promising methodologies related to the effects of toxic substances on the health of aquatic, especially Great Lakes, communities, provide a discussion of the framework in which information on such effects is used. This paper attempts to provide this framework. It summarizes the present status of development of specific water quality objectives for the Great Lakes. More importantly, however, it presents a brief review of the new directions such work is taking in an attempt to develop objectives that serve to aid protection, restoration, and enhancement of, not just Great Lakes water quality, but the environmental quality of the entire Great Lakes ecosystem.

## 2. DEVELOPMENT OF WATER QUALITY OBJECTIVES

The 1972 Great Lakes Water Quality Agreement contained specific objectives for relatively few substances and effects. However, it listed a larger number of substances and effects for which the Parties agreed to consult for the purpose of developing specific objectives. Further, it called for assistance in implementation from the International Joint Commission (IJC), a binational group set up by the Boundary Waters Treaty of 1909 that has as part of its mandate the responsibility to consider questions and matters of difference related to the U.S./Canada boundary waters that are referred to it by the governments. As one of its responsibilities under the Agreement, the IJC is directed to recommend new and modified specific water quality objectives.

To develop the specific objectives called for in the 1972 Agreement, the IJC established two groups. One group, the Scientific Basis for Water Quality Criteria Committee (SBWQC), was composed of scientific experts in such areas as aquatic toxicology, aquatic chemistry, fishery biology, and limnology. The second, the Water Quality Objectives Subcommittee (WQOS), consisted of national, state, and provincial officials with experience and responsibilities related to regulation of water quality. These two groups met jointly and, for most purposes, functioned as one committee (SBWQC/WQOS).

When SBWQC/WQOS first met, it was found that there were major areas of uncertainty and disagreement concerning the basic approach to be used in developing water quality objectives. Thus, the first task of the group was to reach agreement on a general approach to take in the development of objectives. Since that time SBWQC/WQOS and its successor committee have developed objectives for a large number of substances and effects, and the approach adopted has been tested and modifications have been made as required.

During this process a number of general principles or guidelines for the development of objectives have been agreed upon. Although these have not been considered inviolate and they have been disregarded or modified when this seemed justified, these principles have proven very useful as general guides for the group's deliberations. They have assisted in the development of an internally consistent set of objectives that have largely been adopted by the two countries. The most important of these guidelines are presented and briefly discussed in the following list. A more complete discussion of these guidelines and examples of their use in developing objectives are included in Robertson (1982).

1. *The objectives are intended to protect all beneficial uses of the Great Lakes.* For each pollutant or effect being considered, the level at which it starts to interfere with each of the various major beneficial uses of the Great Lakes is de-

termined as well as possible from the information available in the scientific literature. The use that is most sensitive to the pollutant or effect is then designated as the use to be protected, and a level just below (or occasionally just above) the level where that use is first detectably interfered with is established as the objective for that pollutant or effect. Because the Agreement calls for the Great Lakes to be substantially free from substances that adversely affect aquatic life, maintenance of a suitable habitat for aquatic life has been taken as one of the bases for establishing objectives. In fact it has proven to be the most common basis for establishing objectives. However, provision of a public water supply and protection of aesthetic quality for recreational purposes have been found to be the most sensitive uses with regard to a few specific objectives.

2. *The development and adoption of objectives does not preclude the need for other actions to protect beneficial uses.* The objectives have been developed from the point of view of providing protection from the harmful effects of pollutants considered individually. Yet, pollutants usually enter the Great Lakes as components of complex pollutant mixtures and become mixed with additional pollutants after they enter the environment. It is well established that the mixing of pollutants can lead to synergistic or antagonistic effects on the toxicity of the individual substances (e.g., Wong et al., 1978). Thus, it cannot be assumed that harmful effects will be absent just because the objectives for the individual pollutants are met. Also the objectives have been developed based on the assumption that they are to be applied to situations where the limnological conditions and biological communities commonly found in the Great Lakes are present. In local situations, such as restricted harbors and bays, quite different conditions and communities may be found. In such situations there is no certainty that the objectives will still be applicable. Because of these restrictions on the applicability of the objectives, their use should always be accompanied by a complementary biological monitoring program to make certain that unforeseen degradation is not occurring.

3. *The objectives are set at the level of no detectable detrimental effect.* The objectives are set at levels just beyond the limit where detrimental or potentially detrimental effects have been detected. This means that in the common situation where the most sensitive use is protection of aquatic life, the objective is usually set just below the level where subtle, sublethal effects such as metabolic or reproductive changes are first detected. Since the consequences of such changes to populations of organisms in the Great Lakes environment are difficult to ascertain and may well be detrimental, the conservative position of setting the objectives at no-effect levels has been adopted.

4. *Uncertainties concerning the development of the objectives are always resolved in favor of protection.* As discussed above, the objectives are established at the level where reasonable assurance can be given to water resource managers that all beneficial uses will be protected (see, however, guideline 2). Thus, in situations where there are uncertainties concerning how to handle some aspect of objective development, the uncertainties are always resolved in favor of greater protection. For example, it is well established that different chemical species of

certain metals have quite different toxicities (e.g., Andrew, 1976). When developing an objective for these metals, it would clearly be desirable to provide a separate objective for each of the important species based on their differing toxicities. At present, however, there is relatively little information available on such aspects as reliable ways to measure the different species, relative toxicities of these species, and rates of transformation from one to another under various environmental conditions. Given these uncertainties, no attempt has been made to develop objectives for specific metal species. Instead the conservative approach of developing objectives only for allowable total metal concentrations has been adopted. If and when additional research resolves these uncertainties, it may well be possible to develop objectives for each of the important species of the various metals.

5. *All objectives must be scientifically defensible.* The development of the objectives is based entirely on information contained in the published scientific literature. SBWQC/WQOS and its successor committee have evaluated and interpreted the results presented in the pertinent scientific publications and recommended objectives based on these deliberations. A rationale is included with each objective explaining the information and logic, and referencing the publications used in its development. For pollutants where it is judged that there is inadequate information on which to base a valid objective, this fact is reported to the IJC and development of an objective is postponed pending the acquisition of adequate information.

6. *The objectives are not necessarily practical for direct implementation as regulatory standards.* Recommendations for objectives are developed based solely on the available scientific information and relate only to the levels required to protect all beneficial uses. Many important practical questions, such as what would be the social and economical consequences of implementing the recommended objectives as enforced legal standards and are methods available that allow routine measurements at the levels recommended, are excluded from consideration. It is certainly intended that such practical aspects be taken into account if the objectives are considered for adoption as regulatory standards. However, it has been concluded that it is more efficient and effective to have the levels needed to protect all beneficial uses determined on a purely scientific basis for each pollutant before regulatory, economic, and especially political realities are taken into consideration. In other words, the objectives are intended to be just what the word implies, goals for which we should strive if we are to protect all beneficial uses of the Great Lakes. They are not necessarily levels that it is practical or even always desirable to incorporate immediately into law as enforceable standards.

7. *The objectives are intended to be subject to periodic review with revisions as necessary.* The objectives have been developed based on the best information currently available. As more research is carried out and more is learned, the objectives will be reviewed in light of the additional knowledge and revised as appropriate.

Using these guidelines, objectives were developed for a number of pollutants. These objectives primarily take the form of the maximum allowable concentration of a toxic contaminating substance, such as a heavy metal or a synthetic organic, although limits for nontoxic chemicals, such as nutrients, for physical properties, such as temperature and turbidity, and for biological properties, such as microbiological indicators of contamination by pathogens, have also been agreed upon.

## 3. ADOPTION OF WATER QUALITY OBJECTIVES

The 1972 Great Lakes Water Quality Agreement called for Canada and the United States to conduct a comprehensive review of the operation and effectiveness of the Agreement during the fifth year after it came into effect. Such a review was carried out and revisions in the Agreement were agreed upon. This revised document, the Great Lakes Water Quality Agreement of 1978, was signed by the two countries in November 1978 (International Joint Commission, 1978).

Most of the objectives that had been developed by the IJC at the time of the review of the Agreement were included in the 1978 revision, either as new specific objectives or as revisions to existing objectives. Through this process the number

**Table 23.1 Substances and Conditions for Which Specific Objectives Are Included in the Great Lakes Water Quality Agreement of 1978**

| | |
|---|---|
| *Persistent Toxic Substances* | *Nonpersistent Toxic Substances* |
| Aldrin/dieldrin | Diazinon |
| Chlordane | Guthion |
| DDT and metabolites | Parathion |
| Endrin | Other pesticides |
| Heptachlor/heptachlor epoxide | Unspecified substances and complex effluents |
| Lindane | Oil and petrochemicals |
| Methoxychlor | Ammonia |
| Mirex | Hydrogen sulfide |
| Toxaphene | |
| Phthalic acid esters | *Other Substances* |
| Polychlorinated biphenyls | Dissolved oxygen |
| Unspecified organic compounds | pH |
| Arsenic | Nutrients (phosphorus) |
| Cadmium | Tainting substances |
| Chromium | Total dissolved solids |
| Copper | |
| Iron | *Other* |
| Lead | Asbestos |
| Mercury | Temperature |
| Nickel | Settleable and suspended solids and light transmission |
| Selenium | Microbiological conditions |
| Zinc | Radiological conditions |
| Fluoride | |

of objectives went from 13 in the original (1972) Agreement to 41 in the revised (1978) Agreement. Table 23.1 provides a listing of the 41 polluting substances and conditions for which objectives are included in the revised Agreement.

## 4. LIMITATIONS OF THE WATER QUALITY OBJECTIVES

During the process of developing water quality objectives under the 1972 Agreement, it became more and more evident that reliance on such objectives to define the desired state of the Great Lakes ecosystem has several inherent limitations. The development of such objectives is based primarily on the results of experiments in which individuals of one or a few species of biota are exposed under specified conditions to known concentrations of specific chemicals (or, in some cases, to known levels of physical properties) in the laboratory. The effects, if any, on survival, reproduction, or other biological properties at the different concentrations are measured and inferences are drawn as to the likely effects on populations of the same or closely related species in the Great Lakes. Use of objectives based on such toxicity testing can provide a large measure of environmental protection. However, there are a number of questionable assumptions and potential problems involved with basing recommended limits for environmental levels of pollutants on values obtained in this way. The major limitations in such objectives are summarized in the following list.

1. The information on relationships between effects and concentrations of polluting substances is obtained largely in the controlled environment of the laboratory. To apply these results to the field, it is necessary to assume that the laboratory results are representative of what actually occurs in the natural environment. The general applicability of this assumption is unproven (Mount, 1979; Gentile and Schimmel, 1984; Cairns and Buikema, 1984). In some studies where attempts have been made to check it, the laboratory results have been found to be in rather poor agreement with the effects that actually were observed in the field. However, in a number of other such studies good agreement has been found.

2. Only a limited number of species are included in toxicity testing. It must be assumed that no species that are substantially more sensitive than those tested are present in the environment to be protected. Although toxicity testing often attempts to concentrate on the forms that are believed to be the most sensitive to the pollutant under study, there is no assurance that the most sensitive forms can always be determined. Also other considerations, such as economic importance, ease of laboratory culture and maintenance, and ability to serve as a representative of a particular group of organisms, are important in choosing toxicity testing subjects.

3. Chemical pollutants are tested singly, yet in most natural environments, including the Great Lakes, a plethora of chemical substances in both dissolved and particulate form are found. The effects of the tested chemicals can be altered in the real world by synergistic or antagonistic relations with some of these co-oc-

curing substances. It must be assumed that these synergistic and antagonistic effects are unimportant in relation to the Great Lakes objectives.

4. There are hundreds of polluting substances in the Great Lakes, yet for only a few of these is there adequate toxicity information to permit development of a satisfactory objective. It is known that there are many pollutants present that have not yet been adequately identified and measured. Among the pollutants that have been identified, there is a large proportion for which no studies (or only very inadequate studies) have been conducted to determine effects. Further, even among the substances for which there is some information on effects and for which objectives have been developed, there are many for which the available information relates primarily to acute toxicity, and little is known about chronic toxicity, carcinogenicity, or other significant but more subtle, sublethal effects. The overall consequence of these problems is that, for the great majority of potential pollutants in the Great Lakes, no objective, or one based only on inadequate information, has been developed. For all of these pollutants, it must be assumed that their concentrations in the Great Lakes are at levels that present no substantial environmental or health risks or that, where there is an objective, it is at a protective level even though based on inadequate information.

5. Consideration is given only to the direct effects on the species under study. Yet all species occurring in the Great Lakes are part of a complex web of community interactions such as energy and material transfers among the various trophic levels. Thus, effects, even seemingly unimportant ones, on survival or other aspects of the biology of the species under study may have major repercussions on the populations of other components of the community or even on the ecosystem as a whole. It must be assumed that, at the levels specified by the objectives, such indirect effects will be of no more than minor significance.

This list of limitations makes it clear that the use of such water quality objectives to set limits for pollutants in the Great Lakes is not adequate by itself to assure protection of all beneficial uses. Even if all the objectives are met, there can be no guarantee that the Great Lakes ecosystem will always be protected.

## 5. TOWARD ECOSYSTEM OBJECTIVES

By the time of the 1978 revision of the Water Quality Agreement, the difficulties in basing protection solely on water quality and water quality objectives were generally recognized. Thus, the revision places great emphasis not just on controlling levels of chemical pollutants and physical effects, but also in assuring the biological integrity of the waters of the entire Great Lakes Basin ecosystem.

To help further the goal of protecting the ecosystem, SBWQC/WQOS was reconstituted and its terms of reference expanded. The reconstitued group, the Aquatic Ecosystem Objectives Committee (AEOC), was given the expanded charge of formulating ecosystem objectives for the Great Lakes based on a broad, holistic perspective. AEOC devoted considerable thought and discussion to defining the form such objectives should take. The group concluded that they should specify the level or condition desired for certain biologically related properties that could serve as indicators of the overall condition or health of the ecosystem.

It was also concluded that to develop such indicators it was necessary to specify, in more detail than in the revised (1978) Agreement, the desired state of the Great Lakes ecosystem. AEOC considered the general statements in the revised Agreement concerning purposes and goals and found strong emphasis on, and commitment to, restoration and enhancement of water quality in the Great Lakes ecosystem and maintenance of the chemical, physical, and biological integrity of this ecosystem. Based on such general guidance, the committee decided that the spirit of the Agreement could best be furthered by establishing an overall goal calling for the maintenance and restoration of Great Lakes biological communities as similar to those present before human intervention as practicable.

AEOC then considered means to develop specific objectives to serve as indicators of how well this ecosystem quality goal was being met. It established a work group of specialists in the areas of biological indicators and the biology of the Great Lakes to provide expert advice concerning such objectives.

The work group quickly recognized that it would not be possible to develop an indicator objective of this type that would be applicable to the entire Great Lakes System. Because physical and chemical properties vary greatly among the different parts of this system, the same community of organisms is not now, nor was it even before the advent of human intervention, present in all areas. Consequently no indicator objective based on a single species, stock, or community can be expected to be applicable throughout the Great Lakes ecosystem. However, it was also recognized that one rather similar community is present throughout a large part of the offshore waters of the Upper Lakes today and was probably also present in much of the offshore environment of the Lower Lakes before human intervention. Indeed the only major offshore areas where distinctly different communities are believed to have originally occurred are the shallow Western Basin, and probably also the Central Basin, of Lake Erie. The presence or potential presence of this widespread offshore community everywhere but Central and Western Erie was considered to present a promising situation for furthering development of an indicator objective. This ubiquitous community occurs in an environment that may be characterized as relatively cold and nutrient-poor. Accordingly it was decided to develop an indicator objective to help specify the desired state for this cold-water, oligotrophic community.

The work group considered ecological attributes at various hierarchic levels for use as indicators of ecosystem quality. It was decided to use a single integrator organism, preferably a terminal predator, and the status of its various stocks or populations within the Great Lakes as the basis of the initial indicator objective. Attributes, such as particle size conversion efficiency (Borgmann and Whittle, 1983), that provide information at the higher hierarchic levels of community and ecosystem and thus have the advantage of integrating information concerning a number of components of the system were considered. However, it was concluded that an indicator based on a single organism was the preferred approach for the initial indicator objective because there is presently a better theoretical basis and a much larger body of Great Lakes data concerning the use of potential indicator organisms than is available for higher level indicators. It is anticipated that indicators from other than the population level of organization will provide valuable indicators in the future.

Before selecting a specific indicator organism, the work group discussed and reached agreement on a set of criteria to be used to judge the relative suitability of candidate organisms. Although the work group agreed that a suitable indicator need not necessarily meet all the criteria, they felt that the criteria would be helpful in providing an objective basis for selecting organisms to serve as indicators of perturbations in aquatic communities, especially the cold-water, oligotrophic community of the Great Lakes. The criteria are presented in Ryder and Edwards (1985) and are summarized in a somewhat modified form in the following list. They state that a suitable indicator organism will tend to:

- have a broad distribution in the system;
- be easily collected and measured in terms of biomass;
- be indigenous and maintain itself through natural reproduction;
- interact directly with many components of its ecosystem;
- have historical, preferably quantified, information pertaining to its abundance and other critical factors relevant to the state of the organism;
- have well-documented and well-quantified niche dimensions expressed in terms of metabolic and behavioral responses;
- exhibit a gradual response to a variety of human-induced stresses;
- serve as a diagnostic tool for specific stresses of many sorts;
- respond to stresses in a manner that is both identifiable and quantifiable;
- be a suitable species for laboratory investigations;
- be generally recognized as important to humans; and
- serve to indicate aspects of ecosystem quality other than those represented by presently accepted parameters.

A number of organisms were evaluated in relation to these criteria in order to select one to serve as the initial subject for development of the ecosystem indicator concept. It was concluded that the lake trout (*Salvelinus namaycush*) best fit these criteria, and so it was selected to serve as the primary indicator for the Great Lakes cold-water, oligotrophic community.

This native fish was the major terminal predator throughout much of the open water of the upper Great Lakes until the 1940s and 1950s when it was almost completely eliminated in most areas. The major cause of its decline was predation by the sea lamprey (*Petromyzon marinus*), which was inadvertently introduced to the upper four lakes when the Welland Canal was constructed connecting Lake Erie and Lake Ontario (Smith, 1968; Walters *et al.*, 1980). The decline was also assisted by the impacts of overfishing and, to a limited extent, pollution.

Lake trout thrive in pristine oligotrophic environments and are sensitive to many of the environmental stresses that alter these conditions (Marcus et al., 1984). Although sea lamprey populations have been greatly reduced in the Great Lakes recently, self-reproducing populations of lake trout have not become reestablished in most areas. Lake trout bioaccumulate toxic substances such as DDT and PCBs, and there is evidence that the concentrations of such substances in lake trout from the Great Lakes are presently at levels that can cause physiological or toxic effects (Willford, 1984). The difficulty in reestablishing self-reproducing populations may be due to such chemical pollution or to other changes in habitat. Whatever the cause lake trout provide a sensitive indicator for the cold-water, oligotrophic community.

As a terminal predator, lake trout serve as an excellent integrator of degradation throughout the ecosystem. The binational Great Lakes Fishery Commission has, as a major objective, the restoration of self-sustaining populations of this fish in appropriate areas of the Great Lakes; so its use as an objective seems especially important to the international efforts to restore and enhance the Great Lakes ecosystem.

Having selected the lake trout as the primary indicator, the work group then documented and, as much as possible, quantified the critical niche characteristics and habitat requirements of this species (Ryder and Edwards, 1985). These included description of lake trout responses to external stresses, both those of natural origin and those due to pollution and other anthropogenic effects.

This information was then used as the basis for developing criteria to employ in judging the status of the various stocks of lake trout. As a first step this has been done for the lake trout stocks in Lake Superior. Thus, the following objective has been proposed for a healthy Lake Superior ecosystem (Aquatic Ecosystem Objectives Committee, 1985):

> Lake Superior should be maintained as a balanced and stable oligotrophic ecosystem with the lake trout as top aquatic predator of a cold-water community.
> NOTE: In order to determine whether this condition exists, the following criteria should be met:
> (1) the lake trout productivity should be greater than 0.38 kg/ha/yr as determined using mortality rates;
> (2) there should be a stable number of recognizable, self-reproducing stocks;
> (3) the annual harvest of lake trout should not exceed 0.24 kg/ha;
> (4) the harvest of lake trout should be free from contaminants at levels which adversely affect the trout themselves or the quality of the harvested product.

This objective has been recommended to the IJC, and the organization has it under consideration for recommendation to the two Parties for incorporation into the Water Quality Agreement.

Steps are now being taken to expand this approach to other areas and to other indicators. For other oligotrophic, cold-water areas, the lake trout is being considered as the primary indicator. The approach for these areas will be the same as for Lake Superior. However, because the characteristics of the lake trout stocks found in these other areas differ from those in Lake Superior and because the limnological and ecological conditions in these areas are also somewhat different, the criteria will probably have to be altered to fit these particular situations.

In addition to the large offshore areas of the Great Lakes that are, or at least were, occupied by the cold-water, oligotrophic community, there are substantial areas in the Great Lakes where conditions favor a different biological community. For such areas, the lake trout is no longer an appropriate indicator of ecosystem status. A second community of major significance in the Great Lakes is that found in environments that are warmer and more nutrient-enriched than those where the lake trout thrive, i.e., mesotrophic environments. Such mesotrophic environments occur in many nearshore areas and bays such as Saginaw Bay and much of Georgian Bay and Green Bay, as well as the Western and Central Basins of Lake Erie and most of the Great Lakes connecting channels including Lake St. Clair. The work group concluded that the walleye (*Stizostedion vitreum*) is the most promising candidate for use as an indicator in such areas. This fish is, or at least was until decimated by anthropogenic environmental alterations, the major terminal predator in most of these areas and conforms to the general criteria for indicators almost as well as does the lake trout. A second work group, equivalent to the one that developed the use of lake trout as an indicator, has been established to make recommendations for use of the walleye as an indicator for mesotrophic environments.

Efforts are also being initiated by AEOC to explore the formulation of additional indicators for oligotrophic environments. The additional indicators would complement the lake trout indicator and provide information on ecosystem status at trophic levels other than that of the terminal predators and on stresses in addition to those to which the lake trout responds. The amphipod crustacean, *Pontoporeia hoyi,* the benthic macroinvertebrate with by far the greatest biomass in the oligotrophic areas of the Great Lakes, has been judged a promising candidate as a second indicator for oligotrophic environments, and its use for this purpose is presently being investigated.

Thus, substantial progress is being made in going beyond development of traditional water quality objectives to the development of objectives that provide for protection, not just of the water quality of the Great Lakes, but of the entire Great Lakes ecosystem. However, the development of these ecosystem objectives is in its infancy. Given the importance of judging the health of the Great Lakes, not just by the levels of pollutants present, but also by directly assessing the primary concern, that is, the ecosystem itself, there is a clear need for the efforts to further development of indicators of ecosystem health to be continued and expanded.

## ACKNOWLEDGMENTS

I thank Dr. William Strachan, Dr. Eugene Stoermer, and Dr. Marlene Evans for their suggestions for improvement of this paper. Special appreciation is extended to Mr. Richard Ryder for his thorough and very helpful review and comments.

## REFERENCES

Andrew, R. W. (1976). Toxicity relationships to copper forms in natural waters. In P. V. Hodson and D. E. Konasewich (Eds.), *Toxicity to Biota of Metal Forms in Natural Waters.* Great Lakes Research Advisory Board, International Joint Commission, Windsor, Ontario, pp. 127–143.

Aquatic Ecosystem Objectives Committee (1986). *1985 Annual Report.* Great Lakes Science Advisory Board, International Joint Commission, Windsor, Ontario.

Borgmann, U., and Whittle, D. M. (1983). Particle-size-conversion efficiency and contaminant concentrations in Lake Ontario biota. *Can. J. Fish. Aquat. Sci.* **40,** 328–336.

Cairns, J., Jr., and Buikema, A. L. (1984). Verifying predictions of environmental safety and harm. In H. H. White (Ed.), *Concepts in Marine Pollution Measurements.* Maryland Sea Grant, University of Maryland, College Park, pp. 81–111.

Gentile, J. H., and Schimmel, S. C. (1984). Strategies for utilizing laboratory toxicological information in regulatory decisions. In H. H. White (Ed.), *Concepts in Marine Pollution Measurements.* Maryland Sea Grant, University of Maryland, College Park, pp. 57–80.

International Joint Commission (1972). *Great Lakes Water Quality Agreement, with Annexes and Texts and Terms of Reference, Between the United States of America and Canada.* IJC, Washington, DC and Ottawa.

International Joint Commission (1978). *Great Lakes Water Agreement of 1978.* IJC, Washington, DC and Ottawa.

Marcus, M. D., Hubert, W. A., and Anderson, S. H. (1984). Habitat suitability index models: Lake trout (exclusive of the Great Lakes). U.S. Dep. Inter., Fish Wildl. Serv. FWS/OBS-82/10.84, 1–12.

Mount, D. I. (1979). Adequacy of laboratory data for protecting aquatic communities. In K. L. Dickson, A. Maki, and J. Cairns, Jr. (Eds.), *Analyzing the Hazard Evaluation Process.* Am. Fish. Soc., Washington, DC, pp. 112–118.

Robertson, A. (1982). Water quality objectives for the Great Lakes. In *Oceans 82—Conference Record.* Marine Technology Society and IEEE Council on Oceanic Engineering, Washington, DC, pp. 1177–1181.

Ryder, R. A., and Edwards, C. J. (Eds.) (1985). *A Conceptual Approach for the Application of Biological Indicators of Ecosystem Quality in Great Lakes Basin.* Report to Great Lakes Science Advisory Board, International Joint Commission, Windsor, Ontario.

Smith, S. H. (1968). Species succession and fishery exploitation in the Great Lakes. *J. Fish. Res. Board Can.* **25,** 667–693.

Walters, C. J., Steer, G., and Spangler, G. (1980). Responses of lake trout (*Salvelinus namaycush*) to harvesting, stocking, and lamprey predation. *Can. J. Fish. Aquat. Sci.* **37,** 2133–2145.

Willford, W. A. (1984). Contaminants. In R. L. Eshenroder, T. P. Poe, and C. H. Oliver (Eds.),

*Strategies for Rehabilitation of Lake Trout in the Great Lakes: Proceedings of a Conference on Lake Trout Research, August 1983.* Great Lakes Fish. Comm., Ann Arbor, Michigan, Tech. Rept. No. 40, pp. 28–34.

Wong, P. T. S., Chau, Y. K., and Luxon, P. L. (1978). Toxicity of a mixture of metals on freshwater algae. *J. Fish. Res. Board Can.* **35,** 479–481.

# 24

# ECOSYSTEM MANAGEMENT OF PERSISTENT TOXIC CHEMICALS—BACK TO BASICS

***Douglas J. Hallett***

*ELI–Eco Logic Inc., Rockwood, Ontario, Canada N0B 2K0*

## 1. INTRODUCTION

Persistent toxic chemicals such as PCBs, DDT, toxaphene, mirex, chlorobenzenes, and chlorinated dioxins have been found throughout the Great Lakes ecosystem. Residues are evident in fish, herring gulls, food, rainfall, snow, water, and human beings. There is a well-developed data base for PCBs and since they have relatively similar properties to many other persistent organochlorine contaminants, PCBs will serve in this paper as the primary example for most other persistent toxic substances.

The purpose of this chapter is to consider previous management decisions for PCBs as a model for other persistent toxic chemicals. Have the current restrictions been successful, and what further management decisions are necessary?

The theme of the chapter is ''Back to Basics''. It examines the PCB and persistent toxic chemical issue not as a complex jumble containing hundreds of chemicals of unknown toxicity, but as a single-model problem using a persistent chemical type that causes chronic toxic effects. The goal is to promote better management of all persistent chemicals through a thorough basic examination of PCBs.

## 2. THE GREAT LAKES ECOSYSTEM

### 2.1. Sediments

In the Great Lakes, particularly in Lake Ontario, we have observed some dramatic trends in persistent residues. Spatial trends in sediment have indicated that industrial centers on the lower Great Lakes, that is, Lake Erie and Lake Ontario, have contributed vast quantities of mercury, PCBs, and lead to the water.

The sediment record on the Great Lakes clearly shows that the most severe contamination from toxic chemicals of industrial origin occurs either downstream of the major connecting channels or in those harbors and tributaries having intense populations and industries associated with them. The former includes the Niagara River, the Detroit River, the St. Clair River, and the St. Mary's River. Waukegan Harbour, Cleveland Harbour, Hamilton Harbour, the Saginaw, Rouge, and Raisin Rivers in Michigan, and the Humber and Serpent Rivers in Ontario comprise the latter. The toxic chemicals most prevalent in each area are the chemicals associated with the industries adjacent to those water bodies (International Joint Commission, 1978). The PCBs in Waukegan Harbour or the Humber River are associated with direct industrial and municipal discharges. PCBs and chlorinated dioxins are excellent examples of persistent contaminants to follow since the best current scientific data bases are available for these compounds.

PCBs have been measured in sediments of the Great Lakes (Glooschenko et al., 1976; Veith et al., 1977; Eisenreich et al., 1979; Thomas and Frank, 1982) and extensive surveys have been conducted in Lakes Erie and St. Clair (Frank et al., 1977), Lake Ontario (Frank et al., 1979a), Lake Huron (Frank et al., 1979b), Bay of Quinte (Frank et al., 1980a), Lake Michigan (Frank et al., 1981), and Lake Superior (Frank et al., 1980b). Great Lakes harbors with PCB concentrations > 50 $\mu$g/L (Ontario Ministry of the Environment, guideline for open water disposal) are shown in Figure 24.1. These harbor sediments must therefore be regarded as requiring a means of proper disposal. Contaminated harbors appear to be concentrated in the lower highly industrialized Great Lakes. Heavily contaminated lake, river, and harbor sediments will remain as continuing localized sources of PCBs due to sediment resuspension as long as they are present.

## 2.2. The Niagara River

The Niagara River area represents one of the most closely studied areas of high toxic chemical contamination. Temporal trends show a confusing situation. On first glance one can see a dramatic improvement in Lake Ontario residues of PCBs, mirex, and dioxin particularly in wildlife and fish between 1970 and 1980 (Fig. 24.2). However, from 1980 to 1984 the downward trend stopped for all of these residues. On examining the sediment record at the source of Lake Ontario we find a deposition rate for PCBs and mirex from 1950 to 1960 at approximately the present day rate. However, at about 1960 a tremendous pulse of PCBs, mirex, chlorobenzenes, and hexachlorobutadiene is shown in the sediment record. The deposition rate tapers off to 1970 when it plateaus at about the 1950s rate (Durham and Oliver, 1983). If one examines the water in the Niagara River today one still

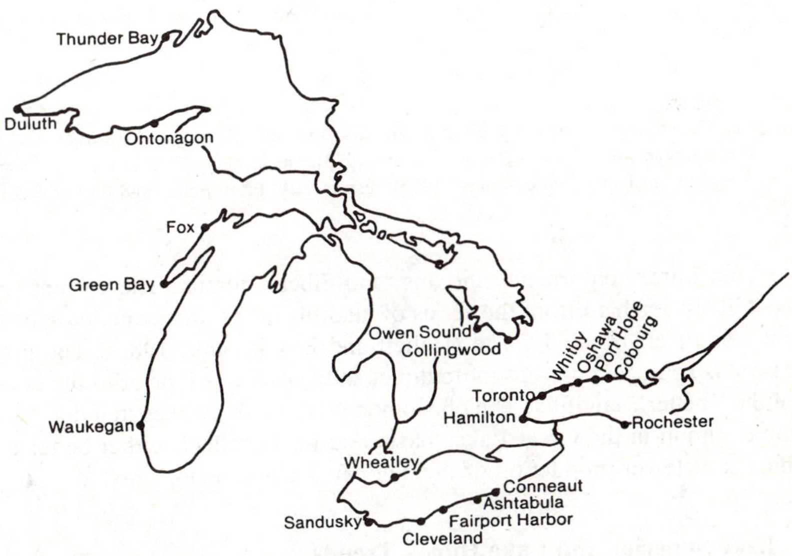


**Figure 24.1.** Great Lakes harbors with PCB concentrations greater than 50 $\mu$g/L. From Thomas and Frank (1982).

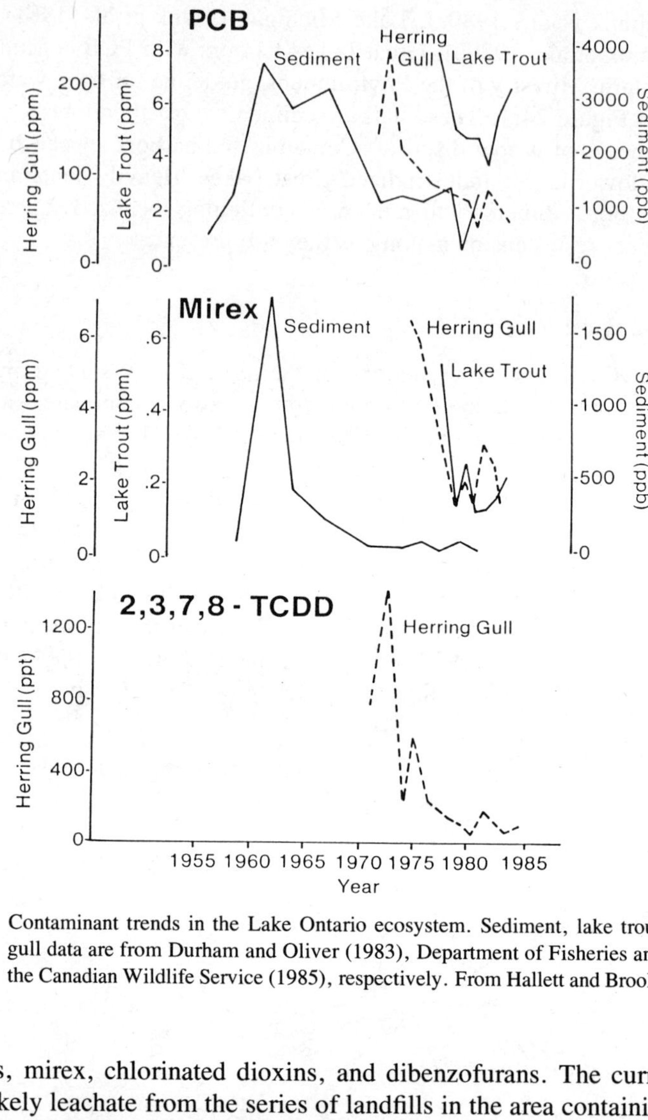


**Figure 24.2.** Contaminant trends in the Lake Ontario ecosystem. Sediment, lake trout, and herring gull data are from Durham and Oliver (1983), Department of Fisheries and Oceans, and the Canadian Wildlife Service (1985), respectively. From Hallett and Brooksbank (1986).

finds PCBs, mirex, chlorinated dioxins, and dibenzofurans. The current sources are most likely leachate from the series of landfills in the area containing over 250 million tons of chlorinated waste (Hallett and Brooksbank, 1986). The previous more serious sources represent more direct sources of waste prior to the construction of the "better" landfills. The phenomenon of rapid declines in industrial areas is quite common in the Great Lakes basin and likely reflects either better control of effluents or lower production of the key persistent contaminants.

## 2.3. Lake Superior and Lake Huron Trends

While levels of PCBs in herring gull eggs of Lake Ontario have decreased over the decade 1974–1984, eggs from Lakes Superior and Huron have not shown the

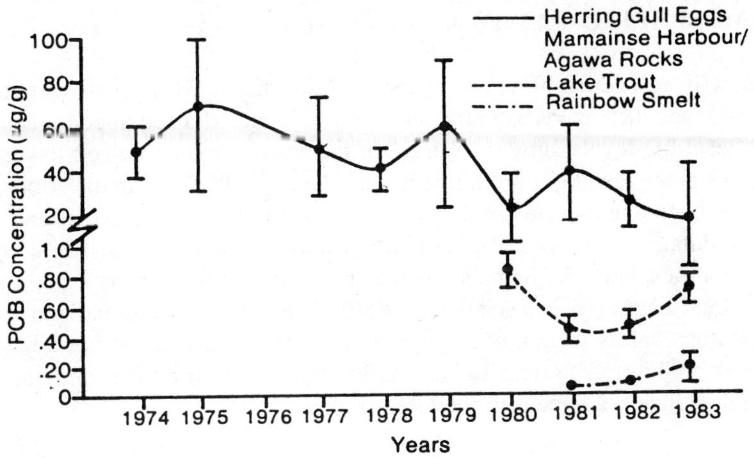


**Figure 24.3.** PCB concentrations in Lake Superior biota. From the International Joint Commission (1985).

same significant decrease in that time (Figs. 24.3 and 24.4). The trend is toward a more uniform level of PCBs throughout the Great Lakes. The difference between mean PCB concentrations in gulls of Lakes Ontario and Superior has narrowed from 110 μg/g in 1975 to only 25 μg/g in 1983. Lake trout from the same two lakes have declined from 6.14 μg/g in 1978 to 4.59 μg/g in 1983. Since only 3 or 4 years of rainbow smelt data are available for most of the lakes, no trend was established. While a general decrease in PCB levels in biota of Lakes Ontario and Huron is seen, lake trout and rainbow smelt from Lake Superior have had an increase in PCB contamination since 1981 indicating increasing deposition of PCBs from unknown sources.

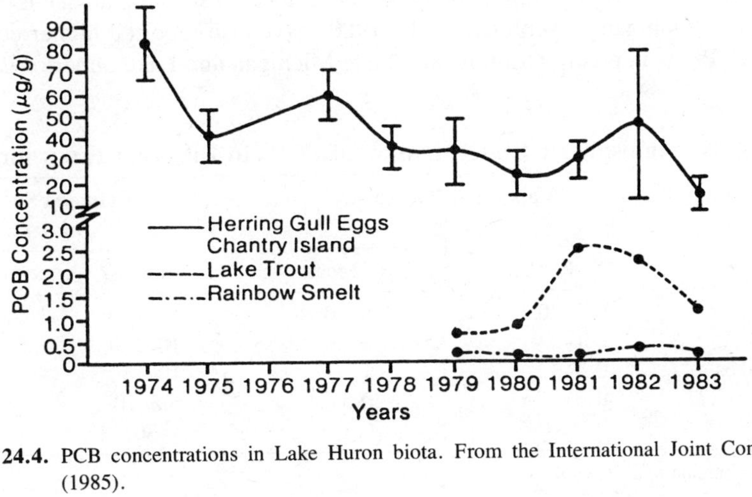


**Figure 24.4.** PCB concentrations in Lake Huron biota. From the International Joint Commission (1985).

## 3. ATMOSPHERIC PATHWAYS

The data with respect to biota and water indicate that residue levels of persistent toxic substances such as PCBs in the relatively nonindustrialized upper Great Lakes are now reaching levels similar to those in the heavily industrialized lower Great Lakes. This is the result of continuing accumulation of PCBs from the atmosphere. There have been a number of studies on the concentrations and deposition of trace organic pollutants from air and precipitation into the Great Lakes. Rural concentrations of atmospheric PCBs have been found to be approximately equal to those over the Great Lakes (Eisenreich et al., 1980). These data are summarized in Table 24.1. "Rural" refers to nonurban, agricultural or forested areas not adjacent to large water bodies and "Great Lakes" refers to atmospheric measurements taken over or immediately adjacent to the Great Lakes.

### 3.1. Atmospheric Deposition

The dry and wet depositions of airborne PCBs to each of the Great Lakes are shown in Figure 24.5 (Eisenreich et al., 1980). Wet deposition was defined as that which occurs during a rain or snow event while all other forms of deposition were termed dry. Dry deposition of airborne PCBs was significantly higher than wet deposition for all of the lakes. Both dry and wet depositions of airborne PCBs were significantly higher in the upper Great Lakes than in the lower Great Lakes region. This is due to the higher surface areas of the upper Great Lakes allowing for a greater catchment area. The lower Great Lakes, however, have many more direct sources of PCBs.

Figure 24.6 details the concentrations of PCBs in rain and snow for areas of the Great Lakes adjacent to Canada (Strachan and Huneault, 1979) and to Lake Michigan (Murphy and Rzeszutko, 1977; Murphy et al., 1980). Significantly higher concentrations of PCBs in rain and snow were found for the area adjacent to Chicago, Lake Michigan. Eisenreich et al. (1980) have also reported higher concentrations of PCBs in precipitation for the Lake Michigan and Lake Superior Basins.

**Table 24.1 Atmospheric Concentration of PCBs in Different Environments**

| | Vapor and Particulate ($ng/m^3$) | | Rain and Snow (ng/L) | |
|---|---|---|---|---|
| Location | Range | Mean | Range | Mean |
| Urban | 0.5–30 | 5–10 | 10–250 | 50 |
| Rural | 0.1–2 | 1 | 10–100 | 15–50 |
| Great Lakes | 0.4–3 | 1 | 10–150 | 20–50 |
| Marine | 0.05–2 | 0.5 | 0.5–10 | 1–5 |
| Remote | 0.02–0.5 | 0.1 | 1–30 | 5 |

*Source.* Eisenreich et al. (1980).

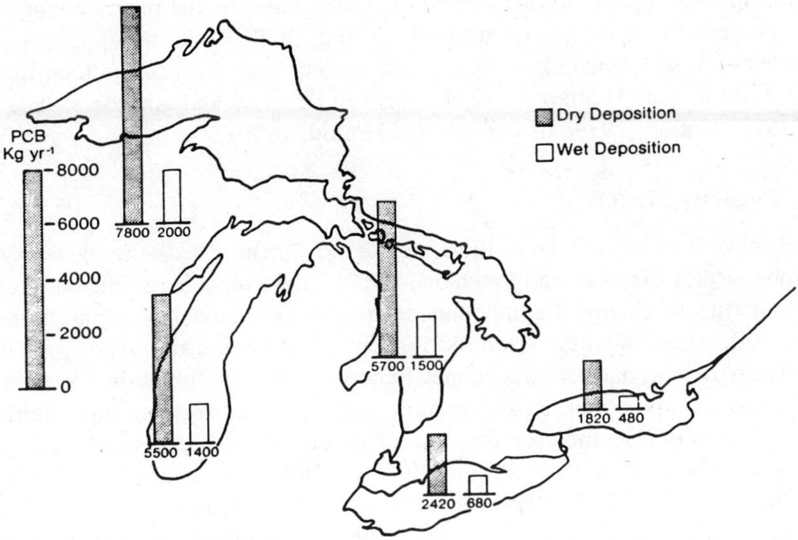


**Figure 24.5.** Dry and wet deposition of airborne PCBs to the Great Lakes. From Eisenreich et al. (1980).

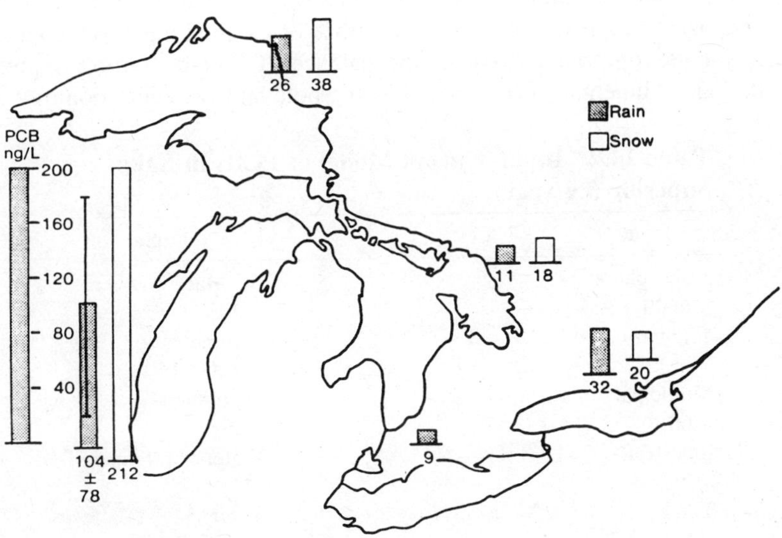


**Figure 24.6.** Atmospheric PCBs in the Great Lakes Basin: rain and snow 1975–1976. From Strachan and Huneault (1979).

Atmospheric deposition of PCBs to the Great Lakes is the primary input pathway. The results of studies of atmospheric flux of PCBs to the upper relatively nonindustrialized Great Lakes suggest that greater than 80% of the total input of PCBs to both lakes is atmospherically derived (Doskey and Andren, 1980; Doskey, 1978; Hollod, 1979; Eisenreich and Hollod, 1980).

### 3.2. Fugacity Models

Models have been developed to illustrate the movement of PCBs in the ecosystem (Murphy, 1985; MacKay and Paterson, 1985). These models use the basic chemical properties of chemical compounds to predict their tendency to stay in water, air, or soil. This tendency is termed fugacity. The chemical's vapor pressure is the primary determinant of its tendency to deposit with precipitation. Lower vapor pressure chemicals such as PCBs are more susceptible to washout in precipitation. Fugacity models also indicate that during precipitation-free periods these compounds volatilize. PCB isomers with three or four chlorine atoms per molecule volatilize at the fastest rate. Thus, cycling of PCBs between air and water is demonstrated. This is of particular concern for the Great Lakes because frequent lower temperatures reduce vapor pressure, which increases the chemicals' tendency to condense and therefore to fall out of the atmosphere as precipitation. Compounds, such as toxaphene, used in warmer climates to the south of the Great Lakes, may preferentially be deposited in cooler areas to the north.

### 3.3. Mass Balance Models

An input–output model for PCB cycling in Lake Superior has been constructed by Hollod (1979) and Eisenreich and Hollod (1980). As seen in Table 24.2, atmospheric deposition represented greater than 85% of the total input to the lake, with tributary inputs comprising most of the balance. Tributary outflow represented about 10% of sedimentary losses based on a whole lake concentration of 2 mg/L.

**Table 24.2 Input–Output Model of PCBs in Lake Superior (kg/year)**

| Input | Output |
|---|---|
| Atmosphere | St. Mary's River |
| 6600–8300 | 140 |
| Tributary | Sedimentation |
| <1311 | 1000–1600 |
| Municipal | Degradation |
| 66 | ? |
| Industrial | Water-Air Transfer |
| 2 | ? |
| Total | Total |
| 8000–9700 | 1140–1740 |

*Source.* Eisenreich et al. (1980)

**Table 24.3 Estimates of Annual Loadings of PCB Substances in Runoff from Selected Urban Areas in the Lower Great Lakes Basin**

| Urban Area | Annual Urban Runoff Volume $\times 10^3 m^3/yr$ | Annual Load of Sediment in Runoff *T*/yr | Loadings in Water (kg/year) | Loadings Attached to Sediment (kg/yr) | Water Plus Sediment Loadings (kg/year) |
|---|---|---|---|---|---|
| Ajax | 2,198 | 706 | 0.04 | 0.87 | 0.92 |
| Belleville | 4,728 | 1482 | 0.10 | 1.84 | 1.93 |
| Brampton | 14,382 | 7,336 | 0.29 | 9.13 | 9.42 |
| Brantford | 9,090 | 3,924 | 0.19 | 4.86 | 5.05 |
| Burlington | 12,728 | 4,542 | 0.26 | 5.63 | 5.89 |
| Cambridge | 8,956 | 2,396 | 0.18 | 2.97 | 3.15 |
| Chatham | 4,481 | 2,230 | 0.09 | 2.76 | 2.85 |
| Cobourg | 1,661 | 362 | 0.03 | 0.45 | 0.48 |
| Cornwall | 3,980 | 1,202 | 0.08 | 1.87 | 1.95 |
| Dundas | 2,351 | 596 | 0.05 | 0.74 | 0.79 |
| Etobicoke | 32,072 | 14,057 | 0.65 | 17.42 | 18.07 |
| Fort Erie | 2,005 | 648 | 0.04 | 0.80 | 0.84 |
| Halton Hills | 3,092 | 1,062 | 0.06 | 1.32 | 1.38 |
| Guelph | 8,554 | 2,462 | 0.17 | 3.05 | 3.22 |
| Hamilton | 30,779 | 18,677 | 0.63 | 23.14 | 23.77 |
| Kingston | 6,826 | 3,381 | 0.14 | 4.19 | 4.33 |
| Kitch-Waterloo | 22,967 | 7,794 | 0.47 | 9.66 | 10.12 |
| Leamington | 1,313 | 393 | 0.03 | 0.49 | 0.51 |
| Lindsay | 1,901 | 439 | 0.04 | 0.54 | 0.58 |
| London | 32,876 | 9,635 | 0.67 | 11.94 | 12.61 |
| Markham | 4,581 | 2,010 | 0.09 | 2.49 | 2.58 |
| Mississauga | 26,916 | 11,134 | 0.55 | 13.79 | 14.34 |
| Newmarket | 3,423 | 1,029 | 0.07 | 1.27 | 1.34 |
| Niagara Falls | 8335 | 3131 | 0.17 | 3.88 | 4.05 |
| Oakville | 8,191 | 2,777 | 0.17 | 3.44 | 3.61 |
| Oshawa | 14,813 | 6,075 | 0.30 | 7.53 | 7.83 |
| Peterborough | 7,518 | 1,835 | 0.15 | 2.27 | 2.43 |
| Pickering | 3,263 | 1,039 | 0.07 | 1.29 | 1.35 |
| Port Colbourne | 2,719 | 584 | 0.06 | 0.72 | 0.78 |
| Richmond Hill | 4,378 | 1,248 | 0.09 | 1.55 | 1.64 |
| St Catharines | 14,296 | 8,931 | 0.29 | 8.59 | 8.88 |
| St. Thomas | 3,507 | 987 | 0.07 | 1.22 | 1.29 |
| Sarnia | 5,900 | 2,935 | 0.12 | 3.64 | 3.76 |
| Scarborough | 41,884 | 21,680 | 0.85 | 26.86 | 27.71 |
| Simcoe | 1,706 | 450 | 0.03 | 0.56 | 0.59 |
| Stratford | 3,934 | 209 | 0.08 | 1.00 | 1.08 |
| Toronto | 41,151 | 19,719 | 0.84 | 24.43 | 25.27 |
| Trenton | 1,894 | 452 | 0.04 | 0.56 | 0.60 |
| Wallaceburg | 1,376 | 401 | 0.03 | 0.50 | 0.52 |
| Welland | 6,019 | 1,573 | 0.12 | 1.95 | 2.07 |
| Whitby | 3,289 | 986 | 0.07 | 1.22 | 1.29 |

**Table 24.3** (*Continued*)

| Urban Area | Annual Urban Runoff Volume $\times 10^3 m^3/yr$ | Annual Load of Sediment in Runoff *T*/yr | Loadings in Water (kg/year) | Loadings Attached to Sediment (kg/yr) | Water Plus Sediment Loadings (kg/year) |
|---|---|---|---|---|---|
| Windsor | 21,827 | 8,735 | 0.45 | 10.82 | 11.27 |
| Woodstock | 3,780 | 829 | 0.08 | 1.03 | 1.10 |
| York | 9,309 | 4,798 | 0.19 | 5.94 | 6.13 |
| York, East | 7,624 | 4,064 | 0.16 | 5.03 | 5.19 |
| York, North | 55,688 | 27,356 | 1.14 | 33.89 | 35.03 |
| Total | 514,262 | 216,918 | 10.5 | 269.1 | 279.6 |

*Source.* Marsalek (1983).

Biological degradation is not considered to be an important loss mechanism since higher molecular weight isomers are resistant to biological degradation (National Academy of Science, 1979). Water-to-air transport of PCBs involves the process of aerosolization and volatilization. These represent a key gap in scientific knowledge and therefore in our full understanding of ecosystem cycling and mass balance equations related to environmental contaminants.

### 3.4. Urban Runoff

Persistent contaminants can also enter the environment through urban runoff. The annual rates of loading of PCBs have been determined in runoff from selected urban areas in the lower Great Lakes Basin (Marsalek, 1983). These results are shown in Table 24.3. Highly industrialized urban centers such as Hamilton, Toronto, and Windsor have the highest estimates of annual loadings of PCBs in urban runoff.

## 4. HUMAN EXPOSURE

Just as wild populations are exposed to persistent toxic chemicals, human populations are exposed to the same chemicals but through different pathways.

### 4.1. Human Concern

Human concern for this exposure stems from the chronic and subchronic effects demonstrated for persistent chemicals such as PCBs or chlorinated dioxins in laboratory animals. Multiple subtle chronic effects that manifest themselves in disease (including terminal illnesses such as cancer, liver sclerosis, and behavioral and neural disorders, and diseases caused as a result of immunosuppression) are very difficult to discern in human populations with statistical rigor.

To find the direct cause of these diseases one must determine both the nature of all the chronic effects or stresses that together caused the disease and also determine all of the most probable variables including toxic chemicals that may have caused the multiples of chronic effects. This investigation cannot generally be carried out until the disease is obvious at a population level. Because of the lag phase between cause and population effect the cause of the disease may have declined or may have been totally eliminated. Therefore, it is important to understand human exposure to persistent toxic chemicals, since these chemicals have the potential to produce human disease in the future or may be contributing to human disease currently. Our knowledge of cancer and most other chronic illnesses is based on studies of the chronic effects that cause such diseases, laboratory experiments on animals or other predictive organisms that prove the cause–effect relationship between exposure to a toxic chemical and the chronic effect, and more expensive multiyear studies of the disease itself. Direct human studies have been impossible except for tragic accidental exposures or isolated occupational problems.

This forms the basis of scientific concern for long-term human exposure to toxic chemicals. Industrial society as a whole has begun to take measures to prevent exposure to toxic substances. By reducing long-term exposure to toxic chemicals, disease manifested from the chronic effects caused by these chemicals will be prevented. Prevention rather than reaction is the best scientific approach to the problem since reaction is always too late.

To prevent human exposure to persistent toxic chemicals we must first understand what toxic chemicals we are exposed to, the basis of human exposure, and the environmental pathways that cause this exposure.

## 4.2. Residues in Great Lakes Populations

Herring gulls and the coho salmon living in Lake Ontario have a common food source, alewives and smelt, which are the next step lower on the food chain. Alewives and smelt are the common source of PCBs, DDE, mirex, photomirex, and chlorobenzenes to the gulls and salmon (Table 24.4) (Norstrom et al., 1978). The same relationship for 2,3,7,8-TCDD can be shown (National Research Council of Canada, 1982). When one examines residues in the human population in Ontario (Table 24.4) one finds exactly the same residues that are found in herring gulls and fish. This demonstrates that humans are exposed to the same persistent chemicals as fish and wildlife from the same region.

Frank et al., 1983 collected data on PCB residues in human adipose tissues from 1976 to 1981 (Fig. 24.7). The data do not show significant differences in residue levels between sexes nor is a trend evident through time, although the sample size used is insufficient to draw any conclusions on trends. Many residues are prevalent in adipose tissue from humans across Canada as shown in Table 24.5 (Mes et al., 1982).

In a study of mother's milk in Sheboygan Wisconsin, women who consumed fish at least twice a month had a PCB content in their breastmilk of 1.26 mg/kg milkfat compared to the breastmilk PCB content of those who ate no fish of 1.07

**Table 24.4 Human, Wildlife, and Fish Residues in Ontario (mg/kg)**

| | *Human* | | | |
|---|---|---|---|---|
| | Adipose Tissue[a] | | | Breast Milk[b] —Fat Basis |
| Compound | Kingston ($n = 12$) | Ottawa ($n = 10$) | Ontario ($n = 47$) | Ontario ($n = 125$) |
| PCB | 3.0 | 2.0 | 2.1 | 0.756 |
| DDT | 3.429 | 2.694 | 4.069 | 0.983 |
| Mirex | 0.027 | 0.011 | 0.02 | — |
| Photomirex | 0.009 | 0.006 | — | — |
| HCB | 0.106 | 0.078 | 0.013 | 0.0184 |
| BHCH | 0.136 | 0.065 | — | — |
| TCDD[c] | $12 \times 10^{-6}$ | $8.6 \times 10^{-6}$ | — | — |
| Heptachlor epoxide | 0.035 | 0.037 | 0.07 | 0.012 |
| Dieldrin | 0.036 | 0.043 | 0.10 | 0.194 |

| | *Wildlife and Fish*[d] | | |
|---|---|---|---|
| Compound | Alewives and Smelt | Coho Salmon | Herring Gull Eggs |
| PCB | 2.21 | 5.7 | 138 |
| DDE | 0.47 | 0.97 | 17.4 |
| Mirex | 0.09 | 0.23 | 4.4 |
| Photomirex | 0.03 | 0.11 | 1.6 |
| HCB | 0.024 | 0.097 | 0.52 |
| BHCH | 0.002 | 0.012 | 0.078 |
| TCDD[e] | $14 \times 10^{-6}$ | $30 \times 10^{-6}$ | $55 \times 10^{-6}$ |
| Heptachlor epoxide | 0.003 | 0.015 | .007 |
| Dieldrin | 0.029 | 0.087 | .060 |

[a]Williams et al. (1984).
[b]R. Frank (personal communication).
[c]Ryan and Williams (1983).
[d]Norstrom et al. (1978)
[e]National Research Council of Canada

mg/kg (Smith, 1984). The mothers' mean PCB blood serum levels were 0.6 $\mu$g/kg for fisheaters and 0.5 $\mu$g/kg for nonfisheaters in the study.

A study of Great Lakes fisheaters showed a significant difference between serum PCB levels of lake trout eaters compared with controls (Humphrey, 1983). This exposed group consumed greater than 10.91 kg/year of fish that ranged from 0.4 to 5.4 mg/kg in PCB content. Serum residue levels from nursing mothers who were not fisheaters ranged from 10 to 23 parts per billion (ppb), whereas residue levels in serum from fish-eating mothers ranged from 41 to 105 ppb.

PCBs have been reported in human milk (Ritcey et al., 1972; Musial et al.,

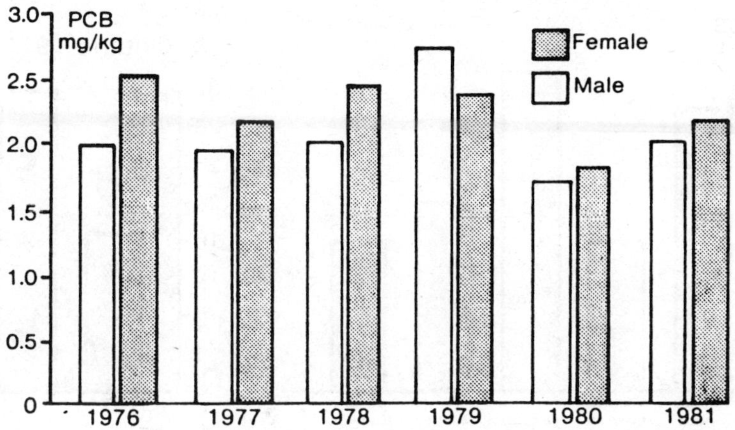


**Figure 24.7.** PCB concentration in human adipose tissue from Ontario 1976–1981. From R. Frank et al. unpublished data.

1974; Mes and Davies, 1979; Frank, R., J. Rasper, H. E. Braun, and M. S. Smout, unpublished data). A nationwide survey by Mes and Davies (1979) of PCBs in the milk of Canadian mothers indicated that the highest average PCB levels were found in Ontario and Western Canada (Fig. 24.8). Frank et al. (unpublished data) analyzed samples of milk and human adipose tissue from Ontario residents

**Table 24.5 Residues in Adipose Tissue from Humans across Canada**

| Compound | Amount[a] (μg/kg wet weight) | Percentage of Samples Containing Residues |
|---|---|---|
| PCB | 907 | 100 |
| Hexachlorobenzene | 62 | 100 |
| BHC (lindane) | 65 | 88 |
| Oxychlordane | 55 | 97 |
| *trans*-Nonachlor | 65 | 99 |
| Heptachlor epoxide | 43 | 100 |
| Dieldrin | 69 | 100 |
| *p,p'*-DDE | 2095 | 100 |
| *o,p'*-DDT | 31 | 63 |
| *p,p'*-TDE | 6 | 26 |
| *p,p'*-DDT | 439 | 100 |

[a]Values are means.
*Source*. Mes and Davies (1979).

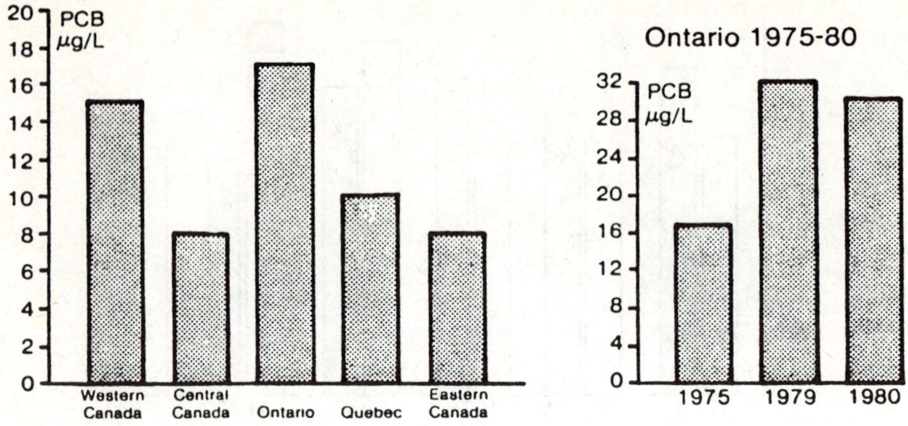


**Figure 24.8.** Arochlor 1260 in mother's milk 1975 Canada-wide. From Mes and Davies (1979).

and found no apparent differences in organochlorine residues between rural and urban populations. They also found no significant differences in human milk samples from year to year (Table 24.6) (Frank et al., 1985). In Ontario, mother's milk was found to contain 0.583 mg/kg PCBs on a milkfat basis, almost 25 times the levels in Ontario cow's milk. Other persistent residues are prevalent in mother's milk (Table 24.4) that parallel the residue patterns found in human body fat.

### 4.3. Human Ecosystem Exposure

Since herring gulls and coho salmon accumulate these residues from their common food supply it seems logical to consider the human exposure routes thoroughly, particularly the food supply.

**Table 24.6 Mean Concentrations and Estimated Exposure to PCBs in Human Milk Samples from Ontario Residents 1971–1985**

| | Concentration | |
|---|---|---|
| Year | Fat (Mean mg/kg) | Whole Milk (Mean mg/kg) |
| 1971–72 | 1.20 | 0.036 |
| 1973–74 | 1.18 | 0.035 |
| 1977 | 0.80 | 0.024 |
| 1978 | 0.066 | 0.002 |
| 1979 | 0.96 | 0.028 |
| 1981 | 0.79 | 0.024 |
| 1983 | 1.24 | 0.037 |
| 1985 | 0.67 | 0.020 |

*Source.* R. Frank (personal communication).

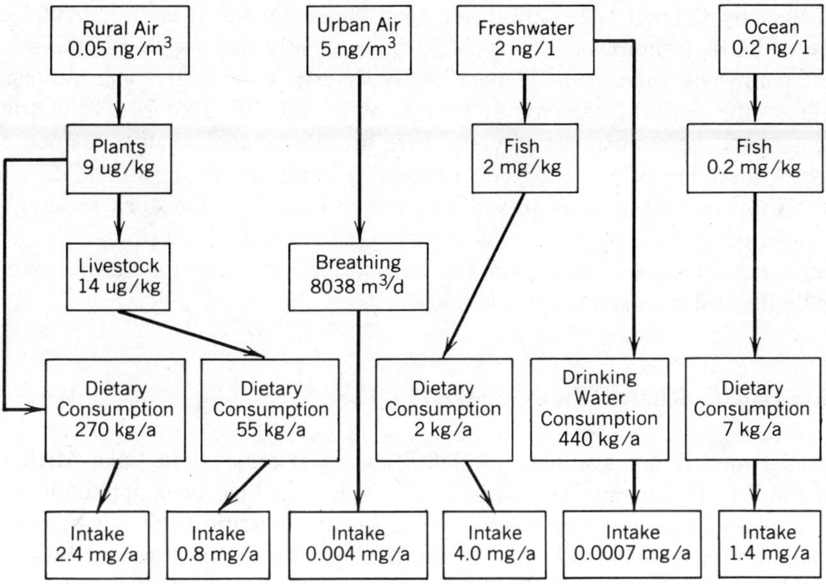


**Figure 24.9.** Pathway evaluation for average exposure to PCBs. From Bennett (1983).

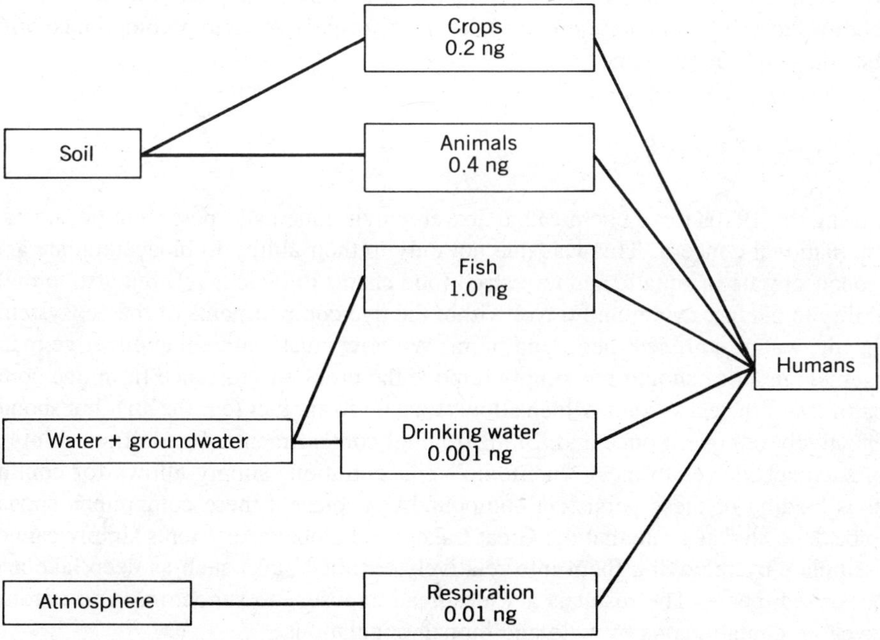


**Figure 24.10.** Fugacity model of annual human exposure for 2,3,7,8-TCDD. From MacKay and Paterson (1985).

A study by Bennett (1983) followed total nonoccupational environmental exposure of PCBs to humans. Figure 24.9 shows clearly that for PCBs, the atmospheric pathway contaminates plants directly, livestock indirectly, and ultimately contributes the major human environmental dosage of PCBs. Exposure from drinking water and the atmosphere directly were inconsequential relative to food exposure. A more recent modeling effort (MacKay et al., 1985) shows that 2,3,7,8-TCDD is transmitted from air to soil and then to plants and livestock resulting in food again being the major route of exposure (Fig. 24.10). The primary basis of human exposure in both cases was the transport via the atmosphere of environmentally derived persistent toxic chemicals.

## 5. LOADING REDUCTION STRATEGIES

Calibrated models are available for DDT and PCB removal in Lake Michigan water and fish (Rogers and Swain, 1983). A PCB model has been applied to forecast the steady-state lake response to various loading conditions. Steady-state conditions imply that time equals infinity and the mass loading is constant. Based on a worst-case situation of a continuous loading of 2400 kg/year (which is less than one-half the present atmospheric loading of PCBs to Lake Michigan), fish species such as coho salmon or lake trout will not achieve PCB concentrations that comply with the present consumption guideline of 2 mg/kg. The PCB loading will have to be reduced to 1520 and 760 kg/year (or to less than one-eighth of the present loading) to bring coho salmon and lake trout, respectively, into compliance with the consumption guideline.

## 6. CONCLUSIONS

During the 1970s toxic chemicals that were environmentally persistent became an international concern. This was due not only to their ability to bioaccumulate and bioconcentrate in aquatic and terrestrial food chains to toxic levels but also to their ability to persist, cycle, and travel within the five compartments of the ecosystem: the air, water, soil, sediment, and biota. We now must realize that effective management practice should not simply remove the problem substance from one compartment of the ecosystem, which allows transfer to another (eg. the air), but should effectively remove it once and for all from all compartments. In addition, dilution of such substances to meet "nontoxic" concentrations simply allows for continuous loading of these persistent compounds. At present these compounds appear to be deposited at a rate that the Great Lakes, and global ecosystems simply cannot assimilate by removing them into relatively terminal sinks such as deep lake and ocean sediments. The result is a widespread exposure to subchronic and chronic levels of contaminants by wild and human populations.

PCBs, which have been studied now for 18 years, should provide guidance and insight in beginning to effectively manage all persistent toxic substances. Loading reduction strategies directed at the major controllable sources seem to be the most scientifically valid option for progress.

## REFERENCES


Bennett, B. G. (1983). Exposure of man to environmental PCBs—an exposure commitment assessment. *Sci. Total Environ.* **29,** 101–111.

Canadian Wildlife Service (1985). *1985 Water Quality Board Report.* Report to the International Joint Commission, Windsor, Ontario.

Doskey, P. V. (1978). Transport of airborne PCBs to Lake Michigan. M. S. Thesis, University of Wisconsin, Madison.

Doskey, P. V., and Andren, A. W. (1980). Modeling the flux of atmospheric PCBs across the air/water interface. *Environ. Sci. Technol.* **15,** 705–711.

Durham, R. W., and Oliver, B. G. (1983). History of Lake Ontario contamination from the Niagara River by sediment radiodating and chlorinated hydrocarbon analysis. *J. Great Lakes Res.* **9,** 160–168.

Eisenreich, S. J., Hollod, G. J., and Johnson, T. C. (1979). Accumulation of polychlorinated biphenyls (PCBs) in surficial Lake Superior sediments. Atmospheric deposition. *Environ. Sci. Technol.* **13,** 569–573.

Eisenreich, S. J., and Hollod, G. J. (1980). Atmospheric input of PCBs to the Great Lakes: Impact on the Lake Superior ecosystem. In S. J. Eisenreich (Ed.), *Atmospheric Input of Pollutants to Natural Waters.* Ann Arbor Sci. Publ., Ann Arbor, MI.

Eisenreich, S. J., Looney, B. B., and Thornton, J. D. (1980). *Assessment of Airborne Organic Contaminants in the Great Lakes Ecosystem,* 1980 Ann. Rep. Great Lakes Science Advisory Board, International Joint Commission, Windsor, Ontario.

Frank, R., Holdrinet, M., Braun, M. E., Thomas, R. L., Kemp, A. L. W., and Jacquet, J. M. (1977). Organochlorine insecticide and PCBs in sediment of Lake St. Clair (1970 and 1974) and Lake Erie (1971). *Sci. Total Environ.* **8,** 205–227.

Frank, R., Thomas, R. L., Holdrinet, M., Kemp, A. L. W., and Braun, H. E. (1979a). Organochlorine insecticides and PCB in surficial sediments (1968) and sediment cores (1976) from Lake Ontario. *J. Great Lakes Res.* **5,** 18–27.

Frank, R., Thomas, R. L., Holdrinet, M., Kemp, A. L. W., Braun, H. E., and Dawson, R. (1979b). Organochlorine insecticides and PCB in the sediments of Lake Huron (1969) and Georgian Bay and North Channel (1973). *Sci. Total Environ.* **13,** 101–117.

Frank, R., Thomas, R. L., Holdrinet, M., and Damiani, V. (1980a). PCB residues in bottom sediments from the Bay of Quinte, Lake Ontario 1972–73. *J. Great Lakes Res.* **6,** 371–376.

Frank, R., Thomas, R. L., Braun, H. E., Rasper, J., and Dawson, R. (1980b). Organochlorine insecticides and PCB in the surficial sediments of Lake Superior (1973). *J. Great Lakes Res.* **6,** 113–120.

Frank, R., Thomas, R. L., Braun, H. E., Gross, D. L., and Davies, T. T. (1981). Organochlorine insecticides and PCB in surficial sediments of Lake Michigan (1975). *J. Great Lakes Res.* **7,** 42–50.

Frank, R., Braun, H. E., and Fleming, G. (1983). Organochlorine and organophosphorus residues in fat of bovine and porcine carcasses marketed in Ontario, Canada from 1969–1981. *J. Food Prot.* **46,** 893–900.

Frank, R., Braun, H. E., Sirons, G. J., Rasper, J., and Ward, G. G. (1985). Organochlorine and organophosphorus insecticides and industrial pollutants in the milk supplies of Ontario -1983. *J. Food Prot.* **48**(6), 499–504.

Glooschenko, W. A., Strachan, W. M. J., and Sampson, R. J. C. (1976). Distribution of pesticides and polychlorinated biphenyls in water, sediments and seston of the Upper Great Lakes 1974. *Pestic. Monit. J.* **10,** 61–67.

Hallett, D. J., and Brooksbank, M. G. (1986). Trends of TCDD and related compounds in the Great Lakes: The Lake Ontario ecosystem. *Chemosphere* **15,** 1405–1416.

Hollod, G. J. (1979). PCBs in the Lake Superior ecosystem: Atmospheric deposition and accumulation in the bottom sediments. Ph.D. Thesis, University of Minnesota, Minneapolis.

Humphrey, H. E. B. (1983). Population studies of PCBs in Michigan residents. In F. M. D'Itri and M. A. Kaman (Eds.), *PCBs: Human and Environmental Hazards.* Ann Arbor Sci. Publ., Ann Arbor, MI, pp. 299–310.

International Joint Commission (1978). *Pollution in the Great Lakes from Land Use Activities.* International Joint Commission, Windsor, Ontario.

International Joint Commission (1982). *Guidelines and register for evaluation of Great Lakes Dredging Projects.* Report of the Dredging Subcommittee to the Water Quality Programs Committee of the Great Lakes Water Quality Board. International Joint Commission, Windsor, Ontario.

International Joint Commission (1985). *1985 Report on Great Lakes Water Quality.* International Joint Commission, Windsor, Ontario.

MacKay, D., and Paterson, S. (1985). *A Fugacity Model of Atmospheric Pathways of Organic Contaminants to the Great Lakes.* Dep. Chem. Eng. Appl. Chem., University of Toronto, Toronto, Ontario.

MacKay, D., Paterson, S., and Cheung, B. (1985). Evaluating the environmental fate of chemicals the fugacity—level III approach as applied to 2,3,7,8, TCDD. Dept. of Chemical Engineering and Applied Chemistry, University of Toronto, Toronto, Ontario. *Chemosphere* **14,**(6/7), 859–863.

Marsalek, J. (1983). *Study of Persistent Toxic Substances in Urban Runoff.* Hydraul. Div., Nat. Water Res. Inst., Burlington, Ontario.

Mes, J., and Davies, D. J. (1979). Presence of polychlorinated biphenyls and organochlorine pesticide residues and absence of polychlorinated terphenyls in Canadian human milk samples. *Bull. Environ. Contam. Toxicol.* **21,** 381–387.

Mes, J., Davies, D. J., and Turton, D. (1982). Polychlorinated biphenyl and other chlorinated hydrocarbon residues in adipose tissue of Canadians. *Bull. Environ. Contam. Toxicol.* **28,** 97–104.

Murphy, T. J. (1985). Atmospheric inputs of chlorinated hydrocarbons to the Great Lakes. In *Toxic Contaminants in the Great Lakes.* Great Lakes Water Quality Board, International Joint Commission, Windsor, Ontario, pp. 52–79.

Murphy, T. J., and Rzeszutko, C. P. (1977). Precipitation inputs of PCBs to Lake Michigan. *J. Great Lakes Res.* **3,** 305–312.

Murphy, T. J., Scjinsky, A., Paolucci, G., and Rzeszutko, C. P. (1980). Inputs of PCB from the atmosphere to Lakes Huron and Michigan. In S. J. Eisenreich (Ed.), *Atmospheric Input of Pollutants to Natural Waters.* Ann Arbor Sci. Publ., Ann Arbor, MI.

Musial, C. J., Hutzinger, O., Zitko, V., and Croker, J. (1974). Presence of PCB, DDE and DDT in human milk in the provinces of New Brunswick and Nova Scotia, Canada. *Bull. Environ. Contam. Toxicol.* **12,** 258–267.

National Academy of Science (NRC) (1979). *Polychlorinated Biphenyls.* NAS, Washington, DC.

National Research Council of Canada (1982). *Polychlorinated Dibenzo-p-dioxins: Criteria for their Effects on Man and His Environment* (NRCC Publ. No. 18574). Associate Committee on Scientific Criteria for Environmental Quality, Ottawa, p. 251.

Norstrom, R. J., Hallett, D. J., and Sonstegard, R. A. (1978). Coho salmon (*Oncorhynchus kisutch*) and herring gulls (*Larus argentatus*) as indicators of organochlorine contamination in Lake Ontario. *J. Fish. Res. Board Can.* **35,** 1401–1409.

Ontario Ministry of the Environment (1983). Data up to and including 1981 in Canada-Ontario Review

Board, Environmental Baseline Report of the Niagara River, November 1981 Update, November 16, 1981 and 1982 data in L. Lind, Troubled waters, May 1982. *Toronto Life*, p. 41.

Ritcey, W. R. G., Savary, G., and McCully, K. A. (1972). Organochlorine insecticide residues in human milk, evaporated milk and some milk substitutes in Canada. *Can. J. Public Health* **63,** 125–132.

Rogers, P. W., and Swain, W. R. (1983). Analysis of polychlorinated biphenyl (PCB) loading trends in Lake Michigan. *J. Great Lakes Res.* **9,** 548–558.

Ryan, J. J., and Williams, D. T. (1983). Analysis of human fat tissue from the Great Lakes area for 2,3,7,8-tetrachlorodibenzo-p-dioxin and -furan residues. *Abstr. Pap.—Chem. Soc., 186th Nat. Meet.* **23**(2), 157–158.

Smith, B. J. (1984). PCB levels in human fluids, Sheboygan case study. M.Sc. Thesis, University of Wisconsin, Madison.

Strachan, W. M. J., and Huneault, H. (1979). PCBs and organochlorine pesticides in Great Lakes precipitation. *J. Great Lakes Res.* **5,** 61–68.

Thomas, R. L., and Frank, R. (1982). PCBs in sediments and fluvial suspended solids in the Great Lakes. *Proc. Int. Workshop PCBs Great Lakes, 1981,* vol. 14, p. 245.

Veith, G. D., Kuehl, D. W., Puglish, F. A., Glass, G. R., and Eaton, J. G. (1977). Residues of PCBs and DDT: In the western Lake Superior ecosystem. *Arch. Environ. Contam. Toxicol.* **5,** 487–499.

Williams, D. T., LeBel, G. L., and Junkins, E. (1984). A comparison of organochlorine residues in human adipose tissue autopsy samples from two Ontario municipalities. *J. Toxicol. Environ. Health* **13,** 19–21.

# PART 3

# SUMMARY AND RECOMMENDATIONS FOR FUTURE RESEARCH

# 25

# PERSISTENT TOXIC SUBSTANCES AND PHYTOPLANKTON IN THE GREAT LAKES

***G-Yull Rhee***

*Wadsworth Center for Labs and Research, New York State Department of Health and School of Public Health Sciences, State University of New York, Albany, New York 12201*

## 1. INTRODUCTION

The health of the phytoplankton community in the aquatic ecosystem is of vital importance, since phytoplankton are the primary producers that form the basis for the food chain. Destabilization of this community thus portends the disruption of the entire ecosystem. The contamination of phytoplankton by toxic pollutants can also ultimately endanger public health when toxicants are transmitted to the upper levels of the food chain, as has been clearly demonstrated for some organic and inorganic contaminants (Biddinger and Gloss, 1984).

Some of the important questions regarding chemical contamination of the Great Lakes are: Have there been changes in phytoplankton communities attributable to chemical pollution? What is the evidence of negative impact? What is the best way to measure the impact of toxic contaminants? Our workshop addressed these questions.

## 2. EVIDENCE OF CHANGES IN PHYTOPLANKTON ASSEMBLAGES

Environmental contamination by pesticides and by organic and inorganic chemicals from industrial discharges has subjected the aquatic ecosystem to new stresses that were never encountered in its evolutionary past. In many cases the effects on phytoplankton are chronic and sublethal, and thus the stress responses are difficult to detect and quantify. It is especially difficult to discern how chemical pollution has changed the Great Lakes phytoplankton community, because historical records of phytoplankton are insufficient and methods have been inconsistent (Vollenweider et al., 1974). These problems are further complicated by the difficulty of separating the effects of the toxic chemicals from the effects of other environmental stresses, such as cultural eutrophication, which have operated simultaneously.

Paleolimnological studies of diatoms preserved in sediments and restudy of old collections from the Great Lakes (Stoermer, 1978) showed clear changes in species distribution. For example, many species, notably the members of the genus *Cyclotella* which had been abundant in the past, were rare in southern Lake Michigan and Lake Ontario. Species indigenous to eutrophic habitats surrounding the Great Lakes, such as *Fragilaria capucina* and *Melosira granulata,* were found in many nearshore waters, suggesting cultural eutrophication as a major cause for their emergence (Stoermer, 1978). The southern basin of Lake Huron also showed a sign of eutrophication with high standing crops (Stevenson, 1985). It is difficult to determine, however, if and how environmental contaminants have modified such changes in species distribution and standing crops.

There is too little information to identify community characteristics that are attributable to pollution by toxic chemicals and that could thus be used as a diagnostic feature. Studies of natural Great Lakes populations experimentally exposed to heavy metal mixtures or sediment elutriates indicated that ultraplankton were more sensitive than larger forms (Munawar et al., 1983) and therefore, it was suggested that the apparent lack of ultraplankton in Lake Ontario near the mouth of the Niagara River might be due to heavy metal contamination (Munawar, 1987). In seawater, on the other hand, chemical pollutants, whether organic or inorganic, favored small flagellates (O'Connors et al., 1978; Menzel, 1980).

## 3. NEGATIVE EFFECTS OF PERSISTENT TOXIC CHEMICALS

### 3.1. Experimental Evidence

Evidence for the detrimental effects of contaminants on phytoplankton and plants in natural waters rests largely on the extrapolation from experimental studies. Bioassays of unialgal cultures (e.g., Butler, 1977; Wright, 1978; Joubert, 1983; Van Coillie et al., 1983; Adams et al., 1985; Wong and Couture, 1986) have demonstrated, among other symptoms, inhibition of growth and/or photosynthesis. When toxic concentrations determined in these experiments, such as $ED_{50}$, were compared with concentrations in natural waters, various degrees of inhibition were indicated for natural assemblages.

In other studies natural populations were exposed to various pollutants in both small-scale (Glooschenko, 1971; Lin and Simmons, 1981; O'Connors et al., 1978; Marshall and Mellinger, 1980; McNaught et al., 1980; Munawar et al., 1983) and mesocosm-scale experiments (Kuiper and Hanstveit, 1984a,b; Wakeham et al., 1983; Menzel, 1980; Thomas et al., 1977). The results suggested that concentrations of these pollutants found in natural waters would not only depress primary production but would also change the species composition, due to the differential sensitivities of assorted species and size classes.

At certain low concentrations of contaminants, enhancement, rather than inhibition, of photosynthesis and growth has been observed (Dunstan et al., 1975; Karydis and Fogg, 1980; McNaught et al., 1980; Lederman and Rhee, 1982b). Whether the effects are inhibitory or stimulatory, any qualitative or quantitative changes in primary producers would engender secondary effects within phytoplankton assemblages as well as on higher trophic levels (Hurlbert, 1975), disrupting the equilibrium of the ecosystem.

### 3.2. Problems in Extrapolating Experimental Results to Natural Waters

Despite the clear demonstration of negative effects by various types of experiments, it is still uncertain how toxic the ambient concentrations of contaminants are to natural populations and to what extent the composition of contemporary phytoplankton communities have been modified by environmental pollutants. The difficulty stems in part from their effects on phytoplankton grazers. The contaminants affect the longevity, reproductive capacity, and respiration of zooplankton, and their filtering and grazing rates (e.g., Geiger and Buikema, 1981; McNaught et al., 1980). Such effects on these grazers would indirectly influence the size and composition of phytoplankton assemblages.

The assessment and prediction of the impact of chemical pollution on phytoplankton communities are further complicated by the fact that phytoplankton can develop resistance to pollutants (Fisher, 1977; Murphy and Belastock, 1980) and can recover from the initial stress over a period of time (e.g., O'Connors et al., 1978; Thomas et al., 1977). Interaction effects among various pollutants in natural waters are also little known. Experimental studies indicate that they can be synergistic (Hutchinson, 1973; Mosser et al., 1974; Wong et al., 1978; Babich and Stotzky, 1983), antagonistic (Hutchinson, 1973; Bartlett et al., 1974), or simply additive (Powers et al., 1975; Christensen and Zielski, 1980). Toxicity is also influenced by environmental factors, such as ionic strength and irradiance (Greene et al., 1975; McNaught et al., 1980).

Bioaccumulation of hydrophotic organic pollutants in algae is by simple physical partitioning. They also partition to particulate matter (e.g., Hiraizumi et al., 1979; Horzempa and Di Toro, 1983) as well as to dissolved organic matter, such as humic material (Herbes, 1977; McCarthy and Jimenez, 1985). Metals and organometallic compounds also behave similarly (P. T. S. Wong, personal communication). Therefore, the extent of bioaccumulation in algae is influenced by the total amount of particulate and dissolved organic matter in the water column. The bioconcentration factor decreases with increasing population density (Harding and Phillips, 1978b; Lederman and Rhee, 1982a; Voice et al., 1983) as the partition coefficient to soil and mineral particles does with increasing concentration of the adsorbents (Hiraizumi et al., 1979; Horzempa and Di Toro, 1983). Probably because of these relationships, the biological effects of hydrophobic compounds, measured as photosynthetic inhibition, were greater at lower population density when the amount of PCBs per volume of culture was held constant in marine diatoms (Harding and Phillips, 1978a) and *Chlorella* (Cole and Plapp, 1974). Conversely, phytoplankton were inhibited by contaminated particulates (Powers et al., 1982; Nau-Ritter et al., 1982), probably due to the redistribution of toxicants by dynamic equilibria among the contaminated particulates, water, and organisms. Therefore the dose–response relationship can greatly vary from place to place. Such variations may not occur if the relationship is expressed on a per biomass basis, but it is difficult, if not impossible, to measure this parameter for natural populations.

There are similar problems in evaluating the effects of heavy metals on Great Lakes phytoplankton. Heavy metals form complexes with dissolved organic ligands, altering their form, activity, and availability (Giesy, 1983; Seritti et al., 1985). Therefore, the toxicity of cadmium varied with qualitative changes in the composition of dissolved organic matter through the alteration of free cadmium activity (Laegreid et al., 1983).

## 4. HOW TO BETTER ASSESS EFFECTS OF TOXIC CHEMICALS

The impact of toxic chemical pollution on Great Lakes phytoplankton is difficult to assess, because it cannot be predicted simply on the basis of concentrations. The impact can be modified by physiological adaptation by organisms at the species level, the sensitivities of individual species at the community level, and effects on phytoplankton grazers at the trophic level. It can also change with the physical and chemical nature of the ecosystem such as the abundance of particulate matter and the presence of other contaminants, humic matter, and other organic ligands. Therefore, the approach to measure the negative impact of toxic pollution can vary from simple to very complex experiments, depending on the the kind of information one seeks.

### 4.1. Parameters to Measure Toxicity

Since metabolic responses to toxic contaminants are diverse, one can also use various physiological parameters to determine toxic effects. However, it is important to choose an appropriate parameter and understand the complexity of physiological responses. Photosynthesis—measured as $O_2$ evolution, $^{14}C$ uptake, or fluorescence yield—is the parameter most frequently used to assess physiological impairment, probably because it is the most important mechanism for primary production. Photosynthesis is also relatively easy to measure. Less often, energy charge and the activities of certain cellular enzymes have also been used (Joubert, 1983; Wong and Couture, 1986). These parameters are useful to determine both short-term and long-term responses.

Toxic effects have also been measured frequently in terms of growth rate, which reflects essentially an integrated long-term response. This parameter is not suitable for a system such as batch enclosures in which growth conditions change rapidly relative to the generation time of organisms, since the growth rate, which requires many generations of growth to measure, cannot be correlated to any definable conditions. This parameter can be used for a system in or approaching a steady state condition such as enclosures with very dilute cell suspensions or continuous cultures.

Whichever parameter is used, it is important to understand the complex cellular responses associated with it. For example, photosynthetic activities can recover or increase after an initial inhibition (Lederman and Rhee, 1982b; Gotham and Rhee, 1982). Thus photosynthetic responses should be measured over a reasonable time period. Alterations in enzyme activity may indicate metabolic inhibition, but they can also indicate the utilization of alternative pathways. The growth rate can initially decrease or increase, followed by the opposite reaction (Lederman and Rhee, 1982b). Therefore, changes in growth rates should also be observed over a period of time like photosynthesis. Cell size may change with growth rates. If this change occurs, the growth-rate responses should be interpreted only in terms of cytokinesis, with no implication for any other cellular events leading to growth.

### 4.2. Bioassays

Extrapolation of the bioassay results to natural assemblages is admittedly hazardous due to the many secondary factors in nature that can modify the toxicity of contaminants as discussed above. However, bioassays are indispensable for the first approximation of toxicity on phytoplankton and are widely used since they are probably the most economical and rapid method for evaluating a wide array of contaminants entering the Great Lakes and differential sensitivities among species. This approach is also valuable for examination of physiological responses and relative toxicity. The design sophistication of bioassay can vary from simple laboratory batch cultures to in situ incubation of natural populations in continuous culture systems (deNoyelles and O'Brien, 1974).

Probably because of its simplicity, the batch-culture system has been utilized most often for bioassays with unialgal cultures or mixed populations despite its many shortcomings (Rhee, 1980; Wong et al., 1983), especially for the study of many toxic chemicals. The well-known problems include progressive nutrient depletion, the accumulation of inhibitory substances, and the artificial patterns in growth phases (lag, logarithmic, and stationary). The ever-changing population density also poses a serious difficulty in investigating chemicals that are not actively transported across the algal membrane. These chemicals accumulate in the cell either by sorption and/or passive transport. Their cellular concentrations are therefore affected by the total area for sorption in the system. In other words, their cellular concentrations would continuously decrease as the population grows, and this decrease would in turn gradually relax the toxic stresses. Observation by Cole and Plapp (1974) and Harding and Phillips (1978a) strongly imply such effects of populations on toxicity.

However, these difficulties can be ameliorated to a large extent by (1) employing very low population densities, (2) keeping at least the initial density constant, (3) completing the experiment within a few generation times, and (4) rigidly standardizing all other culture conditions. The employment of carrier solvents and enclosures with strong sorptive characteristics should be avoided.

Continuous-culture systems, either turbidostats or chemostats, have often been used to overcome many problems of batch cultures (Lederman and Rhee, 1982b; Gotham and Rhee, 1982; Wong et al., 1983). The continuous input and overflow of material resembles in principle the open system of nature. The population can be maintained at a desired level independent of time and the physiological state of the culture can be precisely defined and controlled (Rhee, 1980). Even stable steady states can be obtained in inhibited cultures when two-stage chemostats are utilized (Pirt, 1975). Indeed, understanding the steady state inhibition responses is of critical importance, since in most areas of the Great Lakes, phytoplankton are in a steady state or close to it with respect to contaminants. We know very little of the responses at the metabolic level.

The continuous-culture system is thus an important tool for investigating effects of contaminants on the Great Lakes phytoplankton as well as elucidating underlying mechanisms of toxicity. An understanding of the physiological mechanism of toxicity is of fundamental importance in predicting ecological responses, since they are the basis of phytoplankton responses irrespective of differences in the community composition and geography. The setup and operation requirements of continuous culture, however, are complex and time-consuming. Therefore, this system is not suitable for screening purposes. There are also other drawbacks. For hydrophobic compounds, for example, adsorption to the surface of the tubing between the reservoir and the culture vessel necessitated a direct daily measurement of the initial concentrations in the culture vessel (e.g., Lederman and Rhee, 1982b). For mixed natural assemblages the selective pressure inherent in this system may result in artificial selection of populations.

Species differ in their sensitivity to toxic compounds (e.g., Lederman and Rhee, 1982b; Gotham and Rhee, 1982), and a species sensitive to one chemical may be tolerant of another (Cairns, 1977). Therefore, it is difficult to predict the reactions of natural assemblages with single-species bioassay results. To avoid such difficulties, bioassays are often carried out in situ with natural assemblages in a small batch confinement. The results are a little closer to those expected in natural waters. On the other hand, they are in general specific to the assemblage and by necessity short-term responses. There are other variations in bioassay methods and Wong and Couture (1986), in their excellent review, have discussed them in detail.

### 4.3. Microcosms and Mesocosms

Laboratory microcosms have often been used to simulate natural ecosystems on a small scale. They are made either of natural assemblages with the water of interest, or of standardized components in a chemically defined artificial medium (e.g., Taub et al., 1983; Harrass and Taub, 1985). The standardized microcosm includes algae, zooplankton, bacteria, fungi, and invertebrates in an artificial medium. Sediments, chitin, and cellulose are also added.

On a much larger scale, mesocosms such as MERL (Marine Ecosystem Research Laboratory) on the shore of Narragansett Bay and CEPEX (Controlled Ecosystem Pollution Experiment) in a British Columbia bay can be utilized in the Great Lakes for evaluating toxic contamination effects at the ecosystem level. Since this system can contain many trophic levels, not only the direct effects on phytoplankton populations, but also the secondary effects through altered trophic interactions can be observed. The community metabolism and development of plankton inside such large enclosures follow very closely those of the open waters for relatively long periods with good enclosure replications (Oviatt et al., 1981; Kuiper, 1982). Mesocosms have been especially useful for simulation studies of toxic contamination (e.g., Smith et al., 1982; Reeve et al., 1982; Elmgren and Frithsen, 1982).

These ecosystem approaches also have their own drawbacks. For instance, the results observed in mesocosms or microcosms may be specific to the ecosystem and those of one body of water may not be applicable to another. In these enclosures only the end results of many interactions are observed. Individual reactions can be revealed only with difficulties, if at all. Moreover, a team of specialists in various disciplines is needed for the investigation, and the setup and maintenance requirements of mesocosms are costly.

### 4.4. Recommended Approaches

There is no single approach that can predict the impact of toxic chemicals on Great Lakes phytoplankton at various levels ranging from physiological to ecosystem. An ideal approach is a combination of various methods. For a thorough investigation of selected chemicals, it is ideal to supplement ecosystem level studies of the mesocosm with the examination of physiological responses in batch or continuous cultures at species levels.

In situ bioassays with natural assemblages over a short term (no more than 2–3 days, Barlow et al., 1973) can be used for the cursory assessment of potential species changes in the absence of grazers. Simple and well-defined laboratory bioassays with unialgal cultures are adequate for cellular reactions, screening the numerous toxic contaminants, and comparative studies of different species.

## 5. RESEARCH NEEDS

There is no recommended method for algal toxicity test procedures analogous to the EPA procedure to determine the trophic status of lakes (Miller et al., 1978). The standardization of toxicity test procedures is difficult in part because of the disparate physicochemical behavior of environmental toxicants and poor understanding of their short- and long-term effects and physiological mode of action. However, there is no substitute for the bioassay using a reference species of phytoplankton to study comparative toxicity of various toxicants, and there is urgent need for developing a standard protocol. For such a protocol, the most important are the standardization for the species to be used, their physiological conditions, parameters to be measured, population density, the incubation period, test medium, and the solubilization of toxicants (e.g., without carrier solvents for hydrophobic compounds). Standardization of protocol is also needed for mixed population bioassays in the laboratory and in situ.

There is also a strong need for investigating, using the standard bioassay method, the correlation between algal toxicity and physiochemical properties for a broad range of toxicants such as the relationship found between the toxicity of many hydrocarbons and halogenated hydrocarbons and their solubilities and/or partition coefficients (Hutchinson et al., 1980) or tin compounds (Wong et al., 1982) and their chemical structures. Any such relationship would enable us to easily predict the potential effects of a multitude of chemicals. Mathematical models based on such correlation could be especially helpful in estimating the effects of chemicals entering the aquatic environment, although such models would not provide information on the effects of photo- or biodegradation products, which are often more toxic to phytoplankton than are the parent compounds (McNaught et al., 1980).

It is also important to investigate whether there are any community characteristics indicative of toxic contamination. This may include the identification of resistant and susceptable species or taxonomic groups. Physiological investigations of steady state inhibition responses are also essential. Most of the previous investigations were carried out in transient states while external concentrations were continuously changing. The responses observed were thus essentially transient or non-steady state ones. In most areas of the Great Lakes, however, the time scale of changes in the concentration of toxicants is very long relative to phytoplankton generation time. Therefore, phytoplankton may be considered to be in a steady or pseudosteady state with respect to toxic contaminants and it is important to understand steady state responses.

There is also a need to study the interactive effects of various toxicants and phase partitioning of contaminants between phytoplankton and the aqueous medium. Phase partitioning is in part influenced by the particulate concentrations as well as dissolved organic substances in water and, therefore, phytoplankton toxicity may vary with the seasonal fluctuations of productivity if the input of toxicant remains constant.

## ACKNOWLEDGMENT

I would like to express my appreciation to the following workshop participants for their contribution to the workshop discussions and comments on this manuscript: Frank deNoyelles, Ecology and Systematics, University of Kansas, Lawrence, Kansas; Linda Sicko Goad, Great Lakes Research Division, University of Michigan, Ann Arbor, Michigan; John H. Hartig, International Joint Commission, Windsor, Ontario; Russell G. Kreis Jr., Department of Geology, University of Minnesota, Duluth, Minnesota; Michael Lewis, Environmental Safety Department, Proctor and Gamble Co., Cincinnati, Ohio; Donald C. McNaught, Minnesota Sea Grant Institute, University of Minnesota, St. Paul, Minnesota; Eugene P. Stoermer, Great Lakes Research Division, University of Michigan, Ann Arbor, Michigan; Frieda B. Taub, School of Fisheries, University of Washington, Seattle, Washington; Marc L. Tuchman, Pesticides and Toxic Substances Branch, U.S. Environmental Protection Agency, Chicago, Illinois; and Paul T. S. Wong, Great Lakes Fisheries Research Branch, Canada Centre for Inland Waters, Burlington, Ontario.

## REFERENCES


Adams, N., Goulding, K. H., and Dobbs, A. J. (1985). Toxicity of eight water-soluble organic chemicals to *Selenastrum capricornutum:* A study of methods for calculating toxic values using different growth parameters. *Arch. Environ. Contam. Toxicol.* **14,** 333–345.

Babich, H., and Stotzky, G. (1983). Synergism between nickel and copper in their toxicity to microbes: Mediation by pH. *Ecotoxicol. Environ. Saf.* **7,** 576–587.

Barlow, J. P., Schaffner, W. R., deNoyelles, F., Jr., and Peterson, B. (1973). Continuous flow nutrient bioassays with natural phytoplankton populations. In G. E. Glass (Ed.), *Bioassay Techniques and Environmental Chemistry.* Ann Arbor Sci. Publ., Ann Arbor, MI, pp. 299–319.

Bartlett, L., Rabe, F. W., and Funk, W. H. (1974). Effects of copper, zinc, and cadmium on *Selenastrum capricornutum. Water Res.* **8,** 179–185.

Biddinger, G. R., and Gloss, S. P. (1984). The importance of trophic transfer in the bioaccumulation of chemical contaminants in aquatic ecosystems. *Residue Rev.* **91,** 103–145.

Butler, G. L. (1977). Algae and pesticides. *Residue Rev.* **66,** 19–62.

Cairns, J., Jr. (1977). Quantification of biological assessment of pollution. *J. Water Pollut. Control Fed.* **36,** 650–653.

Christensen, E. R., and Zielski, P. A. (1980). Toxicity of arsenic and PCB to a green alga (*Chlamydomonas*). *Bull. Environ. Contam. Toxicol.* **25,** 43–48.

Cole, D. R., and Plapp, F. W., Jr. (1974). Inhibition of growth and photosynthesis in *Chlorella pyrenoidosa* by polychlorinated biphenyl and several insecticides. *Environ. Entomol.* **3,** 217–220.

deNoyelles, F., and O'Brien, W. J. (1974). The in situ chemostat—a self-contained continuous culturing and water sampling system. *Limnol. Oceanogr.* **19,** 326–331.

Dunstan, W. M., Atkinson, L. P., and Natoli, J. (1975). Stimulation and inhibition of phytoplankton growth by low molecular weight hydrocarbons. *Mar. Biol. (Berlin)* **31,** 305–310.

Elmgren, R., and Frithsen, J. B. (1982). The use of experimental ecosystems for evaluating the environmental impact of pollutants. A comparison of an oil spill in the Baltic Sea and two long-term, low-level oil addition experiments in mesocosms. In G. D. Grice and M. R. Reeves (Eds.), *Marine Mesocosms—Biological and Chemical Research in Experimental Ecosystems.* Springer-Verlag, New York, pp. 153–166.

Fisher, N. S. (1977). On differential sensitivity of estuarine and open ocean diatoms to exotic chemical stress. *Am. Nat.* **111,** 895.

Geiger, J. G., and Buikema, A. L., Jr. (1981). Oxygen consumption and filtering rate of *Daphnia pulex* after exposure to water soluble fractions of naphthalene, phenanthrene, No. 2 fuel oil and coal-tar creosote. *Bull. Environ. Contam. Toxicol.* **27,** 783–789.

Giesy, J. P., Jr. (1983). Metal binding capacity of soft, acid, organic-rich waters. *Toxicol. Environ. Chem.* **6,** 203–224.

Glooschenko, W. A. (1971). The effect of DDT and dieldrin upon carbon-14 uptake by *in situ* phytoplankton in Lakes Erie and Ontario. Proc. 14th Conf. Great Lakes Res. pp. 219–222.

Gotham, I. J., and Rhee, G-Y. (1982). Effects of a hexachlorobiphenyl and a pentachlorophenol on growth and photosynthesis of phytoplankton. *J. Great Lakes Res.* **8,** 328–335.

Greene, J. C., Miller, W. E., Shiroyama, T., and Merwin, E. (1975). Toxicity to the green alga *Selenastrum capricornutum* as a function of phosphorus and ionic strength. In *Biostimulation and Nutrient Assessment Workshop* U.S. Environ. Prot. Agency, EPA 660/3-75-034 pp. 28–43. Washington, DC.

Harding, L. W., Jr., and Phillips, J. H., Jr. (1978b). Polychlorinated biphenyl (PCB) uptake by marine phytoplankton. *Mar. Biol. (Berlin)* **49,** 103–111.

Harding, L. W., Jr., and Phillips, J. H., Jr. (1978a). Polychlorinated biphenyls (PCB) effects on marine phytoplankton photosynthesis and cell division. *Mar. Biol. (Berlin)* **49,** 93–101.

Harrass, M. C., and Taub, F. B. (1985). Comparison of laboratory microcosms and field responses to copper. In T. P. Boyle (Ed.), *Validation and Predictability of Laboratory Methods for the Assessment of Fate and Effects of Environmental Contaminants in Aquatic Ecosystems.* ASTM Special Tech. Publ. 865, Society for Testing and Materials, Philadelphia, Pa., pp. 57–74.

Herbes, S. E. (1977). Partitioning of polycyclic aromatic hydrocarbons between dissolved and particulate phases in natural waters. *Water Res.* **11,** 493–496.

Hiraizumi, Y., Takahashi, M., and Nishimura, H. (1979). Adsorption of polychlorinated biphenyl onto sea bed sediment, marine phytoplankton, and other adsorbing agents. *Environ. Sci. Technol.* **13,** 580–589.

Horzempa, L. M., and DiToro, D. M. (1983). PCB partitioning in sediment-water systems: The effect of sediment concentration. *J. Environ. Qual.* **12,** 373–380.

Hurlbert, S. H. (1975). Secondary effects of pesticides on aquatic ecosystems. *Residue Rev.* **59,** 81–148.

Hutchinson, T. C. (1973). Comparative studies of the toxicity of heavy metals to phytoplankton and their synergistic interactions. *Water Pollut. Res. Can.* **8,** 69–90.

Hutchinson, T. C., Hellebust, J. A., Tam, D., Mackay, D., Mascarenhas, R. A., and Shiu, W. Y. (1980). The correlation of the toxicity to algae of hydrocarbons and halogenated hydrocarbons with their physical-chemical properties. In B. K. Afghan and D. Mackay (Eds.), *Hydrocarbons and Halogenated Hydrocarbons in the Aquatic Environment.* Plenum, New York, pp. 577–586.

Joubert, G. (1983). Detailed method for quantitative toxicity measurements using the green algae *Selenastrum capricornutum.* In J. O. Nriagu (Ed.), *Aquatic Toxicology.* Wiley, New York, pp. 467–486.

Karydis, M., and Fogg, G. E. (1980). Physiological effects of hydrocarbons on the marine diatom *Cyclotella cryptica. Microb. Ecol.* **6,** 281–290.

Kuiper, J. (1982). Ecotoxicological experiments with marine plankton communities in plastic bags. In G. D. Grice and M. R. Reeve (Eds.), *Marine Mesocosms—Biological and Chemical Research in Experimental Ecosystems.* Springer-Verlag, New York, pp. 181–194.

Kuiper, J., and Hanstveit, A. O. (1984a). Fate and effects of 4-chlorophenol and 2,4-dichlorophenol in marine plankton communities in experimental enclosures. *Ecotoxicol. Environ. Saf.* **8,** 15–33.

Kuiper, J., and Hanstveit, A. O. (1984b). Fate and effects of 3,4-dichloroaniline (DCA) in marine plankton communities in experimental enclosures. *Ecotoxicol. Environ. Saf.* **8,** 34–54.

Laegreid, M., Alstad, J., Klaveness, D., and Selp, H. M. (1983). Seasonal variation of cadmium toxicity toward the alga *Selenastrum capricornutum* Printz in two lakes with different humus content. *Environ. Sci. Technol.* **17,** 357–361.

Lederman, T. C., and Rhee, G-Y. (1982a). Bioconcentration of a hexachlorobiphenyl in Great Lakes planktonic algae. *Can. J. Fish. Aquat. Sci.* **39,** 380–387.

Lederman, T. C., and Rhee, G-Y. (1982b). Influence of a hexachlorobiphenyl in Great Lakes phytoplankton in continuous culture. *Can. J. Fish. Aquat. Sci.* **39,** 388–394.

Lin, C. K., and Simmons, M. S. (1981). Effects of pentachlorobiphenyl on growth of nutrient enriched phytoplankton from Lake Michigan. *J. Great Lakes Res.* **7,** 481–485.

Marshall, J. S., and Mellinger, D. L. (1980). Dynamics of cadmium-stressed plankton communities. *Can. J. Fish. Aquat. Sci.* **37,** 403–414.

McCarthy, J. F., and Jimenez, B. D. (1985). Interactions between polycyclic aromatic hydrocarbons and dissolved humic material: Binding and dissociation. *Environ. Sci. Technol.* **19,** 1072–1076.

McNaught, D. C., Griesmer, D., Buzzard, M., and Kennedy, M. (1980). *Inhibition of Productivity by PCB's and Natural Products in Saginaw Bay, Lake Huron.* U.S. Environ. Prot. Agency, Environ. Res. Lab., Duluth, MN.

Menzel, D. W. (1980). Applying results derived from experimental microcosms to the study of natural pelagic marine ecosystems. In J. P. Giesy (Ed.), *Microcosms in Ecological Research* (DOE Symp. Ser. 52 CONF- 781101). Nat. Tech. Inf. Serv., Springfield, VA, pp. 742–751.

Miller, W. E., Greene, J. C., and Shiroyama, T. (1978). The *Selenastrum capricornutum* Printz algal assay bottle test. U.S. Environ. Prot. Agency Rep. EPA-600/9-78-018.

Mosser, J. L., Teng, T.-C., Walther, W. G., and Wurster, C. F. (1974). Interactions of PCBs, DDT and DDE in a marine diatom. *Bull. Environ. Contam. Toxicol.* **12,** 665–668.

Munawar, M. and Munawar, I. F. (1987). Phytoplankton bioassays for evaluating toxicity of in situ sediment contaminants. *Hydrobiologia* **149,** 87–105.

Munawar, M., Mudroch, A., Munawar, I. F., and Thomas, R. L. (1983). The impact of sediment-associated contaminants from the Niagara River mouth on various size assemblages of phytoplankton. *J. Great Lakes Res.* **9,** 303–313.

Murphy, L. S., and Belastock, R. A. (1980). The effect of environmental origin on the response of marine diatoms to chemical stress. *Limnol. Oceanogr.* **25,** 160–165.

Nau-Ritter, G. M., Wurster, C. F., and Rowland, R. G. (1982). Polychlorinated biphenyls (PCB) desorbed from clay particles inhibit photosynthesis by natural phytoplankton communities. *Environ. Pollut., Ser. A* **28,** 177–182.

O'Connors, H. B., Jr., Wurster, C. F., Powers, C. D., Biggs, D. C., and Rowland, R. G. (1978). Polychlorinated biphenyls may alter marine trophic pathways by reducing phytoplankton size and production. *Science* **201,** 737–739.

Oviatt, C., Buckley, B., and Nixon, S. (1981). Annual phytoplankton metabolism in Narragansett Bay calculated from survey field measurements and microcosm observations. *Estuaries* **4,** 167–175.

Pirt, S. J. (1975). *Principles of Microbe and Cell Cultivation.* Wiley, New York.

Powers, C. D., Rowland, R. G., Michaels, R. R., Fisher, N. S., and Wurster, C. F. (1975). The toxicity of DDE to a marine dinoflagellate. *Environ. Pollut.* **9,** 253–262.

Powers, C. D., Nau-Ritter, G. M., Rowland, R. G., and Wurster, C. F. (1982). Field and laboratory studies of the toxicity to phytoplankton of polychlorinated biphenyls (PCBs) desorbed from fine clays and natural suspended particulates. *J. Great Lakes Res.* **8,** 350–357.

Reeve, M. R., Grice, G. D., and Harris, R. P. (1982). The CEPEX approach and its implications for future studies in plankton ecology. In G. D. Grice and M. R. Reeve (Eds.), *Marine Mesocosms—Biological and Chemical Research in Experimental Ecosystems.* Springer-Verlag, New York, pp. 389–398.

Rhee, G.-Y. (1980). Continuous culture in phytoplankton ecology. In M. R. Droop and H. W. Jannasch (Eds.), *Advances in Aquatic Microbiology,* vol. 2. Academic Press, New York. pp. 151–203.

Seritti, A., Pellegrini, D., Barghiniani, C., Ferrara, R., Morelli, E., and Orsini, F. (1985). Influence of phytoplankton on complexing capacity of seawater. *Toxicol. Environ. Chem.* **10,** 299–305.

Smith, W., Gibson, V. R., and Grassle, J. F. (1982). Replication in controlled marine systems: Presenting the evidence. In G. D. Grice and M. R. Reeve (Eds.), *Marine Mesocosms—Biological and Chemical Research in Experimental Ecosystems.* Springer-Verlag, New York, pp. 217–226.

Stevenson, R. J. (1985). Phytoplankton composition, abundance and distribution in Lake Huron. U.S. Environ. Prot. Agency Rep. EPA/905/3-85/004.

Stoermer, E. F. (1978). Phytoplankton assemblages as indicators of water quality in the Laurentian Great Lakes. *Trans. Am. Microsc. Soc.* **97,** 2–16.

Taub, F. B., Read, P. L., Kindig, A. C., Harrass, M. C., Hartmann, H. J., Conquest, L. L., Hardy, F. J., and Munro, P. T. (1983). Demonstration of the ecological effects of streptomycin and malathion on synthetic aquatic microcosms. In W. E. Bishop, R. D. Cardwell, and B. B. Heidolph, Eds., *Aquatic Toxicology and Hazard Assessment:* Sixth Symposium, ASTM Spec. Tech. Pub. 802, American Society for Testing and Materials, Philadelphia, Pa., pp. 5–25.

Thomas, W. H., Holm-Hansen, O., Leibert, D. L. R., Azam, F., Hodson, R., and Takahashi, M. (1977). Effects of copper on phytoplankton standing crop and productivity: Controlled ecosystem pollution experiment. *Bull. Mar. Sci.* **27,** 34–43.

Van Coillie, R., Couture, P., and Visser, S. A. (1983). Use of algae in aquatic ecotoxicology. In J. O. Nriagu (Ed.), *Aquatic Toxicology.* Wiley, New York, pp. 487–502.

Voice, T. C., Rice, C. P., and Weber, W. J., Jr. (1983). Effect of solids concentration on the sorptive partitioning of hydrophobic pollutants in aquatic systems. *Environ. Sci. Technol.* **9,** 513–518.

Vollenweider, R. A., Munawar, M., and Stadelmann, P. (1974). A comparative review of phytoplankton and primary production in the Laurentian Great Lakes. *J. Fish. Res. Board Can.* **31,** 739–762.

Wakeham, S. G., Davis, A. C., and Karas, J. L. (1983). Mesocosm experiments to determine the fate and persistence of volatile organic compounds in coastal seawater. *Environ. Sci. Technol.* **17,** 611–617.

Wong, P. T. S., and Couture, P. (1986). Toxicity screening using phytoplankton. In B. J. Dutka and G. Bitton (Eds.), *Toxicity Testing Using Microorganisms.* CRC Press, Boca Raton, FL, Vol. 2, pp. 79–100.

Wong, P. T. S., Chau, Y. K., and Luxton, P. L. (1978). Toxicity of a mixture of metals in freshwater algae. *J. Fish. Res. Board Can.* **35,** 479–481.

Wong, P. T. S., Chau, Y. K., Kramar, O., and Bengert, G. A. (1982). Structure-toxicity relationship of tin compounds on algae. *Can. J. Fish. Aquat. Sci.* **39,** 483–488.

Wong, P. T. S., Chau, Y. K., and Patel, D. (1983). The use of algal batch and continuous culture techniques in metal toxicity study. In J. O. Nriagu (Ed.), *Aquatic Toxicology.* Wiley, New York, pp. 449–466.

Wright, S. J. L. (1978). Interactions of pesticides with micro-algae. In I. R. Hill and S. J. L. Wright (Eds.), *Pesticide Microbiology.* Academic Press, New York, pp. 535–602.

# 26

# PERSISTENT TOXIC SUBSTANCES AND ZOOPLANKTON IN THE GREAT LAKES

*John E. Gannon*

*International Joint Commission, Windsor, Ontario, Canada*



## 1. INTRODUCTION

The Great Lakes zooplankton community consists of euplanktonic (''true'' planktonic organisms spending all or most of their life history in the water column) and meroplanktonic species (organisms diurnally or seasonally spending part of their life history in the plankton and part in the benthos). Protozoa, Rotifera, and two microcrustacean groups, the Cladocera and Copepoda, are the prominent euplankton in the Great Lakes where they form the critically important intermediate trophic level between phytoplankton and fish. The most numerically and ecologically important meroplanktonic organisms are the oppossum shrimp, *Mysis,* and the deep water amphipod, *Pontoporeia,* who actively migrate into the water column at night and return to or near bottom during the day. Along with the euplankton, these organisms are important links in the food chain, supporting a diverse fish community. Fish larvae are seasonally meroplanktonic and a host of benthic groups (e.g., Ostracoda, Chironomidae) are also meroplanktonic, seasonally or diurnally, especially near the sediment–water interface.

Zooplankton of the Great Lakes have been investigated for over 100 years. Early qualitative studies focused on taxonomy and distribution; early quantitative investigations were prompted by the importance of the zooplankton as a food source for fish. Only in the last several decades has there been a concerted effort to discern the effects of pollution and changes in the balance and composition of planktivorous and predaceous fish populations on zooplankton community composition, structure, and function.

This chapter summarizes our state of knowledge on the effects of pollution on Great Lakes zooplankton and explores methodologies to best measure such effects. For purposes of this workshop, discussions were limited to the euplanktonic cladocerans, copepods, and rotifers because they are best known in terms of ecology and toxicology; long-term data sets, although limited, are also available for these groups of organisms.

## 2. HOW HAS THE ZOOPLANKTON CHANGED?

Limited data indicate that long-term changes in zooplankton composition and abundance have occurred in the Great Lakes. However, cause and effect are difficult to determine because of the paucity of compatible, long-term data sets.

The early studies on Great Lakes zooplankton were conducted mostly in the nearshore waters and were qualitative. Early attempts at quantification are difficult to compare with more recent studies because of uncertainties in taxonomic interpretation and because a myriad of field and laboratory sampling gear has been employed over the years. Moreover, short-term fluctuations in zooplankton abundance and patchiness in distribution can mask long-term trends. In spite of these shortcomings, there is evidence of shifts in the presence or absence of certain species and changes in relative abundance (Davis, 1966; McNaught and Buzzard, 1973; Wells, 1970; Watson, 1974; Gannon and Stemberger, 1978; Gannon, 1981; Evans and Jude, 1986; Evans and McNaught, Chapter 3).

Types of anthropogenic stress on the Great Lakes ecosystem include harvesting of renewable resources, introduction of exotic fish species, physical restructuring of the environment, and many kinds of pollutant discharges. These stresses variously affect the nutrient pool and primary productivity (with subsequent effects on other trophic levels), size distribution, species diversity, and other community structure measures, and system retrogression (Rapport et al., 1985). In spite of the multiplicity of potential causes of change in the zooplankton community, most studies on Great Lakes zooplankton relate changes in community composition to effects of eutrophication. Indicator species, relative abundances, and multivariate statistical analyses of community composition have been widely used to determine the effects of eutrophication on zooplankton, especially along nutrient gradients from eutrophic nearshore waters to comparatively oligotrophic offshore waters (Watson, 1974; Gannon, 1981; Evans and McNaught, Chapter 3). In addition, it has been well known for the past two decades that planktivorous fish can affect zooplankton species and composition by their size-selective feeding habits (Brooks and Dodson, 1965). However, only recently has the extent of influence on the zooplankton been realized in the Great Lakes. In Lake Michigan, for example, recent declines in the population of the alewife, *Alosa pseudoharengus,* an exotic planktivore, apparently have caused substantial species and size shifts in the zooplankton community. A large *Daphnia,* apparently new to the lake, is now predominant in the community and appears to be a highly efficient grazer on phytoplankton, thereby causing an increase in water clarity (Scavia et al., 1986). Management of water quality through food web manipulation and control is currently being debated (Carpenter et al., 1985; Kitchell et al., in press).

## 3. WHAT ROLE HAVE POLLUTANTS PLAYED IN THESE CHANGES?

The effects of toxic contaminants on Great Lakes zooplankton are much more poorly understood (Fitchko, 1986). Some species of zooplankton may be more sensitive to specific toxicants, resulting in spatial or temporal differences in composition and community structure (Gannon and Stemberger, 1978). Mixtures of toxic substances may act synergistically or antagonistically on zooplankton and toxic substances and eutrophication variables may also act synergistically or antagonistically, thereby further complicating interpretation of effects on zooplankton. Most zooplankton studies have been conducted in the offshore waters of the Great Lakes where toxic contaminants are most dilute and effects on zooplankton most difficult to discern. Few studies have been conducted in rivers, harbors, embayments, and connecting channels where concentrations of toxic contaminants are higher. Moreover, functional assays for determining chronic stress on zooplankton have been conducted only in recent years (Evans and McNaught, Chapter 3).

The role of toxic substances in affecting zooplankton community structure in the Great Lakes remains largely unknown. Eutrophication and size-selective fish predation are major confounding factors masking the effects of toxic contaminants. Although continued monitoring of zooplankton populations is recommended, it is readily apparent that further emphasis on functional bioassays is warranted to better elucidate the role toxic pollutants have played in causing the observed changes in Great Lakes zooplankton.

# 4. HOW CAN CHANGES IN ZOOPLANKTON BEST BE MEASURED?

Roop and Hunsaker (1985) classify biomonitoring methods into two broad categories: ecological surveys and toxicity tests. Ecological surveys include qualitative analyses with emphasis on indicator species and integrated assessment using numerical indices, multivariate statistical approaches, and other quantitative measures. Toxicity tests include single-species and multiple-species approaches for detection of acute (lethal) and chronic (sublethal) effects. Chronic tests can be designed to determine growth inhibition, effects on reproduction, and changes in metabolism, behavior, and so forth. Moreover, elements of both ecological survey and toxicity testing can be built into experiments using microcosms and mesocosms in situ or under conditions closely resembling field situations in the laboratory. The challenge is to choose the method or methods most sensitive to demonstrating effects on zooplankton. It must be realized that, in some instances, another group of organisms besides zooplankton should be emphasized in surveys or testing because of their greater sensitivity in demonstrating effects due to certain pollutants. In other instances, zooplankton studies in concert with other biota and with supporting physicochemical data must be employed to best discern effects on the biological community.

## 4.1. Ecological Surveys

### *4.1.1. Long-Term Monitoring*

Davis (1969) aptly stated that we "need to broaden and deepen our Great Lakes investigations so that future generations of investigators will not curse us for leaving them so little information." It is recommended that zooplankton be included in long-term monitoring programs in the Great Lakes and that the long-term program be jealously protected from deletion as resource management priorities shift over the years. As was mentioned before, the Great Lakes has been subjected to a multiplicity of factors potentially influencing zooplankton. Although it is difficult to make one-to-one casual relationships between a pollutant and its impact on zooplankton from field data alone, baseline and benchmark surveys are exceedingly valuable in documenting change. Once change is detected, further studies often of a more experimental nature can be initiated. Consistent sampling even at a few offshore stations representative of the more homogeneous offshore waters will be exceedingly valuable. Often viewed as mundane and routine at the time of sampling, the value of long-term monitoring in the interpretation of ecological change increases substantially with the length of the time series (Gannon et al., 1986).

Besides the atmospheric pathway, pollutants emanate from municipal, industrial, and agricultural point and nonpoint sources in nearshore waters. Effects on zooplankton arguably will be most evident nearest such sources where contaminants are most concentrated. The Great Lakes Water Quality Board of the International Joint Commission has identified 42 Areas of Concern in the Great Lakes, the harbors, river mouths, embayments, and connecting channels where pollution is most severe (International Joint Commission, 1985). Of the 42 sites, 39 are on the list because of identified toxic contaminant problems. It is recommended that these Areas of Concern receive more emphasis in biological community structure analyses, including zooplankton. In addition, sampling from water intakes, where representativencss of nearshore or offshore conditions is determinable, can be a cost-effective method of obtaining long-term data (Evans and Flath, 1984).

#### *4.1.2. Specimen Banking*

Zooplankton samples with value in monitoring long-term change must be saved for future reference. Nearly all of the older Great Lakes zooplankton collections have been lost or destroyed. The inability to verify older taxonomic interpretations has been a major deterrent in our ability to document long-term changes in zooplankton. Specimen banking of zooplankton, and other biota as well, has been previously recommended for retrospective monitoring of toxic contaminants and for taxonomic reanalysis (Beeton and Gannon, 1984), but a mechanism for establishing such a specimen bank has yet to be implemented in the Great Lakes.

### 4.2. Toxicity Tests

Zooplankton have played an important role in the development of the field of toxicology. The single-species bioassay, most often employing the cladoceran, *Daphnia magna,* has been used for decades in dose–response tests on effects of various pollutants on a representative aquatic organism. Interpretations from such laboratory studies, however, are difficult to translate into the mixed effluents and complex biological community composition of the natural environment. These concerns and increasing interest in the effects of pollutants on the ecosystem, the focus of primary management concern, has led to the development of the new field of ecotoxicology (Butler, 1978; Connell and Miller, 1984; Levin and Kimball, 1984). The ecotoxicological approach involves recognition of physicochemical properties of the pollutants, their biogeochemical pathways and fluxes (distribution, transport, and transformation), exposure and uptake, organismal response (lethal and sublethal toxicities, bioaccumulation, biotransformation, and food chain transfer), and population, community, and ecosystem response (modification of structure and function) (Connell, 1986).

From an ecotoxicological perspective, there is a wide spectrum of methods that potentially can be employed to measure the effects of toxic contaminants on biota. As mentioned above, zooplankton field data alone is probably not going to be useful in predicting or measuring toxic effects, but in combination with data from zooplankton toxicity tests and with data from other biological groups such ecological survey data may have greater significance. Toxicity tests can range from measures of single-species response to one chemical to measures of community response to mixed effluents. Single-species tests are recommended, especially if they can be conducted in the field or under situations closely simulating field conditions. The *Ceriodaphnia* bioassay (Mount and Norberg, 1984; Mount, Chapter 11; McNaught and Mount, 1985) appears to be particularly versatile since it can measure lethal and chronic effects using point-source effluents or mixed effluents in ambient water. It can also be employed using elutriate or pore water to test the effects of sediment-related pollutants. Multispecies tests are often suggested because of the different sensitivities to various pollutants expressed by different organisms (Cairns, 1984). Regardless of the test chosen, careful consideration should be given to what is an appropriate control in the experiment.

The microcrustaceans, Cladocera and Copepoda, receive the most attention by zooplankton investigators because they are relatively large and easy to culture, sample, and test and because of their importance in food web dynamics. Indeed, the continued use of *Ceriodaphnia* and mixed species in dose–response tests is recommended. In addition to acute response tests, physiological (feeding and reproduction) and behavioral impairment responses offer potential in discerning chronic effects of toxic contaminants (Evans and McNaught, Chapter 3). However, investigators are encouraged to also work with other zooplankton groups. Rotifers have been shown to be highly sensitive to toxicants and worthy of further ecotoxicological attention (Halbach, 1984). Furthermore, the importance of Protozoa in planktonic food webs is becoming recognized (Porter et al., 1985) and their use in ecotoxicological studies warrants attention.

### 4.3. Microcosms and Mesocosms

The extrapolation of laboratory tests to the field is still a concern expressed by many investigators. Such concern has prompted the use of microcosms and mesocosms to more effectively assess ecosystem response to pollutants and other stresses. Although developed most extensively by marine researchers (Hobbie and Wakeham, Chapter 15; Sullivan and Ritacco, Chapter 16), microcosms and mesocosms are now being successfully applied in freshwater environments (Taub and Crow, 1980; Sanders, 1985). This approach to date has received little emphasis in the Great Lakes for toxicological investigations. It is recommended that tests for toxic effects in the Great Lakes begin with micro/mesocosm experiments with natural assemblages of organisms. Then the organisms determined to be sensitive should be further tested in functional, field-oriented laboratory experiments in an attempt to arrive at specific cause/effect mechanisms. To implement this recommendation, a mesocosm facility (e.g., similar to the Marine Estuarine Research Laboratory (MERL) at Narragansett Bay) is required in the Great Lakes to investigate the effects of toxic contaminants on food web processes and to test hypotheses.

### 4.4. Interdisciplinary Research

It is readily apparent that understanding the extent and effects of toxic contaminants in the Great Lakes cannot be resolved by zooplanktologists looking only at zooplankton data. A strong emphasis in ecotoxicology needs to be given to relations between trophic levels, including human health concerns. Considerable rhetoric is being given to the need for an ecosystem approach to resolve resource management problems (Christie et al., 1986), involving more interdisciplinary, holistic research. As an operational concept, however, the ecosystem approach is poorly defined and rarely achieved (Norstrom, 1986). The toxic contaminants issue demands an interdisciplinary research approach for effective resolution. As retrospectively noted by Gilbertson (1985), our focus of attention on eutrophication during the 1960s and 1970s was likely unwarranted given the actual but unperceived severity of toxic contaminants occurring at that time. As we look to the future, zooplankton research should not focus entirely on toxic contaminants; rather, better understanding of the role zooplankton play in the total context of ecosystem function should dictate research direction and multidisciplinary cooperation.

## ACKNOWLEDGMENTS

The contributions of the workshop participants are gratefully acknowledged: Marlene Evans and Susan Hendricks (University of Michigan), Donald McNaught (University of Minnesota), Andrew Robertson (National Oceanic and Atmospheric Administration), Keith Solomon (University of Guelph), Barbara Sullivan (University of Rhode Island), and Frieda Taub (University of Washington). Student Rapporteur: Scott Bridgeman (University of Minnesota).

## REFERENCES

Beeton, A. M., and Gannon, J. E. (1984). The importance of specimen banking to monitoring the St. Lawrence Great Lakes. In R. A. Lewis, N. Stein, and C. W. Lewis (Eds.), *Environmental Specimen Banking and Monitoring as Related to Banking.* Martinus Nijhoff, Boston, MA, pp. 143–163.

Brooks, J. L., and Dodson, S. I. (1965). Predation, body size and composition of plankton. *Science* **150,** 28–35.

Butler, G. C. (1978). *Principles of Ecotoxicology (Scope 12).* Wiley, New York.

Cairns, J., Jr. (1984). Editorial: Multispecies toxicity testing. *Environ. Toxicol. Chem.* **3,** 1–3.

Carpenter, S. R., Kitchell, J. F., and Hodgson, J. R. (1985). Cascading trophic interactions and lake ecosystem productivity. *BioScience* **35,** 634–639.

Christie, W. J., Becker, M., Cowden, J. W., and Vallentyne, J. R. (1986). Managing the Great Lakes basin as a home. *J. Great Lakes Res.* **12,** 2–17.

Connell, D. (1986). Ecotoxicology: A new approach to understanding hazardous chemicals in the environment. *Search* **17,** 27–31.

Connell, D. W., and Miller, G. J. (1984). *Chemistry and Ecotoxicology of Pollution.* Wiley, New York.

Davis, C. C. (1966). Plankton studies in the largest Great Lakes in the world with special reference to the St. Lawrence Great Lakes of North America. Publ. 14th Great Lakes Res. Div., Univ. Mich. pp. 1–36.

Davis, C. C. (1969). Plants in Lakes Erie and Ontario and changes of their numbers and kinds. *Bull. Buffalo Soc. Nat. Sci.* **25,** 18–44.

Evans, M. S., and Flath, L. E. (1984). Intakes as sampling locations for investigating long-term trends in zooplankton populations. *Can. J. Fish. Aquat. Sci.* **41,** 1513–1518.

Evans, M. S., and Jude, D. J. (1986). Recent shifts in *Daphnia* community structure in southeastern Lake Michigan: A comparison of the inshore and offshore regions. *Limnol. Oceanogr.* **31,** 56–67.

Evans, M. S., and McNaught, D. C. (1988). The effects of toxic substances on zooplankton populations: A Great Lakes perspective. In M. S. Evans (Ed.). *Toxic Contaminants and Ecosystem Health: A Great Lakes Focus.* John Wiley, New York.

Fitchko, J. (1986). A Literature Review of the Effects of Persistent Toxic Substances on Great Lakes Biota. Report to the Health of Aquatic Communities Task Force, Science Advisory Board, International Joint Commission, Windsor, Ontario.

Gannon, J. E. (1981). Zooplankton of the North American Great Lakes. *Verh. Int. Ver. Limnol.* **21,** 1725–1733.

Gannon, J. E., and Stemberger, R. S. (1978). Zooplankton (especially crustaceans and rotifers) as indicators of water quality. *Trans. Am. Microsc. Soc.* **97,** 16–35.

Gannon, J. E., Edwards, C. J., Reynoldson, T. B., and Hartig, J. H. (1986). Indicator approaches used in the Great Lakes International Surveillance Program. In *Monitoring Strategies Symposium,* Oceans 86 Conf. Proc., Vol. 3. Marine Tech. Soc., Washington, DC.

Gilbertson, M. (1985). The Niagara labyrinth—the human ecology of producing organochlorine chemicals. *Can. J. Fish. Aquat. Sci.* **42,** 1681–1692.

Halbach, U. (1984). Populations dynamics of rotifers and its consequences for ecotoxicology. *Hydrobiologia* **109,** 79–96.

Hobbie, J. E., and Wakeham, S. G. (1988). Mesocosms: A tool for studying microbes and effects of behavior of toxic materials. In M. S. Evans (Ed.). Toxic Contaminants and Ecosystem Health: A Great Lakes Focus. John Wiley, New York.

International Joint Commission (1985). *1985 Report on Great Lakes Water Quality.* Great Lakes Water Quality Board, IJC, Windsor, Ontario.

Kitchell, J. F., Evans, M. S., Scavia, D., and Crowder, L. B. (1988). Regulation of water quality in Lake Michigan: Report of the food web workshop. *J. Great Lakes Res.* **14** (in press).

Levin, S. A., and Kimball, K. D., eds. (1984). New perspectives in ecotoxicology. *Environ. Manage.* **8,** 375–442.

McNaught, D. C., and Buzzard, M. (1973). Changes in zooplankton populations in Lake Ontario (1939–1972). Proc. 16th Conf. Great Lakes Res. pp. 76–86.

McNaught, D. C., and Mount, D. I. (1985). Appropriate durations and measures for *Ceriodaphnia* toxicity tests. *ASTM Spec. Tech. Publ.* **STP 891,** 375–381.

Mount, D. I. (1988). Ambient toxicity to assess biological impacts. In M. S. Evans (Ed.). *Toxic Contaminants and Ecosystem Health: A Great Lakes Focus.* John Wiley, New York.

Mount, D. I., and Norberg, T. (1984). A seven-day life-cycle Cladocera toxicity test. *Environ. Toxicol. Chem.* **3,** 425–434.

Norstrom, R. J. (1986). Editorial: Multidisciplinary *vs.* interdisciplinary research. *J. Great Lakes Res.* **12,** 1.

Porter, K. G., Sherr, K. G., Sherr, E. B., Pace, M., and Sanders, R. W. (1985). Protozoa in planktonic food webs. *J. Protozool.* **32,** 409–415.

Rapport, D. J., Regier, H. A., and Hutchinson, T. C. (1985). Ecosystem behaviour under stress. *Am. Nat.* **125,** 617–640.

Roop, R. D., and Hunsaker, C. T. (1985). Biomonitoring for toxics control in NPDES permitting. *J. Water Pollut. Control Fed.* **57,** 271–277.

Sanders, F. S. (1985). Use of large enclosures for perturbation experiments in lentic ecosystems: A review. *Environ. Monit. Assess.* **5,** 55–99.

Scavia, D., Fahnenstiel, G. L., Evans, M. S., Jude, D. J., and Lehman, J. J. (1986). Influence of salmonine predation and weather on long-term water quality trends in Lake Michigan. *Can. J. Fish. Aquat. Sci.* **43,** 435–443.

Sullivan, B. K., and Ritacco, P. J. (1988). Effects of nutrients and copper on copepod population dynamics: A mesocosm study, In M. S. Evans, *Toxic Contaminants and Ecosystem Health: A Great Lakes Focus.* John Wiley, New York.

Taub, F. B., and Crow, M. E. (1980). Synthesizing aquatic microcosms. *DOE Symp. Ser.* **52,** 69–104.

Watson, N. H. F. (1974). Zooplankton of the St. Lawrence Great Lakes—Species composition, distribution and abundance. *J. Fish. Res. Board Can.* **31,** 783–794.

Wells, L. (1970). Effects of alewife predation on zooplankton populations in Lake Michigan. *Limnol. Oceanogr.* **15,** 556–565.

27

# PERSISTENT TOXIC SUBSTANCES AND ZOOBENTHOS IN THE GREAT LAKES

*David S. White*

*Great Lakes Research Division, University of Michigan, Ann Arbor, Michigan 48109*

## 1. INTRODUCTION

This chapter summarizes the discussions and recommendations from a workshop on the effects of persistent toxic substances on benthos in the Great Lakes. The workshop was a discussion of three interrelated questions: (1) How has the benthos, as a biotic group, changed over time? (2) What evidence is there that toxic chemicals have played a role in these changes? (3) How can one better measure changes/effects in the benthos and define/describe the role of toxic substances in these changes? The focus of discussions was on scientific considerations (operational methodologies) and the strengths and weaknesses of various methods applicable to large lakes rather than to the broader aspects of risk management. There is no intent here to extensively review the general literature on benthos and toxics in the Great Lakes. For such reviews, the reader is referred to Nalepa and Landrum (Chapter 4) and Fitchko (1986).

### 1.1. Benthos and the Benthic Environment

Before addressing the specific questions, it was necessary to define the scope of the terms ''benthos'' and the ''benthic environment.'' Benthos was defined as the community of lake bottom dwelling organisms living either in or on the substratum. The primary energy source for the community is the deposition of allochthonous and autochthonous particulate organics, although few general concepts of benthic energy flow and community metabolism in the Great Lakes have been published (Johnson and Brinkhurst, 1971). By definition, the benthos includes microbenthos (bacteria, protozoa, etc.), microphytobenthos (algae, particularly blue-greens and diatoms), macrophytobenthos (rooted aquatic plants), several phyla of meiozoobenthos (nematodes, tardigrades, copepods, rotifers, etc.), larger invertebrate macrozoobenthos (annelids, mollusks, crustaceans, insects, etc.), and certain species or life stages of fish. A complication in defining the benthos comes in the transitory nature of many taxa. Certain species of crustaceans, such as *Mysis,* and fish, particularly the larvae, may be benthic only during portions of their life cycles or over portions of seasonal or diurnal periods.

The depth of the substratum is from the nephloid layer to the greatest depths penetrated by the more deeply burrowing organisms, and the substrata range from bedrock to fine shifting sands to high depositional areas with flocculent organics. The physical environment is a continuum from shallow river mouths, bays, and beaches to the deep profundal zone. With rare exceptions, at least the top few centimeters of sediment in the open Great Lakes are well oxygenated.

The above definitions, or lack of precise definitions, helped to create a framework for discussions. Because the workshop participants primarily dealt with invertebrates (macrozoobenthos or more simply the zoobenthos) in their own research, there is a distinct bias in this paper. Though properly within the benthos, benthic algae, rooted plants, benthic ''zooplankters,'' and fish were left to the other workshops.

### 1.2. Some Relationships between the Zoobenthos Community and the Total Lake System

Other questions arising during preliminary discussions included (1) are the benthos, particularly the invertebrate zoobenthos, a reasonable indicator of the health of an aquatic community, and (2) is this a group of organisms about which there are or should be major concerns? There seem to be several advantages in the use of zoobenthos as indicators of environmental health, which make them probably the best community for field studies in areas of concern. The life spans of many invertebrate taxa (up to a year or more for many species) are usually long enough to determine time-weighted chronic effects but short enough to observe community structure changes within a reasonable period of seasons or years. A second advantage comes in the sedentary habits of most zoobenthos, in contrast with zooplankton and phytoplankton, which have short residence times, and with fish, which are highly mobile.

The sedimentary environment is a repository for virtually all types of toxic substances, even those more volatile in nature, with organic compounds and some metals being sorbed to deposited natural organic matter such as humic and fulvic acids and with other metals being sorbed to deposited clays. Bioturbation processes by the zoobenthos (Robbins, 1982) continually return even once buried compounds to the sediment–water interface making them available for re-release or resuspension into the water column (Karickhoff and Morris, 1985). Additionally, many of the larger zoobenthos, directly exposed to toxic substances in the sediments, are important in the food chain as prey for larger water column invertebrates or fish (Scott and Crossman, 1973).

## 2. HOW THE ZOOBENTHOS HAS CHANGED

### 2.1. Examples of Suspected Changes

Before considering how the zoobenthos has changed, there is the preliminary question of "have there been changes?" There are several examples of changes in zoobenthos abundance, species composition, community structure, and increased teratogenic effects. However, most citations refer to a relatively limited number of publications and types of sites. The literature concerning changes in the open Great Lakes is sparce. Foremost among the studies documenting changes are the more classical publications on "pollution" surveys of Green Bay (Howmiller and Beeton, 1970), Lake Michigan (Robertson and Alley, 1966), Saginaw Bay (Brinkhurst, 1967; Schneider *et al.*, 1969), Western Lake Erie (Carr and Hiltunen, 1965; Hiltunen, 1969), and Lakes Erie and Ontario (Brinkhurst, 1967). Each of the above is primarily descriptive, providing comparisons of changes over a number of years.

There is a greater body of literature associating changes in zoobenthos populations with a variety of suspected environmental alterations in the nearshore regions, particularly in the bays and connecting channels. Some examples are Cook and Veal (1968) for Port Hope, Johnson and Matheson (1968) for Hamilton Bay, Kraft (1979), Kraft and Sypinewski (1981), and Malueg et al. (1984a,b) for the Keweenaw Waterway, Nalepa and Thomas (1976) for Lake Ontario, and Lauritsen et al. (1985) for Lake Michigan.

## 2.2. Problems in Interpreting Suspected Changes

When zoobenthos population changes have been documented, rarely has there been a total elimination of the fauna, particularly zoobenthos, even under the worst of conditions. Rather, most changes are seen in the community composition (usually a reduction in the rarer taxa and an increase in the number of so-called tolerant species), diversity (decreasing), and either a distinct increase or reduction in the total numbers and biomass of organisms (Mozley and Howmiller, 1977). Interpretation of observed changes is difficult because little is known of the environmental requirements even for a few species in the many major zoobenthic taxa (Table 27.1). This is extremely important where closely related species may strongly differ in tolerances or reactions to toxic substances (Resh and Unzicker, 1975).

The oligochaetes have provided some of the better examples of how species composition can change, at least under levels of nutrient enrichment. This can lead to reasonable speculations on the present trophic status of another area in question. For example, nutrient enrichment tends to decrease oligochaete community diversity while increasing density and biomass (Mozley and Howmiller, 1977). Several oligochaete species have received enough attention through association with a range of lake conditions to create some useful trophic indices (Howmiller and Scott, 1977; Mozley and Howmiller, 1977; Lauritsen et al., 1985); however, the indices use only a relatively small portion of the total population and number of species present and then only mature specimens of most of the indicator species can be identified. The distribution of species provides little information on mechanisms behind trophic associations. Such mechanisms eventually must be approached through bioassay and biochemistry (Chapman and Long, 1983; Graney et al., 1986). Some oligochaete species appear to react differently to toxic substances than to nutrient enrichment (Chapman *et al.*, 1982a,b,c), and we have only a rudimentary knowledge of the effects of heavy metals, pesticides, and so forth, on changes in their distribution and community structure.

Other taxa for which there is some understanding of trophic status and change in the fauna include pelecypods (Clarke, 1979; Mackie et al., 1980) and chironomids (Brinkhurst et al., 1968; Saether, 1979; Winnell and White, 1985). Of the major zoobenthic taxa, chironomids eventually may provide the best information on how the benthic environment has changed over longer periods of time. Larval head capsules, like diatoms, are preserved in the sediments and can be identified from cores. Surprisingly few cores have been taken, however, where correlations were made between time, sediment geochemistry, chironomid density, and species composition.

The study by Warwick (1980, Chapter 14) in the Bay of Quinte is probably the best example of the potential of sediment cores. Warwick was able to deter-

**Table 27.1 Present Status of Knowledge for Selected Benthic Taxa in the Laurentian Great Lakes[a]**

| Taxon | Level of Knowledge | | | |
|---|---|---|---|---|
| | Systematics | Life History | Ecology | Water Quality |
| Bacteria | Poor | Poor | Poor | Poor |
| Protozoa | Fair | Poor | Fair to Poor | Poor |
| Benthic plants | | | | |
| Diatoms | Good | Poor | Fair to poor | Fair to poor |
| Blue-greens and greens | Fair | Poor | Poor | Fair to poor |
| Macrophytes | Good | Fair | Fair | Poor |
| Nematoda | Fair | Poor | Poor | Poor |
| Tardigrada | Poor | Poor | Poor | Poor |
| Hydra | Fair | Poor | Poor | Poor |
| Sponges | Fair | Poor | Poor | Poor |
| Turbellaria | Fair | Poor | Poor | Poor |
| Rotatoria | Good | Fair | Fair | Poor |
| Annelida | | | | |
| Polychaeta | Good | Good | Poor | Poor |
| Oligochaete | Good | Fair | Fair | Good |
| Hirudinea | Good | Fair | Fair | Poor |
| Molluska | | | | |
| Pisidiidae | Good | Fair | Fair | Fair |
| Unionidae | Good | Fair | Fair | Fair |
| Gastropoda | Good | Fair | Poor | Poor |
| Acarina | Poor | Poor | Poor | Poor |
| Crustacea | | | | |
| Ostracoda | Fair | Poor | Poor | Poor |
| Copepoda | Good | Good | Good | Fair |
| Harpacticoida | Fair | Poor | Poor | Poor |
| Cladocera | Good | Good | Good | Fair |
| Isopoda | Fair | Poor | Poor | Poor |
| Amphipoda | Fair | Fair | Fair | Fair |
| Mysidacea | Good | Good | Good | Fair |
| Decapoda | Good | Fair | Fair | Poor |
| Insecta | | | | |
| Chironomidae | Good | Fair | Poor | Good |
| *Hexagenia* | Good | Good | Good | Good |
| Other insects | Good to poor | Good to poor | Good to poor | Poor |

[a]The three levels are relative (none could be considered excellent) and reflect knowledge for the entire taxon. In many of taxa listed as poor or fair, a few species are well known; however, the majority of species are not. Relation to water quality includes both nutrient enrichment and toxic substances.

mine patterns over a 2800-year span noting not only a reduction from 152 species to 18 but also a correlation between an increase in head capsule deformities and recent records of chemical input to the bay. Thus, cores would seem to be a greatly under-used tool for comparing present species distributions (or levels of deformities) with patterns from the more distant past, particularly where sedimentation rates are low. The use of cores as a means of establishing baselines from which changes can be extrapolated should receive more emphasis in understanding change. Further, cores may be important in assessing the bioaccumulation potential of in-place pollutants. Cores may be less valuable than actual surveys in documenting zoobenthos changes in the recent past (i.e., the last 20–30 years) because of resolution difficulties and the influence of bioturbation (Robbins, 1982). The application of trophic indices and levels of deformities, while useful in interpretations of general environmental surveys, again do not, by themselves, provide answers to relationships between cause and effect.

Reliable interpretation of "how the zoobenthos has changed" is difficult because we often know very little of past distributions and little of the mechanisms of change. Further, we are only beginning to understand natural cycles, species interactions, the influence of subtle natural changes in the limnology of the lakes, and regional differences. Problems in interpretation are compounded by inconsistent levels of identification and by the accuracy and uncertainty in sampling methodologies, which, as Brinkhurst (1974) points out, may severely limit retrieval of any hard information from past surveys. In many early studies cited above that are available for baseline data, there was relatively loose use of the term "pollution" without differentiation between the effects of nutrient enrichment (eutrophication) and toxic substances. Thus, only in very dramatic cases can we feel confident that observed changes have resulted directly from human influence and not from natural events.

## 3. EVIDENCE THAT TOXIC SUBSTANCES HAVE PLAYED A ROLE IN CHANGES IN THE BENTHOS

At this time, there is little evidence that toxic substances (heavy metals, pesticides, etc.) have directly affected the distribution or abundance of zoobenthic organisms. This is because when a toxic substance has been suspected of affecting populations, rarely have laboratory bioassays been run to corroborate field data, separating the effects of nutrients, organic xenobiontics, and other potentially toxic substances. Virtually all of the published field data on the effects of toxic substances comes from studies with heavy metals. As early as the late 1950s, Surber (1959) proposed that elevated levels of chromium, cyanide and copper affected zoobenthic populations in the Clinton River, Michigan. Similar field examples with heavy metals were published for mercury (Magnuson et al., 1976) and lead (Borgmann et al., 1978). Each of these studies had limitations because of potential effects of cutrophication. In onc of thc fcw studies using laboratory bioassay in conjunction with field data, Malueg et al. (1984a,b) showed that elevated levels of metals were probably responsible for alterations in the zoobenthic populations in the Keweenaw area of Lake Superior.

Surveys in the St. Mary's, St. Clair, and Detroit Rivers indirectly have shown the possibility of pesticides, phenols, oils, napthalene, cyanide, and so on (as well as heavy metals) affecting distributions of zoobenthos (Cook and Veal, 1968; Hamdy et al., 1978; Hiltunen and Schloesser, 1983; Thornley and Hamdy, 1984; Thornley, 1985). Virtually all of these studies rely on field survey data alone to describe changes in distributions and numbers of oligochaetes, chironomids, and the Mayfly, *Hexagenia.* Unfortunately, each of these study areas is influenced by so many potential toxicants that correlations between zoobenthos distribution/abundance and specific toxicants are extremely difficult, even where extensive chemical characterization has been done.

Although still without extensive laboratory validation, there are some good data on deformities in chironomid head capsules which seem more related to toxic substances than to eutrophication alone (Brinkhurst et al., 1968; Cook and Veal, 1968; Hamilton and Saether, 1971; Hare and Carter, 1976; Warwick, 1980, 1988; Crowther and Luoma, 1984). This area of research is quite promising, and further research should be encouraged, particularly the laboratory verification.

Several studies have approached the effect of toxic substances from the standpoint of laboratory bioassays. Specific examples for the Great Lakes include chromium (Batac-Catalan and White, 1983) and PCBs (White, 1987), dissolved solids (summarized by Pringle et al., 1981), and deposited or redeposited "toxic" sediments (Gannon and Beeton, 1969; Prater and Anderson, 1977; Hoke and Prater, 1980; Laskowski-Hoke and Prater, 1984). Although each of these papers reports an effect on a variety of zoobenthic organisms, data either were not translatable to in situ conditions or there was little field validation. Laboratory bioassays should be supplemented, where possible, with on-site testing.

With the rare exceptions of a few studies of heavy metals, there is little evidence directly relating changes in individual zoobenthic organisms, populations, or communities with specific toxic substances. An answer to the question of toxic substances playing roles in observed changes seems most limited by the lack of studies designed to correlate laboratory bioassays with field observations. We are still left with no answer to a greater question—is it even possible to correlate laboratory bioassay with field data?

## 4. TOWARD BETTER MEASUREMENT OF CHANGES/EFFECTS IN THE BENTHOS AND DEFINITION/DESCRIPTION OF THE ROLE OF TOXIC SUBSTANCES IN THESE CHANGES

A primary step must be a better understanding of the taxonomy, life history, and ecology of the species present (Table 27.1). Many advances were made in taxonomy in the past 10–15 years; however, there still are only a few major groups where our ability to identify all the individuals could be considered good. Detailed information on life histories and ecology is available for only a fraction of the Great Lakes species, probably fewer than 10%.

Paleolimnological studies of zoobenthos in the Great Lakes are few. Studies of this kind, when applied to all the lakes, should prove invaluable, not only in giving historic perspective on the distribution and abundance of species but also in examining natural species and community fluctuations.

It is important to recognize that improvements in the measurement and detection of interactions between toxic substance and zoobenthos must be in conjunction with studies of sediment geology, chemistry, fisheries biology, and physical and biological water column limnology. What is required from each specific area will depend upon whether the best approach is at the community level, species level, or the organism level and whether endpoints are at the acute, lethal, or chronic or biochemical (including genetic) levels. Ultimately, the determination of the health of an entire system should be the goal of environmental studies; however, examination of whole systems are at best impractical, and have rarely met with success in pointing out interrelated cause and effect mechanisms. Ideally, one would like to have unlimited funds for environmental monitoring and for creation of appropriate bioassays to test against in situ experiments that eventually would lead to models of whole community structure. There are examples for the Great Lakes where considerable effort and money have been spent on integrated studies of community structure, but we have yet to reach that seemingly elusive endpoint. If we have not achieved desired ends, it is not so much from physical restraints as from divergent individual concepts of how such studies should proceed.

Gaining wider use are approaches that examine the impact of toxicants at the molecular or genetic level. Tests such as the widely used Ames test and alteration of macromolecular patterns (e.g., amino acid ratios, Graney et al., 1986) show an individual organism's reactions to toxic stress. While giving us clues to potential mechanisms behind population changes, molecular alterations have yet to be directly linked to actual alterations in population structure or function.

Thus it would seem obvious that the ideal overall approach to the effects of toxic substances would combine biochemical, behavioral, and acute toxicity bioassay with studies of indigenous benthic lake community changes—a statistically designed, integrated approach. The integrated approach must begin with environmental monitoring, including chemical and biological surveys, to provide adequate baseline data and detect changes beyond normal population fluctuations. If a problem is identified, appropriate laboratory bioassays are run using test species either from the environment or more "standard" sensitive test organisms. Careful choice of test organisms could reduce costs. Multispecies or food chain bioassays would be a logical next step once effects are found at the single-species level. Biochemical and behavioral bioassays should reflect environmental concentrations of the potential toxic substance and should be run in conjunction with the more standard $LD_{50}$ or $LC_{50}$ acute, lethal tests.

The choice of the types of bioassays, whether in lab or in situ, is quite important and should be given considerable thought before implementation. Well-designed experiments will save time and money, as well as give appropriate answers. There is a wealth of information in the Great Lakes literature on bioassay methodology (deletion bioassay, tiered sets of bioassays, etc.) and test organisms (*Hexagenia, Chironomus, Tubifex,* etc.) which should be reviewed (see Fitchko, 1986; Nalepa and Landrum, this volume). It must be kept in mind, however, that just because a bioassay technique or species has been widely used does not necessarily make it appropriate. More important is determining if the bioassay and its endpoints (levels of toxicity, deformity, biochemical change, behavioral alteration) will answer the question being asked. In many cases, this will mean the creation of new bioassay techniques. There remains a need to improve and expand culture techniques that will allow us to increase the number of zoobenthic species available for bioassay.

Studies of toxic substance–zoobenthos interactions in marine environments have not been much more successful than in the Great Lakes, but some have come closer to the ideal, particularly when aimed at a single source or type of contaminant. Chapman and Long (1983) comment upon the integrated approach drawing from what has been learned in Puget Sound. Their paper mirrors much of the discussion and concerns expressed in this workshop. It seems most appropriate here to quote from their ideas on what is involved in the integrated approach, particularly once it has been established that a pollutant or toxic substance is present (also see Fig. 4.5, Nalepa and Landrum, in this volume):

(1) Chemical analysis of bulk sediments;
(2) Infaunal analyses of same sediment samples;
(3) Chemical analyses of benthic or demersal biota;
(4) Sediment bioassays with indigenous (if available) species in which acute toxicity, sublethal effects and genotoxicity are measured; and,
(5) Statistical analyses to determine the extent and significance of correlations among chemical and biological data sets.

## 5. SUMMARY

1. To the question, has the zoobenthos changed in the Great Lakes, it can be concluded that there have been changes, not so much documented in the open lakes as in the bays, connecting channels, and other shallower portions of the lakes. As to how the zoobenthos has changed, we are limited by knowledge of past distributions, taxonomy, ecology, and community structure and function.

2. There is little evidence that toxic substances have directly affected the distribution, abundance, behavior, physiology, or genetics of zoobenthos in the Great Lakes.

3. The integrated approach combining chronic and acute toxicity bioassays with studies of indigenous benthic community changes appears to have the greatest promise for evaluating the effects of persistent toxic substances on Great Lakes zoobenthic populations.

## ACKNOWLEDGMENTS

I would like to express my appreciation to the following people who participated in the discussions and in the review of this paper: Gordon Brown, University of Minnesota, Minneapolis, Minnesota; Jerry Fitchko, IEC Beak Consultants, Mississauga, Ontario; Robert Graney, Michigan State University, East Lansing, Michigan; Janet Lamberson, U.S. EPA, Marine Science Center, Newport, Oregon; Ian Orchard, Environment Canada, Toronto, Ontario; Thomas Nalepa, Great Lakes Environmental Research Laboratory, NOAA, Ann Arbor, Michigan; Trever Reynoldson, International Joint Commission, Windsor, Ontario; Richard Sparks, Illinois Natural History Survey, Havana, Illinois; and William Warwick, National Hydrology Research Center, Saskatoon, Saskatchewan. Contribution number 475 of the Great Lakes Research Division, The University of Michigan.

## REFERENCES

Batac-Catalan, Z., and White, D. S. (1983). Effect of chromium on larval Chironomidae as determined by the optical-fiber light interruption biomonitoring system. *ASTM Spec. Tech. Publ.* **STP 803,** 469–481.

Borgmann, U., Kramar, O., and Loveridge, C. (1978). Rates of mortality, growth and biomass production of *Limnaea palustris* during chronic exposure to lead. *J. Fish. Res. Board Can.* **35,** 1109–1115.

Brinkhurst, R. O. (1967). The distribution of aquatic oligochaetes in Saginaw Bay, Lake Huron. *Limnol. Oceanogr.* **12,** 137–143.

Brinkhurst, R. O. (1974). *The Benthos of Lakes.* Macmillan, London.

Brinkhurst, R. O., Hamilton, A. L., and Herrington, H. B. (1968). Components of the bottom fauna of the St. Lawrence Great Lakes. Univ. Toronto, Great Lakes Inst. Prelim. Rep. 33, 1–50.

Carr, J. F., and Hiltunen, J. K. (1965). Changes in the bottom fauna of western Lake Erie from 1930–1961. *Limnol. Oceanogr.* **10,** 551–569.

Chapman, P. M., and Long, E. R. (1983). The use of bioassays as part of a comprehensive approach to marine pollution assessment. *Mar. Pollut. Bull.* **14,** 81–84.

Chapman, P. M., Farrell, M. A., and Brinkhurst, R. O. (1982a). Relative tolerances of selected aquatic oligochaetes to individual pollutants and environmental factors. *Aquat. Toxicol.* **2,** 47–67.

Chapman, P. M., Farrell, M. A., and Brinkhurst, R. O. (1982b). Relative tolerances of selected aquatic oligochaetes to combinations of pollutants and environmental factors. *Aquat. Toxicol.* **2,** 69–78.

Chapman, P. M., Farrell, M. A., and Brinkhurst, R. O. (1982c). Effects of species interactions on the survival and respiration of *Limnodrilus hoffmeisteri* and *Tubifex tubifex* (Oligochaeta, Tubificidae) exposed to various pollutants and environmental factors. *Water Res.* **16,** 1405–1408.

Clarke, A. H. (1979). Sphaeriidae as indicators of trophic lake stages. *Nautilus* **94,** 178–184.

Cook, W., and Veal, D. (1968). A preliminary biological survey of Port Hope Harbour. Report to the Ministry of the Environment for Ontario, pp. 1–11.

Crowther, R. A., and Luoma, M. E. (1984). Pattern recognition techniques to determine benthic and community responses to industrial input. *Verh. In. Ver. Limnol.* **22,** 2226–2231.

Fitchko, J. (1986). *A Literature Review of the Effects of Persistent Toxic Substances on Great Lakes Biota.* Report of the Health of Aquatic Communities Task Force to the Great Lakes Science Advisory Board, International Joint Commission, Windsor, Ontario.

Gannon, J. E., and Beeton, A. M. (1969). Studies on the effects of dredged materials from selected Great Lakes harbors on plankton and benthos. Spec. Rep., 8, Univ. Wis.-Milwaukee, Cent. Great Lakes Stud., pp. 1–82.

Graney, R. L., Keilty, T. J., and Giesy, J. P. (1986). Free amino acid pools of five species of freshwater oligochaetes. *Can. J. Fish. Aquat. Sci.* **43,** 600–607.

Hamdy, Y., Kinkead, J. D., and Griffiths, M. (1978). St. Mary's River water quality investigations 1973–74. Report to the Ministry of the Environment for Ontario, pp. 1–53.

Hamilton, A. L., and Saether, O. A. (1971). The occurrence of characteristic deformities in the chironomid larvae of several Canadian lakes. *Can. Entomol.* **103,** 363–368.

Hare, L., and Carter, J. C. H. (1976). The distribution of *Chironomus* (s.s.?) *cucini* (*salinaris* group) larvae (Diptera:Chironomidae) in Parry Sound, Georgian Bay with particular reference to structural deformities. *Can. J. Zool.* **54,** 2129–2134.

Hiltunen, J. K. (1969). Distribution of oligochaetes in western Lake Erie. *Limnol. Oceanogr.* **14,** 260–264.

Hiltunen, J. K., and Schloesser, D. W. (1983). The occurrence of oil and the distribution of *Hexagenia* (Ephemeroptera:Ephemeridae) nymphs in the St. Mary's River, Michigan and Ontario. *Freshwater Invertebr. Biol.* **2,** 199–203.

Hoke, R. A., and Prater, B. A. (1980). Relationship of percent mortality of four species of aquatic biota from 96-hour sediment bioassays of five Lake Michigan harbors and elutriate chemistry of the sediments. *Bull. Environ. Contam. Toxicol.* **25,** 394–399.

Howmiller, R. P., and Beeton, A. M. (1970). The oligochaete fauna of Green Bay, Lake Michigan. Proc. 13 Conf. Great Lakes Res. pp. 15–46.

Howmiller, R. P., and Scott, M. A. (1977). An environmental index based on relative abundance of oligochaete species. *J. Water Pollut. Control Fed.* **49,** 809–815.

Johnson, M. G., and Brinkhurst, R. O. (1971). Benthic community metabolism in Bay of Quinte and Lake Ontario. *J. Fish. Res. Board Can.* **28,** 1715–1725.

Johnson, M. G., and Matheson, D. H. (1968). Macroinvertebrate communities of the sediments of Hamilton Bay and adjacent Lake Ontario. *Limnol. Oceanogr.* **13,** 98–111.

Karickhoff, S. W., and Morris, K. R. (1985). Impact of tubificid oligochaetes on pollutant transport in bottom sediments. *Environ. Sci. Technol.* **19,** 51–56.

Kraft, K. J. (1979). *Pontoporeia* distribution along the Keweenaw shore of Lake Superior affected by copper tailings. *J. Great Lakes Res.* **5,** 28–35.

Kraft, K. J., and Sypinewski, R. H. (1981). The effect of sediment copper on the distribution of benthic macroinvertebrates in the Keweenaw Waterway. *J. Great Lakes Res.* **7,** 258–263.

Laskowski-Hoke, R. A., and Prater, B. A. (1984). Multivariate statistical analyses of 96-hour sediment bioassay and chemistry. *Bull. Environ. Contam. Toxicol.* **33,** 400–409.

Lauritsen, D. D., Mozley, S. C., and White, D. S. (1985). Distribution of oligochaetes in Lake Michigan and comments on their use as indices of pollution. *J. Great Lakes Res.* **11,** 67–76.

Mackie, G. L., White, D. S., and Zdeba, T. A. (1980). *A Guide to the Freshwater Mollusks of the Laurentian Great Lakes with Special Emphasis on the Genus* Pisidium. U.S. Environ. Prot. Agency EPA-600/3-80-068, Washington, DC.

Magnuson, J. J., Forbes, A. M., and Hall, R. J. (1976). *An Assessment of the Environmental Effects of Dredged Material Disposal in Lake Superior,* Vol. 3. Mar. Stud. Cent. University of Wisconsin, Madison.

Malueg, K. W., Schuytema, G. S., Krawczyk, D. F., and Gakstatter, J. H. (1984a). Laboratory sediment toxicity tests, sediment chemistry and distribution of benthic macroinvertebrates in sediments from the Keweenaw Waterway, Michigan. *Environ. Toxicol. Chem.* **3,** 233–242.

Malueg, K. W., Schuytema, G. S., Gakstatter, J. H., and Krawczyk, D. F. (1984b). Toxicity of sediments from three metal-contaminated areas. *Environ. Toxicol. Chem.* **3,** 279–291.

Mozley, S. C., and Howmiller, R. P. (1977). *Environmental Status of the Lake Michigan Region,* Vol. 6 (ANL/ES-40). Argonne Natl. Lab., Argonne, IL.

Nalepa, T. F., and Landrum, P. (1988). Benthic invertebrates and contaminant levels in the Great Lakes: effects, fates, and role in cycling. In M. S. Evans (Ed.), *Toxic Contaminants and Ecosystem Health: A Great Lakes Focus.* New York: John Wiley.

Nalepa, T. F., and Thomas, N. A. (1976). Distribution of macrobenthic species in Lake Ontario in relation to sources of pollution and sediment parameters. *J. Great Lakes Res.* **2,** 150–163.

Prater, B. L., and Anderson, M. A. (1977). A 96-hour sediment bioassay of Duluth and Superior Harbor basins (Minnesota) using *Hexagenia limbata, Asellus communis, Daphnia magna,* and *Pimephales promelas* as test organisms. *Bull. Environ. Contam. Toxicol.* **18,** 159–169.

Pringle, C. M., White, D. S., Rice, C. P., and Tuchman, M. J. (1981). *The Biological Effects of Chloride and Sulfate with Special Emphasis on the Laurentian Great Lakes,* Publ. No. 20. Great Lakes Res. Div., University of Michigan, Ann Arbor.

Resh, V. H., and Unzicker, J. D. (1975). Water quality monitoring and aquatic organisms: the importance of species identification. *J. Water. Pollut. Control Fed.* **46,** 9–19.

Robbins, J. A. (1982). Stratigraphic and dynamic effects of sediment reworking by Great Lakes zoobenthos. *Hydrobiologia* **92,** 611–622.

Robertson, A., and Alley, W. P. (1966). A comparative study of Lake Michigan macrobenthos. *Limnol. Oceanogr.* **11,** 576–583.

Saether, O. A. (1979). Chironomid communities as water quality indicators. *Holoarctic Ecol.* **2,** 65–74.

Schneider, J. C., Hooper, F. F., and Beeton, A. M. (1969). The distribution and abundance of benthic fauna in Saginaw Bay, Lake Huron. Proc. 12th Conf. Great Lakes Res., pp. 80–90.

Scott, W. B., and Crossman, E. J. (1973). *Freshwater Fishes of Canada,* Bull. No. 84. Fish. Res. Board Can., Ottawa.

Surber, E. W. (1959). *Cricotopus bicinctus,* an midgefly resistant to electroplating wastes. *Trans. Am. Fish. Soc.* **88,** 111–116.

Thornley, S. (1985). Macrozoobenthos of the Detroit and St. Clair rivers with comparisons to neighbouring waters. *J. Great Lakes Res.* **11,** 290–296.

Thornley, S., and Hamdy, Y. (1984). An assessment of the bottom fauna and sediments of the Detroit River. Report to the Ministry of the Environment for Ontario, pp. 1–48.

Warwick, W. F. (1980). Paleolimnology of the Bay of Quinte, Lake Ontario: 2800 years of cultural influence. *Can. Bull. Fish. Aquat. Sci.* **206,** 1–117.

Warwick, W. F. (1988). Morphological deformities in Chironomidae (Diptera) larvae as biological indicators of toxic stress. In M. S. Evans (Ed.), *Toxic Contaminants and Ecosystem Health: A Great Lakes Focus.* New York, John Wiley.

White, D. S. (1987). Effect of PCB on lake sediment reworking by *Stylodrilus heringianus* (Oligochaeta; Lumbriculidae). *Hydrobiologia* (in press).

Winnell, M. H., and White, D. S. (1985). Trophic status of southeastern Lake Michigan based on the Chironomidae (Diptera). *J. Great Lakes Res.* **11,** 540–548.

# 28

# PERSISTENT TOXIC SUBSTANCES AND THE HEALTH OF FISH COMMUNITIES IN THE GREAT LAKES*

***Wayne A. Willford*†**

*U. S. Fish and Wildlife Service, Ann Arbor, Michigan 48105*

**1. Introduction**
**2. How has your biotic group changed?**
**3. What evidence is there that toxic chemicals have played a role in these changes?**
3.1. For open-lake fish populations
3.2. For nearshore fish populations
**4. How can you better measure changes/effects in your biotic group and define/describe the role of toxic substances in these changes?**
**References**

*National Fisheries Center–Great Lakes Contribution 662.
†Current address: U.S. Environmental Protection Agency, Great Lakes National Program Office, 230 South Dearborn Street, Chicago, IL 60604.

## 1. INTRODUCTION

Of all the changes that have occurred in the Great Lakes, those associated with changes in fish populations and the fisheries they support are perhaps the most commonly noticed and discussed. This is particularly true in the case of toxic substances due to frequent biomagnification of the chemicals through the aquatic food chain and measurement of highly elevated concentrations in fish tissue. As a result, fish are the medium in which problems with toxic substances are initially detected and they have become the medium most routinely used to monitor the levels and trends of toxic substances in the aquatic system. The findings have, in turn, caused the closure of fisheries and the issuance of consumption advisories for certain species and areas of the Great Lakes along with increased public awareness of the problem.

Less well understood, however, is the role, if any, of toxic substances in bringing about changes in the productivity and status of fishery resources. At the "Workshop on Persistent Toxic Substances and the Health of the Aquatic Community," the Fish Biota Discussion Subgroup considered the issue of toxic substances by responding to the three questions addressed to it by the Workshop Organizing Committee as described earlier in the introduction to these proceedings. In responding to these questions, the Subgroup members took into account the earlier, formal presentations made during the Symposium, the published literature applicable to the subject area, and personal knowledge gained from research on Great Lakes fisheries or contaminants.

The major discussions, conclusions, and recommendations of the Subgroup summarized here were developed during these deliberations and are presented under the major headings of the questions asked of the Subgroup.

## 2. HOW HAS YOUR BIOTIC GROUP CHANGED?

As summarized by Hartman (Chapter 5) fish communities of the Great Lakes have changed dramatically since 1900. The changes have been well documented in the literature (e.g., Berst and Spangler, 1973; Christie, 1973; Hartman, 1973; Lawrie and Rahrer, 1973; Wells and McLain, 1973). Although the changes differed in degree and type, depending on the lake, the end result has been a general decline in the highly valued, native fish species such as lake trout (*Salvelinus namaycush*), whitefish (*Coregonus clupeaformis*), lake herring (*Coregonus artedii*), ciscoes (*Coregonus spp.*), walleye (*Stizostedion vitreum vitreum*), and yellow perch (*Perca flavescens*). Among the most dramatic of the changes were the extinction of blue pike (*Stizostedion vitreum glaucum*) in Lakes Erie and Ontario, the near extinction of lake trout in all but Lake Superior, and the loss of several coregonid species throughout the lakes. The decline in high-valued species was frequently accompanied or followed by an increase in lower valued species such as common carp (*Cyprinus carpio*), suckers (*Catostomus spp.*), rainbow smelt (*Osmerus mordax*), and alewife (*Alosa pseudoharengus*).

The rapid spread of nonnative rainbow smelt and alewife populations following their inadvertent introduction into the Great Lakes brought about the stocking of Pacific salmon (*Oncorhynchus kisutch*) and the reintroduction of Atlantic salmon (*Salmo salar*) in an effort to control populations of these forage species. The stocked salmon not only helped to control the smelt and alewife but further support highly successful recreational fisheries. Efforts to rehabilitate lake trout by extensive stocking have resulted in the presence of relatively large lake trout populations throughout their former range in the Great Lakes. Unfortunately, stocked lake trout have failed to reproduce in significant numbers in any of the lakes except Lake Superior and their populations are currently maintained by continued extensive stocking in each of the lakes. Stocked salmon and lake trout have further been shown to frequently accumulate residues of persistent toxic substances in excess of established food guidelines. The presence of these residues has resulted in consumption advisories in several of the Great Lakes and concern about the possible effects of toxic substances on lake trout reproduction.

## 3. WHAT EVIDENCE IS THERE THAT TOXIC CHEMICALS HAVE PLAYED A ROLE IN THESE CHANGES?

The changes and declines in populations of desirable fish species in the Great Lakes mentioned above vary between species and lake and are believed to have been caused by a variety of factors including overfishing, predation by sea lamprey (*Petromyzon marinus*), competition from smelt and alewife, introgressive hybridization, deforestation, siltation and damming of tributaries, draining of the Great Black Swamp, eutrophication, and the introduction of toxic substances. With such a long list of factors operating independently and collectively, the identification of the major causative factor in the decline of a specific species has always been problematic.

In regard to the specific question of the role of toxic chemicals, the Subgroup was unable to link any of the reported changes in biota directly to toxic chemicals. However, the consensus was that several of the more recent changes (i.e., post-industrialization) have some degree of contaminant interaction, but that current contaminant monitoring and fishery assessment programs lack the focus or sensitivity required to demonstrate those interactions. Nevertheless, several changes in the open-lake and nearshore fish populations are thought to be associated with toxic chemicals. These changes are briefly summarized here.

### 3.1. For Open-Lake Fish Populations

1. The presence of toxic chemical residues in fish at levels exceeding food guidelines has reduced the marketability of certain species and, in turn, influenced fisheries (both negatively and positively) and ultimately fish populations. For example, the discovery of mercury residues in walleye in excess of existing food

guidelines in 1970 resulted in major restrictions on their harvest from Lake St. Clair and western Lake Erie. At that time, walleye populations in this region of the Great Lakes were severely depleted due, at least in part, to heavy exploitation. During the decade following harvest restrictions, populations rebounded dramatically and the walleye resource in the region now equals or exceeds historic levels. On the negative side, however, the restrictions on walleye harvest during the 1970s were accompanied by a shift in the fishery from walleye to yellow perch and further aggravated the problem of declining yellow perch populations in Lake Erie (Nepszy, 1977).

2. There have been recurrent problems with egg viability or early-life survival of several species of salmonids stocked into the Great Lakes. Such problems have been reported in four important species: coho salmon (Johnson and Pecor, 1969; Willford et al., 1969), chinook salmon (Giesy et al., 1986; Skea et al., 1985), lake trout (Mac et al., 1985; Willford et al., 1981), and steelhead trout (Skea et al., 1985). In each incidence there has been at least indirect evidence of contaminant association.

3. Induction of mixed function oxidases (MFOs) has been shown to be occurring in certain fish from various areas of the Great Lakes (Melancon et al., Chapter 10). Although the significance of this induction to the fish is not well understood, laboratory studies have shown that MFO induction can result from exposure to a variety of chemical contaminants (Addison, 1984). Observations of inductions in feral fish provides direct evidence that the fish are responding to contaminant exposure.

4. Coho salmon from certain areas of the Great Lakes have been shown to be hyperthyroid (Leatherland and Sonstegard, 1984). Although the exact cause of this hyperthyroidism is unclear, an association with contaminants has been suggested.

### 3.2. For Nearshore Fish Populations

1. Certain harbors, tributaries, and connecting channels of the Great Lakes have been shown to contain fish populations with an elevated incidence of tumors, for example, external and internal tumors in brown bullheads (Baumann et al., 1982) and in suckers and carp (Sonstegard and Leatherland, 1984). Carcinogenic chemicals in sediments from the contaminated areas have been implicated as the cause.

2. Major fish kills and associated population changes have occurred at several nearshore areas due to chemical spills or leaks. Among the areas of past fish kills discussed by the Subgroup were the Spanish River and Redrock areas of Ontario.

3. There have been major changes in the benthic and planktonic populations in contaminated harbors and tributaries of the Great Lakes. These changes are assumed to have caused concommitant effects on fish populations through changes in food availability and loss of spawning habitat (Francis et al., 1979).

Considering the above listed changes in open-lake and nearshore fish populations, the Subgroup concluded that although there are relatively few examples of toxic chemicals being directly linked to population changes, there is considerable circumstantial evidence that fish populations have been and likely continue to be effected by contaminants. The perception therefore exists that there is a direct link between toxic chemicals and population changes in the Great Lakes but that appropriate searches for this link have not been made. The proof of that perception awaits the development of research, monitoring, and assessment programs with sufficient focus and sensitivity to identify and quantify the linkages.

## 4. HOW CAN YOU BETTER MEASURE CHANGES/EFFECTS IN YOUR BIOTIC GROUP AND DEFINE/DESCRIBE THE ROLE OF TOXIC SUBSTANCES IN THESE CHANGES?

The Fish Biota Discussion Subgroup concluded that there were at least three major impediments to solving the complex problem of measuring changes in fish populations and ascribing causes, including toxics:

1. The lack of an integrated, coordinated ecosystem approach to resource assessment.
2. The lack of centralized data storage and summarization of monitoring results that would allow interpretation of commonalities or significance of changes at various ecosystem levels.
3. Inadequate interaction and communication between fish population biologists and toxicologists with the result that each group tends to interpret their results and findings within their own discipline.

The current impediments, as perceived by the Subgroup, thus centered on aspects of policy, institutional arrangements, and communication rather than on the scientific or methodological aspects of measuring cause–effect relationships. This is not to say that scientific or methodological aspects are not a problem; instead there is a recognition that the problem of measuring changes and ascribing causes in heavily used fisheries undergoing simultaneous influence from multiple biotic and abiotic stressors is complex. Resolution of the problem clearly requires well coordinated, integrated, and communicated efforts by several agencies and scientific disciplines. To improve resource assessment and measures of causality for observed changes, the Subgroup recommended seven action items:

1. Development of fully integrated and coordinated monitoring programs (biotic and abiotic) at several ecosystem levels that will aid in the interpretation of changes within individual trophic levels.

2. Development by the International Joint Commission and the Great Lakes Fishery Commission of a centralized system for data storage and summarization

of monitoring results to allow interpretation of commonalities and significances of changes at various ecosystem levels.

3. Development of more formal linkages and communication between population dynamicists and toxicologists, with increased involvement of toxicologists in identified population problems.

4. Implementation of more routine measurement of fish health indicators during population assessment activities. The impetus must come from the management level to force development of such measurements within field programs. The measurements can occur either as part of routine, assessment activities or through development of a special team. In addition to ongoing measurements of recruitment, growth, and year-class strength, the need is for measurements of the incidence of abnormalities (e.g., tumors), disease, and parasites; egg viability; and early-life survival. Because reproduction is the major physiological and population bottleneck, special emphasis must be placed on measuring reproductive success and early-life survival both in the field and under controlled conditions.

5. Continuation of research on clinical and biochemical measurements of stress and on the mechanisms of toxic action in fish. Included is the immediate application of available clinical measurements to known population problems. The Subgroup anticipates that no single method will satisfy the need, but rather that a suite of methods will be needed depending on the type of population change observed and the types of chemicals or other stresses involved.

6. Continuation of chemical residue monitoring programs in fish populations but in a more streamlined and coordinated fashion than in the past.

7. Selection of additional indicator organisms for use in more effectively measuring changes in nearshore and benthic communities. Human-induced changes, particularly those involving toxic substances, are most easily detected in nearshore areas.

## ACKNOWLEDGMENTS

This report represents a summary of discussions and conclusions of the Fish Biota Discussion Subgroup at the "Workshop on Persistent Toxic Substances and the Health of the Aquatic Community." Without the outstanding contributions of the following discussants, this report would not have been possible: Victor Cairns, Canada Department of Fisheries and Oceans; Larry Crowder, North Carolina State University; John Giesy, Michigan State University; Wilbur Hartman, U.S. Fish and Wildlife Service; Michael Mac, U.S. Fish and Wildlife Service; Mark Melancon, Mount Sinai Medical Center; and John Sprague, University of Guelph. Rapporteur for the Subgroup was Raymond Newman, University of Minnesota.

## REFERENCES

Addison, R. F. (1984). Hepatic mixed function oxidase (MFO) induction in fish as a possible biological monitoring system. In V. W. Cairns, P. V. Hodson, and J. O. Nriagu (Eds.), *Contaminant Effects on Fisheries*. John Wiley, New York, pp. 51–60.

Baumann, P. C., Smith, W. D., and Ribick, M. (1982). Hepatic tumor rates and polynuclear aromatic hydrocarbon levels in two populations of brown bullheads (*Ictalurus nebulosis*). In M. Cooke and A. J. Dennis (Eds.), *Physical and Biological Chemistry*. Battelle, Columbus, OH, pp. 93–102.

Berst, A. H., and Spangler, G. R. (1973). Lake Huron—The ecology of the fish community and man's effects on it. Great Lakes Fish. Comm. Tech. Rep. 21, 1–41.

Christie, W. J. (1973). A review of the changes in the fish species composition of Lake Ontario. Great Lakes Fish. Comm. Tech. Rep. 23, 1–55.

Francis, G. R., Magnuson, J. J., Regier, H. A., and Talhelm, D. R. (1979). Rehabilitating Great Lakes ecosystems. Great Lakes Fish. Comm. Tech. Rep. 37, 26–27.

Giesy, J. P., Newsted, J., and Garling, D. L. (1986). Relationships between chlorinated hydrocarbon concentrations and rearing mortality of chinook salmon (*Oncorhyncus tshawytscha*) eggs from Lake Michigan. *J. Great Lakes Res.* **12**(1), 82–98.

Hartman, W. L. (1988). Historical changes in the major fish resources of the Great Lakes. In M. S. Evans (Ed.), *Toxic Contaminants and Ecosystem Health: A Great Lakes Focus*. John Wiley, New York.

Johnson, H. E., and Pecor, C. (1969). Coho salmon mortality and DDT in Lake Michigan. Trans. North Am. Wildl. Nat. Resour. Conf. 34, 159–166.

Lawrie, A. H., and Rahrer, J. F. (1973). Lake Superior—A case history of the lake and its fisheries. Great Lakes Fish. Comm. Tech. Rep. 19, 1–69.

Leatherland, J. F., and Sonstegard, R. A. (1984). Pathobiological responses of feral teleosts to environmental stressors: Interlake studies on the physiology of Great Lakes salmon. In V. W. Cairns, P. V. Hodson, and J. O. Nriagu (Eds.), *Contaminant Effects on Fisheries*. John Wiley, New York, pp. 115–149.

Mac, M. J., Edsall, C. C., and Seelye, J. G. (1985). Survival of lake trout eggs and fry reared in water from the upper Great Lakes. *J. Great Lakes Res.* **11**(4), 520–529.

Melancon, M. J., Binder, R. L., and Lech, J. J. (1988). Environmental induction of monooxygenase activity in fish. In M. S. Evans (Ed.), *Toxic Contaminants and Ecosystem Health: A Great Lakes Focus*. John Wiley, New York.

Nepszy, S. J. (1977). Changes in percid populations and species interactions in Lake Erie. *J. Fish. Res. Board Can.* **34,** 1861–1868.

Skea, J. C., Symula, J., and Niccoli, J. (1985). Separating starvation losses from other early feeding fry mortality in steelhead trout (*Salmo gairdneri*), chinook salmon (*Oncorhyncus tshawytscha*) and lake trout (*Salvelinus namaycush*). *Bull. Environ. Contam. Toxicol.* **35,** 82–91.

Sonstegard, R. A., and Leatherland, J. F. (1984). Comparative epidemiology: The use of fishes in assessing carcinogenic contaminants. In V. W. Cairns, P. V. Hodson, and J. O. Nriagu (Eds.), *Contaminant Effects on Fisheries*. John Wiley, New York, pp. 223–231.

Wells, L., and McLain, A. L. (1973). Lake Michigan—Man's effects on native fish stocks and other biota. Great Lakes Fish. Comm. Tech. Rep. 20, 1–55.

Willford, W. A., Sills, J. B., and Whealdon, E. W. (1969). Chlorinated hydrocarbons in the young of Lake Michigan coho salmon. *Prog. Fish-Cult.* **31**(4), 220.

Willford, W. A., Bergstedt, R. A., Berlin, W. H., Foster, N. R., Hesselberg, R. J., Mac, M. J., Passino, D. R. M., Reinert, R. E., and Rottiers, D. V. (1981). Introduction and Summary. U.S. Fish Wildl. Serv. Tech. Pap. 105, pp. 1–7.

# 29

# PERSISTENT TOXIC SUBSTANCES AND BIRDS AND MAMMALS IN THE GREAT LAKES

*H. J. Harris*

*University of Wisconsin Green Bay, Green Bay, Wisconsin 54302*



## 1. INTRODUCTION

Two groups of warm-blooded vertebrates inhabit Great Lakes ecosystems: birds and mammals. Most are terrestrial creatures rather than aquatic, but many are dependant upon and interact with the Great Lakes in many ways. Many birds and a few mammals (including humans) may be terminal predators in the aquatic food chain. Because of their terminal position, they become integrators of the aquatic food chain and a repository for microcontaminants. In addition to their shared trophic position with humans, birds and other mammal species have significant economic as well as aesthetic value. Consequently, an assessment of the effects of toxic chemicals on the health of the Great Lakes ecosystems would be incomplete without consideration of birds and mammals. This chapter summarizes the results of a one-day workshop held in conjunction with a symposium on the effects of toxic substances in the Great Lakes.

The Birds and Mammals Task Group (BMTG) was asked to address the same three questions as were task groups on other biotic components:

- How have bird and mammal populations on the Great Lakes changed?
- What evidence is there that toxic chemicals, particularly persistent toxic substances, have played a role in these changes?
- How can changes and effects in birds and mammals be better measured to more clearly identify and define the role of toxic substances in these changes?

While our focus is on changes in bird and mammal populations owing to contamination by persistent toxic substances, it seems best to first address what is known about changes in these populations in general.

## 2. WHAT DO WE KNOW IN GENERAL ABOUT BIRD AND MAMMAL POPULATION CHANGES ON THE GREAT LAKES?

To our knowledge, there are no comprehensive reviews of how bird and mammal populations in the Great Lakes Basin have changed over the past 50 years. More to the point, our immediate concern is with bird and mammal populations that are linked intimately to the Great Lakes as an aquatic ecosystem. That link, of course, is through the aquatic food chain with the potential for biomagnification of chloroorganic compounds. This general phenomenon has been recognized and documented for over 20 years (Keith, 1966; Meeks, 1968). Regardless, little has been done over the past two decades to systematically census aquatic dependent bird and mammal populations of the Great Lakes. It was not until 1976 that the U.S. Fish and Wildlife Service attempted to survey colonial nesting bird populations of the U.S. Great Lakes in any comprehensive manner (Scharf et al., 1979). The Canadian Fish and Wildlife Service has conducted a somewhat more intensive survey over a 5-year period.

Comparatively much less is known about mammal populations on the Great Lakes than bird populations. There are, of course, many fewer fish eating mammals than bird species. Otter (*Lutra canadensis*) and mink (*Mustela vison*) are the most likely species to be affected by toxic chemicals because of their food habits. Racoons (*Procyon loter*) may be at risk, but likely less so than otter or mink. In addition, unlike birds, mammals do not occur in large colonies. They are relatively dispersed in their environment and not readily observable. Their young are often hidden in burrows or cavities. Censusing mammal populations is much more difficult than censusing bird populations, particularly colonial nesting birds. Thus, it is not surprising that more information exists on birds on the Great Lakes than on mammals. Consequently, the remainder of this report will focus primarily on birds rather than mammals.

It is obvious that virtually nothing can be concluded about impacts of toxic substances on populations if there is no assessment of the performance or measure of change in the populations themselves. The BMTG observed that avian populations on the Great Lakes are expected to change over time as the rule not the exception. Population characteristics may vary in a number of ways. Sex ratios and age structure may shift over time, geographic distributions may change, age-specific natality and mortality rates may be altered. These changes are part of the "dynamics" of a population that ultimately affect the number of individuals making up the population.

Numbers of breeding pairs or individuals in local populations change for a wide variety of reasons, most of which are not even remotely connected to microcontaminants and toxicity. Populations may change numerically from natural causes on a large geographic base, lakewide or perhaps even systemwide if the causative factor is operating on a lakewide or systemwide scale. Water level change is one systemwide phenomenon that can impact some species positively and others negatively. For example, Forster's tern (*Sterna Foresteri*) populations are altered by changes in Great Lakes water levels because water levels affect the vegetation of coastal marshes and thus the condition of the nesting habitat of the tern (Harris and Trick, 1979; Harris et al., 1981; Trick, 1982; Scharf and Shugart, 1984).

Another obligate piscivore, the double-crested cormorant (*Phlacrocorax auritus*) has experienced marked population fluctuations on the Great Lakes over the past 35 years. Anderson and Hamerström (1967) concluded from historical evidence that the cormorant prior to 1950 was a common migrant and relatively common breeding bird in Wisconsin. From 1950 to 1967, they assessed the population to "have declined greatly in number." They attributed these declines to habitat loss and human disturbance. They did not rule out the possibility of pesticide contamination. Interestingly, Weseloh et al. (1983) report that double-crested cormorants nesting on the Great Lakes, and in particular on Lake Huron in 1972-73, had eggs that were more highly contaminated with DDE, PCBs, and mercury than cormorant eggs from anywhere else in Canada. On Lake Michigan and Green Bay nesting cormorants declined through the late 1970s and then underwent a remarkable recovery (Matteson, 1985; Ludwig, 1984). Water levels on the Great Lakes fluctuated cyclically over 5 ft during this 35-year period.

In conjunction with lake level fluctuations some species may find a temporary improvement in habitat conditions followed by a decline in suitability as natural successional processes ensue. For example, common terns (*Sterna hirundo*) may experience an increase in production due to increased availability of habitat following a decline in water levels and exposure of bare sand and gravel substrate or development of an artificial island only to decline as vegetation takes over. Shugart and Scharf (1983) identified a decline in the number of common terns from colonies of the northern Great Lakes. While they questioned the biological significance of the demise of individual colonies, they did identify fluctuating Great Lakes water levels and possibly increases in ring-billed gulls (*Larus delewarensis*) as factors reducing the amount of suitable nesting habitat for the common tern. The reproductive success of the common tern is influenced by a variety of other factors as well (Morris et al., 1976).

Lakewide changes other than water level fluctuations have been implicated in affecting Great Lakes birds. Ludwig (1965) points to the changing ecology of the fishery in the decline and rebound of the Caspian tern (*Sterna caspia*) populations on the Great Lakes. Specifically, he suggests that the Caspian population of the period 1922–1953 was supported by a steady decline in nine species of lake herring leading to chick starvation and lower recruitment. Subsequently, alewives became available to the terns to feed chicks and resulted in the increased fledging rate from 1957 on.

The most recent and possibly the most dramatic change in an avian population on the Great Lakes is that reported by Blokpoel and Tessier (1986) for the ring-billed gull. These investigators estimated that the nesting population between 1976 and 1984 increased at an average annual growth rate of 11%. They also suggest that because the ring-bills are so highly adaptable, their numbers are likely to continue to increase.

Factors most frequently identified in the literature as causing declines in populations of colony nesting birds are fluctuating water levels, habitat change, human disturbance, and predation. How then are we to factor out the impact of toxics on avian populations in the Great Lakes? First, as we have already noted, we must have some handle on avian populations. Are they increasing or decreasing? Over what period of time and for what species? This is no simple problem, particularly when considering the number of bird species that might be at risk because of their food habits.

Nor is it sufficient to simply keep track of numbers of individuals or breeding pairs over time. As Ludwig (1979) has clearly demonstrated, United States Caspian tern colonies on the Great Lakes in the 1960s and 1970s were being supported by recruitment from Canadian colonies. The net result was that both U.S. and Canadian colonies appeared to be increasing slowly but in actuality the U.S. colonies were experiencing reproductive impairment. Consequently, while breeding pair counts are necessary for whole-population analysis, these are not sufficient data to ascertain the well-being of a particular subpopulation. Only mark–recapture studies of banded birds can supply these needed data.

## 3. WHAT EVIDENCE LINKS CHEMICALS AND POPULATIONS CHANGES?

The BMTG quickly concluded that there is in fact very little evidence to implicate toxic substances as having affected bird and mammal species at the population level, that is, across the geographic range of the species. On the other hand, there is strong evidence that persistent toxic substances have played a role in impairing the reproductive success of local populations (demes) and/or particular colonies of several avian species. Similarly, wild mammals may be affected. This conclusion may be due more to a dearth of information than to any fundamental difference in the susceptibility of the two orders (see Gilbertson, this volume).

Circumstantial evidence suggests that two species of semiaquatic mammals may be affected by contaminants in Great Lakes ecosystems. These two mammals are the mink and river otter.

It has been known for well over a decade that mink are extremely sensitive to the effects of PCBs (Aulerich et al., 1973; Platonow and Karstad, 1973). The established link between PCB-contaminated fish and toxic effects in mink was demonstrated by Hornshaw et al. (1983) as late as 1981. While no clear population data exists for mink bordering the Great Lakes generally, wild New York mink within 5 miles of Lake Ontario have been found to accumulate high body burdens of PCB and organochlorine insecticides unique to the lake. Concentrations of PCBs in these mink are similar to those that impair reproduction in ranch-raised mink, suggesting that the association between high body burdens of PCBs and survival of mink near Lake Ontario should be examined (Robert Foley, N.Y. State Department of Conservation, personal communication, 1986). Elsewhere on the Great Lakes, in 1983 a dead mink was found on the west shore of Green Bay and was subsequently analyzed for PCBs. The mink's liver was found to contain 5.7 ppm PCBs (USFWS, unpublished data). This level is sufficient to be of diagnostic significance given the previous research on PCB liver concentrations found in experimentally exposed dead mink (Ringer et al., 1972; Platonow and Karstad, 1973).

The river otter is a highly prized furbearer of Wisconsin and elsewhere. The Wisconsin population harvest data is suggestive of a void in river otters close to Green Bay (Kohn and Ashbrenner, 1984). Apparently, the same lack of harvest in Michigan portions of Green Bay environs is also occurring (John Stuht, Michigan Department of Natural Resources, personal communication, 1986). The Green Bay west shore tributaries and shorelands have sufficient natural habitat where river otter would be expected to occur and be occasionally trapped. Trapping effort, however, is not known. Regardless, the river otter may be nearly as sensitive to the effects of PCBs as the mink. At least three studies have been published that are suggestive of otter population declines where PCB contamination has been documented. These are the Columbia River Basin, Oregon (Henny et al., 1980), and in Great Britain (Mason et al., 1986) as well as Sweden (Olsson et al., 1981). PCB levels in otters from areas of declining populations have been generally higher than the levels that are known to cause adverse reproductive effects in mink.

In contrast to these few (piscivorous or partially piscivorous) mammalian species, there are many avian species that rely heavily on fish and could be subject to bioconcentration of persistent toxic substances.

Some species of gulls and terns are almost entirely piscivorous, such as Forster's tern and the common tern, while others are less so, such as the herring gull and the ring-billed gull. The double-crested cormorant is a species completely dependent upon fish as a food source. Some waterfowl, particularly mergansers (*Tribe mergini*), or fish-eating ducks are obvious candidates for accumulating persistent toxic substances. Other species of waterfowl, such as Scaup (*Aythya* sp.), old squaw (*Clanqula hyemalis*), and ruddy ducks (*Oxyura jamacicensis*), that rely heavily on benthic macrofauna when vegetation is not abudant could also be involved in a toxic food chain phenomena. Wetland birds such as the great blue heron (*Ardea herodias*), green-backed heron (*Butorides striatus*), and the black-crowned night heron (*Nycticorax nycticorax*), which have a more varied diet but are tied to aquatic systems, are also candidates. Within the raptors, bald eagles (*Haliaeetus leucocephulus*) and osprey (*Pandion haliaetus*) are also among the fish-eating birds associated with portions of the Great Lakes. All of the species have two things in common: they frequent the Great Lakes, and they feed partially or entirely on fish or other aquatic animal life. It should come as little surprise then that what evidence that does exist for chlorinated hydrocarbon toxicity in wild vertebrates is found primarily among birds.

The BMTG was charged with reviewing the evidence that links toxic chemicals and bird and mammal populations in the Great Lakes. We were greatly aided in this task through the excellent review paper of Gilbertson (Chapter 6) titled "Epidemics in Birds and Mammals Caused by Chemicals in the Great Lakes" and by Fitchko's (1986) literature review of the effects of persistent toxic substances on Great Lakes biota. The BMTG categorized the demonstrable impacts expected when population changes are due to persistent toxic substances and identified additional existing studies not addressed by Gilbertson (this volume).

Demonstrable impacts on avian populations may be direct, including reproductive impairment, altered adult survival, and impaired health, or they may be indirect. Reproductive impairment has occurred through declines in egg production, embryo toxicity, eggshell thinning, deformities, altered parental behavior, aberrant behavior of recently hatched young, histopathology, and biochemical dysfunction (see review by Blus, 1982). Changes in adult survival and health may be manifested through direct mortality and abnormal pair bonding (female–female pairing) as discussed by Conover (1984). It is possible, although not documented, that a decline in a prey species due to chloroorganic toxicity could alter interspecific competition between avian predators dependent on that forage resource. This domino effect in the food chain is particularly difficult to substantiate.

Of the array of demonstrable impacts outlined above, six have been documented in avian populations in the Great Lakes: (1) embryo toxicity, (2) eggshell thinning, (3) congenital abnormalities, (4) histopathology, (5) adult mortality, and (6) changes in adult behavior. Gilbertson's paper (Chapter 6) addresses these six factors and the evidence that links them to population changes of different species of birds in the Great Lakes. Here we would like only to add documentation not found in the Gilbertson review.

In the upper Great Lakes states, 9 out of 13 species of fish-eating birds were found to have sustained statistically significant decreases in eggshell thickness post 1946 (Faber and Hickey, 1973). DDE, dieldrin, and PCBs were implicated in different groups of species in relation to the eggshell thinning observed. Anderson and Hickey (1972) utilized archived museum eggs to demonstrate significant eggshell thinning in a number of Great Lakes birds including the herring gull during the period when the use of DDT was prevalent.

Additionally, embryo toxicity, histopathology, parental behavior, and deformities have been associated with impaired reproduction of Forster's terns from colonies located on Green Bay, Lake Michigan (Harris et al., 1985). Levels of 2,3,7,8-tetrachlorodibenzo-*p*-dioxin (TCDD) and a series of structurally related PCB congeners that are isosteric with TCDD (similar structural toxicity) were found, most of which were significantly higher in eggs from the suspected contaminated colony than from eggs in a colony removed from the potential source of contamination. The remarkable agreement of findings among independent laboratory studies of birds, Lake Ontario herring gulls (Gilbertson and Fox, 1977; Gilbertson, 1983) and Green Bay Forster's terns strongly suggests a common etiology. One further aspect of the Forster's tern study should be mentioned here. To better ascertain the relative contribution of 2,3,7,8-TCDD and the numerous isosteres found in the eggs to the toxic response observed in the study, a procedure for normalizing the concentrations of the various isosteres to reflect their individual potencies was developed. Normalizing the isostere concentrations based on their individual aryl hydrocarbon hydroxylase induction capabilities (Sawyer and Safe, 1982) relative to 2,3,7,8-TCDD allowed reporting a relative contribution in a common unit of measure, TCDD equivalents. This approach helps explain the overall biological response of the contaminated population which could not necessarily be attributed to any single toxicant.

A new occurrence of congenital anomalies observed in 1983 and/or 1984 in double-crested cormorants and herring gulls in the Great Lakes has recently been reported by Weseloh et al. (1985). Approximately 1700 herring gull chicks from four colonies and 8300 young double-crested cormorants from 33 colonies were examined for deformities. No anomalies were found in herring gull chicks, but 36 cormorants with anomalies were found. The incidence of anomalies ranged from zero on Lakes Erie, Superior, and Lake-of-the Woods (0 to 2265) to 0.95% (or 31 of 3249) from the Green Bay/Door County peninsula area of Lake Michigan, Wisconsin. The authors suspect toxic chemicals as a cause. Recent preliminary histopathological examination of double-crested cormorant chick tissues indicates that one of the following may be linked to contaminants: liver, thyroids, kidneys, spleen, and lungs (T. Kubiak, personal communication). The need for "control" specimens is paramount in ascertaining the degree of abnormalities from these preliminary findings.

High levels of PCBs have also been recorded in eggs of herring gulls, ring-billed gulls, common terns, and double-crested cormorants from Green Bay (Heinz et al., 1985). Moccia et al. (1986) have found abnormal thyroids in herring gull adults collected from lower Green Bay in 1983 and have suggested that contaminants are the cause of this condition.

Heinz et al. (1984, p. 159) earlier measured organochlorine residues in four species of birds taken from the Sheboygan River, a tributary to Lake Michigan. They concluded "if Great Blue Herons (*Ardea herodias*), Belted Kingfishers (*Ceryle alcyon*), Solitary Sandpipers (*Tringa solitaria*) or Spotted Sandpipers (*Actitis macularia*) were as sensitive to reproductive effects from PCBs as are chickens, their reproduction would have been impaired by the levels they carried." Without reproductive data from the field, there is no way to tell if reproduction was adversely affected in these species.

While there are numerous studies that implicate persistent toxic substances in reproductive failure and/or impairment in fish-eating colonial nesting birds in the Great Lakes (including Vermeer and Peakall, 1977a, b; Gilbertson, Chapter 6), much of the currently available data are circumstantial. Regardless, it is the opinion of the BMTG that the evidence clearly associates persistent toxic chlorinated hydrocarbons to avian population changes at the local population level. While a direct cause and effect relationship cannot be made with any particular chemical(s) and any particular population, the weight of the evidence must not be ignored or discounted. The prudent course of action is to build on what we know in such a manner that the scope and extent of the problem becomes more clearly defined.

## 4. HOW CAN WE RESPOND TO THE PROBLEM?

It seems unlikely that enough funds or personnel can be made available to monitor all avian species that may be at risk to toxic contaminants in the Great Lakes with the intensity necessary to accumulate meaningful data. Consequently, a selection of special data is necessary. The BMTG suggests that two types of data bases be established in conjunction with the selected species. One data base should be a multispecies data base (perhaps 6 species) designed to measure gross population trends, and track breeding and migration trends over time. The second data base should be developed for 2 or 3 key indicator species which would be targets for intensive etiological and epidemiological studies. The BMTG recommended a species selection process to select indicator or key species for which data bases would be developed. "Decision criteria" should be listed on the vertical axis, and "candidate species" on the horizontal axis of a decision matrix. Simple cross-tabulation will reveal which candidate species fit the most (or most "critical") decision criteria.

The BMTG *suggests* that decision criteria should include

- represent a feeding guild
- represent a wide distribution
- reflect conveniently the impact of toxins irrespective of other environmental perturbations
- demonstrate sublethal, chronic impacts of toxicants
- promote ease of censusing population
- represent an indigenous species
- serve directly or indirectly all seven levels of testing for toxins (biochemical, histological, physiological, behavioral, population, community, and ecosystem)
- reveal multigenerational impacts

- represent a top carnivore
- make available long-term data
- involve interspecific interactions
- represent a threatened species

The BMTG purposefully left the list open ended and assigned no "weighting factors" to the criteria, although members thought that the list could be improved upon by the formation of a more broadly based task force or reference group, hopefully, in the near future.

On the horizontal axis of the matrix examples of candidate species are:

Common tern
Forster's tern
Caspian tern
Black terns
Double-crested cormorant
Herring gull
Ring-billed gull
Black-crowned night heron
Great blue heron
Old squaw
Greater scaup
Lesser scaup
Red-breasted merganser
Bald eagle
Osprey

The selection of key or target species is a critical step in the establishment of the data bases referred to earlier. It should be undertaken (possibly as an IJC sponsored workshop or assigned to the Aquatic Ecosystem Objectives Committee) at the earliest possible date.

The intent of the BMTG recommendation for the development of two distinct data base forms is to provide a means of establishing an organized systemwide censusing procedure in addition to the separate but related research-oriented programs focusing on only two or three species.

The establishment of a Systemwide Avian Census (SAC) on the Great Lakes might well take advantage of the fact that many people watch birds. The number of professional and skilled amateur bird-watchers in the Great Lakes region suggests a potentially powerful resource for "ecosystem surveillance." What is needed in this regard is a description of what observations to make and the establishment of a network for compiling and screening the information. It is possible to envision a professional group (such as the Colonial Water Bird Group) collaborating with federal, state, or provincial government agencies to establish monitoring priorities and procedures for colonial waterbirds and noncolonial species. Carefully selected training sessions for both agencies and amateur bird-watchers could lead to the establishment of a network or monitoring system that could be cost-effective in providing reliable data for trend analysis. Thus, an ecologically based and citizenry supported system might provide a useful and significant monitor of the ecosystem and its relative "health" in the Great Lakes.

While the establishment of an avian census is important, it cannot in and of itself provide a factual basis that demonstrates the relation between toxic chemicals in the environment and their impact on avian populations. Intensive research efforts are called for in this regard and of necessity must focus on only two or three species. What is needed is an interactive system that would serve the four aspects of toxics testing as described by J. Cairns (Chapter 1): screening, prediction, monitoring, and validation. The "predictive" aspects of toxics testing would be addressed, in part, by learning more about toxic mechanisms via study of biochemical and physiological processes. This will require the establishment of "model species" and consequently the establishment of captive populations for experimental purposes. This effort must be coupled with monitoring efforts of the "model species" in the field. Nothing less than a multifaceted interdisciplinary effort based on sound epidemiology and etiology, as concluded by Gilbertson (this volume), will be adequate. For example, numerous studies reporting residue levels in eggs and/or birds have not provided the reproductive, histopathological, or biochemical data that are necessary to link exposure with effects. The interdisciplinary approach required is inherent in the study of ring-billed gulls in southern Ontario by Sileo et al. (1977). Field monitoring of the "model species" will also require that an "off-site reference base" be established in conjunction with the population at risk. A first-cut prototype of this kind of approach is suggested in Harris et al. (1985), which built on the pioneering work of the herring gull monitoring program in Canada.

In addition to the laboratory and field components suggested above, the BMTG recommends the establishment of a specimen bank. Archiving egg, tissue, and in some cases whole carcasses may be necessary (Gannon, 1982). Regardless, standards and instructions for preparing, preserving, recording, transporting, and archiving materials and specimens are needed.

Our brief outline of a more effective response to the problem may appear somewhat austere. It is the opinion of the BMTG, however, that the prescribed response is no more forbidding than the problem itself. Grave problems do not call for abstemious measures.

We may add that our recommendations for birds apply to "at-risk" mammals as well. Birds and mammals are key "sensing sites" that can serve as convenient monitors with which to gauge the levels of contamination of the lake ecosystem as a whole (MacKay, 1985). They are more closely related to humans and perhaps in that sense closer to the inevitable ultimate concern of toxic contamination—human health and well being.

## 5. HOW CAN WE PROCEED?

The BMTG was not charged with structuring a course of action to implement these recommendations. To do so may seem presumptuous on our part. On the other hand, not to do so leaves the recommendations little more than an academic exercise. Since not all BMTG members were academicians and those who were felt some urgency in this matter, all chose to make further suggestions.

Comparatively, it seemed to the BMTG that bird and mammal populations have had far less consideration on the Great Lakes than other vertebrate forms (fish) or even invertebrates. Birds in particular cross over national, state, and provincial governmental jurisdictions; they are a shared resource of the Great Lakes system, as are fish. It is logical then that an international Birds and Mammals Reference Group be established within the construct of the International Joint Commission Mandate. This group should have its own funding base as do other reference groups and be assigned a lifetime with periodic reexamination for possible renewal. It should be charged with examining environmental problems specific to bird and mammal species or populations that are inherent in, but not specifically defined or covered, by international agreements or treaties.

In the last analysis, the BMTG believes there has been and is a serious problem of toxic contamination at specific locales in Great Lakes ecosystems. Further, we believe there is sufficient evidence to warrant directing more attention and resources to the impact of persistent toxic substances on bird and mammal populations of the Great Lakes than has been done thus far.

## ACKNOWLEDGMENTS

Members of the Birds and Mammals Task Group (BMTG) contributed materially to this paper. Task Group members were R. T. Dumke, Wisconsin Department of Natural Resources, Madison; R. A. Faber, Biology Department, St. Mary's College, Winnona, Minnesota; D. J. Hallett, Eco Logic Inc., Rockwood, Ontario; T. J. Kubiak, U.S. Fish and Wildlife Service, Green Bay, Wisconsin; J. P. Ludwig, Ecological Research Services, Boyne City, Michigan; D. Blockstein (assisting). Special thanks to Tim Kubiak for helping with revisions following review. Thanks also to Cathi Hinricks and Diane Bubnik for typing this manuscript.

## REFERENCES

Anderson, D. W., and Hamerström, F. (1967). The recent status of Wisconsin cormorants. *Passenger Pigeon* **29,** 5–15.

Anderson, D. W., and Hickey, J. J. (1972). Eggshell changes in certain North American birds. *Proc. Int. Ornithol. Congr.* **15,** 514–540.

Aulerich, R. J., Ringer, R. K., and Iwamoto, S. (1973). Reproductive failure and mortality in mink fed on Great Lakes fish. *J. Reprod. Fertil. Suppl.* **19,** 365–376.

Blokpoel, H., and Tessier, G. D. (1986). The Ring-billed Gull in Ontario: A review of a new problem species. Occas. Pap. 57 Can. Wildl. Serv., pp. 1–34.

Blus, L. J. (1982). Further interpretation of the relation of organochlorine residues in Brown Pelican eggs to reproductive success. *Environ. Pollut.* **28,** 15–33.

Cairns, J., Jr. (1988). Politics, economics, science: going beyond disciplinary boundaries to protect aquatic ecosystems. In M. S. Evans (Ed.). *Toxic Contaminants and Ecosystem Health: A Great Lakes Focus.* John Wiley, New York.

Conover, M. R. (1984). Occurrence of supranormal clutches in the Laridae. *Wilson Bull.* **96,** 249–267.

Faber, R. A., and Hickey, J. J. (1973). Eggshell thinning, chlorinated hydrocarbons and mercury in inland aquatic bird eggs, 1969 and 1970. *Pestic. Monit. J.* **7,** 27–36.

Fitchko, J. (1986). Literature review of the effects of persistent toxic substances on Great Lakes biota. Report of the Health of Aquatic Communities Task Force to the Science Advisory Board, International Joint Commission, 256 pp.

Gannon, J. E. (1982). Specimen banking—an important step toward improving water quality monitoring in large lakes. *J. Great Lakes Res.* **8,** 591.

Gilbertson, M. (1983). Etiology of chick edema disease in Herring Gulls in the Lower Great Lakes. *Chemosphere* **12,** 357–370.

Gilbertson, M. (1988). Epidemics in birds and mammals caused by chemicals in the Great Lakes. In M. S. Evans (ed.), *Toxic Contaminants and Ecosystem Health: A Great Lakes Focus.* John Wiley, New York.

Gilbertson, M., and Fox, G. A. (1977). Pollutant-associated embryonic mortality of Great Lakes Herring Gulls. *Environ. Pollut.* **12,** 211–216.

Harris, H. J., and Trick, J. A. (1979). *Status and Nesting Ecology of the Forster's Tern in Wisconsin.* Report submitted to Wisconsin Dept. of Natural Resources, Madison.

Harris, H. J., Fewless, G., Milligan, M., and Johnson, W. (1981). Recovery processes and habitat quality in freshwater coastal marsh following a natural disturbance. Proc. Midwest Conf. Wetland Values Manage., pp. 363–379.

Harris, H. J., Kubiak, T. J., and Trick, J. (1985). *Microcontaminants and Reproductive Impairment of the Forster's Tern on Green Bay,* Final Report to U.S. Fish and Wildlife Service, Univ. Wis. Sea Grant Inst., Wis. Dep. Nat. Resour. and Green Bay Metropolitan Sewerage District. Sea Grant Inst., University of Wisconsin, Green Bay.

Heinz, G. H., Swineford, D. M., and Katsma, D. E. (1984). High PCB residues in birds from the Sheboygan River, Wisconsin. *Environ. Monit. Assess.* **4,** 155–161.

Heinz, G. H., Erdman, T. C., Hazeltine, S. D., and Stafford, C. (1985). Contaminant levels in colonial waterbirds from Green Bay and Lake Michigan 1975-80. *Environ. Monit. Assess.* **5,** 223–236.

Henny, C. J., Blus, L. J., Gregory, S. V. and Safford, C. J. (1980). PCBs and organochlorine pesticides in wild mink and river otters from Oregon. In J. A. Chapman and D. Pursley, eds. Worldwide Furbearer Conf. Proc., Frostburg, Md. pp. 1963–80.

Hornshaw, T. C., Aulerich, R. J., and Johnson, H. E. (1983). Feeding Great Lakes fish to mink: Effects on mink and accumulation and elimination of PCBs by mink. *J. Toxicol. Environ. Health* **11,** 933–946.

Keith, J. O. (1966). Insecticide contaminations in wetland habitats and their effects on fish-eating birds. *J. Appl. Ecol.* **3** Suppl., 71–85.

Kohn, B. E. and Ashbrenner, J. E. (1984). Harvest and population status of river otter in Wisconsin. Wisconsin Department of Natural Resources Research Report 129. Madison, Wisconsin.

Ludwig, J. P. (1965). Biology and structure of the Caspian Tern (*Hydropogne caspia*) population of the Great Lakes from 1896-1964. *J. Ornithol. Invest.* **36,** 217–233.

Ludwig, J. P. (1979). Present status of the Caspian Tern population of the Great Lakes. *Mich. Acad.* **12,** 69–77.

Ludwig, J. P. (1984). Decline, resurgence and population dynamics of Michigan and Great Lakes Double-crested Cormorants. *Jack Pine Warbler* **62,** 91–102.

MacKay, D. (1985). Editorial: Sensing sites and toxic models. *J. Great Lakes Res.* **11,** 83.

Mason, C. F., Ford, T. C. and Last, N. I. (1986). Organochlorine residues in British otters. *Bull. Environ. Contam. Toxicol.* **36,** 656–661.

Matteson, S. W. (1985). *Population Status of the Double-crested Cormorant in Wisconsin,* Memo. Bureau of Endangered Resources, Wis. Dep. Nat. Resour., Madison.

Meeks, R. L. (1968). The accumulation of $^{36}Cl$ ring-labled DDT in a freshwater marsh. *J. Wildl. Manage.* **32,** 376–398.

Moccia, R. D., Fox, G. A., and Britton, A. (1986). A quantitative assessment of thyroid histopathology of herring gulls (*Larus argentatus*) from the Great Lakes and a hypothesis on the causal role of environmental contaminants. *J. Wildl. Dis.* **22,** 60–70.

Morris, R. D., Hunter, R. A., and McElamn, J. F. (1976). Factors affecting reproductive success of Common Tern (*Sterna hirundo*) colonies on the lower Great Lakes during summer of 1972. *Can. J. Zool.* **54,** 1850–1862.

Olsson, M., Reutergardh, L. and Sandegren, F. (1981). Var a uttern? *Sueriges Natur.* **6/8,** 234–240.

Platonow, N. S., and Karstad, L. K. (1973). Dietary effect of polychlorinated biphenyls on mink. *Can. J. Comp. Med.* **37,** 391–400.

Ringer, R. K., Aulerich, R. J., and Zabik, M. (1972). Effects of dietary polychlorinated biphenyls on growth and reproduction of mink. Abstr. Am. Chem. Soc. Natl. Meet., 164th, Vol. 12, pp. 149–154.

Sawyer, T., and Safe, S. (1982). PCB isomers and congeners: Induction of aryl hydrocarbon hydroxylase and erythoxyresorufin o-deethylase enzyme activities in rat hepatoma cells. *Toxicol. Lett.* **13,** 87–94.

Scharf, W. C., and Shugart, G. W. (1984). Distribution and phenology of nesting Forster's Terns in eastern Lake Huron and Lake St. Clair. *Wilson Bull.* **96,** 306–309.

Scharf, W. C., Chamberlin, M. E., Erdman, T. C., and Shugart, G. W. (1979). Nesting and migration areas of birds of the U.S. Great Lakes. U.S. Fish Wildl. Serv. Rep. FWS/OBS-77/22, 1–113.

Shugart, G. W., and Scharf, W. C. (1983). Common Terns in the Northern Great Lakes. Current status and population trends. *J. Field Ornithol.* **54,** 160–169.

Sileo, L., Karstad, L., Frank, R., Holdrinet, M. V. H., Addison, E., and Braun, H. E. (1977). Organochlorine poisoning of Ring-billed Gulls in southern Ontario. *J. Wildl. Dis.* **13,** 313–322.

Trick, J. A. (1982). Reproductive success and foraging success as related to populations of Forster's Terns (*Sterna fosteri*) on Green Bay, Lake Michigan. M.S. Thesis, University of Wisconsin, Green Bay.

Vermeer, K., and Peakall, D. B. (1977a). Environmental contaminants and the future of fish-eating birds in Canada. *In* T. Mosquin and C. Suchol (eds.), *Canada's Threatened Species and Habitats,* Proc. Symp. Can. Threatened Species Habitats. Can. Nat. Fed. World Wildl. Fund (Can.), Ottawa.

Vermeer, K., and Peakall, D. B. (1977b). Toxic chemicals in Canadian fish-eating birds. *Mar. Pollut. Bull.* **8,** 205–210.

Weseloh, D. V., Teeple, S. M., and Gilbertson, M. (1983). Double-crested Cormorants of the Great Lakes: Egg-laying parameters, reproductive failure, and contaminant residues in eggs, Lake Huron 1972–1973. *Can. J. Zool.* **61,** 427–436.

Weseloh, D. V., Struger, J., and Kubiak, T. J. (1985). The occurrence of congenital anomalies in Double-crested Cormorants and Herring Gulls in the Great Lakes, 1983–1984. Proc. 28th Conf. Great Lakes Res.

# 30

# MODELING AND STATISTICAL APPROACHES FOR DETERMINING THE FATE AND EFFECTS OF TOXICANTS IN NATURAL WATERS

***Thomas D. Fontaine III***

*Great Lakes Environmental Research Laboratory, National Oceanic and Atmospheric Administration, Ann Arbor, Michigan 48105*

# 1. INTRODUCTION

## 1.1. The Challenge

For nearly two decades, aquatic scientists have tried to predict the fate and biological effects of excessive nutrient loading to natural waters. This challenge continues even though our understanding of nutrient effects has advanced greatly through modeling, field, and laboratory studies. A more recent challenge for aquatic scientists is predicting and understanding the fate and effects of toxic contaminants in natural waters. From all scientific viewpoints, this problem is possibly several orders of magnitude more difficult than that posed by nutrients. First, the sheer number of toxic compounds that requires consideration in modeling, field, and laboratory studies is overwhelming in comparison with the number of nutrients usually considered. Second, effluents and receiving waters rarely contain just a single toxic contaminant. More likely to be encountered are contaminant mixtures that vary in composition over time and space. Unfortunately, it is improbable that predictive toxicological cause and effect relationships will soon be developed for all possible mixtures. Third, analytical and experimental techniques used in understanding toxic contaminants require considerably more time and money than nutrient studies. Combined, these problems present a formidable challenge to those who wish to predict, understand, and regulate the behavior of toxic contaminants in the Great Lakes.

## 1.2. Great Lakes Contaminant Questions

Questions often asked about toxic contaminants in the Great Lakes include:

1. What are the contaminant loads to the system and where do they come from? Once input, where will the contaminants go and how long will they stay?
2. What are the risks to human or ecological health from past, present, and projected loads of contaminants? How is risk to be defined?
3. What contaminant loading rates are permissible? What loads will lead to undesirable contaminant concentrations in water or fish?
4. To what extent will proposed contaminant load reduction measures decrease contaminant concentrations in water, sediment, and biota? Which measures will be most cost-effective?
5. Will the most significant effects of contaminants on aquatic organisms be through direct toxic effects, or indirectly, through changes in the ecosystem?

6. Are the effects of multiple contaminants on ecosystems additive?
7. How can a distinction be made between changes in ecosystem structure and function that are caused by contaminants and those that result from other causes (e.g., natural variability of weather)?
8. How will fisheries management and nutrient load reduction strategies affect the fate and effects of contaminants?

Modeling, field, and laboratory studies will be required to answer these and other questions concerning toxic contaminants in the Great Lakes.

The intent of this chapter is to impart the views of the Modeling/Statistical Approaches workshop panel concerning approaches that may be helpful in addressing these questions in the Great Lakes. Where appropriate, brief discussions of literature have been included that are pertinent to the panel's recommendations. The panel's recommendations were not limited to modeling or statistical techniques. Suggestions were also made for laboratory and field studies that would aid modeling and statistical endeavors. Recommendations of the panel are shown in italics.

## 2. MODELING AND STATISTICAL APPROACHES

### 2.1. The Dilemma

Models should be developed or used on the basis of their adequacy to answer a particular question. However, time, funding, and data constraints often bring about a dilemma for modelers and statisticians. That is, questions that need to be answered immediately (e.g., those about contaminant fate and effects) are often those that require the most data (usually unavailable) and the most detailed, process-oriented models. In these instances, a choice can be made between scaling questions back to less specific ones that can be answered quickly with simple, empirical models, or, committing resources and relaxing time constraints so that the original question can be pursued adequately. In the former case, management decisions based upon simplified questions and models may prove to be inappropriate. In the latter case, waiting until the original question can be answered could be disasterous because of inaction. Decision makers should be aware of the uncertainties and assumptions of each approach.

### 2.2. Statistical versus Process-Oriented Models

Statistical and process-oriented modeling approaches are available for addressing questions about the fate and effects of contaminants in aquatic environments. The distinction between these types of models is that the statistical models are strictly empirical and make no attempt to use biological, chemical, or physical theory to define a functional relationship. Process models represent working hypotheses of how a system is causally linked and through simulation can be used to investigate the relative importance of the various processes that control the behavior of the simulated system. As such, process models can provide a framework for identifying needed field measurements and experimental studies. A process-oriented model may employ empirical submodels to describe a process that is not well understood or requires too much detail to describe theoretically. Both types of models can be used for predictions, but process-oriented models have more potential for transferability because they are theoretically based. Statistical models may result in better prediction for the system for which they are developed, but may not be as transferable. Clear examples of where process-oriented or statistical approaches become somewhat indistinguishable are reviewed by Reckhow and Chapra (1983).

An excellent example of how statistical and process-oriented modeling approaches have been helpful in Great Lakes problem solving is described by Scavia and Chapra (1977). Predictions of whole-lake phosphorus concentrations from coarse-resolution, empirical phosphorus models of Chapra (1975) and Dillon and Rigler (1974) differed little from those of the finer resolution, process-oriented Lake Ontario model of Scavia et al. (1976). The empirical models enabled questions about permissible loading rates to be answered quickly, thus providing immediate guidance to policy makers. The process-oriented model provided much needed insight into how ecosystem processes and components might change as a result of changes in phosphorus loading. *We therefore recommend that the development of statistical and process-oriented models continue but that the limitations of each approach be clearly communicated to the intended user. We also recommend that existing and future models be collated and compared to identify relative strengths, weaknesses, and applicability of computer code for use in Great Lakes applications. Based on these comparisons, a central repository of proven, well-documented computer codes (in particular, subroutines that describe a particular process, e.g., volatilization) should be established. Electronic access to this repository would facilitate rapid development of state-of-the-art Great Lakes contaminant models.*

## 2.3. Mass Balance Models

For a finite period of time, a contaminant mass balance may be constructed for a parcel of water by measuring or estimating the right side of Equation 1:

$$[\text{Accumulation}] = [\text{Inputs}] - [\text{Outputs}] \pm [\text{Reactions}] \tag{1}$$

A positive accumulation indicates that the system is a sink for the contaminant being studied. A negative accumulation indicates that the contaminant is exported, transformed, volatilized, or otherwise lost from the system at a rate greater than it is input or created. Inputs can be internal (resuspension or leaching of an in-place pollutant) or external (from point or diffuse sources). By constructing a mass balance it is possible to determine the most effective mode of action (remedial or waste load allocation) for minimizing contaminant inputs since the relative importance of loads is known. *Despite the simplicity and usefulness of the mass balance approach it will be impossible, given current funding levels, to construct a mass balance for most Great Lakes contaminants. Therefore, contaminant candidates for a mass balance must be judiciously selected. Candidates should be representative of groups of compounds that have similar properties and behavior in aqueous systems. By understanding the behavior of representative compounds, the behavior of others can be estimated.*

## 2.4. Fate and Exposure Models

The ultimate goal of contaminant models is to predict the impacts of contaminants on ecosystem structure and function. Humans are part of this ecosystem. Assessing impact requires defining (1) the fate of the contaminant (where does it go, when does it get there, how long does it stay, and what is its concentration and chemical form?), (2) the exposure of organisms to the contaminant (how often does the organism come in contact with the contaminant and what is the duration of contact?), and (3) the effect of the contaminant on the organism (what chronic or acute effects are observable, or, is the organism impacted by contaminant-induced changes in its supporting foodweb?). Most contaminant models address points 1 and, to a lesser degree, 2. These models are termed "fate" or "fate and exposure" models, respectively. To varying degrees, these models include empirical or causal expressions that describe how ecological components affect the fate of contaminants. However, it is implicitly assumed that the contaminant has no effect on ecosystem structure and function. When combined with food web models, fate and exposure models are particularly useful for predicting contaminant body burdens that result from chronic, nonlethal exposures to contaminants. These models are useful for determining if and when contaminant threshold limits might be exceeded, such as those set for fish that are to be consumed by humans. These models are also useful in that they give decision makers a tool for evaluating the effects of remedial options or regulatory actions on contaminant body burdens. Examples of fate or fate and exposure models are given in Burns et al. (1982), Bartell et al. (1983), Brown et al. (1982), Eadie (1981), Kalmaz and Kalmaz (1979), Park et al. (1980), Rodgers (1983), Rodgers and Swain (1983), Thomann (1978, 1979), Thomann and Connolly (1984), and Thomann et al. (1974). Useful reviews of fundamental processes that require consideration in such models are provided by Bedient et al. (1982), Burns (1982), Chapra and Reckhow (1983), Di Toro et al. (1982), Mauersberger (1983), O'Conner and St. John (1982), Jorgensen (1983), and Somlyody (1982). *We recommend the continued development or enhancement of fate and exposure models for those contaminants that exist or are likely to exist at concentrations that will not alter ecosystem structure and function. Contaminants that exist or are likely to exist at concentrations that will cause changes in ecosystem structure and function will require models that explicitly include contaminant effects.*

## 2.5. Combined Fate and Effects Models

Few models include equations for predicting direct or indirect contaminant effects because of the uncertainties involved in accurately representing the causal linkages between contaminants and their effects. Nevertheless, the ability to predict contaminant effects on ecosystems is greatly needed, and in part, hinges on the development of such models. The impetus for developing fate and effects models is based on the concern that contaminant concentrations can become great enough to alter ecosystem structure and function. If contaminant levels reach such critical thresholds, fate and exposure models cannot predict changes in ecosystem structure and function because these models do not have pathways for describing effects. To construct a combined fate and effects model it is necessary to model the trophic interactions of food webs and the interaction of the contaminant within it. In addition, it is necessary to model relevant aspects of the contaminant's water chemistry, as well as its uptake, depuration, and possible transformation by organisms in the food web. Furthermore, if the contaminant exerts acute or chronic toxic effects then the model must account for age class-specific toxicant-induced changes in gamete and biomass production, respiration, and mortality rates.

There is a regrettable lack of data relating changes in the internal processes that regulate the growth rate of plants (photosynthesis, respiration, nutrient uptake and release) and animals (consumption, assimilation, egestion, excretion, respiration, reproduction) to contaminant body burden. Also lacking are verified process models for predicting the effects of contaminant mixtures on aquatic organisms; Calamari and Alabaster (1980) do, however, provide a theoretical discussion relating to this issue. *We therefore recommend that laboratory and field experiments be conducted that will relate changes in an organism's physiological and behavioral processes to its contaminant body burden. Studies of contaminant fate and effects should be designed within the context of a model. If the modeler is involved from the beginning, the most useful data for model development and verification can be collected. When possible, modelers should perform uncertainty or error analyses (see Bartell et al. 1983) on models to aid the experimentalist in setting research priorities.*

Understanding the fate and effects of contaminants relies not only upon understanding the interactions of contaminants with ecosystem processes and components, but also on understanding ecosystem functioning in the absence of the contaminant. Thus, development of toxic contaminant fate and effects models has been viewed as a process in which past ecosystem models (e.g., Park et al., 1974) are expanded to include additional equations that describe toxicant–ecosystem interactions (e.g., Park et al., 1980). Few models have been developed that include toxic contaminant effects. In this regard, work by Breck and Bartell (Chapter 21) on organic contaminants is unique. Similar work by Fontaine and D'Angelo (In prep.) addresses the fate and effect of heavy metals in pelagic systems. *Although the development of combined fate and effects models is in its infancy, we encourage that this work and related field and laboratory experiments continue. Also, because the ultimate goal of many studies is to predict the effects of contaminants on humans, we recommend support for longterm epidemiological studies of human populations that have been significantly exposed to contaminants in the Great Lakes region.*

## 2.6. Complex Effluents: Fate, Exposure, and Effects Modeling

Bergman and Meyer (1982) state that "at some level of [effluent] complexity ... physical modeling approaches [e.g., controlled dosing of artificial ponds or streams with complex effluents] become more reliable and cost-effective than mathematical models for identifying compounds that may pose a persistance or bioaccumulation problem." [my brackets]. The level of effluent complexity where physical and not mathematical modeling approaches will be most useful is a matter of opinion. Nevertheless, it is noteworthy that modeling approaches other than mathematical ones are viable alternatives for understanding the fate and effects of complex effluents. For industries whose wastes vary little in composition or concentration, the physical approach recommended by Bergman and Meyer (1982) may be useful. Indeed, Di Toro et al. (Chapter 20) use an innovative combined modeling and bioassay approach to predict effects of complex effluents. They test the assumption that contaminant impacts on receiving water biota can be modeled without having to specifically characterize the toxicants that make up an effluent. Their approach is analogous to the measurement of biochemical oxygen demand (BOD), which itself does not depend on specific characterization of the media, yet leads to meaningful information concerning the stream's capacity for assimilating complex wastes. While the usefulness of Di Toro's technique is limited by (1) having to conduct bioassays and (2) the assumption that all contaminants flow downstream in the same proportion and physical–chemical form as originally discharged, their approach shows promise for some systems. A clear advantage of Di Toro's approach is that the fate and effects of each toxic compound in a complex effluent need not be modeled.

If a process-oriented modeling approach is to be developed for predicting the fate and effects of multiple contaminants, Bergman and Meyer (1982) warn that these models should account for the fact that contaminants can significantly affect each other's fate and effects. This point has been demonstrated in models of multiple heavy metal fate and transport by Chapman (1982) and Fontaine (1984). Unfortunately, process-oriented models of multiple organic contaminants and their fate and interactions within aquatic ecosystems remain uncommon. However, laboratory experiments by Landrum (1982) and Landrum et al. (1985) demonstrated that the bioavailability of contaminants to the ubiquitous Great Lakes amphipod *Pontoporeia hoyi* is affected by the presence of other contaminants and naturally occurring dissolved organic compounds. *Because interactions among organic contaminants have been shown to significantly affect contaminant fate in the Great Lakes, such interactions should be accounted for in models where necessary. We recommend that more laboratory studies be conducted to determine how contaminant toxicity is affected by the presence of other contaminants and components of the Great Lakes ecosystem.*

### 2.7. Calibration and Verification of Models

Calibration and verification of Great Lakes contaminant models will be difficult. This is because collecting and analyzing necessary and sufficient contaminant data is expensive and time-consuming. Calibration and verification requires reliable load estimates and concentrations of contaminants associated with water, biota, and settled and suspended sediments. The temporal and spatial scale of the model will dictate locations and how often measurements are needed of contaminant concentrations in loads, sediments, and biota. *We recommend that a study be carried out to determine the minimal amount of data (over both time and space) that is necessary and sufficient for calibrating and verifying Great Lakes contaminant models. In addition, this data set should contain sufficient information for detecting contaminant trends and emerging contamination problems in the Great Lakes.*

### 2.8. Statistical Analysis of Data

The usefulness of statistical techniques does not end with the formulation of predictive, empirical relationships. Techniques such as cluster analysis, ordination, and canonical correlation can be used to detect temporal and spatial differences in populations and whether these differences correlate with concentrations of toxic contaminants. Provided that there are sufficient data to conduct a meaningful analysis, these types of analyses can provide scientists with empirical evidence about the relative importance of natural versus contaminant-induced causes of observed population changes (Esterby, Chapter 22). *From the onset, statisticians should be participants in studies of contaminant fate and effects. Doing so should improve chances that necessary and sufficient data will be gathered for determining the relative importance of natural and contaminant-induced population changes. Existing data sets should be examined, particularly those from microcosm studies of contaminant effects, because they may already contain sufficient data for differentiating the effects of contaminants from the effects of natural variability on populations. In addition, modelers should closely examine data from microcosm studies to see if they can be utilized in formulating, calibrating, and validating models of contaminant fate and effects.*

## 3. MANAGEMENT CONSIDERATIONS

### 3.1. Model Uncertainty

As the number of state variables, coefficients, loads, and other attributes of a model increases, prediction uncertainty may also increase. An example of when this might occur is when contaminant process equations are included in ecosystem models to form a fate and effects model. Further increases in prediction uncertainty might occur as the number of contaminants considered in the model increases. As the length of time over which predictions are to be useful increases, predictions would probably become less and less certain (Fedra, 1981). *Therefore, the level of prediction uncertainty at which models become uninformative is a major issue that needs resolution in this and all other fields of modeling.*

Managers are often presented with model output that shows the response of a state variable (e.g., phosphorus concentrations) in relation to average, high, and low values of a forcing function (e.g., phosphorus loading rates). The procedure that produces this type of output is called a sensitivity analysis. Sensitivity analyses are useful for examining the effects of different levels of forcing functions, coefficients, or initial conditions on levels of state variables. This type of procedure was used in the past to generate information for setting allowable phosphorus loading rates to the Great Lakes. For the issue of Great Lakes eutrophication, the question could have been posed: Will phosphorus concentrations that resulted from simulations with average, high, and low loading rates be statistically different from each other? A sensitivity analysis cannot answer this question since a mean response plus a measure of variance is required to determine differences among means. An uncertainty analysis, on the other hand, would produce output phosphorus concentrations in the form of a probability distribution with known mean and variance. A probability distribution is generated because inputs (coefficients, initial conditions, loads, etc.) are also defined as probabilities in order to reflect their natural variability. Using an uncertainty analysis approach, Fontaine and Lesht (1987) show no statistical differences in Great Lakes phosphorus concentrations that result from several different load reduction plans. The importance of such results to a manager should be apparent: if the results from several management options are indistinguishable, there is no reason to expend any more funds than the least expensive option requires. *An uncertainty analysis approach should be used to evaluate Great Lakes contaminant management strategies to determine if changes in contaminant concentrations or effects that will result from strategies can be statistically distinguished from each other.*

## 3.2. Risk Analysis

Risk can be defined as the probability of an undesirable event occurring. Such events in the Great Lakes might include exceedence of water quality standards or exceedence of acceptable levels of contaminants in fish for human consumption. Because uncertainty analyses compute probability distributions for state variables such as contaminant concentrations in water and fish, the risk of exceeding a standard is easily computed. For instance, Fontaine and Lesht (1987) show that the risk of exceeding desired phosphorus concentrations in the eastern basin of Lake Erie is between 0 and 92% depending on the load reduction plan chosen. Breck and Bartell (Chapter 21) and O'Neill et al. (1982) provide examples of how uncertainty analyses can be used to assess contaminant-induced risks to aquatic organisms. *We recommend that risk analysis techniques be used for determining the relative risks of contaminants and other perturbations to Great Lakes biota and regional human populations.*

### 3.3. Multiple-Objective Management: Potential Synergisms

Reduction of toxic substance loads is one of many Great Lakes objectives. Additional objectives include reduction of nutrient loading and managing of fish populations through stocking and harvesting practices. It is reasonable to assume that both nutrient and fisheries management practices could alter the fate and effects of contaminants. This is because either management practice can alter food web relationships, change the rate of existing food web biomass transfers, and change the dynamics of particles to which contaminants adsorb. *We therefore recommend that modeling studies are needed to define how the fate and effects of contaminants will be affected by fisheries management and nutrient load reduction strategies.*

### 3.4. Multiple-Objective Management: Cost Effectiveness

Great Lakes decision makers are confronted with multiple goals, often conflicting, from which they must try to achieve fair and cost-effective policies. Management goals include enhancement of commercial and sports fisheries, attainment of desired water quality, rational consumptive use and diversions of lake water, encouragement of recreation and tourism, and sensible use of the lakes for transportation, mining, hydropower, and waste disposal. Modeling techniques are available that can simultaneously consider multiple goals (Haimes, 1977; Haimes and Allee, 1984) and will produce an optimal plan that maximizes the probability of achieving these goals for the least expense. *A multiobjective optimization analysis of Great Lakes goals, including those concerning toxic contaminants, should be prepared for Great Lakes decision makers. The results of this type of analysis will help decision makers maximize the probability of achieving desired goals for the least cost. In addition, decision makers would be able to clearly evaluate how the pursuit of one goal may enhance or adversely affect the pursuit of another goal.*

## 4. FINAL RECOMMENDATION

Toxic substance contamination of the Great Lakes is a problem that will not disappear without the concerted efforts of scientists, politicians, federal and state agencies, and a well-informed public. Good intentions and rhetoric alone are not sufficient to manage the problem. *An influx of financial support from all sources, state, federal, or private, is needed to address the regulatory, surveillance, and research needs that have been identified in this chapter.* If we remain unable to calculate contaminant mass balances it will continue to be difficult to make informed and cost-effective decisions about waste load allocations. If we remain unable to collect necessary and sufficient data it will be difficult to detect in situ environmental effects of chronic exposure to low concentrations of toxic contaminants. Without necessary and sufficient data, models cannot be calibrated or verified; predictive abilities will remain fuzzy instead of becoming well focused. At some point financial support must be matched against the seriousness of impacts that toxic substances may inflict upon the Great Lakes and its surrounding population and economy.

## ACKNOWLEDGMENTS

This chapter was based in part on discussions among attendees of the Modeling/Statistical Approaches Workshop which was part of the International Joint Commission sponsored symposium ''Persistent Toxic Substances and the Health of Aquatic Communities.'' Workshop participants were Dr. Steven M. Bartell and Dr. James Breck, Oak Ridge National Laboratory; Dr. Sylvia Esterby, National Water Research Institute; Dr. Thomas D. Fontaine III (Workshop Chairman), NOAA-Great Lakes Environmental Research Laboratory; Dr. Kenneth Rose, Martin Marietta Environmental Systems; and Dr. Dominic DiToro, Manhattan College. The views expressed in this article do not necessarily represent a consensus of opinion of the workshop participants. I thank Joel Baker for taking excellent notes on the workshop discussion. The comments of Peter F. Landrum, Greg A. Lang, and three anonymous reviewers helped improve the manuscript. All involved with the workshop thank Dr. Marlene S. Evans and Dr. John E. Gannon for organizing this important symposium. This is contribution No. 528 from the Great Lakes Environmental Research Laboratory, National Oceanic and Atmospheric Administration.

## REFERENCES

Bartell, S. M., Gardner, R. H., O'Neill, R. V., and Giddings, J. M. (1983). Error analysis of predicted fate of anthracene in a simulated pond. *Environ. Toxicol. Chem.* **2,** 19–28.

Bedient, P. B., Springer, N. K., Cook, C. J., and Tomson, M. B. (1982). Modeling chemical reactions and transport in groundwater systems: A review. In K. L. Dickson, A. W. Maki, and J. Cairns, Jr. (Eds.), *Modeling the Fate of Chemicals in the Aquatic Environment.* Ann Arbor Sci. Publ., Ann Arbor, MI, pp. 101–126.

Bergman, H. L., and Meyer, J. S. (1982). Complex effluent fate modeling. In K. L. Dickson, A. W. Maki, and J. Cairns, Jr. (Eds.), *Modeling the Fate of Chemicals in the Aquatic Environment.* Ann Arbor Sci. Publ., Ann Arbor, MI, pp. 101–126.

Breck, J. E., and Bartell, S. M. (1988). Approaches to modeling the fate and effects of toxicants in pelagic systems. In M. S. Evans (Ed.), *Toxic Contaminants and Ecosystem Health: A Great Lakes Focus,* John Wiley, New York.

Brown, M. P., McLaughlin, J. J. A., O'Conner, J. M., and Wyman, K. (1982). A mathematical model of PCB bioaccumulation in plankton. *Ecol. Modell.* **15,** 29–47.

Burns, L. A. (1982). Identification and evaluation of fundamental transport and transformation process models. In K. L. Dickson, A. W. Maki, and J. Cairns, Jr. (Eds.), *Modeling the Fate of Chemicals in the Aquatic Environment.* Ann Arbor Sci. Publ., Ann Arbor, MI, pp. 101–126.

Burns, L. A., Cline, D. M., and Lassiter, R. R. (1982). *Exposure Analysis Modeling System (EXAMS): User Manual and System Documentation* (EPA-600/3-82-023). U.S. Environ. Prot. Agency, Washington, DC.

Calamari, D., and Alabaster, J. S. (1980). An approach to theoretical models in evaluating the effects of mixtures of toxicants in the aquatic environment. *Chemosphere* **9,** 533–538.

Chapman, B. M. (1982). Numerical simulation of the transport and speciation of nonconservative chemical reactants in rivers. *Water Resour. Res.* **18,** 155–167.

Chapra, S. C. (1975). Comment on 'An empirical method of estimating the retention of phosphorus in lakes' by W. B. Kirchner and P. J. Dillon. *Water Resour. Res.* **11,** 1033–1034.

Chapra, S. C., and Reckhow, K. H. (1983). *Engineering Approaches for Lake Management,* Vol. 2. Ann Arbor Sci. Publ., Ann Arbor, MI.

Dillon, P. J., and Rigler, F. H. (1974). The phosphorus-chlorophyll relationship in lakes. *Limnol. Oceanogr.* **19,** 767–773.

Di Toro, D. M., Hallden, J. A., and Plafkin, J. L. (1988). Modeling *Ceriodaphnia* toxicity in the Naugatuck River using additivity and independent action. In M. S. Evans (Ed.), *Toxic Contaminants and Ecosystem Health: A Great Lakes Focus.* John Wiley, New York.

DiToro, D. M., O'Connor, D. J., Thomann, R. V., and St. John, J. P. (1982). Simplified model of the fate of partitioning chemicals in lakes and streams. In K. L. Dickson, A. W. Maki, and J. Cairns, Jr. (Eds.), *Modeling the Fate of Chemicals in the Aquatic Environment.* Ann Arbor Sci. Publ., Ann Arbor, MI, pp. 165–190.

Eadie, B. J. (1981). *An Equilibrium Model for the Partitioning of Synthetic Organic Compounds: Formulation and Calibration* NOAA Tech. Memo. ERL GLERL-35. Great Lakes Environ. Res. Lab., Ann Arbor, MI.

Esterby, S. R. (1988). Statistical methods for studying biotic responses to persistent toxic substances. In M. S. Evans (Ed.), *Toxic Contaminants and Ecosystem Health: A Great Lakes Focus.* John Wiley, New York.

Fedra, K. (1981). Estimating model prediction accuracy: A stochastic approach to ecosystem modeling. In D. M. DuBois (Ed.), *Progress in Ecological Engineering and Management by Mathematical Modelling.* Editions Cebedoc, Liège, Belgium, pp. 87–105.

Fontaine, T. D., III (1984). A non-equilibrium approach to modeling metal speciation in acid, aquatic systems: Theory and process equations. *Ecol. Modell.* **21,** 287–313.

Fontaine, T. D., III, and Lesht, B. M. (1987). Contaminant management strategies for the Great Lakes: Optimal solutions under uncertain conditions. *J. Great Lakes Res.* **13,** 178–192.

Haimes, Y. Y. (1977). *Hierarchical Analysis of Water Resources Systems.* McGraw-Hill, New York.

Haimes, Y. Y., and Allee, D. J. (1984). *Multiobjective Analysis in Water Resources.* Am. Soc. Civ. Eng., New York.

Jorgensen, S. E. (1983). Modeling the distribution and effect of toxic substances in rivers and lakes. In G. T. Orlob (Ed.), *Mathematical Modeling of Water Quality: Streams, Lakes and Reservoirs.* Wiley, New York, pp. 395–424.

Kalmaz, E. V., and Kalmaz, G. D. (1979). Transport, distribution and toxic effects of polychlorinated biphenyls in ecosystems: Review. *Ecol. Modell.* **6,** 223–251.

Landrum, P. F. (1982). The effect of co-contaminants on the bioavailability of polycyclic aromatic hydrocarbons to *Pontoporeia hoyi*. In M. W. Cooke and A. J. Dennis (Eds.), *Polynuclear Aromatic Hydrocarbons.* Battelle, Columbus, OH, pp. 731–743.

Landrum, P. F., Reinhold, M. D., Nihart, S. R., and Eadie, B. J. (1985). Predicting the bioavailability of organic xenobiotics to *Pontoporeia hoyi* in the presence of humic and fulvic materials and natural dissolved organic matter. *Environ. Toxicol. Chem.* **4,** 459–467.

Mauersberger, P. (1983). General principles in deterministic water quality modeling. In G. T. Orlob (Ed.), *Mathematical Modeling of Water Quality: Streams, Lakes and Reservoirs.* Wiley, New York, pp. 42–115.

O'Conner, D. J., and St. John, J. P. (1982). Assessment of modeling the fate of chemicals in the aquatic environment. In K. L. Dickson, A. W. Maki, and J. Cairns, Jr. (Eds.), *Modeling the Fate of Chemicals in the Aquatic Environment.* Ann Arbor Sci. Publ., Ann Arbor, MI, pp. 13–34.

O'Neill, R. V., Gardner, R. H., Barnthouse, L. W., Suter, G. W., Hildebrand, S. G., and Gehrs, C. W. (1982). Ecosystem risk analysis: A new methodology. *Environ. Toxicol. Chem.* **1,** 167–177.

Park, R. A. et al. (24 co-authors). (1974). A generalized model for simulating lake ecosystems. *Simulation* **23,** 33–50.

Park, R. A., Connolly, C. I., Albanese, J. R., Clesceri, L. S., Heitzman, G. W., Herbrandson, H. H., Indyke, B. H., Loehe, J. R., Ross, S., Sharma, D. D., and Shuster, W. W. (1980). *Modeling Transport and Behavior of Pesticides and Other Toxic Organic Materials in Aquatic Environments,* Cent. Ecol. Model. Rep. No. 7. Rensselaer Polytechnic Institute, Troy, NY.

Reckhow, K. H., and Chapra, S. C. (1983). *Engineering Approaches for Lake Management,* Vol. 1. Ann Arbor Sci. Publ., Ann Arbor, MI.

Rodgers, P. W. (1983). *Modeling Analysis of the Fate and Distribution of Polychlorinated Biphenyls in the Saginaw River and Bay.* Rep. Limno-Tech, Inc., Ann Arbor, MI.

Rodgers, P. W., and Swain, W. R. (1983). Analysis of polychlorinated biphenyl (PCB) loading trends in Lake Michigan. *J. Great Lakes Res.* **9,** 548–558.

Scavia, D., and Chapra, S. C. (1977). Comparison of an ecological model of Lake Ontario and phosphorus loading models. *J. Fish. Res. Board Can.* **34,** 286–290.

Scavia, D., Eadie, B. J., and Robertson, A. (1976). *An Ecological Model for Lake Ontario—Model Formulation and Preliminary Evaluation.* NOAA Tech. Rep. ERL 371-GLERL 12. Great Lakes Environ. Res. Lab., Ann Arbor, MI.

Somlyody, L. (1982). Water-quality modelling: A comparison of transport-oriented and ecology-oriented approaches. *Ecol. Modell.* **17,** 183–207.

Thomann, R. V. (1978). *Size Dependent Model of Hazardous Substances in Aquatic Food Chain.* Rept. EPA-600/3-78-036. U.S. Environ. Prot. Agency, Washington, DC.

Thomann, R. V. (1979). An analysis of PCB in Lake Ontario using a size-dependent food chain model. In D. Scavia and A. Robertson (Eds.), *Perspectives on Lake Ecosystem Modeling.* Ann Arbor Sci. Publ., Ann Arbor, MI, pp. 293–320.

Thomann, R. V., and Connolly, J. P. (1984). *An Age-dependent Model of PCB in a Lake Michigan Food Chain.* Rept. EPA-600/3-84-026. U.S. Environ. Prot. Agency, Washington, DC.

Thomann, R. V., Szumski, D. S., DiToro, D. M., and O'Connor, D. J. (1974). A food chain model of cadmium in western Lake Erie. *Water Res.* **8,** 841–849.

# INDEX